CURRENT TRENDS IN
BAYESIAN METHODOLOGY
WITH APPLICATIONS

CURRENT TRENDS IN BAYESIAN METHODOLOGY WITH APPLICATIONS

Edited by

Satyanshu Kumar Upadhyay

Banaras Hindu University
Varanasi, Uttar Pradesh, India

Umesh Singh

Banaras Hindu University
Varanasi, Uttar Pradesh, India

Dipak K. Dey

University of Connecticut
Storrs, USA

Appaia Loganathan

Manonmaniam Sundaranar University
Tirunelveli, Tamil Nadu, India

CRC Press
Taylor & Francis Group
Boca Raton London New York

CRC Press is an imprint of the
Taylor & Francis Group, an **informa** business

A CHAPMAN & HALL BOOK

CRC Press
Taylor & Francis Group
6000 Broken Sound Parkway NW, Suite 300
Boca Raton, FL 33487-2742

First issued in paperback 2019

© 2015 by Taylor & Francis Group, LLC
CRC Press is an imprint of Taylor & Francis Group, an Informa business

No claim to original U.S. Government works

ISBN-13: 978-1-4822-3511-1 (hbk)
ISBN-13: 978-0-367-37762-5 (pbk)

Visit the Taylor & Francis Web site at
http://www.taylorandfrancis.com

and the CRC Press Web site at
http://www.crcpress.com

Dedicated to my Parents and my Teacher Prof. Manju Pandey

— **Satyanshu K. Upadhyay**

Dedicated to my late Mother Mrs. Siddheswari Devi

— **Umesh Singh**

Dedicated to my late Parents Mr. Debendranath Dey and Mrs. Renuka Dey

— **Dipak Dey**

Dedicated to my Teachers Prof. M. Rajagopalan and Prof. G. Nanjundan

— **Appaia Loganathan**

Contents

List of Figures

List of Tables

Preface

The year 2013 can be considered an important year in the history of statistics, with six important global societies declaring the year as the International Year of Statistics. More than 128 countries participated in the International Year of Statistics. It added to its importance because 2013 was 300th anniversary of Jacob Bernoulli's Ars conjectandi and 250th anniversary of Bayes' theorem. India enthusiastically participated in this declaration by organizing an important Bayesian event in January, 2013 that may be considered the beginning of the celebrations of the International Year of Statistics in India. The 20th birth anniversary of International Society for Bayesian Analysis and 250th birth anniversary of the Bayes theorem in the presence of as many as 10 past, future and present presidents of ISBA was another important event celebrated in 2013 and simultaneously a highly desired publication in the form of the current volume *Current Trends in Bayesian Methodology with Applications* was decided to be brought out. We are happy that our thought has taken the shape of reality.

Bayesian statistics has expanded its coverage enormously in the past two to three decades. The Bayesian methods of reasoning are now applied to a wide variety of scientific, social, and business endeavours including areas such as astronomy, biology, economics, education, engineering, genetics, marketing, medicine, psychology, public health, sports, among many others. There are certain situations where Bayesian statistics appears as the only paradigm that offers viable solutions and this has become possible because of the tremendous development of Bayesian theory, methodology, computation and applications. The subject became the forefront of practical statistics with the advent of high-speed computers and sophisticated computational techniques especially in the form of Markov Chain Monte Carlo methods and sample based approaches. In fact, Bayesian modelling in complex problems freely combines components of different sorts of modelling approaches with structural prior information, unconstrained by whether such model combinations have ever been studied or analysed before.

Bayesian publications have also increased enormously in the last thirty years. Why then another publication on Bayesian statistics? The importance of the present volume may be realized from the fact that although the literature on Bayesian statistics is enormous, we do not find any single book that provides various aspects at one place. The literature is undoubtedly widely scattered, and journal articles rarely provide the conceptual background necessary for non-experts to understand and apply the approaches to their own problems. Moreover, the cost of recent publications is getting so high that researchers find it difficult going through the massive amount of literature.

The present volume consists of thirty chapters. We have topics on biostatistics, econometrics, reliability and risk analysis, spatial statistics, image analysis, shape analysis, Bayesian computation, clustering, uncertainty assessment, applications to high-energy astrophysics, neural networking, fuzzy information, objective Bayesian methodologies, empirical Bayes methods, small area estimation, and a lot more. All the articles focus on Bayesian methodologies but each is self-contained and independent so that the present volume may lessen the competition with research level journals and may simultaneously act as a good reference for the researchers and graduate level students. We have preferred to include chapters giving an overview to the area including some theoretical insights and simultaneously expected to emphasize the work of others, give motivating examples, but omit sophisticated technical details. The present volume can be considered as a physical insignia of the inspiration by several experts in the field, which will be quite helpful for the future researches.

Though there may be topics closely related to each other, there is no need to maintain the sequence for reading the chapters. The sequence in the volume is purely alphabetical according to the last name of the first author and, in no way should be taken to mean any preferential order.

We fail in our duties if we do not express our sincere indebtedness to our referees, who were quite critical and unbiased in giving their opinion. We do realize that in spite of their busy schedules they offered us every support in timely commenting on various manuscripts. Undoubtedly, it is the joint endeavor of the contributors and the referees that emerged in the form of such an important and significant volume for the Bayesian world. We sincerely thank them all; the space constraint restricts us to mention the names individually.

We thankfully acknowledge the support rendered by John Kimmel, Executive Editor, Statistics, CRC Press, Taylor & Francis Group, who always stood behind us and always helped us with several of our unusual queries.

We express our indebtedness to everyone who was associated with us directly or indirectly while the work was in progress. The list is certainly too lengthy to be exhaustive but we would like to give special mention to Anuradha, Asha, Rita, Shakila, Geetika, Vertika, Debosri, Om Shankar, Naganandhini, Sangeetha, Rakesh, Rijji, Reema, Praveen, among others. At last but not the least, we express our thankfulness to Mr. Duvvuri Venu Gopal, Banaras Hindu University, who is credited with the present shape of the volume.

– **The Editors**

Foreword

It is a great pleasure to see a new book published on current aspects of Bayesian Analysis and coming out of India. This wide scope volume reflects very accurately on the present role of Bayesian Analysis in scientific inference, be it by statisticians, computer scientists or data analysts. Indeed, we have witnessed in the past decade a massive adoption of Bayesian techniques by users in need of statistical analyses, partly because it became easier to implement such techniques, partly because both the inclusion of prior beliefs and the production of a posterior distribution that provides a single filter for all inferential questions is a natural and intuitive way to process the latter. As reflected so nicely by the subtitle of Sharon McGrayne's *The Theory That Would Not Die*, the Bayesian approach to inference "cracked the Enigma code, hunted down Russian submarines" and more generally contributed to solve many real life or cognitive problems that did not seem to fit within the traditional patterns of a statistical model. Two hundred and fifty years after Bayes published his note, the field is more diverse than ever, as reflected by the range of topics covered by this new book, from the foundations (with objective Bayes developments) to the implementation by filters and simulation devices, to the new Bayesian methodology (regression and small areas, nonignorable response and factor analysis), to a fantastic array of applications. This display reflects very well on the vitality and appeal of Bayesian Analysis. Furthermore, I note with great pleasure that the new book is edited by distinguished Indian Bayesians, India having always been a provider of fine and dedicated Bayesians. I thus warmly congratulate the editors for putting this exciting volume together and I offer my best wishes to readers about to appreciate the appeal and diversity of Bayesian Analysis.

– **Christian P. Robert**

List of Contributors

John A. D. Aston
Department of Statistics
University of Cambridge, UK
J.Aston@statslab.cam.ac.uk

Goodness C. Aye
Department of Economics
University of Pretoria,
South Africa
goodness.aye@gmail.com

Dipankar Bandyopadhyay
Division of Biostatistics
University of Minnesota, USA
dbandyop@umn.edu

Adrian Barbu
Department of Statistics
Florida State University, USA
abarbu@stat.fsu.edu

José M. Bernardo
Department of Statistics
University of Valencia, Spain
jose.m.bernardo@uv.es

Michael Betancourt
Department of Statistics
University of Warwick, UK
betanalpha@gmail.com

Arnab Bhattacharjee
Department of Economics & Spatial
Econ. and Econometrics Centre
(SEEC), Heriot-Watt University, UK
A.Bhattacharjee@hw.ac.uk

Arnab Bhattacharya
Department of Statistics
Trinity College, Ireland
bhattaca@tcd.ie

Madhuchhanda Bhattacharjee
School of Maths. & Statistics
University of Hyderabad, India
chhanda.bhatta@googlemail.com

Darshan Bryner
Naval Surface Warfare Center
Florida State University, USA
dbryner@stat.fsu.edu

Javier Cano
Department of Statistics and O.R.
University Rey Juan Carlos, Spain
javier.cano@urjc.es

Yun Cao
Statistical Consulting
Toronto, ON, Canada
shelleycao@hotmail.com

Luis M. Castro
Department of statistics
University of Concepción, Chile
luiscastrocepero@gmail.com

Eduardo A. Castro
Department of Social,
Political & Territorial Sciences
University of Aveiro, Portugal

Snigdhansu Chatterjee
School of Statistics
University of Minnesota, USA
chatterjee@stat.umn.edu

S.T. Boris Choy
Discipline of Business Analytics
University of Sydney, Australia
boris.choy@sydney.edu.au

Jyotishka Datta
Department of Statistical Science
Duke University, USA and
SAMSI North Carolina, USA
jd298@stat.duke.edu

Dipak K. Dey
Department of Statistics
University of Connecticut, USA
dipak.dey@uconn.edu

Tanujit Dey
Department of Mathematics
The College of William and Mary
Virginia, USA
tanujit.dey@gmail.com

Liangjing Ding
Department of Scientific Computing
Florida State University, USA
liangjingding@gmail.com

Pami Dua
Department of Economics
Delhi School of Economics
University of Delhi, India
dua@econdse.org

Garland Durham
Orfalea College of Business
California Polytechnic State
University, USA

David A. van Dyk
Department of Mathematics
Imperial College London, UK
d.van-dyk@imperial.ac.uk

Evangelos Evangelou
Department of Mathematical
Sciences, University of Bath, UK
E.Evangelou@bath.ac.uk

Michael Evans
Department of Statistics
University of Toronto, Canada
mevans@utstat.utoronto.ca

Ernest Fokoué
Center for Quality and Applied
Statistics, Rochester Institute of
Technology, USA
epfeqa@rit.edu

Diana M. Galvis
Department of Statistics
IMECC-UNICAMP, Brazil
dianagalvis@uniquindio.edu.co

John Geweke
Economics Discipline Group
School of Business, University of
Technology, Sydney, Australia
John.Geweke@uts.edu.au

Jayanta K. Ghosh
Purdue University, USA and
Indian Statistical Institute
Kolkata, India
jayantag1@gmail.com

Mark Girolami
Department of Statistics
University of Warwick, UK
m.girolami@warwick.ac.uk

Irwin Guttman
Department of Mathematics &
Statistics, SUNY at Buffalo, USA
sttirwin@buffalo.edu

Rangan Gupta
Department of Economics
University of Pretoria, South Africa
rangan.gupta@up.ac.za

Fredrik Gustafsson
Department of Electrical Engineering
Linköping University, Sweden
fredrik@isy.liu.se

Gustaf Hendeby
Department of Electrical
Engineering
Linköping University, Sweden
hendeby@isy.liu.se

Rebecca A. Hubbard
Department of Biostatistics
University of Washington
USA
hubbard.r@ghc.org

Lurdes Y. T. Inoue
Department of Biostatistics
University of Washington,
USA
linoue@u.washington.edu

David Ríos Insua
ICMAT-CSIC, Spain
david.rios@urjc.es

Casey M. Jelsema
National Institute of
Environmental
Health Sciences, USA
casey.jelsema@nih.gov

Xun Jiang
Department of Statistics
University of Connecticut, USA
tonyjiangxun@gmail.com

Adam M. Johansen
Department of Statistics
University of Warwick, UK
a.m.johansen@warwick.ac.uk

Vinay Kashyap
High-Energy Astrophysics Div.
Smithsonian Astrophysica,
Observatory, Cambridge, USA
vkashyap@cfa.harvard.edu

Sebastian Kurtek
Department of Statistics
The Ohio State University, USA
kurtek.1@stat.osu.edu

Amy E. Laird
Department of Public Health &
Preventive Medicine, Oregon
Health & Science University, USA
laird@ohsu.edu

Kwok Wai Lau
CSIRO Computational
Informatics, Perth, Australia

Victor H. Lachos
Department of Statistics
IMECC-UNICAMP, Brazil
hlachos@gmail.com

Rosangela H. Loschi
Statistics Department
Federal University
of Minas Gerais, Brasil
loschi@est.ufmg.br

Simón Lunagómez
Department of Statistics
Harvard University, USA
simon.lgz@gmail.com

Tapabrata Maiti
Department of Statistics and
Probability
Michigan State
University, USA
maiti@stt.msu.edu

José M. Martins
Department of Social,
Political and Territorial Sciences
University of Aveiro, Portugal

Antonietta Mira
Swiss Finance Institute
University of Lugano, Switzerland
antonietta.mira@usi.ch

Sayan Mukherjee
Departments of Statistical Science,
Computer Science & Mathematics
Duke University, USA
sayan@stat.duke.edu

Balgobin Nandram
Department of Mathematical
Sciences Worcester Polytechnic
Institute, USA
balnan@wpi.edu

Yasuhiro Omori
Faculty of Economics
University of Tokyo, Japan
omori@ja2.so-net.ne.jp

Ricardo Ortega
Transports Metropolitans de
Barcelona, Spain
rortegap@tmb.cat

Theodore Papamarkou
Department of Statistics
University of Warwick, UK
t.papamarkou@warwick.ac.uk

Rajib Paul
Department of Statistics
Western Michigan University, USA
rajib.paul@wmich.edu

Michael Pellot
Transports Metropolitans de
Barcelona, Spain
mpellot@tmb.cat

Azizur Rahman
Discipline of Statistics
Charles Sturt University,
Australia
azrahman@csu.edu.au

Vivekananda Roy
Department of Statistics
Iowa State University, USA
vroy@iastate.edu

Saikat Saha
Department of Electrical Engineering
Linköping University, Sweden
saha@isy.liu.se

Cristiano C. Santos
Statistics Department, Federal
University of Minas Gerais, Brasil
cristcarvalhosan@yahoo.com.br

Anuj Srivastava
Florida State University
Tallahassee, Florida, USA
anuj@stat.fsu.edu

David C. Stenning
Department of Statistics
University of California, USA
dstenning@gmail.com

Owat Sunanta
Department of Statistics and
Probability Theory, Vienna
University of Technology, Austria
owat.sunanta@tuwien.ac.at

Satyanshu K. Upadhyay
Department of Statistics &
DST-CIMS
Banaras Hindu University, India
sku@bhu.ac.in

Reinhard Viertl
Department of Statistics
and Probability Theory, Vienna
University of Technology, Austria
R.Viertl@tuwien.ac.at

Toshiaki Watanabe
Institute of Economic Research
Hitotsubashi University, USA
twecon@bd5.so-net.ne.jp

Nuttanan Wichitaksorn
School of Maths. and Statistics
University of Canterbury
New Zealand
nuttanan.wichitaksorn@
canterbury.ac.nz

Simon Wilson
Department of Statistics
Trinity College, Ireland
swilson@tcd.ie

Robert Wolpert
Department of Statistical
Science
Duke University, USA
wolpert@stat.duke.edu

Namkyo Woo
Department of Statistics
Kyungpook National University
Korea
namkyo.woo@gmail.com

Qian Xie
Department of Statistics
Florida State University, USA
qxie@stat.fsu.edu

Huaxin Xu
Economics Discipline Group
School of Business
University of Technology
Sydney, Australia

Yaming Yu
Department of Statistics
University of California, Irvine, USA
yamingy@uci.edu

Zhen Zhang
Department of Statistics &
Probability
Michigan State University, USA

Zhengyuan Zhu
Department of Statistics
Iowa State University, USA
zhuz@iastate.edu

1

Bayesian Inference on the Brain: Bayesian Solutions to Selected Problems in Neuroimaging

John A. D. Aston

University of Cambridge

Adam M. Johansen

University of Warwick

CONTENTS

1.1   Introduction

This chapter summarizes certain recent advances in inferential procedures for brain imaging. These advances involve the use of computationally intensive procedures. A goal of this chapter is to demonstrate that although the computational cost of these methods can be significant, it is feasible to employ them, using commodity-grade hardware, to provide Bayesian solutions to problems of realistic complexity in this domain.

Understanding the structure and the functional capabilities of the human brain is something that has been of profound interest for centuries. However, with recent advances in the ability to image the brain *in-vivo*, major advances in this understanding have been possible. Different neuroimaging (brain imaging) modalities typically have their strengths and weaknesses, usually consisting of trade-offs between spatial and temporal resolution and the invasiveness to the patient against the ability to probe specific or more general neuronal systems. However, the statistical methodologies that can be used to make inferences about differing systems can in many ways benefit from considering these modalities not in isolation but rather as inferential problems on spatio-temporal data sets.

It has long been recognized that (computational) Bayesian methods have the potential to provide excellent solutions to complex inferential problems (see, [71] for an application in reliability modelling, for example; and one need only consult the index of any recent volume on Bayesian inference, such as [70], to see how wide-reaching the influence of Monte Carlo methodology has been in this area). However, it is only recently that it has become feasible to apply these methods to the enormous datasets which arise in neuroscience and to do so sufficiently quickly that it is feasible to conduct the sensitivity analysis and replicated runs that are necessary to provide confidence in real applications. This chapter provides a non-technical introduction to these topics; more detailed explanations can be found within the references.

The subfield of neuroimaging is itself far too large to be exhaustively addressed here, and indeed even Bayesian neuroimaging has a very wide scope of applications including brain decoding [21] and more traditional group level hierarchical neuroimaging analyses [66]. Consequently, we confine ourselves here to three particular areas with which we have some experience and in which Bayesian methods provide good performance; there are of course a great many other problems within this field to which Bayesian techniques either have been or could be successfully applied.

1.2 Background

1.2.1 Neuroscience and its imaging problems

Neuroscience can be loosely defined as the study of the mechanism and function of the brain and nervous system, particularly within human subjects. Neuroimaging can be thought of as the subfield of neuroscience which is concerned with the acquisition of images of the (living) brain with a particular emphasis on images which attempt to illustrate the function of the brain and to show those regions which are related to the task or brain state of interest. It is often useful to obtain sequences of images which show, dynamically, changes in the brain during the course of some stimulus.

The focus of this article is the statistical analysis of data obtained in a variety of dynamic neuroimaging scenarios. The details of the different imaging techniques considered are described in the various application sections which follow, but they share a number of common elements. In all cases, complex indirect measurements of the brain's function are obtained.

In the field of neuroscience, Magnetic Resonance Imaging (MRI; [18]) is an imaging technique in which powerful radio-frequency magnetic fields are used to induce nuclear magnetic resonance within the nuclei of certain atoms within the brain. This resonance leads to a measurable perturbation in the magnetic field surrounding the subject. Functional Magnetic Resonance Imaging (fMRI; [34]) is particularly concerned with the location of oxygen-rich blood within the brain; which under the Blood Oxygen-Level Dependent (BOLD) hypothesis [53] is assumed to be related to regions of high neural activity. This technique is considered in Section 1.3.

Positron Emission Tomography (PET; [57]) is an invasive neuroimaging technique. Typically, a positron-labelled tracer is injected intravenously and the PET camera scans a record of positron emission as the tracer decays. With all events detected by the PET camera, the time course of the tissue concentrations are reconstructed as three-dimension images [38]. While it suffers from a significant disadvantage relative to fMRI, in that injection of radioactive material is required, it does benefit from being applicable to a whole range of neurochemical systems, rather than being restricted to the investigation of either direct or indirect changes of blood flow. The digital image so captured shows the signal integrated over small volume elements (voxels). The Bayesian estimation of model parameters of interest, together with selection or averaging is discussed in Section 1.4.

Magnetoencephalography (MEG; [32]) and Electroencephalography (EEG; [52]) are closely-related non-invasive neuroimaging techniques. Magnetoencephalography is an imaging technique which uses a helmet-shaped array of superconducting sensors to measure the magnetic fields produced by the neural currents in a human brain. The sampling rate of MEG recordings is typically around 1 kHz, which allows observation of dynamics on the millisecond scale, several orders of magnitude faster than the two previously mentioned techniques of fMRI and PET. Among other non-invasive neuroimaging tools, only electroencephalography features a comparable temporal resolution. EEG is generally considered complementary to MEG, due to its different sensitivity to source orientation and depth [10], as it detects changes in electrical activity associated with neuronal responses as opposed to the changes in magnetic fields induced by electrical currents that are detected by MEG (see [52] for a comprehensive introduction to EEG). Note that estimation of the neural currents from the measured electric or magnetic fields is an ill-posed inverse problem [61] for which Bayesian inference provides a natural framework for regularization. We consider one particular task associated with inference for MEG in Section 1.5.

1.2.2   Monte Carlo and sequential Monte Carlo

In many Monte Carlo methods, an integral of interest is approximated by simulating a collection of random variables and using a (appropriately weighted) sample average to estimate an expectation. These procedures can be formally justified by appropriate versions of the law of large numbers and central limit theorem and are widely known and used in statistics. The basic principle is that the expectation of some function φ with respect to μ can be well approximated by the sample average $\frac{1}{n}\sum_{i=1}^{n}\varphi(x^i)$ where $x^1,\ldots,x^n$ is a simple random sample from μ — or the realization of an ergodic μ-invariant Markov chain — and n is *large enough*. An excellent book-length summary is provided by [60].

    We focus here on summarizing a less widely-used class of Monte Carlo methods which are used to address several of the neuroimaging problems addressed below. Sequential Monte Carlo (SMC) methods provide collections of weighted samples appropriate for approximating each of a sequence of distributions of interest in turn. Below, such methods are used to address the computational demands associated with Bayesian solutions to various problems in neuroimaging. These approaches are all based upon the framework of [14] which developed techniques for applying techniques developed in the setting of optimal filtering and dating back to [12, 30] (see [17] for a recent summary) to the approximation of arbitrary sequences of distributions and to providing good approximations of posterior distributions in particular.

    SMC methods are based upon two important steps. The first of these is *importance sampling*, an elementary Monte Carlo technique almost as old as the idea of Monte Carlo itself (see, for example, [29]), that uses the identity

$$\int \varphi(x)\frac{\pi(x)}{\mu(x)}\mu(x)\mathrm{d}x = \int \varphi(x)\pi(x)\mathrm{d}x \qquad (1.1)$$

which formally holds for any density μ which vanishes only where π itself vanishes. This allows expectations with respect to π to be approximated using samples from μ by computing a simple Monte Carlo estimate of the expectation of $\varphi\pi/\mu$ using a sample from μ. This can be conveniently viewed as a weighted average with the *importance weight* π/μ correcting for the use of a different sampling distribution. Some simple conditions must be satisfied in order to ensure that the variance of this random approximation is finite, a simple strategy which guarantees this is to ensure that the tails of μ are heavier than those of π. See, for example, [26] for further details. In fact, one can use importance sampling in settings in which the importance weight, π/μ, is known only up to a normalising constant by estimating the normalising constant itself as the empirical mean of the unnormalised importance weights. This is important in Bayesian inference as the posterior density is typically known only to be proportional to the product of the prior and likelihood with the normalising constant, or evidence, being intractable.

The other key step within any SMC algorithm is *resampling*. The basis of this is that given a sample $x^1, \ldots, x^n$ and associated weights $w^1, \ldots, w^n$, which are normalised such that they sum to unity, one can stochastically replicate some of these samples and eliminate others in order to obtain an unweighted sample, $\bar{x}^1, \ldots, \bar{x}^n$, such that:

$$\mathbb{E}\left[\frac{1}{n}\sum_{i=1}^{n}\varphi(\bar{X}^i)\,\middle|\,x^1, \ldots, x^n, w^1, \ldots, w^n\right] = \sum_{i=1}^{n}w^i\varphi(x^i), \qquad (1.2)$$

provided only that the expected number of times that x_i is reproduced in $\bar{x}_1, \ldots, \bar{x}_n$ is nw_i. See [15] for a comparison of some commonly-used resampling schemes.

Resampling in isolation can only increase the variance of the resulting estimate. Its use in SMC settings is due to the fact that it can stabilize the variance of estimates obtained after subsequent importance sampling steps within a framework known as *Sequential Importance Resampling (SIR)*. Given a sequence of distributions $\tilde{\pi}(\tilde{x}_n)$ on the spaces $\tilde{E}_n = \otimes_{p=1}^{n}E_p$ importance sampling can be carried out iteratively to provide weighted samples appropriate for targetting each target in turn.

More precisely, given a weighted sample suitable for approximating $\tilde{\pi}(\tilde{x}_n)$ one can first resample to obtain an unweighted sample, then extend each sample by sampling X_{n+1} from some Markov kernel, concatenating the existing and new samples to obtain samples $\tilde{x}_{1:n+1}$ on $\tilde{E}_{n+1}$ which can then be importance weighted to approximate $\tilde{\pi}_{n+1}$. Algorithm 1.1 shows the steps in a little more detail. Note that in this algorithm and all others in which simulated variables correspond to successive elements of a vector (as is usually the case when performing inference in time series models) we have introduced a double subscript with the first element corresponding to algorithm iteration and the second to the element within the vector. Whilst this may seem cumbersome it allows for cases in which subsequent algorithmic iterations alter the value of earlier elements of the state (and in even simple SMC algorithms, the resampling step in some sense does exactly this). It should be noted that numerous improvements to this algorithm are possible, especially via the use of adaptive resampling schemes in which low variance resampling is used only in those iterations in which the weights have high variance; see [17] for a summary of such improvements.

Algorithm 1.1 The SIR Algorithm targetting $\{\tilde{\pi}_n\}_{n\geq1}$

Iteration $n = 1$:

Sample $X_{1,1}^1, \ldots, X_{1,1}^N \overset{\text{iid}}{\sim} \mu$

Calculate weights, **for** $i = 1, \ldots, N$: $w(X_{1,1}^i) \propto \tilde{\pi}_1(X_{1,1}^i)/\mu(X_{1,1}^i)$.

Normalise weights, **for** $i = 1, \ldots, N$: $W_1^i = w(X_{1,1}^1)/\sum_{j=1}^N w(X_{1,1}^j)$.

Iteration $n \geq 2$:

Resample $\{X_{n-1,1:n-1}^i, W_{n-1}^i\}_{i=1}^N$ to obtain $\{\bar{X}_{n-1,1:n-1}^i, 1/N\}_{i=1}^n$.

for $i = 1, \ldots, N$: sample $X_{n,n}^i \sim q_n(\cdot|\bar{X}_{n-1,1:n-1}^i)$.

for $i = 1, \ldots, N$: set $X_{n,1:n}^i = (\bar{X}_{n-1,1:n-1}^i, X_{n,n}^i)$.

Calculate weights, **for** $i = 1, \ldots, N$:

$$w(X_{n,1:n}^i) \propto \tilde{\pi}_n(X_{n,1:n}^i)/\tilde{\pi}_{n-1}(X_{n,1:n-1}^i)q_n(X_{n,n}^i|X_{n,1:n-1}^i).$$

Normalise weights, **for** $i = 1, \ldots, N$: $W_n^i = w(X_{n,1:n}^i)/\sum_{j=1}^N w(X_{n,1:n}^j)$.

Although the presentation thus far has been quite abstract, much of the development of SMC methods occurred in the field of time series analysis and in particular of online estimation within (general state space) Hidden Markov Models (HMMs; cf. [7] for a recent overview of these models and associated inference problems). HMMs are a broad class of time series models in which an unobserved Markov process $(X_n)_{n\in\mathbb{N}}$ evolves with some transition kernel $f(x_n|x_{n-1})$ and the observation process $(Y_n)_{n\in\mathbb{N}}$ leads to observations with the property that for any p, given x_p, Y_p is conditionally independent of the remainder of the hidden and observed processes and has density $g(y_n|x_n)$. That is, the joint density of $(X_{1:n}, Y_{1:n})$ is, for any n, simply:

$$p(x_{1:n}, y_{1:n}) = p(x_1)g(y_1|x_1) \prod_{p=2}^{n} f(x_p|x_{p-1})g(y_p|y_{p-1}).$$

Although Bayesian inference for this class of models, in particular the characterization of the posterior distribution of $X_{1:n}$ given $y_{1:n}$ may seem trivial, the unavailability for almost all interesting models of the marginal density of $Y_{1:n}$ means that one cannot compute the posterior density of $X_{1:n}$ analytically and is forced to resort to approximate methods. Noting that $p(x_{1:n}|y_{1:n}) \propto p(x_{1:n}, y_{1:n})$ allows us to apply Algorithm 1.1 directly to the sequence of distributions $\pi_n(x_{1:n}) := p(x_{1:n}|y_{1:n})$. Note that in this particular setting the algorithmic steps coincide with the arrival time of observations; when we deal with time series and this is not the case, we will reserve the symbol n to refer to algorithmic iterations and we will use t to refer to *real* time.

The simplest SMC algorithm applicable to filtering is the *bootstrap* particle filter of [30]. The basis of this approach is to set $q(x_n|x_{1:n-1}) = f(x_n|x_{n-1})$; in order to minimize the conditional variance of the importance weights whilst staying within the framework of Algorithm 1.1, [16] demonstrate that one should employ a locally optimal proposal of the form $q(x_n|x_{n-1}) \propto f(x_n|x_{n-1})g(y_n|x_n)$ (i.e. one should use a proposal which *conditions* upon the observed value of the current observation). The use of Markov chain Monte Carlo (MCMC) steps within SMC was first proposed by [27] who designed a filter in which MCMC was used to replace importance sampling; it's straightforward to combine the two in order to obtain the advantages of both techniques and an example of this is shown in the context of MEG in Section 1.5.

A number of approaches to extending the range of SMC methods have been developed in recent years [9, 45, 50] and to apply them to problems which do not have the structure described above. Most of these methods can be considered within the framework of [14], the key insight of which is that, given any sequence of target distributions $\pi_1, \ldots, \pi_n$ on a sequence of spaces $E_1, \ldots, E_n$, an auxiliary sequence of target distributions can be defined via:

$$\tilde{\pi}_n(\tilde{x}_{n,1:n}) = \pi_n(\tilde{x}_{n,n}) \prod_{p=1}^{n-1} L(\tilde{x}_{n,p+1}, \tilde{x}_{n,p}) \tag{1.3}$$

on the spaces $\tilde{E}_n = \otimes_{p=1}^{n} E_p$. This sequence of distributions trivially admits the target distribution of interest as a marginal distribution and is of the correct form to allow the application of SIR algorithms. In principle, the application of standard SIR algorithms to this sequence of auxiliary distributions will provide weighted samples appropriate for approximating the target distributions of interest. In practice, some care is needed over the design of these algorithms.

There has been some perception that these methods are difficult to implement, but it's our view that this perception is due more to a lack of familiar and, historically, of general software implementations than any real technical difficulties. In recent years, several software libraries have been made available to facilitate the implementation of SMC algorithms: simple and general C++ template libraries including [37] and [75], which makes parallel implementation very easy, and some more specialized packages which are likely to be easier to apply within their particular domains including [47]. It should be noted that the problems that will be introduced in the following sections naturally admit approaches based on SMC, either because they can be modelled explicitly or implicitly with HMMs or because the estimation of the normalising constant is of profound interest in the analysis (or because both are important). As will be shown, this will inherently allow similar SMC approaches to be used to investigate different neuroimaging modalities. Our aim is to demonstrate that Bayesian methods are able to provide good solutions to these problems and that SMC is able to facilitate these solutions in these settings; not to advocate this computational technique for every problem in neuroscience.

1.3 Application: Functional Magnetic Resonance Imaging

MRI is the application of nuclear magnetic resonance technology to the problem of imaging. In the context of neuroscience it is a widely used imaging technique in many contexts. In neuroimaging, the term "MRI" itself is synonymous with structural MRI, where anatomical maps of the brain are produced with the differing magnetic resonance (MR) signatures of grey matter, white matter and cerebrospinal fluid (CSF) enabling segmentation of the brain into differing anatomical structures. Another commonly usage of MRI is diffusion tensor MRI [40], which uses the preferential diffusivity of water along neuronal fibres as opposed to across the fibres to build structural maps of brain connectivity.

In this section, we will concentrate on functional MRI (fMRI). This use of MRI allows the detection of blood oxygenation levels with good spatial (and reasonable temporal) resolution; in combination with the BOLD hypothesis [53], the resulting image sequence can be used to infer neural activity. The BOLD hypothesis takes advantage of the differing magnetic properties of oxy- and deoxy- hemoglobin. As neuronal activity requires energy, changes in the relative concentrations of oxygen usage can be measured using fMRI, thanks to these differing magnetic properties. While this is essentially a measure of blood flow, subject to certain assumptions, this can be taken as a surrogate measure for the neuronal activity itself.

In most fMRI experiments, a known design is used in the experiment, allowing regression models (usually linear time series regression models) to be used for the estimation of parameters associated with the experimental task [74]. By investigating whether each spatial location is positively or negatively associated with the task, statistical parametric maps (as they are known in the neuroimaging literature) can be constructed for the task of interest. Correction methods are needed to account for the inherent multiple comparison issues associated with large numbers of tests (for a detailed overview of statistical methods commonly used in fMRI see [42]). However, in many situations, particularly those associated with psychological experiments, it is hard to be robust against the misspecification of the design, if specification is even possible. This is because reactions to psychological stimuli can typically evoke different responses from different subjects over different time scales. It has therefore been proposed [43] that a more natural framework for the investigation of such experiments might be unknown design regression, which can be framed as a change point problem. It is this approach that will be investigated further in this chapter.

A major issue in any modelling, and one which receives considerable discussion in the statistical analysis of fMRI data, is of course which statistical assumptions are appropriate to make when modelling the data. These include, but are not limited to, the dependence or independence assumptions which are made, the presence of interesting or nuisance covariates and the stationarity or otherwise of time series. In this section, we will provide, through an fMRI example, an indicative investigation of how modifying these modelling assumptions can have a dramatic effect on the conclusions which can be drawn. This will be shown to be especially important in the context of change point analyses for fMRI time series.

1.3.1 Problem

In this section the particular problem of identifying both the number and location of *change points* within fMRI time series is considered. Given a sequence of fMRI observations, the observations associated with each region of the brain can be viewed as a time series. Three questions of interest then arise: how *many* qualitative changes in neural activity occur within each region, *when* do those changes occur, and *how certain* can we be about the answers to these questions.

A little more formally, let $\mathbf{y} = (y_1, y_2, \ldots, y_t)$ denote the observations associated with a single neural region. Here, these observations are modelled as realising a (generalised) hidden Markov model with a small finite hidden state sequence $\mathbf{x} = (x_1, x_2, \ldots, x_t)$ whose values correspond to a regime indicator. Change points in this data series are closely related to changes in the value of the hidden state. For a comprehensive treatment of HMMs, we refer the reader to [7].

A slightly broader class of models than those normally termed HMMs are considered, allowing some finite dependence on the past of both the observation and hidden processes. These models are characterised by the following conditional distributions:

$$Y_s | y_{1:s-1}, x_{1:s} \sim f(y_s | x_{s-r:s}, y_{1:s-1}, \theta) \qquad \text{(Emission)} \quad (1.4)$$
$$p(x_s | x_{1:s-1}, y_{1:s-1}, \theta) = p(x_s | x_{s-1}, \theta) \quad s = 1, \ldots, t \qquad \text{(Transition)}.$$

Given the set of model parameters θ, the observation at time $s = 1, \ldots, n$, y_s has emission density dependent on previous observations $y_{1:s-1}$ and previous r states of the underlying states $x_{s-r}, \ldots, x_{s-1}$. The underlying states are assumed to follow a first order Markov chain (although this assumption can easily be relaxed) and takes values in a finite state space, E_X.

The simplest view of changepoints within this class of models, and one which is often used when working with HMMs, is that changepoints coincide exactly with changes in the value of the latent state. With this definition, saying that a changepoint occurs at time s is equivalent to stating that $x_{s-1} \neq x_s$.

We consider a slightly more general definition; a change point to a regime occurs at time s when the change in the underlying chain persists for at least k time periods. That is $x_{s-1} \neq x_s = \ldots = x_{s+j}$ where $j \geq k-1$, and a change of *regime* occurs only when there is a sustained movement in the underlying chain. This is done because spurious short segments will be of little interest in neuroimaging applications.

1.3.2 Methodology

By treating the problem parametrically within the family of generalised (finite state space) HMMs it is possible to compute, exactly, the posterior distribution over changepoint locations given the model parameters using the techniques of [3, 4]. These techniques rely upon Functional Markov Chain Imbedding (FMCI; [22]), and essentially consider an inhomogeneous Markov chain on an extended space which allows both the posterior distribution over changepoint location and the marginal likelihood $p(y|\theta)$, to be evaluated numerically using simple recursive algorithms.

In order to deal with parameter uncertainty, [48] employed an SMC sampler using a well established *likelihood tempering* strategy. This strategy uses a sequence of target distributions $\pi_n(\theta) \propto p(\theta)p(y_{1:t}|\theta)^{\gamma_n}$ where p denotes the prior distribution over the unknown parameters and γ_n is an increasing sequence with $\gamma_1 = 0$ and $\gamma_{n_{\max}} = 1$ which allows this sequence of artificial target distributions to interpolate smoothly between the prior and posterior distribution.

Algorithm 1.2 shows the steps of this algorithm in detail. In practice, some additional refinements are used, in particular resampling adaptively as required rather than every iteration (cf. [17]). Given the approximation of $p(\theta|y_{1:t})$ provided by this algorithm it is straightforward to approximate the distribution of the number of changepoints and their locations by writing the distribution of the quantity of interest as the marginal of the joint distribution of the quantity of interest and the unknown parameters and using the SMC approximation of the parameter posterior within that joint distribution.

We note that the use of essentially this approach, together with the natural evidence estimation procedure provided by the SMC algorithm, to perform model criticism, comparison or selection was further considered by [49]. Given the output of Algorithm 1.2 one can readily estimate the evidence using:

$$\widehat{p(y_{1:n}|\theta)} = \frac{1}{N}\sum_{i=1}^{N}\tilde{w}_1(\theta_0^i)\prod_{n=2}^{n_{\max}}\frac{1}{N}\sum_{i=1}^{N}\tilde{w}_n(\bar{\theta}_{n-1}^i,\theta_n^i). \tag{1.5}$$

It can be shown that this estimator is, in fact, unbiased [13]. Using estimates of this quantity obtained by running the algorithm for several competing models allows for straightforward Bayesian model comparison. In the context of determining how many regimes are required within a switching autoregressive model of the sort considered here, this strategy is advocated by [49].

One of the benefits of Bayesian methodology in this setting is that it provides a meaningful estimate of posterior uncertainty which is especially important in settings such as this one in which there can be many competing

---

Algorithm 1.2 SMC algorithm for quantifying the uncertainty in change points.

Approximating $p(\theta|y_{1:t})$

for $i = 1, \ldots, N$ Sample $\theta_0^i \sim q_1$.

Compute for each i

$$W_1^i = \frac{w_1(\theta_0^i)}{\sum_{j=1}^N w_1(\theta_0^j)} \qquad \text{where } w_1(\theta_0) = \frac{p(\theta_0)}{q_1(\theta_0)}$$

Resample $\{\theta_0^i, W_0^i\}_{i=1}^N$ to obtain $\{\bar{\theta}_0^i, 1/N\}_{i=1}^N$.

for each $i = 1, \ldots, N$: Sample $\theta_1^i \sim K_1(\bar{\theta}_0^i, \cdot)$ where K_1 is a p-invariant Markov kernel.

for $n = 2, \ldots, n_{\max}$ **do**

    Reweighting:

    for each $i = 1, \ldots, N$ **compute:**

$$\widetilde{w}_n(\bar{\theta}_{n-1}^i, \theta_n^i) = \frac{\pi_n(\bar{\theta}_{n-1}^i)}{\pi_{n-1}(\bar{\theta}_{n-1}^i)} = \frac{p(y_{1:t}|\bar{\theta}_{n-1}^i)^{\gamma_n}}{p(y_{1:t}|\bar{\theta}_{n-1}^i)^{\gamma_{n-1}}}.$$

    and normalise $W_n^i = \widetilde{w}_n(\theta_{n-1}^i, \theta_n^i)/\sum_{j=1}^N \widetilde{w}_n(\theta_{n-1}^j, \theta_n^j)$.

    Resample $\{\theta_{n-1}^i, W_n^i\}_{i=1}^N$ to obtain $\{\bar{\theta}_{n-1}^i, 1/N\}_{i=1}^n$.

    for each $i = 1, \ldots, N$: Sample $\theta_n^i \sim K_n(\bar{\theta}_{n-1}^i, \cdot)$ where K_n is a π_n invariant Markov kernel.

end for

Approximation of posterior:

$$\widehat{p(\theta|y_{1:n})}d\theta = \pi_{n_{\max}}(d\theta) \approx \sum_{i=1}^N \frac{1}{N}\delta_{\theta_{n_{\max}}^i}(d\theta).$$

---

explanations for the observed data which are all almost equally probable; simple maximization techniques fail to distinguish between this scenario and one in which a particular configuration of the latent states is overwhelmingly more probable than any alternatives.

1.3.3 Results

To illustrate the approach we consider a simple fMRI study of anxiety induction from [43] which was analyzed using the SMC approach described here in [48].

The experiment as described by [43]:

> The design was an off-on-off design, with an anxiety-provoking speech preparation task occurring between lower-anxiety resting

periods. Participants were informed that they were to be given two minutes to prepare a seven-minute speech, and that the topic would be revealed to them during scanning. They were told that after the scanning session, they would deliver the speech to a panel of expert judges, though there was "a small chance" that they would be randomly selected not to give the speech.

After the start of fMRI acquisition, participants viewed a fixation cross for 2 min (resting baseline). At the end of this period, participants viewed an instruction slide for 15 s that described the speech topic, which was to speak about "why you are a good friend". The slide instructed participants to be sure to prepare enough for the entire 7 min period. After 2 min of silent preparation, another instruction screen appeared (a "relief" instruction, 15 s duration) that informed participants that they would not have to give the speech. An additional 2 min period of resting baseline followed, which completed the functional run.

For brevity we consider the number of regimes to be known and equal to 2, but there would be little difficult in simultaneously estimating the number of hidden regimes concurrently. We utilize a MS-AR(r) model of the form suggested by [55] for application to (suitably detrended) fMRI data:

$$y_t = \mu_{x_t} + a_t \quad a_t = \phi_1 a_{t-1} + \ldots + \phi_r a_{t-r} + \epsilon_t \quad \epsilon_t \sim N(0, \sigma^2), \quad (1.6)$$

which allows an autoregressive dependence in addition to the latent variable structure.

Note that the a_t sequence is introduced only for notational convenience; the model can be written entirely in terms of the observation and latent state sequences, yielding for the emission density:

$$f(y_t | x_{1:t}, y_{1:t-1}, \theta) = \frac{1}{\sqrt{2\pi\sigma^2}} \exp\left(-\frac{1}{2\sigma^2}\left(a_t - \left(\sum_{j=1}^r \phi_j a_{t-j}\right)\right)\right) \quad (1.7)$$

$$= \frac{1}{\sqrt{2\pi\sigma^2}} \exp\left(-\frac{1}{2\sigma^2}((y_t - \mu_{x_t}) - \left(\sum_{j=1}^r \phi_j \left(y_{t-j} - \mu_{x_{t-j}}\right)\right)\right).$$

Notice that the simple HMM with no autoregressive component is recovered when $r = 0$. A regime change is deemed to have occurred when the latent state persists for five consecutive time periods (i.e. $k = 5$ in the notation of subsection 1.3.1). This corresponds to typical estimates of haemodynamic response lengths [28].

We consider two regions of the brain which are expected to experience significant activity during the course of this experiment: the rostral medial pre-frontal cortex (RMPFC), which is known to be associated with anxiety, and the visual cortex (VC), which is expected to show activity when the task-related instructions are seen, as indicated in [43].

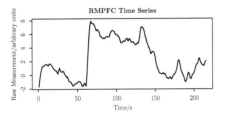
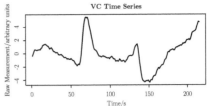

Unprocessed Data

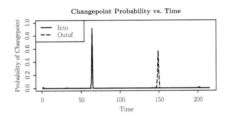
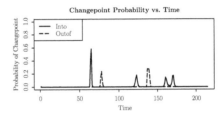

Changepoint probabilities computed with no detrending.

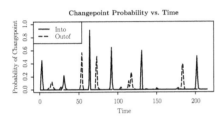
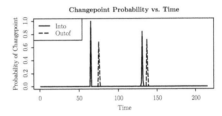

Changepoint probabilities computed with DCB detrending.

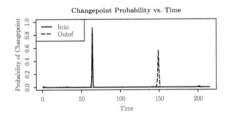
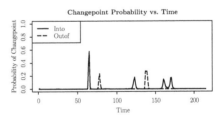

Changepoint probabilities computed with polynomial detrending.

FIGURE 1.1
Estimated changepoint probabilities for both regions using three different approaches to detrending and an independent noise model.

Changepoint probabilities are shown for each region using three different detrending schemes and an AR(0) (i.e., independent) noise model in Figure 1.1 and with an AR(1) noise model in Figure 1.2. In principle, detrending strategies and any other related preprocessing could also be automatically

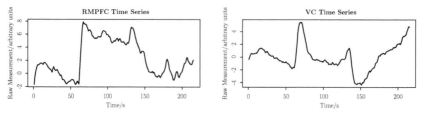

Unprocessed Data

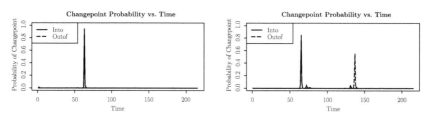

Changepoint probabilities computed with nodetrending.

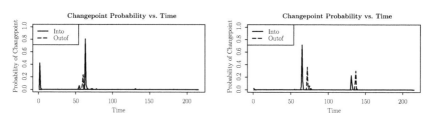

Changepoint probabilities computed with DCB detrending.

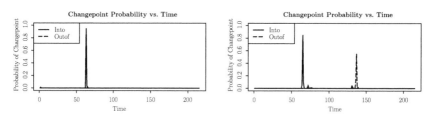

Changepoint probabilities computed with polynomial detrending.

FIGURE 1.2
Estimated changepoint probabilities for both regions using three different approaches to detrending and a first order autoregressive noise model.

selected within a Bayesian framework. However, prior specification in this case is non-trivial and due to confounding between the trend and possible change points, the best explanation of the data might not be a good strategy to consider when the preprocessing is potentially more flexible than the data model.

The detrending strategies are: to do no detrending and to analyze the raw data directly, to employ a low order (cubic) polynomial detrending [74] or to employ a discrete cosine basis (DCB) following [2]. These figures illustrate a number of things. They show that good estimation of the time of changes can be obtained *if* an appropriate detrending approach is considered and that the choice of detrending is appropriate. Change points present in time series can be confounded with slowly varying trends, and as such, care needs to be taken. The figures also show that results are strongly influenced by the choice of detrending procedure and, furthermore, that the use of an AR(1) noise model (with autoregressive parameter fixed at 0.20, as is the default behaviour of the SPM software [2]) can obscure much of the signal found in the data without this noise model.

It is clear that the findings of changepoint analysis in this setting are strongly influenced by preprocessing of the data and noise modelling and so it is extremely important to fully consider the implications of any modelling choices made and to assess the sensitivity of results to these choices when considering such an analysis. That said, the results obtained for the visual cortex with DCB detrending and for the RMPFC with no or polynomial detrending coincide very closely with neuro-physiological expectations illustrating the potential of even such simple models in this setting.

fMRI time series models are still routinely analyzed using known designs. However, hopefully, this chapter has shown that if care is taken in the choice of modelling parameters, it is feasible to determine the underlying design (along with associated uncertainty estimates) from the data rather than pre-specify it. While the analysis here has been undertaken only in a few regions, the increasing speed of algorithms, and the availability of massively parallel hardware offers the realistic possibility that real time change point detection in fMRI may be possible in the not too distant future.

1.4 Application: Positron Emission Tomography

PET is a widely used imaging technique which produces spatially-localised time course data for a very large number of voxels (typically of the order of 250,000). PET is invasive, requiring the introduction of a radionuclide-labelled tracer into the subject's bloodstream and so it is important to obtain the best possible inference from the limited quantity of data which is available.

1.4.1 Problem

There are a number of interesting problems to resolve when using PET to analyze neural activity. Here, we consider the estimation of a particular macro-scopic quantity of interest, the *volume of distribution*; this quantity essentially characterizes the quantity of tracer which would be found at a site in the brain in a steady-state regime obtained in the large time limit in a hypothetical experiment in which the tracer concentration in plasma is maintained at a constant level. In order to do this it is necessary to develop appropriate models for the process and, owing to the complexity of the problem at hand, to consider both parameter estimation and model selection/averaging.

1.4.2   Background

Compartmental models [35] are widely used in the modelling of PET. Such models describe the system of interest, in this case the dynamic behaviour of a radioactive tracer within a human brain, using a number of (real or conceptual) *compartments* between which a substance of interest, in this case the tracer, flows. Linear compartmental models are typically used in PET modelling for several reasons: the number of observations available is often very low and hence identifying more complex models can be difficult or impossible [62]; a single brain image typically contains at least 200,000 voxels and analysis needs to be conducted for every one of these voxels necessitating models for which inference can be performed extremely efficiently; and critically such models have been found to characterize PET data well [39].

Linear compartmental models can be conveniently characterised by a collection of linear differential equations: If we consider a linear m-compartment model and allow $f(t)$ to denote the vector whose i^{th} element corresponds to the concentration in the ith compartment at time t. Let $b(t)$ describe all flow into the system from outside. The ith element of $b(t)$ is the rate of inflow into the ith compartment from the environment. The dynamics of such a model may be written as:

$$\frac{\mathrm{d}f}{\mathrm{d}t}(s) = Af(s) + b(s), \qquad\qquad f(0) = \xi, \qquad\qquad (1.8)$$

where ξ is the vector of initial concentrations and the matrix A characterises the rate of flow between the various compartments (see [31]). The solution to this equation is,

$$f(t) = e^{At}\xi + \int_0^t e^{A(t-s)}b(s)\mathrm{d}s,$$

where the matrix exponential $e^{At} = \sum_{k=0}^{\infty} \frac{(At)^k}{k!}$.

It is possible to attach a loose qualitative interpretation to these equations, that there is an *inflow*, $b(t)$, into the system from outside and that material arriving in the system then flows between its compartments at rates depending upon A, and possibly also out of the system. This is a significant oversimplification in the case of the dynamics of a radioactive tracer entering the brain, and in some instances it's preferable to consider an alternative interpretation of the modelling. It's well known that compartmental models with sufficiently many compartments provide good approximation of essentially arbitrary dynamics. Indeed, in this context [46, p 200], observe that:

The linear combinations of exponential function forms have provided a very rich class of curves to fit to time-concentration data, and compartmental models turn out to be good approximations for many processes.

One can view the compartmental model as no more than a simple, pragmatic approximator of more complex dynamics — although some caution is needed as the small amount of data typically available severely limits the number of compartments which can be identified.

We consider only a particular class of compartmental models widely used for PET modelling, the *tracer input compartmental model*. In this case, it is assumed that in addition to the PET data, a measurement of the concentration of the tracer in plasma is available. The plasma measurement is generally treated as being noise-free as it can be made to a much higher degree of accuracy than the PET measurements themselves.

A plasma input model with m tissue compartments can be written as a set of differential equations,

$$\frac{\mathrm{d}C_T}{\mathrm{d}t}(s) = AC_T(s) + bC_P(s) \qquad C_T(0) = \mathbf{0} \qquad \bar{C}_T(s) = \mathbf{1}^T C_T(t),$$

where $C_T(t)$ is an m-vector of time-activity functions of each tissue compartment, $C_P(t)$ is the plasma time-activity function, i.e., the input function. A is the $m \times m$ state transition matrix, $b = (K_1, 0, \ldots, 0)^T$ is an m-vector, where K_1 is the rate constant of input from the plasma into tissue. The vectors $\mathbf{1}$ and $\mathbf{0}$ correspond to the m-vectors of ones and zeroes, respectively. The matrix A takes the form of a negative semidefinite matrix with non-positive diagonal elements and non-negative off-diagonal elements. The solution to this set of ordinary differential equations (ODEs) is:

$$\bar{C}_T(t) = \int_0^t C_P(t-s)H_{TP}(s)\mathrm{d}s \qquad H_{TP}(t) = \sum_{i=1}^m \phi_i e^{-\theta_i t}, \qquad (1.9)$$

where the ϕ_i and θ_i parameters are determined by the rate constants. The input function $C_P(t)$ is assumed to be nearly continuously measured. The tissue time-activity function $\bar{C}_T(t)$ is measured discretely, leading to measured values of the integral of the signal over each of n consecutive, non-overlapping time intervals ending at time points $t_1, \ldots, t_n$.

The object of principal inferential interest, the aforementioned volume of distribution, is then:

$$V_D := \int_0^\infty H_{TP}(t)\mathrm{d}t = \sum_{i=1}^m \frac{\phi_i}{\theta_i}.$$

In the next section a number of recent Bayesian approaches to estimation of the volume of distribution, together with the characterization of uncertainty, are described.

1.4.3   Modelling & Methodology

The traditional likelihood-based approaches to this problem are difficult to extend to allow for robust noise modelling and unknown model order.

A Bayesian approach to the estimation of the parameters of compartmental models of known model order was proposed in [56], who also provided a comparison with the state of the art in this area.

1.4.3.1   Modelling

In [76], a number of simple modelling improvements were made: considering non-normal noise distributions (much heavier tailed noise is strongly supported by the real data considered below) and developing biologically-motivated prior distributions for the parameters of interest. However, the main focus of that work was extending the approach of [56] to allow for the estimation of model evidence and hence to permit model selection or averaging. The strategy was based around MCMC together with a *careful* implementation of a generalised harmonic mean estimator.

Both [56] and subsequently [76] made use of a model of the form:

$$\bar{C}_T(t_j; \phi_{1:m}, \theta_{1:m}) = \sum_{i=1}^{m} \phi_i \int_0^{t_j} C_P(s) e^{-\theta_i(t_j - s)} ds$$

$$y_j = \bar{C}_T(t_j; \phi_{1:m}, \theta_{1:m}) + \sqrt{\frac{\bar{C}_T(t_j; \phi_{1:m}, \theta_{1:m})}{t_j - t_{j-1}}} \varepsilon_j,$$

where m denotes the number of tissue compartments (1, 2 or 3 in this work), $t_0 = 0$, and $\{\varepsilon_j\}$ are identically independently distributed random variables with mean zero.

In most works it has been assumed that the innovations, ε_j are normally distributed; [76] compares this assumption with one in which t distributions with more than two degrees of freedom are permitted and found that real PET data provided overwhelming support for a heavy-tailed noise distribution and so only that model is considered here. That is, the innovation terms are modelled using zero-mean t random variables, $\varepsilon_j \sim \mathcal{T}(0, \tau, \nu)$, where $\mathcal{T}(0, \tau, \nu)$ denotes the centered Student t distribution with scale τ, and ν degrees of freedom.

The prior distribution for scale τ is an inverse gamma distribution with both parameters equal to 0.001 — a proper approximation to the improper Jeffrey's prior — while that for $1/\nu$ is uniform over interval $[0, 0.5)$, allowing the tails to the likelihood to vary between those of the normal and the much heavier tails of a t distribution with 2 degrees of freedom (following [24]).

Prior distributions for the rate parameters of the three compartment model were obtained from biophysical considerations by [76] (who also considered vague conjugate priors and demonstrated that parameter estimation and model selection results were somewhat insensitive to the precise choices of prior distribution):

$$\phi_1 \sim \mathcal{TN}_{[10^{-5},10^{-2}]}\left(\cdot; 3 \times 10^{-3}, 10^{-3}\right)$$
$$\theta_1|\phi_1 \sim \mathcal{TN}_{[2 \times 10^{-4},10^{-2}]}\left(\cdot; \phi_1/15, 10^{-2}\right)$$
$$\phi_2 \sim \mathcal{TN}_{[10^{-5},10^{-2}]}\left(\cdot; 10^{-3}, 10^{-3}\right)$$
$$\theta_2|\phi_2,\theta_1 \sim \mathcal{TN}_{[\theta_1,6 \times 10^{-2}]}\left(\cdot; \phi_2/4, 10^{-2}\right)$$
$$\phi_3 \sim \mathcal{TN}_{[10^{-5},10^{-2}]}\left(\cdot; 10^{-3}, 10^{-3}\right)$$
$$\theta_1|\phi_3,\theta_2 \sim \mathcal{TN}_{[\theta_2,6 \times 10^{-2}]}\left(\cdot; \phi_3/1, 10^{-2}\right).$$

where the density of the truncated normal distribution is

$$\mathcal{TN}_{[a,b]}\left(x; \mu, \sigma^2\right) := \frac{\mathcal{N}(x; \mu, \sigma^2)\mathbb{I}_{[a,b]}(x)}{\Phi\left(\frac{b-\mu}{\sigma}\right) - \Phi\left(\frac{a-\mu}{\sigma}\right)},$$

with $\mathbb{I}_{[a,b]}$ corresponding to the indicator function on $[a,b]$ and Φ to the standard normal distribution function. For one- and two-compartment models the appropriate subset of these prior distributions are used, ensuring that common priors are used for the shared parameters of nested models.

1.4.3.2 Methodology

The generalised harmonic mean estimator (due to [23]; extending the harmonic mean estimator of [51]) is based around the simple identity, which holds for any probability density, $g(\theta)$:

$$\int_{\theta \in \Theta} g(\theta) \frac{p(\theta|y)p(y)}{p(y|\theta)p(\theta)} d\theta = 1 \Rightarrow p(y) = \left[\int_{\theta \in \Theta} p(\theta|y) \frac{g(\theta)}{p(y|\theta)p(\theta)} d\theta\right]^{-1}.$$

In principle, writing the quantity of interest as the reciprocal of a posterior expectation in this way allows the estimation of $p(y)$ using any sample approximation of the posterior, $\{\theta^i\}_{i=1}^N$ by simply computing:

$$\hat{p}(y) = \left[\frac{1}{N}\sum_{i=1}^N \frac{g(\theta^i)}{p(y|\theta^i)p(\theta^i)}\right]^{-1}.$$

However, this estimator is essentially an importance sampling estimator and will be well-behaved (and have finite variance) only if the tails of g are light compared with those of the true (and typically unknown) posterior density. It will have a small variance only if the dissimilarity between g and the posterior is small. This issue has been the subject of much discussion in the literature [59]. In this application it *is* possible to do this robustly but doing so comes at the cost of both a considerable implementation effort and the use of an adaptive scheme to tune the parameters of the MCMC algorithm as in [76].

    The potential of SMC to provide efficient characterization of the posterior distribution over both model and parameter space in the PET setting was noted by [77]; the approach proposed there corresponds essentially to Algorithm 2 of [78], with Algorithm 4 of that document describing a more sophisticated, adaptive variant of the same general approach which requires little user intervention. This simple algorithm is referred to as SMC2-DS in these papers (the 2 denoting which of three possible approaches to the problem is employed and the DS that the evidence is estimated using Equation 1.5 rather than via an approximation of the path-sampling estimator of [25] which is also possible).

Algorithm 1.3 An SMC approach to evidence estimation for PET compartmental models.

---

for each number of compartments, $k \in \{1, \ldots, k_{\mathbf{max}}\}$**:**

Initialisation: Set $n \leftarrow 0$.

Sample $\theta_0^{(k,i)}, \ldots, \theta_0^{(k,N)} \overset{\text{iid}}{\sim} p(\theta | M_k)$ from the parameter prior.

Set Weight $\{W_0^{(k,i)} = 1/N\}_{i=1}^{N}$.

Iteration: Set $n \leftarrow n + 1$.

for $i = 1, \ldots, N$: Weight $W_n^{(k,i)} \propto W_{n-1}^{(k,i)} p(\mathbf{y} | \theta_{n-1}^{(k,i)}, M_k)^{\gamma_n^k - \gamma_{n-1}^k}$.

Apply resampling if necessary.

for $i = 1, \ldots, N$: Sample $\theta_n^{(k,i)} \sim K_n(\cdot | \theta_{n-1}^{(k,i)})$, a $\pi_{k,n}$-invariant kernel.

Repeat the *Iteration* step *until* $n = n_{max}^k$.

end for

Use Equation 1.5 to estimate $p(\mathbf{y} | M_k)$.

To avoid dwelling unnecessarily on technical details, Algorithm 1.3 shows the simplest form of the algorithm. Here, n_{max}^k describes the number of distributions in the sequence used to deal with model k, which we denote M_k and which corresponds to the k-tissue compartment model in our setting. Tuning parameter γ_n^k corresponds to the nth step in the tempering sequence for model k, i.e., $\pi_{k,n}(\theta) \propto p(\theta | M_k) p(\mathbf{y} | \theta, M_k)^{\gamma_k^n}$. Note that the sequence of distributions employed and the scaling of the proposal distributions can be automated by using the adaptive techniques described in [78] and that this is a major strength of the approach in problems such as that considered here in which it needs to be applied to hundreds of thousands of data series. The adaptation means that a common algorithm can be applied to *all* of these series without tuning on a case-by-case basis and it can then be robust enough to produce good estimation for all of those sequences. Indeed, it's not necessary to specify even the number of intermediate distributions used in each sequence as the adaptive scheme is able to select this (and hence the computational effort) with only one user-specified precision parameter being set. Another significant benefit of the SMC approach is that it can be readily parallelized and can take advantage of modern parallel architectures, see [41] for some illustrations, even those which depend upon SIMD (Same Instruction Multiple Data) operations (in the context of model selection this is studied in some depth by [78]).

1.4.4 Example

We consider data obtained from a PET study using [^{11}C]diprenorphine. The original intention of the study was to quantify opioid receptor concentration in the brain of healthy subjects allowing a baseline to be found for subsequent studies on diseases such as epilepsy which tend to involve changes in brain receptor concentrations or occupancy levels either due to physical lesions within the brain or other chemically relevant differences from controls.

These data have been previously analysed in [36, 56], who focussed on the parameter estimation problem, and by [77] who applied MCMC and generalised harmonic mean estimation to the problem. Another data set from the same study was considered by [78]. Dynamic scans from an [^{11}C]diprenorphine study of healthy subjects, for which an arterial input function was available, were analyzed. [^{11}C]diprenorphine is a tracer that binds to the opioid (pain) receptor system in the brain. The subjects underwent 95 min dynamic [^{11}C]diprenorphine PET baseline scans. Each patient was injected with [^{11}C]diprenorphine with a total activity of 185 MBq. PET scans were acquired in 3D mode on a Siemens/CTI ECAT EXACT3D PET camera, with a linear spatial resolution after image reconstruction of approximately 5 mm. Data reconstruction employed the reprojection algorithm [38] with ramp and Colsher filters cut off at the Nyquist frequency. Reconstructed voxels were 2.096 mm × 2.096mm × 2.43mm. Acquisition was performed in listmode (event-by-event) and scans were rebinned into 32 time frames of increasing duration. Frame-by-frame movement correction was performed on the PET images. Overall this resulted in images of size 128 × 128 × 95 voxels which, when masked to include only brain regions, resulted in ∼ 250,000 separate time series to be analyzed. Each of these time series is analyzed separately within the Bayesian analysis considered here.

We provide here a short summary of the results of this approach while referring the reader to [76] for a comparison of Bayesian and alternative approaches to estimation and to [78] for further details of the sequential Monte Carlo approach and results of the analysis of another data set from the same study.

Figure 1.3 shows three orthogonal slices through the brain, each illustrating the posterior modal model order m obtained for each voxel using the adaptive SMC scheme described. The spatial distribution of the model order can give a crude indication of the types of biological processes occurring at different spatial locations. Figure 1.4 shows the volume of distribution estimated under the most probable model at the same locations. In real applications, it might be more appropriate to consider model averaging, rather than selection, for the purposes of computing parameters of interest; the selection approach is employed here only to simplify exposition.

Both figures show considerable spatial structure: this is not an artefact of the modelling as each voxel was treated independently, but reflects information present within the data itself. There is a non-trivial relationship here

Model order

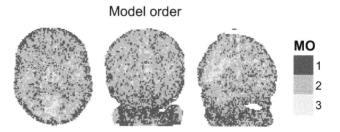

FIGURE 1.3
Modal model order shown as a function of space.

between $\mathbb{E}[V_D]$ and model order. Prior distributions are typically given for micro parameters such as those in the A matrix of (1.8), while the parameter of interest is typically V_D. However, priors on such parameters are much easier to interpret and are therefore the standard in Bayesian PET models. In addition, such specifications are unlikely to cause major issues in interpreting the posterior of V_D. We note, in passing, that similar structure is not readily apparent when traditional likelihood and information-criteria-based methods are employed [76]. Figure 1.4 shows clearly the physical structure of the brain and, perhaps more interestingly, provides a clear indication that the tracer is much more highly concentrated within certain regions of the brain (as would be expected physiologically).

The analysis in this chapter has demonstrated that massive data, in this case hundreds of thousands of voxels, can be analyzed using advanced Monte Carlo techniques. As imaging modalities such as PET become increasingly important in a whole range of clinical applications from cancer studies to psychiatric disorders, robust, automated, computationally tractable Bayesian approaches have the potential to make large contributions to the analysis of such studies.

Volume of distribution

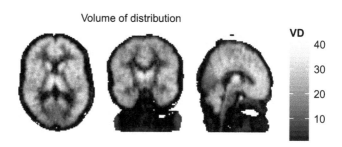

FIGURE 1.4
Posterior mean volume of distribution illustrated as a function of space.

1.5 Application: Magnetoencephalography

Magnetoencephalography enjoys extremely good ($\sim$ 10 ms) temporal resolution and is often used in settings in which the temporal evolution of signals is of interest. The problem of determining the underlying pattern of neural currents from the collection of measured magnetic field strengths is ill-posed and regularized solution of this inverse problem has attracted a great deal of attention.

1.5.1 Problem

This section addresses the specific problem of estimating, online as observations become available, the posterior distribution over current neural currents given the observations received thus far. The approach involves the use of sequential Monte Carlo algorithms within a simple object tracking framework which attempts to make use of a model form the dynamics of the neural currents in order to provide the needed spatio-temporal regularization. Although the methodology discussed has not yet been implemented in real time, doing so is certainly within the range of current computational hardware.

1.5.2 Background

Numerous approaches to the problem of estimating the configuration of neural currents from MEG data have been developed. Broadly speaking, each method uses one of two classes of approximation of the neural currents: *distributed source* models in which currents are treated as varying smoothly throughout the brain or *current dipole* models in which a small number of infinitesimal current dipoles (the closest thing to a point-source of magnetic field which exists) are used to model the current distribution. Both approaches have advantages and disadvantages. See [73] for a detailed discussion.

The simplest approaches produce a single estimate of the *spatial location* of activity averaging over time. Common approaches include Minimum Norm Estimation (MNE; [33], a natural L2-regularized approach) and Minimum Current Estimation (MCE; [72], an L1-regularized approached). These regularized approaches invariably fall into the class of distributed source models.

More recently, approaches which treat the problem explicitly as a dynamic one have been developed. Spatiotemporal regularization approaches attempt to exploit the qualitative observation that activity is expected to be spatially sparse but temporally smooth, typically employing two different norms to provide appropriate regularization in each domain. Works taking such an approach include [54, 68].

Inference under a Bayesian dynamic model which fully describes the evolution of the system has become possible only with the availability of powerful computers. [44] address exact computation of posterior densities under linear Gaussian modelling assumptions imposed upon a (discretised) distributed source model, but such an approach has extremely high computational requirements in addition to substantial modelling restrictions.

A number of papers have considered Monte Carlo approximation of the sequence of filtering distributions using current dipole models (i.e. the distributions over neural current configuration at each time instant given all of the observations received up until that time) using SMC. These include [63, 65] which propose basic particle filters and [6] which attempted to implement an approximation of a so-called Rao-Blackwellized particle filter [1, 8] in which a combination of analytical and Monte Carlo methods are combined by integrating out some unknown variables conditional upon simulated values of the others. In general, this is an excellent idea and one which can significantly reduce the Monte Carlo variance of the resulting estimator; unfortunately, in this case approximations were required to arrive at such an algorithm and these approximations are such that the resulting estimates would not recover the true posterior distribution as the Monte Carlo sample size tends to infinity. All these papers use a model in which current dipoles move around the brain according to a random walk; this is not consistent with the biophysical interpretation of these current dipoles but is required for computational reasons.

Recently, [64] proposed an algorithm which allows for consistent Monte Carlo approximation of the filtering distributions associated with a model in which current dipoles do not move. This approach is discussed in more detail in the following section.

1.5.3 Modelling & Methodology

Multiple object tracking is a difficult problem and has attracted a great deal of attention in the literature ([58] identifies 35 different *types* of multiple object tracking algorithm). The problem within the current dipole model of MEG image reconstruction is especially challenging due to the characteristics of the underlying dynamics.

1.5.3.1 Modelling

As with most tracking problems, the modelling is naturally decomposed into two components: one which describes the underlying dynamics of the hidden system being tracked (in this case, these describe the evolution of the current dipoles within the brain) and a second which describes the relationship between that hidden system and the measurement apparatus. This section commences with a description of the measurement model before returning to a discussion of the dynamic model.

In principle, the magnetic field arising from any latent neural current configuration can be computed directly via Maxwell's equations (or in a steady state regime via the Biot-Savart law). However, doing so can be computationally costly and it is common practice to discretize the brain volume and to compute the influence of three orthogonal current vectors at the centre of each of the resulting N_{grid} volume elements offline in advance and stored in the *leadfield* matrix, G. The forward problem associated with any configuration of current elements located at the centre of those volume elements can then be solved (under the magnetostatic approximation, in which the influence of time varying currents is neglected) by a simple linear superposition.

Such an approximation gives rise to the following measurement model (note that throughout this section parenthetical superscripts index individual dipoles within a collection):

$$b_t = \sum_{i=1}^{N_t} G(r_t^{(i)})q_t^{(i)} + \epsilon_t \qquad (1.10)$$

where b_t is a vector of magnetic currents measured at N_{sensors} detectors at time t, N_t is the number of active dipoles at time t, $q_t^{(i)}$ denotes the ith dipole moment at time t, $r_t^{(i)}$ is the location of dipole i at time t relative to each detector element, $G(r_t^{(i)})$ is the $N_{\text{sensors}} \times 3N_{\text{grid}}$ subset of the leadfield matrix describing the influence of a dipole at location $r_t^{(i)}$ on each sensor and ϵ_t is an additive noise vector.

Assuming that the ϵ_t are independent Gaussian random vectors with covariance Σ_{noise} leads to the likelihood:

$$p(b_t|j_t) = \mathcal{N}\left(b_t; \sum_{i=1}^{N_t} G(r_t^{(i)})q_t^{(i)}, \Sigma_{\text{noise}}\right), \qquad (1.11)$$

where j_t characterizes the underlying population of neural currents and comprises the number, location, strength and direction of current dipoles (i.e. $j_t = (N_t, r_t^{(1:N_t)}, q_t^{(1:N_t)}))$.

Two approaches to modelling the dynamics of the underlying system of current dipoles have been considered in the literature. In both cases, the dynamics are treated as Markovian and the transition kernel is written as a weighted mixture of three components: birth in which an additional dipole becomes active, death in which an existing dipole becomes inactive and a more neutral component in which no dipoles appear or disappear. The difference between the two approaches lies essentially within this third, neutral component.

Given the interpretation of current dipoles as the activity of a fixed neural population, it seems natural to allow only the dipole moment of an existing dipole to change at any given time instant and to keep the location fixed. However, doing this introduces a substantial degree of degeneracy into the transition kernel and such degeneracy can dramatically impede the performance of SMC methods in a filtering setting. Most work on this subject has, therefore, allowed the locations as well as the dipole moments of existing dipoles to evolve according to random walk dynamics (such approaches date back at least as far as [6]); we term this model the *random walk model*. More recent work has demonstrated that it is possible, with care, to implement particle filters for a model in which dipole locations are not permitted to move over time; we term this the *static model*.

1.5.3.2 Methodology

The approach developed in [64] makes use of a number of techniques to obtain substantially better performance than a simple particle filter. These changes are essential to obtain adequate performance with the static model. In particular, it attempts to approximate the locally optimal proposal distribution (cf. [16]), and it employs MCMC steps to improve sample diversity and to allow, to some extent, information provided by the most recent observations to influence estimates associated with the state a few time steps back (unlike a simple SIR algorithm).

In order to avoid entering into a detailed discussion of the technical issues involved in this algorithm, only a summary of the key features of the approach is provided here, noting that full details and a more complete explanation are provided in [64].

As the algorithm is performing particle filtering on a moderately high dimensional space, it is critical to employ as good a proposal distribution as possible in order to obtain importance weights which are as stable as possible (just as the quality of a simple importance-sampling estimator is determined by the quality of the proposal, the performance of a sequential Monte Carlo algorithm can be very heavily influenced by the choice of proposal distribution). In the MEG context, the underlying dynamics can be decomposed as a mixture of three components (birth, death, or steady number of dipoles) and it is natural to employ a proposal of the same form. An approach which [64] found gave a reasonable compromise between proposal quality, computational cost and implementation effort was to employ a mixture distribution. The components of this mixture being the system dynamics for the neutral component, a birth component in which the new dipole is most likely to appear at regions of the brain in which a simple Tychonoff regularized inverse takes large values and a death component which is most likely to eliminate those dipoles which make the smallest contribution to explaining the observed data.

The use of MCMC steps within SMC was first proposed by [27] who gave the approach the name "resample-move" as the original formulation depended upon only those steps. In the MEG context, the use of MCMC moves on the temporal path space allows for data about the location of dipoles to be incorporated retrospectively into estimates. By incorporating such updates into the algorithm, it is (loosely speaking) possible for it to change the estimated location of a dipole *not* by allowing that dipole to move but by updating its estimate to reflect that it was *always* at a different location from that previously estimated. The particular construction used ensures that it remains a consistent (in the number of samples used within the SMC algorithm) estimator of the posterior density associated with the static model. Algorithm 1.4 shows the approach in outline.

Algorithm 1.4 Outline of the Resample-Move algorithm

Sample $J_{0,0}^1, \ldots, J_{0,0}^N \overset{\text{iid}}{\sim} p(j_0)$.

for $i = 1, \ldots, N$: Set $W_0^i = 1/N$.

for $n = 1, \ldots, n_{\mathbf{max}}$:

  for $i = 1, \ldots, N$:

    Set $\tilde{J}_{n,0:n-1}^i = J_{0:n-1}^i$ and sample $\tilde{J}_{n,n}^i$ from $q(J_n|\tilde{J}_{n,0:n-1}^i, b_t)$,

    Compute $\tilde{W}_n^i = \frac{p(b_n|\tilde{J}_{n,n}^i)p(\tilde{J}_{n,n}^i|\tilde{J}_{n,n-1}^i)}{q(\tilde{J}_{n,n}^i|\tilde{J}_{n,n-1}^i, b_n)}$

  end for

  for $i = 1, \ldots, N$: $W_n^i = \tilde{W}_n^i / \sum_j \tilde{W}_n^i$

  Resample $\{\tilde{J}_{n,0:n}^i, W_n^i\}_{i=1}^N$ to obtain $\{\bar{J}_{n,1:n}^i, 1/N\}_{i=1}^n$.

  for $i = 1, \ldots, N$: Sample $J_{n,0:n}^i \sim K_n(\bar{J}_{n,0:n}^i, \cdot)$, a $p(j_{0:n}|b_{1:n})$-invariant Metropolis-Hastings kernel.

end for

1.5.4 Example

This section illustrates the relative behaviour of the static and dynamic dipole models within a particle filtering framework as well as the dynamic Statistical Parametric Mapping (dSPM) method of [11]. dSPM is a well-known method in which Tychonoff regularization is applied independently at each time point, and the so-obtained estimate of the electrical current distribution is standardized by dividing by a location-dependent estimate of the noise variance.

Data from [64], which describes the data acquisition in detail:

> This data comprised measures from a Somatosensory Evoked Field (SEF) mapping experiment. The recordings were performed after informed consent was obtained, and had prior approval by the local ethics committee. Data were acquired with a 306-channel MEG device (Elekta Neuromag Oy, Helsinki, Finland) comprising 204 planar gradiometers and 102 magnetometers in a helmet-shaped array. The left median nerve at wrist was electrically stimulated at the motor threshold with an interstimulus interval randomly varying between 7.0 and 9.0 s. The MEG signals were filtered to 0.1-200 Hz and sampled at 1000 Hz. Electrooculogram (EOG) was used to monitor eye movements that might produce artifacts in the MEG recordings; trials with EOG or MEG exceeding 150 mV or 3 pT/cm, respectively, were excluded and 84 clean trials were averaged. To reduce external interference, signal space separation method [67] was applied to the average. A 3D

digitizer and four head position indicator coils were employed to determine the position of the subject's head within the MEG helmet with respect to anatomical MRIs obtained with a 3-Tesla MRI device (General Electric, Milwaukee, USA).

were used to conduct this comparison. The original work features a more comprehensive comparison.

Figure 1.5 shows estimates of the activity at various points in the time series obtained using three different methods (see [64] for further details of the precise estimators employed and a wider comparison). From this sequence of

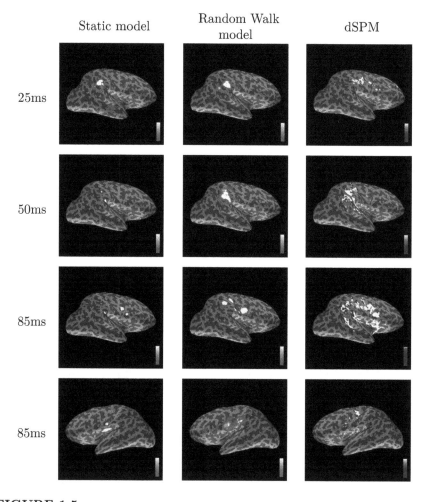

FIGURE 1.5
SEF data. Static Model (left), Random Walk Model (centre) and dSPM estimate (right). Data presented on an inflated brain representation map.

images one can discern some significant features of the methods: both methods based upon the filtering of small numbers of latent current dipoles produced more concentrated estimates of activity location and, in particular, the static model leads to more concentrated estimates than the random walk model (this is not due to the particle system becoming degenerate but to the accumulation of information over time).

It is not possible, of course, to determine how the estimated dynamics of the system behave by looking at these static images. In fact, a major advantage of the static model is that using it eliminates from the estimation non-physical artefacts which are obtained when using the random walk model. In particular, whenever one neural population ceases to be active and another begins at a similar time there are two possible explanations which are compatible with the random walk model: that one dipole has become inactive and another active at similar times *or* that the dipole has *moved* across the brain volume. Only the first of these explanations is biophysically plausible given the interpretation of these current dipoles and only this explanation is compatible with the static model. As such estimated time series obtained from the static model admit much simpler interpretations.

The ability to analyze and produce more scientifically plausible models can only lead to further understanding of the underlying mechanisms of neuronal behaviour. Future research in this area will undoubtedly involve combining complementary imaging modalities, and models such as those proposed above will be needed in order to provide understanding of possible mechanisms to allow this combination.

1.6 Conclusions

Bayesian methodology is at the forefront of neuroimaging analysis and methods, and the recent advances in computational algorithms for high dimensional data can only further enhance its role. This article has outlined only a tiny fraction of the possible applications of Bayesian techniques in Neuroscience. However, they have each demonstrated possible solutions to perceived difficulties in Bayesian modelling. It has been shown that models need not deal with stationary data with known designs, but can, by careful modelling, identify the underlying changes in the system from the data. In addition, the computational cost of Bayesian techniques does not inherently prohibit their usage in even truly massive datasets. Finally, it has been shown that models can be found that are both computationally tractable but which also allow an inherent interpretability to their results.

While in the examples here, the approaches outlined have been based on SMC, it is not the intention of this chapter to imply that this is the only computational methodology suitable for the Bayesian analysis of brain imaging data. Indeed, almost all Bayesian computational approaches currently available have been used successfully in the analysis of neuroimaging data in different contexts. These range from using modelling frameworks which admit analytic solutions [69], to Variational Bayes approaches [20] to MCMC over the whole brain [19], with these three examples being a tiny proportion of the vast literature. Much more important than the particular choice of computation technique is that the model is suitable for the inference problem at hand. This is especially true in the current era of parallel implementations of many computational approaches [41]. However, SMC has been seen above to be a useful approach in a number of situations.

It is hoped that this article has demonstrated some of the power of Bayesian statistics to develop good solutions to real problems in the world of neuroscience and to do so with reasonable implementation effort in a timely fashion. The approaches described in this article are relatively recent and it is clear that there is substantial space for further exploration, particularly with regard to the incorporation of spatial information in the models considered (possibly in similar ways to those specified in structural MRI [19]). In addition, Bayesian methodology even in situations where the likelihood is difficult to specify is a growing area of research with direct application in Neuroscience [5]. It is our view that with the availability of fast computers and the methodology of modern computational statistics, the field of neuroimaging is amenable to substantial methodological advances over the course of the coming years and it is probable that Bayesian methods will be fundamentally involved in many of these advances.

Acknowledgments

Much of the content of this article is based upon several pieces of collaborative research which involved the authors who duly acknowledge the contributions of their collaborators, particularly Dr Christopher F.H. Nam, Dr Alberto Sorrentino (supported by *NIMBLE*, a Marie Curie Intra European Fellowship within the 7th European Community Framework Programme) and Dr. Yan Zhou. Figures 1.1 and 1.2 were adapted from the work of Christopher Nam. Figures 1.3 and 1.4 were produced by Yan Zhou, and Figure 1.5 by Alberto Sorrentino. We are also very grateful for the provision of data from Prof Martin Lindqist (fMRI), Prof Alexander Hammers (PET) and Drs Parkkonen, Pascarella and Sorrentino (MEG). Part of the research summarised here was partially supported by Engineering and Physical Sciences Research Council Grants EP/H016856/1 and EP/I017984/1 as well as the EPSRC/HEFCE CRiSM Grant. Finally, this article was written while both authors were visiting fellows at the Isaac Newton Institute.

Bibliography

[1] C. Andrieu and A. Doucet. Particle filtering for partially observed Gaussian state space models. *Journal of the Royal Statistical Society: Series B (Statistical Methodology)*, 64(4):827–836, 2002.

[2] J. Ashburner, K. Friston, A. P. Holmes, and J. B. Poline. *Statistical Parametric Mapping*. Wellcome Department of Cognitive Neurology, website (http://www.fil.ion.ucl.ac.uk/spm), spm2 edition, 1999.

[3] J. A. D. Aston and D. E. K. Martin. Distributions associated with general runs and patterns in hidden Markov models. *Annals of Applied Statistics*, 1(2), 2007.

[4] J. A. D. Aston, J.-Y. Peng, and D. E. K. Martin. Implied distributions in multiple change point problems. *Statistics and Computing*, 22(4), 2012.

[5] S. Barthelmé and N. Chopin. Expectation propagation for likelihood-free inference. *Journal of the American Statistical Association*, 109(505):315–333, 2014.

[6] C. Campi, A. Pascarella, A. Sorrentino, and M. Piana. A Rao-Blackwellized particle filter for magnetoencephalography. *Inverse Problems*, 24:025023, 2008.

[7] O. Cappé, E. Moulines, and T. Ryden. *Inference in Hidden Markov Models*. Springer Verlag, New York, 2005.

[8] R. Chen and J. S. Liu. Mixture Kalman filters. *Journal of the Royal Statistical Society: Series B (Statistical Methodology)*, 62(3):493–508, 2000.

[9] N. Chopin. A sequential particle filter method for static models. *Biometrika*, 89(3):539–551, 2002.

[10] D. Cohen and B. N. Cuffin. Demonstration of useful differences between magnetoencephalogram and electroencephalogram. *Electroencephalography and Clinical Neurophysiology*, 56:38–51, 1983.

[11] A. Dale, A. K. Liu, B. R. Fischl, R. L. Buckner, J. W. Belliveau, J. D. Lewine, and E. Halgren. Dynamic statistical parametric mapping: Combining fMRI and MEG for high-resolution imaging of cortical activity. *Neuron*, 26:55–67, 2000.

[12] P. Del Moral. Nonlinear filtering using random particles. *Theory of Probability and Its Applications*, 40(4):690–701, 1995.

[13] P. Del Moral. *Feynman-Kac Formulae: Genealogical and Interacting Particle Systems with Applications*. Probability and Its Applications. Springer Verlag, New York, 2004.

[14] P. Del Moral, A. Doucet, and A. Jasra. Sequential Monte Carlo samplers. *Journal of the Royal Statistical Society: Series B (Statistical Methodology)*, 63(3):411–436, 2006.

[15] R. Douc, O. Cappé, and E. Moulines. Comparison of resampling schemes for particle filtering. In *Proceedings of the 4th International Symposium on Image and Signal Processing and Analysis*, volume I, pages 64–69. IEEE, University of Zagreb, Croatia, 2005.

[16] A. Doucet, S. Godsill, and C. Andrieu. On sequential Monte Carlo sampling methods for Bayesian filtering. *Statistics and Computing*, 10(3):197–208, 2000.

[17] A. Doucet and A. M. Johansen. A tutorial on particle filtering and smoothing: Fiteen years later. In D. Crisan and B. Rozovsky, editors, *The Oxford Handbook of Nonlinear Filtering*, pages 656–704. Oxford University Press, Oxford, 2011.

[18] R. R. Edelman and S. Warach. Magnetic resonance imaging. *New England Journal of Medicine*, 328(10):708–716, 1993.

[19] D. Feng, L. Tierney, and V. Magnotta. MRI tissue classification using high-resolution Bayesian hidden Markov normal mixture models. *Journal of the American Statistical Association*, 107(497), 2012.

[20] G. Flandin and W. D. Penny. Bayesian fMRI data analysis with sparse spatial basis function priors. *NeuroImage*, 34(3):1108–1125, 2007.

[21] K. Friston, C. Chu, J. Mourão-Miranda, O. Hulme, G. Rees, W. D. Penny, and J. Ashburner. Bayesian decoding of brain images. *NeuroImage*, 39(1):181–205, 2008.

[22] J. C. Fu and M. V. Koutras. Distribution theory of runs: A Markov chain approach. *Journal of the American Statistical Association*, 89(427):1050–1058, 1994.

[23] A. E. Gelfand and D. K. Dey. Bayesian model choice: Asymptotics and exact calculations. *Journal of the Royal Statistical Society: Series B (Statistical Methodology)*, 56:501–514, 1994.

[24] A. Gelman, J. B. Carlin, H. S. Stern, D. B. Dunson, A. Vehtari, and D. B. Rubin. *Bayesian Data Analysis*. CRC Press, London, 3rd edition, 2013.

[25] A. Gelman and X.-L. Meng. Simulating normalizing constants: From importance sampling to bridge sampling to path sampling. *Statistical Science*, 13(2):163–185, 1998.

[26] J. Geweke. Bayesian inference in econometrics models using Monte Carlo integration. *Econometrica*, 57(6):1317–1339, 1989.

[27] W. R. Gilks and C. Berzuini. Following a moving target – Monte Carlo inference for dynamic Bayesian models. *Journal of the Royal Statistical Society: Series B (Statistical Methodology)*, 63(1):127–146, 2001.

[28] G. H. Glover. Deconvolution of impulse response in event-related BOLD fMRI. *NeuroImage*, 9(4):416–429, 1999.

[29] G. Goertzel. Quota sampling and importance functions in stochastic solution of particle problems. Technical Report, Oak Ridge National Laboratory, USA, 1949.

[30] N. J. Gordon, D. J. Salmond, and A. F. M. Smith. Novel approach to nonlinear/non Gaussian Bayesian state estimation. *IEE Proceedings F-Radar and Signal Processing*, 140(2):107–113, Apr. 1993.

[31] R. N. Gunn, S. R. Gunn, and V. J. Cunningham. Positron emission tomography compartmental models. *Journal of Cerebral Blood Flow & Metabolism*, 21(6):635–52, 2001.

[32] M. Hämäläinen, R. Hari, J. Knuutila, and O. V. Lounasmaa. Magnetoencephalography: Theory, instrumentation and applications to non-invasive studies of the working human brain. *Reviews of Modern Physics*, 65:413–498, 1993.

[33] M. Hämäläinen and R. J. Ilmoniemi. Interpreting magnetic fields of the brain: Minimum norm estimates. *Medical & Biological Engineering & Computing*, 32:35–42, 1994.

[34] S. A. Huettel, A. W. Song, and G. McCarthy. *Functional Magnetic Resonance Imaging*. Sinauer Associates, USA, 2nd edition, 2009.

[35] J. A. Jacquez. *Compartmental Analysis in Biology and Medicine*. University of Michigan Press, 3rd edition, 1996.

[36] C.-R. Jiang, J. A. D. Aston, and J.-L. Wang. Smoothing dynamic positron emission tomography time courses using functional principal components. *NeuroImage*, 47(1):184–93, 2009.

[37] A. M. Johansen. SMCTC: Sequential Monte Carlo in C++. *Journal of Statistical Software*, 30(6):1–41, 2009.

[38] P. E. Kinahan and J. G. Rogers. Analytic 3D image reconstruction using all detected events. *IEEE Transactions on Nuclear Science*, 36(1):964–968, 1989.

[39] A. A. Lammertsma and S. P. Hume. Simplified reference tissue model for PET receptor studies. *NeuroImage*, 4(3):153–158, 1996.

[40] D. Le Bihan. Looking into the functional architecture of the brain with diffusion MRI. *Nature Reviews Neuroscience*, 4(6):469–480, 2003.

[41] A. Lee, C. Yau, M. B. Giles, A. Doucet, and C. C. Holmes. On the utility of graphics cards to perform massively parallel simulation of advanced Monte Carlo methods. *Journal of Computational and Graphical Statistics*, 19(4):769–789, 2010.

[42] M. A. Lindquist. The statistical analysis of fMRI data. *Statistical Science*, 23(4):439–464, 2008.

[43] M. A. Lindquist, T. Waugh, and T. D. Wager. Modeling state-related fMRI activity using change-point theory. *NeuroImage*, 35(3):1125–1141, 2007.

[44] C. J. Long, P. L. Purdon, S. Temeranca, N. U. Desai, M. Hämäläinen, and E. N. Brown. State-space solutions to the dynamic magnetoencephalography inverse problem using high performance computing. *Annals of Applied Statistics*, 5(2B):1207–1228, 2011.

[45] S. N. MacEachern, M. Clyde, and J. S. Liu. Sequential importance sampling for nonparametric Bayes models: The next generation. *Canadian Journal of Statistics*, 27(2):251–267, 1999.

[46] P. Macheras and A. Iliadis. *Modeling in Biopharmaceutics, Pharmacokinetics and Pharmacodynamics*. Springer Verlag, New York, 2005.

[47] L. M. Murray. Bayesian state-space modelling on high-performance hardware using LibBi. Mathematics e-print 1306.3277, arXiv, 2013.

[48] C. F. H. Nam, J. A. D. Aston, and A. M. Johansen. Quantifying the uncertainty in change points. *Journal of Time Series Analysis*, 33(5):807–823, 2012.

[49] C. F. H. Nam, J. A. D. Aston, and A. M. Johansen. Parallel sequential Monte Carlo samplers and estimation of the number of states in a hidden Markov model. *Annals of the Institute of Statistical Mathematics (Tokyo)*, 66(3):553–575, 2014.

[50] R. M. Neal. Annealed importance sampling. *Statistics and Computing*, 11(2):125–139, 2001.

[51] M. A. Newton and A. E. Raftery. Approximate Bayesian inference with the weighted likelihood bootstrap. *Journal of the Royal Statistical Society: Series B (Statistical Methodology)*, 56(1):3–48, 1994.

[52] E. Niedermeyer and F. Lopes da Silva. *Electroencephalography: Basic Principles, Clinical Applications, and Related Fields*. Lippincott Williams & Wilkins, 5th edition, 2012.

[53] S. Ogawa, T. M. Lee, A. S. Nayak, and P. Glynn. Oxygenation-sensitive contrast in magnetic resonance image of rodent brain at high magnetic fields. *Magnetic Resonance in Medicine*, 14(1):68–78, 1990.

[54] W. Ou, M. Hämäläinen, and P. Golland. A distributed spatio-temporal EEG/MEG inverse solver. *NeuroImage*, 44(3):932–946, 2009.

[55] J.-Y. Peng. *Pattern Statistics in Time Series Analysis*. PhD thesis, Department of Computer Science and Information Engineering, College of Electrical Engineering and Computer Science, National Taiwan University, 2008.

[56] J.-Y. Peng, J. A. D. Aston, R. N. Gunn, C.-Y. Liou, and J. Ashburner. Dynamic positron emission tomography data-driven analysis using sparse Bayesian learning. *IEEE Transactions on Medical Imaging*, 27(9):1356–1369, 2008.

[57] M. E. Phelps. Positron emission tomography provides molecular imaging of biological processes. *Proceedings of the National Academy of Science (USA)*, 97(16):9226–9233, 2000.

[58] G. W. Pullford. A taxonomy of multiple target tracking methods. *IEE Proceedings on Radar, Sonar and Nagivation*, 152(5):291–304, 2005.

[59] A. E. Raftery, M. A. Newton, J. M. Satagopan, and P. N. Krivitsky. Estimating the integrated likelihood via posterior simulation using the harmonic mean identity. In J. M. Bernardo et al., editors, *Bayesian Statistics 8*, pages 1–45. Oxford University Press, U.K., 2007.

[60] C. P. Robert and G. Casella. *Monte Carlo Statistical Methods*. Springer Verlag, New York, 2nd edition, 2004.

[61] J. Sarvas. Basic mathematical and electromagnetic concepts of the biomagnetic inverse problem. *Physics in Medicine and Biology*, 32(1):11–22, 1987.

[62] K. Schmidt. Which linear compartmental systems can be analyzed by spectral analysis of PET output data summed over all compartments? *Journal of Cerebral Blood Flow & Metabolism*, 19(5):560–9, 1999.

[63] E. Somersalo, A. Voutilainen, and J. P. Kaipio. Non-stationary magnetoencephalography by Bayesian filtering of dipole models. *Inverse Problems*, 19(5):1047–1063, 2003.

[64] A. Sorrentino, A. M. Johansen, J. A. D. Aston, T. E. Nichols, and W. S. Kendall. Filtering of dynamic and static dipoles in magnetoencephalography. *Annals of Applied Statistics*, 7(2):955–988, 2013.

[65] A. Sorrentino, L. Parkkonen, A. Pascarella, C. Campi, and M. Piana. Dynamical MEG source modeling with multi-target Bayesian filtering. *Human Brain Mapping*, 30(6):1911–1921, 2009.

[66] K. E. Stephan, W. D. Penny, J. Daunizeau, R. J. Moran, and K. J. Friston. Bayesian model selection for group studies. *NeuroImage*, 46(4):1004–1017, 2009.

[67] S. Taulu, M. Kajola, and J. Simola. Suppression of interference and artifacts by the signal space separation method. *Brain Topography*, 16(4):269–275, 2004.

[68] T. S. Tian, J. Z. Huang, H. Shen, and Z. Li. A two-way regulariation method for MEG source reconstruction. *Annals of Applied Statistics*, 6(3):1021–1046, 2012.

[69] F. E. Turkheimer, J. A. D. Aston, M.-C. Asselin, and R. Hinz. Multi–resolution Bayesian regression in PET dynamic studies using wavelets. *NeuroImage*, 32(1):111–121, 2006.

[70] S. K. Upadhyay, U. Singh, and D. K. Dey, editors. *Bayesian Statistics and Its Applications*. Anamaya Publishers, New Delhi, India, & Anshan Ltd., Kent, U.K., 2007.

[71] S. K. Upadhyay, N. Vasishta, and A. F. M. Smith. Bayes inference in life testing and reliability via Markov chain Monte Carlo simulation. *Sankhyā, The Indian Journal of Statistics: Series A*, 63(1):15–40, 2001.

[72] K. Uutela, M. Hämäläinen, and E. Somersalo. Visualization of magnetoencephalographic data using minimum current estimates. *NeuroImage*, 10(2):173–180, 1999.

[73] K. Wendel, O. Väisänen, J. Malmivuo, N. G. Gencer, et al. EEG/MEG source imaging: Methods, challenges, and open issues. *Computational Intelligence and Neuroscience*, 2009:1–12, 2009. doi: 10.1155/2009/656092.

[74] K. J. Worsley, C. Liao, J. A. D. Aston, V. Petre, G. H. Duncan, and A. C. Evans. A general statistical analysis for fMRI data. *NeuroImage*, 15(1):1–15, 2002.

[75] Y. Zhou. vSMC: Parallel sequential Monte Carlo in C++. Mathematics e-print 1306.5583, arXiv, 2013.

[76] Y. Zhou, J. A. D. Aston, and A. M. Johansen. Bayesian model comparison for compartmental models with applications in positron emission tomography. *Journal of Applied Statistics*, 40(5):993–1016, 2013.

[77] Y. Zhou, A. M. Johansen, and J. A. D. Aston. Bayesian model comparison via path-sampling sequential Monte Carlo. In *Proceedings of IEEE Workshop on Statistical Signal Processing*, pages 245–248. IEEE, 2012.

[78] Y. Zhou, A. M. Johansen, and J. A. D. Aston. Towards automatic model comparison: An adaptive sequential Monte Carlo approach. CRiSM Working Paper 13-04, University of Warwick, 2013.

2

Forecasting Indian Macroeconomic Variables Using Medium-Scale VAR Models

Goodness C. Aye

University of Pretoria

Pami Dua

University of Delhi

Rangan Gupta

University of Pretoria

CONTENTS

2.1 Introduction

Prior to the 1990s, the structure of the Indian economy was mostly a closed, regulated and protectionist economy; even after gaining independence. As a result, the performance of various sectors and the aggregate growth rate was not impressive, as the economy experienced a number of obstacles and interruptions. However, during 1990s, the economy had a major structural shift by adopting liberalization and globalization policies to become a market based economy. These policies allowed the economy to grow at an average of over 7 percent per annum since 1994 [9]. Driven by a sterling demographic dividend, diverse nature, continuing structural reforms and further globalisation, the economy is continuously striving for higher growth. Hence, despite the recent slowdown due to the weak global economic environment, the prospects of India are still considered bright in the medium to long-run and it is considered as one of the fastest growing economies in the world. Various sectors including food processing, transportation equipment, petroleum, textiles, software, agriculture, mining, machinery, chemicals, steel, cement, among others, contributed to the economic growth of India. Among these, the agriculture, manufacturing and services sector are three major sources of economic growth over the years [18] and [66]. Over the last decade, the contribution of the services sector to overall growth of the economy is about 65%, while industry and agriculture contribute 27 percent and 8 percent, respectively [41]. However, the economy has slowed down currently, due to the slowing down of the global economy, weighed down by the crisis in the Euro Area and uncertainty about fiscal policy in the USA, while the domestic economy has also been affected by a weak monsoon [41]. Given the growing nature of the Indian economy and uncertainties in economic parameters, it is important to track the movement of vital economic and financial variables in the economy, as these may affect the forecasts of future economic activity.

In general, forecasting the movements of economic and financial variables is of utmost importance for policy makers as they seek to achieve price stability and higher economic growth. Moreover, such forecasts are important to firms and investors, as these help them in making profitable strategies and proper asset allocation decisions. Hence, forecasting may be considered to be at the root of inference in time series analysis [10, 11]. Further, emerging economies' markets are generally characterized by lower levels of liquidity and at the same time by a higher volatility than developed economies' markets [8, 54]. High volatility in these markets is often marked by frequent and erratic changes, which are usually driven by various local events (such as political developments) rather than by the events of global importance [1, 12]. These different features may contribute to different dynamics underlying the macroeconomic variables, making these markets an interesting sphere of research. Therefore, providing accurate forecasts of these variables is very crucial. However, given the difficulty in forecasting financial and economic variables, since the forecast depends on the models used in generating them, it is crucial to evaluate forecasts from different models and select the 'best' based on an objective criterion [36]. Further, [26] argued that in time series models, estimation and inference basically means minimizing of the one-step (or multi-step) forecast errors. Therefore, a model is considered superior to its competitors if it produces smaller forecast errors than its competitors. Therefore, the goal of this study is to evaluate the forecasting performance of 11 econometric models for 15 macroeconomic variables for India. For ease of comparison between the small and medium scale models, the discussion that follows concentrates on the four variables of interest, namely, the industrial production growth, the wholesale price index inflation rate, the 3−month Treasury bill rate and the Indian Rupee-US Dollar exchange rate. We concentrate on these four variables as they tend to be important inputs at various stages of policymaking and in the context of a developing country with an active trade balance, the above series are closely monitored by most interested parties.

There exists a large international literature on forecasting output growth, interest rate, exchange rate and inflation.[1] As far as India is concerned, there are also some notable studies that forecast these variables. On industrial production forecasting, the studies for India include [13, 14, 33, 49, 53, 70, 74] and [84]. Related studies on forecasting inflation include [2, 23, 33, 51, 52, 68] and [84]. For exchange rate, the studies include [28, 37, 55, 62, 65, 67, 69] and [74]. Related studies on interest rates include [32, 34–36] and [64]. While some of these studies have used just one econometric model to produce the relevant forecast, others have compared the forecast performance of two or at most five models, which essentially fall within the category of small comparative analysis. In this current study, we extend the literature by evaluating the performance of 11 models comprising of both small and medium- scale models; namely 3 classical VARs, 4 Bayesian VARs and 4 Bayesian Factor Augmented VARs, which are used to forecast 15 macroeconomic variables for the Indian economy. Hence, we also contribute to previous studies by adding the medium-scale models to the set of forecasting models for economic and financial series in India. Using a large number of models would also provide us a more complete form of analysis. We also provide an ex-ante forecast of these series using the selected 'best' model for each series.

The rest of the chapter is organized as follows: the econometric models that are used to generate the forecasts are described in Section 2.2. The data is described in Section 2.3. In Section 2.4, the results are discussed, while Section 2.5 concludes.

2.2   Econometric Models

2.2.1   VAR and BVAR

The Vector Autoregressive (VAR) model, though 'atheoretical', is particularly useful for forecasting purposes. An unrestricted VAR model, as suggested by [72], can be written as follows:

$$y_t = A_0 + A\left(L\right)y_t + \varepsilon_t \qquad (2.1)$$

where y is a $(n \times 1)$ vector of variables being forecasted; $A\left(L\right)$ is a $(n \times n)$ polynomial matrix in the backshift operator L with lag length p, i.e. $A\left(L\right) =$

---

[1] [16, 19–22, 27, 48, 50] and [80] for a detailed literature review on forecasting real output growth. For forecasting interest rates, see [39, 40, 46, 56, 82] and [88]. For inflation rates forecasting, see [4, 44, 47, 59, 76, 78, 81] and [77]. For forecasting exchange rates, see [5, 24, 25, 29, 43, 86, 87] and literatures cited therein.

$A_1 L + A_2 L^2 + \cdots + A_p L^p$; A_0 is a $(n \times 1)$ vector of constant terms, and ε is a $(n \times 1)$ vector of error terms. In our case, we assume that $\varepsilon \sim N\left(0, \sigma^2 I_n\right)$, where I_n is a $(n \times n)$ identity matrix.

Note that the VAR model generally uses equal lag length for all the variables of the model. One drawback of VAR models is that many parameters need to be estimated, some of which may be insignificant. This problem of overparameterization, resulting in multicollinearity and a loss of degrees of freedom, leads to inefficient estimates and possibly large out-of-sample forecasting errors. One solution, often adapted, is simply to exclude the insignificant lags based on statistical tests. Another approach is to use a near VAR, which specifies an unequal number of lags for the different equations.[2]

However, an alternative approach to overcoming this overparameterization, as described in [31], [63], [75] and [85] is to use a BVAR model. Instead of eliminating longer lags, the Bayesian method imposes restrictions on these coefficients by assuming that they are more likely to be near zero than the coefficients on shorter lags. However, if there are strong effects from less important variables, the data can override this assumption. The restrictions are imposed by specifying normal prior distributions with zero means and small standard deviations for all coefficients with the standard deviation decreasing as the lags increase. Where the data exhibits non-stationary characteristics, the coefficient on the first own lag of a variable has a mean of unity. This procedure is popularly referred to as the ' Minnesota prior' due to its development at the University of Minnesota and the Federal Reserve Bank at Minneapolis, Minnesota.

Formally, as discussed above, the means and variances of the Minnesota prior take the following form:

$$\beta_i \sim N\left(1, \sigma^2_{\beta_i}\right) \quad \text{and} \quad \beta_j \sim N\left(0, \sigma^2_{\beta_j}\right) \tag{2.2}$$

where β_i denotes the coefficients associated with the lagged dependent variables in each equation of the VAR, while β_j represents any other coefficient. In the belief that lagged dependent non-stationary variables are important explanatory variables, the prior means corresponding to them are set to unity, given that, as seen from (2.2), the distribution *apriori* is centered around a random walk. However, for all the other coefficients, β_j's, in a particular equation of the VAR, a prior mean of zero is assigned to suggest that these variables are less important to the model.

The prior variances $\sigma^2_{\beta_i}$ and $\sigma^2_{\beta_j}$, specify uncertainty about the prior means $\bar{\beta}_i = 1$ and $\bar{\beta}_j = 0$, respectively. Because of the overparameterization of the VAR, [31] suggested a formula to generate standard deviations as a function of small numbers of hyperparameters: w, d, and a weighing matrix $f(i, j)$. This approach allows the forecaster to specify individual prior variances for a large number of coefficients based on only a few hyperparameters. The specification

---

[2]The stochastic search variable selection (SSVS) as described in [58], [60] can also be used to do lag length selection in an automatic fashion.

of the standard deviation of the distribution of the prior imposed on variable j in equation i at lag m, for all i, j and m, defined as $\sigma_{i,j,m}$, can be specified as follows:

$$\sigma_{i,j,m} = [w \times g(m) \times f(i,j)] \frac{\hat{\sigma}_i}{\hat{\sigma}_j} \tag{2.3}$$

with $f(i,j) = 1$, if $i = j$ and k_{ij} (interaction variable) otherwise, with ($0 \leq k_{ij} \leq 1$), $g(m) = m^d, d > 0$. Note that $\hat{\sigma}_i$ is the estimated standard error of the univariate autoregression for variable i. The ratio $\hat{\sigma}_i/\hat{\sigma}_j$ scales the variables to account for differences in the units of measurement. The term w indicates the overall tightness and is also the standard deviation of the first own lag, with the prior getting tighter as we reduce the value. The parameter $g(m)$ measures the tightness on lag m with respect to lag 1, and is assumed to have a harmonic shape with a decay factor of d, which tightens the prior on increasing lags. The parameter $f(i,j)$ represents the tightness of variable j in equation i relative to variable i, and by increasing the interaction, i.e., the value of k_{ij}, we can loosen the prior. Given that we have domestic as well as world variables within our dataset, where domestic variables would have minimal, if any, effect on world variables, while world variables influence domestic variables, we use a weighting scheme for k_{ij}. Following [57] and [71], the weight of a world variable in a world equation, as well as a domestic equation, is set at 0.6. The weight of a domestic variable in the other domestic equations is then fixed at 0.1 and its weight in a world equation is set at 0.01. Finally, the weight of the domestic variable in its own equation is 1.0. These weights are in line with Litterman's circle-star approach, where star (world) variables affect both star and circle (domestic) variables, and circle variables primarily influence only other circle variables.[3] We follow [7], [15] and [30] in setting the value of the overall tightness parameter to obtain a desired average fit for the four variables of interest (growth rate of industrial production, WPI inflation rate, three-month Treasury Bill rate and the Indian Rupee to Dollar nominal exchange rate) in the in-sample period (1997 : 04 to 2006 : 12). The optimal value of $w(Fit)(= 0.4273)$ obtained in this fashion is then retained for the entire evaluation period. Note that following [7] and [15], the value of d is set equal to 1.0. Specifically, for a desired *Fit*, w is chosen as follows:

$$w(Fit) = \arg\min_w \left| Fit - \frac{1}{4} \sum_{i=1}^{4} \frac{MSE_i^w}{MSE_i^0} \right| \tag{2.4}$$

where $MSE_i^w = \sqrt{\dfrac{1}{T_0 - p - 1} \sum_{t=p}^{T_0-2} \left(y_{i,t+1|t}^w - y_{i,t+1} \right)^2}$.

Hence, the one-step-ahead mean squared error (MSE) evaluated using the training sample $t = 1, \cdots, T_0 - 1$, with T_0 being the beginning of the sample

---

[3]We experimented with higher and lower interaction values, in comparison to those specified above, to the star variables in both the star and circle equations, but, the rank of the alternative forecasts remained the same.

period and $p (= 8)$ [4] being the order of the VAR. MSE_i^0 is the MSE of variable i with the prior restriction imposed exactly ($w = 0$), while, the baseline Fit is defined as the average relative MSE from an OLS-estimated VAR containing the four variables, i.e.

$$Fit = \frac{1}{4} \sum_{i=1}^{4} \frac{MSE_i^\infty}{MSE_i^0} \qquad (2.5)$$

In addition, we also look at the forecast performances of the large-scale BVAR model for a $Fit = 0.25(w = 0.14), 0.50(w = 0.26)$ and $0.75(w = 0.58)$. Note that the classical VAR attains a highest possible fit of 0.65 for $w = \infty$, where a fit of 0.75 is unattainable. However, the fits of 0.25 and 0.50 are obtained under the values of $w = 0.19$ and 0.41, respectively.

Finally, once the priors have been specified, the BVAR model is estimated using mixed estimation technique [83]. This method involves supplementing the data with prior information on the distribution of the coefficients. The number of observations and degrees of freedom are increased by one in an artificial way, for each restriction imposed on the parameter estimates. The loss of degrees of freedom due to over-parameterization associated with a classical VAR model is, therefore, not a concern in the BVAR. Note, we compute point forecasts using the posterior mean of the parameters.

2.2.2 FAVAR and BFAVAR

This chapter also uses the Dynamic Factor Model (DFM) to extract common components between macroeconomic series. These common components are then used to forecast the four key variables of interest, adding the extracted factors to the 4−variable VAR model to create a FAVAR in the process. We choose a domestic factor (from eleven domestic variables excluding the four of interest) and a world factor (from three world variables). The test [3] for selecting the optimal number of factors also yielded 1 factor each for the domestic and foreign blocks of the dataset.[5] The factors are obtained recursively at each point T in the out-of-sample period. Furthermore, we estimate idiosyncratic component with an $AR(p)$ processes, as suggested by [17].

The DFM expresses individual time series as the sum of two unobserved components: a common component driven by a small number of common factors and an idiosyncratic component for each variable. The DFM extracts the few factors that explain the co-movement of the Indian economy. [45] demonstrated that for a small number of factors relative to the number of variables and a heterogeneous panel, we can recover the factors from present and past observations.

---

[4]The order of the VAR was chosen based on the unanimity of the Akaike Information Criterion (AIC), Final Prediction Error (FPE) criterion and the sequential Likelihood Ratio (LR) criterion.

[5]Details of these results are available upon request from the authors.

For forecasting purposes, we use a 5−variable VAR augmented by extracted common factors using the [79] approach. This approach is similar to the univariate Static and Unrestricted (SU) approach of [17]. We consider a Bayesian FAVAR specification, which augments the FAVAR specification with Bayesian restrictions on lags of the key variables and the two factors based on the prior structure outlined above. Note that the domestic factor is treated like a domestic variable, whereas the world factor is treated like a world variable, when setting up the interaction matrix. The values of w then represent the average relative MSE from the OLS-estimated VAR, which correspond to the priors of $0.25, 0.50$ and 0.75, are $0.747, 0.170, 0.340$ and 2.403, respectively.

2.2.3   Forecast evaluation

We evaluate the forecast performance of the 11 models using the root mean square error (RMSE) relative to a random walk forecasts as the benchmark. A ratio of 1 indicates that the specific model forecasts match the performance of random walk forecasts. A ratio > 1 shows that the random walk forecasts outperform the model forecasts. A ratio < 1 implies that the forecasts from the model outperform the random walk forecasts.

2.3   Data

The dataset contains 15 monthly series of India covering the period of April, 1997 till October, 2011. We use an in-sample of 1997 : 4 till 2006 : 12, with the out-of-sample (2007 : 1 − 2011 : 10) covering the period of slowdown in India following the financial crisis. The variables used are: Indian industrial production growth, Wholesale price index inflation, three−month, one−year, five−year and ten−year Treasury bill rates, $M0, M1, M2$ and $M3$ money growth rates, the Indian Rupee-Dollar exchange rate, government expenditure, three−months and twelve months London Inter Bank Overnight Rates (LIBOR) and the growth rate of world industrial production. Barring the world industrial production and the government expenditure, all data are obtained from the Global Financial Database (GFD). The world industrial production data are obtained from the Organization for Economic Co-operation and Development (OECD) and the government expenditure data is derived from the Controller General of Accounts, India. All series are seasonally adjusted and for the DFM models they are also transformed to ensure that they are covariance stationary. The more powerful DF-GLS test of [42], instead of the popular ADF test, is used to assess the degree of integration of all series. For the DFM models all non-stationary series are made stationary through differencing. The Schwarz information criterion is used in selecting the appropriate lag length in such a way that no serial correlation is left in the stochastic

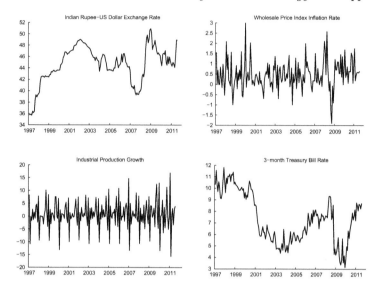

FIGURE 2.1
Historical plots of the four major variables: 1997:04-2011:10.

error term. When there were doubts about the presence of unit root, the KPSS test proposed by [61], with the null hypothesis of stationarity, was applied. All series are standardized to have a mean of zero.

For the BVAR model, non-stationarity is not an issue, since [73] indicates that the likelihood function has the same Gaussian shape regardless of the presence of nonstationarity. Hence, for the sake of comparison amongst the VARs, both classical and Bayesian, we do not take the first difference of the variables, unlike in the DFM. [6] However, following [6] and [15], for the variables in the panel that are characterized by mean-reversion, we set a white-noise prior, i.e. $\overline{\beta}_i = 0$, otherwise, we impose the random walk prior implying that: $\overline{\beta}_i = 1$. Barring the log-levels of the Indian Rupee—Dollar exchange rate and the log-levels of government expenditure, we set $\overline{\beta}_i = 0$, as the remaining 13 variables were all mean reverting.[7] The descriptive statistics of the four variables of interest are presented in Appendix 2, where we note that the deviation of the series from their means is slightly high and they do not appear to be normally distributed, as neither their skewness is zero nor their kurtosis is equal to 3. This is further confirmed by the rejection of the null of normal distribution of the Jarque—Bera test. The plots of the four variables are also presented in Figure 2.1.

---

[6]See [38] for further details.
[7]The list of all the variables and their respective transformation is presented in Appendix 1.

TABLE 2.1
One- to Twelve-Months-Ahead relative RMSE for Industrial production growth (2007:1-2011:10).

Model	Horizon												Average
	1	2	3	4	5	6	7	8	9	10	11	12	
VAR1	0.55	0.81	0.85	0.71	0.62	0.76	0.66	0.59	0.81	0.70	0.61	0.95	0.72
VAR2	0.52	0.74	0.79	0.67	0.59	0.72	0.64	0.56	0.80	0.69	0.59	0.93	0.69
VAR3	0.52	0.78	0.83	0.70	0.61	0.74	0.65	0.57	0.80	0.69	0.60	0.94	0.70
BFAVAR1	0.48	0.70	0.75	0.65	0.57	0.74	0.67	0.58	0.83	0.72	0.64	1.00	0.69
BFAVAR2	0.49	0.70	0.76	0.67	0.59	0.78	0.70	0.60	0.87	0.72	0.62	1.00	0.71
BFAVAR3	0.48	0.69	0.75	0.66	0.58	0.76	0.69	0.60	0.85	0.71	0.62	1.00	0.70
BFAVAR4	0.49	0.70	0.75	0.66	0.58	0.76	0.69	0.60	0.85	0.71	0.62	1.00	0.70
BVAR1	0.51	0.66	0.70	0.60	0.54	0.70	0.64	0.56	0.81	0.69	0.60	0.97	**0.66**
BVAR2	0.52	0.69	0.76	0.67	0.59	0.78	0.70	0.61	0.87	0.72	0.62	1.00	0.71
BVAR3	0.53	0.69	0.75	0.66	0.58	0.77	0.69	0.60	0.86	0.71	0.62	1.00	0.70
BVAR4	0.54	0.71	0.77	0.69	0.59	0.79	0.71	0.61	0.88	0.72	0.62	1.00	0.72

Notes: Entries are RMSE of a specific model relative to the random-walk model.
Average indicates the average of one- to twelve-months-ahead relative RMSEs.
VAR1, VAR2 and VAR3 are small-scale VAR models and correspond to $w = \infty$, $= 0.19$ and 0.41, respectively.
BFAVAR1, BFAVAR2, BFAVAR3 and BFAVAR4 are Bayesian factor-augments VARs and correspond to w=0.75, 0.17, 0.34 and 2.40, respectively.
BVAR1, BVAR2, BVAR3 and BVAR4 are the medium-scale Bayesian VAR models and correspond to w=0.43, 0.14, 0.26 and 0.58, respectively.
Value in bold in the last column relates to the 'best' model.

2.4 Empirical Results

The results of the out-of-sample forecasting performance for the 11 models from 2007 : 01 to 2011 : 10 are presented in Tables 2.1 to 2.4. The set comprises of three classical VARs: VAR1 ($w = \infty$), VAR2 ($w = 0.19$) and VAR3 ($w = 0.41$); four Bayesian VARS: BVAR1 ($w = 0.43$), BVAR2 ($w = 0.14$), BVAR3 ($w = 0.26$) and BVAR4 ($w = 0.58$); and four Bayesian Factor Augmented VARs: BFAVAR1 ($w = 0.75$), BFAVAR2 ($w = 0.17$), BFAVAR3 ($w = 0.34$) and BFAVAR4 ($w = 2.40$). VAR1 is the benchmark classical VAR model. BVAR1 and BFAVAR1 ensure that the in-sample fit of BVARs and BFAVARs are the same as the classical VAR model. Although the VAR, BVARs and BFAVARs are estimated with four, fifteen and six variables, respectively, we report the results for the four variables of interest for ease of comparison.

Table 2.1 reports the one to twelve-month ahead relative RMSEs for industrial production growth. It shows that at all horizons, with the exception of horizon 12, all the 11 models produced a better forecast for industrial production growth than the benchmark random walk model as the RMSEs are less than unity. At horizon 12, the random walk model outperforms the BFAVAR1 to BFAVAR4 and BVAR2 to BVAR4. However, the average RMSE is less than unity for all the models indicating that on average these models perform

TABLE 2.2
One- to Twelve-Months-Ahead relative RMSE for 3-month Treasury bill rate (2007:1-2011:10).

						Horizon							
Model	1	2	3	4	5	6	7	8	9	10	11	12	Average
VAR1	1.13	1.17	1.17	1.12	1.14	1.13	1.13	1.09	1.08	1.08	1.08	1.08	1.12
VAR2	1.07	1.09	1.10	1.06	1.08	1.06	1.07	1.04	1.03	1.03	1.03	1.03	1.06
VAR3	1.11	1.15	1.15	1.10	1.12	1.11	1.11	1.08	1.06	1.06	1.07	1.07	1.10
BFAVAR1	0.99	0.97	0.96	0.93	0.94	0.91	0.89	0.88	0.88	0.89	0.90	0.92	0.92
BFAVAR2	0.94	0.95	0.93	0.89	0.89	0.87	0.86	0.85	0.84	0.82	0.82	0.82	0.87
BFAVAR3	0.94	0.93	0.91	0.86	0.86	0.84	0.83	0.82	0.81	0.80	0.81	0.82	0.85
BFAVAR4	0.95	0.90	0.87	0.85	0.84	0.81	0.80	0.77	0.76	0.75	0.76	0.77	0.82
BVAR1	1.03	0.95	0.91	0.91	0.94	0.95	0.92	0.92	0.96	1.04	1.07	1.10	0.98
BVAR2	0.95	0.93	0.91	0.86	0.87	0.84	0.82	0.79	0.77	0.77	0.76	0.76	0.84
BVAR3	0.97	0.93	0.88	0.84	0.85	0.82	0.79	0.76	0.75	0.75	0.75	0.75	0.82
BVAR4	0.99	0.94	0.87	0.82	0.82	0.79	0.76	0.70	0.68	0.68	0.69	0.71	**0.79**

Note: See notes to Table 2.1.

TABLE 2.3
One- to Twelve-Months-Ahead relative RMSE for Indian Rupee-US Dollar nominal exchange rate (2007:1-2011:10).

						Horizon							
Model	1	2	3	4	5	6	7	8	9	10	11	12	Average
VAR1	1.00	0.98	1.01	1.02	0.99	0.98	0.99	0.99	0.98	0.97	0.95	0.93	0.98
VAR2	0.96	0.95	0.97	0.98	0.95	0.95	0.96	0.95	0.94	0.93	0.92	0.91	0.95
VAR3	0.98	0.97	1.00	1.01	0.98	0.97	0.98	0.98	0.97	0.96	0.94	0.93	0.97
BFAVAR1	0.96	0.97	0.98	0.98	0.97	0.96	0.97	0.97	0.96	0.95	0.94	0.93	0.96
BFAVAR2	0.97	0.97	0.95	0.93	0.90	0.90	0.90	0.89	0.88	0.86	0.85	0.84	**0.90**
BFAVAR3	0.96	0.97	0.96	0.94	0.92	0.91	0.92	0.92	0.90	0.89	0.87	0.87	0.92
BFAVAR4	0.97	0.98	0.98	0.97	0.94	0.94	0.95	0.95	0.93	0.92	0.90	0.89	0.94
BVAR1	0.99	1.04	1.09	1.14	1.17	1.19	1.22	1.25	1.28	1.30	1.30	1.31	1.19
BVAR2	0.99	0.98	0.96	0.95	0.94	0.94	0.93	0.91	0.89	0.87	0.85	0.83	0.92
BVAR3	0.97	0.99	0.98	0.98	0.98	0.98	0.99	0.98	0.96	0.95	0.93	0.91	0.97
BVAR4	0.98	1.00	0.98	0.96	0.95	0.95	0.95	0.95	0.94	0.92	0.91	0.89	0.95

better in forecasting India's industrial production than a simple random walk model. Comparing the performance of the 11 models with one another, Table 2.1 shows that the BFAVAR1 forecasts industrial production better at horizon 1 than the rest of the 10 models, as it has the lowest RMSE. At horizons 2 to 8, the BVAR1 outperforms the other models while at horizons 9 to 12, the VAR2 model outperforms all the others, except at horizon 9 where it produces similar forecasts to the VAR3. So far, it appears that at short and medium term horizons, the BVAR1 model is the best forecasting model for industrial production while over longer term horizon the VAR2 performs better. On average, the BVAR1 outperforms both the random walk benchmark model and

TABLE 2.4

One- to Twelve-Months-Ahead relative RMSE for Wholesale Price Index inflation rate (2007:1-2011:10).

Model	\multicolumn{13}{c}{Horizon}
Model	1	2	3	4	5	6	7	8	9	10	11	12	Average
---	---	---	---	---	---	---	---	---	---	---	---	---	---
VAR1	1.04	0.97	0.96	0.84	0.69	0.66	0.65	0.63	0.71	0.68	0.68	0.77	0.77
VAR2	0.88	0.88	0.90	0.74	0.67	0.64	0.64	0.62	0.67	0.67	0.70	0.79	0.73
VAR3	0.99	0.94	0.94	0.80	0.68	0.65	0.65	0.62	0.70	0.68	0.69	0.77	0.76
BFAVAR1	0.83	0.79	0.85	0.70	0.62	0.61	0.62	0.60	0.66	0.67	0.70	0.78	0.70
BFAVAR2	0.87	0.80	0.87	0.71	0.64	0.63	0.63	0.62	0.65	0.66	0.69	0.77	0.71
BFAVAR3	0.84	0.79	0.85	0.71	0.64	0.62	0.63	0.61	0.65	0.66	0.69	0.76	0.70
BFAVAR4	0.85	0.80	0.85	0.71	0.64	0.62	0.63	0.62	0.65	0.66	0.69	0.77	0.71
BVAR1	0.87	0.83	0.90	0.76	0.68	0.69	0.69	0.65	0.68	0.72	0.76	0.84	0.76
BVAR2	0.80	0.79	0.85	0.70	0.64	0.63	0.63	0.61	0.64	0.65	0.69	0.76	**0.70**
BVAR3	0.80	0.78	0.86	0.73	0.66	0.64	0.64	0.61	0.65	0.66	0.69	0.76	0.71
BVAR4	0.81	0.80	0.89	0.74	0.67	0.65	0.65	0.62	0.65	0.66	0.69	0.76	0.71

Note: For Tables 2.3 and 2.4, see notes to Table 2.1.

the other 10 forecasting models. The RMSE for the BVAR1 model is lower than the RMSE for the random walk model; on average by about 33%.

The out-of-sample forecasting results for the 3−month Treasury bill rate is presented in Table 2.2. The results indicate that the BFAVAR models and the BVAR2 to BVAR3 models outperform the random walk benchmark model at all horizons. However, the random walk model produced a better forecast than all the VAR models at all horizons. In addition, it also outperformed the BVAR1 at horizons 1 and horizons 10 − 12. On average, all other models except the VAR1 to VAR3 have a relative RMSE that is less than unity, indicating that their forecasts for the 3−month Treasury bill rate are better than those of a random walk model. Table 2.2 also shows that the BFAVAR2 and BFAVAR4 outperform the rest of the models at horizons 1 and 2, respectively. The BVAR4 model is the best forecasting model for the 3−month Treasury bill rate at horizons 3 to 12. On average, the results suggest that the movements in India's 3−month Treasury bill rate should be forecast with the BVAR4 model. The RMSE for BVAR4 model is lower than the RMSE for the random walk model by about 21%.

For the Indian Rupee-US Dollar nominal exchange rate, the out-of-sample forecasting performance results are presented in Table 2.3. The results show that in general, the models examined in this study produced more accurate forecasts of the exchange rate than a simple random walk model. The only notable exception is the BVAR1 model. A comparison of the forecasts from the 11 models show that the VAR2, BFAVAR1 and BVAR2 produced more accurate forecasts than other models at horizons 1, 2 and 12, respectively. At horizons 3 to 11, the BFAVAR2 outperforms the other 10 models. Also, the BFAVAR2 outperforms both the random walk benchmark model and the other 10 forecasting models when we take the average of the relative RMSEs for each model. The RMSE for the BFAVAR2 model is lower than the RMSE for the random walk model by about 9%.

The results for the out-of-sample forecasts for the wholesale inflation rate are reported in Table 2.4. The 11 models outperform the simple random walk model at all horizons except at horizon 1 where the random walk model produces better forecast than VAR1 model. There appears to be no model that consistently outperforms the others in forecasting the inflation rate. The forecast capability of a particular model seems to depend very much on the forecast horizon. At horizons 1 and 2, the BVAR3 is the best while at horizon 3, the BFAVAR4 is the best. At horizons 4 to 8, the BFAVAR1 is the best performing model, while at horizons 9 and 10, it is BVAR2. At the longer horizons of 11 and 12, the VAR1 and BVAR4 would appear to be superior. On average, the BVAR2 produced more accurate forecasts of the inflation rate, when compared to other 10 models. Its gain over the random walk model is about 29%.

Having selected the 'best' models for forecasting each series based on the minimum average RMSEs relative to the random walk benchmark model, we now use the respective 'best' models to do an ex-ante forecast of Indian industrial production growth, Wholesale price index inflation, three-month Treasury bill rates and the Indian Rupee-Dollar exchange rate over 2011 : 11 to 2012 : 10. The results are presented in Figure 2.2. For industrial production, the BVAR1 model appears to capture the direction of change or the turning

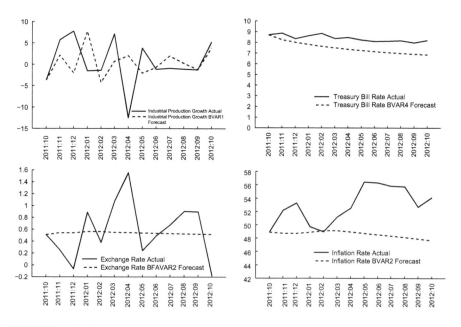

FIGURE 2.2
Ex-ante forecast (2011:11 – 2012:10).

points at the beginning and toward the end of the ex-ante forecast period. For the 3−month Treasury bill rate, the BVAR4 model appears to capture the direction of change perfectly, but the actual and forecast values differ slightly. For the Indian Rupee−US Dollar exchange rate and the inflation rate, neither the BFAVAR2 nor the BVAR2 captures the turning points. Furthermore, the forecast values for these variables are not very close to the actual values. Hence, while these models predict somewhat constant values for exchange and inflation rates, the actual values vary greatly over the period.

2.5   Conclusion

Forecasting the movement of economic and financial variables is of utmost importance for policy makers, that seek to achieve a level of price stability and higher economic growth. Moreover, such forecasts are of importance to firms and investors, as they facilitate more appropriate strategic decision making and better asset allocation. However, given the difficulty in forecasting financial and economic variables, one should evaluate the forecasts of a number of different models at regular periodic intervals. On this basis, the current study examined the forecasting performance of 11 vector autoregressive models, which included 3 standard classical VAR models, 4 Bayesian VAR models and 4 Bayesian Factor Augmented VAR models. These models were used to generate forecasts for 15 macroeconomic variables in India. Using the minimum average RMSE relative to a random walk model, we show that in most cases, the models in this study performed better than a simple random walk model.

It was also noted that the performance of each of the 11 models varies across different horizons. However, in general, the Bayesian VARs and Bayesian Factor Augmented VARs outperform the classical VARs. The superiority of the models may be due to their ability to overcome the problem of overparameterization and loss of degrees of freedom that is encountered in classical VAR estimation. Using the 'best' selected models, we perform an ex-ante forecast for 2011 : 11 to 2012 : 10 period. The ex−ante results suggest that these linear models do not predict the critical effects that affected the variables, following the Global Financial Crisis. Hence, it may be important to consider the use of non-linear frameworks in future research as these can better capture the volatilities and frequent structural breaks that exist in the economy.

Acknowledgment

The authors gratefully acknowledge competent research assistance from Surbhi Badhwar.

Appendix 1: List of all variables used with their respective sources and transformations.

No.	Names of Variables	Source	Transformation
1	Indian industrial production growth	GFD	0 (0)
2	Wholesale price index inflation	GFD	0 (0)
3	Three-month, Treasury bill rate	GFD	0 (0)
4	One-year Treasury bill rate	GFD	0 (0)
5	Five-year Treasury bill rate	GFD	0 (0)
6	Ten-year Treasury bill rate	GFD	0 (0)
7	M0	GFD	1 (1)
8	M1	GFD	1 (1)
9	M2	GFD	1 (1)
10	M3	GFD	1 (1)
11	The Indian Rupee-Dollar exchange rate	GFD	1 (2)
12	Government expenditure	CGA, India	1 (2)
13	Three-months London Inter Bank Overnight Rates	GFD	0 (0)
14	Twelve-months London Inter Bank Overnight Rates	GFD	0 (0)
15	World industrial production	OECD	1 (1)

Note: GFD:Global Financial Database, OECD:Organization for Economic Co-operation and Development, CGA: Controller General of Accounts
Transformation codes for FAVAR: 0: No transformation; 1: Growth rate in percentages
Transformation codes for BVAR in parentheses: 0: No transformation;
1: Growth rate in percentages; 2: Natural logarithms

Appendix 2: Summary Statistics of Selected Indian Macroeconomic Variables.

Statistics	Industrial Production growth	3-month Treasury bill rate	Indian Rupee-US Dollar exchange rate	Wholesale price index inflation rate
Mean	0.70	7.34	44.66	0.46
Maximum	16.89	11.8	50.94	2.98
Minimum	−15.58	3.14	35.71	−1.89
Standard deviation	5.41	2.13	3.19	0.67
Skewness	−0.24	0.19	−0.77	0.35
Kurtosis	3.93	1.98	3.50	4.75
Jarque-Bera	8.16 (0.01)	8.66 (0.01)	19.32 (0.00)	26.13 (0.00)

Note: p-value for Jarque-Bera is in parenthesis.

Bibliography

[1] R. Aggarwal, C. Inclan, and R. Leal. Volatility in emerging stock markets. *Journal of Financial and Quantitative Analysis*, 34(1):33–55, 1999.

[2] A. K. M. S. Alam and S. J. Kamath. Models and forecasts of inflation in a developing economy. *Journal of Economic Studies*, 13(4):3–29, 1986.

[3] L. Alessi, M. Barigozzi, and M. Capasso. Improved penalization for determining the number of factors in approximate factor models. *Statistics & Probability Letters*, 80(23–24):1806–1813, 2010.

[4] A. Atkeson and L. E. Ohanian. Are Phillips curves useful for forecasting inflation? *Federal Reserve Bank of Minneapolis Quarterly Review*, 25(1):2–11, 2001.

[5] G. C. Aye, M. Balcilar, A. Bosch, R. Gupta, and F. Stofberg. The out-of-sample forecasting performance of non-linear models of real exchange rate behaviour: The case of the South African Rand. *European Journal of Comparative Economics*, 10(1):121–148, 2013.

[6] M. Bańbura, D. Giannone, and L. Reichlin. Bayesian VARs with large panels. Centre for Economic Policy Research (CEPR) Discussion Papers 6326, http://ideas.repec.org/p/cpr/ceprdp/6326.html, 2007.

[7] M. Bańbura, D. Giannone, and L. Reichlin. Large Bayesian vector auto regressions. *Journal of Applied Econometrics*, 25(1):71–92, 2010.

[8] J. T. Barkoulas, C. F. Baum, and N. Travlos. Long memory in the Greek stock market. *Applied Financial Economics*, 10(2):177–184, 2000.

[9] K. Basu. The Indian economy: Up to 1991 and since. In K. Basu, editor, *India's Emerging Economy: Performance and Prospects in the 1990s and Beyond*, pages 3–32. MIT Press, Cambridge, 2004.

[10] N. Beck, G. King, and L. Zeng. Improving quantitative studies of international conflict: A conjecture. *American Political Science Review*, 94(1):21–35, 2000.

[11] N. Beck, G. King, and L. Zeng. Theory and evidence in international conflict: A response to de Marchi, Gelpi, and Grynaviski. *American Political Science Review*, 98(2):379–389, 2004.

[12] G. Bekaert and C. R. Harvey. Emerging equity market volatility. *Journal of Financial economics*, 43(1):29–77, 1997.

[13] R. Bhattacharya, R. Pandey, and G. Veronese. Tracking India growth in real time, 2011. Working Paper no. 2011-90, National Institute of Public Finance and Policy, New Delhi.

[14] D. Biswas, S. Singh, and A. Sinha. Forecasting inflation and IIP growth: Bayesian vector autoregressive model. *Reserve Bank of India Occasional Papers*, 31(2):31–48, 2010.

[15] C. Bloor and T. Matheson. Analysing shock transmission in a data-rich environment: A large BVAR for New Zealand. *Empirical Economics*, 39(2):537–558, 2010.

[16] G. Bodo, A. Cividini, and L. F. Signorini. Forecasting the Italian industrial production index in real time. *Journal of Forecasting*, 10(3):285–299, 1991.

[17] J. Boivin and N. Serena. Understanding and comparing factor-based forecasts. *International Journal of Central Banking*, 1(3):117–151, 2005.

[18] B. Bosworth, S. M. Collins, and A. Virmani. Sources of growth in the Indian economy. Working Paper 12901, National Bureau of Economic Research (NBER), February 2007.

[19] G. Bruno and C. Lupi. Forecasting Euro-area industrial production using (mostly) business surveys data. MPRA Paper 42332, University Library of Munich, Germany, Oct 2003.

[20] G. Bruno and C. Lupi. Forecasting industrial production and the early detection of turning points. *Empirical Economics*, 29(3):647–671, 2004.

[21] G. Bulligan, R. Golinelli, and G. Parigi. Forecasting industrial production: The role of information and methods. In B. for International Settlements, editor, *The IFC's Contribution to the 57th ISI Session, Durban, August 2009*, volume 33 of *IFC Bulletins Chapters*, pages 227–235. Bank for International Settlements, June 2010.

[22] G. Bulligan, R. Golinelli, and G. Parigi. Forecasting monthly industrial production in real-time: From single equations to factor-based models. *Empirical Economics*, 39(2):303–336, 2010.

[23] T. Callen and D. Chang. Modeling and forecasting inflation in India. IMF Working Paper 99/119, International Monetary Fund, Sept. 1999.

[24] A. Carriero, G. Kapetanios, and M. Marcellino. Forecasting exchange rates with a large Bayesian VAR. *International Journal of Forecasting*, 25(2):400–417, 2009.

[25] R. H. Clarida, L. Sarno, M. P. Taylor, and G. Valente. The out-of-sample success of term structure models as exchange rate predictors: A step beyond. *Journal of International Economics*, 60(1):61–83, 2003.

[26] M. Clements and D. Hendry. *Forecasting Economic Time Series*. Cambridge University Press, UK, 1998.

[27] M. Costantini. Forecasting the industrial production using alternative factor models and business survey data. *Journal of Applied Statistics*, 40(10):2275–2289, October 2013.

[28] K. Datta and C. K. Mukhopadhyay. RBI forecast vs. GARCH-based ARIMA forecast for Indian Rupee-US Dollar exchange rate: A comparison. *The IUP Journal of Bank Management*, IX(4):7–20, November 2010.

[29] P. De Grauwe and A. Markiewicz. Learning to forecast the exchange rate: Two competing approaches. *Journal of International Money and Finance*, 32:42–76, February 2013.

[30] C. De Mol, D. Giannone, and L. Reichlin. Forecasting using a large number of predictors: Is Bayesian shrinkage a valid alternative to principal components? *Journal of Econometrics*, 146(2):318–328, 2008.

[31] T. Doan, R. Litterman, and C. Sims. Forecasting and conditional projection using realistic prior distributions. *Econometric Reviews*, 3(1):1–100, 1984.

[32] P. Dua. Multiperiod forecasts of interest rates. *Journal of Business & Economic Statistics*, 6(3):381–384, 1988.

[33] P. Dua and L. Kumawat. Modelling and forecasting seasonality in Indian macroeconomic time series. Working Paper 136, 2005.

[34] P. Dua and B. L. Pandit. Interest rate determination in India: Domestic and external factors. *Journal of Policy Modeling*, 24(9):853–875, 2002.

[35] P. Dua, N. Raje, and S. Sahoo. Interest rate modelling and forecasting in India. Department of Economic Analysis and Policy, Reserve Bank of India, 2003.

[36] P. Dua, N. Raje, and S. Sahoo. Forecasting interest rates in India. *Margin: The Journal of Applied Economic Research*, 2(1):1–41, 2008.

[37] P. Dua and R. Ranjan. Modelling and forecasting the exchange rate. In *Exchange Rate Policy and Modelling in India*, chapter 8, pages 170–219. Oxford University Press, India, 2012.

[38] P. Dua and S. C. Ray. A BVAR model for the Connecticut economy. *Journal of Forecasting*, 14(3):167–180, 1995.

[39] G. R. Duffee. Term premia and interest rate forecasts in affine models. *The Journal of Finance*, 57(1):405–443, 2002.

[40] G. R. Duffee. Forecasting interest rates. Economics Working Paper Archive 599, The Johns Hopkins University, Department of Economics, July 2012.

[41] Economic Survey. *Union Budget and Economic Survey.* Government of India, http://indiabudget.nic.in/es2012-13/echap-01.pdf, 2012–2013.

[42] G. Elliott, T. J. Rothenberg, and J. H. Stock. Efficient tests for an autoregressive unit root. *Econometrica*, 64(4):813–836, 1996.

[43] C. Engel. Can the Markov switching model forecast exchange rates? *Journal of International Economics*, 36(1):151–165, 1994.

[44] J. Faust and J. H. Wright. Forecasting inflation. In G. Elliott and A. Timmermann, editors, *Handbook of Economic Forecasting*, volume 2(A), pages 2–56. North Holland, Amsterdam, 2013.

[45] M. Forni, M. Hallin, M. Lippi, and L. Reichlin. The generalized dynamic factor model: One-sided estimation and forecasting. *Journal of the American Statistical Association*, 100(471):830–840, 2005.

[46] C. A. E. Goodhart and C. W. B. Lim. Interest rate forecasts: A pathology. *International Journal of Central Banking*, 7(2):135–171, June 2011.

[47] J. J. J. Groen, R. Paap, and F. Ravazzolo. Real time inflation forecasting in a changing world. *Journal of Business & Economic Statistics*, 31(1):29–44, 2013.

[48] R. Gupta and F. Hartley. The role of asset prices in forecasting inflation and output in South Africa. *Journal of Emerging Market Finance*, 12(3):239–291, 2013.

[49] R. Gupta, Y. Ye, and C. Sako. Financial variables and the out-of-sample forecastability of the growth rate of Indian industrial production. *Technological and Economic Development of Economy*, 20 (Supplement 1):S103–S119, 2014.

[50] H. Hassani, S. Heravi, and A. Zhigljavsky. Forecasting UK industrial production with multivariate singular spectrum analysis. *Journal of Forecasting*, 32(5):395–408, 2013.

[51] A. R. Joshi and D. Acharya. Inflation model for India in the context of open economy. *South Asia Economic Journal*, 12(1):39–59, 2011.

[52] M. Kapur. Inflation forecasting: Issues and challenges in India. *Reserve Bank of India Working Paper*, (01), 2012.

[53] S. Kar and K. Mandal. Banks, stock markets and output: Interactions in the Indian economy. In *12th Annual Global Development Conference*, Columbia, January 2011. GDN.

[54] A. Kasman, S. Kasman, and E. Torun. Dual long memory property in returns and volatility: Evidence from the CEE countries' stock markets. *Emerging Markets Review*, 10(2):122–139, 2009.

[55] M. Khashei and H. Bijari. Exchange rate forecasting better with hybrid artificial neural networks models. *Journal of Mathematical and Computational Science*, 1(1):103–125, 2012.

[56] T.-H. Kim, P. Mizen, and T. Chevapatrakul. Forecasting changes in UK interest rates. *Journal of Forecasting*, 27(1):53–74, 2008.

[57] T. Kinal and J. Ratner. A VAR forecasting model of a regional economy: Its construction and comparative accuracy. *International Regional Science Review*, 10(2):113–126, 1986.

[58] G. Koop and D. Korobilis. Bayesian multivariate time series methods for empirical macroeconomics. *Foundations and Trends in Econometrics*, 3(4):267–358, 2008.

[59] G. Koop and D. Korobilis. Forecasting inflation using dynamic model averaging. *International Economic Review*, 53(3):867–886, 2012.

[60] D. Korobilis. VAR forecasting using Bayesian variable selection. *Journal of Applied Econometrics*, 28(2):204–230, 2013.

[61] D. Kwiatkowski, P. C. B. Phillips, P. Schmidt, and Y. Shin. Testing the null hypothesis of stationarity against the alternative of a unit root: How sure are we that economic time series have a unit root? *Journal of Econometrics*, 54(1):159–178, 1992.

[62] L. Lam, L. Fung, and Ip-wing Yu. Comparing forecast performance of exchange rate models. Working Papers 0808, Hong Kong Monetary Authority, Jun 2008.

[63] R. B. Litterman. Forecasting with Bayesian vector autoregressions – five years of experience. *Journal of Business & Economic Statistics*, 4(1):25–38, 1986.

[64] D. Nachane and J. G. Clavel. Forecasting interest rates: A comparative assessment of some second-generation nonlinear models. *Journal of Applied Statistics*, 35(5):493–514, 2008.

[65] M. K. Newaz. Comparing the performance of time series models for forecasting exchange rate. *BRAC University Journal*, 5(2):55–65, 2008.

[66] P. C. Padhan. Application of ARIMA model for forecasting agricultural productivity in India. *Journal of Agriculture & Social Sciences*, 8(2):50–56, 2012.

[67] Y. Perwej and A. Perwej. Forecasting of Indian rupee INR/US dollar (USD) currency exchange rate using artificial neural network. *International Journal of Computer Science, Engineering and Applications (IJCSEA)*, 2(2):41–52, 2012.

[68] V. N. Pillai. A Markov chain model of inflation in India. *Indian Economic Review*, 37(1):91–116, 2002.

[69] R. P. Pradhan and R. Kumar. Forecasting exchange rate in India: An application of artificial neural network model. *Journal of Mathematics Research*, 2(4):111–117, 2010.

[70] P. Ray and J. S. Chatterjee. The role of asset prices in Indian inflation in recent years: Some conjectures. In B. for International Settlements, editor, *Modelling Aspects of the Inflation Process and the Monetary Transmission Mechanism in Emerging Market Countries*, volume 8, pages 131–150. Bank for International Settlements, April 2001.

[71] G. L. Shoesmith. Co-integration, error correction and improved medium-term regional VAR forecasting. *Journal of Forecasting*, 11(2):91–109, 1992.

[72] C. A. Sims. Macroeconomics and reality. *Econometrica*, 48(1):1–48, 1980.

[73] C. A. Sims, J. H. Stock, and M. W. Watson. Inference in linear time series models with some unit roots. *Econometrica*, 58(1):113–144, 1990.

[74] P. Sinha, S. Gupta, and N. Randev. Modeling & forecasting of macroeconomic variables of India: Before, during & after recession. *Journal of Applied Economic Sciences (JAES)*, 6(1-15):43–60, 2011.

[75] D. E. Spencer. Developing a Bayesian vector autoregression forecasting model. *International Journal of Forecasting*, 9(3):407– 421, 1993.

[76] A. Stella and J. H. Stock. A state-dependent model for inflation forecasting. International Finance Discussion Papers 1062, Board of Governors of the Federal Reserve System (USA), 2012.

[77] J. H. Stock and M. W. Watson. Phillips curve inflation forecasts. In J. Fuhrer, Y. Kodrzycki, J. Little, and G. Olivei, editors, *Understanding Inflation and the Implications for Monetary Policy*. MIT Press, Cambridge.

[78] J. H. Stock and M. W. Watson. Forecasting inflation. *Journal of Monetary Economics*, 44(2):293–335, 1999.

[79] J. H. Stock and M. W. Watson. Forecasting using principal components from a large number of predictors. *Journal of the American Statistical Association*, 97(460):1167–1179, 2002.

[80] J. H. Stock and M. W. Watson. Forecasting output and inflation: The role of asset prices. *Journal of Economic Literature*, 41(3):788–829, 2003.

[81] J. H. Stock and M. W. Watson. Why has U.S. inflation become harder to forecast? *Journal of Money, Credit and Banking*, 39:3–33, 2007.

[82] X. Tang. Analysis of interest rate forecasts from professional forecasters. 2012. http://www4.ncsu.edu/ njtraum/seminar/XiaoyanTang.pdf.

[83] H. Theil. *Principles of Econometrics*. John Wiley & Sons, New York, 1971.

[84] D. D. Thomakos and P. S. Bhattacharya. Forecasting inflation, industrial output and exchange rates: A template study for India. *Indian Economic Review*, 40(2):145–165, 2005.

[85] R. M. Todd. Improving economic forecasting with Bayesian vector autoregression. *Quarterly Review*, (Fall), 1984. Federal Reserve Bank of Minneapolis, USA.

[86] G. A. Vasilakis, K. A. Theofilatos, E. F. Georgopoulos, A. Karathanasopoulos, and S. D. Likothanassis. A genetic programming approach for EUR/USD exchange rate forecasting and trading. *Computational Economics*, 42(4):415–431, 2013.

[87] J. Wang and J. J. Wu. The Taylor rule and forecast intervals for exchange rates. *Journal of Money, Credit and Banking*, 44(1):103–144, 2012.

[88] J. Xiang and X. Zhu. A regime-switching Nelson - Siegel term structure model and interest rate forecasts. *Journal of Financial Econometrics*, 11(3):522–555, 2013.

3

Comparing Proportions: A Modern Solution to a Classical Problem

José M. Bernardo

University of Valencia

CONTENTS

3.1 Introduction

From a Bayesian viewpoint, the final outcome of any problem of inference is the posterior distribution of the vector of interest. Thus, given a probability model $\mathcal{M}_z = \{p(z \mid \omega), z \in \mathcal{Z}, \omega \in \Omega\}$ which is assumed to describe the mechanism which has generated the available data z, *all* that can be said about any function $\theta(\omega) \in \Theta$ of the parameter vector ω is contained in its posterior distribution $p(\theta \mid z)$. This is computed using standard probability theory techniques from the posterior distribution $p(\omega \mid z) \propto p(z \mid \omega) \, p(\omega)$ obtained by Bayes theorem from the assumed prior $p(\boldsymbol{\omega})$. To facilitate the assimilation of the inferential contents of $p(\boldsymbol{\theta} \,|\, \boldsymbol{z})$, one often tries to *summarize* the information contained in this posterior by

1. providing $\boldsymbol{\theta}$ values which, in the light of the data, are likely to be close to its true value (*estimation*), and

2. measuring the compatibility of the data with one or more possible values $\boldsymbol{\theta}_0 \in \boldsymbol{\Theta}$ of the vector of interest which might have been suggested by the research context (*hypothesis testing*).

One would expect that the *same* prior $p(\boldsymbol{\omega})$, whatever its basis, could be used to derive both types of summaries. However, since the pioneering work by Jeffreys [13], Bayesian methods have often made use of two *radically different* types of prior, some for estimation and some for hypothesis testing. It is argued that this is certainly *not necessary*, and probably not convenient, and that a coherent solution to both problems using the same prior is possible within the standard framework of Bayesian decision theory.

Section 3.2 specifies a decision theoretic formulation for point estimation, region estimation and precise hypothesis testing, emphasizes that the results are highly dependent on the choices of both the loss function and the prior distribution, and reviews a set of desiderata for loss functions to be used in stylized non context-specific problems of inference.

Section 3.3 proposes the use of the *average log-likelihood ratio against the null*, abbreviated to *intrinsic logarithmic loss*, as a self-calibrated information-based continuous loss function, which is suggested for general use in precise hypothesis testing.

Section 3.4 applies that methodology to the problem of comparing the parameters of two independent binomial populations, providing a *coherent* set of solutions—using the same prior—to *both* the problem of estimating their ratio, and the problem of testing whether or not data are compatible with the hypothesis that both parameters are equal. This is illustrated with the analysis of results published in the literature on the 2009 RV144 HIV vaccine efficacy trial held in Thailand.

3.2   Integrated Bayesian Analysis

3.2.1   Bayesian inference summaries

Let $\boldsymbol{z}$ be the available data which are assumed to have been generated as one random observation from model $\mathcal{M}_{\boldsymbol{z}} = \{p(\boldsymbol{z} \,|\, \boldsymbol{\omega}), \boldsymbol{z} \in \mathcal{Z}, \boldsymbol{\omega} \in \boldsymbol{\Omega}\}$. Often, but not always, data will consist of a random sample $\boldsymbol{z} = \{\boldsymbol{x}_1, \ldots, \boldsymbol{x}_n\}$ from some distribution $q(\boldsymbol{x} \,|\, \boldsymbol{\omega})$, $\boldsymbol{x} \in \mathcal{X}$; then, $p(\boldsymbol{z} \,|\, \boldsymbol{\omega}) = \prod_{i=1}^{n} q(\boldsymbol{x}_i \,|\, \boldsymbol{\omega})$, and

$\mathcal{Z} = \mathcal{X}^n$. Let $\boldsymbol{\theta}(\boldsymbol{\omega})$ be the vector of interest. Without loss of generality, the model may explicitly be expressed in terms of the quantity of interest $\boldsymbol{\theta}$, so that $\mathcal{M}_{\boldsymbol{z}} = \{p(\boldsymbol{z} \,|\, \boldsymbol{\theta}, \boldsymbol{\lambda}), \boldsymbol{z} \in \mathcal{Z}, \boldsymbol{\theta} \in \boldsymbol{\Theta}, \boldsymbol{\lambda} \in \boldsymbol{\Lambda}\}$, where $\boldsymbol{\lambda}$ is some appropriately chosen nuisance parameter vector. Let $p(\boldsymbol{\theta}, \boldsymbol{\lambda}) = p(\boldsymbol{\lambda} \,|\, \boldsymbol{\theta}) \, p(\boldsymbol{\theta})$ be the assumed prior, and let $p(\boldsymbol{\theta} \,|\, \boldsymbol{z})$ be the corresponding marginal posterior distribution of $\boldsymbol{\theta}$. Appreciation of the inferential contents of $p(\boldsymbol{\theta} \,|\, \boldsymbol{z})$ may be enhanced by providing both point and region estimates of the vector of interest $\boldsymbol{\theta}$, and by declaring whether or not some context suggested specific value $\boldsymbol{\theta}_0$ (or maybe a set of values $\boldsymbol{\Theta}_0$), is (are) compatible with the observed data $\boldsymbol{z}$. A large number of Bayesian estimation and hypothesis testing procedures have been proposed in the literature. It is argued that their construction is better made within a *coherent* decision theoretical framework, making use of the *same* prior distribution in all cases.

Let $\ell\{\boldsymbol{\theta}_0, (\boldsymbol{\theta}, \boldsymbol{\lambda})\}$ describe, as a function of the (unknown) parameter values $(\boldsymbol{\theta}, \boldsymbol{\lambda})$ which have generated the available data, *the loss* to be suffered if, working with model $\mathcal{M}_{\boldsymbol{z}}$, the value $\boldsymbol{\theta}_0$ were used as a proxy for the unknown value of $\boldsymbol{\theta}$. As summarized below, point estimation, region estimation and hypothesis are all appropriately described as specific decision problems using a *common* prior distribution and a *common* loss structure. The results may dramatically depend on the particular choices made for both the prior and the loss function but, given the available data $\boldsymbol{z}$, they all only depend on $\boldsymbol{z}$ through the corresponding posterior expected loss,

$$\bar{\ell}(\boldsymbol{\theta}_0 \,|\, \boldsymbol{z}) = \int_{\boldsymbol{\Theta}} \int_{\boldsymbol{\Lambda}} \ell\{\boldsymbol{\theta}_0, (\boldsymbol{\theta}, \boldsymbol{\lambda})\} \, p(\boldsymbol{\theta}, \boldsymbol{\lambda} \,|\, \boldsymbol{z}) \, d\boldsymbol{\theta} d\boldsymbol{\lambda}.$$

As a function of $\boldsymbol{\theta}_0 \in \boldsymbol{\Theta}$, the expected loss $\bar{\ell}(\boldsymbol{\theta}_0 \,|\, \boldsymbol{z})$ provides a *direct* measure of the relative unacceptability of all possible values of the quantity of interest in the light of the information provided by the data. Together with the marginal posterior distribution $p(\boldsymbol{\theta} \,|\, \boldsymbol{z})$, this provides the basis for an *integrated* coherent Bayesian analysis of the inferential content of the data $\boldsymbol{z}$ with respect to the quantity of interest $\boldsymbol{\theta}$.

3.2.1.1   Point estimation

To choose a point estimate for $\boldsymbol{\theta}$ may be seen as a decision problem where the action space is the class $\boldsymbol{\Theta}$ of all possible $\boldsymbol{\theta}$ values. Foundations of decision theory dictate that the best estimator is that which minimizes the expected loss; this is called the *Bayes estimator* which corresponds to this particular loss:

$$\boldsymbol{\theta}^*(\boldsymbol{z}) = \arg \inf_{\boldsymbol{\theta}_0 \in \boldsymbol{\Theta}} \bar{\ell}(\boldsymbol{\theta}_0 \,|\, \boldsymbol{z}).$$

Conventional examples of loss functions include the ubiquitous quadratic loss $\ell\{\boldsymbol{\theta}_0, (\boldsymbol{\theta}, \boldsymbol{\lambda})\} = (\boldsymbol{\theta}_0 - \boldsymbol{\theta})^t (\boldsymbol{\theta}_0 - \boldsymbol{\theta})$, which yields the posterior expectation as the Bayes estimator, and the zero-one loss on a neighborhood of the true value, which yields the posterior mode as a limiting result.

3.2.1.2   Region estimation

Bayesian region estimation is easily achieved by quoting posterior credible regions. To choose a q-credible region for $\boldsymbol{\theta}$ may be seen as a decision problem where the action space is the class of subsets of $\boldsymbol{\Theta}$ with posterior probability q. Foundations dictate that the best region is that which contains those $\boldsymbol{\theta}$ values with minimum expected loss. A *Bayes q-credible region* $\boldsymbol{\Theta}_q^*(\boldsymbol{z}) \subset \boldsymbol{\Theta}$ is a q-credible region where any value within the region has a smaller posterior expected loss than any value outside the region, so that

$$\forall \boldsymbol{\theta}_i \in \boldsymbol{\Theta}_q^*(\boldsymbol{z}), \ \forall \boldsymbol{\theta}_j \notin \boldsymbol{\Theta}_q^*(\boldsymbol{z}), \quad \bar{\ell}(\boldsymbol{\theta}_i \,|\, \boldsymbol{z}) \le \bar{\ell}(\boldsymbol{\theta}_j \,|\, \boldsymbol{z}).$$

The concept of a Bayes credible region was introduced by Bernardo in [7] under the name of *lower posterior loss* (LPL) credible regions.

  The quadratic loss function yields credible regions which contain those values of $\boldsymbol{\theta}$ closest to the posterior expectation in the Euclidean distance sense. A zero-one loss function leads to highest posterior density (HPD) credible regions.

3.2.1.3   Precise hypothesis testing

Consider a value $\boldsymbol{\theta}_0$ of the vector of interest which deserves special consideration, either because assuming $\boldsymbol{\theta} = \boldsymbol{\theta}_0$ would noticeably simplify the model, or because there are additional context specific arguments suggesting that $\boldsymbol{\theta} = \boldsymbol{\theta}_0$. Intuitively, the value $\boldsymbol{\theta}_0$ should be judged to be *compatible* with the observed data $\boldsymbol{z}$ if its posterior density $p(\boldsymbol{\theta}_0 \,|\, \boldsymbol{z})$ is relatively high. However, a more precise form of conclusion is typically required.

  Formally, testing the hypothesis $H_0 \equiv \{\boldsymbol{\theta} = \boldsymbol{\theta}_0\}$ may be described as a decision problem where the action space $\mathcal{A} = \{a_0, a_1\}$ contains only two elements: to accept (a_0) or to reject (a_1) the hypothesis under scrutiny. Foundations require to specify a loss function $\ell_h\{a_i, (\boldsymbol{\theta}, \boldsymbol{\lambda})\}$ measuring the consequences of accepting or rejecting H_0 as a function of the actual parameter values. By assumption, a_0 means to *act as if H_0 were true*, that is, to work with the submodel $\mathcal{M}_0 = \{p(\boldsymbol{z} \,|\, \boldsymbol{\theta}_0, \boldsymbol{\lambda}_0), \boldsymbol{z} \in \mathcal{Z}, \boldsymbol{\lambda}_0 \in \boldsymbol{\Lambda}\}$, while a_1 means to reject this simplification and to keep working with the full model $\mathcal{M}_{\boldsymbol{z}} = \{p(\boldsymbol{z} \,|\, \boldsymbol{\theta}, \boldsymbol{\lambda}), \boldsymbol{z} \in \mathcal{Z}, \boldsymbol{\theta} \in \boldsymbol{\Theta}, \boldsymbol{\lambda} \in \boldsymbol{\Lambda}\}$. Notice that the full submodel $\mathcal{M}_0$, a class of densities with $\boldsymbol{\theta} = \boldsymbol{\theta}_0$ and $\boldsymbol{\lambda}_0 \in \boldsymbol{\Lambda}$ must be considered, for, in general, $p(\boldsymbol{z} \,|\, \boldsymbol{\theta}_0)$ is not a fully specified probability density. Alternatively, an already established model $\mathcal{M}_0$ may have been embedded into a more general model $\mathcal{M}_{\boldsymbol{z}}$, constructed to include promising departures from $\boldsymbol{\theta} = \boldsymbol{\theta}_0$, and it is required to verify whether presently available data $\boldsymbol{z}$ are compatible with $\boldsymbol{\theta} = \boldsymbol{\theta}_0$, or whether the extension to $\boldsymbol{\theta} \in \boldsymbol{\Theta}$ is really necessary. The optimal action will be to reject the hypothesis if (and only if) the expected posterior loss of accepting (a_0) is larger than that of rejecting (a_1), so that

$$\int_{\boldsymbol{\Theta}} \int_{\boldsymbol{\Lambda}} [\ell_h\{a_0, (\boldsymbol{\theta}, \boldsymbol{\lambda})\} - \ell_h\{a_1, (\boldsymbol{\theta}, \boldsymbol{\lambda})\}] \, p(\boldsymbol{\theta}, \boldsymbol{\lambda} \,|\, \boldsymbol{z}) \, d\boldsymbol{\theta} d\boldsymbol{\lambda} > 0.$$

Hence, only the difference $\Delta \ell_h \{\boldsymbol{\theta}_0, (\boldsymbol{\theta}, \boldsymbol{\lambda})\} = \ell_h \{a_0, (\boldsymbol{\theta}, \boldsymbol{\lambda})\} - \ell_h \{a_1, (\boldsymbol{\theta}, \boldsymbol{\lambda})\}$, which measures the *marginal advantage of rejecting* $H_0 \equiv \{\boldsymbol{\theta} = \boldsymbol{\theta}_0\}$ as a function of the parameter values, must be specified. The hypothesis H_0 should be rejected whenever the expected marginal advantage of rejecting is positive. Without loss of generality, the function $\Delta \ell_h$ may be written in the form

$$\Delta \ell_h \{\boldsymbol{\theta}_0, (\boldsymbol{\theta}, \boldsymbol{\lambda})\} = \ell \{\boldsymbol{\theta}_0, (\boldsymbol{\theta}, \boldsymbol{\lambda})\} - \ell_0$$

where (precisely as in estimation), $\ell \{\boldsymbol{\theta}_0, (\boldsymbol{\theta}, \boldsymbol{\lambda})\}$ describes, as a function of the parameter values which have generated the data, the non-negative loss to be suffered if $\boldsymbol{\theta}_0$ were used as a proxy for $\boldsymbol{\theta}$. Since $\ell \{\boldsymbol{\theta}_0, (\boldsymbol{\theta}_0, \boldsymbol{\lambda})\} = 0$, so that $\Delta \ell_h \{\boldsymbol{\theta}_0, (\boldsymbol{\theta}_0, \boldsymbol{\lambda})\} = -\ell_0$, the constant $\ell_0 > 0$ describes (in the same loss units) the context-dependent non-negative marginal advantage of accepting $\boldsymbol{\theta} = \boldsymbol{\theta}_0$ when it is true. With this formulation, the optimal action is to reject $\boldsymbol{\theta} = \boldsymbol{\theta}_0$ whenever the expected value of $\ell \{\boldsymbol{\theta}_0, (\boldsymbol{\theta}, \boldsymbol{\lambda})\} - \ell_0$ is positive, *i.e.*, whenever $\overline{\ell}(\boldsymbol{\theta}_0 \,|\, \boldsymbol{z})$, the posterior expectation of $\ell \{\boldsymbol{\theta}_0, (\boldsymbol{\theta}, \boldsymbol{\lambda})\}$, is larger than ℓ_0. Thus the solution is found in terms of the *same* expected loss function that was needed for estimation. The *Bayes test criterion* to decide on the compatibility of $\boldsymbol{\theta} = \boldsymbol{\theta}_0$ with available data $\boldsymbol{z}$ is to reject $H_0 \equiv \{\boldsymbol{\theta} = \boldsymbol{\theta}_0\}$ if (and only if), $\overline{\ell}(\boldsymbol{\theta}_0 \,|\, \boldsymbol{z}) > \ell_0$, where ℓ_0 is a context dependent positive constant.

Using the quadratic loss function leads to rejecting a $\boldsymbol{\theta}_0$ value whenever its Euclidean distance to the posterior expectation of $\boldsymbol{\theta}$ is sufficiently large. The use of the (rather naive) zero-one loss function, $\ell \{\boldsymbol{\theta}_0, (\boldsymbol{\theta}, \boldsymbol{\lambda})\} = 0$ if $\boldsymbol{\theta} = \boldsymbol{\theta}_0$, and $\ell \{\boldsymbol{\theta}_0, (\boldsymbol{\theta}, \boldsymbol{\lambda})\} = 1$ otherwise, so that the loss advantage of rejecting $\boldsymbol{\theta}_0$ is a constant whenever $\boldsymbol{\theta} \neq \boldsymbol{\theta}_0$ and zero otherwise, leads to rejecting H_0 if (and only if) $\Pr(\boldsymbol{\theta} = \boldsymbol{\theta}_0 \,|\, \boldsymbol{z}) < p_0$ for some context-dependent p_0. Notice however that, using this particular loss function *requires* the prior probability $\Pr(\boldsymbol{\theta} = \boldsymbol{\theta}_0)$ to be *strictly positive*; if $\boldsymbol{\theta}$ is a continuous parameter this forces the use of a non-regular "sharp" prior, concentrating a positive probability mass at $\boldsymbol{\theta}_0$, which would typically *not* be appropriate for estimation and will obviously depend on the particular $\boldsymbol{\theta}_0$ value to test. Foundations would suggest however that the *same* prior—which is supposed to describe the available knowledge about the parameter values—should be used for *any* aspect of the Bayesian analysis of the problem.

The threshold constant ℓ_0—which is used to decide whether or not the expected loss $\overline{\ell}(\boldsymbol{\theta}_0 \,|\, \boldsymbol{z})$ is too large—is part of the specification of the decision problem, and should be context-dependent. However, as demonstrated below, a judicious choice of the loss function leads to *self-calibrated* expected losses, where the relevant threshold constant has an immediate, operational interpretation.

3.2.2   Continuous invariant loss functions

The formulation above is *totally general*, and may be used with any loss function $\ell \{\boldsymbol{\theta}_0, (\boldsymbol{\theta}, \boldsymbol{\lambda})\}$—which measures the loss to be suffered if a value $\boldsymbol{\theta}_0$ where used as a proxy for the true value $\boldsymbol{\theta}$—and any prior $p(\boldsymbol{\theta}, \boldsymbol{\lambda})$—which describes the available knowledge about the parameter values—to provide solutions for both estimation and hypothesis testing. If both the loss function and the prior distribution are *continuous*, precisely the *same* loss and the *same* prior may be used to obtain a coherent, integrated set of solutions for both estimation and testing, which may all be derived from the *joint* use of the corresponding posterior density $p(\boldsymbol{\theta} \mid \boldsymbol{z})$, and the posterior expected loss function $\overline{\ell}(\boldsymbol{\theta}_0 \mid \boldsymbol{z})$. Moreover the prior used may well be improper, as will typically be the case when an 'objective' analysis is required.

For most conventional loss functions, Bayes estimators are *not* invariant under one to one transformations. For example, the Bayes estimator of a variance under quadratic loss (its posterior expectation), is not the square of the Bayes estimator of the standard deviation. This is rather difficult to justify when, as it is the case in pure inference problems, one merely wishes to report an estimate of some quantity of interest.

Similarly, Bayes credible regions are generally *not* invariant under one to one transformations. Thus, HPD regions in one parameterization—obtained from a zero-one loss function—will *not* transform to HPD regions in another.

Rather more dramatically, Bayes test criteria are generally *not* invariant under one-to-one transformations so that, if $\boldsymbol{\phi}(\boldsymbol{\theta})$ is a one-to-one transformation of $\boldsymbol{\theta}$, rejecting $\boldsymbol{\theta} = \boldsymbol{\theta}_0$ does *not* generally imply rejecting the—logically equivalent—proposition $\boldsymbol{\phi}(\boldsymbol{\theta}) = \boldsymbol{\phi}(\boldsymbol{\theta}_0)$.

Invariant Bayes point estimators, credible regions and test procedures may all be easily obtained by using *invariant* loss functions, so that

$$\ell\{\boldsymbol{\theta}_0, (\boldsymbol{\theta}, \boldsymbol{\lambda})\} = \ell\{\boldsymbol{\phi}(\boldsymbol{\theta}_0), (\boldsymbol{\phi}(\boldsymbol{\theta}), \boldsymbol{\psi}(\boldsymbol{\lambda})\}$$

for any one-to-one transformations $\boldsymbol{\phi}(\boldsymbol{\theta})$ and $\boldsymbol{\psi}(\boldsymbol{\lambda})$ of $\boldsymbol{\theta}$ and $\boldsymbol{\lambda}$, rather than conventional (non-invariant) loss functions such as the quadratic or the zero-one loss functions. A particularly interesting family of invariant loss functions is described below.

Conditional on model $\mathcal{M}_{\boldsymbol{z}} = \{p(\boldsymbol{z} \mid \boldsymbol{\theta}, \boldsymbol{\lambda}), \boldsymbol{z} \in \mathcal{Z}, \boldsymbol{\theta} \in \boldsymbol{\Theta}, \boldsymbol{\lambda} \in \boldsymbol{\Lambda}\}$, the required loss function $\ell\{\boldsymbol{\theta}_0, (\boldsymbol{\theta}, \boldsymbol{\lambda})\}$ should describe, in terms of the unknown parameter values $(\boldsymbol{\theta}, \boldsymbol{\lambda})$ which are assumed to have generated the data, the loss to be suffered if, in work with model $\mathcal{M}_{\boldsymbol{z}}$, the value $\boldsymbol{\theta}_0$ were used as a proxy for $\boldsymbol{\theta}$. It may naively appear that what is needed is just some measure of the discrepancy between $\boldsymbol{\theta}_0$ and $\boldsymbol{\theta}$. However, since all parameterizations are arbitrary, what is really required is some measure of the discrepancy between the *models* labelled by $\boldsymbol{\theta}$ and by $\boldsymbol{\theta}_0$. By construction, such a discrepancy measure will be independent of the particular parameterization used. C. P. Robert [15] coined the word *intrinsic* to refer to these model-based loss functions; by construction, they are always invariant under one-to-one reparameterizations.

3.2.3 The intrinsic logarithmic loss

A particular *intrinsic* loss function with very attractive properties, the *logarithmic intrinsic loss*, is now introduced.

Let $\mathcal{M}_z = \{p(z \,|\, \boldsymbol{\theta}, \boldsymbol{\lambda}), z \in \mathcal{Z}\}$ be the model which is *assumed* to have generated the available data $z \in \mathcal{Z}$, where $\boldsymbol{\theta} \in \boldsymbol{\Theta}$ and $\boldsymbol{\lambda} \in \boldsymbol{\Lambda}$ are both unknown, and consider any other model $\mathcal{M}_0 = \{p(z \,|\, \boldsymbol{\omega}_0), z \in \mathcal{Z}\}$, for some $\boldsymbol{\omega}_0 \in \boldsymbol{\Omega}_0$, with the same or larger support. The Kullback–Leibler [14] directed divergence of the probability density $p(z \,|\, \boldsymbol{\omega}_0)$ *from* the probability density $p(z \,|\, \boldsymbol{\theta}, \boldsymbol{\lambda})$,

$$\kappa\{p_z(\cdot \,|\, \boldsymbol{\omega}_0) \,|\, p_z(\cdot \,|\, \boldsymbol{\theta}, \boldsymbol{\lambda})\} = \int_{\mathcal{Z}} p(z \,|\, \boldsymbol{\theta}, \boldsymbol{\lambda}) \log \frac{p(z \,|\, \boldsymbol{\theta}, \boldsymbol{\lambda})}{p(z \,|\, \boldsymbol{\omega}_0)} \, dz,$$

is the average (under repeated sampling) log-likelihood ratio *against* the alternative model $p(z \,|\, \boldsymbol{\omega}_0)$. This is known to be *nonnegative*, and zero if, and only if, $p(z \,|\, \boldsymbol{\omega}_0) = p(z \,|\, \boldsymbol{\theta}, \boldsymbol{\lambda})$ almost everywhere, and it is *invariant* under one-to-one transformations of either the data z or the parameters $\boldsymbol{\theta}$, $\boldsymbol{\lambda}$ and $\boldsymbol{\omega}_0$. It is also *additive*, in the sense that if $z = \{x_1, \ldots, x_n\}$ is assumed to a random sample from some model, then

$$\kappa\{p_z(\cdot \,|\, \boldsymbol{\omega}_0) \,|\, p_z(\cdot \,|\, \boldsymbol{\theta}, \boldsymbol{\lambda})\} = n \, \kappa\{p_x(\cdot \,|\, \boldsymbol{\omega}_0) \,|\, p_x(\cdot \,|\, \boldsymbol{\theta}, \boldsymbol{\lambda})\}.$$

And it is *invariant* under reduction to sufficient statistics in the sense that, if $t \in \mathcal{T}$ is a sufficient statistic for both $\mathcal{M}_z$ and $\mathcal{M}_0$, then

$$\kappa\{p_z(\cdot \,|\, \boldsymbol{\omega}_0) \,|\, p_z(\cdot \,|\, \boldsymbol{\theta}, \boldsymbol{\lambda})\} = \kappa\{p_t(\cdot \,|\, \boldsymbol{\omega}_0) \,|\, p_t(\cdot \,|\, \boldsymbol{\theta}, \boldsymbol{\lambda})\}.$$

Definition 1. Intrinsic logarithmic loss function. *Let z be the available data, let $\mathcal{M}_z = \{p(z \,|\, \boldsymbol{\theta}, \boldsymbol{\lambda}), z \in \mathcal{Z}\}$ be the model from which the data are assumed to have been generated, and let H_0 be the hypothesis that the data have actually been generated from a member of the family*

$$\mathcal{M}_0 = \{p(z \,|\, \boldsymbol{\omega}_0), \boldsymbol{\omega}_0 \in \boldsymbol{\Omega}_0, z \in \mathcal{Z}_0\}, \quad \mathcal{Z} \subseteq \mathcal{Z}_0$$

The intrinsic logarithmic loss function from assuming H_0 is the minimum average under sampling of the log-likelihood ratio against an element of $\mathcal{M}_0$,

$$\delta\{H_0 \,|\, \boldsymbol{\theta}, \boldsymbol{\lambda}, \mathcal{M}_z\} = \inf_{\boldsymbol{\omega}_0 \in \boldsymbol{\Omega}_0} \kappa\{p_z(\cdot \,|\, \boldsymbol{\omega}_0) \,|\, p_z(\cdot \,|\, \boldsymbol{\theta}, \boldsymbol{\lambda})\}.$$

Notice the complete generality of this definition. It may be used with either discrete or continuous data models (in the discrete case, the integrals will obviously be sums), and with either discrete or continuous parameter spaces, of any dimensionality.

The particular case which obtains when $H_0 \equiv \{\boldsymbol{\theta} = \boldsymbol{\theta}_0\}$, so that

$$\delta\{H_0 \,|\, \boldsymbol{\theta}, \boldsymbol{\lambda}, \mathcal{M}_z\} = \delta_z\{\boldsymbol{\theta}_0 \,|\, \boldsymbol{\theta}, \boldsymbol{\lambda}\} = \inf_{\boldsymbol{\lambda}_0 \in \boldsymbol{\Lambda}_0} \int_{\mathcal{Z}} p(z \,|\, \boldsymbol{\theta}, \boldsymbol{\lambda}) \log \frac{p(z \,|\, \boldsymbol{\theta}, \boldsymbol{\lambda})}{p(z \,|\, \boldsymbol{\theta}_0, \boldsymbol{\lambda}_0)} \, dz,$$

is an appropriate loss function for both point and region estimation of $\boldsymbol{\theta}$, and for testing whether or not a particular $\boldsymbol{\theta}_0$ value is compatible with the observed data.

The intrinsic logarithmic loss function $\delta\{H_0 \mid \boldsymbol{\theta}, \boldsymbol{\lambda}, \mathcal{M}_{\boldsymbol{z}}\}$ formalizes the use of log-likelihood ratios against the null to define a general loss function. With this loss structure, a precise hypothesis H_0 will be rejected if, and only if

$$d(H_0 \mid \boldsymbol{z}) = \int_{\Theta} \int_{\Lambda} \delta\{H_0 \mid \boldsymbol{\theta}, \boldsymbol{\lambda}, \mathcal{M}_{\boldsymbol{z}}\} \, p(\boldsymbol{\theta}, \boldsymbol{\lambda} \mid \boldsymbol{z}) \, d\boldsymbol{\theta} \, d\boldsymbol{\lambda} > \ell_0,$$

that is if, and only if, the posterior expectation of the average log-likelihood ratio loss—which estimates the minimum log-likelihood ratio against H_0—is larger than a suitably chosen constant ℓ_0. In particular, if $\ell_0 = \log[R]$, then H_0 would be rejected whenever, given the observed data, the minimum average likelihood ratio against H_0, may be expected to be larger than about R. Conventional choices for ℓ_0 are $\{\log 20, \log 100, \log 1000\} \approx \{3.0, 4.6, 6.9\}$.

In a multivariate normal model with known covariance matrix the intrinsic logarithmic loss is proportional to the Mahalanobis distance. Thus, if $\boldsymbol{z}$ is a random sample of size n from a k-variate normal distribution $\mathrm{N}(\boldsymbol{x} \mid \boldsymbol{\mu}, \boldsymbol{\Sigma})$,

$$\delta_{\boldsymbol{z}}\{\boldsymbol{\mu}_0 \mid \boldsymbol{\mu}, \boldsymbol{\Sigma}\} = \frac{n}{2}(\boldsymbol{\mu}_0 - \boldsymbol{\mu})^t \boldsymbol{\Sigma}^{-1} (\boldsymbol{\mu}_0 - \boldsymbol{\mu}),$$

which is $n/2$ times the Mahalanobis distance between $\boldsymbol{\mu}_0$ and $\boldsymbol{\mu}$. This result may be used to obtain large-sample approximations to the intrinsic logarithmic loss. In particular, if $\boldsymbol{z}$ is a random sample of size n from the single parameter model $p(x \mid \theta)$, and $\tilde{\theta}_n = \tilde{\theta}_n(\boldsymbol{z})$ is an asymptotically sufficient consistent estimator of θ whose sampling distribution is asymptotically normal with standard deviation $s(\theta)/\sqrt{n}$, then, for large values of n,

$$\delta_{\boldsymbol{z}}\{\theta_0 \mid \theta, \mathcal{M}_{\boldsymbol{z}}\} \approx \frac{n}{2}[\phi(\theta_0) - \phi(\theta)]^2,$$

where $\phi(\theta) = \int^{\theta} s(y)^{-1} dy$ is the corresponding variance stabilization transformation.

As mentioned in the introduction, the results of a decision theoretical structured Bayesian analysis may be very dependent on the particular choice of the loss function, specially in hypothesis testing. In particular, the naive zero-loss function may produce rather different results than those derived from a continuous loss function. We have argued that hypothesis testing should really be done with a continuous, preferably invariant, loss function, and we have suggested a systematic use of the intrinsic logarithmic loss. Other *continuous* loss functions would certainly produce slightly different results, but those would be qualitatively close to those obtained with the logarithmic loss, and would all converge to the same answer as the sample size increases, as the Mahalanobis approximation indicates.

3.3   Intrinsic Reference Analysis

The decision-theoretic procedures described above to derive summaries for Bayesian inference are totally general, so that they may be used with any loss function and any prior distribution. The advantages of using the intrinsic logarithmic loss have been described above: it is invariant under both reparameterization and reduction to sufficient statistics, and—most important—it has a simple operational interpretation in terms of average log-likelihood ratios against the null, so it is self-calibrated in terms of simple log-likelihood ratios.

3.3.1   Intrinsic reference estimation and testing

Foundations indicate that the prior distribution should describe available prior knowledge. In many situations however, either the available prior information is too vague to warrant the effort required to formalize it, or it is too subjective to be useful in scientific communication. An "objective" procedure, where the prior function is intended to describe a situation where there is no relevant information about the quantity of interest, is therefore often required. Objectivity is an emotionally charged word, and it should be explicitly qualified whenever it is used. No statistical analysis is really objective, since both the experimental design and the model assumed have very strong subjective inputs. However, frequentist procedures are often branded as "objective" just because their conclusions are only conditional on the model assumed and the data obtained. Bayesian methods where the prior function is directly derived from the assumed model are objective is this limited, but precise sense. There is a vast literature devoted to the formulation of objective priors. Reference analysis, introduced by Bernardo in [6], and further developed in [1], [2], [3], [4] and references therein, is probably the most popular approach for deriving objective priors.

Reference priors may be numerically obtained (see [3] for details) but, under appropriate regularity conditions, explicit formulae for the reference priors are readily available (see [2] and [11] for details). In particular, if the posterior distribution of θ given a random sample of size n from $p(x \mid \theta)$ is asymptotically normal with standard deviation $s(\tilde{\theta}_n)/\sqrt{n}$, where $\tilde{\theta}_n$ is a consistent estimator of θ, then the reference prior is $\pi(\theta) = s(\theta)^{-1}$. This includes the well-known one-parameter Jeffreys prior

$$\pi(\theta) \propto i(\theta)^{1/2}, \quad i(\theta) = \mathrm{E}_{\boldsymbol{x} \mid \theta}[-\partial^2 \log p(\boldsymbol{z} \mid \theta)/\partial \theta^2],$$

as a particular case.

For objective Bayesian solutions to inferential problems the *combined* use of the *intrinsic logarithmic loss* function and the relevant *reference prior* are recommended. The corresponding Bayes point estimators, Bayes credible regions and Bayes test criteria are respectively referred to as *intrinsic reference estimators*, credible regions or test criteria. The basic ideas were respectively introduced in [9], [8] and [10].

All inference summaries depend on the data only through the expected reference intrinsic loss, $d(\boldsymbol{\theta}_0 \,|\, \boldsymbol{z})$, the expectation of the intrinsic loss with respect to the appropriate joint reference posterior,

$$d(\boldsymbol{\theta}_0 \,|\, \boldsymbol{z}) = \int_{\Theta} \int_{\Lambda} \delta\{\boldsymbol{\theta}_0 \,|\, \boldsymbol{\theta}, \boldsymbol{\lambda}, \mathcal{M}_{\boldsymbol{z}}\} \, \pi(\boldsymbol{\theta}, \boldsymbol{\lambda} \,|\, \boldsymbol{z}) \, d\boldsymbol{\theta} d\boldsymbol{\lambda}.$$

Most other *intrinsic* loss functions (invariant, continuous loss functions which measure the discrepancy between the models rather than the discrepancy between their parameters) would yield qualitatively similar results, but attention will here be confined to the *intrinsic logarithmic* loss defined above, for this is often easily derived, and—most important—it is self-calibrated in terms of easily interpretable log-likelihood ratios against the null.

The following example is intended to illustrate the general procedure:

3.3.2   Example: The normal variance

Let $\boldsymbol{z} = \{x_1, \ldots, x_n\}$ be a random sample from a normal $\mathrm{N}(x \,|\, \mu, \sigma)$ distribution whose variance σ^2 is of interest. Since reference analysis is invariant under one-to-one transformations, one may equivalently work in terms of σ, $\log \sigma$, of any other one-to-one transformation of σ. The Kullback-Leibler discrepancy of $p(\boldsymbol{z} \,|\, \mu_0, \sigma_0)$ from $p(\boldsymbol{z} \,|\, \mu, \sigma)$ is given by

$$
\begin{aligned}
\kappa\{\mathrm{N}_{\boldsymbol{z}}(\cdot \,|\, \mu_0, \sigma_0) \,|\, \mathrm{N}_{\boldsymbol{z}}(\cdot \,|\, \mu, \sigma)\} &= n \int_{\Re} \mathrm{N}(x \,|\, \mu, \sigma) \log \frac{\mathrm{N}(x \,|\, \mu, \sigma)}{\mathrm{N}(x \,|\, \mu_0, \sigma_0)} \, dx \\
&= \frac{n}{2} \left[\log \frac{\sigma_0^2}{\sigma^2} + \frac{\sigma^2}{\sigma_0^2} - 1 + \frac{(\mu - \mu_0)^2}{\sigma_0^2} \right],
\end{aligned}
$$

which is minimized when $\mu_0 = \mu$. Hence, the intrinsic logarithmic loss function for $H_0 \equiv \{\sigma = \sigma_0\}$ is

$$\delta\{H_0 \,|\, \sigma, \mu, \mathcal{M}_{\boldsymbol{z}}\} = \delta_{\boldsymbol{z}}\{\sigma_0 \,|\, \sigma, \mu\} = \frac{n}{2} \left[\log \frac{\sigma_0^2}{\sigma^2} + \frac{\sigma^2}{\sigma_0^2} - 1 \right].$$

Since the normal is a location-scale model, the reference prior is the conventional improper prior $\pi(\mu, \sigma) = \sigma^{-1}$. The corresponding reference posterior density of σ, after a random sample $\boldsymbol{z} = \{x_1, ..., x_n\}$ of size $n \geq 2$ has been observed, which is always proper, is

$$\pi(\sigma \,|\, \boldsymbol{z}) = \pi(\sigma \,|\, n, s^2) = \frac{(ns^2)^{(n-1)/2}}{2^{(n-3)/2}\Gamma[(n-1)/2]} \sigma^{-n} \exp[-\frac{ns^2}{2\sigma^2}],$$

where $s^2 = n^{-1} \sum_{j=1}^{n} (x_j - \bar{x})^2$ is the MLE of σ^2.

The corresponding *reference posterior expected loss* from using σ_0 as a proxy for σ, given a random sample of size n, is

$$d(\sigma_0 \mid z) = \int_0^\infty n \, \delta_z \{\sigma_0 \mid \sigma, \mu\} \, \pi(\sigma \mid z) \, d\sigma$$

$$= \frac{n}{2} \left[\psi\left(\frac{n-1}{2}\right) - 1 + \frac{ns^2}{(n-3)\sigma_0^2} + \log\left(\frac{2\sigma_0^2}{ns^2}\right) \right].$$

By definition, the Bayes point estimator with respect to this loss function, the intrinsic reference estimator of σ is that value of σ_0 which minimizes $d(\sigma_0 \mid z)$; this is found to be $\sigma^*(z) = \sqrt{ns}/\sqrt{n-3}$. Thus, the intrinsic reference estimator of the variance is

$$\sigma^{2*}(z) = \frac{n s^2}{n-3},$$

an estimator already suggested by Stein [16], which is always larger than both the MLE and the conventional unbiased estimator, that respectively divide the sum of squares ns^2 by n and by $n-1$; for small samples, the differences are noticeable. Since intrinsic estimation is consistent under one-to-one reparametrizations, the intrinsic reference estimator of, say, $\log \sigma$ is simply $\log[\sigma^*(z)]$.

As an illustration, a random sample z of size $n = 10$ was simulated from a normal distribution with $\mu = 1$ and $\sigma = 2$, yielding $\bar{x} = 0.951$ and $s = 1.631$. Intrinsic reference analysis of σ is well summarized by two complementary functions: (i) the reference posterior density $\pi(\sigma \mid z)$, and (ii) the expected posterior intrinsic logarithmic loss $d(\sigma_0 \mid z)$ of using σ_0 as a proxy for σ. Figure 3.1 represents both $\pi(\sigma \mid z)$ (upper panel) and $d(\sigma_0 \mid z)$ (lower panel), in the same horizontal scale.

The expected intrinsic logarithmic loss $d(\sigma_0 \mid z)$ is minimized at the reference intrinsic estimate, $\sigma^* = 1.949$, represented in both panels by a black dot. Thus, the intrinsic estimator of the variance is $\sigma^{*2} = 3.80$, which may be compared with the MLE $s^2 = 2.66$, or with the conventional unbiased estimator $\hat{\sigma}^2 = 2.96$.

To test if a particular σ_0 value is supported by the data one simply checks its expected loss. For example, all σ_0 values smaller than 1.29 or larger than 3.46 have an expected intrinsic logarithmic loss larger than $\log(20) \approx 3.0$, and would be rejected if the threshold were set to reject σ_0 values with an expected log-likelihood ratio against the true (unknown) value of σ larger than $\log(20)$, suggesting that the average likelihood under the true model may be expected to be at least 20 times larger than that under any model with $\sigma = \sigma_0$. The corresponding acceptance region, the interval $(1.29, 3.46)$, shaded area in the upper panel) has a posterior probability of 0.92. Since all elements within that region have smaller expected loss than those outside, this is an intrinsic reference 0.92-credible region. By definition, this region is

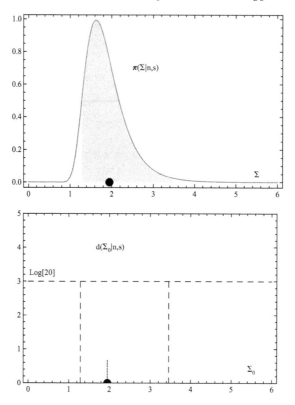

FIGURE 3.1
Intrinsic reference analysis for the standard deviation of a normal distribution, given a random sample of size $n = 5$, with $s = 1.631$.

invariant under transformations; thus, the 0.92 reference intrinsic region for σ^2 is simply $(1.29^2, 3.46^2)$. Notice, that these regions are *not* HPD regions. The intrinsic reference 0.95-credible interval is $(1.24, 3.74)$.

The expected logarithmic intrinsic loss of using $\sigma_0 = 5$ as a proxy for σ is $d(5 \,|\, z) = 5.86 \approx \log[350]$, so that the hypothesis $H_0 \equiv \{\sigma = 5\}$ would typically be rejected on the grounds that the likelihood ratio *against* H_0 may be expected to be at least 350.

3.4 Comparing Binomial Proportions

Consider two random samples of sizes n_1 and n_2 from independent binomial populations with parameters θ_1 and θ_2, respectively yielding r_1 and r_2 successes. Thus, the data are $z = \{(r_1, n_1), (r_2, n_2)\}$, the unknown parameter vector is $\boldsymbol{\theta} = \{\theta_1, \theta_2\}$, and the sampling model is

$$p(\boldsymbol{z} \,|\, \boldsymbol{\theta}) = \mathrm{Bi}(r_1 \,|\, n_1, \theta_1)\,\mathrm{Bi}(r_2 \,|\, n_2, \theta_2) \propto \theta_1^{r_1}(1 - \theta_1)^{n_1 - r_1}\theta_2^{r_2}(1 - \theta_2)^{n_2 - r_2}$$

Interest focuses in comparing θ_1 and θ_2 and, more specifically, in deciding whether or not there is evidence against the hypothesis $H_0 \equiv \{\theta_1 = \theta_2\}$ that the two proportions are actually equal.

3.4.1   Intrinsic logarithmic loss to test equality

The Kullback–Leibler discrepancy of a model $p(\boldsymbol{z} \,|\, \alpha)$ with equal parameters $\theta_1 = \theta_2 = \alpha$, so that $p(\boldsymbol{z} \,|\, \alpha) = \mathrm{Bi}(r_1 \,|\, n_1, \alpha)\,\mathrm{Bi}(r_2 \,|\, n_2, \alpha)$, from the assumed model $p(\boldsymbol{z} \,|\, \boldsymbol{\theta}) = \mathrm{Bi}(r_1 \,|\, n_1, \theta_1)\,\mathrm{Bi}(r_2 \,|\, n_2, \theta_2)$, is the expected value under the true model $p(\boldsymbol{z} \,|\, \boldsymbol{\theta})$ of the corresponding log-likelihood ratio ratio,

$$
\begin{aligned}
\kappa\{p_{\boldsymbol{z}}(\cdot \,|\, \alpha) \,|\, p_{\boldsymbol{z}}(\cdot \,|\, \boldsymbol{\theta})\} \;=\; & \mathrm{E}_{\boldsymbol{z}\,|\,\boldsymbol{\theta}}\left[\log\frac{p(\boldsymbol{z} \,|\, \boldsymbol{\theta})}{p(\boldsymbol{z} \,|\, \alpha)}\right] \\
=\; & n_1 \log(1 - \theta_1) + n_2 \log(1 - \theta_2) \\
& -[n_1(1 - \theta_1) + n_2(1 - \theta_2)]\log[1 - \alpha] \\
& -n_1\theta_1 \log\frac{\alpha\,(1 - \theta_1)}{\theta_1} - n_2\theta_2 \log\frac{\alpha\,(1 - \theta_2)}{\theta_2}\;,
\end{aligned}
$$

which is minimized when

$$\alpha = \frac{n_1\theta_1 + n_2\theta_2}{n_1 + n_2}\;.$$

The intrinsic logarithmic loss function for $H_0 \equiv \{\theta_1 = \theta_2\}$ is

$$\delta\{H_0 \,|\, \theta_1, \theta_2, n_1, n_2\} = \min_{\alpha \in (0,1)} \kappa\{p_{\boldsymbol{z}}(\cdot \,|\, \alpha) \,|\, p_{\boldsymbol{z}}(\cdot \,|\, \boldsymbol{\theta})\},$$

and substitution of the minimizing value of α yields

$$
\begin{aligned}
\delta\{H_0 \,|\, \theta_1, \theta_2, n_1, n_2\} \;=\; & n_1 \log(1 - \theta_1) + n_2 \log(1 - \theta_2) \\
& + n_1\theta_1 \log[\frac{\theta_1}{1 - \theta_1}] + n_2\theta_2 \log[\frac{\theta_2}{1 - \theta_2}] \\
& - (n_1\theta_1 + n_2\theta_2)\log[\frac{n_1\theta_1 + n_2\theta_2}{n_1 + n_2}] \\
& - (n_1(1 - \theta_1) + n_2(1 - \theta_2))\log[1 - \frac{n_1\theta_1 + n_2\theta_2}{n_1 + n_2}],
\end{aligned}
$$

a non-negative function of (θ_1, θ_2) which is zero if, and only if, $\theta_1 = \theta_2$ and reaches a maximum

$$\delta_{max} = (n_1 + n_2)\log(n_1 + n_2) - n_1 \log n_1 - n_2 \log n_2$$

when $(\theta_1, \theta_2) = (0, 1)$ or $(\theta_1, \theta_2) = (1, 0)$, which reduces to $2\,n \log[2]$ when $n_1 = n_2 = n$.

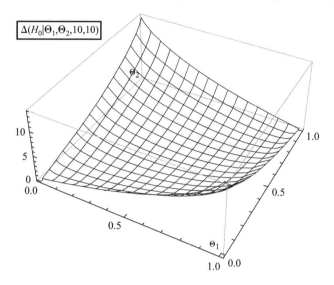

FIGURE 3.2
Intrinsic logarithmic loss function for $H_0 = \{\theta_1 = \theta_2\}$ when $n_1 = n_2 = 10$.

Figure 3.2 represents $\delta\{H_0 \mid \theta_1, \theta_2, n_1, n_2\}$ when $n_1 = n_2 = 10$. The function $\delta\{H_0 \mid \theta_1, \theta_2, n_1, n_2\}$ unambiguously and precisely describes, as a function of θ_1 and θ_2, the discrepancy of the hypothesis $H_0 \equiv \{\theta_1 = \theta_2\}$ from the model $\mathrm{Bi}(r_1 \mid n_1, \theta_1)\,\mathrm{Bi}(r_2 \mid n_2, \theta_2)$.

3.4.2   Reference posterior distributions

Objective solutions to inferences about the possible differences between θ_1 and θ_2 require the use of an objective joint prior $\pi(\theta_1, \theta_2)$. In single parameter problems, the reference prior is uniquely defined, and it is invariant under reparameterization. In multiparameter models however, the reference prior depends on the quantity of interest.

In this problem there are several possible choices for the quantity of interest. A clear option would be the intrinsic logarithmic loss function itself, for this precisely measures the discrepancy of H_0 for the model, and one is interested in checking whether or not this might be zero. Thus, one could define $\phi_0(\theta_1, \theta_2) = \delta\{H_0 \mid \theta_1, \theta_2, n_1, n_2\}$ and proceed to derive the reference prior $\pi_{\phi_0}(\theta_1, \theta_2)$ when ϕ_0 is the quantity of interest; this is a non trivial exercise, but it may be done. However, other options more easily interpretable by the user suggest themselves, as the difference $\phi_1(\theta_1, \theta_2) = \theta_1 - \theta_2$, or the ratio $\phi_2(\theta_1, \theta_2) = \theta_1/\theta_2$.

Indeed, in models with many parameters, there are many situations where one is simultaneously interested in several functions of them, and it would then be useful to have a *single* objective prior which could safely be used to produce reasonable marginal posteriors for all the quantities of interest. Berger,

Bernardo and Sun propose in [5] a criterium to select an *overall* joint prior function which may be considered a good approximate joint reference prior, in the sense that, for all datasets, it may be expected to produce marginal posteriors for all the quantities of interest which are not too different from the relevant reference posteriors. In situations where independent binomial situations are considered this leads to the use of the corresponding reference priors for each of the binomial models considered, which are known to be the relevant (proper) Jeffreys priors, $\pi(\theta_i) = \mathrm{Be}(\theta_i \mid \frac{1}{2}, \frac{1}{2})$. In the case discussed here, this reduces to

$$
\begin{aligned}
\pi(\theta_1, \theta_2) &= \mathrm{Be}(\theta_1 \mid \tfrac{1}{2}, \tfrac{1}{2}) \, \mathrm{Be}(\theta_2 \mid \tfrac{1}{2}, \tfrac{1}{2}) \\
&= \pi^{-2} \theta_1^{-1/2} (1 - \theta_1)^{-1/2} \theta_2^{-1/2} (1 - \theta_2)^{-1/2},
\end{aligned}
$$

and this is the *overall* objective prior suggested to analyze this problem. The corresponding joint reference posterior is

$$
\begin{aligned}
\pi(\theta_1, \theta_2 \mid \boldsymbol{z}) &= \pi(\theta_1, \theta_2 \mid r_1, r_2, n_1, n_2) \\
&= \mathrm{Be}(\theta_1 \mid r_1 + \tfrac{1}{2}, n_1 - r_1 + \tfrac{1}{2}) \, \mathrm{Be}(\theta_2 \mid r_2 + \tfrac{1}{2}, n_2 - r_2 + \tfrac{1}{2}),
\end{aligned}
$$

and the expected logarithmic intrinsic loss is

$$
d(H_0 \mid \boldsymbol{z}) = \int_0^1 \int_0^1 \delta\{H_0 \mid \theta_1, \theta_2, n_1, n_2\} \, \pi(\theta_1, \theta_2 \mid r_1, r_2, n_1, n_2) \, d\theta_1 \, d\theta_2,
$$

which may easily be numerically evaluated.

The numerical results will certainly depend on the particular choice of the prior, but simulations show that practical differences between the results obtained from the alternative reference priors mentioned above are rather negligible even for quite moderate sample sizes.

3.4.3 The RV144 HIV vaccine efficacy trial in Thailand

In 2009, the RV144 randomized, double-blind, efficacy trial in Thailand reported that a prime-boost human immunodeficiency virus (HIV) vaccine regimen conferred about 30% protection against HIV acquisition, but different analyses seemed to give conflicting results, and a heated debate followed as scientists and the broader public struggled with their interpretation; see [12] for a detailed description of the issues involved. The main result concerned individuals in the general population in Thailand, mostly at heterosexual risk, 61% of which were men, randomized within the "intention to treat" population, excluding subjects found to be HIV positive at the time of randomization. The press release reported $r_1 = 51$ infected among $n_1 = 8197$ who have taken the vaccine, to be compared with $r_2 = 74$ infected among $n_2 = 8198$ who have taken a placebo. Using conventional frequentist testing, the two corresponding binomial parameters, θ_1 and θ_2, were said to be significantly different, with

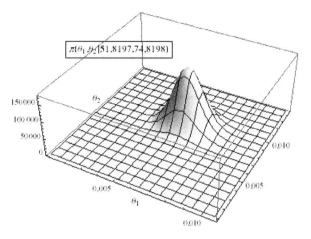

FIGURE 3.3
Joint posterior reference density $\pi(\theta_1, \theta_2 \mid r_1, r_2, n_1, n_2)$, when $r_1 = 51$, $n_1 = 8197$, $r_2 = 74$ and $n_2 = 8198$.

a p-value of 0.04. Morever, the results suggested an *estimated* vaccine efficacy (one minus the relative hazard rate of HIV in the vaccine versus placebo group) of

$$\mathrm{VE}(r_1, n_1, r_2, n_2) = 1 - \frac{r_1/n_1}{r_2/n_2} = 0.31.$$

An objective Bayesian, intrinsic reference analysis of these data will now be provided.

The joint reference posterior distribution for θ_1 and θ_2 which corresponds to the overall reference prior $\pi(\theta_1, \theta_2) = \mathrm{Be}(\theta_1 \mid \frac{1}{2}, \frac{1}{2}) \mathrm{Be}(\theta_2 \mid \frac{1}{2}, \frac{1}{2})$ is $\pi(\theta_1, \theta_2 \mid r_1, r_2, n_1, n_2) = \mathrm{Be}(\theta_1 \mid 51.5, 8146.5) \mathrm{Be}(\theta_2 \mid 74.5, 8124.5)$, represented in Figure 3.3, which has with a unique mode at $(0.0063, 0.0091)$.

The reference posterior probability that the proportion θ_1 of infected among those vaccinated is actually smaller than the proportion θ_2 of infected among those take the placebo is simply

$$\Pr[\theta_1 < \theta_2 \mid \boldsymbol{z}] = \int_0^1 \int_0^{\theta_2} \pi(\theta_1, \theta_2 \mid \boldsymbol{z}) \, d\theta_1 \, d\theta_2 = 0.981,$$

so there seems to be evidence that, indeed, θ_1 may well be smaller than θ_2.

The posterior expectation of the intrinsic logarithmic loss corresponding to the hypothesis $H_0 = \{\theta_1 = \theta_2\}$ that both parameters are actually equal is

$$d(H_0 \mid \boldsymbol{z}) = \int_0^1 \int_0^1 \delta\{H_0 \mid \boldsymbol{\theta}, n_1, n_2\} \pi(\theta_1, \theta_2 \mid \boldsymbol{z}) \, d\theta_1 \, d\theta_2 = 2.624 = \log[13.8].$$

Thus, given the data, it is estimated that the *average* log-likelihood ratio

$$\delta\{H_0 \mid \boldsymbol{\theta}, n_1, n_2\} = \inf_{\boldsymbol{\theta}_0 \in H_0} E_{\boldsymbol{z}\mid\boldsymbol{\theta}}\left[\log \frac{p(\boldsymbol{z}\mid\boldsymbol{\theta})}{p(\boldsymbol{z}\mid\boldsymbol{\theta}_0)}\right]$$

for the model who has generated to data against any model within H_0 is at least $2.624 = \log[13.8]$; hence, the observed data may be expected to be about 14 times more likely under the true model than under a model within H_0. This certainly indicates that there is some evidence of the existence of a difference between the two hazard rates and that the vaccine may well be effective, but the evidence is far from strong.

Standard probability calculus may be used to derive the reference posterior distribution of the actual efficacy of the vaccine,

$$\phi(\theta_1, \theta_2) = 1 - \frac{\theta_1}{\theta_2} ,$$

from the joint reference posterior $\pi(\theta_1, \theta_2 \mid \boldsymbol{z})$ to obtain

$$
\begin{aligned}
\pi(\phi \mid \boldsymbol{z}) &= \int_0^1 \pi(\theta_1, \theta_2 \mid \boldsymbol{z})\, \theta_2\big|_{\theta_1 \to \theta_2(1-\phi)}\, d\theta_2 \\
&= \frac{n_1!\, n_2!\, (r_1 + r_2)!}{\Gamma[r1 + 1/2]\,\Gamma[r2 + 1/2]\,\Gamma[n1 - r1 + 1/2]} (1 - \phi)^{r_1 - 1/2} \\
&\quad \times\ {}_2F_1^R[1/2 - n_1 + r_1, 1 + r_1 + r_2, 3/2 + n_2 + r_1, 1 - \phi],
\end{aligned}
$$

where ${}_2F_1^R(a, b, c, z)$ is the regularized ${}_2F_1$ hypergeometric function.

The posterior density of $\phi(\theta_1, \theta_2) = 1 - (\theta_1/\theta_2)$ which corresponds to the vaccine trial data is represented in the upper panel of Figure 3.4. The lower panel represents $d(\phi_0 \mid \boldsymbol{z})$, the reference posterior expectation of the corresponding intrinsic logarithmic loss,

$$\delta\{\phi_0 \mid \phi, \theta_2, n_1, n_2\} = \inf_{\theta_{20} \in [0,1]} E_{\boldsymbol{z}\mid\phi,\theta_2}\left[\log \frac{p(\boldsymbol{z}\mid\phi,\theta_2)}{p(\boldsymbol{z}\mid\phi_0,\theta_{20})}\right],$$

which is given by

$$d(\phi_0 \mid \boldsymbol{z}) = \int_{-\infty}^{\infty} \int_0^1 \delta\{\phi_0 \mid \phi, \theta_2, n_1, n_2\}\, \pi(\phi, \theta_2 \mid \boldsymbol{z})\, d\theta_2\, d\phi.$$

and precisely describes the loss to be expected (in self-calibrated average log-likelihood ratio terms) if a particular value for the vaccine efficacy ϕ_0 were used as a proxy for the true unknown value of ϕ. This is minimized at $\phi^* = 0.297$ which is therefore the intrinsic reference estimate of the vaccine efficacy (represented by a solid point in both panels). The values within the interval $(-0.071, 0.544)$ have all an expected loss smaller than $\log[20]$ so they could possibly be accepted as proxies for ϕ in that the expected log-likelihood

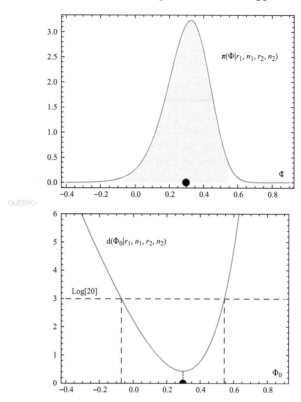

FIGURE 3.4
Reference posterior analysis of the vaccine efficacy $1 - (\theta_1/\theta_2)$ for the RV144 vaccine efficacy trial data.

ratio against them would be smaller than log [20]. Notice that this includes the value $\phi_0 = 0$ of zero efficacy. The region has a reference posterior probability of 0.97, and thus provides a intrinsic reference 0.97-credible interval for ϕ (shaded region in the upper panel). The intrinsic reference 0.95-credible interval is $(-0.009, 0.514)$, which also contains $\phi_0 = 0$.

All these results elaborate on the basic conclusion already provided by the computation of $d(H_0 \mid z) = \log[13.8]$, which indicates some evidence (but not too strong) against the hypothesis H_0 that there is no difference between the two hazard rates.

There is certainly some suggestion of a vaccine efficacy of about 30%, but the true value of the efficacy could really be anywhere between about -1% and 50%, so that—against the "firm conclusion" of an existing difference between the parameters apparently implied by the $p = 0.04$ frequentist significance—more information is really necessary before any final answers could possibly be reached.

3.5 Discussion

As described in Section 3.1, standard Bayesian decision theory provides a unified framework where the problems posed by point estimation, region estimation and hypothesis testing, may all be coherently solved within the same structure. Although the formulation is totally general, there are clear advantages in the use of a *continuous* loss function and a *continuous* prior (which may well be improper) that, for consistency, should both be the same in all those problems.

We have argued that one should preferably make use of *intrinsic* loss functions, for those have the required invariance properties and thus provide solutions which are invariant under reparameterization. Among intrinsic loss functions, there are important arguments to choose that derived from the average log-likelihood ratio against the null, the *intrinsic logarithmic loss*, for this is self-calibrated and has an immediate interpretation.

The *combined* use of the *intrinsic logarithmic loss function* and an *overall reference prior* provides the elements for an *integrated Bayesian reference analysis*, including both the posterior densities of the quantities of interest, and the expected logarithmic intrinsic losses associated to any alternative values. In particular this provides an immediate solution to any precise hypothesis testing problem in terms of the estimated (expected posterior) average log-likelihood ratio against the null.

As illustrated by the analysis of the RV144 HIV vaccine efficacy trial in Thailand, the solutions proposed are not difficult to obtain, they have a simple, very intuitive interpretation, and have far-reaching consequences, which in hypothesis testing problems often contradict conventional statistical practice.

Bibliography

[1] J. O. Berger and J. M. Bernardo. Estimating the product of means: Bayesian analysis with reference priors. *Journal of the American Statistical Association*, 84:200–207, 1989.

[2] J. O. Berger and J. M. Bernardo. On the development of reference priors (with discussion). In J. M. Bernardo, J. O. Berger, A. P. Dawid, and A. F. M. Smith, editors, *Bayesian Statistics 4*, pages 35–60, Oxford, UK, 1992. Oxford University Press.

[3] J. O. Berger, J. M. Bernardo, and D. Sun. The formal definition of reference priors. *The Annals of Statistics*, 37(2):905–938, 2009.

[4] J. O. Berger, J. M. Bernardo, and D. Sun. Objective priors for discrete parameter spaces. *Journal of the American Statistical Association*, 107:636–648, 2012.

[5] J. O. Berger, J. M. Bernardo, and D. Sun. Overall Objective priors. *Bayesian Analysis 10* (with discussion). In press, March 2015.

[6] J. M. Bernardo. Reference posterior distributions for Bayesian inference (with discussion). *Journal of the Royal Statistical Society: Series B (Statistical Methodology)*, 41(2), 1979.

[7] J. M. Bernardo. Intrinsic credible regions: An objective Bayesian approach to interval estimation (with discussion). *Test*, 14:317–384, 2005.

[8] J. M. Bernardo. Reference analysis. In D. K. Dey and C. R. Rao, editors, *Bayesian Thinking: Modeling and Computation, Handbook of Statistics 25*, pages 465–476, Amsterdam, The Netherlands, Elsevier, 2005.

[9] J. M. Bernardo and M. A. Juárez. Intrinsic estimation. In J. M. Bernardo, M. J. Bayarri, J. O. Berger, A. P. Dawid, D. Heckerman, A. F. M. Smith, and M. West, editors, *Bayesian Statistics 7*, pages 465–476, Oxford, UK, Oxford University Press, 2003.

[10] J. M. Bernardo and R. Rueda. Bayesian hypothesis testing: A reference approach. *International Statistical Review*, 70(3):351–372, 2002.

[11] J. M. Bernardo and A. F. M. Smith. *Bayesian Theory*. John Wiley & Sons, 2000.

[12] P. B. Gilber, J. O. Berger, D. Stablein, S. Becker, M. Essex, S. M. Hammer, J. H. Kim, and V. G. DeGruttola. Statistical interpretation of the RV144 HIV vaccine efficacy trial in Thailand: A case study for statistical issues in efficacy trials. *The Journal of Infectious Diseases*, 203:969–975, 2011.

[13] H. Jeffreys. *Theory of Probability* (3rd ed.). Oxford University Press, Oxford, UK, 1961.

[14] S. Kullback and R. A. Leibler. On information and suffiency. *The Annals of Mathematical Statistics*, 22(1):79–86, 1951.

[15] C. P. Robert. Intrinsic losses. *Theory and Decision*, 40(2):191–214, 1996.

[16] C. Stein. Inadmissibility of the usual estimator for the variance of a normal distribution with unknown mean. *Annals of the Institute of Statistical Mathematics*, 16(1):155–160, 1964.

4

Hamiltonian Monte Carlo for Hierarchical Models

Michael Betancourt

University of Warwick

Mark Girolami

University of Warwick

CONTENTS

Many of the most exciting problems in applied statistics involve intricate, typically high-dimensional, models and, at least relative to the model complexity, sparse data. With the data alone unable to identify the model, valid inference in these circumstances requires significant prior information. Such information, however, is not limited to the choice of an explicit prior distribution: it can be encoded in the construction of the model itself.

79

Hierarchical models take this latter approach, associating parameters into exchangeable groups that draw common prior information from shared parent groups. The interactions between the levels in the hierarchy allow the groups to learn from each other without having to sacrifice their unique context, partially pooling the data together to improve inferences. Unfortunately, the same structure that admits powerful modeling also induces formidable pathologies that limit the performance of those inferences.

After reviewing hierarchical models and their pathologies, we will discuss common implementations and show how those pathologies either make the algorithms impractical or limit their effectiveness to an unpleasantly small space of models. We then introduce Hamiltonian Monte Carlo and show how the novel properties of the algorithm can yield much higher performance for general hierarchical models. Finally we conclude with examples which emulate the kind of models ubiquitous in contemporary applications.

4.1   Hierarchical Models

Hierarchical models [6] are defined by the organization of a model's parameters into exchangeable groups, and the resulting conditional independencies between those groups.[1] A one-level hierarchy with parameters (θ, ϕ) and data $\mathcal{D}$, for example, factors as (Figure 4.1)

$$\pi(\theta, \phi | \mathcal{D}) \propto \prod_{i=1}^{n} \pi(\mathcal{D}_i | \theta_i) \, \pi(\theta_i | \phi) \, \pi(\phi). \qquad (4.1)$$

A common example is the one-way normal model,

$$y_i \sim \mathcal{N}\left(\theta_i, \sigma_i^2\right)$$
$$\theta_i \sim \mathcal{N}\left(\mu, \tau^2\right), \text{ for } i = 1, \dots, I, \qquad (4.2)$$

or, in terms of the general notation of (4.1), $\mathcal{D} = (y_i, \sigma_i)$, $\phi = (\mu, \tau)$, and $\theta = (\theta_i)$. To ease exposition we refer to any elements of ϕ as global parameters, and any elements of θ as local parameters, even though such a dichotomy quickly falls apart when considering models with multiple layers.

Unfortunately for practitioners, but perhaps fortunately for pedagogy, the one-level model (4.1) exhibits all of the pathologies typical of hierarchical models. Because the n contributions at the bottom of the hierarchy all depend on the global parameters, a small change in ϕ induces large changes in the density. Consequently, when the data are sparse the density of these models looks like a "funnel", with a region of high density confined within a

---

[1]Not that all parameters have to be grouped into the same hierarchical structure. Models with different hierarchical structures for different parameters are known as *multilevel models*.

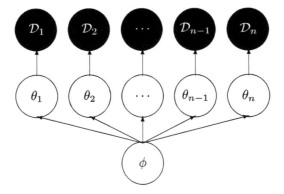

FIGURE 4.1
In hierarchical models "local" parameters, θ, interact via a common dependency on "global" parameters, ϕ. The interactions allow the measured data, $\mathcal{D}$, to inform all of the θ instead of just their immediate parent. More general constructions repeat this structure, either over different sets of parameters or additional layers of hierarchy.

small neighborhood below a region of low density spanning a large volume. The probability *mass* of the two regions, however, is comparable and any successful sampling algorithm must be able to manage the dramatic variations in curvature in order to fully explore the posterior.

For visual illustration, consider the funnel distribution [14] resulting from a one-way normal model with no data, latent mean μ set to zero, and a log-normal prior on the variance $\tau^2 = e^v$,[2]

$$\pi(\theta_1, \ldots, \theta_n, v) \propto \prod_{i=1}^{n} \mathcal{N}\left(\theta_i | 0, (e^{-v/2})^2\right) \mathcal{N}\left(v | 0, 3^2\right).$$

The hierarchical structure induces large correlations between v and each of the θ_i, with the correlation between the parameters strongly varying with position (Figure 4.2). Note that the position-dependence of the correlation ensures that no global correction, such as a rotation and rescaling of the parameters, will simplify the distribution to admit an easier implementation.

---

[2]The exponential relationship between the latent v and the variance τ^2 may appear particularly extreme, but it arises naturally whenever one transforms from a parameter constrained to be positive to an unconstrained parameter more appropriate for sampling.

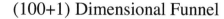

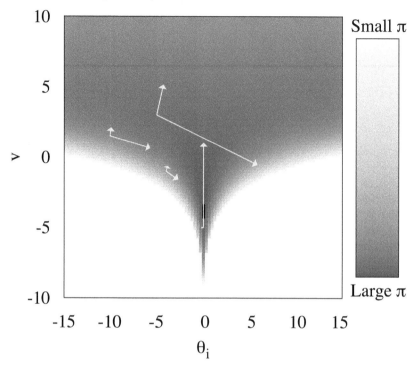

FIGURE 4.2
Typical of hierarchical models, the curvature of the funnel distribution varies strongly with the parameters, taxing most algorithms and limiting their ultimate performance. Here the curvature is represented visually by the eigenvectors of $\sqrt{\left| \partial^2 \log \pi(\theta_1, \ldots, \theta_n, v) / \partial \theta_i \partial \theta_j \right|}$ scaled by their respective eigenvalues, which encode the direction and magnitudes of the local deviation from isotropy.

4.2   Common Implementations of Hierarchical Models

Given the utility of hierarchical models, a variety of implementations have been developed with varying degrees of success. Deterministic algorithms, for example [17, 18, 21], can be quite powerful in a limited scope of models. Here we instead focus on the stochastic Markov Chain Monte Carlo algorithms, in particular the Metropolis and Gibbs samplers, which offer more breadth.

4.2.1 Naïve implementations

Although they are straightforward to implement for many hierarchical models, the performance of algorithms like Random Walk Metropolis and the Gibbs sampler [19] is limited by their incoherent exploration. More technically, these algorithms explore via transitions tuned to the conditional variances of the target distribution. When the target is highly correlated, however, the conditional variances are much smaller than the marginal variances and many transitions are required to explore the entire distribution. Consequently, the samplers devolve into random walks which explore the target distribution extremely slowly.

As we saw above, hierarchical models are highly correlated by construction. As more groups and more levels are added to the hierarchy, the correlations worsen and naïve MCMC implementations quickly become impractical (Figure 4.3).

4.2.2 Efficient implementations

A common means of improving Random Walk Metropolis and the Gibbs sampler is to correct for global correlations, bringing the conditional variances closer to the marginal variances and reducing the undesired random walk behavior. The correlations in hierarchical models, however, are not global but rather local and efficient implementations require more sophistication. To reduce the correlations between successive layers and improve performance, we have to take advantage of the hierarchical structure explicitly. Note that, because this structure is defined in terms of conditional independencies, these strategies tend to be more natural, not to mention more successful, for Gibbs samplers.

One approach is to separate each layer with auxiliary variables [8, 12, 13], for example the one-way normal model (4.2) would become

$$\theta_i = \mu + \xi\eta_i, \ \eta_i \sim \mathcal{N}(0, \sigma_\eta^2),$$

with $\tau = |\xi|\sigma_\eta$. Conditioned on η, the layers become independent and the Gibbs sampler can efficiently explore the target distribution. On the other hand, the multiplicative dependence of the auxiliary variable actually introduces strong correlations into the joint distribution that diminishes the performance of Random Walk Metropolis.

In addition to adding new parameters, the dependence between layers can also be broken by reparameterizing existing parameters. Non-centered parameterizations, for example [16], factor certain dependencies into deterministic transformations between the layers, leaving the actively sampled variables uncorrelated (Figure 4.4). In the one-way normal model (4.2), we would apply both location and scale reparameterizations yielding

$$y_i \sim \mathcal{N}(\vartheta_i\tau + \mu, \sigma_i^2)$$
$$\vartheta_i \sim \mathcal{N}(0, 1),$$

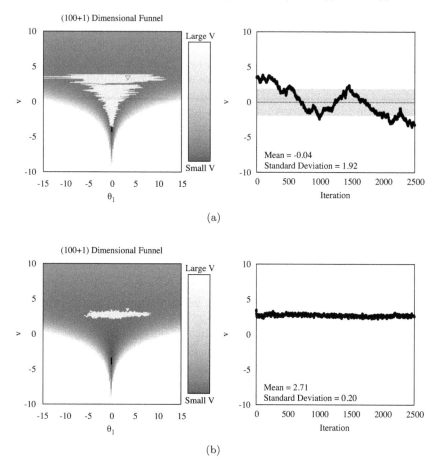

(a)

(b)

FIGURE 4.3
One of the biggest challenges with modern models is not global correlations
but rather local correlations that are resistant to the corrections based on
the global covariance. The funnel distribution, here with $N = 100$, features
strongly varying local correlations but enough symmetry that these correlations cancel globally, so no single correction can compensate for the ineffective
exploration of (a) the Gibbs sampler and (b) Random Walk Metropolis. After 2500 iterations neither chain has explored the marginal distribution of v,
$\pi(v) = \mathcal{N}(v|0, 3^2)$. Note that the Gibbs sampler utilizes a Metropolis-within-
Gibbs scheme as the conditional $\pi(v|\vec{\theta})$ does not have a closed form.

effectively shifting the correlations from the latent parameters to the data.
Provided that the data are not particularly constraining, i.e. the σ_i^2 are large,
the resulting joint distribution is almost isotropic and the performance of both
Random Walk Metropolis and the Gibbs sampler improves. Given sufficient

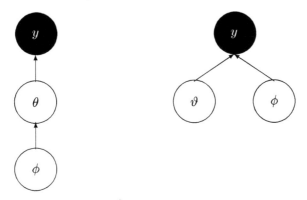

FIGURE 4.4
In one-level hierarchical models with global parameters, ϕ, local parameters, θ, and measured data y, correlations between parameters can be mediated by different parameterizations of the model. Non-centered parameterizations exchange a direct dependence between ϕ and θ for a dependence between ϕ and y; the reparameterized ϑ and ϕ become independent conditioned on the data. When the data are weak these non-centered parameterizations yield simpler posterior geometries.

analytical results the two parameterizations can even be combined within a single algorithm [24].

Unfortunately, the ultimate utility of these efficient implementations is limited to the relatively small class of models where analytic results can be applied. Parameter expansion, for example, requires that the expanded conditional distributions can be found in closed form (although, to be fair, that is also a requirement for any Gibbs sampler in the first place) while non-centered parameterizations are applicable mostly to models where the dependence between layers is given by a generalized linear model. To enable efficient inference without constraining the model, we need to consider more sophisticated Markov Chain Monte Carlo techniques.

4.3   Hamiltonian Monte Carlo for Hierarchical Models

Hamiltonian Monte Carlo [4, 5, 15] utilizes techniques from differential geometry to generate transitions spanning the full marginal variance, eliminating the random walk behavior endemic to Random Walk Metropolis and the Gibbs samplers.[3]

The algorithm introduces auxiliary *momentum* variables, p, to the parameters of the target distribution, q, with the joint density

$$\pi(p, q) = \pi(p|q)\,\pi(q)\,.$$

After specifying the conditional density of the momenta, the joint density defines a Hamiltonian,

$$\begin{aligned}
H(p, q) &= -\log \pi(p, q) \\
&= -\log \pi(p|q) - \log \pi(q) \\
&= T(p|q) + V(q)\,,
\end{aligned}$$

with the *kinetic energy*,

$$T(p|q) \equiv -\log \pi(p|q)\,,$$

and the *potential energy*,

$$V(q) \equiv -\log \pi(q)\,.$$

This Hamiltonian function generates a transition by first sampling the auxiliary momenta,

$$p \sim \pi(p|q)\,,$$

and then evolving the joint system via Hamilton's equations,

$$\begin{aligned}
\frac{dq}{dt} &= +\frac{\partial H}{\partial p} = +\frac{\partial T}{\partial p} \\
\frac{dp}{dt} &= -\frac{\partial H}{\partial q} = -\frac{\partial T}{\partial q} - \frac{\partial V}{\partial q}\,.
\end{aligned}$$

The gradients guide the transitions through regions of high probability and admit the efficient exploration of the entire target distribution. For how long to evolve the system depends on the shape of the target distribution, and the optimal value may vary with position [3]. Dynamically determining the optimal integration time is highly non-trivial as naïve implementations break the detailed balance of the transitions; the No-U-Turn sampler preserves detailed balance by integrating not just forward in time but also backwards [11].

Because the trajectories are able to span the entire marginal variances, the efficient exploration of Hamiltonian Monte Carlo transitions persists as the target distribution becomes correlated, and even if those correlations are largely local, as is typical of hierarchical models.

As the target distribution becomes correlated, even if the correlations are local and vary with position, the trajectories of Hamiltonian Monte Carlo continue to span the full marginal variances of the target distribution and the efficient exploration of the transitions persists. Distributions such as those arising from hierarchical models, however, can introduce pathologies beyond correlations that can pose their own challenges for Hamiltonian Monte Carlo.

---

[3]Hamiltonian Monte Carlo derives from the *Hybrid Monte Carlo* algorithm developed for computing expectations in complicated physical systems [5].

4.3.1 Euclidean Hamiltonian Monte Carlo

The simplest choice of the momenta distribution, and the one almost exclusively seen in contemporary applications, is a Gaussian independent of q,

$$\pi(p|q) = \mathcal{N}(p|0, \Sigma),$$

resulting in a quadratic kinetic energy,

$$T(p, q) = \frac{1}{2}p^T \Sigma^{-1} p.$$

Because the subsequent Hamiltonian also generates dynamics on a Euclidean manifold, we refer to the resulting algorithm at Euclidean Hamiltonian Monte Carlo. Note that the metric, Σ, effectively induces a global rotation and rescaling of the target distribution, although it is often taken to be the identity in practice.

Despite its history of success in difficult applications, Euclidean Hamiltonian Monte Carlo does have two weaknesses that are accentuated in hierarchical models: the introduction of a characteristic length scale and limited variations in density.

4.3.1.1 Characteristic length scale

In practice Hamilton's equations are sufficiently complex to render analytic solutions infeasible; instead the equations must be integrated numerically. Although symplectic integrators provide efficient and accurate numerical solutions [10], they introduce a characteristic length scale via the time discretization, or step size, ϵ.

Typically the step size is tuned to achieve an optimal acceptance probability [1], but such optimality criteria ignore the potential instability of the integrator. In order to prevent the numerical solution from diverging before it can explore the entire distribution, the step size must be tuned to match the curvature. Formally, a stable solution requires [10]

$$\epsilon \sqrt{\lambda_i} < 2$$

for each eigenvalue, λ_i, of the matrix[4]

$$M_{ij} = \left(\Sigma^{-1}\right)_{ik} \frac{\partial^2 V}{\partial q_k \partial q_j}.$$

Moreover, algorithms that adapt the step size to achieve an optimal acceptance probability require a relatively precise and accurate estimate of the

---

[4] Note that the ultimate computational efficiency of Euclidean Hamiltonian Monte Carlo scales with the condition number of M averaged over the target distribution. A well chosen metric reduces the condition number, explaining why a global decorrelation that helps with Random Walk Metropolis and Gibbs sampling is also favorable for Euclidean Hamiltonian Monte Carlo.

global acceptance probability. When the chain has high autocorrelation or overlooks regions of high curvature because of a divergent integrator, however, such estimates are almost impossible to achieve. Consequently, adaptive algorithms can adapt too aggressively to the local neighborhood where the chain was seeded, potentially biasing resulting inferences.

Given the ubiquity of spatially-varying curvature, these pathologies are particularly common to hierarchical models. In order to use adaptive algorithms we recommend relaxing any adaptation criteria to ensure that the Markov chain hasn't been biased by overly assertive adaptation. A particularly robust strategy is to compare inferences, especially for the latent parameters, as adaptation criteria are gradually weakened, selecting a step size only once the inferences have stabilized and divergent transitions are rare (Figure 4.5). At the very least an auxiliary chain should always be run at a smaller step size to ensure consistent inferences.

4.3.1.2   Limited density variations

A more subtle, but no less insidious, vulnerability of Euclidean Hamiltonian Monte Carlo concerns density variations within a transition. In the evolution of the system, the Hamiltonian function,

$$H(p, q) = T(p|q) + V(q),$$

is constant, meaning that any variation in the potential energy must be compensated for by an opposite variation in the kinetic energy. In Euclidean Hamiltonian Monte Carlo, however, the kinetic energy is a χ^2 variate which, in expectation, varies by only half the dimensionality, d, of the target distribution. Consequently the Hamiltonian transitions are limited to

$$\Delta V = \Delta T \sim \frac{d}{2},$$

restraining the density variation within a single transition. Unfortunately the correlations inherent to hierarchical models also induce huge density variations, and for any but the smallest hierarchical model this restriction prevents the transitions from spanning the full marginal variation. Eventually random walk behavior creeps back in and the efficiency of the algorithm plummets (Figure 4.6, 4.7).

Because they remove explicit hierarchical correlations, non-centered parameterizations can also reduce the density variations of hierarchical models and drastically increase the performance of Euclidean Hamiltonian Monte Carlo (Figure 4.8). Note that as in the case of Random Walk Metropolis and Gibbs sampling, the efficacy of the parametrization depends on the relative strength of the data, although when the nominal centered parameterization is best there is often enough data that the partial pooling of hierarchical models isn't needed in the first place.

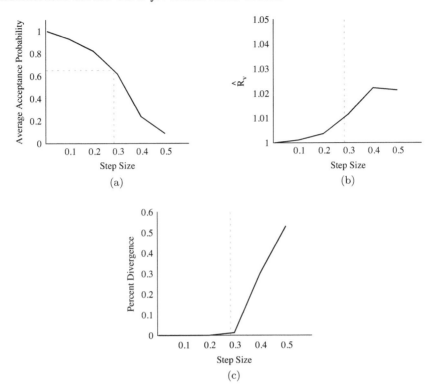

FIGURE 4.5
Careful consideration of any adaptation procedure is crucial for valid inference
in hierarchical models. As the step size of the numerical integrator is decreased
(a) the average acceptance probability increases from the canonically optimal
value of 0.651 but (b) the sampler output converges to a consistent distribution. Indeed, (c) at the canonically optimal value of the average acceptance
probability the integrator begins to diverge. Here consistency is measured
with a modified potential scale reduction statistic, $\hat{R}$, [7, 23] for the latent
parameter v in a $(50 + 1)$-dimensional funnel.

4.3.2 Riemannian Hamiltonian Monte Carlo

Euclidean Hamiltonian Monte Carlo is readily generalized by allowing the
covariance to vary with position

$$\pi(p|q) = \mathcal{N}(p|0, \Sigma(q)),$$

giving,

$$T(p,q) = \frac{1}{2}p^T \Sigma^{-1}(q)\, p - \frac{1}{2} \log |\Sigma(q)|.$$

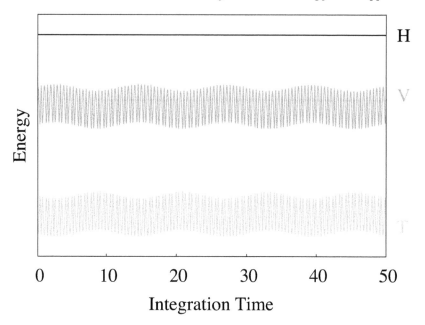

FIGURE 4.6
Because the Hamiltonian, H, is conserved during each trajectory, the variation
in the potential energy, V, is limited to the variation in the kinetic energy, T,
which itself is limited to only $d/2$.

With the Hamiltonian now generating dynamics on a Riemannian manifold
with metric Σ, we follow the convention established above and denote the
resulting algorithm as Riemannian Hamiltonian Monte Carlo [9].

The dynamic metric effectively induces local corrections to the target distribution, and if the metric is well chosen then those corrections can compensate for position-dependent scalings and correlations, not only reducing
the computational burden of simulating the Hamiltonian evolution but also
relieving the sensitivity to the integrator step size.

Note also the appearance of the log determinant term, $\frac{1}{2}\log|\Sigma(q)|$. Nomi-
nally in place to provide the appropriate normalization for the momenta distribution, this term provides a powerful feature to Riemannian Hamiltonian
Monte Carlo, serving as a reservoir that absorbs and then releases energy along
the evolution and potentially allowing much larger variations in the potential
energy.

Of course, the utility of Riemannian Hamiltonian Monte Carlo is dependent on the choice of the metric $\Sigma(q)$. To optimize the position-dependent corrections we want a metric that leaves the target distribution locally isotropic,
motivating a metric resembling the Hessian of the target distribution.

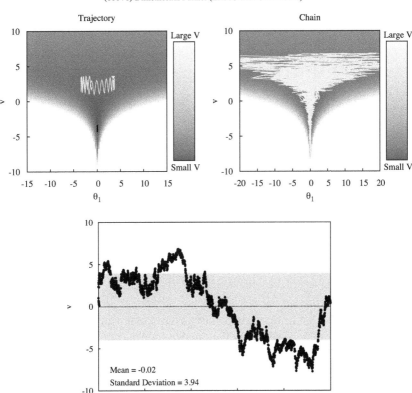

FIGURE 4.7
Limited to moderate potential energy variations, the trajectories of Euclidean HMC, here with a unit metric $\Sigma = \mathbb{I}$, reduce to random walk behavior in hierarchical models. The resulting Markov chain explores more efficiently than Gibbs and Random Walk Metropolis (Figure 4.3), but not efficiently enough to make these models particularly practical.

Unfortunately the Hessian isn't sufficiently well-behaved to serve as a metric itself; in general, it is not even guaranteed to be positive-definite. The Hessian can be manipulated into a well-behaved form, however, by applying the SoftAbs transformation,[5] and the resulting SoftAbs metric [2] admits a generic but efficient Riemannian Hamiltonian Monte Carlo implementation (Figure 4.10, 4.11).

---

[5] Another approach to regularizing the Hessian is with the Fisher-Rao metric from information geometry [9], but this metric is able to regularize only by integrating out exactly the correlations needed for effective corrections, especially in hierarchical models [2].

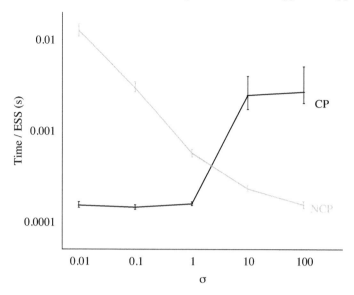

FIGURE 4.8

Depending on the common variance, σ^2, from which the data were generated, the performance of a 10-dimensional one-way normal model (4.2) varies drastically between centered (CP) and non-centered (NCP) parameterizations of the latent parameters, θ_i. As the variance increases and the data become effectively more sparse, the non-centered parameterization yields the most efficient inference and the disparity in performance increases with the dimensionality of the model. The bands denote the quartiles over an ensemble of 50 runs, with each run using Stan [22] configured with a diagonal metric and the No-U-Turn sampler. Both the metric and the step size were adapted during warmup, and care was taken to ensure consistent estimates (Figure 4.9).

4.4   Example

To see the advantage of Hamiltonian Monte Carlo over algorithms that explore with a random walk, consider the one-way normal model (4.2) with 800 latent θ_i and a constant measurement error, $\sigma_i = \sigma$ across all nodes. The latent parameters are taken to be $\mu = 8$ and $\tau = 3$, with the θ_i and y_i randomly sampled in turn with $\sigma = 10$. To this generative likelihood we add weakly-informative priors,

$$\pi(\mu) = \mathcal{N}\left(0, 5^2\right)$$
$$\pi(\tau) = \text{Half-Cauchy}(0, 2.5).$$

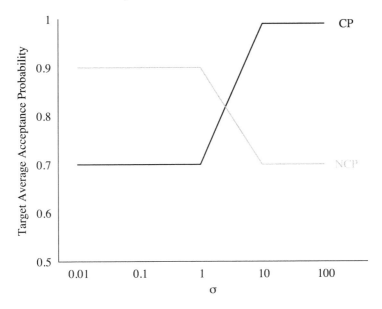

FIGURE 4.9
As noted in subsection 4.3.1.1, care must be taken when using adaptive implementations of Euclidean Hamiltonian Monte Carlo with hierarchical models. For the results in Figure 4.8, the optimal average acceptance probability was relaxed until the estimates of τ stabilized, as measured by the potential scale reduction factor, and divergent transitions did not appear.

All sampling is done on a fully unconstrained space, in particular the latent τ is transformed into $\lambda = \log \tau$.

Noting the results of Figure 4.8, the nominal centered parameterization,

$$y_i \sim \mathcal{N}\left(\theta_i, \sigma_i^2\right)$$
$$\theta_i \sim \mathcal{N}\left(\mu, \tau^2\right), \text{ for } i = 1, \ldots, 800,$$

should yield inferior performance to the non-centered parameterization,

$$y_i \sim \mathcal{N}\left(\tau \vartheta_i + \mu, \sigma_i^2\right) \tag{4.3}$$
$$\vartheta_i \sim \mathcal{N}(0, 1), \text{ for } i = 1, \ldots, 800; \tag{4.4}$$

in order to not overestimate the success of Hamiltonian Monte Carlo we include both. For both parameterizations, we fit Random Walk Metropolis, Metropolis-within-Gibbs,[6] and Euclidean Hamiltonian Monte Carlo with a

---

[6]Because of the non-conjugate prior distributions, the conditionals for this model are not analytic and we must resort to a Metropolis-within-Gibbs scheme as is common in practical applications.

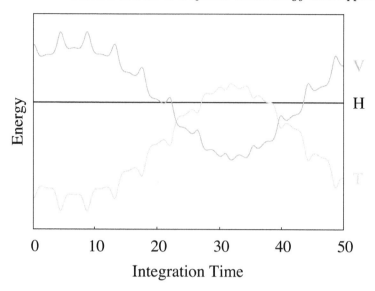

FIGURE 4.10
Although the variation in the potential energy, V, is still limited by the variation in the kinetic energy, T, the introduction of the log determinant term in Riemannian Hamiltonian Monte Carlo allows the kinetic energy sufficiently large variation that the potential is essentially unconstrained in practice.

diagonal metric[7] to the generated data.[8] The step size parameter in each case was tuned to be as large as possible while still yielding consistent estimates with a baseline sample (Figure 4.12).

Although the pathologies of the centered parameterization penalize all three algorithms, Euclidean Hamiltonian Monte Carlo proves to be at least an order-of-magnitude more efficient than both Random Walk Metropolis and the Gibbs sampler. The real power of Hamiltonian Monte Carlo, however, is revealed when those penalties are removed in the non-centered parameterization (Table 4.1).

Most importantly, the advantage of Hamiltonian Monte Carlo scales with increasing dimensionality of the model. In the most complex models that populate the cutting edge of applied statistics, Hamiltonian Monte Carlo is not just the most convenient solution, it is often the only practical solution.

---

[7]Hamiltonian Monte Carlo is able to obtain more accurate estimates of the marginal variances than Random Walk Metropolis and Metropolis-within-Gibbs and, in a friendly gesture, the variance estimates from Hamiltonian Monte Carlo were used to scale the transitions in the competing algorithms.

[8]Riemannian Hamiltonian Monte Carlo was not considered here as its implementation in Stan is still under development.

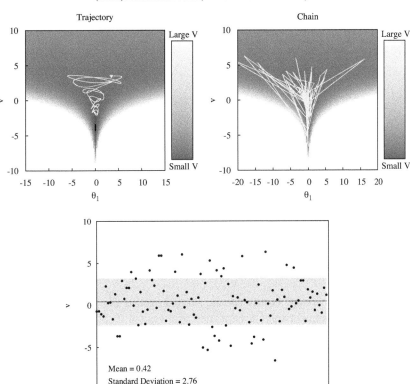

FIGURE 4.11
Without being limited to small variations in the potential energy, Riemannian
Hamiltonian Monte Carlo with the SoftAbs metric admits transitions that
expire the entirety of the funnel distribution, resulting in nearly independent
transitions, and drastically smaller autocorrelations (compare with Figure 4.7,
noting the different number of iterations).

4.5 Conclusion

By utilizing the local curvature of the target distribution, Hamiltonian Monte
Carlo provides the efficient exploration necessary for learning from the complex hierarchical models of interest in applied problems. Whether using Euclidean Hamiltonian Monte Carlo with careful parameterizations or Riemannian Hamiltonian Monte Carlo with the SoftAbs metric, these algorithms admit inference whose performance scales not just with the size of the hierarchy

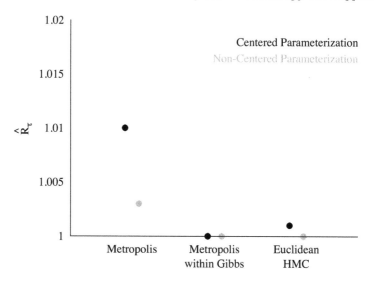

FIGURE 4.12
In order to ensure valid comparisons, each sampling algorithm was optimized
to achieve an optimal target acceptance probability [1, 20], but only so long as
the resulting estimates were consistent with each other; as before, consistency
is quantified with the modified potential scale reduction statistic, $\hat{R}$, [7, 23]
of the latent variance, τ. The clear exception was the centered Random Walk
Metropolis algorithm which proved difficult to tune; because improving consistency would only reduce performance, the sampler was not pushed further.

but also with the complexity of local distributions, even those that may not
be amenable to analytic manipulation.

The immediate drawback of Hamiltonian Monte Carlo is increased difficulty of implementation: not only does the algorithm require non-trivial
tasks such as the the integration of Hamilton's equations, the user must
also specify the derivatives of the target distribution. The inference engine
Stan [22] removes these burdens, providing not only a powerful probabilistic
programming language for specifying the target distribution and a high performance implementation of Hamiltonian evolution but also state-of-the-art
automatic differentiation techniques to compute the derivatives without any
user input. Through Stan, users can build, test, and run hierarchical models
without having to compromise for computational constraints.

TABLE 4.1
Euclidean Hamiltonian Monte Carlo significantly outperforms both Random Walk Metropolis and the Metropolis-within-Gibbs sampler for both parameterizations of the one-way normal model. The difference is particularly striking for the more efficient non-centered parameterization that would be used in practice.

Algorithm	Parameter-ization	Step Size	Average Acceptance Probability	Time (s)	Time/ESS (s)
Metropolis	Centered	$5.00 \cdot 10^{-3}$	0.822	$4.51 \cdot 10^4$	1220
Gibbs	Centered	1.50	0.446	$9.54 \cdot 10^4$	297
EHMC	Centered	$1.91 \cdot 10^{-2}$	0.987	$1.00 \cdot 10^4$	**16.2**
Metropolis	Non-Centered	0.0500	0.461	398	1.44
Gibbs	Non-Centered	2.00	0.496	817	1.95
EHMC	Non-Centered	0.164	0.763	154	$\mathbf{2.94 \cdot 10^{-2}}$

Acknowledgments

We are indebted to Simon Byrne, Bob Carpenter, Michael Epstein, Andrew Gelman, Yair Ghitza, Daniel Lee, Peter Li, Sam Livingstone, and Anne-Marie Lyne for many fruitful discussions as well as invaluable comments on the text. Moreover we thank the reviewers who offered helpful critiques. Michael Betancourt is supported under EPSRC grant EP/J016934/1 and Mark Girolami is an EPSRC Research Fellow under grant EP/J016934/1.

4.6   Appendix: Stan Models

Here we include the Stan models used above.

Funnel

```
transformed data {
  int<lower=0> J;
  J <- 25;
}
parameters {
  real theta[J];
  real v;
}
model {
  v ~ normal(0, 3);
  theta ~ normal(0, exp(v/2));
}
```

Generate One-Way Normal Pseudo-data

```
transformed data {
  real mu;
  real<lower=0> tau;

  real alpha;
  int N;

  mu <- 8;
  tau <- 3;

  alpha <- 10;
  N <- 800;

}

parameters {
  real x;
}

model {
  x ~ normal(0, 1);
}

generated quantities {
  real mu_print;
  real tau_print;

  vector[N] theta;
  vector[N] sigma;
  vector[N] y;

  mu_print <- mu;
  tau_print <- tau;

  for (i in 1:N) {
    theta[i] <- normal_rng(mu, tau);
    sigma[i] <- alpha;
    y[i] <- normal_rng(theta[i], sigma[i]);
  }

}
```

One-Way Normal (Centered)

```
data {
  int<lower=0> J;
  real y[J];
  real<lower=0> sigma[J];
}

parameters {
  real mu;
  real<lower=0> tau;
  real theta[J];
}

model {
  mu ~ normal(0, 5);
  tau ~ cauchy(0, 2.5);
  theta ~ normal(mu, tau);
  y ~ normal(theta, sigma);
}
```

One-Way Normal (Non-Centered)

```
data {
  int<lower=0> J;
  real y[J];
  real<lower=0> sigma[J];
}

parameters {
  real mu;
  real<lower=0> tau;
  real var_theta[J];
}

transformed parameters {
  real theta[J];
  for (j in 1:J) theta[j] <- tau * var_theta[j] + mu;
}

model {
  mu ~ normal(0, 5);
  tau ~ cauchy(0, 2.5);
  var_theta ~ normal(0, 1);
  y ~ normal(theta, sigma);
}
```

Bibliography

[1] A. Beskos, N. Pillai, G. Roberts, J. Sanz-Serna, and S. Andrew. Optimal tuning of the hybrid Monte Carlo algorithm. *Bernoulli*, 19(5A):1501–1534, 2013.

[2] M. Betancourt. A general metric for Riemannian Hamiltonian Monte Carlo. In F. Nielsen and F. Barbaresco, editor, *First International Conference on the Geometric Science of Information*, volume 8085 of *Lecture Notes in Computer Science*, Berlin Heidelberg, Springer Verlag, 2013.

[3] M. Betancourt. Generalizing the No-U-Turn sampler to Riemannian manifolds. *arXiv e-prints*, Apr. 2013.

[4] M. Betancourt and L. C. Stein. The geometry of Hamiltonian Monte Carlo. *arXiv e-prints*, 2011.

[5] S. Duane, A. D. Kennedy, B. J. Pendleton, and D. Roweth. Hybrid Monte Carlo. *Physics Letters B*, 195(2):216 – 222, 1987.

[6] A. Gelman, J. B. Carlin, H. S. Stern, D. B. Dunson, A. Vehtari, and D. B. Rubin. *Bayesian Data Analysis*. CRC Press, London, 3rd edition, 2013.

[7] A. Gelman and D. B. Rubin. Inference from iterative simulation using multiple sequences. *Statistical Science*, 7(4):457–472, 1992.

[8] A. Gelman, D. A. Van Dyk, Z. Huang, and J. W. Boscardin. Using redundant parameterizations to fit hierarchical models. *Journal of Computational and Graphical Statistics*, 17(1):95–122, 2008.

[9] M. Girolami and B. Calderhead. Riemann manifold Langevin and Hamiltonian Monte Carlo methods. *Journal of the Royal Statistical Society: Series B (Statistical Methodology)*, 73(2):123–214, 2011.

[10] E. Hairer, C. Lubich, and G. Wanner. *Geometric Numerical Integration: Structure-Preserving Algorithms for Ordinary Differential Equations*. Springer Verlag, The Netherlands, 2006.

[11] M. D. Hoffman and A. Gelman. The No-U-Turn sampler: Adaptively setting path lengths in Hamiltonian Monte Carlo. *Journal of Machine Learning Research*, 15:1351–1381, April 2014.

[12] C. Liu, D. B. Rubin, and Y. N. Wu. Parameter expansion to accelerate EM: The PX-EM algorithm. *Biometrika*, 85(4):755–770, 1998.

[13] J. S. Liu and Y. N. Wu. Parameter expansion for data augmentation. *Journal of the American Statistical Association*, 94(448):1264–1274, 1999.

[14] R. M. Neal. Slice sampling (with discussion). *Annals of Statistics*, 31(3):705–767, 2003.

[15] R. M. Neal. MCMC using Hamiltonian dynamics. In S. Brooks, A. Gelman, G. L. Jones, and X.-L. Meng, editors, *Handbook of Markov Chain Monte Carlo*, pages 113–162. CRC Press, New York, 2011.

[16] O. Papaspiliopoulos, G. O. Roberts, and M. Sköld. A general framework for the parametrization of hierarchical models. *Statistical Science*, 22(1):59–73, 2007.

[17] J. C. Pinheiro and D. M. Bates. Linear mixed-effects models: Basic concepts and examples. In *Mixed-Effects Models in S and S-PLUS*, pages 3–56. Springer Verlag, New York, 2000.

[18] S. Rabe-Hesketh and A. Skrondal. *Multilevel and Longitudinal Modeling Using Stata*. STATA Press, USA, 2008.

[19] C. P. Robert and G. Casella. *Monte Carlo Statistical Methods*. Springer Verlag, New York, 2nd edition, 2004.

[20] G. O. Roberts, A. Gelman, and W. R. Gilks. Weak convergence and optimal scaling of random walk Metropolis algorithms. *The Annals of Applied Probability*, 7(1):110–120, 1997.

[21] H. Rue, S. Martino, and N. Chopin. Approximate Bayesian inference for latent Gaussian models by using integrated nested Laplace approximations. *Journal of the Royal Statistical Society: Series B (Statistical Methodology)*, 71(2):319–392, 2009.

[22] Stan Development Team. Stan: A C++ library for probability and sampling, version 2.0, 2013.

[23] Stan Development Team. *Stan Modeling Language User's Guide and Reference Manual, Version 2.0*, 2013.

[24] Y. Yu and X.-L. Meng. To center or not to center: That is not the question—An ancillarity–sufficiency interweaving strategy (ASIS) for boosting MCMC efficiency. *Journal of Computational and Graphical Statistics*, 20(3):531–570, 2011.

5

On Bayesian Spatio-Temporal Modeling of Oceanographic Climate Characteristics

Madhuchhanda Bhattacharjee

University of Hyderabad

Snigdhansu Chatterjee

University of Minnesota

CONTENTS

5.1 Introduction

Recent change in the planet's climate conditions is a matter of great concern, owing to its potential devastating effects for all life forms. A thorough study of climate variables should inform us about possible patterns of the climate of this planet, which further goes on to inform and guide policy decisions relating to adaptation and mitigation strategies to counter climate change, better management of resources, risk management, and for a better quality of life for all. Many of these aspects relate to extremes as well as typical values of climate variables. Consequently, it is of interest to obtain the joint distribution of the several climate variables. Owing to the complex dependence patterns possible in such joint distributions, a Bayesian modeling of the data is needed. Moreover, using Bayesian methodologies in the climate field also allows for systematic, coherent and simple treatment of Physics-driven known relations and constraints, combining multiple sources of data of varying size, dimensions and precision.

However, attempting to understand the patterns in data on climate variables presents unique challenges for Bayesian modeling. Note that many of the software tools that are available to the practitioner of Bayesian analysis are tailor-made to handle data supported on two-dimensional space, or for the analysis of a univariate variable of interest, or allow for very limited forms of spatial or temporal dependencies. As examples of such tools, consider several of the contributed packages in the statistical computing platform R, which have been developed primarily for epidemiological, forestry mapping and other application domains.

Unlike typical datasets from these domains, climate data is typically available over three dimensional space and time, on multiple variables, and generally form a non-stationary random field. In this chapter, we present an analysis of such multivariate, multiple-indexed (by three dimensional space as well as time) climate data using the example of Arctic Ocean seawater data. We consider the dataset on global seawater ([25]), accessible from the repository (http://data.giss.nasa.gov) of the Goddard Institute of Space Studies, which will be the focus for the rest of this chapter. We will discuss details of the dataset later.

In this chapter, we demonstrate how to capture the spatial variability of the response variables, allowing for smooth change over space without using specific fixed spatial relational function. For this we employ a technique similar to the *distributed lag model* in economics or *normal dynamic linear models* in clinical trials. In these comparable models, the lag would be considered that of time. Here, we generalize this framework to use the geographical space, and consider a spatial version of a dynamic model. We require additional modifications to account for the circular nature of earth surface and for the resulting cyclic dependence in the data. By adding constraints on the effect parameters we are able to achieve this and retain the model in a DAG (directed acyclic graph) framework.

We consider several climate variables together as the response. These response variables are known to have interdependence in this case. While one option would be to use known explicit functional form is for this, note that these are only approximate relations, and not necessarily shared or common for geographic locations and depths of the sea. Thus we would rather like to keep possibilities to explore such a relationship open and study it a-posteriori. We modeled the response variables jointly as a multivariate normal random vector where the parameters depend on the three dimensional geographic coordinates.

As a component of climate related research, analysis of oceanographic data is less commonly observed compared to that of atmospheric variables. Mention must be made of the GLODAP project ([16]) and research that it generated. A concern for the planet's climate springs from possible acidification of the oceans, see [20], [23], [24] and others. Other studies include [12], [10], and various others. However, we have not come across a comprehensive Bayesian analysis of multiple variables relating to climatic properties of the

Earth's oceans. It is challenging to analyze multiple response data indexed by multiple state variables and potentially having complex non-linear and non-stationary patterns, and in the context of climate data such cases may form a M-open problem, as presented in [4]. In view of this, the chapter potentially contributes in two ways. First, we present a framework that allows for the posterior to "talk" outside the strictures of stylized parametric dependency structures (simple examples of which are temporal autoregression or spatial conditional autoregression) which may be useful for analyzing M-open problems. Second, we demonstrate an applied Bayesian analysis exercise in the context of oceanography that uses such a framework.

5.2   A Description of the Dataset

We obtain the data from the repository http://data.giss.nasa.gov. This dataset includes measurements of temperature, salinity, deuterium, the ratio of the O-18 and O-16 isotopes of oxygen; co-variable information about the depth of the sea, the latitude and longitude, and the month and year in which the data was collected; along with references and notes. It is a compilation of data gathered by various teams of researchers at different points of time and locations. Calibrations are carried out to correct for the difference in standards, techniques and instruments used by these teams, and such corrections are flagged. Missing values are present. Information on the time at which the data is collected is available in terms of month and years. Further technical minutiae relating to the dataset is available in the aforementioned website. Exploratory analysis of an earlier edition of this dataset have been carried case of [5]. A small-area type predictive analysis of this dataset has been presented in [22].

We access the data records from this source that correspond to Latitude values 60 degrees North and higher. We further limit that data to be from 1975 or more recent times. The region from which the data has been gathered is depicted in Figure 5.1. This map has been produced by the aforementioned website.

5.3   A Description of the Methodology

We consider the variables *temperature, salinity*, and the ratio of Oxygen-16 and Oxygen-18 isotopes, called *Oxygen-18* in the sequel. We might have considered modeling for the hydrogen and deuterium isotope ratio, but only 8 records out of a total of 11800 contained values for these; the rest were missing. Consequently the deuterium isotope ratio was not considered as a variable

Location of matching data points

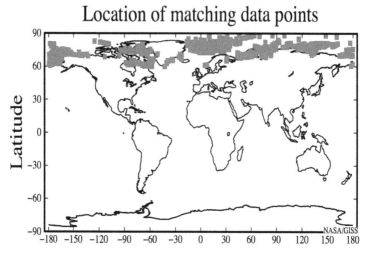

Longitude

FIGURE 5.1
The spatial region from where the data has been gathered. This figure is
generated from the same website from which the data is obtained.

in this study. There were some missing values in each of the other variables
as well, but they are not as overwhelming.

We consider observations from 60° North and above latitudes, which
largely correspond to the Arctic Ocean region. The data is gathered at various depths, at different longitudes, over different months of the year, from
1975 onwards. Since data collection is sparse and uneven outside the sum-
mer months, we restrict our attention to data from July–September only. We
consider spatial blocks of data, in order to model spatial smoothness and co-
herence. In the analysis presented below, we consider four bands of latitude
values (60–70, 70–75, 75–80 and 80–90 North), ten levels for longitude values produced by the following break points (−160, −80, −40, −5, 0, 20, 40,
90, 130, 180), and eleven levels of depths (bins separated at depth values of
0, 5, 15, 30, 50, 100, 150, 300, 500, 1000, 10000). These choices were made
keeping data availability in mind. Note that the latitude, longitude and depth
bands of values have a natural ordering each. Thus, spatial dependence may
be modeled by considering neighboring bands. In view of the circular nature
of longitudes, an additional constraint is imposed.

A generic notation for a data point may be $Y(\ell_1, \ell_2, d, t, j)$, where ℓ_1 specifies the latitude level, ℓ_2 the longitude level, d the depth level, t the time, and
j an oceanic feature (temperature, salinity, O-18 ratio). We might further col-
lapse the first four indices (latitude, longitude, depth and time), and consider
a generic observation $y_i \in \mathbb{R}^3$ as a feature vector indexed by space-time.

For each observed sample point $i = 1, \ldots, n$ (n = 6795 number of samples) and $N (= 3)$ number of response variables, we assume the following model:

$$[y_i | \eta_i, \Theta_i] \sim N_3 (\eta_i, \Theta_i),$$

where η_i is the mean vector $\in \mathbb{R}^N$, and Θ_i is the corresponding precision matrix. Notice that these vary with the observed sample points.

We assume that the mean η_i are affected by the spatial interdependence between the response variables, and not the variance-covariance structures.

In the above, the precision matrices Θ_i are allowed to be different for different observations, with identical distributions for different levels of longitude and depth. Suppose

$$\tau_1 \sim Wishart(\mathbb{I}_N, N),$$

is a $N \times N$ random matrix. Here $\mathbb{I}_N$ is the identity matrix of dimension N. The degrees of freedom parameter is set at N as this corresponds to the least informative proper prior. For observations at a given level of longitude and depth values, we use the prior that the precision matrices Θ_i are distributionally identical copies of the corresponding τ_1 matrices. Thus, geographic locations were assumed to have exchangeable prior distributions which in this case were expressed for the precision matrices as Wishart distributions with prefixed hyper-parameters.

We use a spatially smoothed dynamic linear model for the mean vectors. The response variables are assumed to have individual overall means with variation around them depending upon geographical location given by the triplet latitude, longitude and depth. Initial data exploration suggested that the behavior of the response variables at various depths is possibly non-smooth. Thus smoothness on the remaining 2-dimensional surface were explored and the following model describes possibilities of using lags in both (latitude, longitude) and (longitude).

The additive model for the mean vectors are given by, for $i = 1, \ldots, n$ and $j = 1, 2, 3 (= N)$,

$$\eta_{i,j} = \mu_{0j} + \mu_{lat_i, long_i, depth_i, j}.$$

Here the overall effect vector is modeled with multivariate normal distribution as follows:

$$\mu_0 \sim N_N (\mathbf{0}_{3 \times 1}, \mathbb{I}_N).$$

We have a more complex modeling structure for the space-dependent mean structure, as follows:

$$\mu(k, \ell, m) = \begin{cases} \mathbf{0}, & \text{if } \ell = 1, \text{ for all } k, m, \\ \sim & N_N (\mu(k, l-1, m, 1:N), \tau_{N \times N}) \\ & \text{if } k = 1, \ell = 2, \ldots, 10 \text{ and all } m \\ \sim & N_3 (0.5\mu(k-1, \ell, m) + 0.5\mu(k, l-1, m), \tau_{N \times N}) \\ & \text{for all } m, k = 2, \ldots, 4 \text{ and } \ell = 2, \ldots, 10. \end{cases}$$

The hyper-parameter τ appearing above is given a Wishart distribution in this hierarchical structure.

$$\tau \sim Wishart\left(\mathbb{I}_{N \times N}, N\right).$$

In the above framework, the spatial dependence within a given block of (latitude, longitude, depth)-level is captured by common parameters, while between-block dependencies are captured by shared hyper-parameters. Temporal dependencies are captured by the built-in dependence structure for each longitude and depth block in the precision matrices Θ_i's, and in the shared dependencies in the η_i's. Note that it is guaranteed that we have a proper posterior distribution, since all the priors were chosen to be probability density measures. This model can be enhanced with more complex features. However, our analysis showed that adding more complexity either by more complex mean and dispersion structures, or with additional spatial or temporal dependency measurements, does not enhance the quality of the statistical model. This is at least partially because of the nature of the data, and in keeping with the results of earlier non-Bayesian attempts at analyzing this data, see [5, 22]. In addition, this feature of additional complexity not leading to model improvement strongly suggests that this problem is M-open. A Bayesian analysis problem is considered M-open if the data generating mechanism is not within the collection of models used for analysis, and there is no prior belief on how the data related to the "true model". One of the most important reason for analyzing climate data is predicting future climate patterns, especially in a climate-change regime. It may be noted that prediction presents unique challenges in M-open problems, see [6–8] for detailed discussion on such issues.

5.4   Results

We ran a Markov Chain Monte Carlo procedure according to the model specified above, to generate an approximate posterior distribution of the parameters of interest. We used 2 parallel chains, and a burn-in of 10,000 for each chain, followed by a further sequence of 50,000 iterations, which we thinned by a factor of 10. This allowed us to generate posteriors' summaries based on (2 x 50000 / 10 =) 10000 samples, which are presented below.

In Figure 5.2, we show how the three response variables, (temperature, salinity and O-18 isotope ratio) vary with depth and longitude. Figure 5.3 shows how the elements of the precision matrix, denoting the joint variability of the three responses, vary across depth and longitude. The figures for the corresponding posterior variance matrix over these three responses are given in Figure 5.4. In Figure 5.5, Figure 5.6 and Figure 5.7, we present the posterior distributions of these three responses as various latitude values.

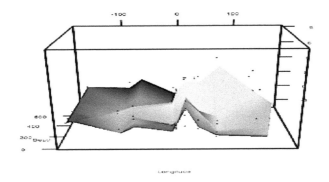

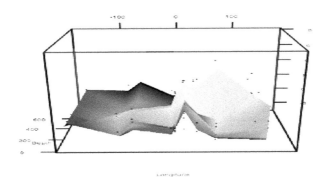

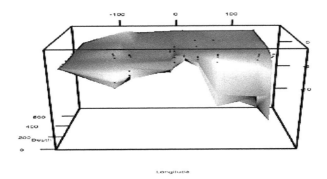

FIGURE 5.2

The pattern of posterior means *temperature* (top), *salinity* (middle) and *O-18 isotope ratio* (bottom) at various longitudes and depths of sea in the Arctic Ocean region.

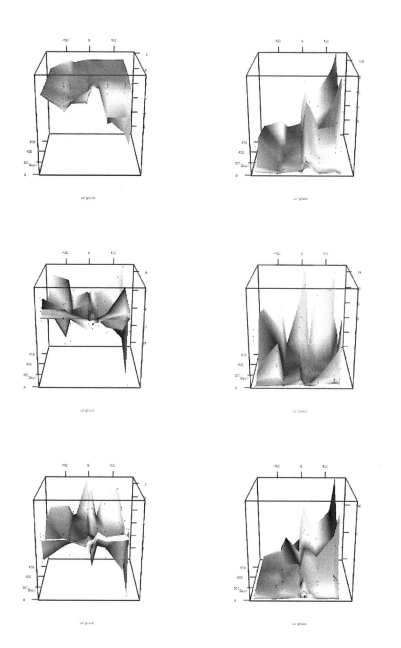

FIGURE 5.3
The top row figures are the (1,1) and (1,2) elements, middle row figures are
the (1,3) and (2,2) elements, and bottom row figures are the (2,3) and (3,3)
elements in the *posterior precision matrix* across *temperature, salinity* and *O-
18 isotope ratio* at various longitudes and depths of sea in the Arctic Ocean
region.

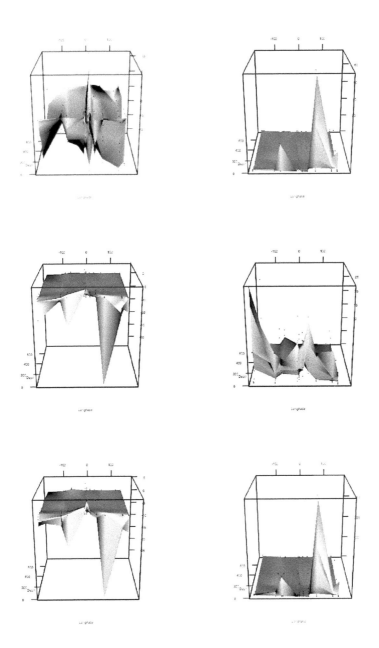

FIGURE 5.4
The top row figures are the (1,1) and (1,2) elements, middle row figures are
the (1,3) and (2,2) elements, and bottom row figures are the (2,3) and (3,3)
elements in the *posterior variance matrix* across *temperature, salinity* and *O-
18 isotope ratio* at various longitudes and depths of sea in the Arctic Ocean
region

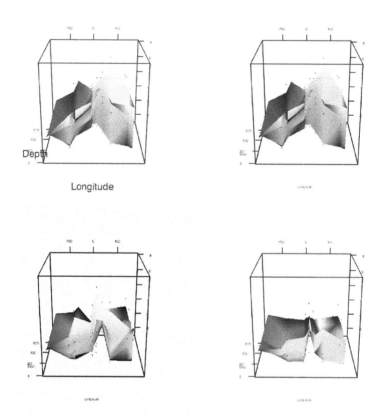

FIGURE 5.5
The top row figures are the patterns in the posterior mean of *temperature* of sea water, at various longitude and depth values, over Latitudes 60-70 North, and 70-75 North. The bottom row contains corresponding figures for Latitudes 75-80 North, and 80-90 North.

Our major conclusion from this quite extensive analysis is that ocean variables are co-dependent, their mean and covariance structures seem to be strongly related to the spatial and temporal frames from which the observations have been gathered. There are naturally some differences between the variables, and some scope of bringing in Physics-guided knowledge among the response variables we studied. The patterns we found in the data suggest that no simple relationship would possibly suffice to explain the nature of any one of the response variables over space and time, or for the nature of co-dependence among the variables themselves. The figures we have presented show that the lack of a simple relationship should not be confused with a lack of a relationship; there is very strong suggestion of a pattern; see Figure 5.2,

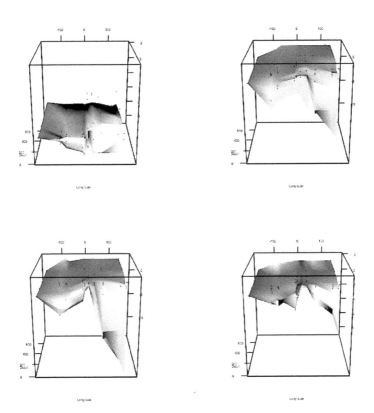

FIGURE 5.6
The top row figures are the patterns in the posterior mean of *salinity* of sea
water, at various longitude and depth values, over Latitudes 60-70 North, and
70-75 North. The bottom row contains corresponding figures for Latitudes
75-80 North, and 80-90 North.

for example. One common theme that emerges from these graphics is that the
posterior is neither simple, nor smooth over space, and standard summary
measures like posterior location or scale parameters may be misleading.

We now present some evidence about the performance of the Markov
Chain Monte Carlo procedure we adopted. In Figure 5.8 we show the posterior histograms of the μ-parameters for the three responses. The precision
τ-parameters have posterior histograms as shown in Figure 5.9. Various diagnostic results, for example convergence graphs, posterior deviance results,
and auto-correlation plots are presented in Figure 5.10.

These results strongly suggest that the extracted samples from the MCMC
runs may resemble a sample from the true posterior distribution of various

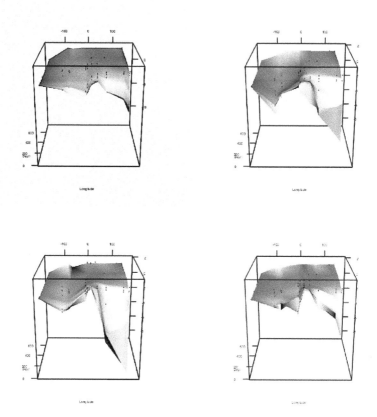

FIGURE 5.7
The top row figures are the patterns in the posterior mean of *Oxygen-18 isotope ratio* of sea water, at various longitude and depth values, over Latitudes 60-70 North, and 70-75 North. The bottom row contains corresponding figures for Latitudes 75-80 North, and 80-90 North.

parameters under consideration. We have performed some robustness studies, whereby our conclusions do not seem to be altered by a choice of hyper-parameter values.

5.5   Discussion

The existing literature on climate modeling is essentially one of modeling nature using knowledge from a variety of scientific disciplines. From a Physics-based perspective, it is typical to consider the variable under study as a

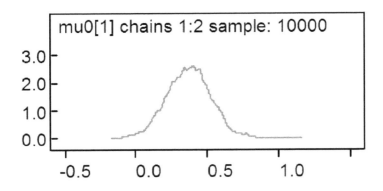

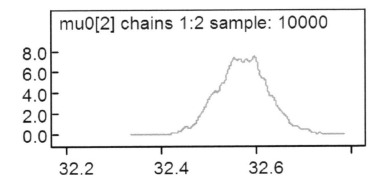

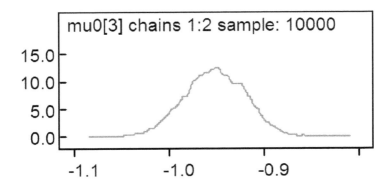

FIGURE 5.8
The histograms depicting the posterior distribution of the μ parameters across *temperature, salinity* and *O-18 isotope ratio*.

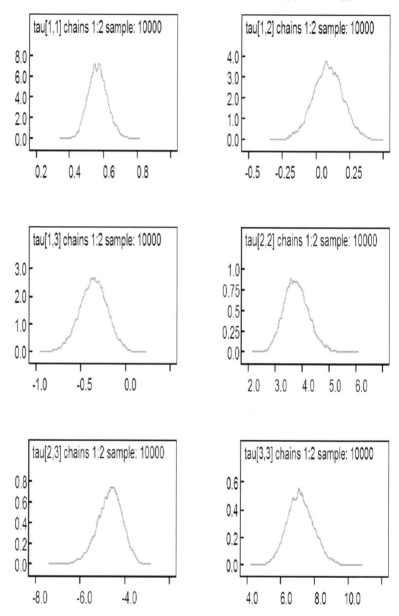

FIGURE 5.9

The top row figures are the histograms of the (1,1) and (1,2) elements, middle row figures are the histograms of the (1,3) and (2,2) elements, and bottom row figures are the histograms of the (2,3) and (3,3) elements depicting the posterior distribution of the *precision matrix* across *temperature*, *salinity* and *O-18 isotope ratio*.

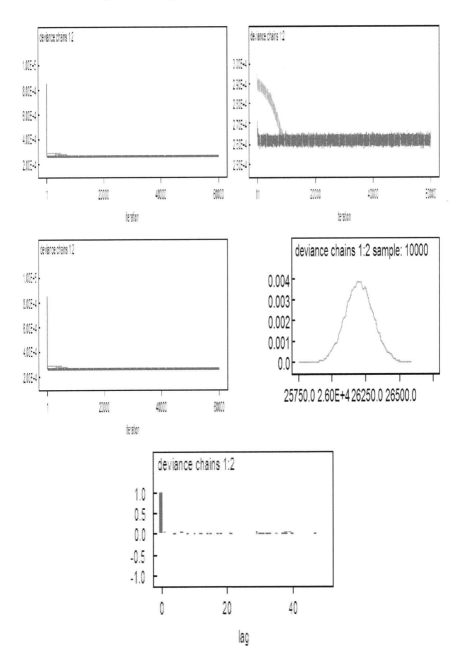

FIGURE 5.10
Graphs denoting convergence properties, deviance, and autocorrelation structures in the MCMC runs.

deterministic, but extremely complex, function of other physical variables. This kind of modeling retains a high degree of fidelity to the true process by which the data is generated. However, it is inevitable that not all features of a system as complex as climate will be measured or retained in various forms of data records. Also, our present state of knowledge about how different physical, atmospheric, geophysical and other variables interplay is limited, as is natural in any scientific discipline. A very partial and incomplete review of natural scientific modeling of climate may be obtained from [14], [21], [9], [11] and several references therein. A Bayesian framework where such Physics-based approach has been considered may be obtained from [15].

Climate models are used for several purposes. Of these, some of the important ones are *detection* of climate change, *attribution* of climate change to a cause, *forecasting* of future climate scenarios. A study of climate in the prehistoric past, using data and proxies based on tree-rings, ice-core samples and other geophysical and fossilized sources, forms the topic of *paleoclimate*, and is useful as a reference for climate as of today and in future. Examples of paleoclimate studies may be found in [19], [18] [1], [27], [13] and other sources.

Climate research has progressed beyond change detection and attribution. Forecasts, and quantifying errors of forecasts relating to future climate scenarios, predicting possible consequences of climate change; combination of outputs from several *global climate models* (GCM) models, and other important studies now form a part of climate research. Several works using Bayesian ideas have contributed to this research, see, for example [3], [17], [2], [26] and references therein. In much of the above cited literature, oceanographic variables have not been considered. This is partially because the atmosphere is better understood than the hydrosphere in Physics, partially because it was felt that modeling the atmosphere was of greater importance for understanding climate change. However, in recent times, the hydrosphere, cyrosphere, and biosphere are being studied with greater vigor.

In this context, the chapter attempts to present a Bayesian analysis of a three-dimensional response on ocean water. We illustrated the complexity of patterns in oceanographic variables, and suggested a possible approach towards understanding the statistical properties of these variables. While a more detailed and thorough study needs to be done, our MCMC results suggest that a Bayesian approach may provide answers to several interesting questions in this domain.

In addition, the possibility of analyzing data from M-complete and M-open problems using the broad framework we adopted here: namely, constructing blocks of indexing variables to reduce or eliminate data sparsity, using least informative proper priors, and assuming minimal structure otherwise, should be explored.

Acknowledgment:

We thank two anonymous referees for their comments, which greatly improved the present chapter. The second author's research was partially funded by NSF grants # SES-0851705 and # IIS-1029711, and research grants from the Institute on the Environment and College of Liberal Arts, University of Minnesota.

Bibliography

[1] M. R. Allen and S. F. B.Tett. Checking for model consistency in optimal fingerprinting. *Climate Dynamics*, 15:419–434, 1999.

[2] L. M. Berliner, R. A. Levine, and D. J. Shea. Bayesian climate change assessment. *Journal of Climate*, 13:3805–3820, 1999.

[3] L. M. Berliner, D. Nychka, and T. Hoar. *Studies in the Atmospheric Sciences*. Springer Verlag, New York, 2000.

[4] J. M. Bernardo and A. F. M. Smith. *Bayesian Theory*. John Wiley, Chichester, U.K., 2000.

[5] S. Chatterjee, Q. Deng, and J. Xu. The statistical evidence of climate change : An analysis of global seawater data. Technical Report 677, School of Statistics, University of Minnesota, 2009.

[6] B. Clarke. Desiderata for a predictive theory of statistics. *Bayesian Analysis*, 5(2):283–318, 2010.

[7] J. L. Clarke, B. Clarke, and C.-W. Yu. Prediction in m-complete problems with limited sample size. *Bayesian Analysis*, 8(3):647–690, 2013.

[8] M. Clyde and E. S. Iversen. *Bayesian Theory and Applications*, chapter Bayesian model averaging in the M-open framework, pages 483–498. Oxford University Press, 2013.

[9] P. M. Cox, R. A. Betts, C. B. Bunton, R. L. H. Essery, P. Rowntree, and J. Smith. The impact of new land surface physics on the GCM simulation of climate and climate sensitivity. *Climate Dynamics*, 15:183–203, 1999.

[10] P. J. Durack, S. E. Wijffels, and R. J. Matear. Ocean salinities reveal strong global water cycle intensification during 1950 to 2000. *Science*, 336(6080):455–458, 2012.

[11] F. Giorgi and L. O. Mearns. Calculation of average, uncertainty range, and reliability of regional climate changes from AOGCM simulations via the "reliability ensemble averaging" (REA) method. *Journal of Climate*, 10:1141–1158, 2002.

[12] V. Gouretski and F. Reseghetti. On depth and temperature biases in bathythermograph data : Development of a new correction scheme based on analysis of a global ocean database. *Deep Sea Research Part I: Oceanographic Research Papers*, 57(6):812–833, 2010.

[13] G. C. Hegerl, T. R. Karl, M. R. Allen, N. L. Bindoff, N. P. Gillett, D. J. Karoly, X. Zhang, and F. W. Zwiers. Climate change detection and attribution: Beyond mean temperature signals. *Journal of Climate*, 19:5058–5077, 2006.

[14] IPCC. *Climate Change 2007 - The Physical Science Basis (Working Group I Contribution to the Fourth Assessment Report of the IPCC)*. Cambridge University Press, U.K., 2007.

[15] M. Kennedy and A. O'Hagan. Bayesian calibration of computer models (with discussion). *Journal of the Royal Statistical Society: Series B (Statistical Methodology)*, 63(3):425–464, 2001.

[16] R. M. Key, A. Kozyr, C. L. Sabine, K. Lee, R. Wanninkhof, J. Bullister, R. A. Freely, F. Millero, C. Mordy, and T.-H. Peng. A global ocean carbon climatology: Results from GLODAP. Technical Report GB4031, Global Biogeochemical Cycles, 18, 2004.

[17] R. A. Levine and L. M. Berliner. Statistical principles for climate change studies. *Journal of Climate*, 12:564–574, 1999.

[18] B. Li, W. D. Nychka, and C. M. Ammann. The "Hockey Stick" and the 1990s: A statistical perspective on reconstructing hemispheric temperatures. *Tellus*, 59:591–598, 2007.

[19] M. E. Mann, R. S. Bradley, and M. K. Hughes. Global-scale temperature patterns and climate forcing over the past six centuries. *Nature*, 392:779–787, 1998.

[20] K. Matsumoto and N. Gruber. How accurate is the estimation of anthropogenic carbon in the ocean? An evaluation of the ΔC^* method. *Global Biogeochemical Cycles*, 19, 2005. GB3014, doi:10.1029/2004GB002397.

[21] L. O. Mearns, M. Hulme, T. R. Carter, R. Leemans, M. Lal, and P. Whetton. *Climate Change 2001: The Scientific Basis. Contribution of Working Group I to the Third Assessment Report of the IPCC*, chapter Climate scenario development, pages 739–768. Cambridge University Press, Cambridge, UK, 2001.

[22] U. Mukherjee and S. Chatterjee. A Fay-Herriot type approach for better prediction in multi-indexed response with application to Arctic seawater data analysis. *Journal of Indian Society for Agricultural Statistics*, 68(4):257–272, 2013.

[23] J. C. Orr and et al. Anthropogenic ocean acidification over the twenty-first century and its impact on calcifying organisms. *Nature*, 437:681–686, 2005.

[24] J. A. Raven and et al. *Ocean Acidification Due to Increasing Atmospheric Carbon Dioxide*. The Royal Society, London, UK, 2004.

[25] G. A. Schmidt, B. G. R., and E. J. Rohling. Global seawater oxygen-18 database. http://data.giss.nasa.gov/o18data/, 1999.

[26] C. Tebaldi, R. W. Smith, D. Nychka, and L. O. Mearns. Quantifying uncertainty in projections of regional climate change: A Bayesian approach to the analysis of multi-model ensembles. *Journal of Climate*, 18(10):1524–1540, 2005.

[27] The International ad hoc detection and attribution group. Detecting and attributing external influences on the climate system: A review of recent advances. *Journal of Climate*, 18(9):1291–1314, 2005.

6

Sequential Bayesian Inference for Dynamic State Space Model Parameters

Arnab Bhattacharya

Heriot-Watt University

Simon Wilson

Trinity College

CONTENTS

6.1 Introduction

Dynamic state-space models [24], consisting of a latent Markov process (state/system process) $X_0, X_1, \ldots$ and noisy observations $Y_1, Y_2, \ldots$ that are conditionally independent, are used in a wide variety of applications, for example wireless networks [9], object tracking [21] and econometrics [7], among many others. The model is specified by an initial distribution $p(x_0|\theta)$, a transition kernel $p(x_t|x_{t-1}, \theta)$ and an observation distribution $p(y_t|x_t, \theta)$. These distributions are defined in terms of a set of K static (e.g. non time-varying) parameters $\theta = (\theta_1, \ldots, \theta_K)$. The joint model to time T is:

$$p(\mathbf{y}_{1:T}, \mathbf{x}_{0:T}, \theta) = \left(\prod_{t=1}^{T} p(y_t|x_t, \theta) p(x_t|x_{t-1}, \theta) \right) \times p(x_0|\theta) p(\theta), \qquad (6.1)$$

123

where $\mathbf{y}_{1:T} = (y_1, \ldots, y_T)$, etc. These models are also known as hidden Markov models [20].

In this chapter, we focus on sequential Bayesian estimation of static parameters (θ) for these models; at time T, we observe y_T and wish to compute the posterior distribution $p(\theta|\mathbf{y}_{1:T})$ for every $T = 1, 2, 3, \ldots$. Further, this is to be done in a setting where online estimation is required, so that there is an issue of trade-off between computation speed and accuracy. The constraint on computation-time means that for some sufficiently large T it becomes infeasible to simply recompute $p(\theta|\mathbf{y}_{1:T})$ "from scratch" by

- Monte Carlo (e.g. MCMC) or,

- functional approximation method (e.g. the integrated nested Laplace approximation) or,

- by some *offline* applications of particle filter [6] adapted to also infer these static parameters [1], or even by

- maximum likelihood based filtering methods like in [10].

We propose a method that accomplishes this for a fairly broad set of dynamic state space models.

As regards the static parameter estimation problem, almost no closed form solutions are available for the posterior of static parameters. This is easy to see that even in linear Gaussian models the following integral is not in closed form [2]:

$$p(\theta|\mathbf{y}_{1:T}) = \int p(\theta, \mathbf{x}_{0:T}|\mathbf{y}_{1:T}) \mathrm{d}\mathbf{x}_{0:T}. \qquad (6.2)$$

[24] shows that conjugate sequential updates of the state and observation variances, as well as for x_0, are available for some specific cases. Noteworthy works specific to online inference on static parameters, applicable for general state space models are found in [14, 23] and [3]. [12] is a good overview of parameter estimation, including both offline approaches and the use of sequential Monte Carlo for online parameter estimation.

The rest of the chapter is organized as follows. Section 6.2 outlines the principle of the method. Section 6.3 describes one of the main issues to be resolved in order to implement the method: approximations to one-step ahead filtering and prediction densities. Section 6.4 illustrates the method and assesses its performance against alternative approaches. Section 6.5 contains some concluding remarks.

6.2 Principle

The basic algorithm explaining our new method is the construction of the posterior for a set of static parameters on a grid of deterministically chosen points, and a subsequent methodology that can update the posterior density sequentially over time. The principle of the proposed method is based on two fundamental theoretical ideas.

The first idea is that many dynamic state space models have a relatively small number of static parameters, so that in principle $p(\theta|\mathbf{y}_{1:T})$ can be computed and stored on a discrete grid of practical size. In a good number of situations, the parameters are time-varying processes themselves; and there are hyper-parameters that are static but unknown. This has been noted as a property of many latent models [22]. It is noted that the transition kernel of some dynamic state space models is itself defined in terms of a set of static parameters (θ_1) e.g. $p(x_t|x_{t-1}, \psi_t, \theta_1)$ and time-varying parameters (ψ_t). The latter may also evolve as a Markov process depending on some hyper-parameters (θ_2) e.g. $p(\psi_{0:T}|\theta_2) = p(\psi_0|\theta_2) \prod_{t=1}^{T} p(\psi_t|\psi_{t-1}, \theta_2)$; for example dynamic linear models with a trend [24]. Without loss of generality, such cases are also incorporated in our problem by considering (x_t, ψ_t) to be the latent process and by denoting the complete set of static parameters and hyper-parameters as θ.

The second significant point to pay attention to is that there exists a useful identity for parameter estimation in latent models. This identity is known as the *basic marginal-likelihood identity* (BMI) as it appeared in [4] and is used for the calculation of marginal likelihood in the original work. In this chapter, the following approach is taken

$$p(\theta|\mathbf{y}_{1:T}) \propto p(\mathbf{y}_{1:T}, \theta)$$
$$= \frac{p(\mathbf{y}_{1:T}, \mathbf{x}_{0:T}, \theta)}{p(\mathbf{x}_{0:T}|\mathbf{y}_{1:T}, \theta)}\bigg|_{\mathbf{x}_{0:T}=\mathbf{x}^*(\theta)}, \tag{6.3}$$

valid for any $\mathbf{x}_{0:T}$ for which $p(\mathbf{x}_{0:T}|\mathbf{y}_{1:T}, \theta) > 0$. Under the assumption that $p(\mathbf{x}_{0:T}|\theta)$ is Gaussian, the above identity forms the basis of the integrated nested Laplace approximation (INLA) of [22]. Here a Gaussian approximation is made for the denominator term, and it is evaluated on a discrete grid of values of θ. The method also includes a way to derive such a grid "intelligently". The value $\mathbf{x}_{0:T} = \mathbf{x}^*(\theta)$ is allowed to be a function of θ and typically $\mathbf{x}^*(\theta) = \arg\max_{\mathbf{x}_{0:T}} p(\mathbf{x}_{0:T}|\mathbf{y}_{1:T}, \theta)$ is used, which is the mean of the Gaussian approximation to $p(\mathbf{x}_{0:T}|\mathbf{y}_{1:T}, \theta)$.

Another useful identity is:

$$p(\theta|\mathbf{y}_{1:T}) \propto p(\theta|\mathbf{y}_{1:T-1}) \, p(y_T|\mathbf{y}_{1:T-1}, \theta). \tag{6.4}$$

In our case, we estimate $p(y_T|\mathbf{y}_{1:T-1},\theta)$ by the following:

$$p(y_T|\mathbf{y}_{1:T-1},\theta) = \left.\frac{p(y_T|x_T,\theta)p(x_T|\mathbf{y}_{1:T-1},\theta)}{p(x_T|\mathbf{y}_{1:T},\theta)}\right|_{x_T=x^*(\theta)} ; \qquad (6.5)$$

as with equation (6.3), we choose $x^*(\theta) = \arg\max_{x_T} p(x_T|\mathbf{y}_{1:T},\theta)$. This identity is clearly useful for sequential estimation and does not suffer from the dimension-increasing problem of equation (6.3), in the sense that its computation does not depend on the whole set of values of $\mathbf{y}$, from 1 to T. Thus from equation (6.4), if prediction and filtering approximations $\tilde{p}(x_T|\mathbf{y}_{1:T-1},\theta)$ and $\tilde{p}(x_T|\mathbf{y}_{1:T},\theta)$ are available, any approximation $\tilde{p}(\theta|\mathbf{y}_{1:T-1})$ at time $T-1$ can be updated:

$$\tilde{p}(\theta|\mathbf{y}_{1:T}) \propto \tilde{p}(\theta|\mathbf{y}_{1:T-1})\,\tilde{p}(y_T|\mathbf{y}_{1:T-1},\theta)$$

$$= \left.\tilde{p}(\theta|\mathbf{y}_{1:T-1})\,p(y_T|x_T,\theta) \times \frac{\tilde{p}(x_T|\mathbf{y}_{1:T-1},\theta)}{\tilde{p}(x_T|\mathbf{y}_{1:T},\theta)}\right|_{x_T=x^*(\theta)}, \qquad (6.6)$$

where $x^*(\theta) = \arg\max_{x_T} \tilde{p}(x_T|\mathbf{y}_{1:T},\theta)$. For θ of low dimension, computing equation (6.6) on a discrete grid offers the potential for fast sequential estimation.

This suggests the following sequential estimation algorithm when approximate prediction and filtering distributions are available, which we call SINLA or Sequential INLA. Initially, $p(\theta|\mathbf{y}_{1:T})$ is approximated by INLA because it is both fast and accurate and produces a discrete grid Θ_T over which $p(\theta|\mathbf{y}_{1:T})$ is computed, for each T. At some time T_{INLA}, this will prove to be too slow to compute, and from then on the sequential update of equation (6.6) will be used.

The main issue that remains to be addressed in order to implement this algorithm is the form of the approximations $\tilde{p}(x_T|\mathbf{y}_{1:T-1},\theta)$ and $\tilde{p}(x_T|\mathbf{y}_{1:T},\theta)$. It is addressed in the next section.

6.3   Predicting and Filtering Density Approximations

For the Kalman filter (where $p(y_t|x_t,\theta)$, $p(x_0|\theta)$ and $p(x_t|x_{t-1},\theta)$ are linear and Gaussian), the prediction and filtering distributions are Gaussian, equation (6.4) can be computed exactly, and the Gaussian approximation is also exact. The means and variances of these Gaussians are sequentially updated [18]. All that we need to store are the means and variances of the prediction and filtering distributions for each θ in the grid; from this $p(\theta|\mathbf{y}_{1:T})$ can be computed.

An equivalent definition of equation (6.1), and one that is useful in describing some aspects of the approximations that we propose, is the general state-space representation:

$$y_t = f(x_t, u_t, v_t, \theta); \tag{6.7}$$
$$x_t = g(x_{t-1}, w_t, \theta), \tag{6.8}$$

where v_t and w_t are observation and state process errors, and u_t are (possibly non-existent) exogenous variables. The likelihood $p(y_t|x_t, \theta)$ is specified by f and v_t, while the transition density $p(x_t|x_{t-1}, \theta)$ is specified by g and w_t. Two of several different algorithms found in the literature are listed here.

6.3.1 Basic approximations

When either the linear or Gaussian property does not hold, two extensions of the Kalman filter can be computed quickly.

Extended KF

The extended Kalman filter was one of the first generalisations of the Kalman filter to nonlinear models [17]. In its basic form, it linearises a nonlinear model (i.e. uses the first term in a Taylor expansion of the function) to create a Kalman filter (e.g. Gaussian) approximation to the filtering and prediction densities [8]. Hence the prediction and filtering distribution approximations are Gaussian and make use of the fast sequential updating of the first two moments.

Unscented KF

The unscented Kalman filter also produces Gaussian approximations to the filtering and prediction densities but avoids linearising by approximately propagating the means and covariances through the nonlinear functions of equations (6.7) and (6.8) [11]. It tends to be more accurate than the extended Kalman filter, more so for strongly nonlinear models.

The nonlinearity in the model is propagated deterministically through a small set of points, known as sigma points. Weights are associated with each point, and estimates of the mean and variance of the Gaussian approximations to $p(x_t|\mathbf{y}_{1:t-1}, \theta)$ and $p(x_t|\mathbf{y}_{1:t}, \theta)$ are made as weighted means and variances of these points. The method is computationally fast as it only requires the propagation of these points, and then the computation of means and variances of the approximation, for each observation.

6.4 Examples: Linear Dynamic Model

Our method is implemented on a dynamic linear model and subsequently compared to INLA (an offline method), and a particle filter method which is developed for online inference of static parameters (online Bayesian sequential Monte Carlo or BSMC) by [14]). A dynamic linear model with Gaussian errors is chosen since it is the easiest form of state space model, and would be a good starting point to test our method. Average performance is measured across many replications of simulated data. To keep the computation-time comparison fair, all methods were implemented in R [19]. T_{INLA} is usually set at some value up till which INLA produces results extremely quickly. For this example, it was found to be around 20, hence $T_{\text{INLA}} = 20$. The model performances were compared using two different measures, as described below:

Mahalanobis distance (MD): This is used as a measure to judge the accuracy of the estimates of the parameters in the model in a multivariate parameter space setting [15].

computation-time: The time to compute the posterior approximation is also recorded.

The statistical model in this example has been assumed to be of the form:

$$y_t = x_{t-1}\mathbf{1} + \eta_t \tag{6.9}$$
$$x_t = \phi x_{t-1} + \epsilon_t \tag{6.10}$$

where $\epsilon_t \sim N(0, \sigma_{state}^2)$, $\mathbf{1}$ is a vector of n 1's and $\eta_t \sim \mathcal{MVN}(\mathbf{0}, \Sigma)$; with $\dim(\eta_t) = n$ and $\dim(\epsilon_t) = 1$. The covariance matrix Σ is assumed to be dependent on a single unknown parameter:

$$\Sigma = \sigma_{obs}\Sigma^*.$$

The entries of Σ^* are of the following type:

$$\Sigma_{ij} = \begin{cases} 1 & \text{if } i = j, \\ \exp(-rd(i,j)) & \text{if } i \neq j, \end{cases}$$

where $r > 0$ and $d(i,j)$ is some measure of distance betweens nodes i and j. Σ defines the well-known Gaussian spatial process [16].

Data has been generated by fixing the values at $\phi = 0.7, \sigma_{obs}^2 = 1$ and $\sigma_{state}^2 = 1$, and it is assumed that dimension of observation process, $n = 3$. Further, the form of Σ^* is known and fixed. Instead of variance parameters, precision parameters are used which are denoted as ρ_{obs} and ρ_{state} respectively.

The Kalman filter has been used for optimal filtering at each step, since the system and observation equations are linear. The above simulation has been replicated 10 times for SINLA and 5 times each for INLA and BSMC (as these methods were very slow). For SINLA (and also INLA), the AR parameter ϕ has a normal prior with mean 0.1 and s.d. 1, truncated at -0.99 and 0.99. Both ρ_{state} and ρ_{obs} have a gamma prior with both the shape and rate parameters being 3. For BSMC, stronger informative priors were provided. The prior for ϕ is normal with mean 0.5 and s.d. 1, again truncated at -0.99 and 0.99; whereas for the gamma priors, both the parameters are now set at 1.

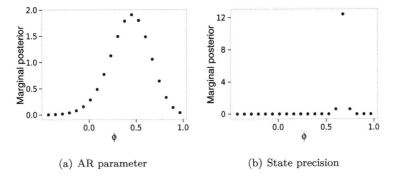

(a) AR parameter   (b) State precision

FIGURE 6.1
Figure (a) is the marginal posterior density of the AR parameter (ϕ) as constructed by INLA at time $T = 20$. Figure (b) shows the marginal posterior of
the same parameter at time $T = 1000$, as updated using SINLA. Note in Figure (b) that several grid points can be dropped as they are not contributing
to the region of high density, which in turn also shows that adding new grid
points will help.

Two plots are provided exhibiting the performance of SINLA. Figure 6.1
contains the posterior of AR parameter at time $T = 20$, as computed by INLA;
and at $T = 1000$, as computed using SINLA. It is obvious from Figure 6.1
that several grid points are not contributing to the support of the marginal
posterior and can be dropped, while new grid points need to be added at
regions of high density. Trace plots having the approximate mode of each of
the parameters along with the approximate 95% probability bounds are shown
in a second plot, Figure 6.2. From this plot, one can ascertain that behavior of
the traces degrades over time; in the sense that modes of the parameters are
tending toward the edge of the support for the grid. This hints at the fact that
the grid needs to shift over time; i.e. for $T > T_{\text{INLA}}$ fast alternative methods
of updating the grid must be determined.

Table 6.1 has been constructed to compare the accuracy and computation-
time of our algorithm with BSMC and offline Bayesian inference using INLA.
The comparison is done for $T = 500$ and $T = 1000$, and is based on point estimates, Mahalanobis distance and computation-times for each of the methods.

INLA, as one can expect, is computationally the most expensive algorithm
while producing very accurate outputs. The particle filter has been implemented using the R package pomp [13]. The computation-time of the SMC is
much higher than the new algorithm; which of course is partially dependent
on the number of particles used. For this example, 10000 particles have been
used, with the intention of achieving greater accuracy. But degeneracy in all
the examples caused the output to be extremely inaccurate. No Mahalanobis
distance values are reported for BSMC as they are extremely high, caused
possibly by inflation of variance due to degeneracy, something which is well

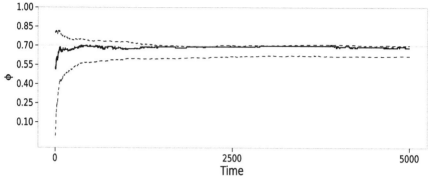

(a) AR parameter

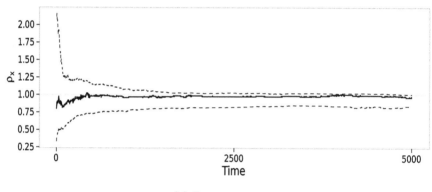

(b) State precision

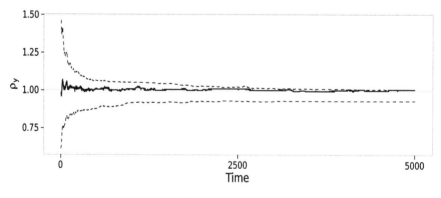

(c) Observation precision

FIGURE 6.2

Plots (a), (b) and (c) represent trace plots showing trajectories of the averaged approximate mode and approximate 95% probability bounds of the posteriors of ϕ, ρ_{state} and ρ_{obs} respectively. The light grey line displays the true parameter value.

TABLE 6.1
Table containing point estimates of the parameters and measures of accuracy and computation-time for the three algorithms, namely SINLA, BSMC and INLA. The true values of the parameters are $\theta = (0.7, 1, 1)$.

Methods	T = 500			T = 1000		
	Estimates	MD	Time(s)	Estimates	MD	Time(s)
SINLA	0.68, 0.98, 1.01	1.81	73.63	0.67, 0.96, 1.01	1.94	136.71
BSMC	0.72, 1.66, 0.86	⋆	2230.11	0.69, 1.68, 0.81	⋆	4437.83
INLA	0.7, 0.99, 1.00	0.71	3358.67	0.7, 1.00, 1.00	0.79	23631.54

⋆ MD for BSMC are not reported as they are extremely large.

known for SMC [5]. The point estimates also show the relative inaccuracy of BSMC compared to the other two methods. An interesting feature is the increase in Mahalanobis distance value for SINLA, as T moves from 500 to 1000. It can possibly be justified by the decrease in the standard error of the parameters caused because the support rested on only a few grid points, as can be seen in Figure 6.1.

6.5 Concluding Remarks

A method of fast sequential parameter estimation for dynamic state space models has been proposed and compared to two alternatives: the integrated nested Laplace approximation, and a particle filter. In all the examples that we consider, INLA proved the most accurate but the slowest. Our method achieved much better accuracy than the particle filter and also proved to be much faster than both the algorithms. It is also worth noting that our method also suffers from issues of degeneracy, albeit slower than BSMC.

The principal disadvantages of the approach, in its current form are the following. The constant grid, computed using INLA does not cover the true support of the posterior over time. There is a need to develop a deterministic grid shifting algorithm, i.e. one that dynamically adds or drops specifically chosen grid points over time. Another crucial disadvantage of this method is that it is restricted to models with a relatively small number of fixed parameters.

Finally there is a need to develop asymptotic properties related to the convergence of the filter. While some idea of the accuracy of posterior of $\theta|\mathbf{Y}_{1:t}$ seems to be directly related to the dimension of the latent process as shown by [22] for INLA, it needs to be extended to a sequential setting. We feel that the consistency properties of our filter are completely dependent on that of the state filtering mechanism. But this needs to be looked into as future work.

Bibliography

[1] C. Andrieu, A. Doucet, and R. Holenstein. Particle Markov chain Monte Carlo methods. *Journal of the Royal Statistical Society: Series B (Statistical Methodology)*, 72(3):269–342, 2010.

[2] C. Andrieu, A. Doucet, and V. B. Tadic. On-line parameter estimation in general state-space models. In IEEE, editor, *Proceedings of the 44th IEEE Conference on Decision and Control*, pages 332–337, 2005.

[3] C. M. Carvalho, M. S. Johannes, H. F. Lopes, and N. G. Polson. Particle learning and smoothing. *Statistical Science*, 25(1):88–106, 2010.

[4] S. Chib. Marginal likelihood from the Gibbs output. *Journal of the American Statistical Association*, 90(432):1313–1321, 1995.

[5] A. Doucet and A. M. Johansen. A tutorial on particle filtering and smoothing: Fifteen years later. In D. Crisan and B. Rozovsky, editors, *The Oxford Handbook of Nonlinear Filtering*, pages 656–704. Oxford University Press, Oxford, 2011.

[6] N. J. Gordon, D. J. Salmond, and A. F. M. Smith. Novel approach to nonlinear/non-Gaussian Bayesian state estimation. *IEE Proceedings F-Radar and Signal Processing*, 140(2):107–113, Apr. 1993.

[7] J. D. Hamilton. State-space models. In R. F. Engle and D. McFadden, editors, *Handbook of Econometrics*, volume 4, chapter 50, pages 3039–3080. Elsevier, 1986.

[8] S. Haykin, editor. *Kalman Filtering and Neural Networks*. John Wiley & Sons, New York, 2001.

[9] S. Haykin, K. Huber, and Z. Chen. Bayesian sequential state estimation for mimo wireless communications. *Proceedings of the IEEE*, 92(3):439–454, 2004.

[10] E. L. Ionides, A. Bhadra, Y. Atchadé, and A. King. Iterated filtering. *Annals of Statistics*, 39(3):1776–1802, 2011.

[11] S. J. Julier and J. K. Uhlmann. A new extension of the Kalman filter to nonlinear systems. In *Proceedings of AeroSense: The 11th International Symposium on Aerospace/Defense Sensing, Simulation and Controls, Orlando, Florida*, pages 182–193, 1997.

[12] N. Kantas, A. Doucet, S. S. Singh, and J. M. Maciejowski. An overview of sequential Monte Carlo methods for parameter estimation in general state-space models. In *Proceedings of the IFAC Symposium on System Identification (SySid) Meeting*, 2009.

[13] A. A. King, E. L. Ionides, C. M. Bretó, S. E., B. Kendall, H. Wearing, M. J. Ferrari, M. Lavine, and D. C. Reuman. *pomp: Statistical inference for partially observed Markov processes (R Package)*, 2010.

[14] J. Liu and M. West. Combined parameter and state estimation in simulation-based filtering. In D. Freitas and N. J. Gordon, editors, *Sequential Monte Carlo Methods in Practice*. Springer Verlag, New York, 2000.

[15] P. C. Mahalanobis. On the generalised distance in Statistics. *Proceedings of the National Institute of Sciences of India*, 2(1):49–55, 1936.

[16] B. Matérn. *Spatial Variation*. Springer Verlag, Berlin, 2nd edition, 1986.

[17] B. A. McElhoe. An assessment of the navigation and course corrections for a manned flyby of Mars or Venus. *IEEE Transactions on Aerospace and Electronic Systems*, AES-2(4):613–623, 1966.

[18] R. J. Meinhold and N. D. Singpurwalla. Understanding the Kalman filter. *The American Statistician*, 37(2):123–127, 1983.

[19] R Development Core Team. *R: A language and environment for statistical computing*. R Foundation for Statistical Computing, Vienna, Austria, 2013. ISBN 3-900051-07-0.

[20] L. R. Rabiner. A tutorial on hidden Markov models and selected applications in speech recognition. In *Proceedings of the IEEE*, volume 77(2), pages 257–286, February 1989.

[21] B. Ristic, S. Arulampalam, and N. Gordon. *Beyond the Kalman Filter: Particle Filters for Tracking Applications*. Artech House, USA, 2004.

[22] H. Rue, S. Martino, and N. Chopin. Approximate Bayesian inference for latent Gaussian models by using integrated nested Laplace approximations. *Journal of the Royal Statistical Society: Series B (Statistical Methodology)*, 71(2):319–392, 2009.

[23] G. Storvik. Particle filters for state-space models with the presence of unknown static parameters. *IEEE Transactions on Signal Processing*, 50(2):281–289, 2002.

[24] M. West and J. Harrison. *Bayesian Forecasting and Dynamic Models*. Springer Verlag, New York, 2nd edition, 1997.

7

Bayesian Active Contours with Affine-Invariant Elastic Shape Prior

Darshan Bryner

Florida State University

Anuj Srivastava

Florida State University

CONTENTS

7.1 Introduction

An object of interest in an image can be characterized to some extent by the shape of its external boundary. It is therefore important to develop procedures for boundary extraction in problems of detection, tracking, and classification of objects in images. Active contour algorithms have become an important tool in image segmentation for object detection [4, 5, 9]. As segmentation algorithms become more sophisticated, they are tested in more difficult imaging environments of real-world scenarios where images do not have enough contrast to provide sharp boundaries, there is some occlusion of the target, or there exists target-like clutter or noise. Thus, it is of increasing importance that boundary extraction algorithms make use of prior knowledge about the expected target class in order to help compensate for the lack of clear data. This is accomplished by influencing the contour evolution in part with a *shape prior*, a statistical model derived from a set of known training shapes, in a *Bayesian active contour* approach [3, 7, 13, 18].

Most of the past Bayesian segmentation methods use a shape prior designed to be invariant to similarity transformations of translation, rotation, and global scaling. However, in situations when the image plane of a camera is not parallel to the plane containing the defining part of the shape, perspective effects can transform the observed shapes in a more complicated manner than what can be modeled by the similarity group alone. The affine group is commonly used to approximate such shape deformations, and thus it is our goal to develop a segmentation algorithm that uses a shape prior built from affine-invariant shape statistics. In other words, *training and test shapes can be at random affine transformations from each other and the segmentation results will be invariant to those transformations.* Thus, our segmentation will be robust not only to poor image quality but also to perspective skews due to different viewing angles, either in test or training.

7.1.1   Past work on prior-driven active contours

There are two broad categories of active contour methods: *parametric* methods that evolve an explicitly defined parameterized curve, and *geometric* methods that evolve implicitly defined zero-level sets of higher-dimensional functions. Due to the popularity and versatility of geometric methods, pioneered by the works of [4, 14] among others, most Bayesian methods have been applied in the geometric realm and follow the ideas presented in Leventon *et al.* [13], which uses PCA of level-set functions to form a shape prior. Tsai *et al.* in [18] incorporate a similar shape prior in an improved level-set segmentation framework given by Chan and Vese [5]. Others, e.g. [11], improve on Leventon's Gaussian shape prior by applying non-parametric density estimation techniques in $\mathbb{L}^2$ space. Yezzi and Soatto in [20] propose a shape prior based on an average shape that is invariant to any finite-dimensional group transformation, which includes the affine group.

There have been a few Bayesian active contour models that take a parametric approach. One such example, [8] makes use of "landmark-based" shape analysis [6] to impose a shape prior based on the statistics of similarity-invariant point sets. Recent advancements in the modeling of shapes as continuous curves, given by elastic shape analysis, e.g. [17], have allowed for more accurate and parsimonious shape models compared to those of [6, 11, 13, 18, 20]. Elastic shape analysis offers the important advantage of simultaneous registration and deformation of curves with an optimal combination of stretching and bending. Joshi *et al.* [7] create a shape prior from an intrinsic density on elastic shape space, but the method uses an older, more computationally expensive representation. The work of Bryner *et al* [3] incorporates recent simplifications for elastic shape analysis provided in [17] for a computational speed-up. The works of [7] and [3] only formulate an intrinsic similarity-invariant shape prior rather than allowing for an affine-invariant, elastic shape model, which has been partially developed in [2]. *In summary, we are unaware of the existence of any Bayesian active contour model that incorporates fully intrinsic, affine-invariant shape statistics of parametric curves.*

7.1.2   Our approach and contributions

Our goal is to develop a method for *representing, modeling, and incorporating prior information* about shapes of closed curves, invariant to affine transformation and re-parameterization, in a parametric boundary extraction algorithm. Using the mathematical representation presented in [2] on affine-invariant elastic shape, we develop an intrinsic statistical model on the space of canonical, or affine-standardized, closed curves that will serve as a shape prior for the segmentation. Due to elastic matching of curves, this shape model captures the underlying shape variation of a shape class more accurately and leads to a more parsimonious shape model than its extrinsic counterparts often used in previous geometric Bayesian contour models. Furthermore, an invariance to affine transformation allows the model to be robust to perspective skews.

With respect to many state-of-the-art Bayesian contour models that use intrinsic shape statistics (e.g. [3, 7]), we make one further key advancement to their approach aside from our novel, affine-invariant shape prior term. In our work we compute a true gradient descent flow for energy minimization [19]; that is, for each energy functional in the active contour model, we compute its gradient with respect to the same ($\mathbb{L}^2$) metric. Previous methods tend to mix gradients by computing the shape prior energy gradient with respect to the intrinsic shape metric and the remaining energy gradients with respect to the $\mathbb{L}^2$ metric. We show that a true gradient descent flow allows for the contour to optimally fit both shape and image data. In summary, the main contributions of this chapter are: (1) develop an intrinsic, elastic affine-invariant shape model of planar curves, (2) incorporate this statistical shape model in driving parametric active contours via true gradient descent, and (3) demonstrate this framework using segmentation of targets subject to noise or occlusion as well as perspective skew.

The organization of the rest of the chapter is as follows. Section 7.2 reviews the affine-invariant, elastic shape analysis method from [2], and then develops the algorithmic tools to build intrinsic statistical shape models on the affine shape space. Section 7.3 describes the Bayesian active contour model, focusing on computation of the shape prior energy gradient with respect to the $\mathbb{L}^2$ metric. Section 7.4 shows a variety of experimental results that showcase the effectiveness of using our affine shape prior with $\mathbb{L}^2$ gradient compared to other methods. Section 7.5 is the conclusion and summary of future efforts.

7.2   Affine-Invariant, Elastic Shape Statistics

Here, we summarize the affine-invariant, elastic shape analysis method presented in [2], and we then develop the procedures necessary to form intrinsic statistical models on such a shape space.

7.2.1   Affine-invariant, elastic shape analysis

Let $\beta \in \mathcal{B}$ where $\mathcal{B}$ is the set of all closed, parameterized, absolutely continuous curves. The action of the orientation preserving affine group $G_a = GL_+(2) \ltimes \mathbb{R}^2$ on β results in the orbits $[\beta] = \{A\beta + b | A \in GL_+(2), b \in \mathbb{R}^2\}$, where the group $GL_+(2)$ represents all 2×2 invertible matrices with positive determinant. Let Γ be the set of all re-parameterizations of the form $\gamma : \mathbb{S}^1 \to \mathbb{S}^1$ such that γ is a diffeomorphism. We wish to analyze the space of all equivalence classes $\mathcal{B}/(G_a \times \Gamma)$, yet, with respect to the standard $\mathbb{L}^2$ metric, the group $G_a \times \Gamma$ does not act on $\mathcal{B}$ by isometries. In other words, for two elements $\beta_1, \beta_2 \in \mathcal{B}$ and for an arbitrary $g \in (G_a \times \Gamma)$, we do not have the distance preserving property of $\|\beta_1 - \beta_2\| = \|(\beta_1, g) - (\beta_2, g)\|$, where $(\cdot, g)$ represents the group action of g on an element in $\mathcal{B}$. Thus, it is not possible to impose a proper distance between equivalence classes and perform affine-invariant shape analysis of curves with this metric in this representation space. The solution proposed in [2] is to define a space M/G_0 with the following properties: (a) $M \subset \mathcal{B}$ and G_0 is a subgroup of $G_a \times \Gamma$, (b) there exists a bijection between M/G_0 and $\mathcal{B}/(G_a \times \Gamma)$, and (c) G_0 acts by isometries on M with respect to the chosen metric. Thus, a proper statistical analysis on M/G_0 is possible and represents, implicitly, a statistical analysis on $\mathcal{B}/(G_a \times \Gamma)$. The space M/G_0 is called a *section* of affine orbits.

The section is defined in the following manner. Let $L_\beta = \int_0^1 |\dot{\beta}(t)| dt$ be the length of the curve β, where $|\cdot|$ is Euclidean 2-norm. (Please contrast it from $\|\cdot\|$ which is used to denote the $\mathbb{L}^2$-norm of a curve or function.) The *centroid* of β is defined as $C_\beta = \frac{1}{L_\beta} \int_0^1 \beta(t)|\dot{\beta}(t)| dt \in \mathbb{R}^2$. The *covariance* of β is defined as $\Sigma_\beta = \frac{1}{L_\beta} \int_0^1 (\beta(t) - C_\beta)(\beta(t) - C_\beta)^T |\dot{\beta}(t)| dt \in \mathbb{R}^{2 \times 2}$. It can be shown that for any $\beta \in \mathcal{B}$ there exists a canonical, or standard, element $\beta_0 \in [\beta]$ that satisfies the following three conditions: (1) $L_{\beta_0} = 1$, (2) $C_{\beta_0} = 0$, and (3) $\Sigma_{\beta_0} \propto I$. Furthermore, for any two curves $\beta^{(1)}$ and $\beta^{(2)}$ within an affine transformation of each other, the corresponding standard elements, $\beta_0^{(1)}$ and $\beta_0^{(2)}$, are related by a rotation and a re-parameterization, i.e. $\beta_0^{(2)} = O(\beta_0^{(1)} \circ \gamma)$, where $O \in SO(2)$ and $\gamma \in \Gamma$. If $\mathcal{B}_0$ is the space of all such affine-standardized curves, then the quotient space $\mathcal{B}_0/(SO(2) \times \Gamma)$ satisfies properties (a) and (b) of a section as defined in the previous paragraph. Property (c) is not yet satisfied since the group Γ does not act on $\mathcal{B}_0$ by isometries with respect to the $\mathbb{L}^2$ metric.

In order to achieve the isometry property (c), we make the following transformation as given by [17]. Define $q(t) = \dot{\beta}(t)/\sqrt{|\dot{\beta}(t)|}$ as the square-root velocity function (SRVF) of β. The action of a $\gamma \in \Gamma$ on q is given by $(q, \gamma) = (q \circ \gamma)\sqrt{\dot{\gamma}}$, and now the group Γ acts by isometries on the space of SRVF's with respect to the $\mathbb{L}^2$ metric. Therefore, if $\mathcal{Q}_0$ is the set of all SRVF's of $\mathcal{B}_0$, the space $\mathcal{Q}_0/(SO(2) \times \Gamma)$ is in one-to-one correspondence with $\mathcal{B}_0/(SO(2) \times \Gamma)$ and satisfies all three properties of a section. Furthermore, the paper [17] shows that the $\mathbb{L}^2$ metric of SRVF's is equivalent to the elastic metric of curves. Hence, performing statistical shape analysis on $\mathcal{Q}_0/(SO(2) \times \Gamma)$ is equivalent to that of an affine-invariant analysis on $\mathcal{B}/(G_a \times \Gamma)$, and it is in fact an elastic shape analysis framework with respect to the standard $\mathbb{L}^2$ metric. We denote $\mathcal{S} = \mathcal{Q}_0/(SO(2) \times \Gamma)$ as affine-invariant, elastic shape space.

7.2.2 Statistical modeling

In order to build intrinsic statistical models on $\mathcal{S}$, one must first develop a set of algorithmic tools that use the geometry of $\mathcal{Q}_0$ to compute pertinent statistical quantities. Bryner *et al* [2] describe geometry of $\mathcal{Q}_0$, and we will not repeat it here. Instead, we discuss the algorithms and subroutines necessary to compute an intrinsic sample mean and covariance on $\mathcal{S}$. These two statistics are then used to define a Gaussian-type probability model of elastic shapes invariant to affine transformation and re-parameterization. This probability model is in turn used as a shape prior in our Bayesian active contour framework, described later.

Statistical modeling on $\mathcal{S}$ requires the use of two algorithms, one for sample mean calculation and one for covariance calculation. Both of these algorithms make use of five subroutines that each requires as input a basis for $N_q(\mathcal{Q}_0)$, the normal space of $\mathcal{Q}_0$ at any point q. The five subroutines are (1) Projection onto the manifold, (2) Projection onto a tangent space, (3) Parallel translation of a tangent vector, (4) Exponential mapping, and (5) Inverse exponential mapping. An expression for the basis of $N_q(\mathcal{Q}_0)$ as well as descriptions of Subroutines (1)–(3) appear in [2]. To summarize, Subroutine 1 takes any arbitrary function in the space $\mathbb{L}^2([0, 1], \mathbb{R}^2)$ and projects it to $\mathcal{S}$; Subroutine 2 takes an arbitrary function in $\mathbb{L}^2([0, 1], \mathbb{R}^2)$, given $[q] \in \mathcal{S}$, and projects it to the tangent space $T_{[q]}(\mathcal{S})$; and Subroutine 3 takes a vector $v \in T_{[q_1]}(\mathcal{S})$ and parallel translates it to $T_{[q_2]}(\mathcal{S})$, given $[q_1], [q_2] \in \mathcal{S}$. Now, we provide Subroutines 4 and 5.

Given an element $[q] \in \mathcal{S}$ and the shooting vector $v \in T_{[q]}(\mathcal{S})$, the exponential mapping computes a point $[p] = \exp_{[q]}(v)$ in $\mathcal{S}$ that represents the point that is reached by traveling along a constant-speed geodesic starting at $[q]$, and with the initial velocity v. Subroutines 4 and 5 make use of the fact that $\mathcal{Q}_0 \subset \mathbb{S}^\infty$, the unit hypersphere in $\mathbb{L}^2$ space, which has well-known analytical formulas for the exponential and inverse exponential mappings.

Subroutine 4 – Exponential Map: *Given $[q] \in \mathcal{S}$, $v \in T_{[q]}(\mathcal{S})$, an integer n, and $\epsilon > 0$,*

1. *If $\|v\| < \epsilon$, return $[p] = [q]$, else*

2. *Let $\delta = 1/n$. For $i = 1, ..., n$*

   (a) *Compute $q_{adv} = \exp_q(\delta v)$ on $\mathbb{S}^\infty$ via the formula $q_{adv} = \cos(\delta\|v\|)q + \sin(\delta\|v\|)\frac{v}{\|v\|}$.*

   (b) *Project $[q_{adv}]$ to $\mathcal{S}$ using Subroutine 1.*

   (c) *Parallel translate v from $T_{[q]}(\mathcal{S})$ to $T_{[q_{adv}]}(\mathcal{S})$ using Subroutine 3. Let $q = q_{adv}$.*

3. *Return $[p] = [q]$.*

The opposite of the exponential map, the inverse exponential map, computes a vector $v = \exp_{[q]}^{-1}([p])$ in $T_{[q]}(\mathcal{S})$ that represents the shooting vector that satisfies $[p] = \exp_{[q]}(v)$ given $[p], [q] \in \mathcal{S}$. Since we wish to compute statistics modulo rotation and re-parameterization on the quotient space $\mathcal{S}$, when computing this shooting vector v, we must select either q or p and optimally rotate and re-parameterize it to the other. This registration is accomplished using a combination of Procrustes rigid body alignment and dynamic programming (see [17]).

Subroutine 5 – Inverse Exponential Map: *Given $[q], [p] \in \mathcal{S}$ and $\epsilon > 0$,*

1. *Optimally rotate/register WLOG q to p.*

2. *Compute the arclength $\theta = \cos^{-1}(\langle q, p \rangle)$.*

3. *If $\theta < \epsilon$, let $v = 0$, else compute $v = \exp_q^{-1}(p)$ in $T_q(\mathbb{S}^\infty)$ via the formula $v = \frac{\theta}{\sin(\theta)}(q - p\cos(\theta))$.*

4. *Project v to $T_{[q]}(\mathcal{S})$ via Subroutine 2.*

Now we are ready to present the algorithms to compute the mean $[\mu]$ and the covariance K of a set of n shapes $\{[q_i]\}$ in $\mathcal{S}$. A popular intrinsic mean calculation is the Karcher mean, which is defined as $[\mu] = \arg\min_{[q] \in \mathcal{S}} \sum_{i=1}^n d_{\mathcal{S}}([q], [q_i])^2$, where $d_{\mathcal{S}}(\cdot, \cdot)$ is the geodesic distance on shape space. An iterative algorithm to find the Karcher mean of a set of shapes is outlined below. The general idea is to update the current estimate $[\mu_j]$ in the direction of the average shooting vector from $[\mu_j]$ to each of data points $\{[q_i]\}$.

Algorithm (Karcher Mean): *Let $[\mu_0] \in \mathcal{S}$ be an initial estimate of the mean of $\{[q_i]\}$, e.g. let $[\mu_0] = [q_1]$. Set $j = 0$.*

1. *For each $i = 1, ..., n$, register/rotate q_i to μ_j, and compute $v_i = \exp^{-1}_{[\mu_j]}([q_i])$ using Subroutine 5.*

2. *Compute the average direction $\bar{v} = \frac{1}{n} \sum_{i=1}^{n} v_i$.*

3. *If $\|\bar{v}\|$ is small, stop. Else, update $[\mu_j]$ by $[\mu_{j+1}] = \exp_{[\mu_j]}(\delta \bar{v})$ via Subroutine 4, where $\delta \approx 0.5$.*

4. *Set $j = j + 1$ and return to step 1.*

Once we have found a Karcher mean $[\mu]$, we obtain the Karcher covariance matrix via $K = \frac{1}{n-1} \sum_{i=1}^{n} v_i v_i^T$, where the v_i's are shooting vectors from μ to the respective q_i's, each optimally rotated and registered to μ. While in theory $v : [0, 1] \to \mathbb{R}^2$ is a vector valued function, in practice it is computed using T equally spaced samples on the interval $[0, 1]$. Therefore, $v \in \mathbb{R}^{2 \times T}$ or re-arranged to be $\mathbb{R}^{1 \times 2T}$ and K is a $2T \times 2T$ covariance matrix.

Algorithm (Karcher Covariance): *Given a set of shapes $\{[q_i]\}$ and its Karcher mean $[\mu]$,*

1. *For $i = 1, ..., n$, register/rotate q_i to μ, and calculate the shooting vector $v_i = \exp^{-1}_{[\mu]}([q_i])$ using Subroutine 5.*

2. *Compute $K = \frac{1}{n-1} \sum_{i=1}^{n} v_i v_i^T$.*

Now that we have the tools to compute a sample mean and covariance of data on $\mathcal{S}$, we can speak of defining a probability density function from the shape class $(\{[q_i]\}, \mu, K)$. There are many densities one can define on $\mathcal{S}$ from $(\{[q_i]\}, \mu, K)$, but for this research we only consider a *truncated wrapped-normal density* [12], which is formed as follows. First, obtain the singular value decomposition of K as $[U, S, V] = \text{svd}(K)$, and let U_m be the m-dimensional principal subspace of $T_{[\mu]}(\mathcal{S})$ defined as the first m columns of U. The truncated wrapped-normal density is given as

$$f_m([q]) = \frac{1}{Z} e^{-\frac{1}{2}\left(v_\parallel^T S_m^{-1} v_\parallel + \|v_\perp\|^2 / \delta^2\right)} J_{[\mu]} \mathbf{1}_{\|v\| < \pi}, \tag{7.1}$$

where $v = \exp^{-1}_{[\mu]}([q])$, $v_\parallel = U_m^T v$ is the projection of v into U_m, $v_\perp = v - U_m v_\parallel$, S_m is the diagonal matrix containing the first m singular values, $J_{[\mu]}$ is the Jacobian of the exponential mapping, and Z is the normalizing constant. The scalar value δ is chosen to be less than the smallest singular value in S_m. In other words, this density is defined as a multivariate Gaussian density on $U_m \subset T_{[\mu]}(\mathcal{S})$ wrapped onto the manifold via the exponential mapping.

After computing the density in Eqn. 7.1, it is rather straightforward to randomly sample from it. Since it is not possible to visualize the density function itself, in order to illustrate the shape variation explained in the density, we show a number of random samples instead. Fig. 7.1 displays four shape models in this manner formed from the same training data – 20 crown shapes

Data	Mean	Rando Samples

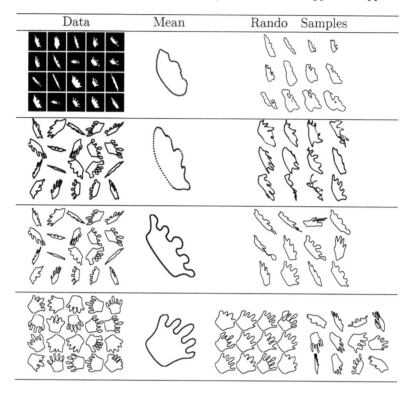

FIGURE 7.1

Shape models from top to bottom: [13], [6], [17], and our elastic affine-invariant model.

from the well-known MPEG-7 database [1] – in different shape spaces. Applied to each shape is a random orientation-preserving affine transformation, and the resulting curves are then sampled to have uniform speed parameterization. The shape spaces from top to bottom are the level-set similarity-invariant space from [13], landmark similarity-invariant space from [6], elastic similarity-invariant space [17], and elastic affine-invariant space. The random samples in the last case are displayed in two stages. The left set of samples are those created in $\mathcal{S}$, while the right set are the same samples but with a random orientation-preserving affine transformation re-applied to them. In this manner we separate the shape variability and affine variability when forming random samples.

From Fig. 7.1, one can see that the ability of the elastic shape models to describe the underlying shape variation in a complicated shape class is superior to the non-elastic ones [6, 13]. The mean shapes in the non-elastic models wash out the defining crown features, while the elastic models retain the five points of the crown. Furthermore, the random samples from [6, 13] are not representative of the training data. The elastic affine model is more accurate

TABLE 7.1
Top 5 singular values in the affine-invariant shape model from Figure 7.1.

Index	1	2	3	4	5
Singular Value	4.86	1.72	1.02	0.79	0.61
Cumulative Energy Percentage	0.42	0.57	0.66	0.73	0.78

and parsimonious than the elastic similarity model since affine variability and shape variability are separated. Thus, our elastic affine model is the superior choice in general for a shape prior in a Bayesian active contour model. All models shown in Fig. 7.1 use the top $m = 5$ principal components of variation, and Table 7.1 shows these singular values of the elastic affine model along with their cumulative energy percentage.

7.3 Bayesian Active Contour Model

7.3.1 Model definition

The problem of boundary extraction in our active contour model can be posed as a maximum *a posteriori* (MAP) estimation via the energy minimization

$$\hat{\beta} = \underset{\beta \in \mathcal{B}}{\operatorname{argmin}} \left(\lambda_1 E_{image}(\beta) + \lambda_2 E_{smooth}(\beta) + \lambda_3 E_{prior}(\beta) \right), \qquad (7.2)$$

where the λ_i's are user-defined constant weights. The terms E_{image} and E_{smooth} are common energy terms based on given image pixel data and the smoothness of the evolving boundary curve, respectively, while the term E_{prior} is the novel term representing the shape prior energy functional built from a wrapped-normal density on $\mathcal{S}$. If E_{total} represents the weighted sum of all the the energy functionals in the model, the active contour evolution is given by the gradient descent iteration $\beta_{n+1} = \beta_n - \epsilon \nabla E_{total}$, where $\epsilon > 0$ is a step size selected to maintain numerical stability.

Our research focuses on the influence of E_{prior} on active contour evolution, and thus the formulation of E_{image} and E_{smooth} are less important for our overall exposition. For this reason we have selected the relatively simple and well-known E_{image} term based on "region competition" [21] and E_{smooth} term based on curvature smoothing. The important point to consider here is that E_{image} and E_{smooth} are functionals of the curve β, and their gradients ∇E_{image} and ∇E_{smooth} are taken with respect to the $\mathbb{L}^2$ metric of curves. Thus, as noted in [19], in order for our active contour evolution to be a true gradient descent flow, we must take the gradient ∇E_{prior} also with respect to the $\mathbb{L}^2$ metric of curves. For completeness, though, we include descriptions of our formulation of E_{image} and E_{smooth} as well as our novel E_{prior}.

7.3.2   Computing E_{image} and its gradient

Our image energy term is based on the work presented in [15, 21]. The active contour model assumes the existence of a set of training images with their corresponding ground truth segmentations, and we use this prior knowledge to form E_{image} in the following manner. Let $I(x)$ be the pixel information in a training image with domain Ω. For our application the region-based approach assumes that any image domain is partitioned into two regions, Ω_{in} and Ω_{out}, representing the target and the background, respectively. First, obtain *a priori* a sample of target pixel values from $I(\Omega_{in})$ and a sample of background pixel values from $I(\Omega_{out})$. Then estimate two probability densities, $p_{in}(I)$ and $p_{out}(I)$, from the normalized histograms of these sample target and background pixel intensity values, respectively. One can collect as much pixel data from various training images as needed to form accurate density functions. These two learned densities will be used to calculate E_{image} during an active contour evolution on a test image.

Paragios *et al.* [15] show that for a closed curve β defining the two regions Ω_{in} and Ω_{out}, the minimization of the following energy functional is equivalent to the maximization of the image-based *a posteriori* segmentation probability.

$$E_{image}(\beta) = -\int_{\Omega_{in}} \log(p_{in}(I(x)))dx - \int_{\Omega_{out}} \log(p_{out}(I(x)))dx. \qquad (7.3)$$

Using functional differentiation and Green's Theorem, as shown in [21], we calculate the gradient of the image energy at a point $\beta(t)$ as

$$\nabla E_{image}(\beta)(t) = -\log\left(\frac{p_{in}(I(\beta(t)))}{p_{out}(I(\beta(t)))}\right)\mathbf{n}(t), \qquad (7.4)$$

where $\mathbf{n}(t)$ is the outward unit normal vector to the curve β. Define $\ell_\beta(t) \equiv \log\left(\frac{p_{in}(I(\beta(t)))}{p_{out}(I(\beta(t)))}\right)$. Notice that contour evolution according to the negative gradient will be along the outward normal direction if $\ell_\beta(t) > 0$ and along the inward normal direction if $\ell_\beta(t) < 0$. This evolution will therefore push any part of the contour more likely lying in the target out toward its most likely boundary, and it will pull any part lying outside of the target in toward its most likely boundary.

7.3.3   Computing E_{smooth} and its gradient

A common feature of all active contour models is a smoothness penalty that prevents the evolving contour from becoming too jagged. We follow the well-known approach proposed in [10] and [4], which is based on the idea of Euclidean heat flow. Define the smoothing energy functional as

$$E_{smooth}(\beta) = \int_0^1 |\dot{\beta}(t)|dt, \qquad (7.5)$$

which is equal to the length of the curve and is naturally invariant to any re-parameterization. It is shown in [10] that the gradient of E_{smooth} is given by the Euclidean heat flow equation

$$\nabla E_{smooth}(\beta)(t) = \kappa_\beta(t)\mathbf{n}(t), \tag{7.6}$$

where $\kappa_\beta(t)$ is the curvature of $\beta(t)$. It has been shown that this penalty on a curve's length automatically leads to smoothing of a curve by forcing the curve to become convex over time. Eventually, the curve evolves to a circle and shrinks to a point as the evolution time goes to infinity.

7.3.4   Computing E_{prior} and its gradient

Given a prior shape class $(\{[q_i]\}, [\mu], K)$ in $\mathcal{S}$, $E_{prior} : \mathcal{B} \to \mathbb{R}$ is defined as such:

$$E_{prior}(\beta) = \frac{1}{2}v^T (U_m S_m^{-1} U_m^T)v + \frac{1}{2\delta^2}\|v - U_m U_m^T v\|^2, \tag{7.7}$$

where all terms are defined in the description of Eqn. 7.1. Thus, minimizing E_{prior} corresponds to maximizing the log-likelihood of the prior shape density. The global minimizer of this functional is the curve representation of the mean $[\mu]$. Note that $v = \exp_{[\mu]}^{-1}([q])$, where q is the SRVF of the standardized curve $\beta_0 \in [\beta]$. Even though the calculation of E_{prior} is based on an elastic shape distance between SRVF's, we ultimately treat it as any black box functional on $\mathbb{L}^2$ space. A numerical technique to approximate the gradient of such a functional is given as follows. Select an orthonormal basis for $T_\beta(\mathcal{B})$, say $\{b_i, i = 1, 2, ...\}$. The standard Fourier basis functions that are periodic on $[0, 1]$ serve as a basis for this $\mathbb{L}^2$ space. After truncating to the first N basis elements for practical implementation, a first order numerical approximation is given as

$$\nabla E_{prior}(\beta) \approx \sum_{i=1}^{N} \frac{E_{prior}(\beta + \epsilon b_i) - E_{prior}(\beta)}{\epsilon}b_i, \tag{7.8}$$

where $\epsilon > 0$ is sufficiently small. The full algorithm for computing $\nabla E_{prior}(\beta)$ is given below.

Algorithm (∇E_{prior} Calculation): *Given a curve $\beta \in \mathcal{B}$,*

1. *Standardize β to $\mathcal{B}_0$ and convert to SRVF representation to obtain $[q] \in \mathcal{S}$.*

2. *Register/rotate μ to q to obtain $\mu^* = O^*(\mu, \gamma^*)$, and calculate $v = \exp_{[\mu]}^{-1}([q])$ via Subroutine 5. Calculate $E_{prior}(\beta)$ via Eqn. 7.7.*

3. *For each $i = 1, ..., N$,*

   (a) Compute q_i, the SRVF of $\beta + \epsilon b_i$.

(b) *Register/rotate μ to q_i using O^* and γ^* from step 2, and approximate $v = \exp^{-1}_{[\mu]}([q_i])$ via Subroutine 5. Calculate $E_{prior}(\beta + \epsilon b_i)$ via Eqn. 7.7.*

4. *Compute $\nabla E_{prior}(\beta)$ via Eqn. 7.8.*

Note that in step (3b) above, we compute an approximation of the term $E_{prior}(\beta + \epsilon b_i)$ since we do not require for each i an optimization over $SO(2) \times \Gamma$. This would be quite expensive computationally. The approximation is valid since we assume ϵ small enough, *i.e.* a small enough perturbation of β, that the values O^* and γ^* obtained from optimizing μ to q can be used for optimizing μ to q_i.

This completes our calculation of the shape prior gradient with respect to the $\mathbb{L}^2$ metric. Our formulation advances the works of [3, 7] because in each of these papers, the authors compute ∇E_{prior} with respect to the elastic metric on shape space, which is inconsistent with the remaining energy functional gradients. In particular, they formulate ∇E_{prior} as the shooting vector Av on the tangent space of the shape manifold, where $v = \exp^{-1}_{[\mu]}([q])$ and $A = U_m S_m^{-1} U_m^T + (I - U_m U_m^T)/\delta^2$. Then, they numerically approximate the vector field on the curve resulting from this shooting vector. While this formulation provides an analytical and precise expression for the shape prior gradient on shape space, the summation with ∇E_{image} and ∇E_{smooth} does not result in a true gradient of E_{total}.

Each of the three gradients $\nabla E_{image}(\beta)$, $\nabla E_{smooth}(\beta)$, and $\nabla E_{prior}(\beta)$ are defined as a vector field along the parameterized curve β. That is, for every $t \in [0, 1]$, there are three gradient vectors associated to the point $\beta(t)$. The vector field $\nabla E_{total}(\beta)$ is therefore the sum, weighted by the coefficients λ_i, of these three vector fields. The equation for gradient descent evolution of β according to the functional E_{total} is therefore given by

$$\beta_{n+1} = \beta_n - \lambda_1 \nabla E_{image}(\beta_n) - \lambda_2 \nabla E_{smooth}(\beta_n) - \lambda_3 \nabla E_{prior}(\beta_n), \quad (7.9)$$

where the subscript $n = 0, 1, \ldots$ represents the index of the discretized contour evolution time. Note that in this formula, we omit any numerical time stepping factor since it can be absorbed into the selection of the λ_i's. All experimental results in the following section reflect an evolution according to Eqn. 7.9.

7.4   Experimental Results

Here, we evaluate the segmentation performance of multiple active contour models on various datasets. The contour models use all possible combinations of the following tools to formulate E_{prior} and its gradient: similarity-invariant shape statistics (from [17]), affine-invariant shape statistics (developed here as an extension of [2]), elastic gradient (from [7]), and $\mathbb{L}^2$ gradient (developed here). The E_{prior} scenarios are thus given as (1) no shape prior, (2) similarity-invariant with elastic gradient, (3) similarity-invariant with $\mathbb{L}^2$ gradient, (4) affine-invariant with elastic gradient, and (5) affine-invariant with $\mathbb{L}^2$ gradient. Scenario (2) is exactly the model presented in [3], while scenarios (3)–(5) are novel to this work. It is our proposition that scenario (5) will yield the best segmentation results due to its robustness to perspective effects as well as consistency of gradients.

In order to evaluate the accuracy of any segmentation result, we compare the converged contour to the associated ground truth curve via two metrics: $d_{geod}(\cdot, \cdot)$ and $d_{bin}(\cdot, \cdot)$. The distance d_{geod} is the geodesic distance on similarity-invariant, elastic shape space from [17]. The distance d_{bin} is a binary image metric that measures the area of non-overlapping regions and is defined in the following manner. If $\hat{B}$ is the binary image obtained by the segmentation and B is the ground truth binary image, the binary image distance is defined as $d_{bin}(\hat{B}, B) = \text{area}(\hat{B} \cup B - \hat{B} \cap B)/\text{area}(\hat{B} \cup B)$. The values of these two metrics together show how accurately our segmentation result matches the correct shape as well as the correct location, orientation, and scale in the image. We have that $d_{geod} \in [0, \pi/2]$ and $d_{bin} \in [0, 1]$, and in each case a lower distance value corresponds to greater accuracy to ground truth.

The average computational cost for 100 iterations of each of the five scenarios, as computed in MATLAB on a 2.8 gHz processor, are given in seconds as (1) 0.367, (2) 20.8, (3) 26.0, (4) 38.9, and (5) 38.9. The computations all use 100 discrete sample points along the curve. The affine cases are more complex than the similarity cases due to the necessity to standardize the active contour at each iteration. Most results shown in this section converged within 100–200 iterations with an initialization fairly close to the true boundary; thus, the computational complexity remains in the realm of practicality. Now, we present our segmentation results.

7.4.1 Multiview curve database (MCD)

The MCD [22] has been constructed from the widely used MPEG-7 shape database. Here, a number of shapes were selected from the MPEG-7 database and printed on white paper as binary images, where the region enclosed by the shape was colored black. Variations of each shape were recorded by photographing the printed shapes under seven different camera angles. Although the MCD consists of only the extracted boundary curves from these binary images, we simply recreate the binary images from these curves and use them as test images for segmentation. Conveniently, in this manner the ground truth curves for segmentation are precisely the MCD curves themselves. Since each shape in the MCD comes from one shape class in the MPEG-7 database, we use that shape class to build a shape prior for segmentation. Thus, we construct an experiment where the test image is of a different perspective than the training shapes and show the necessity of an affine-invariant shape prior

TABLE 7.2
Averaged segmentation results of occluded MCD shapes.

Prior	Crown (# 1)	Fountain (# 2)
(1) None	$(0.61, 0.17)$	$(0.60, 0.19)$
(2) El. Sim	$(0.37, 0.23)$	$(0.31, 0.30)$
(3) $\mathbb{L}^2$ Sim	$(0.40, 0.14)$	$(0.23, 0.12)$
(4) El. Aff	$(0.55, 0.16)$	$(0.33, 0.20)$
(5) $\mathbb{L}^2$ Aff	$\mathbf{(0.35, 0.10)}$	$\mathbf{(0.17, 0.067)}$

for accurate segmentation. Furthermore, to increase the difficulty of segmentation, we introduce some occlusion to each skewed test image and segment under the aforementioned five E_{prior} scenarios.

The experiment is as follows. First, we select two shape classes in the MCD – "crown" and "fountain" – on which to perform the segmentation. Then, for each of the two shape classes, we create two prior shape models, a similarity and an affine-invariant model, from 20 MPEG-7 shapes belonging to each respective shape class. Next, we create the test images for segmentation by introducing occlusion to each of the 7 perspective variations in each of the MCD binary images. Finally, we segment all test images under each of the five E_{prior} scenarios, and average the results over each shape class. Fig. 7.2 shows the segmentation results from two test images, and for each case we show eight images. From left to right, top to bottom, the eight images are as follows: the original image under centered camera view, the skewed and occluded test image, segmentation under scenarios (1)–(5) respectively, and the ground truth segmentation. Table 7.2 lists the values (d_{geod}, d_{bin}) for each of the five scenarios averaged over the seven different test images (camera angles) for each shape. The segmentation is best in both instances under scenario (5), the affine prior with $\mathbb{L}^2$ gradient.

The average segmentation results are best in each shape class under scenario (5), the affine prior with $\mathbb{L}^2$ gradient, with respect to both performance metrics. Scenario (5) is a true gradient descent method, which allows the segmentation to simultaneously fit the shape data and the image data in an optimal fashion, and has an affine-invariant prior, which accounts for the perspective variation seen across each shape class. Performance issues under scenarios (2) and (4), the segmentation with mixed gradients, become apparent when compared visually to the results of (3) and (5). When comparing (2) to (3), the similarity prior scenarios, it is clear that the shape data overtakes the image data in the segmentation of scenario (2), while in scenario (3) the image data and the shape data seem to have the same level of influence on the result. When comparing the results from scenario (4) to (5), we see that in scenario (4) the active contour does not land directly on the boundaries defined by the image data. It appears that the contour evolution in (4) struggles to find the correct affine transformation of the mean that best minimizes the

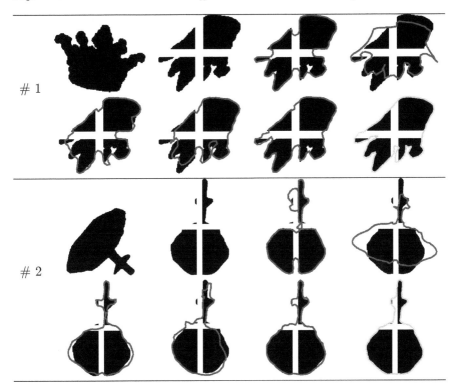

FIGURE 7.2
Segmentation results for occluded versions of the MCD shapes "crown" (camera angle 5) and "fountain" (camera angle 4).

total energy. The image data does not provide enough influence to overtake and correct the shape data in these cases. This issue is solved by the use of a true gradient descent flow, where in (3) and (5) we see that the contour respects both the shape and the image data equally. For this reason in scenario (3), the average segmentation is quite close to ground truth even though it uses a similarity-invariant prior rather than an affine-invariant prior. The affine-invariant prior in (5) clearly is closest to ground truth, though, yielding the best segmentation results.

7.4.2 Leaf segmentation

Another useful application of our affine-invariant contour algorithm is the segmentation of leaves in images. The shape of a leaf is distinctive and indicative of the type of tree that it comes from, and moreover, since a leaf is a planar object, perspective effects of imaged leaves can be approximated by an affine transformation. In the upper-left image of Fig. 7.3 we see the

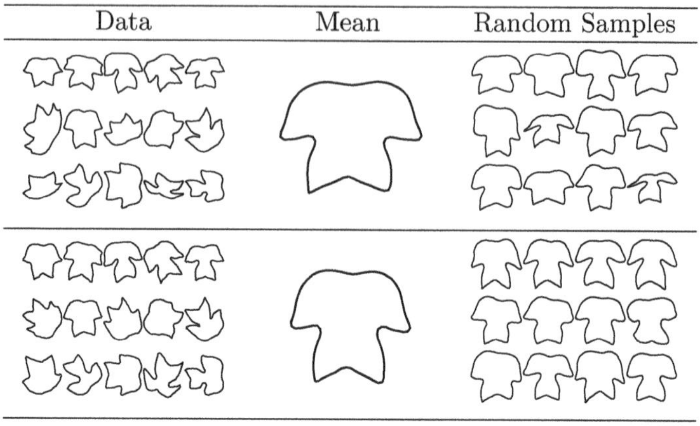

Data	Mean	Random Samples

FIGURE 7.3
Elastic shape models of the tulip poplar leaf. Top: Similarity-invariant. Bottom: Affine-invariant.

ground truth boundary curves of 15 tulip poplar leaves that were extracted from images found in a Google Image search. Notice that in addition to the inherent leaf shape variability, this data additionally exhibits an approximate affine variability due to imaging from different camera angles. Fig. 7.3 shows the resulting similarity and affine invariant elastic shape models, and one can see from the random samples that the affine-invariant model eliminates the perspective variability while the similarity-invariant model does not.

Using the statistical models in Fig. 7.3 to build E_{prior}, we segment two test images of tulip poplar leaves in a somewhat noisy background with target-like clutter. Fig. 7.4 from left to right and top to bottom shows the test image and the segmentation results from scenarios (1)–(5) respectively. We forgo showing the ground truth segmentation because visually it is rather apparent. As in Table 7.2, Table 7.3 lists the pair (d_{geod}, d_{bin}) comparing the segmentation result to ground truth in each of the 5 scenarios for both of the test images. Again as predicted, scenario (5), the curve evolution using the $\mathbb{L}^2$ gradient of the affine-invariant E_{prior}, yields the best results, optimally fitting the shape data and the image data.

7.4.3 SAS shadow segmentation

Here, we segment a dataset of imagery collected beyond the visible spectrum, where we segment the shadows of a cylinder target in underwater synthetic aperture sonar (SAS) imagery. Segmentation is typically difficult in the synthetic aperture imaging modalities due to background noise, clutter, and imaging artifacts. The SAS images were created from the Shallow Water Acoustics Toolkit (SWAT), a program developed by the Naval Surface Warfare Center Panama City Division (NSWC PCD) that synthesizes SAS imagery of

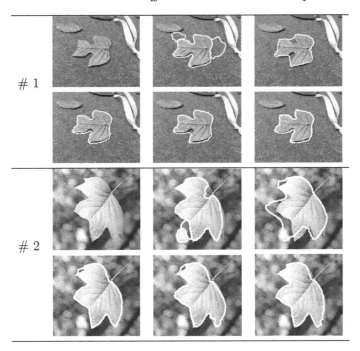

FIGURE 7.4
Segmentation results for tulip poplar leaves.

various targets in seabed environments [16]. The SWAT simulator is considered accurate to reality and is widely used to test automatic target detection and recognition algorithms in place of real SAS data. This particular dataset consists of imagery of the same cylinder target at different aspect angles and ranges, which yields shadow signatures that exhibit a shape variability that can be modeled by an affine transformation. Fig. 7.5 shows an example of five images in the SAS dataset, and Fig. 7.6 shows the similarity and affine invariant shape models from 10 training shapes. Notice that much of the shape variability is removed in an affine-invariant framework.

TABLE 7.3
Segmentation results of the tulip poplar leaves.

Prior	Leaf 1	Leaf 2
(1) None	$(0.83, 0.32)$	$(0.52, 0.22)$
(2) El. Sim	$(0.35, 0.13)$	$(0.28, 0.25)$
(3) $\mathbb{L}^2$ Sim	$(0.23, 0.11)$	$(0.26, 0.13)$
(4) El. Aff	$(0.22, 0.11)$	$(0.29, 0.14)$
(5) $\mathbb{L}^2$ Aff	$\mathbf{(0.18, 0.096)}$	$\mathbf{(0.21, 0.12)}$

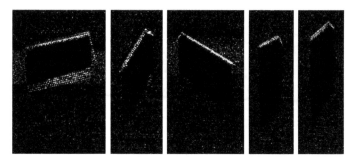

FIGURE 7.5
Example images from the PC SWAT cylinder database.

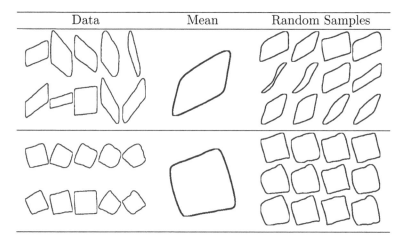

FIGURE 7.6
Elastic shape models of the SAS cylinder shadow. Top: Similarity-invariant. Bottom: Affine-invariant.

Assuming that the ground truth shadow boundary curves are available, we perform a cross-validation experiment for each of the five scenarios. In each cross-validation iteration, we select 10 images at random for training, form the shape prior density on the appropriate shape space from the corresponding ground truth curves, and segment the remaining 90 test images with the influence of that shape prior. After each segmentation we calculate the values (d_{geod}, d_{bin}) to ground truth. The averages of these two distance values across all cross-validation iterations in each scenario are as follows: (1) $(0.28, 0.14)$, (2) $(0.21, 0.12)$, (3) $(0.17, 0.099)$, (4) $(0.18, 0.11)$, and **(5) (0.17, 0.090)**.

Fig. 7.7 shows segmentation results from one of the test images. The six segmentations shown from left to right and top to bottom are from scenarios (1)–(5), respectively, and ground truth. Although in many instances in the cross validation experiment, segmentation with scenarios (1)–(4) yields

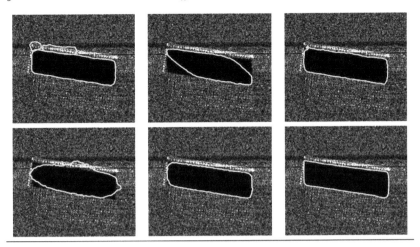

FIGURE 7.7
Segmentation results for the SAS database.

perfectly acceptable results, there yet remain a few cases where it fails. Segmentation failure in scenario (1) is often attributed to the active contour looping over itself after hanging onto background noise or artifacts from the SAS image-forming algorithms. In scenario (2), the segmentation fails to yield good results due to the affine variability of the test images. Here, the active contour tends toward the mean similarity shape, which is at one particular affine transformation of the training data. Segmentation with (4) fails often as a result of the segmentation flowing towards a different affine skew than ground truth. Segmentation with (3) and (5) corrects the issues in (2) and (4) due to their true gradient flow, as described in the MCD segmentation experiment. The results in these cases are rather similar, with scenario (5) yielding the best results.

7.5   Conclusion

We present a Bayesian active contour model for image segmentation that improves the state-of-the-art in two key aspects: (1) we use a shape prior term based on intrinsic affine-invariant, elastic shape statistics, and (2) we perform a true gradient descent flow to minimize the total energy functional. Elastic shape analysis allows us to build shape models that more accurately capture the underlying variation of complicated shape classes when compared to common extrinsic methods used in geometric active contours. Furthermore, an affine-invariant shape model is robust to perspective skew, allowing us to accurately segment when either test or training images are taken with respect to different camera angles. By computing the gradient of each of the three energy functionals – E_{image}, E_{smooth}, and E_{prior} – with respect to the same ($\mathbb{L}^2$) metric, the active contour evolution becomes a true gradient descent flow along E_{total}. With such a flow, segmentation results in an optimal fitting of both image and shape data.

Bibliography

[1] M. Bober. Mpeg-7 visual shape descriptors. *IEEE Transactions on Circuits and Systems for Video Technology*, 11(6):716–719, Jun. 2001.

[2] D. Bryner, E. Klassen, and A. Srivastava. Affine-invariant, elastic shape analysis of planar contours. *CVPR*, pages 390–397, 2012.

[3] D. Bryner, A. Srivastava, and Q. Huynh. Elastic shapes models for improving segmentation of object boundaries in synthetic aperture sonar images. *Computer Vision and Image Understanding*, 117(12):1695–1710, 2013.

[4] V. Caselles, R. Kimmel, and G. Sapiro. Geodesic active contours. *International Journal of Computer Vision*, 22(1):61–79, Feb. 1997.

[5] T. F. Chan and L. A. Vese. Active contours without edges. *IEEE Transactions on Image Processing*, 10(2):266–277, Feb 2001.

[6] I. L. Dryden and K. V. Mardia. *Statistical Shape Analysis*. John Wiley & Sons, Chichester, U.K., 1998.

[7] S. H. Joshi and A. Srivastava. Intrinsic Bayesian active contours for extraction of object boundaries in images. *International Journal of Computer Vision*, 81(3):331–355, Mar. 2009.

[8] M. Kamandar and S. A. Seyedin. Procrustes - based shape prior for parametric active contours. In *International Conference on Machine Vision*, pages 135–140, Dec. 2007.

[9] M. Kass, A. Witkin, and D. Terzopoulos. Snakes: Active contour models. *International Journal of Computer Vision*, 1(4):321–331, 1988.

[10] S. Kichenassamy, A. Kumar, P. Olver, A. Tannenbaum, and A. Yezzi. Gradient flows and geometric active contour models. In *IEEE International Conference on Computer Vision*, pages 810–815, Jun. 1995.

[11] J. Kim, M. Çetin, and A. S. Willsky. Nonparametric shape priors for active contour-based image segmentation. *Signal Processing*, 87(12):3021–3044, Dec. 2007.

[12] S. Kurtek, A. Srivastava, E. Klassen, and Z. Ding. Statistical modeling of curves using shapes and related features. *Journal of American Statistical Association*, 107(499):1152–1165, 2012.

[13] M. E. Leventon, W. E. L. Grimson, and O. Faugeras. Statistical shape influence in geodesic active contours. In *Computer Vision and Pattern Recognition*, volume 1, pages 316–323, 2000.

[14] S. Osher and J. A. Sethian. Fronts propagating with curvature dependent speed: Algorithms based on Hamilton-Jacobi formulations. *Journal of Computational Physics*, 79(1):12–49, 1988.

[15] N. Paragios and R. Deriche. Geodesic active regions for supervised texture segmentation. In *IEEE International Conference on Computer Vision*, volume 2, pages 926–932, 1999.

[16] G. Sammelmann, J. Christoff, and J. Lathrop. Synthetic images of proud targets. *MTS/IEEE Oceans Conference*, pages 1–6, Sept. 2006.

[17] A. Srivastava, E. Klassen, S. H. Joshi, and I. H. Jermyn. Shape analysis of elastic curves in Euclidean spaces. *IEEE Transactions on Pattern Analysis and Machine Intelligence*, 33(7):1415–1428, July 2011.

[18] A. Tsai, A. Yezzi, Jr., W. Wells, C. Tempany, D. Tucker, A. Fan, W. E. Grimson, and A. Willsky. A shape-based approach to the segmentation of medical imagery using level sets. *IEEE Transactions on Medical Imaging*, 22(2):137–154, 2003.

[19] A. Yezzi and A. Mennucci. Conformal metrics and true "gradient flows" for curves. In *10th IEEE International Conference on Computer Vision (ICCV)*, volume 1, pages 913–919, 2005.

[20] A. Yezzi and S. Soatto. Deformotion: Deforming motion, shape average and the joint registration and approximation of structures in images. *International Journal of Computer Vision*, 53(2):153–167, 2003.

[21] S. C. Zhu, T. S. Lee, and A. L. Yuille. Region competition: Unifying snakes, region growing, energy/Bayes/MDL for multi-band image segmentation. In *Fifth International Conference on Computer Vision*, pages 416–423. IEEE, Jun. 1995.

[22] M. Zuliani, S. Bhagavathy, B. S. Manjunath, and C. S. Kenney. Affine invariant curve matching. In *International Conference on Image Processing (ICIP'04)*. IEEE, 2004.

8

Bayesian Semiparametric Longitudinal Data Modeling Using NI Densities

Luis M. Castro

University of Concepción

Victor H. Lachos

IMECC-UNICAMP

Diana M. Galvis

IMECC-UNICAMP

Dipankar Bandyopadhyay

University of Minnesota

CONTENTS

8.1   Introduction

Longitudinal data abounds in bio-statistical research, leading to exploration of a wide variety of statistical models with varying complexity. Linear mixed effects (LME) models [see e.g. 18, 31, 32] are routinely used to analyze these data, allowing researchers to capture correlations between responses that exhibit multivariate, clustered, multilevel, spatially-referenced and various other data structures. The LME model for continuous responses assumes normal distributions for the between-subject random effects and the within-subject random errors. However, this may lack robustness in parameter estimation under departures from normality (namely, heavy tails) and/or outliers [24]. To deal with this issue, some proposals in the literature consider replacing the normality assumption with a more flexible class of distributions. For example, [24] proposed a multivariate Student-t LME model in the presence of outliers. [20] and [21] developed some additional tools for the t-LME model from a Bayesian perspective. [28] advocated the use of a subclass of elliptical distributions, called normal/independent (NI) distributions [22], and adopted a Bayesian framework to carry out posterior analysis for heavy-tailed LME (NI-LME) models. [1, 2] proposed extensions of the normal LME to deal with both asymmetry and outliers, with the LME as a particular case.

In addition, in longitudinal studies, the response-covariate relationship throughout the entire longitudinal profile may not remain linear for certain covariates, such as time, and utilizing a simple parametric framework can be too restrictive to capture the evolution of the response over time. To alleviate this, semi-parametric extensions [15] to the elliptical class of densities (henceforth, E-SME models) have been proposed. After model fitting, it is imperative to check model assumptions through sensitivity analysis to detect potential extremes, or influential observations, which might distort inference. Following the celebrated work of [7], case-deletion and local influence diagnostics have been widely applied to a variety of regression models. Diagnostic assessments are relatively rare in nonparametric and semiparametric mixed models. From a frequentist (classical) perspective, [38] developed local influence measures in semiparametric mixed and partially linear models under Gaussian errors. Recently, [15] extended those to E-SME models.

In this chapter, we propose robust modeling of the SME model using the NI class of distributions, such that the NI semiparametric linear mixed-effects (NI-SME) model is defined. The non-linear effect of time is modeled nonparametrically using penalized splines, hence our model is semiparametric. We utilize a fully Bayesian approach that utilizes relevant Markov chain Monte Carlo (MCMC) steps [5]. Although the term robustness is very broad, in this chapter, robustness is achieved with respect to Bayesian parameter estimation. The NI class attributes varying weights to each subject stochastically, (such as lower weight for outliers), to control for the influence of outlying observations on the overall inference. As a consequence, developing diagnostic measures for outlier detection based on q-divergences [10, 23] is immediate. The NI class preserves the normality, such that members of the NI class tend to the normal density under suitable conditions, for example, as the degrees of freedom goes to ∞, Student-t approaches normality.

The rest of the chapter proceeds as follows. In Section 8.2, after a brief introduction to the NI density, the NI-SME model is defined. In Section 8.3, we present the Bayesian inferential framework for the proposed NI-SME model, including choices of priors, model comparison criteria, etc. Also, we develop the Bayesian case influence diagnostics based on the q-divergence measures. The method is illustrated using a motivating dataset on HIV/AIDS infected patients in Section 8.4. Section 8.5 compares finite sample performance of our method with other 'normality' based alternatives, and also the efficiency in accommodating outliers, using simulation studies. Finally, Section 8.6 concludes with some viable future research directions.

8.2 Semiparametric Linear Mixed-Effects Model

In this section, we introduce the NI-SME model. Following [24], this model assumes that the random effects and error terms belongs to the NI class. For a self-contained proposal, we start with some background on the NI class as proposed in [19].

8.2.1 Normal/independent distributions

An element of the NI family [19, 22] is defined as the distribution of the p-variate random vector

$$\mathbf{Y} = \boldsymbol{\mu} + U^{-1/2}\mathbf{Z}, \tag{8.1}$$

where $\boldsymbol{\mu}$ is a location vector, $\mathbf{Z}$ is a normal random vector with mean vector $\mathbf{0}$, variance-covariance matrix $\boldsymbol{\Sigma}$, and U is a mixing positive random variable with cumulative distribution function (*cdf*) $H(u|\boldsymbol{\nu})$, and probability density function (*pdf*) $h(u|\boldsymbol{\nu})$, independent of $\mathbf{Z}$, where $\boldsymbol{\nu}$ is a scalar or parameter vector indexing the distribution of U. Note that given U, $\mathbf{Y}$ follows a multivariate normal distribution with mean vector $\boldsymbol{\mu}$ and variance-covariance matrix $u^{-1}\boldsymbol{\Sigma}$. As the reader can see, the NI distributions are scale mixtures of the normal distribution, where the distribution of the scale factor U is the mixing distribution. Hence, the *pdf* of $\mathbf{Y}$ is given by $\mathrm{NI}(\mathbf{y}|\boldsymbol{\mu}, \boldsymbol{\Sigma}, \boldsymbol{\nu}) = \int_0^\infty \phi_p(\mathbf{y}; \boldsymbol{\mu}, u^{-1}\boldsymbol{\Sigma})dH(u|\boldsymbol{\nu})$, where $\phi_p(\cdot; \boldsymbol{\mu}, \boldsymbol{\Sigma})$ stands for the *pdf* of the p-variate normal distribution with mean vector $\boldsymbol{\mu}$ and covariate matrix $\boldsymbol{\Sigma}$. We use the notation $\mathrm{NI}_p(\boldsymbol{\mu}, \boldsymbol{\Sigma}, H)$ when $\mathbf{Y}$ is a member of the NI class. Three scale mixtures of multivariate normal distribution are commonly used for robust estimation, namely, the multivariate Student-t, multivariate slash and multivariate contaminated-normal distributions (see subsection 8.7.1, Appendix A, for more details).

8.2.2   Model specification

Following [15], we propose the following semiparametric mixed-effects model where the random terms are assumed to follow a NI distribution within the class defined in (8.1). This is defined as:

$$\mathbf{y}_i \;\; = \;\; \mathbf{X}_i \boldsymbol{\beta} + \mathbf{Z}_i \mathbf{b}_i + \mathbf{N}_i \mathbf{f}_i + \boldsymbol{\epsilon}_i, \tag{8.2}$$

with the assumption that

$$\begin{pmatrix} \mathbf{b}_i \\ \boldsymbol{\epsilon}_i \end{pmatrix} \overset{ind.}{\sim} \mathrm{NI}_{q+n_i} \left(\begin{pmatrix} \mathbf{0} \\ \mathbf{0} \end{pmatrix}, \begin{pmatrix} \mathbf{D} & \mathbf{0} \\ \mathbf{0} & \sigma_e^2 \mathbf{I}_{n_i} \end{pmatrix}, H \right), \;\; i = 1, \ldots, n, \tag{8.3}$$

where the subscript i is the subject index; $\mathbf{f}_i = (f(t_1^0), \ldots, f(t_{r_i}^0))^\top$ is an $r_i \times 1$ vector with $t_1^0, \ldots, t_{r_i}^0$ being the distinct and ordered values of t_{ij}, with $f(\cdot)$ a smooth function of time t_{ij}; $\mathbf{N}_i$ is an $(n_i \times r_i)$ incidence matrix whose (j, s)-th element equals the indicator function $I(t_{ij} = t_s^0)$ for $j = 1, \ldots, n_i$ and $s = 1, \ldots, r_i$; $\mathbf{y}_i = (y_{i1}, \ldots, y_{in_i})^\top$ is a $n_i \times 1$ vector of observed continuous responses for subject i; $\mathbf{X}_i$ is the $n_i \times p$ design matrix corresponding to the $p \times 1$ vector of fixed-effects $\boldsymbol{\beta}$; $\mathbf{Z}_i$ is the $n_i \times q$ design matrix corresponding to the $q \times 1$ vector of random effects $\mathbf{b}_i$ and $\boldsymbol{\epsilon}_i$ is the $n_i \times 1$ vector of random errors. The dispersion matrix $\mathbf{D} = \mathbf{D}(\boldsymbol{\gamma})$ models between-subjects variability, and depends on the unknown and reduced parameter vector $\boldsymbol{\gamma}$.

Using the definition of a NI random vector and (8.3), it follows that, marginally,

$$\mathbf{b}_i \overset{ind}{\sim} \mathrm{NI}_q(0, \mathbf{D}, H) \quad \text{and} \quad \boldsymbol{\epsilon}_i \overset{ind.}{\sim} \mathrm{NI}_{n_i}(\mathbf{0}, \sigma_e^2 \mathbf{I}_{n_i}, H), \quad i = 1, \ldots, n \tag{8.4}$$

and they are uncorrelated, since $Cov(\mathbf{b}_i, \boldsymbol{\epsilon}_i) = \mathrm{E}\{\mathbf{b}_i \boldsymbol{\epsilon}_i^\top\} = \mathrm{E}\{\mathrm{E}\{\mathbf{b}_i \boldsymbol{\epsilon}_i^\top | U_i\}\} = 0$. Note that, in the absence of the nonparametric function $\mathbf{N}_i \mathbf{f}_i$, $i = 1, \ldots, n$, in the model (8.2)-(8.3), the NI-SME model reduces to the NI-LME model proposed by [28], and in the absence of the random effects term $\mathbf{b}_i$ (other terms in 8.2 remaining intact), the NI-SME model reduces to the well-known partial linear model of [14]. Finally, when $\boldsymbol{\beta} = \mathbf{0}$, the NI-SME model reduces to the nonparametric mixed model developed by [34].

It follows from (8.2) and (8.4) that our proposition can be formulated in terms of a flexible hierarchical representation as follows:

$$\mathbf{Y}_i | \mathbf{b}_i, U_i = u_i \overset{ind.}{\sim} N_{n_i}(\mathbf{X}_i \boldsymbol{\beta} + \mathbf{Z}_i \mathbf{b}_i + \mathbf{N}_i \mathbf{f}_i, u_i^{-1} \sigma_e^2 \mathbf{I}_{n_i}), \tag{8.5}$$

$$\mathbf{b}_i | U_i = u_i \overset{ind.}{\sim} N_q(0, u_i^{-1} \mathbf{D}), \tag{8.6}$$

$$U_i \overset{ind.}{\sim} H(\cdot; \boldsymbol{\nu}). \tag{8.7}$$

Let $\boldsymbol{\Omega} = (\boldsymbol{\beta}^{\top}, \sigma_e^2, \boldsymbol{\gamma}^{\top}, \boldsymbol{\nu}^{\top})^{\top}$ be the parameter vector of interest. Then, the marginal model induced by $\mathbf{y}_i$, is given by

$$\mathbf{y}_i \sim \mathrm{NI}_{n_i}(\mathbf{X}_i\boldsymbol{\beta} + \mathbf{N}_i\mathbf{f}_i, \mathbf{V}_i, H),$$

where $\mathbf{V}_i = \sigma_e^2\mathbf{I}_{ni} + \mathbf{Z}_i\mathbf{D}\mathbf{Z}_i^{\top}$ is an $n_i \times n_i$ covariance matrix. Therefore, the likelihood function of $\boldsymbol{\Omega}$, given the observed data $\mathbf{y} = (\mathbf{y}_1^{\top}, \ldots, \mathbf{y}_n^{\top})^{\top}$, is given by

$$L(\boldsymbol{\Omega}|\mathbf{y}) = \prod_{i=1}^{n} \mathrm{NI}_{n_i}(\mathbf{y}_i|\mathbf{X}_i\boldsymbol{\beta} + \mathbf{N}_i\mathbf{f}_i, \mathbf{V}_i, \boldsymbol{\nu}),$$

where $\mathrm{NI}_{ni}(\mathbf{y}_i|\boldsymbol{\mu}, \boldsymbol{\Sigma}, \boldsymbol{\nu})$ is the *pdf* of a NI distribution with location vector $\boldsymbol{\mu}$, covariance matrix $\boldsymbol{\Sigma}$, and the mixture parameter $\boldsymbol{\nu}$. This likelihood function will be also useful for Bayesian model selection and to develop case-deletion influence diagnostics based on q-divergence measures [see, 17].

8.3 Bayesian Approach

Although one can use standard algorithms for maximum-likelihood (ML) inference, such as, the maximum penalized likelihood estimates [see 14, 15], there are difficult challenges for likelihood-based inference, such as finding standard errors in multidimensional settings, and invalidity of the asymptotic theory of MLE under small and moderate samples. Hence, in this chapter, we choose a Bayesian route primarily for computational simplicity. Our approach relies on MCMC algorithms to obtain posterior inference for the parameters with easy and straightforward implementation in conventional software like WinBUGS, OpenBUGS or JAGS.

8.3.1 Modeling the non-parametric part

Often in AIDS studies, the viral load trajectories exhibit a complex evolution process (see Figure 8.1, Panel a), and it might be difficult to specify a known functional form to it. In this context, previous research [14, 15] has demonstrated that the relation between the response variable and some covariates (say, time) are often non-linear, and may not be modelled parametrically. Although (response) transformations and/or adding quadratic terms can be used to handle nonlinearities [29], these are not universal, and their specifications require a good deal of expertise. Hence, keeping the functional form of these response-covariate relationships completely unspecified (except some smoothness conditions) is justified, leading to the SME model.

In this chapter, we use penalized splines to approximate the non-parametric functions. As a smoothing technique, penalized splines are very popular [see 29], and are easy to implement using conventional software like WinBUGS. Following [9], the connection with the mixed model representation

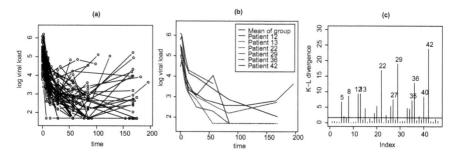

FIGURE 8.1
ACTG 315 data. (a) Viral loads in $\log_{10}$ scale. (b) Individual profiles of some influential observations under the normal SME model. (c) KL distance under the normal SME model.

is achieved by using the cubic spline basis to represent $f(t_{ij})$ as $\alpha_0 + \alpha_1 t_{ij} + \alpha_2 t_{ij}^2 + \alpha_3 t_{ij}^3 + \sum_{s=1}^{K} c_s |t_{ij} - \kappa_s|_+^3$. Thus, the non-parametric component $\mathbf{N}_i \mathbf{f}_i$ from the model defined in (8.2) can be expressed in matrix form as

$$\mathbf{g}_i = \mathbf{N}_i \mathbf{f}_i = \mathbf{T}_i \boldsymbol{\alpha} + \mathbf{K}_i \mathbf{c}, \tag{8.8}$$

where $\mathbf{T}_i = (1, \mathbf{t}_i, \mathbf{t}_i^2, \mathbf{t}_i^3)$ is an $n_i \times 4$ matrix associated with the fixed effects components of the spline function $\boldsymbol{\alpha} = (\alpha_0, \ldots, \alpha_3)^\top$, with $\mathbf{t}_i = (t_{i1}, \ldots, t_{in_i})^\top$; $\mathbf{K}_i$ is an $n_i \times K$ matrix corresponding to the random splines' coefficients $\mathbf{c} = (c_1, \ldots, c_K)^\top$, whose lj-element is $|t_{il} - \kappa_j|_+^3$ with $l = 1, \ldots, n_i$, $j = 1, \ldots, K$, $|a|_+$ a function equal to a if $a > 0$ and zero otherwise and fixed knots $\kappa_1 \leq \kappa_2 \leq \ldots \leq \kappa_K$, typically placed at quantiles of the distribution of unique values of the covariate t_{ij}. Here, we choose radial basis functions instead of B-splines, leading to a considerably less computation time by specifying a simple knot sequence [29]. Note that B-splines can sometimes lead to numerical instability when a large number of knots is considered and the penalty parameter is small.

To choose the dimension K, we follow the recommendations of [29] such that the actual choice of K and the location of knots have little influence on the resulting penalized fit as long as K is large. The value of K is chosen between 5 and 35 to ensure enough flexibility. Note that a large value of K induces under-smoothing, whereas a small value of K over-smooths the estimates. However, there are computational advantages to keeping the number of knots relatively low. Moreover, as the number of basis functions increases, the regression problem becomes more ill-conditioned, making the numerical computation less stable.

Knot placement remains a delicate problem in nonparametric regression as well. It is customary in the nonparametric splines literature to assign the interior knots on the order statistics, or to assume they are equally spaced. For example, [29] (subsection 5.5.3) proposed to choose the knots $\kappa_1, \ldots, \kappa_K$ as $\kappa_k = \left(\frac{k+1}{K+2} \right) th$ sample quantile of the unique x_i, where $K =$

min(0.25 number of unique x_i, 35) and x_i is the covariate that is modeled non-parametrically. This well-known procedure in nonparametric regression reduces the computational cost substantially, and avoids having to solve a difficult problem of optimizing the knot positions. As mentioned in [33], any other procedure needs to consider the fact that changes in the knot positions might cause considerable change in the function f. The random coefficient of the penalized spline functions c_s, $s = 1, \ldots, K$, is assumed to be normally distributed (*i.i.d.*) with zero-mean and variance σ_c^2. Note that the Bayesian penalized splines allow one to assess the effects of uncertainty in the smoothing parameters upon a smooth fit, and consider a simultaneous estimation of smooth functions and smoothing parameters.

8.3.2 Prior distributions

From (8.5)–(8.8), the complete-data likelihood function $L(\boldsymbol{\theta}|\mathbf{y}, \mathbf{b}, \mathbf{u})$ is proportional to

$$\prod_{i=1}^{n} \phi_{n_i}(\mathbf{y}_i; \mathbf{X}_i\boldsymbol{\beta} + \mathbf{Z}_i\mathbf{b}_i + \mathbf{g}_i, \sigma_e^2 u_i^{-1}\mathbf{I}_{n_i})\phi_q(\mathbf{b}_i; \mathbf{0}, u_i^{-1}\mathbf{D})h(u_i|\boldsymbol{\nu}), \qquad (8.9)$$

where $\boldsymbol{\theta} = (\boldsymbol{\beta}^\top, \sigma_e^2, \boldsymbol{\gamma}^\top, \boldsymbol{\nu}^\top, \boldsymbol{\alpha}^\top, \sigma_c^2)^\top$, $\mathbf{y} = (\mathbf{y}_1^\top, \ldots, \mathbf{y}_n^\top)^\top$, $\mathbf{b} = (\mathbf{b}_1^\top, \ldots, \mathbf{b}_n^\top)^\top$, and $\mathbf{u} = (u_1, \ldots, u_n)^\top$.

Now, to complete the Bayesian specification, we need to consider prior distributions of all the unknown parameters given in $\boldsymbol{\theta}$. A popular choice to ensure posterior propriety in the LME models setting is to consider proper (but diffuse) conditionally conjugate priors [13, 37]. In general, we have $\boldsymbol{\beta} \sim N_p(\boldsymbol{\beta}_0, \mathbf{S}_\beta)$, $\sigma_e^2 \sim \text{IGamma}(\tau_o/2, T_o/2)$ and $\mathbf{D} \sim \text{IWish}_q(\mathbf{M}_o, l)$, where IGamma$(a, b)$ is the inverse gamma distribution with mean $b/(a-1)$, $a > 1$, and IWish$_q(\mathbf{M}, l)$ is the inverse Wishart distribution with mean $\mathbf{M}/(l-q-1)$, $l > q + 1$, where $\mathbf{M}$ is a $q \times q$ known positive definite matrix. For the specific NI models in subsection 8.7.1, Appendix A, the prior for $\boldsymbol{\nu}$ was chosen accordingly as follows [see 16, 17, 27]:

(i) *Student-t model:* Here $\nu \sim \text{TExp}(\gamma/2; (2, \infty))$, *i.e.*, the degrees of freedom parameter ν has a truncated exponential prior distribution in the interval $(2, \infty)$. This truncation point was chosen to assure finite variance.

(ii) *Slash model:* A Gamma(a, b) distribution with small positive values of a and b ($b << a$) is adopted as a prior distribution for ν, primarily to ensure conjugacy.

(iii) *Contaminated normal model:* A Beta(ν_0, ν_1) distribution is used as a prior for ν, and an independent Beta(ρ_0, ρ_1) is adopted as prior for ρ to ensure conjugacy.

For the nonparametric functions $f(\cdot)$, we have unknown parameters for fixed and random effects components. For the fixed effects, we consider a multivariate normal distribution with mean $\mathbf{0}$ and variance $\boldsymbol{\Sigma}_\alpha$ as a prior distribution for $\boldsymbol{\alpha} = (\alpha_0, \alpha_1, \alpha_2, \alpha_3)$. For the variances of the P-spline random coefficients, we consider $\sigma_c^2 \sim \text{IGamma}(a, b)$. Finally, the likelihood function in (8.9) is combined with the prior specifications to derive posterior inference. This procedure is implemented using the Gibbs sampling algorithm. By considering the hierarchical representation (8.5)-(8.7) of the NI-SME model and the prior specifications, the full conditional distributions required for the MCMC algorithm are straightforward, and are presented in subsection 8.7.2, Appendix B.

In this work, the MCMC simulation sampling was implemented using the WinBUGS software. The program codes are available from the first author on request. Convergence of the MCMC samples was assessed using standard tools, such as the trace plots, ACF plots, as well as Gelman-Rubin convergence diagnostics. After discarding the initial 50000 burn-in samples, 10000 more samples (with thinning of 20) were used to derive inference. After fitting these models, we also use formal Bayesian model selection techniques to choose the model that produces the best fit.

8.3.3 Model comparison criteria

We use the conditional predictive ordinate (CPO) [11] for our model selection derived from the posterior predictive distribution (ppd), and summarize these CPOs via the log pseudo-marginal likelihood (LPML) statistic [5]. Larger values of LPML indicate better fit. Owing to the instability of the harmonic-mean identity used for CPO computations [26], we consider a more pragmatic route and compute the CPO (and LPML) statistics using 500 non-overlapping blocks of the Markov chain, each of size 2000, post-convergence (i.e., after discarding the initial burn-in samples), and report the expected LPML computed over the 500 blocks. In addition, we also apply the expected Akaike information criterion (EAIC), the expected Bayesian (or Schwarz) information criterion (EBIC) [5], and the DIC$_3$ [6] criteria. The DIC$_3$ was used as an alternative to the usual DIC [30] because of the ease of computation directly from the MCMC output, and also due to the mixture modeling framework. All these criteria abide by the 'lower is better' law, i.e., the model producing the lowest value gets selected.

8.3.4 Case influence diagnostics

In addition, as a direct byproduct from the MCMC output, some influence diagnostic measures are developed to study the impact of outliers on mainly the fixed effects parameters due to data perturbation schemes based on case-deletion statistics [8], and the q-divergence measures [10, 35] between posterior distributions. We consider three choices of these divergences, namely, the Kullback-Leibler (KL) divergence, the J-distance (symmetric version of the KL divergence), and the L_1-distance. Using the calibration method of

[23], we obtained the cut-off values as 0.98, 1.61 and 2.25 for the L_1, KL and J-distances, respectively. They were subsequently used to identify influential observations in the ACTG 315 dataset.

8.4   Application: ACTG 315 Data Revisited

In this section, we apply our proposed semi-parametric linear mixed-effects model to the motivating ACTG 315 protocol HIV1-RNA viral loads data previously analyzed by [36]. In HIV-AIDS research, it is hypothesized that the relationship between the viral load and the time of treatment within an antiviral regimen is nonlinear, whereas the relationship between the viral loads and certain immunologic responses such as CD4+ and CD8+ cell counts is linear. Since the viral load is recorded for patients at specific time points, mixed-effects models are typically used.

   The ACTG 315 data consider 46 HIV-1 infected patients treated with a potent antiretroviral drug cocktail based on the protease inhibitor ritonavir and reverse transcriptase inhibitor drugs (zidovudine and lamivudine). The aim of this antiretroviral regimen is to partially restore immunity in people with moderately advanced HIV disease. Viral load was measured on days 0, 2, 7, 10, 14, 21, 28, 56, 84, 168 and 196 after start of treatment. The dataset has 361 observations. Immunologic markers known as CD4+ and CD8+ cell counts were also measured along with viral load. Since one of our motivations is to investigate the relationship between virologic and immunologic responses in AIDS clinical trials, we consider the standardized version of CD4+ cell count as a covariate for the parametric part of the model, whereas the time of treatment is modeled using splines. The CD8+ cell counts were not considered in the analysis.

   Our NI-SME model with random intercepts b_i is given as

$$y_{ij} = \beta_1 \text{CD4+} + f_i(t_{ij}) + b_i + \epsilon_{ij} \tag{8.10}$$

where y_{ij} denotes the $\log_{10}$ transformation of the viral load for the i-th subject at time t_{ij} ($i = 1, 2, \ldots, 46$; $j = 1, 2, \ldots, n_i$), $f_i(t_{ij})$ is a smooth function, b_i is the random effect for the i-th patient, and ϵ_{ij} are random errors. Following (8.2), one may express (8.10) in matrix form as

$$\mathbf{y}_i = \mathbf{X}_i \boldsymbol{\beta} + \mathbf{N}_i \mathbf{f}_i + \mathbf{Z}_i \mathbf{b}_i + \boldsymbol{\epsilon}_i, \tag{8.11}$$

where $\mathbf{y}_i$ is an $(n_i \times 1)$ vector of responses for the i-th patient, $\mathbf{X}_i = (\text{CD4+}_{i1}, \ldots, \text{CD4+}_{in_i})^\top$ where CD4+_{ij} indicates a summary of the unobserved CD4+ values up to time t_{ij}, $\mathbf{N}_i = \mathbf{I}_{n_i}$ is the identity matrix of order n_i, $\mathbf{f}_i$ is a $(n_i \times 1)$ vector whose components are the function $f(\cdot)$ evaluated at the times in the set $\mathbf{t}_i = \{t_{i1}, \ldots, t_{in_i}\}$, $\mathbf{Z}_i = \mathbf{1}_{n_i}$, with $\mathbf{1}_{n_i}$, an $(n_i \times 1)$ vector of ones, $\mathbf{b}_i = b_i$ the random intercept, and $\boldsymbol{\epsilon}_i = (\epsilon_{i1}, \ldots, \epsilon_{in_i})^\top$ represents the within-subject random error.

Figure 8.1 (Panel a) presents the individual viral load profiles as a function of time. Clearly, the viral load changes with time in a nonlinear manner. Moreover, a variation in the intercept across individuals is also observed. Figure 8.1 (Panel b) presents the fitted mean profile of the viral loads via. a normal semiparametric mixed-effects model, and the trajectories of six outlying subjects. Interestingly, the mean profile decreases initially, but increases at higher time-points. The KL distances for these observations (and other outliers) are presented in Figure 8.1 (Panel c). It is immediate that normality-based analysis for this dataset might be inadequate. With our goal to provide robust inference, our analysis of this dataset considers other heavy-tailed members of the NI class of density, namely, the Student-t, slash and contaminated normal distributions. We compare the results with those obtained under the normal SME model.

For choice of priors, we have, $\beta_1 \sim N(0, 10^3)$, $\alpha_l \sim N(0, 10^3)$, $l = 1, 2, 3$, $\sigma_e^2 \sim \text{IGamma}(0.1, 0.01)$, $\gamma = \sigma_b^2 \sim \text{IGamma}(0.1, 0.01)$. In addition, $\nu \sim \text{TExp}(0.3, (2, \infty))$ for the Student-t model, $\nu \sim \text{Gamma}(0.1, 0.05)$, for the slash model and $\nu \sim \text{Beta}(1, 1)$ and $\rho \sim \text{Beta}(1, 1)$, for the contaminated normal model. We generated three parallel independent MCMC runs of size 100,000 with widely dispersed initial values for each parameter, where the first 50,000 iterations (burn-in samples) were discarded to compute the posterior estimates. To eliminate potential problems due to auto-correlation, we considered a spacing of size 20. The convergence of the MCMC chains was monitored via trace plots, auto-correlation plots and Gelman-Rubin $\hat{R}$ diagnostics [4]. Following [12], we considered a sensitivity analysis on the routine use of the inverse-gamma prior on the variance components, and found the results to be fairly robust under different choices of priors.

Table 8.1 presents the LPML, DIC$_3$, EAIC and EBIC values after fitting the various subclasses of the NI-SME model. Note that all of these indicate that the SME models with heavy tails perform significantly better than the normal one, with the Student-t SME outperforming the rest. In addition, by considering the nonparametric route, the performance of the best fitting Student-t model was uniformly better than the corresponding parametric counterpart with DIC$_3$ = 2994.425, EAIC = 2995.839 and EBIC = 3003.154. Table 8.2 reports the posterior mean, standard deviations (SD), and 95%

TABLE 8.1
ACTG 315 data. Comparison between NI-SME models using various Bayesian model selection criteria.

	LPML	DIC$_3$	EAIC	EBIC
Normal	−978.641	1799.091	1574.723	1582.038
Student-t	**−664.046**	**1321.203**	**1291.934**	**1301.694**
Slash	−696.763	1382.818	1294.55	1303.694
Contaminated Normal	−847.415	1567.851	1408.237	1419.209

TABLE 8.2
ACTG 315 data. Posterior estimates under the selected NI-SME models.

Model	Parameter	Mean	SD	2.5%	97.5%
Normal	CD4+	−0.159	0.049	−0.258	−0.065
	σ_ϵ	0.259	0.022	0.219	0.306
	σ_b	0.262	0.067	0.161	0.423
Student-t	CD4+	−0.099	0.049	−0.197	−0.006
	σ_ϵ	0.183	0.027	0.134	0.240
	σ_b	0.227	0.064	0.129	0.381
	ν	6.007	2.187	2.990	11.376
Slash	CD4+	−0.113	0.048	−0.209	−0.018
	σ_ϵ	0.133	0.027	0.091	0.193
	σ_b	0.157	0.046	0.086	0.265
	ν	1.971	0.731	1.159	3.970
Cont. Normal	CD4+	−0.107	0.052	−0.210	−0.005
	σ_ϵ	0.114	0.030	0.067	0.183
	σ_b	0.134	0.048	0.063	0.249
	ν	0.574	0.141	0.263	0.823
	ρ	0.311	0.078	0.198	0.476

credible intervals of the model parameters after fitting all of the NI-SME models. The posterior estimates of CD4+ from the four models are negative, with the estimates from the three NI-SME models (with heavy tails) quite close. The 95% posterior credible intervals do not include 0 for all models. Hence, we can conclude that the viral loads and the CD4+ counts are negatively correlated. The estimates of the between-subject variance σ_b and within-subject scale parameter σ_e for the NI-SME models with heavy tails are slightly smaller compared to the normal one. Finally, for the Student-t and slash SME models, the estimated value of ν is small, indicating the inadequacy of the normality assumptions in this dataset.

Using the q-divergence measures (see Figures 8.2 and 8.3), we notice that under the normal SME model, observations 5, 8, 12, 13, 22, 27, 29, 35, 36, 40 and 42 have a larger $d_q(\cdot)$ than in the Student-t and slash SME models. Consequently, the SME models with heavy tails attenuate the effect of these cases on the posterior estimates, making these models a robust alternative for the SME model with influential or outlying observations.

8.5 Simulation Studies

In this section, we conduct two simulation studies to illustrate the finite sample performance of our proposed Bayesian methods.

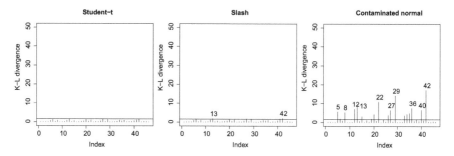

FIGURE 8.2
ACTG 315 data. KL distance under the Student-t, slash and contaminated normal SME models.

The goals of the first simulation study are: (i) to investigate, using empirical mean square error (MSE) and empirical bias, the consequences on parametric inference when the normality assumption is inappropriate; and (ii) to determine whether the model comparison measures, e.g., LPML, DIC$_3$, EAIC and EBIC select the best fitting model for the simulated data. We consider the following SME model:

$$y_{ij} = \beta_1 x_{1ij} + \beta_2 x_{2ij} + f_i(t_{ij}) + b_i + \epsilon_{ij}, \quad i = 1, \ldots, n, \quad j = 1, \ldots, 9 \, (8.12)$$

where $t_{ij} = \{2, 3, 4, 5, 6, 7, 8, 10, 12\}$ for all i. The random effect b_i and the error term $\epsilon_i = (\epsilon_{i1}, \ldots, \epsilon_{i9})^\top$ are non-correlated with

$$\begin{pmatrix} b_i \\ \epsilon_i \end{pmatrix} \overset{ind.}{\sim} t_{1+9} \left(\begin{pmatrix} 0 \\ 0 \end{pmatrix}, \begin{pmatrix} \sigma_b^2 & \mathbf{0} \\ \mathbf{0} & \sigma_e^2 \mathbf{I}_9 \end{pmatrix}, \nu \right), \quad i = 1, \ldots, n, \qquad (8.13)$$

where $t(\mu, \Sigma, \nu)$ represents the location-scale Student-t distribution with ν degrees of freedom (see section 8.7.1, Appendix A). We set $\beta_1 = -0.5$, $\beta_2 = 1$, $f_i(t_{ij}) = 2\sin(2\pi\, t_{ij}/12)$, $\sigma_b^2 = 3$, $\sigma_e^2 = 2$ and $\nu = 4$. We consider a Monte Carlo scheme with 200 iterations under different sample sizes $n = 50, 100$ and 200, for each of the members of the NI family. With our parameter vector $\theta = \{\beta_1, \beta_2, \sigma_b^2, \sigma_e^2, f(\cdot)\}$ and θ_s an element of θ, we compute

$$\text{RelBias}(\widehat{\theta}_s) = \frac{1}{200}\sum_{i=1}^{200}\left(\widehat{\theta}_s^{(i)}/\theta_s - 1\right) \quad \text{and} \quad \text{MSE}(\widehat{\theta}_s) = \frac{1}{200}\sum_{i=1}^{200}\left(\widehat{\theta}_s^{(i)} - \theta\right)^2,$$

where $\widehat{\theta}_s^{(i)}$ is the posterior estimate of the parameter of interest θ_s at the i-th iteration.

From Table 8.3, we observe that the NI-SME models with heavy tails have the smallest RelBias and MSE for the parameters β_1 and β_2 for all sample sizes considered. A similar conclusion is obtained for the estimation of the nonparametric functions $f(t_1), \ldots, f(t_9)$, and for the within-subject

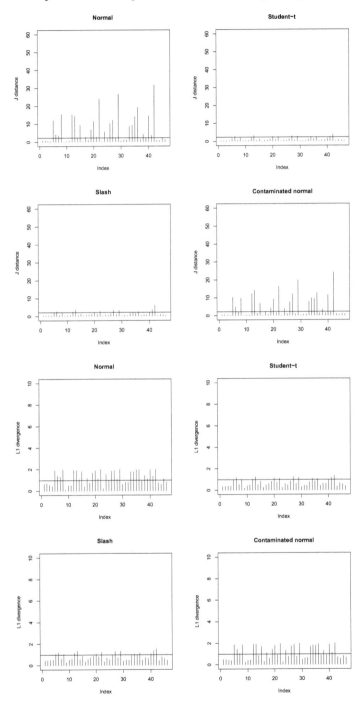

FIGURE 8.3
ACTG 315 data. J and L_1 distances under the normal, Student-t, slash and contaminated normal SME models.

TABLE 8.3

Simulated data: Relative bias and mean squared error (MSE) of the parameter estimates comparing NI-SME models based on 200 simulated datasets for various sample sizes.

Parameter	Normal model					
	RelBias			MSE		
	n=50	n=100	n=200	n=50	n=100	n=200
β_1	-0.086	-0.030	-0.026	0.109	0.055	0.032
β_2	-0.004	-0.013	-0.008	0.129	0.065	0.028
$f(t_1)$	-0.013	-0.004	-0.006	0.470	0.211	0.125
$f(t_2)$	-0.008	-0.003	0.004	0.442	0.184	0.121
$f(t_3)$	-0.007	-0.007	-0.003	0.454	0.186	0.123
$f(t_4)$	-0.014	-0.008	-0.010	0.460	0.187	0.126
$f(t_5)$	-0.021	-0.005	-0.010	0.461	0.190	0.123
$f(t_6)$	-0.014	-0.002	0.001	0.446	0.195	0.121
$f(t_7)$	-0.004	-0.004	0.014	0.433	0.193	0.127
$f(t_8)$	-0.030	-0.026	-0.011	0.477	0.187	0.141
$f(t_9)$	-0.013	0.001	0.000	0.474	0.212	0.123
σ_b^2	1.357	0.964	1.086	128.575	14.213	13.769
σ_e^2	0.980	0.991	0.978	6.462	4.889	4.163
	Student-t model					
β_1	-0.041	0.013	-0.014	0.078	0.032	0.019
β_2	0.014	-0.009	0.001	0.070	0.036	0.016
$f(t_1)$	0.002	0.011	0.002	0.282	0.130	0.086
$f(t_2)$	-0.007	0.002	-0.004	0.278	0.121	0.076
$f(t_3)$	-0.012	-0.001	-0.006	0.290	0.124	0.077
$f(t_4)$	-0.011	0.001	-0.004	0.278	0.125	0.075
$f(t_5)$	-0.005	0.006	-0.001	0.257	0.128	0.074
$f(t_6)$	0.003	0.012	0.001	0.247	0.133	0.076
$f(t_7)$	0.006	0.018	0.004	0.249	0.136	0.079
$f(t_8)$	-0.029	-0.005	-0.011	0.307	0.126	0.090
$f(t_9)$	-0.005	0.009	-0.000	0.273	0.130	0.077
σ_b^2	0.091	0.025	0.038	0.783	0.309	0.168
σ_e^2	0.051	0.015	0.009	0.116	0.038	0.020
ν	0.255	0.110	0.043	4.593	1.112	0.334
	Slash model					
β_1	-0.030	0.007	-0.013	0.076	0.032	0.020
β_2	0.018	-0.010	0.000	0.071	0.037	0.016
$f(t_1)$	0.007	0.010	0.003	0.288	0.133	0.089
$f(t_2)$	-0.011	-0.000	-0.006	0.283	0.128	0.081
$f(t_3)$	-0.013	0.000	-0.005	0.290	0.132	0.079
$f(t_4)$	-0.009	0.002	-0.002	0.278	0.126	0.078
$f(t_5)$	-0.003	0.005	-0.001	0.257	0.128	0.079
$f(t_6)$	0.004	0.009	-0.000	0.246	0.134	0.079
$f(t_7)$	0.009	0.017	0.006	0.252	0.139	0.081
$f(t_8)$	-0.026	-0.002	-0.008	0.295	0.129	0.091
$f(t_9)$	-0.007	0.007	-0.000	0.278	0.135	0.080
σ_b^2	-0.336	-0.395	-0.401	1.354	1.534	1.507
σ_e^2	-0.359	-0.401	-0.417	0.580	0.665	0.707
	Contaminated normal model					
β_1	-0.036	-0.002	-0.013	0.083	0.035	0.023
β_2	0.013	-0.011	-0.003	0.077	0.039	0.017
$f(t_1)$	0.006	0.008	0.003	0.320	0.146	0.100
$f(t_2)$	-0.011	-0.001	-0.005	0.321	0.134	0.092
$f(t_3)$	-0.014	-0.001	-0.004	0.330	0.136	0.092
$f(t_4)$	-0.009	0.001	-0.001	0.314	0.136	0.091
$f(t_5)$	-0.002	0.003	-0.000	0.294	0.139	0.090
$f(t_6)$	0.004	0.006	0.001	0.281	0.144	0.090
$f(t_7)$	0.008	0.014	0.008	0.281	0.149	0.094
$f(t_8)$	-0.028	-0.005	-0.007	0.330	0.139	0.104
$f(t_9)$	-0.007	0.004	0.001	0.323	0.150	0.094
σ_b^2	0.006	-0.022	0.012	0.917	0.562	0.281
σ_e^2	-0.030	-0.026	-0.022	0.242	0.139	0.063

TABLE 8.4
Simulated data: Monte Carlo estimates of the various model comparison measures after fitting the NI-SME models.

n	Model	LPML	DIC$_3$	EAIC	EBIC
50	Normal	-1014.73	2027.39	2011.00	2018.64
	Student-t	$\mathbf{-960.15}$	**1919.91**	**1910.26**	**1919.82**
	Slash	-967.46	1934.49	1920.66	1930.22
	Cont. normal	-967.53	1932.32	1919.89	1931.36
100	Normal	-2029.52	4056.55	4036.10	4046.52
	Student-t	$\mathbf{-1911.04}$	**3821.91**	**3811.75**	**3824.78**
	Slash	-1925.73	3851.23	3836.35	3849.38
	Cont. normal	-1928.13	3853.80	3838.41	3854.04
200	Normal	-4061.71	8122.84	8102.06	8115.26
	Student-t	$\mathbf{-3825.60}$	**7651.12**	**7640.54**	**7657.04**
	Slash	-3855.24	7710.40	7694.89	7711.38
	Cont. normal	-3862.31	7722.73	7704.01	7723.80

and between-subject variances. These quantities tend to zero with increasing sample size. Note that the Student-t SME model also detects the heavy-tailed feature of the simulated data (the RelBias and MSE for ν are small).

In Table 8.4, we present the Monte Carlo averages (MC LPML, MC DIC$_3$, MC EAIC, and MC EBIC) of the various model comparison measures mentioned earlier. All these measures favor the Student-t-SME model for our (true) simulated data. This fact demonstrates the capacity of these Bayesian selection methods to detect an obvious departure from normality.

In the next simulation study, we examine the efficiency of our proposed models in detecting and accommodating atypical data, and also assess the sensitivity of the Bayesian estimates in the presence of these outliers. We considered a case-weight perturbation scheme, using the following normal SME model,

$$y_{ij} = \beta_1 x_{1ij} + \beta_2 x_{2ij} + f_i(t_{ij}) + b_i + \epsilon_{ij}, \tag{8.14}$$

for $t_{ij} = \{1, 3, 5, 6, 7, 9, 10, 11, 14, 15\}$ for all i, where $\beta_1 = -1.5$, $\beta_2 = 0.9$, $f_i(t_{ij}) = 2\log(2\pi t_{ij}) + 10$, $\sigma_b^2 = 0.5$ and $\sigma_e^2 = 1$. The random effect b_i and the error term $\epsilon_i = (\epsilon_{i1}, \ldots, \epsilon_{i10})^\top$ are normally distributed as follows

$$\begin{pmatrix} b_i \\ \epsilon_i \end{pmatrix} \overset{ind.}{\sim} \mathrm{N}_{1+10} \left(\begin{pmatrix} 0 \\ \mathbf{0} \end{pmatrix}, \begin{pmatrix} \sigma_b^2 & \mathbf{0} \\ \mathbf{0} & \sigma_e^2 \mathbf{I}_{10} \end{pmatrix} \right), \quad i = 1, \ldots, n. \tag{8.15}$$

Using a sample of size 50, we perturbed five cases, namely, 24, 26, 28, 31 and 45. The perturbation scheme is as follows: $\tilde{y}_{kj} = y_{kj} + 2S_y$, for $k = 26, 31, 45$ and all j, and $\tilde{y}_{\ell j} = y_{\ell j} - 2S_y$, for $\ell = 24, 28$ and all j, where S_y is the standard deviation of the simulated sample. For the sake of brevity, we

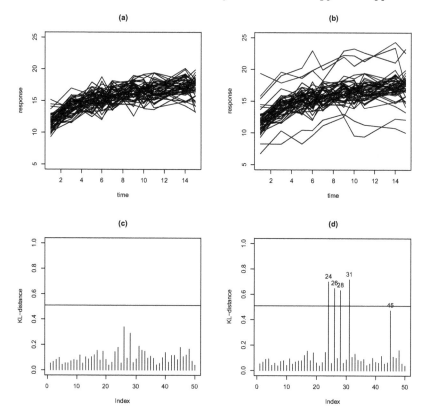

FIGURE 8.4
Simulated data. (a) Unperturbed individual profiles. (b) Perturbed individual
profiles. (c) KL distance under the normal SME model for the unperturbed
data. (d) KL distance under the normal SME model for the perturbed data.

only consider the KL distance in our analysis. Figure 8.4 (Panel a) presents
the unperturbed simulated profiles as a function of time, while Figure 8.4
(Panel b) displays the response profiles along with the perturbed ones. The
respective KL distances under the normal SME model for these observations
are sketched in Figure 8.4 (Panels c and d).

Table 8.5 reports the estimates of the LPML, DIC_3, EAIC and EBIC for
each perturbed version of the original data set, under different NI-SME models. All these criteria indicate that the NI models with heavy tails present a
better fit than the normal one. From Table 8.6, we find that the posterior inferences are sensitive to the perturbation of the selected cases, specifically, the
between-subjects variance is influenced under the normality assumption. As
expected, the estimates of the fixed effects β_1 and β_2, and within-subject variance σ^2 under the NI distributions are closer to the corresponding true values,
as compared to the normal one. Hence, we can conclude that the posterior

TABLE 8.5
Comparison among the NI-SME models using various Bayesian model selection criteria for the simulated perturbed data.

	LPML	DIC$_3$	EAIC	EBIC
Normal	−825.040	1648.709	1633.926	1641.574
Student-t	−820.530	1640.442	1626.098	1635.658
Slash	−819.987	1639.293	1624.751	1634.311
Contaminated Normal	−815.688	1630.89	1619.538	1631.01

TABLE 8.6
Posterior estimates for the unperturbed and perturbed simulated data sets under the NI-SME model.

Data	Model	Parameter	Mean	SD	2.5%	97.5%
Unperturbed	Normal	β_1	−1.405	0.171	−1.736	−1.068
		β_2	0.786	0.173	0.444	1.123
		σ_ϵ	1.114	0.075	0.977	1.266
		σ_b	0.447	0.119	0.262	0.729
Perturbed	Normal	β_1	−1.426	0.173	−1.766	−1.093
		β_2	0.761	0.175	0.427	1.111
		σ_ϵ	1.110	0.075	0.974	1.271
		σ_b	3.041	0.676	1.997	4.574
	Student-t	β_1	−1.487	0.170	−1.809	−1.150
		β_2	0.742	0.177	0.394	1.081
		σ_ϵ	1.002	0.096	0.824	1.203
		σ_b	1.770	0.517	0.947	2.956
		ν	15.266	7.818	6.094	35.986
	Slash	β_1	−1.476	0.167	−1.799	−1.143
		β_2	0.736	0.172	0.383	1.075
		σ_ϵ	0.720	0.085	0.564	0.903
		σ_b	1.147	0.397	0.590	2.118
		ν	2.592	0.855	1.467	4.692
	Cont. Normal	β_1	−1.473	0.168	−1.810	−1.148
		β_2	0.743	0.175	0.401	1.093
		σ_ϵ	0.909	0.074	0.772	1.061
		σ_b	1.239	0.379	0.697	2.178
		ν	0.172	0.089	0.064	0.423
		ρ	0.316	0.090	0.184	0.525

parameter estimates are highly sensitive in the presence of atypical observations for normality based models. Figure 8.5 presents the KL distances under the heavy-tailed NI-SME models, showing that the effect of these perturbations was attenuated for the heavy-tailed alternatives.

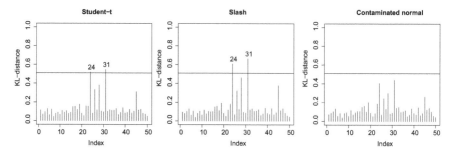

FIGURE 8.5
KL distances for the simulated perturbed data under the Student-t, slash and
contaminated normal SME models.

8.6   Conclusions

This article proposed a Bayesian implementation of a robust alternative to
the usual SME models by replacing the Gaussian assumptions of the random
terms by the class of normal/independent (NI) distributions. Four specific
cases were studied, namely, the normal, Student-t, slash, and contaminated
normal distributions. The method was applied to an interesting and widely
analyzed HIV viral load dataset to illustrate how the procedures developed
can be used to evaluate model assumptions, detect outliers and obtain robust
parameter estimates. Simulation studies revealed improvement in efficiency
and accuracy of fixed effects parameter estimates under conditions where normality assumptions are questionable. The models considered in this article can
be fitted using standard available software packages, like R and WinBUGS/JAGS
making our approach quite powerful and accessible to practitioners in the field.
Our current proposition only handles thickness in tails, because typically in
HIV studies, the viral loads are log-transformed to remove effects of skewness.
However, the non-normal features might still persist even in log-transformed
responses, and can be attributed to (additional) skewness and tail-thickness in
the random components. In the context of SME models, one might take a parametric route, via the skew-normal model [1, 2] or skew-normal/independent
(SNI) family [3]. Other semiparametric, or fully nonparametric alternatives
based on Dirichlet processes are certainly possible, but they come with additional complexities in the estimation framework. In addition, HIV viral loads
are often left-censored [16], and covariates like CD4+ might be measured with
error [36]. Currently, our framework excludes these complications, yet all these
remain viable directions to pursue in the future.

Acknowledgment

Luis M. Castro acknowledges funding support from FONDECYT (Grant 1130233) from the Chilean government and Grant 2012/19445-0 from FAPESP–Brazil. The research of Victor H. Lachos was supported by CNPq-Brazil (Grant 305054/2011-2) and by FAPESP-Brazil (Grant 2011/17400-6) from the Brazilian government. Diana M. Galvis acknowledges support from CAPES/CNPq - IEL Nacional Brasil. Bandyopadhyay acknowledges support from the US National Institutes of Health grants UL1TR000114 (CTSA award), and P30 CA77598 (Biostatistics and Bioinformatics Core shared resource of the University of Minnesota Masonic Cancer Center).

8.7 Appendix

8.7.1 Appendix A: Densities of some specific NI distributions

The multivariate Student-t distribution, denoted by $t_p(\boldsymbol{\mu}, \boldsymbol{\Sigma}, \nu)$, where ν is the degree of freedom parameter, can be derived from the mixture model (8.1), where U is distributed as Gamma$(\nu/2, \nu/2)$, with $\nu > 0$. The *pdf* of $\mathbf{Y}$ takes the form:

$$T(\mathbf{y}|\boldsymbol{\mu}, \boldsymbol{\Sigma}, \nu) = \frac{\Gamma(\frac{p+\nu}{2})}{\Gamma(\frac{\nu}{2})\pi^{p/2}} \nu^{-p/2}|\boldsymbol{\Sigma}|^{-1/2} \left(1 + \frac{d}{\nu}\right)^{-(p+\nu)/2}, \quad \mathbf{y} \in \mathbb{R}^p,$$

where $\Gamma(\cdot)$ is the standard gamma function, and $d = (\mathbf{y} - \boldsymbol{\mu})^\top \boldsymbol{\Sigma}^{-1}(\mathbf{y} - \boldsymbol{\mu})$ is the Mahalanobis distance. $\nu = 1$ leads to the multivariate Cauchy distribution, whereas $\nu \longrightarrow \infty$ leads to the multivariate normal distribution.

The multivariate slash distribution, denoted by $\mathrm{SL}_p(\boldsymbol{\mu}, \boldsymbol{\Sigma}, \nu)$, arises when the distribution of U is Beta$(\nu, 1)$, with $u \in (0, 1)$ and $\nu > 0$. Its *pdf* is given by

$$\mathrm{SL}(\mathbf{y}|\boldsymbol{\mu}, \boldsymbol{\Sigma}; \nu) = \nu \int_0^1 u^{\nu-1} \phi_p(\mathbf{y}; \boldsymbol{\mu}, u^{-1}\boldsymbol{\Sigma}) du, \quad \mathbf{y} \in \mathbb{R}^p,$$

and can be evaluated using the R function `integrate` [25]. When $\nu \longrightarrow \infty$, we have the multivariate normal distribution.

The multivariate contaminated normal distribution is denoted by $\mathrm{CN}_p(\boldsymbol{\mu}, \boldsymbol{\Sigma}, \nu, \rho)$, where $\nu, \rho \in (0, 1)$. Here, U is a discrete random variable taking one of two possible values, with probability function given by

$$h(u|\boldsymbol{\nu}) = \nu \mathbb{I}_{\{\rho\}}(u) + (1 - \nu)\mathbb{I}_{\{1\}}(u),$$

where $\boldsymbol{\nu} = (\nu, \rho)^\top$ and $\mathbb{I}_{\{\tau\}}(\cdot)$ is the indicator function of the set $\{\tau\}$. The associated density is

$$\mathrm{CN}(\mathbf{y}|\boldsymbol{\mu}, \boldsymbol{\Sigma}, \boldsymbol{\nu}) = \nu\phi_p(\mathbf{y}; \boldsymbol{\mu}, \rho^{-1}\boldsymbol{\Sigma}) + (1 - \nu)\phi_p(\mathbf{y}; \boldsymbol{\mu}, \boldsymbol{\Sigma}).$$

The multivariate contaminated normal distribution reduces to the multivariate normal when $\rho \longrightarrow 1$, or $\nu \longrightarrow 0$.

8.7.2   Appendix B: Conditional posterior distributions

Let $\mathbf{y} = (\mathbf{y}_1, \ldots \mathbf{y}_n)$, $\mathbf{b} = (\mathbf{b}_1, \ldots, \mathbf{b}_n)^\top$, $\mathbf{u} = (u_1, \ldots, u_n)^\top$. Given $\mathbf{u}$, the conditional posterior distributions are given as follows:

$\pi(\boldsymbol{\beta}|\mathbf{y}, \mathbf{b}, \mathbf{u}, \boldsymbol{\theta}_{(-\beta)})$ is a multivariate normal distribution *i.e.*, $\mathrm{N}_p(\mathbf{A}_\beta^{-1}\mathbf{d}_\beta, \mathbf{A}_\beta^{-1})$, where $\mathbf{A}_\beta = \mathbf{S}_\beta + \sum_{i=1}^n \frac{u_i\mathbf{X}_i^\top\mathbf{X}_i}{\sigma_e^2}$, $\mathbf{d}_\beta = \mathbf{S}_\beta^{-1} + \sum_{i=1}^n \frac{u_i\mathbf{X}_i^\top\mathbf{y}_{i_\beta}}{\sigma_e^2}$ and $\mathbf{y}_{i_\beta} = \mathbf{y}_i - \mathbf{Z}_i\mathbf{b}_i - \mathbf{T}_i\boldsymbol{\alpha} - \mathbf{K}_i\mathbf{c}$.

$\pi(\sigma_e^2|\mathbf{y}, \mathbf{b}, \mathbf{u}, \boldsymbol{\theta}_{(-\sigma_e^2)})$ is an inverse gamma distribution *i.e.*, $\mathrm{IGamma}\left(\frac{\tau_0+N}{2}, \frac{T_0+\sum_{i=1}^n u_i\lambda_i}{2}\right)$, where $\lambda_i = (\mathbf{y}_i - \mathbf{X}_i\boldsymbol{\beta} - \mathbf{Z}_i\mathbf{b}_i - \mathbf{T}_i\boldsymbol{\alpha} - \mathbf{K}_i\mathbf{c})^\top(\mathbf{y}_i - \mathbf{X}_i\boldsymbol{\beta} - \mathbf{Z}_i\mathbf{b}_i - \mathbf{T}_i\boldsymbol{\alpha} - \mathbf{K}_i\mathbf{c})$ and $N = \sum_{i=1}^n n_i$.

$\pi(\mathbf{b}_i|\mathbf{y}, \mathbf{u}_i, \boldsymbol{\theta})$ is a multivariate normal distribution *i.e.*, $\mathrm{N}_q(\mathbf{A}_b^{-1}\mathbf{d}_b, \mathbf{A}_b^{-1})$, where $\mathbf{A}_b = \mathbf{D}^{-1} + \sum_{i=1}^n \frac{u_i\mathbf{Z}_i^\top\mathbf{Z}_i}{\sigma_e^2}$, $\mathbf{d}_b = \sum_{i=1}^n \frac{u_i\mathbf{Z}_i^\top\mathbf{y}_{i_b}}{\sigma_e^2}$ and $\mathbf{y}_{i_b} = \mathbf{y}_i - \mathbf{X}_i\boldsymbol{\beta} - \mathbf{T}_i\boldsymbol{\alpha} - K_i\mathbf{c}$.

$\pi(\mathbf{D}|\mathbf{y}, \mathbf{b}, \mathbf{u}, \boldsymbol{\theta}_{(-\gamma)})$ is an inverse Wishart distribution *i.e.*, $\mathrm{IWishart}(M_o + \sum_{i=1}^n u_i\mathbf{b}_i\mathbf{b}_i^\top, l + n)$.

$\pi(\boldsymbol{\alpha}|\mathbf{y}, \mathbf{b}, \mathbf{u}, \boldsymbol{\theta})$ is a multivariate normal distribution *i.e.*, $\mathrm{N}_4(\mathbf{A}_\alpha^{-1}\mathbf{d}_\alpha, \mathbf{A}_\alpha^{-1})$, where $\mathbf{A}_\alpha = \Sigma_\alpha^{-1} + \sum_{i=1}^n \frac{u_i\mathbf{T}_i^\top\mathbf{T}_i}{\sigma_e^2}$, $\mathbf{d}_\alpha = \sum_{i=1}^n \frac{u_i\mathbf{T}_i^\top\mathbf{y}_{i_\alpha}}{\sigma_e^2}$ and $\mathbf{y}_{i_\alpha} = \mathbf{y}_i - \mathbf{X}_i\boldsymbol{\beta} - \mathbf{Z}_i\mathbf{b}_i - \mathbf{K}_i\mathbf{c}$.

$\pi(\mathbf{c}|\mathbf{y}, \mathbf{b}, \mathbf{u}, \boldsymbol{\theta})$ is a multivariate normal distribution *i.e.*, $\mathrm{N}_K(\mathbf{A}_c^{-1}\mathbf{d}_c, \mathbf{A}_c^{-1})$, where $\mathbf{A}_c = \mathbf{I}_K + \sum_{i=1}^n \frac{u_i\mathbf{K}_i^\top\mathbf{K}_i}{\sigma_e^2}$, $\mathbf{d}_c = \sum_{i=1}^n \frac{u_i\mathbf{K}_i^\top\mathbf{y}_{i_c}}{\sigma_e^2}$ and $\mathbf{y}_{i_c} = \mathbf{y}_i - \mathbf{X}_i\boldsymbol{\beta} - \mathbf{Z}_i\mathbf{b}_i - \mathbf{T}_i\boldsymbol{\alpha}$.

$\pi(\sigma_c^2|\mathbf{y}, \mathbf{b}, \mathbf{u}, \boldsymbol{\theta}_{(-\sigma_c^2)})$ is an inverse gamma distribution *i.e.*,

$$\mathrm{IGamma}\left((a + 3/2)K - 1, bK + \sum_{s=1}^K \frac{c_s^2}{2}\right).$$

$\pi(u_i|\mathbf{y}_i, \mathbf{b}_i, \boldsymbol{\theta})$ corresponds to:

- a gamma distribution in the Student-t case *i.e.*, $\text{Gamma}\left(\frac{\nu+n_i}{2}, \frac{\nu}{2} + \frac{\lambda_i}{2\sigma_e^2}\right)$;

- a truncated gamma distribution in the slash case *i.e.*,
$\text{TGamma}\left(\nu + \frac{n_i}{2}, \frac{\lambda_i}{2\sigma_e^2}, (0,1)\right)$; and

- a discrete distribution taken one of two states in the contaminated normal case *i.e.*,

$$\pi(u_i|\mathbf{y}_i, \mathbf{b}_i, \boldsymbol{\theta}) = \left\{ \begin{array}{ll} \frac{\eta_i}{\xi_i + \eta_i} & \text{if } u_i = \rho \\ \frac{\xi_i}{\xi_i + \eta_i} & \text{if } u_i = 1 \end{array} \right.$$

where $\eta_i = \nu\rho^{n_i/2}\exp\left\{-\frac{\rho\lambda_i}{2\sigma_e^2}\right\}$ and $\xi_i = \nu\exp\left\{-\frac{\lambda_i}{2\sigma_e^2}\right\}$.

$\pi(\boldsymbol{\nu}|\mathbf{y}, \mathbf{b}, \mathbf{u}, \boldsymbol{\theta})$ is:

- in the Student-t case, proportional to

$$(\nu/2)^{n\nu/2}\left(\Gamma(\nu/2)\right)^n \exp\left(-\frac{\nu}{2}\left[\sum_{i=1}^{n}(u_i - log(u_i)) + \gamma\right]\right) \mathbb{I}_{\{(2,\infty)\}}(\nu)$$

- a Gamma distribution *i.e.*, $\text{Gamma}(a+n, b - \sum_{i=1}^{n}\log u_i)$ in the slash case;

- and given by,

$$\pi(\nu|\mathbf{y}, \mathbf{b}, \mathbf{u}, \boldsymbol{\theta}_{(-\nu)}) \equiv \text{Beta}\left(\sum_{i=1}^{n}\mathbb{I}_{\{\rho\}}(u_i) + \nu_0, n - \sum_{i=1}^{n}\mathbb{I}_{\{\rho\}}(u_i) + \nu_1\right),$$

$$\pi(\rho|\mathbf{y}, \mathbf{b}, \mathbf{u}, \boldsymbol{\theta}_{(-\rho)}) \propto \nu^{\sum_{i=1}^{n}\mathbb{I}_{\{\rho\}}(u_i)}(1-\nu)^{n-\sum_{i=1}^{n}\mathbb{I}_{\{\rho\}}(u_i)}\rho^{\rho_0 - 1}(1-\rho)^{\rho_1 - 1};$$

in the contaminated normal case.

Bibliography

[1] R. B. Arellano-Valle, H. Bolfarine, and V. H. Lachos. Skew-normal linear mixed models. *Journal of Data Science*, 3:415–438, 2005.

[2] R. B. Arellano-Valle, H. Bolfarine, and V. H. Lachos. Bayesian inference for skew-normal linear mixed models. *Journal of Applied Statistics*, 34(6):663–682, 2007.

[3] D. Bandyopadhyay, V. H. Lachos, L. M. Castro, and D. K. Dey. Skew-normal/independent linear mixed models for censored responses with applications to HIV viral loads. *Biometrical Journal*, 54(3):405–425, 2012.

[4] S. P. Brooks and A. Gelman. General methods for monitoring convergence of iterative simulations. *Journal of Computational and Graphical Statistics*, 7(4):434–455, 1998.

[5] B. P. Carlin and T. A. Louis. *Bayesian Methods for Data Analysis*. Chapman & Hall/CRC, New York, 3rd edition, 2008.

[6] G. Celeux, F. Forbes, C. P. Robert, and D. M. Titterington. Deviance information criteria for missing data models. *Bayesian Analysis*, 1(4):651–673, 2006.

[7] R. D. Cook. Assessment of local influence. *Journal of the Royal Statistical Society: Series B (Statistical Methodology)*, 48:133–169, 1986.

[8] R. D. Cook and S. Weisberg. *Residuals and Influence in Regression*. Chapman & Hall/CRC, Boca Raton, FL, 1982.

[9] C. M. Crainiceanu, D. Ruppert, and M. P. Wand. Bayesian analysis for penalized spline regression using WinBUGS. *Journal of Statistical Software*, 14:1–24, 2005.

[10] I. Csiszár. Information-type measures of difference of probability distributions and indirect observations. *Studia Scientiarum Mathematicarum Hungarica*, 2:299–318, 1967.

[11] S. Geisser and W. F. Eddy. A predictive approach to model selection. *Journal of the American Statistical Association*, 74(365):153–160, 1979.

[12] A. Gelman, J. B. Carlin, H. S. Stern, D. B. Dunson, A. Vehtari, and D. B. Rubin. *Bayesian Data Analysis*. CRC Press, London, 3rd edition, 2013.

[13] J. P. Hobert and G. Casella. The effect of improper priors on Gibbs sampling in hierarchical linear mixed models. *Journal of the American Statistical Association*, 91(436):1461–1473, 1996.

[14] G. Ibacache-Pulgar, G. Paula, and F. Cysneiros. Semiparametric additive models under symmetric distributions. *TEST*, 22(1):103–121, 2013.

[15] G. Ibacache-Pulgar, G. A. Paula, and M. Galea. Influence diagnostics for elliptical semiparametric mixed models. *Statistical Modelling*, 12(2):165–193, 2012.

[16] V. H. Lachos, D. Bandyopadhyay, and D. K. Dey. Linear and nonlinear mixed-effects models for censored HIV viral loads using normal/independent distributions. *Biometrics*, 67(4):1594–1604, 2011.

[17] V. H. Lachos, L. M. Castro, and D. K. Dey. Bayesian inference in nonlinear mixed-effects models using normal independent distributions. *Computational Statistics & Data Analysis*, 64:237–252, 2013.

[18] N. M. Laird and J. H. Ware. Random effects models for longitudinal data. *Biometrics*, 38(4):963–974, 1982.

[19] K. L. Lange and J. S. Sinsheimer. Normal/independent distributions and their applications in robust regression. *Journal of Computational and Graphical Statistics*, 2(2):175–198, 1993.

[20] T. I. Lin and J. C. Lee. A robust approach to *t* linear mixed models applied to multiple sclerosis data. *Statistics in Medicine*, 25(8):1397–1412, 2006.

[21] T. I. Lin and J. C. Lee. Bayesian analysis of hierarchical linear mixed modeling using the multivariate *t* distribution. *Journal of Statistical Planning and Inference*, 137(2):484–495, 2007.

[22] C. Liu. Bayesian robust multivariate linear regression with incomplete data. *Journal of the American Statistical Association*, 91(435):1219–1227, 1996.

[23] F. Peng and D. K. Dey. Bayesian analysis of outlier problems using divergence measures. *The Canadian Journal of Statistics*, 23(2):199–213, 1995.

[24] J. C. Pinheiro, C. H. Liu, and Y. N. Wu. Efficient algorithms for robust estimation in linear mixed-effects models using a multivariate t-distribution. *Journal of Computational and Graphical Statistics*, 10(2):249–276, 2001.

[25] R Development Core Team. *R: A language and environment for statistical computing*. R Foundation for Statistical Computing, Vienna, Austria, 2013. ISBN 3-900051-07-0.

[26] A. E. Raftery, M. A. Newton, J. M. Satagopan, and P. N. Krivitsky. Estimating the integrated likelihood via posterior simulation using the harmonic mean identity. In J. M. Bernardo et al., editors, *Bayesian Statistics 8*, pages 1–45. Oxford University Press, U.K., 2007.

[27] G. Rosa, D. Gianola, and C. Padovani. Bayesian longitudinal data analysis with mixed models and thick-tailed distributions using MCMC. *Journal of Applied Statistics*, 31(7):855–873, 2004.

[28] G. J. M. Rosa, C. R. Padovani, and D. Gianola. Robust linear mixed models with normal/independent distributions and Bayesian MCMC implementation. *Biometrical Journal*, 45(5):573–590, 2003.

[29] D. Ruppert, M. P. Wand, and R. J. Carroll. *Semiparametric Regression*, volume 12. Cambridge University Press, New York, 2003.

[30] D. J. Spiegelhalter, N. G. Best, B. P. Carlin, and A. Van Der Linde. Bayesian measures of model complexity and fit. *Journal of the Royal Statistical Society: Series B (Statistical Methodology)*, 64(4):583–639, 2002.

[31] G. Verbeke and E. Lessafre. The linear mixed model. A critical investigation in the context of longitudinal data. In T. Gregoire, editor, *Proceedings of the Nantucket Conference on Modelling Longitudinal and Spatially Correlated Data: Methods, Applications, and Future Directions*, Lecture Notes in Statistics 122, pages 89–99, New York, 1997. Springer Verlag.

[32] G. Verbeke and G. Molenberghs. *Linear Mixed Models for Longitudinal Data*. Springer Verlag, New York, 2000.

[33] G. Wahba. Constrained regularization for ill posed linear operator equations, with applications in meteorology and medicine. In S. S. Gupta and J. O. Berger, editors, *Statistical Decision Theory and Related Topics III, Vol. 2*, pages 383–418. Academic Press, New York.

[34] Y. Wang. Mixed effects smoothing spline analysis of variance. *Journal of the Royal Statistical Society: Series B (Statistical Methodology)*, 60(1):159–174, 1998.

[35] R. Weiss. An approach to Bayesian sensitivity analysis. *Journal of the Royal Statistical Society: Series B (Statistical Methodology)*, 58(4):739–750, 1996.

[36] L. Wu. A joint model for nonlinear mixed-effects models with censoring and covariates measured with error, with application to AIDS studies. *Journal of the American Statistical Association*, 97(460):955–964, 2002.

[37] Y. Zhao, J. Staudenmayer, B. A. Coull, and M. P. Wand. General design Bayesian generalized linear mixed models. *Statistical Science*, 21(1):35–51, 2006.

[38] Z. Zhu, X. He, and W. Fung. Local influence analysis for penalized Gaussian likelihood estimators in partially linear models. *Scandinavian Journal of Statistics*, 30(4):767–780, 2003.

9

Bayesian Factor Analysis Based on Concentration

Yun Cao

Statistical Consulting

Michael Evans

University of Toronto

Irwin Guttman

SUNY at Buffalo

CONTENTS

9.1   Introduction

Applications of factor analysis arise in contexts where there are p dependent response variables and we believe that the dependencies among these variables can be explained by q latent variables where $q \ll p$. By a latent variable we mean that the values of the variable are not observed. A factor analysis is a methodology for identifying a value of q that accounts for a meaningful fraction of the variation in the observed responses, provides estimates of the dependencies of the response variables on the latent variables and also estimates of the values of the latent variables. The latent variables may or may not have meaning in light of the application.

Factor analysis is known to be a difficult problem. This difficulty arises due to some complicated geometry associated with correlation matrices and occurs irrespective of whether one takes a frequentist or a Bayesian

181

point-of-view. Still there are many practical contexts where factor analysis models seem to be just the right thing to use and so they are often employed.

We have several purposes here. First we want to provide a careful discussion of the basic sampling model and we do this in Section 9.2. For a variety of reasons we recommend that a factor analysis be approached as a Bayesian problem, namely, we will consider the problem where a prior is used in addition to the sampling model. We provide an explicit methodology for eliciting an appropriate proper prior. It is of some importance that the prior used be elicited as the methods currently advocated for a Bayesian factor analysis typically involve default choices of priors on the distributions of latent variables. We take the position that there is little to no justification for using default priors, whether proper or improper, in Bayesian analyses. This section also contains a survey of various methods put forward in the literature for carrying out a Bayesian factor analysis and we give our reasons for developing a somewhat different methodology. In particular, much simpler simulation algorithms are available with the elicited prior we propose.

We base our inferences on relative beliefs as discussed in [2]. In general, suppose we have a statistical model $\{f_\theta : \theta \in \Theta\}$ for data $x \in \mathcal{X}$, a proper prior π for θ, and we are interested in making inferences about a quantity $\psi = \Psi(\theta)$. The relative belief ratio $RB(\psi_0) = \pi_\Psi(\psi_0 \mid x)/\pi_\Psi(\psi_0)$, where π_Ψ and $\pi_\Psi(\cdot \mid x)$ are the prior and posterior densities of Ψ, is a measure of whether beliefs have increased (evidence for) or decreased (evidence against) concerning the truth of ψ_0 from *a priori* to *a posteriori*. A calibration of the relative belief ratio $RB(\psi_0)$ is given by

$$\Pi_\Psi(RB(\psi) \leq RB(\psi_0) \mid x), \tag{9.1}$$

which is the posterior probability that the true value of ψ has a relative belief ratio, and thus evidence in favor, no greater than the hypothesized value ψ_0. So, for example, we have strong evidence against the hypothesis $H_0 : \Psi(\theta) = \psi_0$ when $RB(\psi_0) < 1$ and (9.1) is small and have strong evidence in favor of H_0 when $RB(\psi_0) > 1$ and (9.1) is large. Also, we estimate ψ by the value $\arg\sup RB(\psi)$ as this value of ψ has the greatest evidence in its favor. An assessment of the accuracy of this estimate is obtained by selecting $\gamma \in (0,1)$ and looking at the size of the credible region given by the γ-relative belief region for ψ, namely, $C_\gamma(x) = \{\psi : RB(\psi) \geq c_\gamma(x)\}$ where $c_\gamma(x) = \inf\{k : \Pi_\Psi(RB(\psi) > k \mid x) \leq \gamma\}$.

In Section 9.3 we discuss the prescription and computation of relevant Ψ functions for factor analysis. This involves the *method of concentration* where we measure the extent to which a probability measure P on a set $\mathcal{X}$ concentrates about a set $C \subset \mathcal{X}$. The most obvious measure of concentration is $E_P(d(x, C))$ where $d(x, C) = \inf\{d(x, y) : y \in C\}$ for some distance measure $d(x, y)$ on $\mathcal{X}$. In the factor analysis context, the set $\mathcal{X}$ corresponds to the set of correlation matrices, C is a subset of correlation matrices and d is Frobenius distance. In particular, the hypothesis that q factors suffices to specify a correlation matrix, corresponds to a subset H_0^q of the set of all correlation matrices. We then assess the hypothesis H_0^q by computing the relative belief ratio that compares the posterior and prior concentrations about H_0^q. The method of concentration for hypothesis assessment has been previously used for inference in [7], [8] and [9]. Section 9.4 is concerned with inference for the factor analysis problem and we apply this in several problems. Proofs of all stated results are presented in the Appendix 9.6.

9.2 The Model, Prior and Posterior

Suppose that $\mathbf{y} \in R^p$ has unknown mean $\mu \in R^p$ and variance $\mathbf{\Sigma} \in R^{p \times p}$. A factor model corresponds to $\mathbf{\Sigma}$ possessing a particular structure, namely,

$$\mathbf{\Sigma} = \mathbf{\Gamma}_q \mathbf{\Gamma}_q' + \mathbf{\Lambda} \tag{9.2}$$

where $0 \leq q \leq p$, $\mathbf{\Gamma}_q \in R^{p \times q}$ is of rank q and $\mathbf{\Lambda}$ is diagonal with nonnegative entries. We refer to $\mathbf{\Gamma}_q$ as the *factor loading* matrix and the diagonal elements of $\mathbf{\Lambda}$ as the *residual variances*. The structure (9.2) arises from the existence of latent factors $\mathbf{f} \in R^q$, having distribution with mean $\mathbf{0}$ and variance $\mathbf{I}$, and unique variables $\mathbf{e} \in R^p$, uncorrelated with $\mathbf{f}$ and having mean $\mathbf{0}$ and variance $\mathbf{\Lambda}$, such that $\mathbf{y} = \mu + \mathbf{\Gamma}_q \mathbf{f} + \mathbf{e}$. Note that when $q = p$, then we can take $\mathbf{\Gamma}_q = \mathbf{\Sigma}^{1/2}$ (the symmetric square root of $\mathbf{\Sigma}$) and $\mathbf{\Lambda} = 0$ and so (9.2) is always correct for some q. When $q = 0$, then $\mathbf{\Gamma}_q = 0$ and the response variables are independent. The point of a factor analysis is to identify the smallest q such that (9.2) holds and further choose $\mathbf{\Gamma}_q$ to help in providing interpretations for the latent factors.

For any orthogonal matrix $\mathbf{Q} \in R^{q \times q}$, (9.2) implies that $\mathbf{\Sigma} = \mathbf{\Gamma}_q \mathbf{\Gamma}_q' + \mathbf{\Lambda} = (\mathbf{\Gamma}_q \mathbf{Q})(\mathbf{\Gamma}_q \mathbf{Q})' + \mathbf{\Lambda}$ and so $\mathbf{\Gamma}_q$ is not unique. We partially avoid the nonuniqueness by noting that $\mathbf{\Gamma}_q$ can be written uniquely as $\mathbf{\Gamma}_q = \mathbf{T}\mathbf{Q}$ where $\mathbf{Q} \in R^{q \times q}$ is orthogonal and $\mathbf{T} \in R^{p \times q}$ is lower triangular with nonnegative diagonal elements. Accordingly, we require that $\mathbf{\Gamma}_q$ be lower triangular with nonnegative diagonal elements hereafter, effectively taking $\mathbf{Q} = \mathbf{I}$, although we will weaken this restriction slightly for the computations. After $\mathbf{\Gamma}_q$ is estimated, rotations can be applied as is usual in factor analysis.

We assume that $\mathbf{e} \sim \mathbf{N}_p(\mathbf{0}, \mathbf{\Lambda})$ independent of $\mathbf{f} \sim \mathbf{N}_q(\mathbf{0}, \mathbf{I})$. After observing a sample $\mathbf{Y} = (\mathbf{y}_1 \cdots \mathbf{y}_n)$, our goal is to make inferences about the values of q and the parameters $(\mu, \mathbf{\Lambda}, \mathbf{\Gamma}_q)$. It is common in the literature to treat μ as known and given by $\bar{\mathbf{y}}$ but we don't do this, as it is generally not appropriate.

With these assumptions, likelihood methods can be used for factor analysis but these are known to suffer from computational challenges. In fact, without some fairly arbitrary restrictions being placed on the estimates, it can happen that software does not produce sensible answers. This is discussed in [18]. We use Bayesian methodology for this problem and, while this does not entirely avoid computational challenges, the use of prior information can be helpful in this regard.

For our purposes it will be convenient to consider a parameterization other than $(\mu, \mathbf{\Lambda}, \mathbf{\Gamma}_q)$. Let $\mathbf{\Sigma} = \Delta^{1/2} \mathcal{R} \Delta^{1/2}$ where $\Delta = \mathrm{diag}(\delta_1, \ldots, \delta_p)$ gives the variances of the observed variables and $\mathcal{R}$ is the correlation matrix. Then we write $\Delta^{-1/2}(\mathbf{y} - \mu) = \mathbf{\Gamma}_q \mathbf{f} + \mathbf{e}$ where $\mathbf{e}$ and $\mathbf{f}$ are as before but now $\mathbf{\Gamma}_q$ is lower triangular with nonnegative diagonal elements and $\sum_{k=1}^{\min(i,q)} \gamma_{ik}^2 + \lambda_i = 1$ for $i = 1, \ldots, p$. So $\sum_{k=1}^{\min(i,q)} \gamma_{ik}^2 \leq 1$ for $i = 1, \ldots, p$ and the λ_i values are now determined by the values in $\mathbf{\Gamma}_q$. This leads to the equation

$$\mathcal{R} = \mathbf{\Gamma}_q \mathbf{\Gamma}_q' + \mathbf{\Lambda}. \tag{9.3}$$

We let H_0^q denote the set of all correlation matrices satisfying (9.3). Note that the role of $\mathbf{\Gamma}_q$ and $\mathbf{\Lambda}$ is different in (9.3) than in (9.2).

A typical Bayesian factor analysis proceeds as follows. Let π_q^{factor} be the prior probability that $\mathcal{R} \in H_0^q$, i.e., there are q common factors, and let π_q be a prior on $(\mu, \mathbf{\Lambda}, \mathbf{\Gamma}_q)$ for $q = 0, \ldots, p$. After observing $\mathbf{Y}$ we compute the posterior probabilities $\pi_q^{\mathrm{factor}}(\mathbf{Y})$ that $\mathcal{R} \in H_0^q$ and select the q that maximizes this posterior probability. The posterior $\pi_q(\cdot \mid \mathbf{Y})$ is then available for inferences about $(\mu, \mathbf{\Lambda}, \mathbf{\Gamma}_q)$. This approach to Bayesian factor analysis is thoroughly described in [16]. Also, relevant material can be found in [14], [1], [19], [15], [20], [17], [6], [10], [11], and [4].

There are several difficulties associated with this approach. Perhaps the most significant is the need to specify $p+1$ prior distributions and prior model probabilities. This requires that, for the model with q factors, we have information about the relevant $\mathbf{\Gamma}_q$, as specified in a $q[p - (q - 1)/2]$ dimensional distribution, and we need this for $q = 0, \ldots, p$. This involves a demanding amount of elicitation. Furthermore, it is generally unrealistic to imagine that we have information of any kind about latent variables. So placing a prior on parameters of the distributions of such variables does not seem feasible. To avoid the elicitation problem it is common to place default improper priors on these quantities to represent a lack of information. Even if the subsequent posterior is proper, this leads to ambiguities concerning the proper interpretation of posterior model probabilities and Bayes factors due to the dependence of these quantities on arbitrary constants multiplying improper priors, see the discussion in Section 6.7 of [12]. In essence, there is little in the way of foundational support for using such priors. It is also known that choosing very diffuse proper priors can lead to bias in favor of hypotheses being true, as in the Jeffreys–Lindley paradox, and so this should be avoided. Simply writing down a complicated model and choosing some kind of default prior on the parameters cannot be viewed as an appropriate approach to a statistical analysis.

Undoubtedly the correct way to assign priors in a statistical analysis is through elicitation. It is reasonable to say that effectively solving a statistical problem using Bayesian methods requires the specification of a practically realistic elicitation process. So we need a methodology for the factor analysis problem that allows the elicitation of proper priors. The approach to the

implementation of a Bayesian factor analysis adopted here only requires that
we place a prior on (μ, Σ), namely, the parameters of the distribution of
the variables we are actually observing. Given that we are measuring these
variables, it almost goes without saying that we know something about their
location and scaling. Therefore, eliciting priors on the means μ_i and variances
σ_{ii} is certainly feasible, although eliciting priors on the covariances σ_{ij} seems
more challenging. In fact, this latter concern effectively rules out using the
conjugate prior on (μ, Σ) given by $\mu \mid \Sigma \sim N_p(\mu_0, \sigma_0^2 \Sigma), \Sigma^{-1} \sim W_p(k_0, \mathbf{A}_0)$,
where W_p denotes the Wishart distribution on $p \times p$ matrices and μ_0, σ_0^2, k_0
and $\mathbf{A}_0$ are hyperparameters. While posterior computations with this prior
are straightforward, it is not clear how to specify an elicitation procedure for
the hyperparameters.

Suppose, however, we write $\Sigma = \Delta^{1/2} \mathcal{R} \Delta^{1/2}$ where $\Delta = \mathrm{diag}(\delta_1, \ldots, \delta_p)$,
and specify a proper prior as

$$
\begin{aligned}
\mu \mid \Sigma &\sim N_p(\mu_0, \sigma_0^2 \Sigma), \mathcal{R} \sim \mathrm{uniform}(C^p), \\
\delta_i^{-1} &\sim \mathrm{gamma}(\alpha_{0i}, \beta_{0i}) \text{ for } i = 1, \ldots, p,
\end{aligned}
\tag{9.4}
$$

where C^p is the space of all $p \times p$ correlation matrices, and σ_0^2 and $(\mu_{0i}, \alpha_{0i}, \beta_{0i})$
are hyperparameters with μ_{0i} the i-th element of μ_0. We note that C^p is
a compact set and so the $\mathrm{uniform}(C^p)$ prior is proper. Elicitation of (9.4)
requires a total of $3p+1$ quantities and we avoid eliciting priors for correlations.
The hyperparameters all relate to the location and scaling of the response
variables and so elicitation is relatively straightforward, as we will show.

The use of a uniform prior on $\mathcal{R}$ has received some attention in [3], but it
is not commonly used. The reason for this is undoubtedly the difficulties with
posterior computations but we will show that there is a relatively straightforward and effective solution. Due to the symmetry of this distribution, all
the correlations have the same marginal prior symmetric about 0 and, as such,
have mean 0.

We now consider eliciting the hyperparameters σ_0^2 and $(\mu_{0i}, \alpha_{0i}, \beta_{0i})$ for
(9.4). To start, we specify an interval (m_{1i}, m_{2i}) such that we are virtually
certain this contains the true value of μ_i, where μ_i is the i-th element of μ.
Putting $\mu_{0i} = (m_{1i} + m_{2i})/2$, which specifies a hyperparameter, we interpret
this to mean that, given δ_i,

$$
p \le \Phi\left(\frac{m_{2i} - \mu_{0i}}{\sigma_0 \delta_i^{1/2}}\right) - \Phi\left(\frac{m_{1i} - \mu_{0i}}{\sigma_0 \delta_i^{1/2}}\right) = 2\Phi\left(\frac{m_{2i} - m_{1i}}{2\sigma_0 \delta_i^{1/2}}\right) - 1
\tag{9.5}
$$

where $p \in (0, 1)$ is chosen close to 1. For example, we could take $p = 0.999$
but other values could be used to interpret the phrase 'virtual certainty'. Now
(9.5) implies that

$$
\delta_i^{1/2} \le (m_{2i} - m_{1i})/2\sigma_0 z_{(1+p)/2}
\tag{9.6}
$$

where $z_{(1+p)/2} = \Phi^{-1}((1+p)/2)$. The inequality (9.6) only uses information
about the location of a measurement of y_i.

An interval that will contain an observed value of y_i with virtual certainty, given that we know μ_i and δ_i, is $\mu_i \pm \delta_i^{1/2} z_{(1+p)/2}$. Let s_{1i} and s_{2i} be lower and upper bounds respectively on the half-length of this interval. The values s_{1i} and s_{2i} represent what we know about the scaling of y_i and we choose these so that s_{1i} is not excessively small or s_{2i} excessively large. These choices imply that

$$s_{2i}^{-2} z_{(1+p)/2}^2 \leq \delta_i^{-1} \leq s_{1i}^{-2} z_{(1+p)/2}^2. \tag{9.7}$$

Therefore, $\sigma_0 = \max\{(m_{2i} - m_{1i})/2s_{2i} : i = 1, \ldots, p\}$ is compatible with (9.6) and (9.7), which specifies another hyperparameter. It is perhaps logical for $(m_{2i} - m_{1i})/2s_{2i}$ to be constant but this is not necessary.

Let $G(\alpha_{0i}, \beta_{0i}, \cdot)$ denote the gamma$(\alpha_{0i}, \beta_{0i})$ cdf, using the rate parameterization, so that $G(\alpha, \beta, x) = G(\alpha, 1, \beta x)$. Therefore, (9.7) contains δ_i^{-1} with virtual certainty, when α_{0i}, β_{0i} satisfy $G^{-1}(\alpha_{0i}, \beta_{0i}, (1+p)/2) = s_{1i}^{-2} z_{(1+p)/2}^2$, $G^{-1}(\alpha_{0i}, \beta_{0i}, (1-p)/2) = s_{2i}^{-2} z_{(1+p)/2}^2$, or equivalently

$$G(\alpha_{0i}, 1, \beta_{0i} s_{1i}^{-2} z_{(1+p)/2}^2) = (1+p)/2, \tag{9.8}$$

$$G(\alpha_{0i}, 1, \beta_{0i} s_{2i}^{-2} z_{(1+p)/2}^2) = (1-p)/2. \tag{9.9}$$

It is a simple matter to solve these equations iteratively for $(\alpha_{0i}, \beta_{0i})$. For this, choose an initial value for α_{0i} and, using (9.8), solve for x satisfying $G(\alpha_{0i}, 1, x) = (1+p)/2$, which implies $\beta_{0i} = x/s_{1i}^{-2} z_{(1+p)/2}^2$. If the left-side of (9.9) is less (greater) than $(1-p)/2$, then decrease (increase) the value of α_{0i} and solve for a new value of x as just described. Continue iterating this process until satisfactory convergence is attained.

We illustrate this process with a simple numerical example.

Example 1. For simplicity let us suppose that the response variables under consideration are all similar so that they all have essentially the same location and scaling. Therefore, we will use common values for the m_{ij} and s_{ij} although it is straightforward to have these values depend on the variable. Suppose that, based on background information about the measured variables, we know with virtual certainty ($p = 0.999$) that each mean value will lie in the interval $(1, 10)$. So we have $\mu_{0i} = 5.5$ for each i and $\delta^{1/2} \leq (10 - 1)/2\sigma_0 z_{(1+p)/2} = 1.3677/\sigma_0$ since $z_{(1+p)/2} = 3.2905$. Also, based on background information, we assume that $s_1 = 0.5$ and $s_2 = 2.0$ so that $0.5 \leq \delta^{1/2} z_{(1+p)/2} \leq 2.0$ with virtual certainty. Therefore, we get the inequalities for δ^{-1} given by (9.7) with $s_1^{-2} z_{(1+p)/2}^2 = 43.3102$ and $s_2^{-2} z_{(1+p)/2}^2 = 2.7069$. Also $\sigma_0 = (10 - 1)/2(2.0) = 2.25$.

To start the iteration using (9.8) and (9.9) we choose $\alpha_0 = 1$ and obtain $x = G^{-1}(1, 1, (1+p)/2) = 7.6009$ so $\beta_0 = x/s_1^{-2} z_{(1+p)/2}^2 = 0.1755$ which gives $G(1, 1, 2.7069) = 0.3782$. As this is greater than $(1-p)/2 = 0.0005$, we increase the value of α_0 to $\alpha_0 = 2$, repeat the above steps and continue increasing α_0 until we get a value of $G(\alpha_0, 1, \beta_0 s_2^{-2} z_{(1+p)/2}^2) < (1-p)/2$. When this is achieved we use the bisection algorithm to find the solution to (9.9). The results of this process are presented in Table 9.1 and we see that it converges

TABLE 9.1
The results of the iterative procedure to determine α_0 and β_0 in Example 1.

i	α_0	β_0	$G(\alpha_0, 1, \beta_0 s_2^{-2} z_{(1+p)/2}^2)$
1	1.0000	0.1755	0.3782
2	2.0000	0.2309	0.1302
3	8.0000	0.4769	0.0001
4	5.0000	0.3627	0.0034
5	6.5000	0.4211	0.0005
6	5.7500	0.3923	0.0013
7	6.1250	0.4068	0.0008
8	6.3125	0.4140	0.0006
9	6.4063	0.4176	0.0005
10	6.4531	0.4193	0.0005

quite quickly to give the values of $\alpha_0 = 6.4531$ and $\beta_0 = 0.4193$. It is easy to write a short routine in R to carry out these computations.

When we took the criterion of virtual certainty to be $p = 0.99$, and used the same values for the m_{ij} and s_{ij}, we obtained, again after 10 iterations, $\alpha_0 = 4.0391$ and $\beta_0 = 0.4161$. The two priors for $\delta = 1/\sigma^2$ are similar although the prior obtained with $p = 0.999$ has more mass near 0 indicating larger possible values for the variances. Of course, it is not sensible to use the same intervals for different choices of p as these should be longer the greater p is. Notice that changing the criterion for virtual certainty does not affect the conditional prior on the location variables unless we change the m_{ij} and s_{ij}.

■

Obtaining values from the prior (9.4) requires generating $\mathcal{R} \sim$ uniform(C^p). For this there is an algorithm due to [13], called the onion method. We have the following result for the posterior.

Proposition 1. Suppose that $\mathbf{Y} \in R^{n \times p}$ is a sample from a $N_p(\mu, \Sigma)$ distribution, $\mu \mid \Sigma \sim N_p(\mu_0, \sigma_0^2 \Sigma)$ and let π_1 denote a prior on Δ and π_2 a prior on $\mathcal{R}$ with Δ and $\mathcal{R}$ a priori independent. Then the conditional posterior distribution of μ is $\mu \mid \Sigma, \mathbf{Y} \sim N_p((\sigma_0^{-2} + n)^{-1}(\sigma_0^{-2}\mu_0 + n\bar{y}), (\sigma_0^{-2} + n)^{-1}\Sigma)$ and the posterior density of $\mathcal{J} = \Sigma^{-1}$ is proportional to a $W_p(A^{-1}(\mathbf{Y}), n - p + 1)$ density times $k(\mathcal{J}) = \pi_1(\text{diag}(\mathcal{J}^{-1}))\pi_2((\text{diag}(\mathcal{J}^{-1}))^{-1/2}\mathcal{J}^{-1}(\text{diag}(\mathcal{J}^{-1}))^{-1/2}) \times |\text{diag}(\mathcal{J}^{-1})|^{-(p-1)/2}$ where $\mathbf{A}(\mathbf{Y}) = (n-1)\mathbf{S} + n(n\sigma_0^2 + 1)^{-1}(\bar{y} - \mu_0)(\bar{y} - \mu_0)'$.

It is clear that generating from the posterior of Proposition 1 is not easy. Also a Gibbs sampler for this posterior is not readily available. We note from Proposition 1 that the posterior density of $\mathcal{J}$ factors, as a Wishart density times a function of $\mathcal{J}^{-1} = \Sigma$, that does not involve the data $\mathbf{Y}$. This function only depends on the prior π_1 on Δ and the prior π_2 on $\mathcal{R}$. Depending on how we choose π_1 and π_2, this can lead to a simple importance sampling algorithm for approximating integrals with respect to the posterior. For our importance sampler we will use the $W_p(\mathbf{A}^{-1}(\mathbf{Y}), n - p + 1)$ distribution. This is typically the 'hard' part of the density as it contains all the dependence on the data.

When we use the prior given by (9.4) we have the following result.

Corollary 2. If π_2 is the uniform prior on $\mathcal{R}$, and π_1 is the product of inverse gamma$(\alpha_{0i}, \beta_{0i})$ densities, then $k(\mathcal{J}) = \prod_{i=1}^{p} \sigma_{ii}^{-\alpha_{0i}-(p+1)/2} \exp\{-\beta_{0i}/\sigma_{ii}\}$.

Note that $k(\mathcal{J})$ is a fairly simple function of the σ_{ii}. It is easy to see from this result that other choices could be made for π_1 that would also lead to simple importance sampling algorithms.

9.3   Concentration for Factor Analysis Models

To assess the hypothesis H_0^q using the method of concentration, we need to compute the distance of a correlation matrix $\mathcal{R}$ from H_0^q. It seems natural to use Frobenius distance which, for correlation matrices $\mathcal{R}^{(1)}$ and $\mathcal{R}^{(2)}$, is given by $d(\mathcal{R}^{(1)}, \mathcal{R}^{(2)}) = 2 \sum_{i<j} (\xi_{ij}^{(1)} - \xi_{ij}^{(2)})^2$. Since $H_0^q = \{\mathcal{R} : d(\mathcal{R}, H_0^q) = 0\}$, we need to compute $d(\mathcal{R}, H_0^q)$ for an arbitrary correlation matrix $\mathcal{R}$ and ultimately compute the relative belief ratio when $d(\mathcal{R}, H_0^q) = 0$. Computing $d(\mathcal{R}, H_0^q)$ requires minimizing, as a function of Γ_q,

$$\sum_{i<j} \left(\xi_{ij} - \sum_{k=1}^{\min(i,q)} \gamma_{ik}\gamma_{jk} \right)^2 \tag{9.10}$$

where $\gamma_{ii} \geq 0$ for $i = 1, \ldots, q$ and $\sum_{k=1}^{\min(i,q)} \gamma_{ik}^2 \leq 1$ for $i = 1, \ldots, p$. Note that (9.10) is invariant under multiplying a column of Γ_q by -1. Therefore, we can consider minimizing (9.10) subject to the simpler constraint that the i-th row of Γ_q lies in the unit ball $B^{\min(i,q)}(\mathbf{0})$ centered at the origin in $R^{\min(i,q)}$. To minimize (9.10) as a function of Γ_q, we have to find the point in $B^{\min(1,q)}(\mathbf{0}) \times \cdots \times B^{\min(p,q)}(\mathbf{0})$ where this minimum is attained. Since (9.10) is continuous on this compact set, an absolute minimum exists.

When $q = 0$ then, as $\Gamma_0 = 0$, the minimum of (9.10) is $\sum_{i<j} \xi_{ij}^2$. For other values of q we need to proceed iteratively. Consider the case when $q = 1$, so we need to minimize $\sum_{i<j} (\xi_{ij} - \gamma_{i1}\gamma_{j1})^2$. First we start with $(\gamma_{11}(0), \ldots, \gamma_{p1}(0)) \in B^1(\mathbf{0}) \times \cdots \times B^1(\mathbf{0}) = [-1, 1]^p$. We can write

$$\sum_{i<j} (\xi_{ij} - \gamma_{i1}(0)\gamma_{j1}(0))^2 = \left(\sum_{j=2}^{p} \gamma_{j1}^2(0) \right) \gamma_{11}^2(0) - 2 \left(\sum_{j=2}^{p} \xi_{1j}\gamma_{j1}(0) \right) \gamma_{11}(0) + c$$

where c is a constant as a function of $\gamma_{11}(0)$. If $\sum_{j=2}^{p} \gamma_{j1}^2(0) \neq 0$, then this quadratic in $\gamma_{11}(0)$ is minimized by $\gamma_{11}(0)$ equal to

$$\sum_{j=2}^{p} \xi_{1j}\gamma_{j1}(0) / \sum_{j=2}^{p} \gamma_{j1}^2(0). \tag{9.11}$$

If (9.11) is not in $[-1,1]$, then the quadratic is minimized over this interval by setting $\gamma_{11}(0) = -1$, when (9.11) is less than -1, or setting $\gamma_{11}(0) = 1$, when (9.11) is greater than 1. If $\sum_{j=2}^{p} \gamma_{j1}^2(0) = 0$, then $\sum_{j=2}^{p} \xi_{1j}\gamma_{j1}(0) = 0$, and there is no dependence on $\gamma_{11}(0)$ so we set $\gamma_{11}(1) = \gamma_{11}(0)$. So $\gamma_{11}(0)$ is replaced by a value $\gamma_{11}(1)$ that minimizes the quadratic over $[-1,1]$. After updating γ_{11} we use the same argument to replace $\gamma_{21}(0)$ by the value $\gamma_{21}(1)$ and continue cycling through the variables in this way. We call this algorithm *constrained univariate quadratic iteration.*

We see immediately that at each step of the iteration, the value of (9.10) never increases. Since (9.10) is bounded below by 0, the iteration converges to a minimum value. The convergence is typically very fast. This minimum value is not necessarily the absolute minimum as it depends on the starting value. Accordingly, we proceed as follows. We select m i.i.d. starting points from the uniform distribution on $B^1(\mathbf{0}) \times \cdots \times B^1(\mathbf{0})$ and compute the m minima $d_1, \ldots, d_m$ via this iterative procedure applied to each starting value. We then estimate $d(\mathcal{R}, H_0^1)$ by $d_{(1):m}$, i.e., the smallest order statistic. The values $d_1, \ldots, d_m$ comprise an i.i.d. sample from a distribution with compact support in R^1 and so we have that $d_{(1):m}$ converges in probability to the absolute minimum of (9.10). Computational experience indicates that there are typically a small number of local minima for a given $\mathcal{R}$ and often there is only one found. So this represents an efficient method for computing $d(\mathcal{R}, H_0^1)$.

The same iterative procedure works for general q where we generate the starting values uniformly in $B^{\min(1,q)}(\mathbf{0}) \times \cdots \times B^{\min(p,q)}(\mathbf{0})$.

Proposition 2. Constrained univariate quadratic iteration with $\boldsymbol{\Gamma}_q(0) \in B^{\min(1,q)}(\mathbf{0}) \times \cdots \times B^{\min(p,q)}(\mathbf{0})$, always gives a nonincreasing sequence of values of (9.10) and as such converges.

The convergence of $\boldsymbol{\Gamma}_q(k)\boldsymbol{\Gamma}_q'(k) + \boldsymbol{\Lambda}(k)$ is not necessary for the assessment of H_0^q but in our experience this always occurs. In fact, the $\boldsymbol{\Gamma}_q(k)$ sequence typically converges to a point in $B^{\min(1,q)}(\mathbf{0}) \times \cdots \times B^{\min(p,q)}(\mathbf{0})$ but this depends on the starting value $\boldsymbol{\Gamma}_q(0)$. Since $B^{\min(1,q)}(\mathbf{0}) \times \cdots \times B^{\min(p,q)}(\mathbf{0})$ is compact, this sequence always has a convergent subsequence but we have not been able to prove that there is only one limit point.

We consider some examples of using the algorithm.

Example 2. First consider a correlation matrix of the form $\mathcal{R} = \boldsymbol{\Gamma}_1\boldsymbol{\Gamma}_1' + \boldsymbol{\Lambda}$, namely, we have a 1-factor model and suppose we want to compute $d(\mathcal{R}, H_0^1)$. The minimization algorithm should converge to the actual distance 0. Suppose we take $p = 6$ and $\boldsymbol{\Gamma}_1 = \mathbf{1}_6 = (1,1,1,1,1,1)'$, so $\boldsymbol{\Lambda}$ is a matrix of zeros. To assess the performance of the minimization, we generated 10^3 starting values $\boldsymbol{\Gamma}_1(0)$ uniformly in $B_1^1(\mathbf{0}) \times \cdots \times B_1^1(\mathbf{0})$ and applied the algorithm to each case. For the stopping rule, we iterated until the squared distance between two successive $\Gamma_1(i)$ was less than 10^{-5}. In this case all the minimum distances were equal to 0. The mean number of iterations required to stop was 2.884 and the maximum number was 4, so convergence was very fast.

Now suppose the correlation matrix possesses a two factor structure, namely, $\mathcal{R} = \Gamma_2\Gamma_2' + \Lambda$ and we want to compute $d(\mathcal{R}, H_0^2)$. The exact distance in this case is again 0. Suppose $p = 8$ and

$$\Gamma_2 = \left(\begin{array}{cccccccc} 0.8 & 0 & 0.8 & 0.8 & 0 & 0 & 0.8 & 0.6 \\ 0 & 0.8 & 0 & 0 & 0.8 & 0.8 & 0 & 0.6 \end{array} \right)'$$

so Λ is not the zero matrix in this case. We used 10^3 starting values generated uniformly from $B^1(\mathbf{0}) \times B^2(\mathbf{0}) \times \cdots \times B^2(\mathbf{0})$ and stopped iterating when the precision 10^{-5} was reached. For a two factor model, the iterative process takes longer with a mean number of 20 and a maximum of 47 iterations, but still converges very quickly. The minimum distance obtained was $d_{(1):m} = 8.678564 \times 10^{-7}$ while the maximum was equal to 1.459. To 4 decimal places there were 999 instances of the distance equaling 0. So we really found only 2 minima with the absolute minimum equaling 0. ∎

In practice we do not need anything like 10^3 starting values to compute the absolute minimum. It is the case, however, that as q increases the number of iterations required to achieve a given precision increases. We can offset this, however, with a generalization of the algorithm where we proceed by iterating on the rows of Γ_q rather than individual entries, see [5].

In the following section we present examples of computing the prior and posterior densities of $d(\mathcal{R}, H_0^q)$ for various q. We estimate the prior density π_d based on a sample obtained by generating $\mathcal{R} \sim \text{Uniform}(C^p)$ using the onion method and then use constrained univariate quadratic iteration to obtain $d(\mathcal{R}, H_0^q)$. To obtain the posterior density $\pi_d(\cdot \mid x)$ we use importance sampling as described in Proposition 1 and Corollary 2. A numerical problem sometimes arises in computing $\pi_d(0 \mid x)/\pi_d(0)$ when both terms in the ratio are close to 0. We then use $\pi_d(d_\alpha \mid x)/\pi_d(d_\alpha)$ as an approximation to the limit $\lim_{\epsilon \searrow 0} \pi_d(\epsilon \mid x)/\pi_d(\epsilon)$ where d_α is the α-th prior quantile of $d(\mathcal{R}, H_0^q)$ with α small, e.g., $\alpha = 0.01$. This procedure was found to work well in a wide variety of examples.

9.4 Inferences and Examples

For inferences we first assess the hypothesis H_0^0 based on the relative belief ratio for $d(\mathcal{R}, H_0^0)$ at $d(\mathcal{R}, H_0^0) = 0$. If we obtain evidence against H_0^0, then we proceed to assess H_0^1, etc. We stop at the q-th step when we first obtain evidence in favor of H_0^q. This approach seems natural as our goal is to find the smallest number of factors required to explain the observed correlations.

Suppose then that we have selected the factor model with q factors. This says that the true correlation matrix satisfies (9.3) and we must estimate $\mathcal{R}$ so that our estimate is in H_0^q. Naturally we want to maximize the relative belief ratio $RB(\mathcal{R})$ for $\mathcal{R} \in H_0^q$ but this is computationally difficult as the set is not convex. Note that this also rules out using the conditional expectation given that $\mathcal{R} \in H_0^q$, as we are not guaranteed that this value is in H_0^q. We can, however, use the following approximation. Let $\hat{\mathcal{R}}$ denote the plug-in MLE of $\mathcal{R}$, i.e., the estimate obtained from the MLE of Σ given by $(n-1)n^{-1}\mathbf{S}$. Now let $\hat{\mathcal{R}}_q$ denote the point in H_0^q that is closest to $\hat{\mathcal{R}}$ and note that this can be computed using our minimization algorithm. We take $\hat{\mathcal{R}}_q$ as our estimate of $\mathcal{R}$. Certainly, when H_0^q is true, then $\hat{\mathcal{R}}$ converges to a point in H_0^q and so $\hat{\mathcal{R}}_q$ will also converge to the true value. Furthermore, we can quantify the uncertainty in $\hat{\mathcal{R}}_q$ by looking at the posterior distribution of $d(\mathcal{R}, \hat{\mathcal{R}}_q)$ and comparing this to its prior to assess how much the data have increased our belief in the estimate.

Corresponding to $\hat{\mathcal{R}}_q$ there is a $\hat{\Gamma}_q$, obtained using the minimization algorithm, and so we could use this as an estimate of the factor loadings. There is no guarantee, however, that there is only one lower triangular matrix of factor loadings satisfying $\hat{\mathcal{R}}_q = \hat{\Gamma}_q \hat{\Gamma}_q' + \hat{\Lambda}$, where $\hat{\Lambda}$ is determined from $\hat{\Gamma}_q$ as previously described. We can also always postmultiply $\hat{\Gamma}_q$ by a $q \times q$ orthogonal matrix to obtain more interpretable loadings.

We now consider implementing inferences in several examples.

Example 3. *Simulated Data.*

We first consider simulating the data $\mathbf{Y}$ from a distribution with correlation matrix $\mathcal{R} \in H_0^2$ where $\Gamma_2 \in R^{8 \times 2}$ is as given in Example 2. The entries in the correlation matrix above the main diagonal are given in Table 9.2.

TABLE 9.2
The true correlation matrix (above diagonal) in Example 3.

0.00	0.64	0.64	0.00	0.00	0.64	0.48
	0.00	0.00	0.64	0.64	0.00	0.48
		0.64	0.00	0.00	0.64	0.48
			0.00	0.00	0.64	0.48
				0.64	0.00	0.48
					0.00	0.48
						0.48

The prior chosen was as given in (9.4) with $(\alpha_{0i}, \beta_{0i}) = (1,1)$ for $i = 1, \ldots, p$, $\mu_0 = \mathbf{0}$ and $\sigma_0^2 = 1$. We generated a sample of $n = 200$ values from the $N_8(\mathbf{0}, \Gamma_2 \Gamma_2' + \Lambda)$ distribution. In Figure 9.1 we have plotted the prior and posterior densities for the $q = 0, 1, 2$ cases. We see that the posterior becomes increasingly concentrated near 0 as q increases. The relative belief ratio when $q = 0$ is effectively 0 and (9.1) is also effectively 0, so we have strong evidence against H_0^0. For H_0^1, the relative belief ratio is also very close to 0 and (9.1) equals 1.7×10^{-5}, so again we have strong evidence against H_0^1. For H_0^2, the relative belief ratio is very large and (9.1) is effectively 1, so we have strong evidence in favor of H_0^2 which is of course correct.

Given that we accept H_0^2 we estimated $\boldsymbol{\Gamma}_2$ and obtained

$$
\hat{\boldsymbol{\Gamma}}_2 = \begin{pmatrix} 0.804 & 0.046 & 0.784 & 0.755 & 0.068 & 0.064 & 0.774 & 0.717 \\ 0.000 & 0.797 & 0.076 & 0.067 & 0.783 & 0.785 & -0.021 & 0.528 \end{pmatrix}'.
$$

This gives $\hat{\mathcal{R}}_2$ as shown in Table 9.3 which is close to $\mathcal{R}$ provided in Table 9.2.

To assess the accuracy of the importance sampling, we estimated the coefficient of variation of the estimator of the normalizing constant for the posterior. For a Monte Carlo sample of size N, this is given by $CV = N^{-1}(N \sum_{i=1}^{N} w_{*i}^2 - 1)$ where $w_{*i} = k(\mathcal{J}_i)/\sum_{j=1}^{N} k(\mathcal{J}_i)$. A value of $\sum_{i=1}^{N} w_{*i}^2$ close to $1/N$ indicates that the importance sampling is working while a value near 1 indicates a failure. In this case, $N = 10^5$ and we obtained the values in Table 9.4. So the importance sampling is working very well as was found to be the case in a number of examples discussed in [5]. ∎

Example 4. *Currency Exchange Data*

We now consider a data set involving monthly international exchange rates ($n = 144$) available in [21]. These time series are the monthly changes in exchange rates in British pounds of the following $p = 6$ currencies: US dollar (US), Canadian dollar (CAN), Japanese yen (JAP), French franc (FRA), Italian lira (ITA) and the German Deutschmark (GER). The data span the period from January of 1975 to December of 1986 inclusive. For this data the correlation matrix is given by Table 9.5.

In [21], it was determined that up to three factors are needed. Although the dimension has been reduced to half, it offers no simplification based on a degrees of freedom argument. In other words, the factor model contains as many parameters as $\boldsymbol{\Sigma}$. Also, the maximum likelihood approach failed to converge for this data set.

In [17] a reversible jump MCMC was developed to handle the change in dimension as q changes. Very diffuse proper priors were chosen and it was concluded that two factors are needed to explain the correlations.

Our analysis is based on the prior given by (9.4). As noted, this greatly simplifies the elicitation of the prior as a uniform prior is specified for the correlation matrix. For the $N_6(\mu_0, \sigma_0^2 \Sigma)$ prior we chose the hyperparameters to equal $\mu_0 = (1.8, 2.1, 430.6, 10.1, 1984, 12.4)'$, where the individual entries are the sample means of these quantities, and also set $\sigma_0^2 = 100$ to reflect ignorance about the location parameters. We note that all other Bayesian factor analyses that we are aware of, have effectively set $\sigma_0^2 = 0$ so that all the prior probability for μ is concentrated at the sample means. This is not necessary with our approach. For the scaling parameters we used $(\alpha_{0i}, \beta_{0i}) = (2.2, 0.1)$ for $i = 1, \ldots, 6$. These choices are the same as those made in [17] and are selected here for comparison purposes. In an actual application we

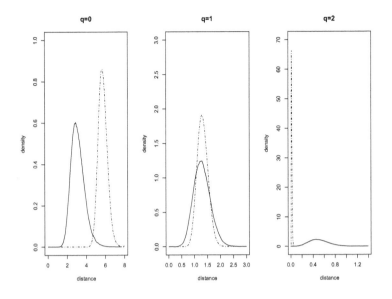

FIGURE 9.1
The prior (—) and the posterior (- -) densities of the distance for $q = 0, 1$ and 2 in Example 3.

TABLE 9.3
The estimated correlation matrix (above diagonal) in Example 3 when $q = 2$.

0.0370	0.630	0.607	0.054	0.052	0.622	0.576
	0.097	0.088	0.627	0.628	0.019	0.454
		0.597	0.112	0.110	0.605	0.602
			0.103	0.101	0.583	0.576
				0.618	0.036	0.462
					0.033	0.460
						0.544

TABLE 9.4
Sums of squared normalized importance sampling weights and coefficient of variation in Example 3.

	$\sum_{i=1}^{N} w_{*i}^2$	CV
$q = 0$	1.94×10^{-5}	9.38×10^{-6}
$q = 1$	1.78×10^{-5}	7.76×10^{-6}
$q = 2$	1.78×10^{-5}	7.80×10^{-6}

TABLE 9.5

Sample correlations in Example 4.

	CAN	JAP	FRA	ITA	GER
US	0.858	0.801	−0.453	−0.501	0.148
CAN		0.429	−0.144	−0.075	0.191
JAP			−0.446	−0.652	0.043
FRA				0.922	0.068
ITA					0.067

would want to elicit the values of μ_0, σ_0^2 and the $(\alpha_{0i}, \beta_{0i})$ based on historical knowledge of the exchange rates.

First we assess H_0^0. The posterior and prior densities of $d(\mathcal{R}, H_0^0)$ are plotted in Figure 9.2. We see that the posterior is much less concentrated near 0 than the prior. The relative belief ratio for H_0^0 is effectively 0 and (9.1) equals 0 to four decimal places. So we have strong evidence against H_0^0 and conclude that $q > 0$, i.e., an independence model does not hold.

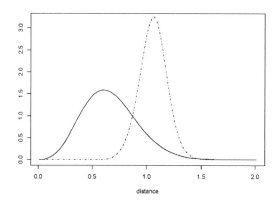

FIGURE 9.2

The prior (—) and the posterior (− −) densities of $d(\mathcal{R}, H_0^0)$ in Example 4.

We next assess H_0^1. The posterior and prior density comparison is presented in Figure 9.3. This figure also shows that the posterior is much less concentrated near 0 than the prior. The relative belief ratio for H_0^1 is also effectively 0 and (9.1) is 0.0001. Thus we have strong evidence against H_0^1 and conclude that $q > 1$, i.e., a model with more than 1 factor is needed.

We now proceed to assess H_0^2 and plot the prior and posterior densities in Figure 9.4. This plot shows that posterior concentrates much more near 0 than the prior does. The relative belief ratio for H_0^2 is approximated by $\pi(d_\alpha|\mathbf{X})/\pi(d_\alpha) = 17.2$, where $d_\alpha = 0.018$ with $\alpha = 0.01$, and (9.1) equals

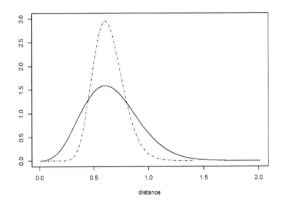

FIGURE 9.3
The prior (—) and the posterior (– –) densities of $d(\mathcal{R}, H_0^1)$ in Example 4.

0.467. Therefore, we have moderate evidence in favor of H_0^2 and conclude that a 2-factor model is reasonable, which agrees with [17].

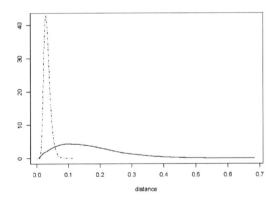

FIGURE 9.4
The prior (—) and the posterior (– –) densities of $d(\mathcal{R}, H_0^2)$ in Example 4.

We now estimate $\mathcal{R}$, given that $\mathcal{R} \in H_0^2$. For Γ_2 we obtained the estimate

$$\widehat{\Gamma}_2 = \begin{pmatrix} 1.0 & 0.82345 & 0.72684 & -0.45423 & -0.50842 & 0.14552 \\ 0 & 0.37541 & -0.27513 & 0.74464 & 0.86111 & 0.17547 \end{pmatrix}'.$$

This leads to the estimate $\widehat{\mathcal{R}}_2$ given by the entries of Table 9.6. We note that these are all close to the values in Table 9.5 so the two factor model seems

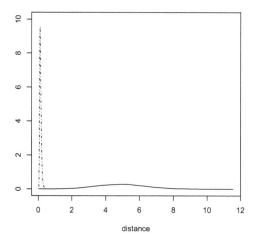

FIGURE 9.5
Plots of prior (—) and posterior (- -) densities of distance to $\hat{\mathcal{R}}_2$ in Example 4.

appropriate. Figure 9.5 plots the prior and posterior densities of the distance from $\hat{\mathcal{R}}_q$ which shows a huge increase in belief for the estimate.

TABLE 9.6
Estimated correlations in Example 4 based on H_0^2.

	CAN	JAP	FRA	ITA	GER
US	0.823	0.727	−0.454	−0.508	0.146
CAN		0.495	−0.094	−0.095	0.186
JAP			−0.535	−0.606	0.057
FRA				0.872	0.065
ITA					0.077

From this analysis we see that we have strong evidence that the five currencies are related by two common variables. We can examine the entries in $\hat{\Gamma}_2$ to see if meaning can be assigned to these variables. The first latent variable would appear to be a contrast between {CAN, JAP} and {FRA, ITA} while the second latent variable has a less obvious interpretation. ∎

9.5 Conclusions

We have developed a Bayesian approach to factor analysis which has several attractive features. In particular, we need only place a prior on the full model parameter (μ, Σ) as opposed to requiring priors on parameters associated with unobservable latent variables. The computational approach allows for the use of a uniform prior on the correlation matrix and we have provided a convenient elicitation algorithm for location and scaling parameters. The methodology is seen to work well in a variety of examples. Also see [5] for more examples.

Various aspects of this problem can be worked on for improvements. For example, the difficult part of the computations, due to the geometry of the sets H_0^q, is the minimization step. Techniques from algebraic geometry may prove useful in obtaining more efficient algorithms. The difficulties caused by the geometry of the sets H_0^q are intrinsic to factor analysis itself. Certainly the computational burden rises as the dimension and the number of factors rise. The usefulness of factor analysis lies in applications where relatively few factors are found to summarize the dependencies among manifest variables. For such problems, the methodology developed here is effective.

9.6 Appendix

Proof of Proposition 1 and Corollary 2 The Jacobian of the transformation $(\Delta, \mathcal{R}) \to \Sigma$ is $|\text{diag}(\Sigma)|^{-(p-1)/2}$ and so the joint prior on (μ, Σ) is proportional to

$$|\Sigma|^{-1/2} \exp\left\{-\frac{1}{2\sigma_0^2}(\mu - \mu_0)'\Sigma^{-1}(\mu - \mu_0)\right\} \pi_1((\text{diag}(\Sigma))^{1/2}) \times$$
$$\pi_2((\text{diag}(\Sigma))^{-1/2}\Sigma(\text{diag}(\Sigma))^{-1/2})|\text{diag}(\Sigma)|^{-(p-1)/2}.$$

Making the change of variable $\Sigma \to \mathcal{J} = \Sigma^{-1}$, which has Jacobian $|\Sigma|^{p+1}$, the prior density of $(\mu, \mathcal{J})$, where $\text{etr}(A) = \exp(\text{tr}(A))$ for matrix A, is

$$|\mathcal{J}|^{-(p+1/2)}\text{etr}\left\{-\frac{1}{2\sigma_0^2}\mathcal{J}(\mu - \mu_0)(\mu - \mu_0)'\right\} \pi_1(\text{diag}(\mathcal{J}^{-1})) \times$$
$$\pi_2((\text{diag}(\mathcal{J}^{-1}))^{-1/2}\mathcal{J}^{-1}(\text{diag}(\mathcal{J}^{-1}))^{-1/2})|\text{diag}(\mathcal{J}^{-1})|^{-(p-1)/2}.$$

The likelihood function is proportional (up to a constant function of the data and parameters) to $|\Sigma|^{-n/2}\text{etr}\{-(n/2)\Sigma^{-1}(\bar{y} - \mu)(\bar{y} - \mu)' - (n-1)/2)\mathbf{S}\Sigma^{-1}\}$, so the joint posterior of $(\mu, \mathcal{J})$ is similarly proportional to

$$|\mathcal{J}|^{n/2-(p+1/2)} \operatorname{etr} \left\{ \begin{array}{c} -\frac{1}{2\sigma_0^2} \mathcal{J}(\mu - \mu_0)(\mu - \mu_0)' - \frac{n}{2} \mathcal{J}(\bar{\mathbf{y}} - \mu)(\bar{\mathbf{y}} - \mu)' \\ -\frac{(n-1)}{2} \mathbf{S} \mathcal{J} \end{array} \right\} \times$$

$$\pi_1(\operatorname{diag}(\mathcal{J}^{-1})) \pi_2((\operatorname{diag}(\mathcal{J}^{-1}))^{-1/2} \mathcal{J}^{-1} (\operatorname{diag}(\mathcal{J}^{-1}))^{-1/2}) |\operatorname{diag}(\mathcal{J}^{-1})|^{-(p-1)/2}.$$

Now we have that

$$\sigma_0^{-2} (\mu - \mu_0)' \mathcal{J}(\mu - \mu_0) + n(\mu - \bar{\mathbf{y}})' \mathcal{J}(\mu - \bar{\mathbf{y}})$$

$$= (\sigma_0^{-2} + n) \left(\mu - \frac{\sigma_0^{-2}\mu_0 + n\bar{\mathbf{y}}}{\sigma_0^{-2} + n} \right)' \mathcal{J} \left(\mu - \frac{\sigma_0^{-2}\mu_0 + n\bar{\mathbf{y}}}{\sigma_0^{-2} + n} \right)$$

$$+ (\sigma_0^2 + n^{-1})^{-1} (\mu_0 - \bar{\mathbf{y}})' \mathcal{J}(\mu_0 - \bar{\mathbf{y}})$$

Integrating out μ gives that the posterior of $\mathcal{J}$ is proportional (up to a constant function of the data and parameters) to

$$|\mathcal{J}|^{n/2-p} \operatorname{etr} \left\{ -\frac{A(\mathbf{Y})\mathcal{J}}{2} \right\} \pi_1(\operatorname{diag}(\mathcal{J}^{-1})) \times$$

$$\pi_2((\operatorname{diag}\mathcal{J}^{-1})^{-1/2} \mathcal{J}^{-1} (\operatorname{diag}\mathcal{J}^{-1})^{-1/2}) |\operatorname{diag}(\mathcal{J}^{-1})|^{-(p-1)/2}$$

where $A(\mathbf{Y}) = (n-1)\mathbf{S} + (\sigma_0^2 + 1/n)^{-1} (\bar{\mathbf{y}} - \mu_0)(\bar{\mathbf{y}} - \mu_0)'$.

For Corollary 2, put $\sigma_{ii} = \delta_i$ and then $k(\mathcal{J}) = \prod_{i=1}^p (1/\delta_i)^{\alpha_{0i}+1} \exp(-\beta_{0i}/\delta_i)(\delta_i)^{-(p-1)/2} = \prod_{i=1}^p (\sigma_{ii})^{-\alpha_{0i}-(p+1)/2} \exp(-\beta_i/\sigma_{ii})$.

Proof of Proposition 2 We note first that if $ax^2 + bx + c$ is such that $a > 0$, then the minimum of the quadratic occurs at $-b/2a$. If $-b/2a$ is not in $[l, u]$ then the minimum over the interval occurs at l or u.

Note that when $r < s$, then $\gamma_{rs} = 0$ and so we need only consider the case $r \geq s$. We see that γ_{rs} occurs in terms of (9.11) only when $i = r$ or $j = r$, so we can write (9.11) as

$$\sum_{j=r+1}^p \left(\sigma_{rj} - \sum_{k=1}^{\min(r,q)} \gamma_{rk}\gamma_{jk} \right)^2 + \sum_{i=1}^{r-1} \left(\sigma_{ir} - \sum_{k=1}^{\min(i,q)} \gamma_{ik}\gamma_{rk} \right)^2 + c \quad (A1)$$

where c is some constant not involving γ_{rs}. Now when $s = r \leq q$, then the second term in (A1) does not involve γ_{rr} and so we need only consider the first term. As a function of γ_{rr}, the first term can be written as

$$\left(\sum_{j=r+1}^p \gamma_{jr}^2 \right) \gamma_{rr}^2 - 2 \left[\sum_{j=r+1}^p \left(\sigma_{rj} - \sum_{k=1}^{r-1} \gamma_{rk}\gamma_{jk} \right) \gamma_{jr} \right] \gamma_{rr} + c \quad (A2)$$

where c is a constant. Note that if $\sum_{j=r+1}^p \gamma_{jr}^2 = 0$, then the coefficient of γ_{rr}

in (A2) is also 0. When $s < r$, then $s \leq q$ and (A1) equals

$$
\left(\sum_{j=\max(r+1,s)}^{p} \gamma_{js}^2 + \sum_{i=s}^{r-1} \gamma_{is}^2 \right) \gamma_{rs}^2 -
$$

$$
2 \left[\begin{array}{c} \sum_{j=\max(r+1,s+1)}^{p} \left(\sigma_{rj} - \sum_{k=1,k\neq s}^{\min(r,q)} \gamma_{rk}\gamma_{jk} \right) \gamma_{js} + \\ \sum_{i=s}^{r-1} \left(\sigma_{ir} - \sum_{k=1,k\neq s}^{\min(i,q)} \gamma_{ik}\gamma_{rk} \right) \gamma_{is} \end{array} \right] \gamma_{rs} + c \qquad (A3)
$$

where c is a constant. Again if the coefficient of γ_{rs}^2 in (A3) is 0, then the coefficient of γ_{rs} is 0. Now for $1 \leq r \leq q$, then

$$
\gamma_{rr} \in \left[-\left(\sigma_{rr} - \sum_{k=1}^{r-1} \gamma_{rk}^2 \right)^{1/2}, \left(\sigma_{rr} - \sum_{k=1}^{r-1} \gamma_{rk}^2 \right)^{1/2} \right] \qquad (A4)
$$

and otherwise

$$
\gamma_{rs} \in \left[-\left(\sigma_{rr} - \sum_{k=1,k\neq s}^{\min(r,q)} \gamma_{rk}^2 \right)^{1/2}, \left(\sigma_{rr} - \sum_{k=1,k\neq s}^{\min(r,q)} \gamma_{rk}^2 \right)^{1/2} \right]. \qquad (A5)
$$

Now just as in the $q = 1$ case we can select a starting $\Gamma_q(0)$ and then iterate using (A2) and (A3), to minimize each quadratic over the relevant interval as determined by (A4) and (A5). So we might start with $\gamma_{11}(0)$ replacing it by $\gamma_{11}(1)$, and then, with this new value for γ_{11}, replace $\gamma_{21}(0)$ by $\gamma_{21}(1)$, etc. At each step, the distance (9.11) does not increase and so the sequence converges to a minimum.

Bibliography

[1] D. J. Bartholomew. *Latent Variable Models and Factor Analysis*. Charles Griffin, London, 2004.

[2] Z. Baskurt and M. Evans. Hypothesis assessment and inequalities for Bayes factors and relative belief ratios. *Bayesian Analysis*, 8(3):569–590, 2013.

[3] J. Bernard, R. McCulloch, and X. L. Meng. Modeling covariance matrices in terms of standard deviations and correlations with application to shrinkage. *Statistica Sinica*, 10:1281–1311, 2000.

[4] A. Bhattacharya and D. B. Dunson. Sparse Bayesian infinite factor models. *Biometrika*, 98:291–306, 2011.

[5] Y. Cao. *A Bayesian Approach to Factor Analysis via Comparing Prior and Posterior Concentration.* Ph. D. thesis, University of Toronto, 2010.

[6] C. M. Carvalho, J. Chang, J. Lucas, Q. Wang, J. Nevins, and M. West. High-dimensional sparse factor modelling: Applications in gene expression genomics. *Journal of the American Statistical Association,* 103(4):1438–1456, 2008.

[7] M. Evans, Z. Gilula, and I. Guttman. Computational issues in the Bayesian analysis of categorical data: Loglinear and Goodman's RC model. *Statistica Sinica,* 3:391–406, 1993.

[8] M. Evans, Z. Gilula, and I. Guttman. Conversion of ordinal attitudinal scales: An inferential Bayesian approach. *Quantitative Marketing and Economics,* 10:283–304, 2012.

[9] M. Evans, Z. Gilula, I. Guttman, and T. Swartz. Bayesian analysis of stochastically ordered distributions of categorical variables. *Journal of the American Statistical Association,* 92(437):208–214, 1997.

[10] J. Ghosh and D. B. Dunson. Bayesian model selection in factor analytic models. In D. B. Dunson, editor, *Random Effect and Latent Variable Model Selection.* Springer Verlag, New York, 2008.

[11] J. Ghosh and D. B. Dunson. Default prior distributions and efficient posterior computation in Bayesian factor analysis. *Journal of Computational and Graphical Statistics,* 18(2):306–320, 2009.

[12] J. K. Ghosh, M. Delampady, and T. Samanta. *An Introduction to Bayesian Analysis: Theory and Methods.* Springer Verlag, New York, 2006.

[13] S. Ghosh and S. G. Henderson. Behavior of the NORTA method for correlated random vector generation as the dimension increases. *ACM Transactions on Modeling and Computer Simulation,* 13:276–294, 2003.

[14] G. Kaufman and S. J. Press. Bayesian factor analysis. *Sloan School of Management, MIT,* Working Paper 662-73, 1973.

[15] S. E. Lee and S. J. Press. Robustness of Bayesian factor analysis estimates. *Communications in Statistics–Theory and Methods,* 27(8):1871–1893, 1998.

[16] S.-Y. Lee. *Structural Equation Modeling, A Bayesian Approach.* John Wiley & Sons, New York, 2007.

[17] H. F. Lopes and M. West. Bayesian model assessment in factor analysis. *Statistica Sinica,* 14(1):41–67, 2004.

[18] S. J. Press. *Applied Multivariate Analysis: Using Bayesian and Frequentist Methods of Inference.* Robert E. Krieger Publishing Co., Malabar, Florida, 1982.

[19] S. J. Press and K. Shigemasu. Bayesian inference in factor analysis. In L. J. Gleser, M. D. Perlman, S. J. Press, and A. R. Sampson, editors, *Contributions to Probability and Statistics: Essays in Honor of Ingram Olkin.* Springer Verlag, New York, 1989.

[20] D. B. Rowe. *Multivariate Bayesian Statistics: Models for Source Separation and Signal Unmixing.* Chapman and Hall/CRC, Florida, USA, 2003.

[21] M. West and J. Harrison. *Bayesian Forecasting and Dynamic Models.* Springer Verlag, New York, 2nd edition, 1997.

10

Regional Fertility Data Analysis: A Small Area Bayesian Approach

Eduardo A. Castro

University of Aveiro

Zhen Zhang

Michigan State University

Arnab Bhattacharjee

Heriot-Watt University

José M. Martins

University of Aveiro

Tapabrata Maiti

Michigan State University

CONTENTS

10.1   Introduction

We use small area Bayesian statistics to develop a model for age-specific fertility rates. The model is then used to estimate age-specific fertility rates and total fertility rates at the regional NUTS III area level for Portugal. The chapter makes important contributions to small area Bayesian statistics in a spatial domain focusing on estimation of fertility rates. The estimates obtained are useful for demographic policy in Portugal.

Micro-demographic variables such as mortality, fertility and migrations by age group are not only the main drivers of demographic dynamics but are also key elements used to describe the behaviour of populations, both at national and regional levels. Both regional population forecasting and inter-regional comparisons to support policy making require the accurate data on such variables, and are strongly sensitive to the quality of estimates. In turn, data quality depends essentially on the survey and data collection processes, on the size of the analysed population and on the forecast time horizon under study. Quality issues are particularly a matter of great concern in the analyses of regional data, with small populations and relatively rare events such as births, deaths and migrations in specific age groups. This is the basic problem addressed within the literature on small area statistics as well as its application to demographic estimates and projections; see, for example, [17], [18], [28], [30] and [32] as some classic studies reflecting the importance of these issues in the area of demography.

The remainder of the chapter is organised as follows. In Section 10.2, we start with a short overview of the literature on small areas statistics applied to demography. Particular emphasis is placed on estimation where instability arises from the size of the analysed units rather than from poor quality surveys. Section 10.3 presents a spatial random effects Poisson model for fertility outcomes, together with priors, Bayesian implementation and statistics for model comparison. This is followed (Section 10.4) by Bayesian inferences on a spatial clustering model to estimate an adjacency matrix. In Section 10.5, the above methods are applied to the estimation of age group fertility rates for Portuguese NUTS III regions. This model assumes that each particular fertility rate is a random variable with a Poisson distribution, which in turn is defined by the combination of age group fixed effects with spatially structured random effects described by a Conditional Auto-Regressive (CAR) model. Fertility rates are thus estimated using a hierarchical Bayesian methodology and a Markov Chain Monte Carlo (MCMC) computational procedure. The results show that estimates for small regions and low fertility age groups are considerably different from the observed values. Such age groups are important for several types of demographic policies, independently of their contribution to the total number of births (TNB). However, this contribution is expected to have an increasing importance in ageing populations, where the older fertile age groups are relatively large. Therefore, the chapter ends with a comparison between TNB estimates and observed values, followed by concluding comments (Section 10.6).

10.2   Theoretical and Methodological Background

We start with a short review of the literature on small area statistics in demography. This is followed, by way of motivation for the current research, by a discussion of the challenges in applying existing methods to estimation of regional age-specific fertility rates in Portugal.

10.2.1   Smoothing techniques applied to demographic analysis

Small area estimation (SAE) under frequentist approach have been developed for cross-classified counts by borrowing strength from related characteristics to enhance the quality for cases with small sample size or sparse observations. For example, [34] developed a family of log-linear structural modes for the estimation of small area cross-classification, with application to the labour force characteristics. [21] proposed a multinomial logit mixed model with random area effects for modeling the employment data, while [20] adopted independent random effects on the categories of the multinomial responses. [13] derived a multivariate SAE model to borrow strength not only from areas but also from the correlations between multiple response variables. In order to obtain more efficient estimates for sub-national domains, smoothing techniques aimed either at correcting inaccurate data or at finding substitutes for missing values are attracting a growing interest in the fields of demography and planning; see, for example, [1], [5], [14], [26] and [29]. Such techniques estimate demographic variables either by using covariates, or data for individual units assumed to have a similar behaviour. In the first case, missing or inaccurate data is estimated assuming that the variables under analysis vary according to fixed effects related to a set of explanatory variables (percentage of rural population, per capita GDP, etc.). These indirect estimation methods are widely used by demographers, mainly for population predictions for developing countries ([2], [24], [25]). In the second case, smoothing is performed by using any combination of demographic data for other age groups and total population in the same region and year ([35], [27]), for the same age group and region in different years [23], or for areas expected to be similar, either because they are geographical neighbours or because they have similar socio-economic patterns ([3], [23]).

Even with good quality data, smoothing is essential when estimates are unstable and show excessive variation over space or population cohorts. This can occur either because the demographic behaviour is affected by qualitative changes or because the analysed areas or cohorts are too small. A good example of the first case is the rapid change in fertility rates, with heterogenous effects in different age groups and regions, which occurred in Portugal in the final decades of the previous century; this evidence is also in line with observed variation across other developed countries [16]. The second case, which is the main focus of this chapter, corresponds to structural instability, independent of data accuracy and transitional changes, and is particularly important in regional statistics where the number of occurrences is reduced, because both the base population and the per capita frequency of such occurrences are small. This structural instability cannot always be corrected for by the use of non-demographic covariates, particularly when these other covariates are affected by the same sort of problems as well. Therefore, the appropriate alternative is the application of shrinkage techniques where each particular estimate is improved by using information concerning the same variable, either in related observational units [3] or in the overall set of estimates [10].

Borrowing strength from related observational units or overall averages is done by assuming that the variable under analysis has a given probability distribution which produces the individual observations and is described by a set of parameters θ, which in turn can be generated by specific distributions described by hyper-parameters. Shrinkage, the correction of observed values taking into account the distribution from which they are generated, can be implemented through various Bayesian methods, namely: i) Hierarchical Bayesian approaches, applied to demography by [6] and [12], where the model is described at two or more levels, the first typically at the individual unit level highlighting heterogeneity across these individuals, and the second at a broader regional or cohort level explaining the reasons for such heterogeneity; and ii) Empirical Bayesian approaches, applied by [3] and [10], where the complete definition of the priors is substituted by inferences provided by observed data. See, for example, [8] for a full description of both approaches.

When probability distributions are related to a spatial structure, a common case in demographic variables, this spatial structure must be adequately described by a model. Typically, either CAR [4] or the spatial autoregressive model (SAR) ([33]) describes such spatial dependence. Observationally, these two models are very similar [31] even if the interpretation of the models is somewhat different. SAR models were adopted for example by [6] and [12]. In our work we use a CAR dependence structure, which is better suited for the adopted estimation methodology. First, it offers interpretation in terms of conditional distributions of fertility rates across different regions. Second, the CAR spatial model is better suited to Bayesian modeling, and better adapted for interpretation within a Bayesian model. This general overview of the application of small area estimation techniques to demography is the basis for the presentation, in the following subsection, of our methodology for empirical analysis.

10.2.2   Estimation of fertility rates for Portuguese NUTS III regions

Now, we turn to the context of estimating fertility rates for Portuguese regions. Mainland Portugal is divided in 28 NUTS III regions with 2011 census population varying between 2.04 million inhabitants (Greater Lisbon) and slightly more than 40,000 inhabitants (Pinhal Interior Sul). The relevant population (women between 15 and 49 years old) varies from 743,000 in Greater Lisbon to 1,200 in Pinhal Interior Sul. In the smaller regions, the annual numbers of births for some age groups are very small; in an extreme case, the figure for Pinhal Interior Sul, 45-49 cohort, is zero births. This renders the estimates of the corresponding fertility rates unstable, thus justifying adoption of small area estimation techniques. We apply these techniques to estimate the fertility rates for 28 regions (indexed by i), 7 quinquennial age groups (denoted by j: 15-19, 20-24, 25-29, 30-34, 35-39, 40-44, and 45-49 years old) and time t. The simultaneous consideration of temporal and spatial dimensions is crucial for forecasting. This is beyond the scope of the current chapter and retained for future work. In this chapter, we focus on spatial cross-section estimates and comparisons, setting t fixed at the year 2009.

As referred to above, the adopted estimation technique would ideally use a selected set of covariates defining how fertility is affected by factors such as economic and social well-being or cultural attitudes driving the willingness to produce and raise children, factors and attitudes which are expected to explain a substantial part of heterogeneity in child-bearing behaviour. However, Portugal has been subject to a rapid process of demographic transition, which changed drastically the fertility rates of all age groups and its spatial distribution [9]. Births are now concentrated in the age groups between 25 and 35 years old, being very low in the extreme age groups, while the highest fertilities moved from the rural areas and northern regions to the urban regions. This means that the spatial effect of the covariates changed over time, following a process yet to be stabilised, which renders use of cross-section covariates difficult.

In the absence of suitable covariates, we assume that each particular value $\hat{Y}_{ijt}$ (the observed fertility rate) is a singular event determined by a given probability distribution, and subject to a statistical error term with both a structural and an idiosyncratic element. The structural element, in turn, is the outcome of fixed effects (the age group specific fertility rate) and random effects arising from spatially structured geographical heterogeneity. In other words, it is assumed that the error term in each individual observation reflects heterogeneity and has a pattern of autocorrelation with the neighbour regions defined by a CAR model. We assume the Poisson distribution as the chosen model to describe the observed pattern of fertility rate variation. Though the binomial distribution is an alternative option often adopted in demographic studies (see, for example, [6]), it is not recommended in our particular case, since we deal with events with very low frequency.

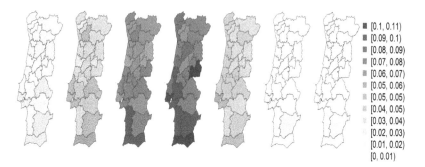

FIGURE 10.1
Observed Fertility Rate by quinquennial age group j ($j = 1, \ldots, 7$, ages 15 to 49 years old).

The importance of the simultaneous consideration of age effects and spatially autocorrelated heterogeneity is highlighted by the highly significant values of the Moran's I measure of spatial autocorrelation [22] index for the regional values of fertility rates in six of the seven age groups (except $j = 7$). The spatial patterns of observed fertility rates across the territory of Portugal for the different age groups (Figure 10.1) also shows the same evidence. At the same time, fertility rates across the different age groups are also considerably different; hence the disregard of age fixed effects would generate serious specification errors.

The parameters defining the Poisson distribution and the respective hyperparameters which define the fixed effects and the CAR model are estimated using the Hierarchical Bayesian methodology and a MCMC posterior sampling algorithm. The MCMC is implemented by running 3 chains each with 6,000 iterations. In addition to the estimation of fertility rates for all the age groups it is useful to estimate the total number of births for each region and to check how much they differ from the observed values. For this we run a Monte Carlo simulation for each region where all the vectors of fertility rates obtained for the iterations are multiplied by the vector of female population for each age group, providing thereby estimates of the total number of births. A full description of the statistical model is presented in the following section.

10.3 Spatial Random-Effects Poisson Model

10.3.1 Model specification

Suppose we have data containing count response Y_{ij} observed for $i = 1, 2, \cdots, N$ sites and $j = 1, 2, \cdots, J$ groups. Let E_{ij} be the expected value, or the population size. Consider the Spatial Random-effect Poisson model (SRP)

$$Y_{ij} \sim \text{Poisson}(E_{ij}\xi_{ij}) \tag{10.1}$$

$$\log(\xi_{ij}) = \eta_{ij} \sim \mathcal{N}(\mu + \alpha_i + \theta_j, \ \delta^2) \tag{10.2}$$

where we treat the group effect $\boldsymbol{\theta}_g = (\theta_1, \theta_2, \cdots, \theta_J)'$ as fixed with $\theta_1 \equiv 0$ since the grand mean μ is included, and assume the site-specific random effect $\boldsymbol{\alpha} = (\alpha_1, \cdots, \alpha_N)'$ admits the CAR structure

$$\boldsymbol{\alpha} \sim \mathcal{N}(0, \ \tau^2 \boldsymbol{D}(\gamma)), \quad \boldsymbol{D}(\gamma) = (I - \gamma \boldsymbol{MW})^{-1} \boldsymbol{M}. \tag{10.3}$$

$\boldsymbol{W}$ is the adjacency matrix with $w_{ij} = 1$ if site i and site j are neighbors and 0 otherwise, with $w_{ii} \equiv 0$ for $1 \leq i, j \leq N$. For each site i, define $w_{i+} \equiv \sum_{j=1}^{N} w_{ij}$ to be the sum that represents the total number of neighbors of site i. $\boldsymbol{M}$ is the diagonal matrix with $m_{ii} = 1/w_{i+}$.

Let $\boldsymbol{\eta} = (\boldsymbol{\eta}_1', \cdots, \boldsymbol{\eta}_J')$ where each $\boldsymbol{\eta}_j$ is the $N \times 1$ response log-rates for j-th group. We employ $\boldsymbol{1}_{m,n}$ to denote a $m \times n$ matrix with all entries 1, and I_m a $m \times m$ identity matrix. Accordingly, letting $\boldsymbol{\theta} = (\mu, \theta_2, \cdots, \theta_J)'$ with the corresponding design matrix $\boldsymbol{X}$, and $\boldsymbol{Z} = \boldsymbol{1}_{J,1} \otimes I_N$, we can write the model as

$$\boldsymbol{\eta} \sim \mathcal{N}\left(\boldsymbol{X\theta} + \boldsymbol{Z\alpha}, \ \delta^2 I_{NJ}\right) \tag{10.4}$$

To perform Bayesian inference for the parameters of interests, we need to specify their prior distributions. In practice, when the prior knowledge on the parameters is not available, we choose the following non-informative priors to represent the lack of the prior information, and allow the posterior estimates to be more data driven. More specifically, we choose the uniform prior distributions for θ_j's, γ, $\log \delta^2$ and $\log \tau^2$ over their respective supports, which are equivalent to the forms

$$\pi_0(\theta_j) \propto 1, \quad \pi_2(\gamma) = \text{Uniform}(\lambda_N^{-1}, \lambda_1^{-1}), \quad \pi_1(\delta^2) \propto \delta^{-2}, \quad \pi_1(\tau^2) \propto \tau^{-2} \tag{10.5}$$

where $\lambda_1 \geq \lambda_2 \geq \cdots \geq \lambda_N$ are ordered eigenvalues of $\boldsymbol{MW}$. The condition $\gamma \in (\lambda_N^{-1}, 1)$ is required to ensure positive definiteness of $\boldsymbol{D}(\gamma)$. The upper limit of the interval is one since the row sum of $\boldsymbol{MW}$ is one and $\lambda_N^{-1} < 0$ since the trace of $\boldsymbol{MW}$ is zero.

10.3.2 Implementation

We obtain the posterior samples of the parameters $\{\boldsymbol{\theta}, \boldsymbol{\alpha}, \delta^2, \tau^2, \gamma\}$ via Gibbs Sampler. Under the choice of priors in (10.5), the full conditional distributions given the data and the remaining parameters for fixed-effects and random-effects are, respectively

$$\pi(\boldsymbol{\theta}|\boldsymbol{\eta}, \boldsymbol{\alpha}, \delta^2) = \mathcal{N}(\mu_{\boldsymbol{\theta}}, \Sigma_{\boldsymbol{\theta}}) \begin{cases} \Sigma_{\boldsymbol{\theta}} = \delta^2 \left(\boldsymbol{X}^T \boldsymbol{X}\right)^{-1} \\ \mu_{\boldsymbol{\theta}} = \left(\boldsymbol{X}^T \boldsymbol{X}\right)^{-1} \boldsymbol{X}^T \left(\boldsymbol{\eta} - \boldsymbol{Z\alpha}\right) \end{cases} \tag{10.6}$$

$$\pi(\boldsymbol{\alpha}|\boldsymbol{\eta},\boldsymbol{\theta},\delta^2,\tau^2,\gamma) = \mathcal{N}(\mu_{\boldsymbol{\alpha}},\Sigma_{\boldsymbol{\alpha}}) \begin{cases} \Sigma_{\boldsymbol{\alpha}} = \left(\delta^{-2}JI_N + \tau^{-2}\boldsymbol{D}(\gamma)^{-1}\right)^{-1} \\ \mu_{\boldsymbol{\alpha}} = \delta^{-2}\Sigma_{\boldsymbol{\alpha}}\boldsymbol{Z}^T\left(\boldsymbol{\eta}-\boldsymbol{X}\boldsymbol{\theta}\right) \end{cases} \quad (10.7)$$

The conditional distributions of variance components and spatial dependence are

$$\begin{aligned} \pi(\delta^2 \mid \boldsymbol{\eta},\boldsymbol{\theta},\boldsymbol{\alpha}) &= \texttt{igamma}\left(NJ/2, \ \boldsymbol{\epsilon}^T\boldsymbol{\epsilon}/2\right), \text{with } \boldsymbol{\epsilon} = \boldsymbol{\eta} - \boldsymbol{X}\boldsymbol{\theta} - \boldsymbol{Z}\boldsymbol{\alpha} \quad &(10.8) \\ \pi(\tau^2 \mid \boldsymbol{\alpha},\gamma) &= \texttt{igamma}\left(N/2, \ \boldsymbol{\alpha}^T\boldsymbol{D}(\gamma)^{-1}\boldsymbol{\alpha}/2\right) &(10.9) \\ \pi(\gamma \mid \boldsymbol{\alpha},\tau^2) &\propto |\boldsymbol{D}(\gamma)|^{-1/2}\exp\{\gamma\boldsymbol{\alpha}^T\boldsymbol{W}\boldsymbol{\alpha}/(2\tau^2)\}\cdot\boldsymbol{I}(\gamma\in(\lambda_N^{-1},\lambda_1^{-1})) \end{aligned}$$

$$(10.10)$$

where $\texttt{igamma}(a,b)$ denotes the Inverse Gamma density with shape a and scale b, i.e., $p(x) \propto x^{-a-1}\exp(-b/x)$. Note that the full conditional distribution of γ is not known but can be sampled by numerically evaluating the density over its support.

Finally, for each region $i = 1, 2, \cdots, N$ and age group $j = 1, 2, \cdots, J$, we sample the response log-rates η_{ij} from its full conditional distribution, which has the logarithm of the density in the form

$$\begin{aligned} \log\pi(\eta_{ij} \mid Y_{ij},\mu,\alpha_i,\theta_j,\delta^2) &= K - \eta_{ij}^2/(2\delta^2) + \eta_{ij}(Y_{ij}+(\mu+\alpha_i+\theta_j)/\delta^2) \\ &\quad - E_{ij}\exp\{\eta_{ij}\} \end{aligned}$$

$$(10.11)$$

where K does not involve η_{ij}.

10.3.3 Model comparisons

To compare models and validate the importance of the spatial models, we use Deviance Information Criterion (DIC) for mixed-effects model, DIC_4 [11] based on the complete likelihood

$$\begin{aligned} \text{DIC}_4 &= -4\mathbb{E}_{\boldsymbol{\theta},\boldsymbol{\alpha}}[\log f(\boldsymbol{Y},\boldsymbol{\alpha}|\boldsymbol{\theta})|\boldsymbol{Y}] + 2\mathbb{E}_{\boldsymbol{\alpha}}[\log f(\boldsymbol{Y},\boldsymbol{\alpha}|\mathbb{E}_{\boldsymbol{\theta}}[\boldsymbol{\theta}|\boldsymbol{Y},\boldsymbol{\alpha}])|\boldsymbol{Y}] \\ &\triangleq -4\mathbb{E}_1 + 2\mathbb{E}_2, \end{aligned}$$

and, to evaluate the second conditional expectation $\mathbb{E}_2$, we need the information of $\mathbb{E}_{\boldsymbol{\theta}}[\boldsymbol{\theta}|\boldsymbol{Y},\boldsymbol{\alpha}]$ which can be evaluated by sampling $\boldsymbol{\theta}$ for each posterior sample of $\boldsymbol{\alpha}$ and obtain the mean. We also report $\overline{D(\boldsymbol{\theta})} = -2\mathbb{E}_1$ as the posterior expected value of the joint deviance, and $p_{D4} = \overline{D(\boldsymbol{\theta})} + 2\mathbb{E}_2$ as a measure of model dimensionality. Generally, a smaller DIC_4 indicates a better model.

10.4 Spatial Clustering Model for Estimating the Adjacency Matrix

The adjacency matrix $\boldsymbol{W}$ can play an important role in measuring the spatial dependence for improving the fit of the model in the previous section. In addition to the observed adjacency of regions on a map, it can be also estimated from the variables of interests in the data. The method below is based on a spatial clustering model that we use to estimate our adjacency matrix .

10.4.1   Model specification

The spatial dependence γ is modeled to improve the predictive power, and can rely on the definition of the adjacency matrix $\boldsymbol{W}$. For small area data where sampling unit sites are represented as polygons, a natural choice is to define $w_{ij} = 1$ if site i and site j share overlapping boundary, and 0 otherwise. However it is possible to estimate $\boldsymbol{W}$ from the data, using spatial clustering techniques, and this can potentially further improve the prediction. Specifically, consider a spatial clustering configuration $\varpi = (d, G_d)$, where d is the number of clusters, $G_d = (g_1, \cdots, g_d)$ are the cluster centers, and the memberships of the N sites are determined according to minimal distance criterion based on a distance measure $D(i,j)$ between each pair of sites. We choose $D(i,j)$ to be the smallest number of boundaries crosses in going from site i to site j.

Suppose under the partition $\cup_{r=1}^d C_r$ of the set $\{1, 2, ..., N\}$, introduced by the clustering configuration ϖ, each cluster C_r has n_r member sites and $\sum_{r=1}^d n_r = N$. We consider the Spatial Clustering Poisson Model (SCPM) for each $i \in C_r$, $r = 1, \cdots, d$, that

$$Y_{ij} \sim \text{Poisson}(E_{ij}\xi_{ij}) \tag{10.12}$$
$$\log(\xi_{ij}) = \eta_{ij} \sim \mathcal{N}(\mu_r + \theta_j, \ \delta^2) \tag{10.13}$$

with $\theta_1 \equiv 0$. Comparing with the SRP in (10.1), since the adjacency matrix is considered as unobserved and the spatial dependence mechanism is unknown, we cluster the $\mu_i = \mu + \alpha_i$ into μ_r that is shared by all members in cluster C_r. Letting $\boldsymbol{\mu} = (\mu_1 \cdots, \mu_d)$ and $\boldsymbol{\theta} = (\theta_2, \cdots, \theta_J)$, the parameters associated with SCPM are $\{\varpi, \boldsymbol{\mu}, \boldsymbol{\theta}, \boldsymbol{\eta}, \delta^2\}$.

Next, we consider prior elicitation. For the spatial clustering configuration ϖ, a prior model considered by [19] is $\pi(\varpi) = \pi(G_d|d)\pi(d|\kappa)\pi(\kappa)$, with $\pi(d|\kappa) \propto (1 - \kappa)^{d-1}$ for $d = 1, \cdots, N_0 < N$ and $\pi(\kappa) \sim \text{Uniform } [0,1]$. Here, the hyperparameter κ controls the penalty of model complexity for large d. The prior for d receives power decay from this elicitation and when κ is larger, then d is more likely to be smaller, while when it is fixed at a very small value, d is almost uniformly distributed over $\{1, 2, \cdots, N_0\}$ to indicate weak prior information. However, when the clustering parameters have high dimensionality, a power decay can be insufficient.

Instead, we consider a exponential decay: $\pi(d|\kappa) \propto \exp(-\kappa d)$ and when κ is fixed to be (*# of parameters under* ϖ) $\log(NJ)/2$, it is closely related to the Bayesian Information Criterion (BIC). To allow the posterior of d to be more data driven, we fix κ at 0 such that the prior for d is non-informative and all possible d's receive equal probability. Conditional on d, we again consider the non-informative prior for centers $\pi(G_d|d) = (N-d)!/N!$ where the $\binom{N}{d}$ $d!$ possible G_d's receive equal probability of being the cluster centers. We then elicit non-informative prior for the remaining parameters for $r = 1, \cdots, d$ and $j = 2, \cdots, J$ as:

$$\pi_0(\mu_r|\varpi) \propto 1, \quad \pi_1(\theta_j) \propto 1, \quad \pi_2(\delta^2) \propto \delta^{-2}.$$

10.4.2   Implementation

To obtain posterior samples for $\{\varpi, \boldsymbol{\mu}, \boldsymbol{\theta}, \boldsymbol{\eta}, \delta^2\}$, the main challenge is that the model dimensionality changes when the clustering configuration ϖ is updated. We therefore construct the reversible jump MCMC ([15]) updating procedure for $(\varpi, \boldsymbol{\mu}, \delta^2)$ given $(\boldsymbol{\theta}, \boldsymbol{\eta})$, and then update $(\boldsymbol{\theta}, \boldsymbol{\eta})$ from the full conditional distribution.

The steps of the algorithm are described as follows.

(i) Update $(\varpi, \boldsymbol{\mu}, \delta^2)$: Let the $n_r J$-vector $\Delta_r = (\Delta_{ij})_{i \in C_r, 1 \leq j \leq J}$ with $\Delta_{ij} = \eta_{ij} - \theta_j$, and the NJ-vector $\boldsymbol{\Delta} = (\Delta_r)_{1 \leq r \leq d}$. Also let $\bar{\Delta}_r$ be the first moment of Δ_r. From the current state $(\varpi, \boldsymbol{\mu}, \delta^2)$, we propose a new state $(\varpi^*, \boldsymbol{\mu}^*, \delta^{2*})$ by first proposing a new clustering configuration ϖ^* from a certain proposal function $g(\cdot|\varpi)$, and then proposing $(\boldsymbol{\mu}^*, \delta^{2*})$ given ϖ^*. More specifically, we define the auxiliary variable $U = \boldsymbol{\psi}^* = (\boldsymbol{\mu}^*, \delta^{2*})$ and let $\boldsymbol{\psi} = (\boldsymbol{\mu}, \delta^2) = U^*$, the corresponding invertible map $\boldsymbol{q} : (\boldsymbol{\psi}, U)$ to $(\boldsymbol{\psi}^*, U^*)$ is one-to-one with Jacobian 1. We propose $U \sim h(\cdot|\varpi, \boldsymbol{\mu}, \varpi^*)$ from a certain proposal density function h. The probability of accepting the proposed state is

$$\min\left\{1, \frac{g(\varpi|\varpi^*)}{g(\varpi^*|\varpi)} \times \frac{\pi(\varpi^*, \boldsymbol{\psi}^*|\boldsymbol{\Delta})}{\pi(\varpi, \boldsymbol{\psi}|\boldsymbol{\Delta})} \times \frac{h(u^*|\varpi^*, \boldsymbol{\psi}^*, \varpi)}{h(u|\varpi, \boldsymbol{\psi}, \varpi^*)} \times 1\right\} \quad (10.14)$$

We specify h to be the full conditional density of $\boldsymbol{\psi}^*$ as the proposal, i.e., $h(u|\varpi, \boldsymbol{\mu}, \varpi^*) = \pi(\boldsymbol{\mu}^*, \delta^{2*}|\varpi^*, \boldsymbol{\Delta}) = \pi(\delta^{2*}|\varpi^*, \boldsymbol{\Delta}) \times \pi(\boldsymbol{\mu}^*|\varpi^*, \delta^{2*}, \boldsymbol{\Delta})$. Specifically, we first generate δ^{2*} from

$$\pi(\delta^{2*}|\varpi^*, \boldsymbol{\Delta}) \propto \pi_2(\delta^{2*}) \int \pi(\boldsymbol{\Delta}|\boldsymbol{\mu}, \delta^{2*}) \pi_0(\boldsymbol{\mu}) \, d\boldsymbol{\mu} \quad (10.15)$$

which is an inverse-Gamma density with shape $(N-d)J/2$ and scale $\sum_{r=1}^{d} \sum_{i \in C_r, j} (\Delta_{ij} - \bar{\Delta}_r)^2/2$. Next, we generate $\boldsymbol{\mu}^*$ from the full conditional distribution given δ^{2*}

$$\pi(\boldsymbol{\mu}_r^*|\varpi^*, \delta^{2*}, \Delta_r) \sim \mathcal{N}\left(\bar{\Delta}_r, (n_r J)^{-1}\delta^{2*}\right), \quad r = 1, 2, \cdots, d \quad (10.16)$$

Under this choice of proposal density, letting $m(\boldsymbol{\Delta})$ be the normalizing constant which is finite by the posterior propriety that is evident from the following context, we can substitute

$$\pi(\varpi, \boldsymbol{\psi}|\boldsymbol{\Delta}) = \pi(\boldsymbol{\Delta}|\varpi, \boldsymbol{\mu}, \delta^2) \pi(\boldsymbol{\mu}, \delta^2|\varpi, \boldsymbol{\Delta}) \pi(\varpi)/m(\boldsymbol{\Delta})$$

into the Metropolis-Hasting ratio in (10.14) which becomes

$$\frac{g(\varpi|\varpi^*)}{g(\varpi^*|\varpi)} \times \frac{\pi(\boldsymbol{\Delta}|\varpi^*, \boldsymbol{\mu}^*, \delta^{2*}) \, \pi(\boldsymbol{\mu}^*, \delta^{2*}|\varpi^*) \, \pi(\boldsymbol{\mu}, \delta^2|\varpi, \boldsymbol{\Delta})}{\pi(\boldsymbol{\Delta}|\varpi, \boldsymbol{\mu}, \delta^2) \, \pi(\boldsymbol{\mu}, \delta^2|\varpi) \, \pi(\boldsymbol{\mu}^*, \delta^{2*}|\varpi^*, \boldsymbol{\Delta})} \times \frac{\pi(\varpi^*)}{\pi(\varpi)} \quad (10.17)$$

Using the fact that

$$\pi(\boldsymbol{\Delta}|\varpi) = \frac{\pi(\boldsymbol{\Delta}|\varpi, \boldsymbol{\mu}, \delta^2) \, \pi(\boldsymbol{\mu}, \delta^2|\varpi)}{\pi(\boldsymbol{\mu}, \delta^2|\varpi, \boldsymbol{\Delta})},$$

the ratio (10.17) reduces to

$$\frac{g(\varpi|\varpi^*)}{g(\varpi^*|\varpi)} \times \frac{\pi(\boldsymbol{\Delta}|\varpi^*)}{\pi(\boldsymbol{\Delta}|\varpi)} \times \frac{\pi(\varpi^*)}{\pi(\varpi)} \quad (10.18)$$

When the prior and proposal density of ϖ are non-informative, the acceptance rate is dominated by the marginal likelihood ratio $r = \pi(\boldsymbol{\Delta}|\varpi^*)/\pi(\boldsymbol{\Delta}|\varpi)$ with

$$\pi(\boldsymbol{\Delta}|\varpi) = \int \left(\prod_{r=1}^{d} \int \pi(\boldsymbol{\Delta}_r|\delta^2, \mu_r) \, \pi_0(\mu_r|\varpi) \, \pi_2(\delta^2) \, d\mu_r \right) d\delta^2 \quad (10.19)$$

Letting $n^* = (N - d)J/2$, we can simplify it into

$$\log(\pi(\boldsymbol{\Delta}|\varpi)) = -n^* \log(\pi) - \frac{1}{2} \sum_{r=1}^{d} \log(n_r J) + \log \Gamma(n^*)$$

$$-n^* \log \sum_{r=1}^{d} \sum_{i \in C_r, j} (\Delta_{ij} - \bar{\Delta}_r)^2 \quad (10.20)$$

The posterior propriety for $\boldsymbol{\Delta}$ immediately follows that $\pi(\boldsymbol{\Delta}|\varpi) < \infty$ for $d < N$ and the collection of all ϖ's is finite.

The efficiency of the algorithm can heavily rely upon reasonable choices of $g(\varpi^*|\varpi)$ such that the proposed ϖ^* behaves similarly, or falls in the neighborhood of the current ϖ in the space of all possible clustering configurations. The difficulty in specifying g also lies in the varying dimensions (in terms of the number of clusters d) that are involved in exploring the clustering configurations that resemble ϖ, and one typically construct different move types with different prescribed probabilities for specifying g. We choose similar proposal function g as described in [19]. Specifically, at each iteration, we consider proposing a new ϖ^* using one of the 3 steps according to the different move types of d, i.e., $d \to d + 1$ for Growth step, $d \to d - 1$ for Merge step, and $d \to d$ for Shift step, with the prescribed proposal probabilities P(Growth) = P(Merge) = 0.4 and P(Shift) = 0.2 that are similar to the choice in [19]. The motivation underlying the choice of the prescribed probabilities, which gives higher chance of proposing a new clustering configuration with a different number of clusters, d, is to fully explore the uncertainty associated with d. Next, to keep this chapter self-contained, we give the details of the three steps in g, along with the respective acceptance probability under our choice of the prior densities and h, as follows:

1. Growth step $(d, G_d) \rightarrow (d+1, G^*_{d+1})$: We create a new cluster. We first draw a random variable uniformly distributed on the $N-d$ non-center sites, to determine the new cluster C^* with center g^*. Secondly, we draw another random variable r uniformly distributed on $\{1, \cdots, d+1\}$ to determine the position of g^* in G^*_{d+1}. The n^*_r sites that have minimal distance from g^* automatically enter C^*. In this case,

$$\frac{g(\varpi|\varpi^*)}{g(\varpi^*|\varpi)} = \frac{\text{P(Merge)}(N-d)}{\text{P(Growth)}}, \qquad \frac{\pi(\varpi^*)}{\pi(\varpi)} = \frac{\exp(-(d+1)\kappa)(N-d-1)!}{\exp(-d\kappa)(N-d)!}$$

and the acceptance probability is $\min\{1, \exp(-\kappa) \times \dfrac{\text{P(Merge)}}{\text{P(Growth)}} \times r\}$.

2. Merge step $(d+1, G_{d+1}) \rightarrow (d, G^*_d)$: We delete one existing cluster and merge its members into other existing clusters. First, generate a random variable uniformly r distributed on $\{1, \cdots, d+1\}$, which determines the cluster C_r with center g_r to be removed and all its members merging into one of the remaining clusters by the minimal distance criterion. The acceptance probability is the reciprocal of that in Growth step.

3. Shift Step: $(d, G_d) \rightarrow (d, G^*_d)$: We adopt a shift step for moving one cluster center to its non-center neighborhood when d is invariant. For each site s, we define all sites that are directly connected (i.e., not via third site) to it by latitude or longitude as its neighbors. Among d current cluster centers there are $n(G_d)$ cluster centers that have at least one non-center neighbors. Draw $r \sim \text{Uniform}\{1, \cdots, n(G_d)\}$ to obtain one such cluster center g_r with $m(g_r)$ non-center neighbors. Secondly, draw l from $\{1, \cdots, m(g_r)\}$ uniformly. The l-th non-center neighbor becomes the new cluster center g^*_r that replaces g_r in G_d. The acceptance probability is $\min\{1, \dfrac{n(G_d)m(g_r)}{n(G^*_d)m(g^*_r)} \times r\}$

(ii) Update (θ, η): The full conditional distribution of θ_j is

$$\mathcal{N}\left(\sum_{r=1}^{d} \sum_{i \in C_r} (\eta_{ij} - \mu_r)/N, \ \delta^2/N\right)$$

for each $j = 2, \cdots, J$. Finally, to update η_{ij} for $i \in C_r$ and $j = 1, 2, \cdots, J$, we have the log-density of the full conditional distribution of η_{ij}

$$\log \pi(\eta_{ij} \mid Y_{ij}, \mu_r, \theta_j, \delta^2) = K - \eta^2_{ij}/(2\delta^2) + \eta_{ij}(Y_{ij} + (\mu_r + \theta_j)/\delta^2) - E_{ij} \exp\{\eta_{ij}\}$$

where K does not involve η_{ij}.

10.4.3   Posterior inference

To estimate the adjacency matrix $\boldsymbol{W}$, once we obtain posterior samples with size B, consider $w_b(i,j) = 1$ if site i and j share the same clustering membership in the b-th sample, and 0 otherwise. Let $w_{ij} = \sum_{b=1}^{B} w_b(i,j)/B \in [0,1]$ and $w_{ii} \equiv 0$ be the estimated adjacency between site i and j. We can then fit SRP with the estimated $\hat{\boldsymbol{W}} = (w_{ij})_{1 \leq i,j \leq N}$. To further obtain a central clustering configuration ϖ^* from the posterior samples $\varpi^1, \cdots, \varpi^B$, consider the dissimilarity measure $Diss(i,j) = 1 - w_{ij}$ based on which an agglomerative clustering algorithm is performed with number of clusters as the posterior mode of d.

10.5   Application and Posterior Inference

To apply the aforementioned methods to the fertility data in Portugal, we treat the age group $j = 1$ as the baseline group and compare the remaining 6 groups, $j = 2, \cdots, J$ to its mean level. SRP with $\gamma = 0$ (non-spatial) and γ updated (spatial) are separately implemented by running 3 Monte Carlo Markov Chains , each with $6,000$ iterations. The convergence is well committed after $5,000$ iterations with the Potential Scale Reduction Factor $\sqrt{R} < 1.2$ for all parameters ([7]). The final $1,000$ samples for each chain are then used as the posterior samples.

Next, we also fit SCPM using 4 MCMC runs with a total of $55,000$ iterations, and sample every 5-th iteration of the last $5,000$ samples after convergence, and obtain a total of $4,000$ posterior samples for analysis. The convergence is validated using the criterion $\sqrt{R} < 1.2$ by monitoring $(\delta^2, \boldsymbol{\theta}, \boldsymbol{\eta})$, the Gaussian likelihood $f(\boldsymbol{\eta}|\boldsymbol{\mu}, \boldsymbol{\theta}, \delta^2)$ and the marginal likelihood $\pi(\boldsymbol{\Delta}|\varpi)$ in (10.20). The posterior distribution of d turns out to be highly concentrated on $d = 2$ with probability 0.9985, and the central clustering configuration based on $Diss(i,j)$ is shown in Figure 10.2. It indicates that after taking the fixed effects on age groups $j = 2, \cdots, J$ into account, the mean level of the response log-rates highly suggests two clusters that can be interpreted as the north and south regions. This is consistent with the observed fertility rates in Figure 10.1 which shows the south regions have higher fertility rates compared to the north regions, particularly for the reference age group $j = 1$. Consequently, the estimated adjacency matrix based on the spatial connectivity under the clustering configuration ϖ at each iteration of the posterior samples for SCPM, is almost partitioned into two blocks, with members in each block share a high value of adjacency (close to 1). We then re-fit SRP using the estimated $\hat{\boldsymbol{W}}$ as the third SRP implementation for comparisons.

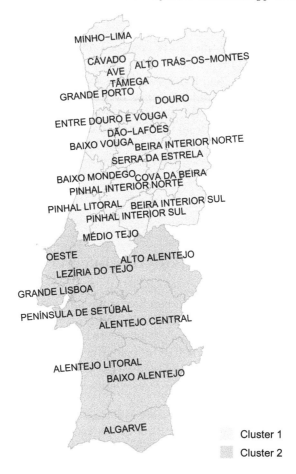

FIGURE 10.2
The central clustering configuration under SCP fit from 4,000 posterior samples.

The parameter estimates and model assessment for SRP with non-spatial case, spatial case with the natural adjacency matrix W, and spatial case with estimated adjacency matrix $\hat{W}$ from SCPM, are summarized in Table 10.1.

From the output, it can be observed that the spatial dependence parameter γ is highly significant with credible interval away from 0. The spatial case of SRP with estimated $\hat{W}$ from SCPM turns out to have the smallest DIC_4, and hence is preferable. Consequently, the histograms of the 3,000 posterior samples of parameters from the most favorable model are shown in Figure 10.3. We also plot the estimated mean fertility rate $\hat{\xi}_{ij}$'s by groups and regions in Figure 10.4, and the comparison with observed $\xi_{ij} = Y_{ij}/E_{ij}$ in Figure 10.5.

TABLE 10.1
Posterior inference for SRP, with posterior mean, $(2.5\%, 97.5\%)$th quantile from the total 3000 posterior samples.

Parameter	Non-spatial	Spatial with W	Spatial with $\hat{W}$
intercept : μ	$-4.413\,(-4.496, -4.333)$	$-4.390\,(-4.548, -4.244)$	$-4.397\,(-4.565, -4.242)$
$j = 2 : \theta_2$	$1.049\,(0.965, 1.134)$	$1.048\,(0.963, 1.134)$	$1.048\,(0.962, 1.134)$
$j = 3 : \theta_3$	$1.719\,(1.634, 1.804)$	$1.720\,(1.637, 1.803)$	$1.719\,(1.636, 1.803)$
$j = 4 : \theta_4$	$1.888\,(1.807, 1.970)$	$1.887\,(1.806, 1.974)$	$1.886\,(1.805, 1.974)$
$j = 5 : \theta_5$	$1.067\,(0.983, 1.153)$	$1.067\,(0.984, 1.154)$	$1.067\,(0.984, 1.154)$
$j = 6 : \theta_6$	$-0.644\,(-0.735, -0.549)$	$-0.644\,(-0.743, -0.547)$	$-0.644\,(-0.743, -0.546)$
$j = 7 : \theta_7$	$-3.479\,(-3.654, -3.289)$	$-3.477\,(-3.674, -3.275)$	$-3.476\,(-3.673, -3.274)$
δ^2	$0.017\,(0.012, 0.022)$	$0.016\,(0.012, 0.022)$	$0.016\,(0.012, 0.022)$
τ^2	$0.065\,(0.032, 0.122)$	$0.036\,(0.016, 0.074)$	$0.101\,(0.043, 0.213)$
γ	0	$0.864\,(0.453, 0.993)$	$0.924\,(0.689, 0.996)$
$\overline{D(\theta)}$	1201.505	1186.047	1176.368
p_{D4}	8.994	9.175	9.103
DIC_4	1210.499	1195.222	1185.471

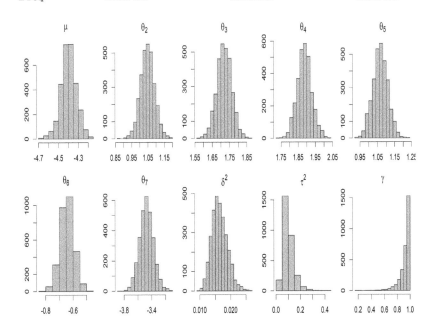

FIGURE 10.3
Histogram of the posterior samples for SRP with estimated $\hat{W}$ from SCPM.

10.6 Final Results and Conclusions

Figure 10.4 and 10.5 show that fertility in Portugal is heavily concentrated between 25 and 35 years old age groups, the corresponding fertility rates being

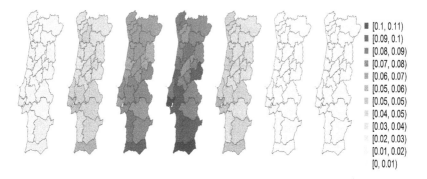

FIGURE 10.4
Estimated Fertility Rate by quinquennial age group j ($j = 1, \ldots, 7$, ages 15 to 49 years old).

higher and spatially homogeneous. This is a typical situation of societies where mothers plan to have one or two children when they simultaneously have more stable economic conditions and are still young. Conversely, the extreme groups have a marginal contribution to population reproduction, showing low and spatially heterogeneous rates with visible differences between observed and estimated values; in more extreme cases such as Baixo Alentejo ($j = 1$), Pinhal Interior Sul ($j = 6$) and several regions ($j = 7$), observed values are outside 95% confidence intervals, typifying erratic observations arising from unstable distributions. The coverage rates of the confidence intervals in Figure 10.5 are, respectively, 92.9%, 100%, 100%, 100%, 100%, 89.3% and 50% for the age group $j = 1, 2, \cdots, 7$, indicating a satisfactory performance for major age groups with high fertility rates, and the overall coverage rate is 90.3% which is generally consistent with the false discovery rate associated with testing at 5% level for multiple tests that are potentially dependent. Comparing the maps of observed and estimated values we can conclude that the model produced a visible shrinkage of fertility heterogeneity, with changes occurring even in the central age groups. The weaker shrinkage in central groups is due both to higher fertility levels and to lower regional heterogeneity.

Another remarkable feature of Portuguese fertility landscape is the incidence of higher rates both in rural and poorer northern regions (Tâmega) and the more urbanized and richer areas around Lisbon (Península de Setubal, Grande Lisboa). This means that the demographic transition, moving higher fertility from north to south and from rural to urban places, is still occurring. As we argued above, this hinders the use of covariates to effectively explain spatial heterogeneity in fertility and justifies our methodological approach.

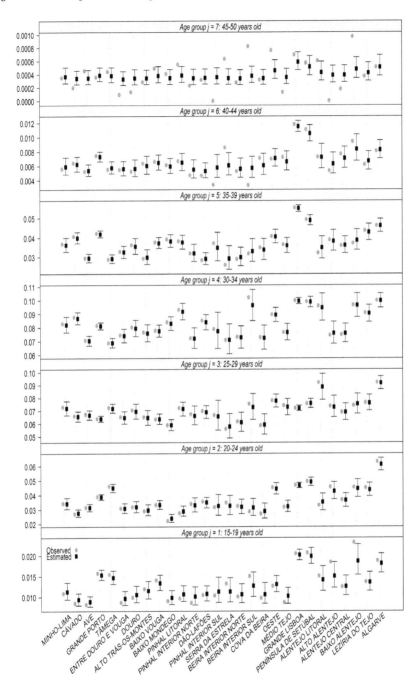

FIGURE 10.5
Observed vs. Estimated Fertility Rate by quinquenial age group.

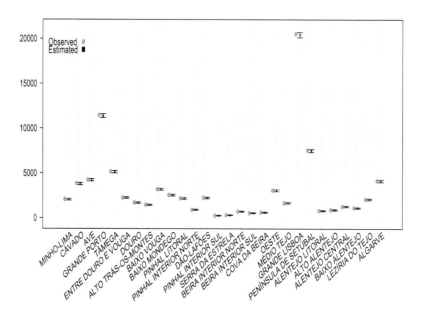

FIGURE 10.6
Observed vs. Estimated total number of births.

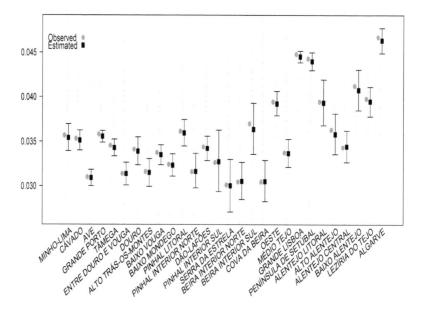

FIGURE 10.7
Observed vs. Estimated total fertility rates.

Finally, it is interesting to see how variation in age specific fertility groups generates variation in the total number of births and total fertility rates. The observed and estimated total number of births and fertility rates are shown in Figure 10.6 and Figure 10.7, respectively. The close correspondence between observed and estimated total number of births across all regions highlights the fit achieved by our modeling approach. Likewise the mean estimated fertility rates in each region is also in close alignment with the corresponding observed rates, even if the prediction intervals indicate substantial uncertainty and spatial variation.

In summary, our proposed model and methodology are effective in the estimation of age-specific regional fertility rates, using small area estimation methods to smooth out large instability and capture the spatial structure in an appropriate way. The estimates obtained thereby are in good alignment with observed values, at the same time as they are robust to sampling fluctuations due to small sample sizes in some regions and age groups.

Acknowledgment

The authors acknowledge financial support provided by the research project DEMOSPIN (PTDC/CS-DEM/100530/2008), from the Portuguese Foundation for Science and Technology, the Operational Programme 'Thematic Factors of Competitiveness' of the EU Community Support Framework, and from the European Regional Development Fund. The usual disclaimer applies.

Bibliography

[1] L. Alkema, A. Raftery, P. Gerland, S. Clark, and F. Pelletier. Estimating the total fertility rate from multiple imperfect data sources and assessing its uncertaninty. Technical Report 89, Center for Statistics and the Social Sciences, University of Washington, 2008.

[2] L. Alkema, A. Raftery, P. Gerland, S. Clark, F. Pelletier, T. Buettner, and G. Heilig. Probabilistic projections of the total fertility rate for all countries. *Demography*, 48:815–839, 2011.

[3] R. Assunção, C. Schmertmann, J. Potter, and S. Cavenaghi. Empirical Bayes estimation of demographic schedules for small areas. *Demography*, 42(3):537–558, 2005.

[4] J. E. Besag. Spatial interaction and the statistical analysis of lattice systems (with discussion). *Journal of the Royal Statistical Society: Series B (Statistical Methodology)*, 36(2):192–236, 1974.

[5] J. Bongaarts and R. Bulatao. Completing the demographic transition. *Population and Development Review*, 25(3):515–529, 1999.

[6] R. Borgoni and F. Billari. Bayesian spatial analysis of demographic survey data. *Demographic Research*, 8(3):61–92, 2003.

[7] S. P. Brooks and A. Gelman. General methods for monitoring convergence of iterative simulations. *Journal of Computational and Graphical Statistics*, 7(4):434–455, 1998.

[8] B. P. Carlin and T. A. Louis. *Bayes and Empirical Bayes Methods for Data Analysis*. Chapman and Hall/CRC, Boca Raton, FL, 2nd edition, 2000.

[9] E. A. Castro, M. C. S. Gomes, C. J. Silva, and J. M. Martins. An inter-regional migration model applied to Portugese data. In *European Population Conference 2012*, Stockholm, http://epc2012.princeton.edu/papers/120824, Sweden, 2012.

[10] S. M. Cavenaghi, J. E. Potter, C. P. Schmertmann, and R. Assunção. Estimating total fertility rates for small areas in Brazil. In *Population Association of America 2004 Annual Meeting*, http://paa2004.princeton.edu/papers/42194, Boston, MA, 2004.

[11] G. Celeux, F. Forbes, C. P. Robert, and D. M. Titterington. Deviance information criteria for missing data models. *Bayesian Analysis*, 1(4):651–673, 2006.

[12] F. Divino, V. Egidi, and M. A. Salvatore. Geographical mortality patterns in Italy: A Bayesian analysis. *Demographic Research*, 20(18):435–466, 2009.

[13] M. R. Ferrante and C. Trivisano. Small area estimation of the number of firms' recruits by using multivariate models for count data. *Survey Methodology*, 36(2):171–180, 2010.

[14] P. Festy and F. Prioux. An evaluation of the fertility and family survey project. Technical Report, http://www.unece.org/leadmin/DAM/pau/docs/s/FFS 2000 Prog EvalReprt.pdf, United Nations, New York and Geneva, 2002.

[15] P. J. Green. Reversible jump Markov chain Monte Carlo computation and Bayesian model determination. *Biometrika*, 82(4):711–732, 1995.

[16] T. W. Guinnane. The historical fertility transition: A guide for economists. *Journal of Economic Literature*, 49(3):589–614, 2011.

[17] A. M. Isserman. The accuracy of population projections for subcounty areas. *Journal of the American Institute of Planners*, 43:247–259, 1977.

[18] N. Keyfitz. The limits of population forecasting. *Population and Development Review*, 7:579–593, 1981.

[19] L. Knorr-Held and G. Raßer. Bayesian detection of clusters and discontinuities in disease maps. *Biometrics*, 56(1):13–21, 2000.

[20] E. López-Vizcaíno, M. J. Lombardía, and D. Morales. Multinomial-based small area estimation of labour force indicators. *Statistical Modelling*, 13(2):153–178, 2013.

[21] I. Molina, A. Saei, and M. J. Lombardía. Small area estimates of labour force participation under a multinomial logit mixed model. *Journal of the Royal Statistical Society: Series A (Statistics in Society)*, 170(4):975–1000, 2007.

[22] P. A. P. Moran. Notes on continuous stochastic phenomena. *Biometrika*, 37(1):17–23, 1950.

[23] J. Potter, C. Schmertmann, R. Assunção, and S.Cavenaghi. Mapping the timing, pace and scale of the fertility transition in Brazil. *Population and Development Review*, 36(2):283–307, 2010.

[24] J. Potter, C. Schmertmann, and S. Cavenaghi. Fertility and development: Evidence from Brazil. *Demography*, 39(4):739–761, 2002.

[25] A. E. Raftery. Bayesian model selection in social research. *Sociological Methodology*, 25:111–163, 1995.

[26] A. E. Raftery, L. Alkema, P. Gerland, S. Clark, F. Pelletier, T. Buettner, G. Heilig, N. Li, and H. Ševčíková. White paper: Probabilistic projections of the total fertility rate for all countries for the 2010 world population prospects. Technical Report, http://www.un.org/esa/population/meetings/EGMFertility2009/P16 Raftery.pdf, United Nations, 2009.

[27] R. Retherford, N. Ogawa, R. Matsukura, and H. Eini-Zinab. Multivariate analysis of parity progression-based measures of the total fertility rate and its components. *Demography*, 47(1):97–124, 2010.

[28] R. C. Schmitt and A. H. Crosetti. Accuracy of the ratio method for forecasting city population. *Land Economics*, 27(4):346–348, 1951.

[29] S. Smith and T. Sincich. Stability over time in the distribution of population forecast errors. *Demography*, 25(3):461–474, 1988.

[30] S. K. Smith. Tests of forecast accuracy and bias for county population projections. *Journal of the American Statistical Association*, 82(400):991–1003, 1987.

[31] M. M. Wall. A close look at the spatial structure implied by the CAR and SAR models. *Journal of Statistical Planning and Inference*, 121:311–324, 2004.

[32] H. R. White. Empirical study of the accuracy of selected methods of projecting state populations. *Journal of the American Statistical Association*, 49(267):480–498, 1954.

[33] P. Whittle. On stationary processes in the plane. *Biometrika*, 41(3-4):434–449, 1954.

[34] L.-C. Zhang and R. L. Chambers. Small area estimates for cross-classifications. *Journal of the Royal Statistical Society: Series B (Statistical Methodology)*, 66(2):479–496, 2004.

[35] Z. Zhao and B. Guo. An algorithm for determination of age-specific fertility rate from initial age structure and total population. *Journal of Systems Science and Complexity*, 25(5):833–844, 2012.

11

In Search of Optimal Objective Priors for Model Selection and Estimation

Jyotishka Datta

Duke University

Jayanta K. Ghosh

Purdue University and Indian Statistical Institute

CONTENTS

11.1 Introduction

In a very interesting and basic paper, [3] argue that it makes sense to talk of an optimal objective prior in model selection. They make this plausible by providing a set of definitive properties for such a prior and then producing a prior which indeed has all these properties. Just to give a flavor of these properties, we give a few of our favourites. These are the more usual sense of the posterior increasingly putting most of its mass on the true model, i.e. Model Selection Consistency (Criterion 2), Information Consistency (Criterion

3), Predictive Matching Criteria (Criterion 5), and Invariance under certain groups of transformations (Criterion 7).

[3] then go on to apply these ideas to get an optimal prior for linear models, which remains one of the most important and difficult problems for all classical statisticians and objective Bayesians, henceforth referred to as just 'Bayesians'. Of course the notion of optimality of a prior makes no sense in subjective Bayes analysis.

The use of the word optimality, so rare, even if not completely absent in the Bayesian literature, does give rise to technical expectations beyond what we find in the wonderful paper by [3]. The purpose of this modest note is to supplement [3] by requiring that an optimal prior should lead to some sort of optimality of performance in common Bayesian senses. Most likely [3] also have this in mind even though they don't spell it out.

There have been another set of priors that we will refer to collectively as the "global-local" shrinkage priors. These priors include the Horseshoe prior ([9]), Hypergeometric Inverted-Beta ([24]), the Generalized Double Pareto ([2]), and the Three Parameter Beta priors ([1]). These new priors are called the "global-local" shrinkage priors after [8, 9, 25, 26] because they shrink the noise observations but leave the tails unshrunk. These priors have also been claimed to be optimal for their performance in sparse signal recovery problems. Section 11.3 introduces these innovative priors, proposed recently by [1, 2, 9], for possible comparison with the optimal prior of [3].

The shrinkage priors have some similarities with the Information Consistency requirement in [3] which in turn goes back to [19], but was popularized by [4]. It suggests that in model selection, or for testing a sharp null about normal mean, the normal prior does not have the right tail, scale mixtures of normal, e.g. Cauchy-priors are to be preferred.

We elucidate the notion of optimality in the context of multiple testing. [5] introduced the notion of 'Asymptotic Bayes Optimality under Sparsity' (ABOS) and provided conditions under which the Benjamini-Hochberg procedure is ABOS for a two-groups model. To our knowledge, this is the only notion of optimality for multiple testing. Under the asymptotic framework of [5], [12] have proved the following theorem on the optimality of the Horseshoe prior.

Theorem 11.1.1 ([12]). *If the "global shrinkage parameter" τ of the Horseshoe prior is chosen to be of the same order as the proportion of signals p, then the natural decision rule induced by the Horseshoe prior attains the risk of the Bayes oracle up to $O(1)$ with a constant close to the constant in the oracle.*

Lasso is a famous penalized model selection rule for linear models which is often used as a gold-standard in classical statistics. It also appears in some of the new Bayesian work that we discuss later. We now know a lot of deep facts relating to (frequentist) optimality of Lasso and its modifications for linear models, thanks to the deep work of [7]. Some of these results and the conditions are discussed in Section 11.4.

We summarize the contents of this note below:

We discuss the framework for the optimal priors for variable selection in Section 11.2. We describe the "global-local" shrinkage priors and their optimality properties in Section 11.3. In Section 11.4, we formally define the Lasso and discuss some of the optimality properties that makes it an attractive candidate for both model selection and estimation. In Section 11.5, we compare the performance of the explicit optimal prior in [3] with that of the Lasso and the other shrinkage priors with respect to three criteria: accuracy of estimation, out-of-bag prediction, and variable selection.

In Section 11.5, we show numerically that the 'Robust' prior does better variable selection and the Horseshoe prior does better estimation for $p < n$ case. As we discuss in Section 11.2, the Robust prior is restricted to the small p case, and the only fully Bayesian model selection tool for $p > n$ problem is the generalized g-prior by [22]. It remains to be seen how the two Bayesian approaches differ when the model dimension p exceeds the sample size n. We hope to return to a comparison between [22] and Horseshoe for a large p, small n regression problem in the future. In Section 11.6, we compare the variable selection performances of the Horseshoe prior and the Robust prior for the US Crime dataset.

Finally, we discuss the goal of this particular Bayesian Analysis. *Is it to be Model Selection, or the closely related topic of Variable Selection, or estimation of the parameters, e.g. the regression coefficients?*

11.2   The Optimal Prior

We discuss below the framework for the optimal prior for variable selection by [3]. In [3]'s formulation, the simplest model M_0 contains k_0 pre-selected covariates and the ith model M_i contains $k_0 + k_i$ variables where,

$$M_0 : \mathbf{Y} \sim N_n(\mathbf{X}_0\boldsymbol{\beta}_0, \mathbf{I}) \tag{11.1}$$
$$M_i : \mathbf{Y} \sim N_n(\mathbf{X}_0\boldsymbol{\beta}_0 + \mathbf{X}_i\boldsymbol{\beta}_i, \mathbf{I}) \tag{11.2}$$

The prior proposed by [3] that satisfies the desiderata is given by:

$$\pi_i^R(\boldsymbol{\beta}_i|\boldsymbol{\beta}_0, \sigma) = \int_0^\infty \mathcal{N}_{k_i}(\boldsymbol{\beta}_i|\mathbf{0}, g\sigma^2(\mathbf{V}_i^t\mathbf{V}_i)^{-1})p_n(g)dg \tag{11.3}$$

where, $\mathbf{V}_i = (\mathbf{I}_n - \mathbf{X}_0(\mathbf{X}_0^t\mathbf{X}_0)^{-1}\mathbf{X}_0^t)\mathbf{X}_i$. For the "robust" prior $p_n(g)$ is given by:

$$p_n^R(g) = a(\rho_i(b+n))^a(g+b)^{-(a+1)}I(g > \rho_i(b+n) - b) \tag{11.4}$$

where $\mathbf{I}(.)$ is the indicator function. Here, the parameters a, b and ρ_i are chosen to achieve optimal properties for model selection. [3] recommends $a = \frac{1}{2}, b = 1, \rho_i = (k_0 + k_i)^{-1}$. We refer the readers to Section 3.4 of [3] for a detailed discussion on this. Henceforth we refer to the prior simply as the "Robust" prior after [3]. Please note that the optimal or the "robust" prior of [3] requires $p < n$ as the covariance matrix for the prior on $\boldsymbol{\beta}$ requires $\mathbf{X}_i$ to have full-rank. This requirement is present for most of the popular Bayesian approaches for variable selection, such as the standard g-prior by [29], the [30] Cauchy prior and more recent hyper-g prior and mixture of g-priors by [21] and [11]. A modification of the standard g-prior for use in Bayesian variable selection in the large p, small n case was proposed in [22], who orthogonalize the $\boldsymbol{\beta}$ vector and use the singular values of the design matrix instead.

11.3   Shrinkage Priors

We provide a brief survey of the recently proposed shrinkage priors starting with the horseshoe prior due to [8, 9, 26] . As mentioned in the introduction, it is now clear that for a single location parameter, one should use a prior with Cauchy-like tails and ability to shrink near its center. The Horseshoe prior is a new prior in the same spirit, with a strong shrinkage near zero. The Horseshoe prior for a linear model is given as follows:

$$\mathbf{Y} \sim \mathcal{N}(\mathbf{X}\boldsymbol{\beta}, \sigma^2\mathbf{I})$$
$$\beta_i \sim \mathcal{N}(0, \sigma^2\lambda_i^2\tau^2)$$
$$\lambda_i \sim C^+(0, 1)$$
$$\tau \sim C^+(0, 1);$$

where $C^+(0, 1)$ is a standard half-Cauchy distribution on $\mathcal{R}^+$. [8] call λ_i's the local shrinkage parameter and τ the global shrinkage parameter.

[12] show that the natural multiple testing rule induced by the horseshoe prior attains the risk of the Bayes oracle up to $O(1)$ when the global shrinkage parameter is suitably tuned to handle the sparsity in the data. [12] also show that τ is numerically estimable from the data and the Full Bayes estimate of τ seems to adapt to the sparsity level (*vide* Fig. 3 in their paper). On the other hand the empirical Bayes estimate of τ can lead to degeneracy, a phenomenon noted in [6, 20] and noted and discussed in detail in [27].

For several location parameters as well as for regression problems, an important new prior is the generalized double pareto (GDP) due to [2]. It appears more complex than the Horseshoe, but has an explicit form as required by [3] for making the MCMC relatively simple and more importantly, it has tails like the Cauchy-distribution and shrinks near its center. The GDP density with parameters ξ and α is given by:

$$f(\beta|\xi, \alpha) = \frac{1}{2\xi} \left(1 + \frac{|\beta|}{\alpha\xi}\right)^{-(\alpha+1)} \qquad \beta \in \Re \qquad (11.5)$$

In the context of linear models, [2] also argued that the generalized pareto distribution can induce a sparsity favoring penalty in regularized least squares, thus allowing for analysis of frequentist optimality properties.

Another very important class of new priors for the β_j's in regression problems is proposed in [1]. They propose a new three-parameter Beta distribution and use it as a mixing distribution for the parameter λ_j of the $\mathcal{N}(0, \lambda_j\sigma^2)$ - prior for β_j. It seems that one would need two additional hyper-parameters for a shrinkage prior, one for controlling the behaviour near zero and the other for a heavy tail. In reality, some of the proposed shrinkage priors have several hyper-parameters. It would be interesting to study if they help in fine-tuning or if they are redundant.

It is a well-known fact after [9] that a continuous shrinkage prior with "good" estimation properties should have both a spike at zero and very heavy tails. In fact, a shrinkage prior with tails like a Laplace distribution would over-shrink even the large observations. To illustrate this further, we have plotted the prior density on a single regression coefficient β for the shrinkage priors Horseshoe and Generalized Double Pareto vis-a-vis the Robust prior in Fig.11.1. We also plot the densities for Laplace and Cauchy prior for comparing the heaviness in the tails.

We plot the density of Robust prior for the simpler case when the common parameter $\beta_0 = 0$ for $a = \frac{1}{2}, b = 1$ and $\rho_i = \frac{1}{2}$. This choice of hyper-parameters corresponds to the hyper-g/n prior proposed in [21]. Fig.11.1(a) displays the behaviour of the priors under consideration near zero and Fig.11.1(b) displays their behavior in the tails. As expected, all the shrinkage priors have a peak near $\beta = 0$ but only the Horseshoe prior is unbounded. By Fig. 11.1(a) the order of peakedness near zero would be $HS > GDP = Laplace > Robust > Cauchy$. Fig. 11.1(b) shows that all the shrinkage priors as well as the Robust prior have heavier tails than the Laplace distribution. By Fig. 11.1(b) the order of tail heaviness is as follows: $Robust > GDP > Cauchy > HS > Laplace$.

11.4 The Lasso Based Priors

To fix ideas, we consider a standard linear model

$$\mathbf{Y} = \mathbf{X}\beta + \epsilon \qquad (11.6)$$

where $\mathbf{Y}$ is an $n \times 1$ vector of observations, $\mathbf{X}$ is an $n \times p$ non-random design matrix, β is the $p \times 1$ vector of unknown parameters and ϵ is the vector of

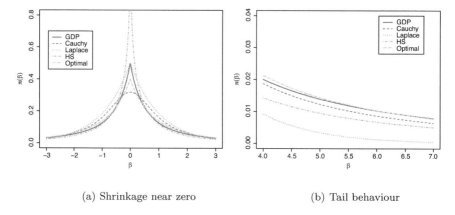

(a) Shrinkage near zero   (b) Tail behaviour

FIGURE 11.1
Probability density functions for generalized double Pareto, Cauchy, Laplace,
Horseshoe and Robust prior.

n i.i.d. $\mathcal{N}(0, \sigma^2)$ r.v.s. For convenience in presentation, σ^2 is assumed known.
The parameter dimension p may be larger than n, i.e. $p > n$ in which case we
have a high-dimensional problem, but we may also have $p < n$. Various details
in the Lasso based analysis depend on whether $p > n$ or $p < n$. The latter
case is relatively standard and usual least squares estimates are known to be
consistent as p and n tend to infinity. The case $p > n$ is the high-dimensional
case, for which most of the results mentioned below are new and taken from
[7]. The likelihood is, up to a constant multiplier,

$$L(\boldsymbol{\beta}) = \exp(-\frac{1}{2\sigma^2}\|\mathbf{Y} - \mathbf{X}\boldsymbol{\beta}\|^2) \tag{11.7}$$

and the Lasso based estimate of $\boldsymbol{\beta}$ is the value of $\boldsymbol{\beta}$ that maxmizes the penalized log-likelihood, namely,

$$\log L(\boldsymbol{\beta}) - \lambda \sum_{j=1}^{p} |\beta_j| \tag{11.8}$$

It is a standard fact that a penalty may be interpreted as a prior. Our Lasso
induced prior is simply the double exponential

$$\pi_L(\boldsymbol{\beta}) = \exp(-\lambda \sum_{j=1}^{p} |\beta_j|) \tag{11.9}$$

Clearly, we get the same estimate of $\boldsymbol{\beta}$ as the classical Lasso estimate if we
maximize $\pi_L(\boldsymbol{\beta})L(\boldsymbol{\beta})$ and get the posterior mode or the highest posterior
density (HPD) estimate of $\boldsymbol{\beta}$. The posterior mode, being the same as the
classical Lasso based estimate, will have all the good properties of Lasso proved

by [7]. Of course, λ will have to be chosen by minimizing the squared error risk found by cross-validation. The simplest case, as expected, is prediction. Then the above procedure will lead to optimal prediction and in principle can be compared with other competing procedures for which the squared error risk can be obtained by cross-validation.

[7] also have an Oracle inequality (*vide* Equation (2.8) as well as Theorem (6.1)), which shows that up to $O(\log(p))$ and a "compatibility" constant ϕ_0^2, the mean squared prediction error is of the same order as if one knew the truth. (Here 'truth' means knowledge of the active set $S_0 = \{j : \beta_j^0 \neq 0\}$.) But, unfortunately, the compatibility constant will be known only in simulated models. They have extremely good theorems also on rates of convergence of consistent point estimates of $\boldsymbol{\beta}$, and optimality results on variable selection. As expected, good variable selection is both most difficult to define in a satisfactory way or to achieve in practice.

As we discussed before, the posterior mode of $\boldsymbol{\beta}$ under the double exponential prior will have all the above optimal properties of the Lasso estimate of $\boldsymbol{\beta}$. Unfortunately, the same is not expected to hold for the posterior mean, which is the Bayes estimate under squared error loss for the following reasons. Thanks to the pioneering work on shrinkage of observations and the Horseshoe prior by [8, 9, 26] and some theory developed by [12], the following has become clear. A good prior should shrink relatively small observations but basically leave alone, i.e. not shrink, the observations near the tails.

Various people, including [12] have noted that this is not the case with the double exponential, which shrinks at both tails of the distribution. It is to be expected that this will prevent optimality of these priors for vector valued parameters also. A major new contribution on this topic is [25].

Before we close this section it is worth pointing out that an important additional fact, related to variable selection, is that the naive estimate $\hat{S}(\lambda) = \{j; \hat{\beta}_j^{Lasso}(\lambda) \neq 0, j = 1, \ldots, p_n\}$ satisfies $P(\hat{S}(\lambda) = S_0) \to 1$ as $n \to \infty$ under a certain condition called "the neighbourhood stability condition" on the design matrix and "beta-min" condition on the coeffcient vector [1](*vide* Theorems 1, 2 in [23] and Section 2.6.1 in [7]). The neighbourhood stability condition is equivalent to the "irrepresentable" condition of [32, 33]. In particular, the probability of Lasso selecting the true model is almost 1 when $n_\infty > 0.2$ and it is almost zero when $n_\infty < -0.3$. Later, in Section 11.5, we examine how badly the irrepresentability condition affects the variable screening properties of the Robust prior and the Horseshoe prior based model selection.

---

[1] The "beta-min" condition is requiring that the non-zero coefficients are sufficinetly large, i.e. the non-zero β_j^0s satisfy $\inf_{j \in S_0^c} |\beta_j^0| >> \sqrt{s_0 \log(p)/n}$. For further discussion on this, please refer to Ch. 7 in [7].

11.5 Numerical Results

11.5.1 Effect of the irrepresentability condition

As discussed in Section 11.4, the "beta-min" condition along with the neighborhood stability or the equivalent irrepresentability condition is critical for consistent variable selection with Lasso. A natural question is to check how this condition affects the two Bayesian procedures, *viz.*, the Robust prior and the Horseshoe prior. To examine the effect of irreprsentable condition on the three approaches, we follow the example in [32] and [31] which shows the relationship between probability of selecting the true model and the irrepresentable condition number defined as

$$n_\infty = 1 - \left\lVert \hat{\Sigma}_{p-s_0,s_0} \hat{\Sigma}_{s_0,s_0}^{-1} sign(\beta_{S_0}) \right\rVert_\infty \qquad (11.10)$$

We followed the example in [32] and simulated data with $n = 100, p = 32$ and $s_0 = 5$ with the sparse coefficient vector $\beta_{S_0}^* = (7,4,2,1,1)^T$, σ^2 was set to 0.1 to obey the asymptotic properties of the Lasso. [32] first draw the covariance matrix Σ from $Wishart(p, \mathbf{I}_p)$ and then generate design matrix $\mathbf{X}$ from $\mathcal{N}(0, \Sigma)$. [32] showed that the irrepresentability condition may not hold for such a design matrix. In fact, in our simulation studies the n_∞'s for the 100 simulated designs were between $[-1.02, 0.36]$, with 66 of them being lower than zero and 34 of them being higher than zero. We expect the Lasso to perform well when $n_\infty > 0$ and poorly when $n_\infty < 0$. We generate $n = 100$ design matrices and for each design, 100 simulations were conducted by generating the noise vector from $\mathcal{N}(0, \sigma^2 \mathbf{I})$.

Figure 11.2 shows the percentage of correctly selected model as a function of the irrepresentable condition number, n_∞ for Lasso, the Robust prior, and the Horseshoe prior. As expected, Lasso's variable selection performance is crucially dependent on the irrepresentability condition but the Robust prior almost always recovers the true sparse β vector irrespective of n_∞, with the Horseshoe performing slightly worse.

11.5.2 Performance comparison

In this section, we compare the three variable selection tools, *viz.*, the Robust prior, the Horseshoe prior and the Lasso in terms of three different criteria. We describe the criteria briefly as follows.

1. First we compare the misclassification probability associated with the three procedures in subsection 11.5.3. We also plot the 'posterior' inclusion probabilities for the two Bayesian procedures (Horseshoe and Robust prior) for a simulated dataset. In this connection, it may be noted that the 'posterior' inclusion probability is defined as the conditional probability that a variable is selected given the data, i.e. $P(\beta_j \neq 0| \mathbf{y})$.

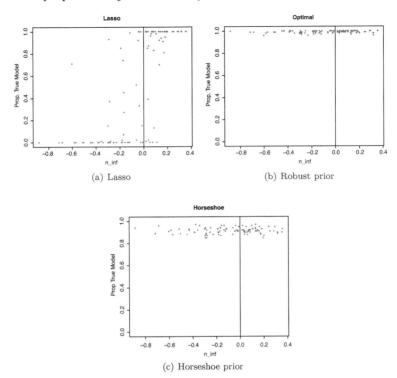

(a) Lasso   (b) Robust prior

(c) Horseshoe prior

FIGURE 11.2
Probability of selecting true model as a function of the 'irrepresentability condition number' for Lasso, Robust prior and the Horseshoe prior.

2. We compare the relative estimation accuracy (and predictive performance for the given data) for Lasso and the two priors. We also calculate what [7] call 'average squared error loss' (see Equation (2.7)-(2.8) in [7]) for all the contenders in this chapter (subsection 11.5.4).

3. Finally, we compare the 'out-of-sample' prediction error[2] for all the methods (subsection 11.5.5).

We follow the following hierarchical model as described in [25]'s simulation scheme to generate the datasets.

$$\mathbf{Y}_{n\times 1} \sim \mathcal{N}(\mathbf{X}_{n\times p}\boldsymbol{\beta}_{p\times 1}, \mathbf{I}) \tag{11.11}$$

$$\boldsymbol{\beta} \sim \omega t_3 + (1-\omega)\delta_0 \tag{11.12}$$

$$\omega \sim Be(1, \frac{1}{4}) \tag{11.13}$$

---

[2]In the present context, 'Out-of-Sample' prediction pertains to prediction based on un-observed explanatory variables.

Here t_3 denotes the t-distribution with 3 degrees of freedom, and δ_0 denotes the degenerate distribution at zero. The last hierarchy implies that 80% of the elements of $\boldsymbol{\beta} = (\beta_1, \beta_2, \ldots, \beta_p)$ are zero on an average. For the numerical examples, we have fixed the number of observations n at 60, and the dimension of $\boldsymbol{\beta}$, p at 40. The entries of the design matrix $\mathbf{X}$ are i.i.d standard Gaussian random variables. We have compared the performance of the new "robust" prior proposed in [3] with that of the Horseshoe prior and Lasso for datasets generated using the aforementioned scheme.

We have used the R-package `BayesVarSel` by [15] for the Robust prior of [3] and the R-package `monomvn` by [17] for implementing the analysis based on Horseshoe prior of [9]. The `glmnet`-package is used for carrying out Lasso with a 10-fold cross-validation. For calculating the inclusion probabilities for both the Robust and the Horseshoe model, the respective packages use a Full Bayes approach with Gibbs sampling. We have run the MCMC samplers for 1000 iterations and discarded the first 200 samples as burn-in.

11.5.3 Variable selection via inclusion probability

Fig. 11.3 shows the inclusion probabilities $\omega_j = P(\beta_j \neq 0|\mathbf{y})$ for the Robust prior and the Horseshoe prior. The black diamonds and the grey filled circles denote the true non-zero β_j's and the true zero β_j's respectively. An interesting technical fact is that the inference based on the Horseshoe prior helps us identify the inclusion probabilities as approximations to the conditional probability that a parameter is significant given the data. The exact computation requires introduction of the two-group model and involves much more intensive calculation. For a full discussion of the one-group approximation to the two-group model, see [12, 25].

In Bayesian model selection literature, it is common to report the highest probability density model or the median probability density model as the single model chosen by the decision rule. Alternatively, one can also select only those variables for which the marginal inclusion probability is greater than $1/2$.

Fig. 11.3 shows the posterior inclusion probabilities for three different choices of the design matrix, uncorrelated (leftmost column), equicorrelated with $\rho = 0.2$ (middle column) and $\rho = 0.9$ (rightmost column). Table 11.1 shows the number of misclassification errors committed by the Robust and the Horseshoe prior for each design matrix. It seems that the performance of both the priors deteriorates in the presence of dependence between the predictor variables, but the degree of deterioration seems more for the Horseshoe prior than the Robust prior (see both Table 11.1 and Fig. 11.3).

One can also see from Figure 11.3 that the dichotomy between the inclusion probabilities for the zero and non-zero components of the $\boldsymbol{\beta}$ vector is more enhanced for the "robust" prior, leading to a possibly better model selection performance. However, one may not draw a firm conclusion without further comparison in many examples.

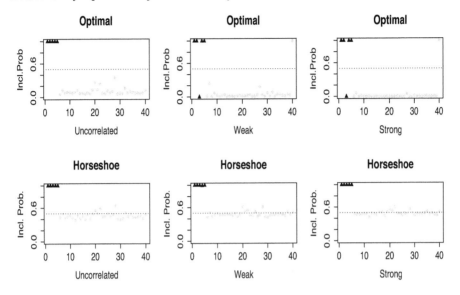

FIGURE 11.3
Posterior inclusion probabilities due to the Robust prior and the Horseshoe prior for different correlation structures of the design matrix. The black diamonds indicate the true non-zero β_j's and the grey filled circles indicate the true zero β_j's.

11.5.4 Estimation accuracy

To compare the estimation accuracies of the three different model selection procedures, we compute two different quantities. The first is the squared error in estimating β, calculated as the squared norm difference between the true and the estimated coefficient vector, $\left\|\hat{\beta} - \beta_0\right\|_2^2$. The second quantity is the error in estimating the underlying regression function, and is calculated as $(\hat{\beta} - \beta_0)'\Sigma_X(\hat{\beta} - \beta_0)$, where Σ_X is given by $\frac{1}{n}X^TX$ for a fixed design and by the covariance of X for a random design. This quantity is called the average squared error loss, by noting that $X(\hat{\beta} - \beta_0) = \hat{Y} - Y_0$. One can show that, under suitable sparsity conditions on β_0 and "compatibility" conditions on

TABLE 11.1
Number of misclassified β_j's for the Robust prior and the Horseshoe prior for different degress of correlation in the design matrix.

	Orthogonal	Weakly Correlated	Strongly Correlated
Horseshoe	6	8	9
Robust	0	2	2

TABLE 11.2
Comparison of average squared estimation error (and their standard error) for
different model selection procedures.

	Lasso (10-fold CV)	Horseshoe	Robust
$\|\hat{\beta} - \beta_0\|_2^2$	0.680 (1.646)	0.443 (1.370)	0.611 (1.502)
$(\hat{\beta} - \beta_0)' \frac{1}{n} \mathbf{X}^T \mathbf{X}(\hat{\beta} - \beta_0)$	0.406 (0.281)	0.241 (0.194)	0.325 (0.148)

the design matrix $\mathbf{X}$, the Lasso is consistent for estimation in the following
sense.

$$(\hat{\beta} - \beta_0)' \Sigma_X (\hat{\beta} - \beta_0) = o_p(1) \quad (n \to \infty) \tag{11.14}$$

For a detailed proof and more information about the conditions, see Chapters
2 and 6 of [7]. Table 11.2 displays both the errors averaged over 100 datasets
simulated from the scheme displayed in (11.12) and (11.13). It seems that the
Horseshoe prior is doing better than the others in terms of both the measures
of estimation accuracy and Lasso seems to be the worst, with the Robust prior
falling somewhere in between.

11.5.5 Out-of-sample prediction error

In many of the high-throughput problems, the goal of a 'model selection' process is to find a model with the best predictive performance for an yet-unseen
explanatory variable. In this subsection, we compare the 'out-of-sample' predictive performances of the Lasso, the Horseshoe and the Robust prior. In
order to do that, we generate one dataset from the scheme (11.12) and (11.13)
and estimate the coefficient vector β using all the competing methods. The
true coefficient β_0 has $s_0 = 4$ non-zero coefficients, their values being equal
to $\{-0.659, 0.381, -1.200, 1.517\}$. We calculate the relative prediction accura-
cies $\frac{\|X(\hat{\beta} - \beta_0)\|^2}{\|X\beta_0\|^2}$ for 100 new design matrices $\mathbf{X}$. Table 11.3 shows the average
relative prediction accuracies for the three procedures.

TABLE 11.3
Comparison of average relative prediction error (and their standard error) for
different model selection procedures.

	Lasso (10-fold CV)	Horseshoe	Robust
$E\left(\frac{\|X(\hat{\beta} - \beta_0)\|^2}{\|X\beta_0\|^2}\right)$	0.850(1.026)	0.243(1.026)	0.300(0.411)

11.6   Examples with Real Data

In this section, we show the small sample performance of the Robust prior compared to a few popular model selection tools, such as BIC, Benchmark prior (BRIC, due to [14]), Empirical Bayes (due to [10, 16]). A similar study appears in [21], where the mixture g-priors are compared with other candidates for two real datasets, *viz.*, the US Crime data and the ground ozone data. Both the datasets have been used extensively in literature for evaluating and benchmarking the model selection performances of competing procedures.

11.6.1   US crime data

The US crime data contains an aggregate measure for crime rates for 47 states in the United States in 1960. The original data comes from [13] and [28] where the aim was to explain crime rate as a function of economic decision process and probability of punishment. The $p = 15$ regressors in the crime data are given in Table 11.4. This data has been used as an illustrative example in Bayesian linear model literature, see for example Raftery et al.(1997), [14, 21].

We continue along the lines of [21] who investigate how the mixture of g-priors affect 'variable importance'. In Table 11.6, we show the marginal inclusion probabilties for the 15 variables in the US Crime data set for the Robust and the Horseshoe prior, compared with the model selection methods

TABLE 11.4
Description of the variables in US Crime dataset.

Description	Name
Percentage of males aged 14âĂŞ24.	AGE
Indicator variable for a Southern state.	S
Mean years of schooling	Ed
Police expenditure in 1960.	Ex0
Police expenditure in 1959.	Ex1
Labour force participation rate	LF
Number of males per 1000 females	M
State population	N
Number of non-whites per 1000 people	NW
Unemployment rate of urban males 14âĂŞ24.	U1
Unemployment rate of urban males 35âĂŞ39	U2
Gross Domestic Product per head.	W
Income inequality	X
Probability of imprisonment.	prison
Average time served in state prisons	time

TABLE 11.5

Description of the model selection methods used for the US Crime example.

BRIC	g prior with $g = \max(n, p^2)$
HG-n	Hyper-g prior with a = n
HG3	Hyper-g prior with a = 3
HG4	Hyper g-prior with a = 4
EB-L	Local EB estimate of g in g-prior
EB-G	Global EB estimate of g in g-prior
ZS-N	Base Model in Bayes factor taken as the Null model,
ZS-F	Base Model in Bayes factor taken as the Full model,
BIC	Bayesian Information Criterion
HS	Horseshoe Prior
Robust	Robust prior

used in [21]. The methods being compared are listed in Table 11.5. As Table 11.6 shows, the marginal inclusion probabilities for the Robust prior are very similar to the fully Bayesian mixture of g-priors as well as the Zellner-Siow null based apporach and the Empirical Bayes approaches EB-global and EB-local. However, the Horseshoe prior leads to higher marginal inclusion probabilities, and as a result the median probability model for Horseshoe will be the least parsimonious and will include three more variables compared to the fully Bayesian and Empirical Bayes methods. [21] also pointed out that BRIC leads to the maximum shrinkage of the marginal inclusion probabilities and lead to the most parsimonious model.

TABLE 11.6

Marginal Inclusion Probabilities for $p = 15$ variables in US Crime dataset.

	BRIC	HG-n	HG3	HG4	EB-L	EB-G	ZS-N	ZS-F	BIC	HS	Robust
log(AGE)	0.75	0.85	0.84	0.84	0.85	0.86	0.85	0.88	0.91	0.82	0.83
S	0.15	0.27	0.29	0.31	0.29	0.29	0.27	0.36	0.23	0.46	0.30
log(Ed)	0.95	0.97	0.97	0.96	0.97	0.97	0.97	0.97	0.99	0.93	0.96
log(Ex0)	0.66	0.66	0.66	0.66	0.67	0.67	0.67	0.68	0.69	0.77	0.66
log(Ex1)	0.39	0.45	0.47	0.47	0.46	0.46	0.45	0.5	0.4	0.69	0.47
log(LF)	0.08	0.2	0.23	0.24	0.22	0.21	0.2	0.3	0.16	0.44	0.23
log(M)	0.09	0.2	0.23	0.24	0.22	0.22	0.2	0.3	0.17	0.48	0.23
log(N)	0.23	0.37	0.39	0.39	0.39	0.38	0.37	0.46	0.36	0.51	0.38
log(NW)	0.51	0.69	0.69	0.68	0.7	0.7	0.69	0.75	0.78	0.77	0.67
log(U1)	0.11	0.25	0.27	0.28	0.27	0.27	0.25	0.35	0.23	0.44	0.27
log(U2)	0.45	0.61	0.61	0.61	0.62	0.62	0.61	0.68	0.7	0.59	0.59
log(W)	0.18	0.35	0.38	0.39	0.38	0.38	0.36	0.47	0.36	0.52	0.37
log(X)	0.99	1	0.99	0.99	1	1	1	0.99	1	0.97	0.99
log(prison)	0.78	0.89	0.89	0.89	0.9	0.9	0.9	0.92	0.95	0.83	0.88
log(time)	0.19	0.37	0.38	0.39	0.39	0.38	0.37	0.47	0.41	0.49	0.38

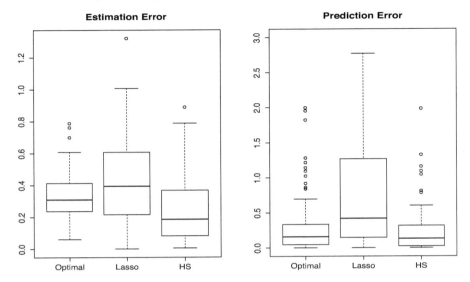

FIGURE 11.4
Boxplots for the average squared error loss,$(\hat{\beta} - \beta_0)' \frac{1}{n} \mathbf{X}^T \mathbf{X}(\hat{\beta} - \beta_0)$ and the relative prediction error $E\left(\frac{\|X(\hat{\beta} - \beta_0)\|^2}{\|X\beta_0\|^2}\right)$ for the Lasso, the Horseshoe and the Robust prior.

11.7 Discussion

(A) The notion of optimality in performance is somewhat ambiguous. Strictly speaking, classical decision theory has only one such notion — a minimax estimator. Most objective Bayesians would think that is too restrictive. In the context of the present note, we have chosen optimality to mean similar performance as the shrinkage priors, which shrink small signals but basically leaves alone the large signals. Even in this definition, "small" and "large" are neither well-defined nor based on a consensus.

A practical alternative could be make a list of problems and require that the chosen procedure achieves a specified result in that problem. This would be something like Donoho's examples for comparing minimax estimators.

(B) An important point that emerges from our numerical comparison of different priors is that some priors do some things well, but none does all things well. For example, the Robust prior does good model selection but the shrinkage priors estimate better.

This point has been noted and discussed well in [18]. It appears to us that this is a problem involving more than one loss function, recently brought to our notice by Yaacov Ritov and Peter Bickel at the Joint Statistical Meeting, 2013. It also seems clear that an optimal objective prior would be different for these two loss functions. Our personal view is that one should decide what is the more important loss function; in most cases it will be model selection, but in principle it could be otherwise too. We favor choosing the Robust prior for the more important model selection problem and use it for both problems. It is important to point out that both in the problems of [18] and the examples that we have seen, the loss in optimality in one of the two problems due to lack of selection of the optimal prior for that problem, does not lead to significant loss of optimality. Given a single prior, an optimal decision rule for weighted additive losses would not be difficult in principle.

To further illustrate this point, we compare the average squared error loss and the relative prediction accuracies for the three procedures by the means of boxplots of the errors in Fig. 11.4. It seems that a shrinkage prior like the Horseshoe might be the best choice when it comes to estimation or prediction but the loss of accuracy by using the Robust prior is not significant.

Acknowledgment

We thank the anonymous referees for their suggestions, which led to a better presentation of our manuscript. This material was based upon work partially supported by the National Science Foundation under Grant DMS-1127914 to the Statistical and Applied Mathematical Sciences Institute. Any opinions, findings, and conclusions or recommendations expressed in this material are those of the author(s) and do not necessarily reflect the views of the National Science Foundation.

Bibliography

[1] A. Armagan, D. B. Dunson, and M. Clyde. Generalized beta mixtures of Gaussians. In *NIPS*, pages 523–531, 2011.

[2] A. Armagan, D. B. Dunson, and J. Lee. Generalized double Pareto shrinkage. *Statistica Sinica*, 23(1):119, 2013.

[3] M. J. Bayarri, J. O. Berger, A. Forte, and G. García-Donato. Criteria for Bayesian model choice with application to variable selection. *The Annals of Statistics*, 40(3):1550–1577, 2012.

[4] J. O. Berger and L. R. Pericchi. Objective Bayesian methods for model selection: Introduction and comparison. In *Model selection*, volume 38 of *IMS Lecture Notes Monogr. Ser.*, pages 135–207. Institute of Mathematical Statistics, Beachwood, OH, 2001. With discussion by J. K. Ghosh, Tapas Samanta, and Fulvio De Santis, and a rejoinder by the authors.

[5] M. Bogdan, A. Chakrabarti, F. Frommlet, and J. K. Ghosh. Asymptotic Bayes-optimality under sparsity of some multiple testing procedures. *The Annals of Statistics*, 39(3):1551–1579, 2011.

[6] M. Bogdan, J. K. Ghosh, and S. T. Tokdar. A comparison of the Benjamini-Hochberg procedure with some Bayesian rules for multiple testing. In *Beyond parametrics in Interdisciplinary Research: Festschrift in Honor of Professor Pranab K. Sen*, volume 1 of *Institute of Mathematical Statistics Collection*, pages 211–230. 2008.

[7] P. Bühlmann and S. van de Geer. *Statistics for High-Dimensional Data: Methods, Theory and Applications*. Springer Verlag, Heidelberg, 2011.

[8] C. M. Carvalho, N. G. Polson, and J. G. Scott. Handling sparsity via the horseshoe. *Journal of Machine Learning Research W&CP*, 5(73-80), 2009.

[9] C. M. Carvalho, N. G. Polson, and J. G. Scott. The horseshoe estimator for sparse signals. *Biometrika*, 97(2):465–480, 2010.

[10] M. Clyde and E. I. George. Model uncertainty. *Statistical Science*, 19(1):81–94, 2004.

[11] W. Cui and E. I. George. Empirical Bayes vs. fully Bayes variable selection. *Journal of Statistical Planning and Inference*, 138(4):888–900, 2008.

[12] J. Datta and J. K. Ghosh. Asymptotic properties of Bayes risk for the horseshoe prior. *Bayesian Analysis*, 8(1):111–131, 2013.

[13] I. Ehrlich. Participation in illegitimate activities: A theoretical and empirical investigation. *The Journal of Political Economy*, pages 521–565, 1973.

[14] C. Fernández, E. Ley, and M. F. J. Steel. Benchmark priors for Bayesian model averaging. *Journal of Econometrics*, 100(2):381–427, 2001.

[15] G. Garcıa-Donato and A. Forte. *BayesVarSel: Bayesian Variable Selection in Linear Models*, 2014. R Package version 1.5.1.

[16] E. I. George and D. P. Foster. Calibration and empirical Bayes variable selection. *Biometrika*, 87(4):731–747, 2000.

[17] R. B. Gramacy. monomvn: Estimation for multivariate normal and student-t data with monotone missingness. *R Package Version*, pages 1–8, 2010.

[18] P. R. Hahn and C. M. Carvalho. Decoupling shrinkage and selection in Bayesian linear models: A posterior summary perspective, 2014. Working Paper.

[19] H. Jeffreys. *The Theory of Probability*. Oxford University Press, Oxford, UK, 3rd edition, 1998.

[20] I. M. Johnstone and B. W. Silverman. Needles and straw in haystacks: Empirical Bayes estimates of possibly sparse sequences. *The Annals of Statistics*, 32(4):1594–1649, 2004.

[21] F. Liang, R. Paulo, G. Molina, M. A. Clyde, and J. O. Berger. Mixtures of g priors for Bayesian variable selection. *Journal of the American Statistical Association*, 103(481):410–423, 2008.

[22] Y. Maruyama and E. I. George. Fully Bayes factors with a generalized g-prior. *The Annals of Statistics*, 39(5):2740–2765, 2011.

[23] N. Meinshausen and P. Bühlmann. High-dimensional graphs and variable selection with the Lasso. *The Annals of Statistics*, 34(3):1436–1462, 2006.

[24] N. G. Polson and J. G. Scott. Large-scale simultaneous testing with hypergeometric inverted-beta priors. *arXiv preprint: arXiv:1010.5223*, 2010.

[25] N. G. Polson and J. G. Scott. Local shrinkage rules, Lévy processes, and regularized regression. *Journal of the Royal Statistical Society: Series B (Statistical Methodology)*, 74(2):287–311, 2012.

[26] N. G. Polson and J. G. Scott. On the half-Cauchy prior for a global scale parameter. *Bayesian Analysis*, 7(2):1–16, 2012.

[27] J. G. Scott and J. O. Berger. An exploration of aspects of Bayesian multiple testing. *Journal of Statistical Planning and Inference*, 136(7):2144–2162, 2006.

[28] W. Vandaele. Participation in illegitimate activities: Ehrlich revisited, 1960. *(ICPSR 8677)*, 1987.

[29] A. Zellner. On assessing prior distributions and Bayesian regression analysis with g-prior distributions. In P. K. Goel and A. Zellner, editors, *Bayesian Inference and Decision Techniques: Essays in Honor of Bruno de Finetti*, pages 233–243. Elsevier Science Ltd, Amsterdam, North-Holland, June 1986.

[30] A. Zellner and A. Siow. Posterior odds ratios for selected regression hypotheses. *Trabajos de estadística y de investigación operativa*, 31(1):585–603, 1980.

[31] J. Zhang, X. J. Jeng, and H. Liu. Some two-step procedures for variable selection in high-dimensional linear regression. *arXiv preprint: arXiv:0810.1644*, 2008.

[32] P. Zhao and B. Yu. On model selection consistency of Lasso. *The Journal of Machine Learning Research*, 7:2541–2563, Dec. 2006.

[33] H. Zou. The adaptive Lasso and its oracle properties. *Journal of the American Statistical Association*, 101(476):1418–1429, 2006.

12

Bayesian Variable Selection for Predictively Optimal Regression

Tanujit Dey

The College of William and Mary

Ernest Fokoué

Rochester Institute of Technology

CONTENTS

12.1 Introduction

We are given a training set $\mathcal{D} = \{(\mathbf{x}_i, y_i), i = 1, \cdots, n : \mathbf{x}_i \in \mathcal{X} \subset \mathbb{R}^p, y_i \in \mathbb{R}\}$, where the y_i's are realizations of $Y_i = f^*(\mathbf{x}_i) + \epsilon_i$, with the ϵ_i's representing the noise terms, herein assumed to be independently normally distributed with mean 0 and constant variance σ^2. Our goal is to use the information contained in the data $\mathcal{D}$ to build an estimator $\hat{f}$ of the true unknown function f^* that achieves the smallest mean squared error all over $\mathcal{X} \times \mathcal{Y}$, or more specifically.

$$f^* = \underset{f}{\arg\min}\, R(f) = \underset{f}{\arg\min}\, \mathbb{E}[\ell(Y, f(X))]$$

where

$$R(f) = \mathbb{E}[(Y - f(X))^2] = \int_{\mathcal{X} \times \mathcal{Y}} (y - f(\mathbf{x}))^2 p_{XY}(\mathbf{x}, y) d\mathbf{x} dy$$

is the expected loss over all pairs of the cross space $\mathcal{X} \times \mathcal{Y}$. In practice, the theoretical risk $R(f)$ is unavailable, because the distribution of the data is unknown. We therefore seek the best/optimal according to the empirical standard, namely

$$\hat{f}^* = \underset{f}{\arg\min}\, \widehat{R}(f) = \underset{f}{\arg\min}\, \left\{ \frac{1}{n} \sum_{i=1}^{n} (Y_i - f(X_i))^2 \right\}$$

Since it is impossible to search all possible functions, it is usually crucial to choose the "right" function space. In this article, we will focus on the traditional linear regression model class as our function class. Our goal in this article is to compare the predictive performances of various methods of model selection and model averaging when the function class under consideration is the traditional multiple linear regression model (12.1). To perform our predictive comparisons, we use the test set $\{(\mathbf{x}_i^*, \mathbf{y}_i^*),\ i = 1, \cdots, m\}$ to compute the test error, also referred to here as the Empirical Predictive Mean Squared Error (EPMSE) given by

$$\text{EPMSE}(\hat{f}) = \frac{1}{m} \sum_{i=1}^{m} \left(\mathbf{y}_i^* - \hat{f}(\mathbf{x}_i^*) \right)^2.$$

We repeatedly (say $R = 500$ times) split the provided dataset $\mathcal{D} = \{\mathbf{z}_i = (\mathbf{x}_i, \mathbf{y}_i),\ i = 1, \cdots, n\}$ into a training and a test set, and compute the EPMSE at each replication, and then use the average EPMSE as our score for comparing different model selection and model averaging techniques. Set R, the number of replications, then choose π, the percentage of training, and compute $\mathit{l} = \lceil n\pi \rceil$.

- For $r = 1$ to R

    - Draw l items without replacement from $\mathcal{D}$ and form $\mathcal{T}_r$
    - Train $\hat{f}^{(r)}(\cdot)$ based on the l items in $\mathcal{T}_r$
    - Predict $\hat{f}^{(r)}(\mathbf{x}_i)$ for the m items in $\mathcal{V}_r = \mathcal{D} \backslash \mathcal{T}_r$

- Calculate $\widehat{\text{EPMSE}}(\hat{f}^{(r)}) = \dfrac{1}{m} \displaystyle\sum_{\mathbf{z}_i \in \mathcal{V}_r} \left(\mathbf{y}_i - \hat{f}^{(r)}(\mathbf{x}_i) \right)^2.$

- End

- Compute the average EPMSE for $\hat{f}$, namely

$$\text{average}\{\text{EPMSE}(\hat{f})\} = \frac{1}{R} \sum_{r=1}^{R} \widehat{\text{EPMSE}}(\hat{f}^{(r)}).$$

For any given dataset and a collection $\{\hat{f}_j, \ j = 1, \cdots, J\}$ of estimators of f used for predicting the response given $\mathbf{x}$, the estimator $\hat{f}_j$ will be declared the best of the J estimators (at least with respect to the task at hand) if

$$\text{average}\{\text{EPMSE}(\hat{f}_j)\} < \text{average}\{\text{EPMSE}(\hat{f}_k)\}, \ \ k = 1, \cdots, J, \ k \neq j.$$

We now consider various approaches to predicting the response in a regression setting, and we compare them on qualitatively and quantitatively different datasets.

12.2 Approaches to Response Prediction in Regression

Modern statistical learning and data mining are filled with thousands of studies where the main statistical task revolves around estimation and prediction based on the traditional multiple linear regression (MLR) model given by

$$M_f : \quad \boldsymbol{Y} = \alpha \mathbf{1}_n + \boldsymbol{X}\boldsymbol{\beta} + \boldsymbol{\epsilon}, \tag{12.1}$$

where $\boldsymbol{\beta} = (\beta_1, \beta_2, \cdots, \beta_p)^\top$, $\boldsymbol{Y} = (\mathrm{y}_1, \mathrm{y}_2, \cdots, \mathrm{y}_n)^\top$, $\boldsymbol{\epsilon} = (\epsilon_1, \epsilon_2, \cdots, \epsilon_n)^\top \sim$ $\text{MVN}(0, \sigma^2 I)$, the design matrix $\boldsymbol{X}$ is an $n \times p$ matrix, and $\mathbf{1}_n = (1, 1, \cdots, 1)^\top$ is a $n \times 1$ dimensional vector of 1's. We shall refer to (12.1) as the *full model.*

12.2.1 Basic elements and assumptions

We assume that many of the β_j's are essentially zero, so that the intrinsic rank of the design matrix $\boldsymbol{X}$ is a number $q \in \mathbb{N}$ with $q \ll p$. Many data mining problems do exhibit such a characteristic of rank deficiency, mainly because variables are typically picked up as they are available, and therefore will turn out to be either noise variables (no relationship with the response) or redundant variables. A basic result in the linear model analysis shows that when $\boldsymbol{X}$ is rank deficient, the ordinary least squares estimator

$$\hat{\beta}^{(\text{OLS})} = (\boldsymbol{X}^\top \boldsymbol{X})^{-1} \boldsymbol{X}^\top \boldsymbol{Y} \tag{12.2}$$

of $\boldsymbol{\beta}$ will tend to exhibit a high (inflated) variance, thereby corrupting all predictions and inferences with the computed model. It is therefore crucial to determine (if possible) the intrinsic model that generated the data, i.e. the model made up of only the most significant and non redundant variables. For many decades, both frequentist and Bayesian statisticians have contributed substantially to this topic of *variable selection*. In elementary statistical regression analysis courses, the method of choice for variable selection has been overwhelmingly frequentist with the *stepwise regression heuristic* occupying a prominent place, and *best subsets selection* occasionally used whenever possible. While a heuristic like stepwise regression does provide a workable approach to variable selection, it is not a principled method, and does have the extra limitation of not providing any measure of variable importance. Besides, when the number of variables p is larger than the sample size n (a setting now known as large p small n or short fat data), the stepwise regression heuristic cannot be used because the submodels cannot even be built, let alone compared. In recent years, both frequentists and Bayesians have developed new methods for handling some of the most formidable variable selection tasks, many of which arose from the statistical learning and data mining community. The regularization framework has allowed the variable selection problem to be formulated and solved as penalized least squares problem, with the penalty function suitably chosen (devised) to induce reduction of the dimensionality of the parameter space and therefore selection of the most significant variable. One of the earliest uses of the regularization framework in statistical learning could be attributed to both [15] and [14], who introduced the seminal paper treating the theory and application of ridge regression, a device aimed at addressing the rank deficiency of the design matrix. Under the ridge regression framework, the corresponding parameter estimate is of the form

$$\hat{\boldsymbol{\beta}} = \arg\min_{\beta \in \mathbb{R}^p} \left\{ \frac{1}{n} \|\boldsymbol{y} - \boldsymbol{X}\boldsymbol{\beta}\|_2^2 + \lambda \|\boldsymbol{\beta}\|_2^2 \right\}. \tag{12.3}$$

While ridge regression achieves shrinkage, it does not have the ability to achieve selection, because its ℓ_2 norm based penalty function is inherently non-sparsity inducing. [24] provides the seminal paper that formalized the Least Absolute Shrinkage and Selection Operator (LASSO), which, as the name suggests achieves both shrinkage and variable selection. Despite its inherent computational and mathematical challenges, along with the fact that it does not yield a unique solution to the estimation problem, the phenomenal success of the LASSO caused it to be widely used and extended by a wide variety of authors. The LASSO corresponds to the use of the ℓ_1 norm as the penalty function, and is often written in regularization format as

$$\hat{\beta} = \arg\min_{\beta \in I\!\!R^p} \left\{ \frac{1}{n} \|y - X\beta\|_2^2 + \lambda \sum_{j=1}^{p} |\beta_j| \right\} \tag{12.4}$$

or as a constrained optimization problem

$$\hat{\beta} = \arg\min_{\beta \in I\!\!R^p} \left\{ \frac{1}{n} \|y - X\beta\|_2^2 \right\} \quad \texttt{subject to} \quad \sum_{j=1}^{p} |\beta_j| < \tau. \tag{12.5}$$

The popular Least Angle Regression [7] is one of the most direct extensions of LASSO. Motivated by the need to further extend the LASSO, [28] proposed the Elastic Net regularizer to achieve both variable selection and variable clustering, that has also become very popular with data mining practitioners. [13] provide a detailed account of some of the modern approaches to model selection in general and variable selection in particular. Another popular and very substantial adaptation of the use of the ℓ_1 norm as a device for achieving variable selection is captured by the so-called Dantzig selector of [3] who provide an extension of the LASSO capable of handling regression analysis problems when p is much larger than n. [2] provide a very illuminating discussion of the breakthrough offered by [3]. Alongside the above frequentist contributions to variable selection, various Bayesian statisticians have contributed a wealth of scholarly research work covering both the traditional setting of variable selection where n is much larger than p and the now popular and more tricky short fat data context where p is much larger than n. The vast majority of recent Bayesian contributions to variable selection have concentrated on the use of conjugate prior, with the typical choice of prior on β being a Gaussian prior of the form

$$\beta | \sigma^2, W \sim \texttt{MVN}\left(0, \sigma^2 W^{-1}\right), \tag{12.6}$$

where W is the prior precision matrix. Of course, the use of a zero mean prior expresses the assumption of many insignificant coefficients. The LASSO approach to variable selection mentioned earlier, can be formulated in the Bayesian framework, using the prior

$$p(\beta_j|\lambda) = \frac{1}{2\lambda} \exp\left(-\frac{1}{\lambda}|\beta_j|\right), \tag{12.7}$$

which corresponds to a Bayesian formulation of a regularized approach to variable selection. While this regularized framework approach to variable selection by soft thresholding has enjoyed tremendous success and popularity, the vast majority of Bayesians have adopted an approach to variable selection that uses hard thresholding and a lot of model space search. In the most commonly used Bayesian formulation, the prior density of β does not need to have any sparsity inducing properties, whereas in the regularization framework, the success of the selection depends on the prior density's sparsity inducing properties. In both cases, as one would expect, the computation can quickly become prohibitive. To the best of our knowledge, one of the early papers that formalized the currently used approach to Bayesian selection is [11], who provide one of the earliest thorough details of the use of the g-prior of [26] in Bayesian Variable Selection. It is important to note however that [12] provided an even earlier detailed account on the computational implementation of Bayesian Variable Selection *via* the Gibbs Sampling. Arguably, one of the greatest appeals of the *g*-prior lies in the fact that it allows the computation of the *marginal likelihood* of each submodel in closed-form, and thereby the computation of the crucially important *posterior inclusion probability* of each variable.

One of the key building blocks of the Bayesian variable selection machinery is the use of a vector of indicator variables. With the p original predictor variables, there are $2^p - 1$ non empty models each corresponding to a subset of the provided variables. We shall use a vector $\boldsymbol{\gamma} = (\gamma_1, \gamma_2, \cdots, \gamma_p)^\top$ to denote the index of a given model, with each γ_j being an indicator of the variable's presence in the model under consideration, namely

$$\gamma_j = \begin{cases} 1 & \texttt{If variable } X_j \texttt{ appears in the model} \\ 0 & \texttt{Otherwise} \end{cases}$$

Clearly, $\boldsymbol{\gamma} = (1, 1, \cdots, 1)^\top$ corresponds to the *full model* M_f, while $\boldsymbol{\gamma} = (0, 0, \cdots, 0)^\top$ corresponds to the empty model also referred to as the *null model*, and given by

$$M_0 : \quad \boldsymbol{Y} = \alpha \mathbf{1}_n + \boldsymbol{\epsilon}. \tag{12.8}$$

Equipped with this index, $p_{\boldsymbol{\gamma}} = \sum_{j=1}^p \gamma_j$ is the number of predictor variables in model $M_{\boldsymbol{\gamma}}$, and $\boldsymbol{\beta}_{\boldsymbol{\gamma}}$ is the subset of $\boldsymbol{\beta}$ made up of only the β_j's picked up by $\boldsymbol{\gamma}$. Finally, $\boldsymbol{X}_{\boldsymbol{\gamma}}$ is the submatrix of $\boldsymbol{X}$ whose columns are only those $p_{\boldsymbol{\gamma}}$ columns of $\boldsymbol{X}$ picked up by $\boldsymbol{\gamma}$, so that $\boldsymbol{X}_{\boldsymbol{\gamma}}$ is really an $n \times p_{\boldsymbol{\gamma}}$ matrix, and the corresponding model $M_{\boldsymbol{\gamma}}$ is given by

$$M_{\boldsymbol{\gamma}} : \quad \boldsymbol{Y} = \alpha \mathbf{1}_n + \boldsymbol{X}_{\boldsymbol{\gamma}} \boldsymbol{\beta}_{\boldsymbol{\gamma}} + \boldsymbol{\epsilon}. \tag{12.9}$$

Without loss of generality, we will assume from now on that the columns of $\boldsymbol{X}$ are standardized in such a way that $\mathbf{x}_j^\top \mathbf{1}_n = 0$ and $\mathbf{x}_j^\top \mathbf{x}_j / n = 1$, for $j = 1, \cdots, p$.

12.2.2 Methods of response prediction

For any given model $M_{\boldsymbol{\gamma}}$ indexed by $\boldsymbol{\gamma}$, the OLS estimate of the regression coefficient is given by

$$\hat{\boldsymbol{\beta}}_{\boldsymbol{\gamma}}^{\texttt{(OLS)}} = \hat{\boldsymbol{\beta}}_{\boldsymbol{\gamma}} = (\boldsymbol{X}_{\boldsymbol{\gamma}}^\top \boldsymbol{X}_{\boldsymbol{\gamma}})^{-1} \boldsymbol{X}_{\boldsymbol{\gamma}}^\top \boldsymbol{Y}, \tag{12.10}$$

while the corresponding Bayesian estimate is given by

$$\tilde{\beta}_{\gamma}^{(\text{Bayes})} = \tilde{\beta}_{\gamma} = \mathbb{E}[\beta_{\gamma}|M_{\gamma}, Y] = \int \beta_{\gamma} p(\beta|M_{\gamma}, Y) d\beta_{\gamma} \qquad (12.11)$$

Least squares prediction of Y: It is easy to see that the frequentist estimated average response at x is given by

$$\hat{Y}_{\gamma}^* = \hat{f}_{\gamma}^{(\text{OLS})}(\mathbf{x}) = \sum_{j=1}^{p} I(\gamma_j = 1) \mathbf{x}_j \hat{\beta}_{\gamma_j} = \mathbf{x}^{\top} V_{\gamma} \hat{\beta}_{\gamma} \qquad (12.12)$$

where $\hat{\beta}_{\gamma} = (\hat{\beta}_{\gamma_1}, \cdots, \hat{\beta}_{\gamma_p})^{\top}$ with $\hat{\beta}_{\gamma_j} = 0$ if $\gamma_j = 0$, V_{γ} is the $p \times p_{\gamma}$ matrix such that for $\mathbf{x} \in I\!R^p$, the vector $\mathbf{x}^{\top} V_{\gamma}$ is a p_{γ}-dimensional subvector of $\mathbf{x}$ corresponding to $\gamma_j \neq 0$ (nonzero coordinates of γ). For example, if $p = 5$ and we have $\gamma = (1, 0, 1, 0, 1)$, then

$$V_{\gamma} = \begin{bmatrix} 1 & 0 & 0 \\ 0 & 0 & 0 \\ 0 & 1 & 0 \\ 0 & 0 & 0 \\ 0 & 0 & 1 \end{bmatrix} \quad \text{and} \quad \mathbf{x}^{\top} V_{\gamma} = [\mathbf{x}_1, \mathbf{x}_2, \mathbf{x}_3, \mathbf{x}_4, \mathbf{x}_5] \begin{bmatrix} 1 & 0 & 0 \\ 0 & 0 & 0 \\ 0 & 1 & 0 \\ 0 & 0 & 0 \\ 0 & 0 & 1 \end{bmatrix} = \begin{bmatrix} \mathbf{x}_1 \\ \mathbf{x}_3 \\ \mathbf{x}_5 \end{bmatrix}$$

Generic Bayesian prediction of Y: From a Bayesian perspective, if model M_{γ} is selected, then the predictor of the response Y given $\mathbf{x}$ is given by

$$\hat{Y}_{\gamma} = \hat{f}_{\gamma}^{(\text{Bayes})}(\mathbf{x}) = \mathbf{x}^{\top} V_{\gamma} \mathbb{E}[\beta_{\gamma}|M_{\gamma}, Y] = \mathbf{x}^{\top} V_{\gamma} \tilde{\beta}_{\gamma} = \sum_{j=1}^{p} I(\gamma_j = 1) \mathbf{x}_j \tilde{\beta}_{\gamma_j} \quad (12.13)$$

where $\tilde{\beta}_{\gamma} = (\tilde{\beta}_{\gamma_1}, \cdots, \tilde{\beta}_{\gamma_p})^{\top}$ with $\tilde{\beta}_{\gamma_j} = 0$ if $\gamma_j = 0$.

Optimal Prediction of Y: A well-known theoretical result, see [1], establishes that the optimal Bayesian predictor of Y is well known to be the model averaging predictor.

$$\begin{aligned} \tilde{Y} &= \hat{f}_{\gamma}^{(\text{BMA})}(\mathbf{x}) = \mathbf{x}^{\top} \sum_{\gamma \in \Gamma} p(M_{\gamma}|Y) V_{\gamma} \tilde{\beta}_{\gamma} \\ &= \sum_{\gamma \in \Gamma} \sum_{j=1}^{p} I(\gamma_j = 1) p(M_{\gamma}|Y) \mathbf{x}_j \hat{\beta}_{\gamma_j} \end{aligned} \qquad (12.14)$$

where $\Gamma = \{0, 1\}^p$ is the entire model space, made up of the 2^p vectors of p indicators, and $p(M_{\gamma}|Y)$, the posterior probability of M_{γ} given Y is given by

$$\begin{aligned} p(M_{\gamma}|Y) &= \frac{p(M_{\gamma}) m_{\gamma}(Y)}{\sum_{\gamma} p(M_{\gamma}) m_{\gamma}(Y)} = \frac{p(Y|M_{\gamma}) p(M_{\gamma})}{p(Y)} = \frac{p(Y|M_{\gamma}) p(M_{\gamma})}{\sum_{j=1}^{2^p} p(Y|M_j) p(M_j)} \\ &= \frac{p(M_{\gamma}) \mathbf{BF}_{\gamma:0}}{\sum_{\gamma} p(M_{\gamma}) \mathbf{BF}_{\gamma:0}} \end{aligned} \qquad (12.15)$$

where $m_{\boldsymbol{\gamma}}(\boldsymbol{Y})$ is the marginal density of Y under the model $M_{\boldsymbol{\gamma}}$ given by

$$m_{\boldsymbol{\gamma}}(\boldsymbol{Y}) = p(\boldsymbol{Y}|M_{\boldsymbol{\gamma}}) = \int p(\boldsymbol{Y}|\boldsymbol{X}_{\boldsymbol{\gamma}}, \boldsymbol{\beta}_{\boldsymbol{\gamma}}, \sigma^2)p(\boldsymbol{\beta}_{\boldsymbol{\gamma}}, \sigma^2)d\boldsymbol{\beta}_{\boldsymbol{\gamma}}d\sigma^2.$$

and the Bayes Factor $\mathbf{BF}_{\boldsymbol{\gamma}:0}$ given by

$$\mathbf{BF}_{\boldsymbol{\gamma}:0} = \frac{m_{\boldsymbol{\gamma}}(\boldsymbol{Y})}{m_0(\boldsymbol{Y})}.$$

Median Probability Model predictor of Y: [1] defined the median probability model index vector $\boldsymbol{\gamma}^{(\mathtt{med})} = (\gamma_1^{(\mathtt{med})}, \cdots, \gamma_p^{(\mathtt{med})})^{\top}$ where

$$\gamma_j^{(\mathtt{med})} = \begin{cases} 1 & \text{if } p_j = \mathtt{PIP}_j \geq \frac{1}{2} \\ 0 & \text{otherwise} \end{cases}$$

with the posterior inclusion probability $p_j = \mathtt{PIP}_j$ given by

$$p_j = \mathtt{PIP}_j = \Pr[\gamma_j = 1|\boldsymbol{Y}] = \sum_{\boldsymbol{\gamma} \in \boldsymbol{\Gamma}} I(\gamma_j = 1)p(M_{\boldsymbol{\gamma}}|\boldsymbol{Y}),$$

In other words, the median probability model is simply the model made up of variables appearing in at least half of the models in model space. The median probability model introduced and developed in [1] seeks to achieve optimal prediction along with consistent model selection. It has the main limitation that the model does not always exist. [9] provides a remedy to this limitation by suggesting an approach for optimal predictive atom selection in the general basis function expansion framework. If one decides to choose the median probability model $M_{\boldsymbol{\gamma}^{(\mathtt{med})}}$, then

$$\begin{aligned} \tilde{Y}_{\boldsymbol{\gamma}^{(\mathtt{med})}} &= \hat{f}_{\boldsymbol{\gamma}^{(\mathtt{med})}}^{(\mathtt{Bayes})}(\mathbf{x}) = \mathbf{x}^{\top} V_{\boldsymbol{\gamma}^{(\mathtt{med})}} \mathbb{E}[\boldsymbol{\beta}_{\boldsymbol{\gamma}^{(\mathtt{med})}}|M_{\boldsymbol{\gamma}^{(\mathtt{med})}}, \boldsymbol{Y}] \\ &= \mathbf{x}^{\top} V_{\boldsymbol{\gamma}^{(\mathtt{med})}} \tilde{\boldsymbol{\beta}}_{\boldsymbol{\gamma}^{(\mathtt{med})}} = \sum_{j=1}^{p} I(\gamma_j^{(\mathtt{med})} = 1)\mathbf{x}_j \tilde{\beta}_{\gamma_j^{(\mathtt{med})}} \end{aligned} \quad (12.16)$$

Highest Posterior Probability Model predictor of Y: An alternative to the median probability model is the so-called highest posterior model (HPM), whose model index vector is given by

$$\boldsymbol{\gamma}^{(\mathtt{HPM})} = \operatorname*{argmax}_{\boldsymbol{\gamma} \in \boldsymbol{\Gamma}} \{p(M_{\boldsymbol{\gamma}}|\boldsymbol{Y})\}.$$

In other words, the HPM model is the model with the highest posterior probability.

BIC predictor of Y: One may also consider the BIC model defined by

$$\boldsymbol{\gamma}^{(\mathtt{BIC})} = \operatorname*{argmin}_{\boldsymbol{\gamma} \in \boldsymbol{\Gamma}} \{\mathtt{BIC}(M_{\boldsymbol{\gamma}})\} = \operatorname*{argmax}_{\boldsymbol{\gamma} \in \boldsymbol{\Gamma}} \{e^{-\mathtt{BIC}(M_{\boldsymbol{\gamma}})}\}.$$

where

$$\texttt{BIC}(M_\gamma) = -2\log\widehat{L(M_\gamma)} + p\log(n)$$

and $\widehat{L(M_\gamma)}$ is the estimated maximum likelihood for model M_γ.

AIC predictor of Y: Closely related to the BIC score function is the AIC score function

$$\texttt{AIC}(M_\gamma) = -2\log\widehat{L(M_\gamma)} + 2p$$

which can also be used in the assessment of predictive performance of estimating functions.

LASSO predictor of Y: Given the immense popularity of the LASSO, it is important to compare its predictive performance to those of the other methods. We define

$$\hat{f}^{(\texttt{LASSO})}(\mathbf{x}) = \mathbf{x}^\top \mathbf{V}_{\gamma^{(\texttt{LASSO})}} \hat{\boldsymbol{\beta}}^{(\texttt{LASSO})}$$

where

$$\hat{\boldsymbol{\beta}}^{(\texttt{LASSO})} = \arg\min_{\boldsymbol{\beta}\in I\!R^p} \left\{ \frac{1}{n}\|\mathbf{y} - \mathbf{X}\boldsymbol{\beta}\|_2^2 + \lambda\sum_{j=1}^{p}|\beta_j| \right\},$$

and $\gamma_j^{(\texttt{LASSO})} = (\gamma_1^{(\texttt{LASSO})}, \cdots, \gamma_p^{(\texttt{LASSO})})$ with

$$\gamma_j^{(\texttt{LASSO})} = \begin{cases} 1 & \text{if} \quad \hat{\beta}_j^{(\texttt{LASSO})} \neq 0 \\ 0 & \text{otherwise} \end{cases}$$

12.3 Bayesian Approaches to Variable Selection

Throughout the previous section, the predictions referred to were based on the knowledge of the model index indicator vector γ. In this section, we look at both the frequentist and Bayesian approaches to estimating γ from the data. As will become clear in the subsequent paragraph, finding γ corresponds to searching the model space, a task that can become computationally prohibitive when the dimension p of the input space becomes larger. This article does not intend to go into any details as far as model space search in concerned. We shall simply use some of the existing results and refer the reader to the literature for greater coverage on the topic. Many authors have contributed detailed reviews of Bayesian Variable Selection, and among them [25] provide somewhat of an overview, while [23] give a full review. [5] dedicate the entirety of their Chapter 10 to modern approaches to variable selection, with some highlights on the Bayesian approach. More recently, [4] provides an excellent review of the most recent advances in Bayesian variable selection, contrasting both frequentist and Bayesian contributions to the field, and touching on contributions as recent as the Dantzig selector [3].

12.3.1   Bayesian variable selection with Zellner's g-prior

Bayesian Variable Selection using Zellner's g-prior ([26]) has received considerable attention partly because it allows the derivation of closed-form expressions needed in model space search, and also because it tends to yield good results. One of the most recent accounts is given by [19] with greater details on aspects of model space search and explorations of various choices of g. The basic idea of the Zellner g-prior is that, for a model indexed by $\boldsymbol{\gamma}$, one specifies for $\boldsymbol{\beta}_{\boldsymbol{\gamma}}$ a Gaussian prior of the form $\boldsymbol{\beta}_{\boldsymbol{\gamma}}|\boldsymbol{\gamma}, g, \sigma^2 \sim \texttt{MVN}\left(\mathbf{0}, g\sigma^2 (\mathbf{X}_{\boldsymbol{\gamma}}^{\top} \mathbf{X}_{\boldsymbol{\gamma}})^{-1}\right)$ where $g > 0$ is a constant for which different authors have suggested various schemes for setting its value. In fact, one such author suggests setting $g = \max(n, p^2)$, but the literature on the topic is filled with many other choices of g. Now, with a g-prior on $\boldsymbol{\beta}_{\boldsymbol{\gamma}}$, the joint prior structure is given by $p(\boldsymbol{\beta}_{\boldsymbol{\gamma}}, \boldsymbol{\gamma}, \sigma^2 | \pi, g) = p(\boldsymbol{\beta}_{\boldsymbol{\gamma}}|\boldsymbol{\gamma}, g, \sigma^2)p(\boldsymbol{\gamma}|\pi)p(\sigma^2)$, where

$$p(\boldsymbol{\beta}_{\boldsymbol{\gamma}}|\boldsymbol{\gamma}, g, \sigma^2) = \texttt{MVN}\left(\mathbf{0}, g\sigma^2 (\mathbf{X}_{\boldsymbol{\gamma}}^{\top} \mathbf{X}_{\boldsymbol{\gamma}})^{-1}\right)$$

and the prior on $\boldsymbol{\gamma}$ is either uniform, i.e $p(\boldsymbol{\gamma}) = 2^{-p}$ or binomial, i.e. $p(\boldsymbol{\gamma}|\pi) = \pi^{p_{\boldsymbol{\gamma}}}(1-\pi)^{p-p_{\boldsymbol{\gamma}}}$. One common choice of the prior on σ^2 is $p(\sigma) \propto \sigma^{-1}$, although other choices like the Gamma distribution are plausible. It is straightforward to show that the Bayesian estimator of $\boldsymbol{\beta}_{\boldsymbol{\gamma}}$ is given by

$$\tilde{\boldsymbol{\beta}}_{\boldsymbol{\gamma}} = \mathbb{E}[\boldsymbol{\beta}_{\boldsymbol{\gamma}}|M_{\boldsymbol{\gamma}}, g, \mathbf{Y}] = \frac{g}{1+g}\hat{\boldsymbol{\beta}}_{\boldsymbol{\gamma}}^{(\texttt{OLS})},$$

which means that under the g-prior, the Bayesian estimate is a shrunken version of the least squares estimate. It is also easy to see that

$$\lim_{g \to \infty} \mathbb{E}[\boldsymbol{\beta}_{\boldsymbol{\gamma}}|\mathbf{Y}, g, M_{\boldsymbol{\gamma}}] = \lim_{g \to \infty}\left\{\frac{g}{1+g}\hat{\boldsymbol{\beta}}_{\boldsymbol{\gamma}}^{(\texttt{OLS})}\right\} = \hat{\boldsymbol{\beta}}_{\boldsymbol{\gamma}}^{(\texttt{OLS})}.$$

[27] proposed the use of a Cauchy prior for $\boldsymbol{\beta}_{\boldsymbol{\gamma}}$, which [19] reformulated as a mixture of g-priors with an inverse gamma prior $\texttt{InvGamma}(1/2, n/2)$ of g, specifically

$$\pi(g) = \frac{(n/2)^{1/2}}{\Gamma(1/2)}g^{-3/2}e^{-n/(2g)},$$

and

$$\pi(\boldsymbol{\beta}_{\boldsymbol{\gamma}}|\sigma^2) \propto \int N\left(\boldsymbol{\beta}_{\boldsymbol{\gamma}}|\mathbf{0}, \sigma^2 g(\mathbf{X}_{\boldsymbol{\gamma}}^{\top} \mathbf{X}_{\boldsymbol{\gamma}})^{-1}\right)\pi(g)dg.$$

[19] also suggested and deeply explored the so-called hyper-g prior as an alternative to [27] as

$$\pi(g) = \frac{a-2}{2}(1+g)^{-a/2}, \quad g > 0.$$

An extension of this prior used to address the consistency of the estimator is referred to by [19] as the hyper-g/n prior and defined as

$$\pi(g) = \frac{a-2}{2n} \left(1 + \frac{g}{n}\right)^{-a/2},$$

which yields the marginal likelihood

$$m_\gamma(Y) \propto \left(\frac{g}{1+g}\right)^{p_\gamma/2} \left(\frac{1}{1+g} Y^\top P_\gamma Y + \frac{g}{1+g} Y^\top (I_n - J_n)Y\right)^{-\frac{n-1}{2}}$$

for which a closed-form expression is found to be

$$m_\gamma(Y) = \frac{\Gamma((n-1)/2)}{\sqrt{\pi}^{(n-1)} \sqrt{n}} \left[Y^\top (I_n - J_n)Y\right]^{-\frac{n-1}{2}} \frac{(1+g)^{(n-p_\gamma-1)/2}}{[1 + g(1 - R_\gamma^2)]^{(n-1)/2}}.$$

The corresponding Bayes Factor is given by

$$\mathbf{BF}_{\gamma:0} = (1+g)^{(n-p_\gamma-1)/2} \left[1 + g(1 - R_\gamma^2)\right]^{-(n-1)/2}$$

where the multiple coefficient of determination R_γ^2 for M_γ is given by

$$1 - R_\gamma^2 = \frac{Y^\top (I_n - H_\gamma)Y}{Y^\top (I_n - J_n)Y}$$

with $H_\gamma = X_\gamma (X_\gamma^\top X_\gamma)^{-1} X_\gamma^\top$ and $P_\gamma = I_n - H_\gamma$, I_n is the identity matrix, and J_n is the n-dimensional matrix with all entries equal to $1/n$. It is remarkable to see that despite not being an inherently sparsity inducing prior, the Zellner's g-prior, by providing tractable marginal likelihood functions, allows computationally efficient model space searches and thereby practically useful variable selection in the Bayesian framework. For a model indexed by γ, the marginal likelihood is $m_\gamma(Y)$, and the appeal of the g-prior comes from the fact that with it, the marginal likelihood can be written in closed form. [8] give some recommendations on how to choose g in the g-prior context. In fact, the R package BMS offers a wide variety of options on how to set the g-prior hyperparameter g. It touches on the recommendations made by [10], [8], and [19]. Model-based choices of g include $g_\gamma = \max(0, F_\gamma - 1)$ where

$$F_\gamma = \frac{R_\gamma^2 / p_\gamma}{(1 - R_\gamma^2)/(n - p_\gamma - 1)}$$

and the hyper-g prior $\frac{g}{1+g} \sim \mathtt{beta}(1, a/2 - 1)$ where $2 < a \le 4$.

12.3.2 Spike and slab approach to variable selection

The spike and slab prior in the MLR model adhere to the following Bayesian hierarchy:

$$
\begin{aligned}
(Y_i|X_i, \boldsymbol{\beta}, \sigma^2) &\stackrel{\text{ind}}{\sim} N(X_i^t\boldsymbol{\beta}, \sigma^2), \quad i = 1, \dots, n, \\
(\boldsymbol{\beta}|\boldsymbol{\gamma}) &\sim N(\mathbf{0}, \boldsymbol{\Gamma}), \\
\boldsymbol{\gamma} &\sim \pi(d\boldsymbol{\gamma}), \\
\sigma^2 &\sim \mu(d\sigma^2),
\end{aligned}
\tag{12.17}
$$

where $\mathbf{0}$ is a p-dimensional vector of zero, $\boldsymbol{\Gamma}$ is the $p \times p$ diagonal matrix, π is the prior measure for $\boldsymbol{\Gamma}$ and μ is the prior measure for σ^2. [18] and later [22] pioneered the *spike and slab prior* in the context of variable selection in the MLR model. The terminology *spike and slab* denotes the prior for $\boldsymbol{\beta}$ in the Bayesian hierarchy. It turns out that it assigns a two-point mixture prior for β_j, $j = 1, 2, \dots, p$ comprises of a uniform flat distribution (the slab) and a degenerate distribution at zero (the spike). The main idea of using this special type of prior is to zero out β_j which are truly zero as contributing *via* their posterior mean very small. [12] proposed an alternative version of spike and slab prior using zero-one latent variables; each β_j in the Bayesian hierarchy entertains a scale mixture of two normal distributions as

$$
\beta_j|\gamma_j \sim (1 - \mathcal{I}_j)N(0, \tau_j^2) + \mathcal{I}_j N(0, c_j\tau_j^2), \quad j = 1, \dots, p.
$$

In general, τ_j is chosen to be a small value and $c_j > 0$ to be some comfortably large value. This results in getting variables with larger hypervariances corresponding to latent variables $\gamma_j = 1$, producing larger posterior values of β_js. The reverse happens while $\gamma_j = 0$. In this hierarchy, it is also customary to assign prior for the latent variable $\mathcal{I}_j$. A common practice is to use an indifference or uniform prior by assuming $\mathcal{I}_j$s as independent Bernoulli(w_j) random variables with $0 < w_j < 1$. Then the prior $\pi(d\boldsymbol{\gamma})$ in (12.17) reduces to the following

$$
\begin{aligned}
(\gamma_j|c_j, \tau_j^2, \mathcal{I}_j) &\stackrel{\text{ind}}{\sim} (1 - \mathcal{I}_j)\delta_{\tau_j^2}(\cdot) + \mathcal{I}_j\delta_{c_j\tau_j^2}(\cdot), \quad j = 1, 2, \dots, p, \\
(\mathcal{I}_j|w_j) &\stackrel{\text{ind}}{\sim} (1 - w_j)\delta_0(\cdot) + w_j\delta_1(\cdot),
\end{aligned}
\tag{12.18}
$$

where $\delta_t(\cdot)$ denotes a discrete measure at the value t. [17] proposed a rescaled version of spike and slab model in modifications to [12] as follows:

$$
\begin{aligned}
(Y_i^*|X_i, \boldsymbol{\beta}, \sigma^2) &\stackrel{\text{ind}}{\sim} N(X_i^t\boldsymbol{\beta}, n\sigma^2), \quad i = 1, \dots, n, \\
(\beta_j|\mathcal{I}_j, \tau_j^2) &\stackrel{\text{ind}}{\sim} N(0, \mathcal{I}_j\tau_j^2), \quad i = 1, \dots, p, \\
(\mathcal{I}_j|v_0, w) &\stackrel{\text{iid}}{\sim} (1 - w)\delta_{v_0}(\cdot) + w\delta_1(\cdot), \\
(\tau_j^{-2}|s_1, s_2) &\stackrel{\text{iid}}{\sim} \text{Gamma}(c_1, c_2), \\
w &\sim \text{Uniform}(0, 1), \\
\sigma^2 &\sim \text{InvGamma}(s_1, s_2).
\end{aligned}
\tag{12.19}
$$

In model (12.19), the two-point mixture prior in [12] has been replaced by a continuous bimodal prior to get rid of selecting hyper-parameters of the improper priors. Besides, a major breakthrough improvised via rescaling the response Y_i by $Y_i^* = \hat{\sigma}^{-1} n^{1/2} Y_i$, where $\hat{\sigma}^2$ is an unbiased estimator for σ^2 based on the full model. [17] pointed out that besides the flexibility of the spike and slab priors, as in the case of other priors, it has non-enforceable effect on the posterior as compared to the increase in the sample size. So there should be some adjustment required in the prior structure to avoid this issue. They have showed that rescaling the response helps to avoid this issue. For further details, we refer readers to [17]. [6] modified the rescaled spike and slab model by using a two-component mixture prior to incorporate the selective shrinkage property in prior specification. Besides this feature, it has been shown that because of this choice of prior, the resulting posterior mean approximately equals to the the restricted least squares estimates. So the outputs from the Bayesian hierarchy can be used for further frequentist model selection procedures. [20] used different choices of spike and slab priors to compare the performance of posterior inclusion probabilities based on MCMC sampling. [16] have written and presented an R package implementing the Spike and Slab approach to Bayesian Variable Selection and Prediction, essentially implementing the concepts, methods and results contained in [17].

12.4 Computational Demonstrations

Among the many R packages implementing Bayesian variable selection, the most noteworthy are BMS, spikeslab. The package lars, does not profess to be Bayesian but implements Efron's Least Angle Regression [7] and offers an implementation of the LASSO [24], which is set in the regularization framework but admits a Bayesian formulation using the Laplace distribution as the prior on the regression coefficient. For our comparison of the methods presented in this article, we use several datasets with distinct qualitative and quantitative properties. The packages contained in the Bayesian CRAN view of R offer not only functions for performing Bayesian variable selection and Bayesian model averaging but also contain a wide variety of datasets.

12.4.1 Analysis of the Diabetes dataset

Our first dataset is the diabetes data from [16] and [17]. The original version of this data has $n = 442$ observations on $p = 64$ variables. In the spikeslab package, [16] considered a p larger than n setup by expanding the data to become $n = 442$ and $p = 2064$ so that the 2000 additional variables are in reality just noise.

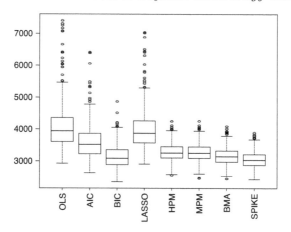

FIGURE 12.1
Boxplot of test errors for the Diabetes dataset with interactions.

The boxplot in Figure 12.1 shows the predictive performances of the methods explored. Each boxplot is constructed based on $R = 500$ random realizations of the test error. For this dataset, the spike and slab method yields the lowest average test error, while the worst method is the LASSO.

12.4.2   Analysis of the Boston housing dataset

The Boston housing dataset has $n = 506$ and $p = 13$. This dataset has been used as a benchmark dataset in statistics and machine learning communities for many years; available in the R package MASS. Despite not being of a particularly high dimension (p is only 13 variables), this dataset has been deemed rich in various aspects that make it interesting in evaluating regression methods.

Figure 12.2 shows a disappointing performance of all the methods based on the g-prior on the dataset. The spike and slab does well on this dataset, so does the basic OLS. It is interesting to further investigate as to why the methods based on the g-prior are predictively inferior to the other methods in this dataset. Could the standardization of the dataset lead to better predictive performances as a result of better model space searches?

12.4.3   Analysis of the Prostate cancer dataset

Another dataset used in our comparison is the prostate cancer dataset which is small in size compared to the previous dataset with $n = 97$ and $p = 8$; available in the R package lasso2.

The boxplot in Figure 12.3 shows the predictive performances of the methods explored. For this dataset, no methods emerge as predictively superior.

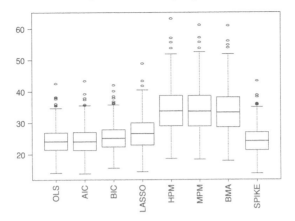

FIGURE 12.2
Boxplot of test errors for the Boston Housing dataset.

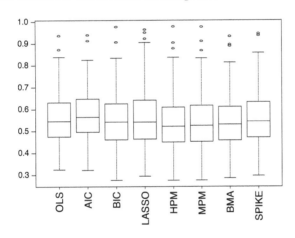

FIGURE 12.3
Boxplot of test errors for the Prostate cancer dataset.

12.4.4 Analysis of the Prestige dataset

The Prestige dataset used here comes from the R package car. This dataset has $n = 102$ rows and we use only the $p = 4$ numeric variables (excluding the categorical one). Clearly, this dataset is dimensionally smaller than the previous ones.

Figure 12.4 shows that all methods yield almost the same predictive performances except the LASSO, which is below par when compared to the rest. It is interesting to explore this dataset which causes the LASSO method to under perform a good deal in comparison to the other methods.

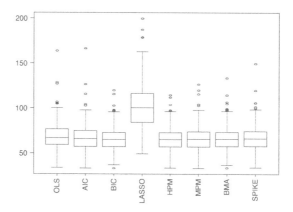

FIGURE 12.4
Boxplot of test errors for the Prestige dataset.

12.4.5   Analysis of the Cars dataset

The cars dataset comes from the R package datasets. The cars dataset has $n = 80$ and $p = 4$. Here we try to predict the gas mileage (MPG) of a car given four characteristics, namely, VOL, HP, SP and WT. We spot from Figure 12.5 that no method emerges as predictively superior for this dataset.

12.4.6   Analysis of the Hald Cement dataset

The Hald Cement dataset comes from the R package MPV. This dataset with $n = 13$ and $p = 4$ is clearly the smallest dataset considered here. However, with the presence of strong multicollinearity, it is exciting to probe how different methods perform predictively on it.

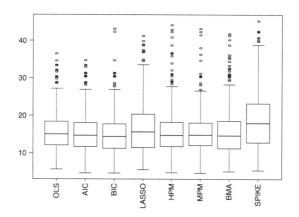

FIGURE 12.5
Boxplot of test errors for the Cars dataset.

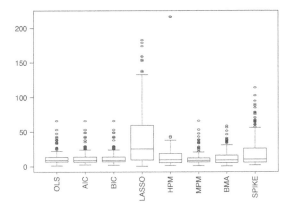

FIGURE 12.6
Boxplot of test errors for the Hald Cement dataset.

12.4.7 Analysis of the Attitude dataset

The attitude dataset has $n = 30$ and $p = 6$ and contains some levels of multicollinearity and many noise variables; available in the R package datasets. Figure 12.7 shows that HPM gives better performance than all other methods.

12.4.8 Simulated example with interactions

The dataset in this example is simulated data with different setup on the level of correlation among the variables, and the ratio n and p. In this simulated example, the true function is

$$f(\mathbf{x}) = 1 + 2\mathbf{x}_3 + \mathbf{x}_7 + 3\mathbf{x}_9 - 2\mathbf{x}_1\mathbf{x}_7 - 1.5\mathbf{x}_2\mathbf{x}_6$$

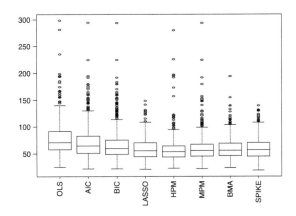

FIGURE 12.7
Boxplot of test errors for the Attitude dataset.

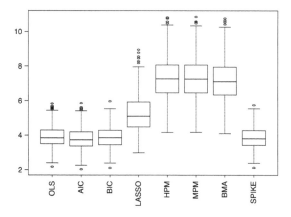

FIGURE 12.8
Boxplot of test errors for the simulated dataset with interactions.

where $X \sim \mathtt{MVN}(\mathbf{1}_9, \Sigma_\rho)$ and $\epsilon \sim \mathbf{N}(0, 2^2)$. It turns out that the presence of interactions leads to various forms of difficulties for different methods. In particular, we simulate data by defining $\rho \in [0, 1)$ and $\tau = 1$, then generate our predictor variables using a multivariate normal distribution with zero mean and the following covariance matrix.

$$\Sigma = \tau \begin{bmatrix} 1 & \rho & \rho^2 & \cdots & \rho^{p-1} \\ \rho & 1 & \rho & \cdots & \rho^{p-2} \\ \vdots & \vdots & \vdots & \ddots & \vdots \\ \rho^{p-1} & \rho^{p-2} & \rho^{p-3} & \cdots & 1 \end{bmatrix}.$$

Figures 12.8 and 12.9 are the boxplots representing the test error performance of several methods considered in the simulated example. In both situations, surprisingly OLS performs better than other methods.

12.4.9 Final predictive comparison tables

We have considered several datasets arising from diverse research problems. Several frequentist and Bayesian techniques have been used to analyze these datasets. Tables 12.1 and 12.2 represent the predictive performances of the methods on previously considered datasets. It is interesting to note that no method outperforms the other uniformly across these datasets. This phenomenon is referred to in the data mining and machine learning community as the *"No free lunch theorem"* [5]. Surprisingly, the predictive performance of simple methods like OLS sometimes is better than sophisticated methods like LASSO or spike and slab model.

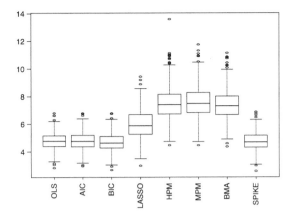

FIGURE 12.9
Boxplot of test errors for the simulated dataset with interactions and a correlation of $\rho = 0.72$.

TABLE 12.1
Test errors over $R = 500$ hold-out subsamples, with the training set representing $2/3$ of the data. The best performances are shown in boldface while the worst are *italic*.

	OLS	AIC	BIC	LASSO
Cement	11.48	11.26	11.28	*38.94*
Prestige	69.41	67.85	**66.00**	*99.65*
Attitude	*77.39*	71.36	66.48	60.69
Prostate	0.56	*0.57*	0.55	0.55
Boston	24.38	**24.34**	25.41	26.81
Diabetes (B)	**30.42**	30.32	30.58	30.73
Diabetes (I)	*40.52*	35.76	31.38	39.47
Cars	15.40	15.15	**14.98**	16.32
Simulated (I)	5.09	5.01	**4.93**	5.95
Simulated (I) ($\rho = 0.72$)	4.76	4.77	**4.68**	5.97

12.5   Conclusion and Discussion

We have considered a realistic comparison of the predictive performances of some of the common methods of variable selection and model selection in the context of the traditional multiple linear regression model. Our comparisons involve a mixture of both frequentist and Bayesian methods, under various scenarios of inherent data structure. Our results reveal that there is no such thing as a universally superior technique that would outperform all other techniques

TABLE 12.2

Test errors over $R = 500$ hold-out subsamples, with the training set representing 2/3 of the data. The best performances are shown in boldface while the worst are *italic*. Here we are comparing Bayesian based methods along with the LASSO. Interestingly, BMA never comes out as the worst.

	LASSO	HPM	MPM	BMA	SPIKE
Cement	*38.94*	12.82	**10.78**	12.28	19.66
Prestige	*99.65*	66.11	66.80	66.57	67.10
Attitude	60.69	**56.34**	59.49	59.66	60.08
Prostate	0.55	**0.54**	**0.54**	**0.54**	0.55
Boston	26.81	*34.29*	34.21	33.83	24.41
Diabetes (B)	30.73	31.05	*31.24*	30.66	**30.42**
Diabetes (I)	39.47	32.81	32.71	31.57	**30.52**
Cars	15.40	15.45	15.36	15.47	*18.70*
Simulated (I)	5.95	*7.88*	7.85	7.70	5.02
Simulated (I) ($\rho = 0.72$)	5.97	7.50	*7.58*	7.38	4.73

all the time regardless of the dataset at hand. However from a predictive optimality perspective, model averaging seem to emerge as the best in the sense that it always produces a relatively low average test error, even though it is not the lowest. As expected, model space search turns out to be computationally intensive, and can become extremely slow when, p, the number of explanatory variables increase. A useful and appealing area of future research is the continued involvement in the development of more computationally efficient model space search methods. The LASSO seems to substantially underperform predictively as the correlation among predictor variables gets larger, especially when the design is random as opposed to the traditionally assumed fixed design. Also noteworthy is the fact that all the methods based on the g-prior for variable selection appear to under perform when there are interactions among the variables. Interestingly, the spike and slab method consistently yields good predictive performances when there are interactions. It would be interesting to analyze using all the currently available choices of the hyperparameter g of Zellner's g-prior. It is reassuring that at present there exists many readily available computer packages, in R that allow practitioners to perform variable selection for a variety of real life scenarios. In the traditional setting where n is much larger than p, i.e., the number of observations is more than the number of predictor variables, both frequentist and Bayesian methods achieve comparable/similar performances. In both frequentist and Bayesian frameworks, handling variable selection when p is much larger than n must be done with extra care and caution. [21] introduces the generalized g-prior and provides a thorough treatment of linear regression when p is larger than n, offering both theory and computational insights.

Bibliography

[1] M. M. Barbieri and J. O. Berger. Optimal predictive model selection. *Annals of Statistics*, 32(3):870–897, 2004.

[2] T. T. Cai and J. Lv. Discussion: The Dantzig selector: Statistical estimation when p is much larger than n. *Annals of Statistics*, 35(6):2365–2369, 2007.

[3] E. Candes and T. Tao. The Dantzig selector: Statistical estimation when p is much larger than n. *Annals of Statistics*, 35(6):2313–2351, 2007.

[4] G. Celeux, M. El Anbari, J.-M. Marin, and C. P. Robert. Regularization in regression: Comparing Bayesian and frequentist methods in a poorly informative situation. *Bayesian Analysis*, 7(2):477–502, 2012.

[5] B. Clarke, E. Fokoue, and H. Zhang. *Principles and Theory for Data Mining and Machine Learning*. Springer Verlag, New York, 1st edition, 2009.

[6] T. Dey. A bimodal spike and slab model for variable selection and model exploration. *Journal of Data Science*, 10:363–383, 2012.

[7] B. Efron, T. Hastie, I. Johnstone, and R. Tibshirani. Least angle regression (with discussion). *Annals of Statistics*, 32:407–499, 2004.

[8] C. Fernández, E. Ley, and M. F. J. Steel. Benchmark priors for Bayesian model averaging. *Journal of Econometrics*, 100(2):381–427, 2001.

[9] E. Fokoué. Estimation of atom prevalence for optimal prediction. *Contemporary Mathematics (The American Mathematical Society)*, 443:103–129, 2007.

[10] D. P. Foster and E. I. George. The risk inflation factor for multiple regression. *Annals of Statistics*, 22:1947–1975, 1994.

[11] E. I. George and D. P. Foster. Calibration and empirical Bayes variable selection. *Biometrika*, 87(4):731–747, 2000.

[12] E. I. George and R. E. McCulloch. Variable selection via Gibbs sampling. *Journal of the American Statistical Association*, 88(423):881–889, 1993.

[13] T. Hastie, R. Tibshirani, and J. Friedman. *Elements of Statistical Learning*. Springer Verlag, New York, 2001.

[14] A. E. Hoerl and R. W. Kennard. Ridge regression: Applications to nonorthogonal problems. *Technometrics*, 12(1):69–82, 1970.

[15] A. E. Hoerl and R. W. Kennard. Ridge regression: Biased estimation for nonorthogonal problems. *Technometrics*, 12(1):55–67, 1970.

[16] H. Ishwaran, U. B. Kogalur, and J. S. Rao. spikeslab: Prediction and variable selection using spike and slab regression. *The R Journal*, 2(2):68–73, December 2010.

[17] H. Ishwaran and J. S. Rao. Spike and slab variable selection: Frequentist and Bayesian strategies. *Annals of Statistics*, 33(2):730–773, 2005.

[18] F. B. Lempers. *Posterior Probabilities of Alternative Linear Models*. Rotterdam University Press, Rotterdam, 1971.

[19] F. Liang, R. Paulo, G. Molina, M. A. Clyde, and J. O. Berger. Mixtures of g priors for Bayesian variable selection. *Journal of the American Statistical Association*, 103(481):410–423, 2008.

[20] G. Malsiner-Walli and H. Wagner. Comparing spike and slab priors for Bayesian variable selection. *Austrian Journal of Statistics*, 40(4):241–264, 2011.

[21] Y. Maruyama and E. I. George. Fully Bayes factors with a generalized g-prior. *The Annals of Statistics*, 39(5):2740–2765, 2011.

[22] T. J. Mitchell and J. J. Beauchamp. Bayesian variable selection in linear regression. *Journal of the American Statistical Association*, 83(404):1023–1032, 1988.

[23] R. B. O'Hara and M. J. Sillanpää. A review of Bayesian variable selection methods: What, how and which. *Bayesian Analysis*, 4(1):85–117, 2009.

[24] R. Tibshirani. Regression shrinkage and selection via the lasso. *Journal of the Royal Statistical Society: Series B (Statistical Methodology)*, 58(1):267–288, 1996.

[25] A. Yardimci and A. Erar. Bayesian variable selection in linear regression and a comparison. *Hacettepe Journal of Mathematics and Statistics*, 31:63–76, 2002.

[26] A. Zellner. *An Introduction to Bayesian Inference in Econometrics*. John Wiley & Sons, New York, 1971.

[27] A. Zellner and A. Siow. Posterior odds ratios for selected regression hypotheses. *Proceedings of the First International Meeting held in Valencia (Spain), (eds. J. M. Bernardo, M. H. DeGroot, D. V. Lindley and A. F. M. Smith)*, pages 585–603, 1980.

[28] H. Zou and T. Hastie. Regularization and variable selection via the elastic net. *Journal of the Royal Statistical Society: Series B (Statistical Methodology)*, 67(2):301–320, 2005.

13

Scalable Subspace Clustering with Application to Motion Segmentation

Liangjing Ding

Florida State University

Adrian Barbu

Florida State University

CONTENTS

13.1 Introduction

Subspace clustering is the problem of grouping an unlabeled set of points into a number of clusters corresponding to subspaces of the ambient space. This problem has applications in unsupervised learning and computer vision. One of the computer vision applications is motion segmentation, where a number of feature point trajectories need to be grouped into a small number of clusters according to their common motion model. The feature point trajectories are obtained by detecting a number of feature points using an interest point

267

detector and tracking them through many frames using a feature point tracker or an optical flow algorithm.

A common approach in the state of the art sparse motion segmentation methods (see [6, 12, 13, 24] and [27]) is to project the feature trajectories to a lower dimensional space and use a subspace clustering method based on spectral clustering to group the projected points and obtain the motion segmentation.

Even though these methods obtain very good results on standard benchmark datasets, the spectral clustering algorithm requires expensive computation of eigenvectors and eigenvalues on an $N \times N$ dense matrix where N is the number of data points. In this manner, the computation time for these subspace clustering/motion segmentation methods scales as $O(N^3)$, so it can become prohibitive for large problems (e.g. $N = 10^5 - 10^6$).

This chapter proposes a completely different approach to subspace clustering, based on the Swendsen–Wang Cut (SWC) algorithm [2] and brings the following contributions:

• The subspace clustering problem is formulated as Maximum A Posteriori (MAP) optimization problem in a Bayesian framework with Ising/Potts prior [16] and likelihood based on a linear subspace model.

• The optimization problem is solved using the Swendsen–Wang Cuts (SWC) algorithm and simulated annealing. The SWC algorithm needs a weighted graph to propose good data-driven clusters for label switching. This graph is constructed as a k-NN graph from an affinity matrix.

• The computation complexity of the SWC algorithm is evaluated and observed to scale as $O(N^2)$, making the proposed approach more scalable than spectral clustering (an $O(N^3)$ algorithm) to large scale problems.

• Motion segmentation experiments on the Hopkins 155 dataset are conducted and the performance of the proposed algorithm is compared with the state of the art methods. The SWC obtains an error less than twice the error of the state of the art methods. The experiments obtain an observed scaling of about $O(N^{1.3})$ for the SWC and about $O(N^{2.5})$ for the spectral clustering.

Compared to other statistical methods [7, 14, 21], the proposed SWC method does not require a good initialization, which can be hard to obtain.

Overall, the proposed method provides a new perspective to solve the subspace clustering problem, and demonstrates the power of the Swendsen-Wang Cuts algorithm in clustering problems. While it does not obtain a better average error, it scales better to large datasets, both theoretically as $O(N^2)$ and practically as $O(N^{1.3})$.

13.2   Subspace Clustering by Spectral Clustering

Given a set of points $\{\boldsymbol{x}_1, ..., \boldsymbol{x}_N\} \in \mathbb{R}^D$, the subspace clustering problem is to group the points into a number of clusters corresponding to linear subspaces of $\mathbb{R}^D$. The problem is illustrated in Figure 13.1, left, showing two linear subspaces and a number of outliers in $\mathbb{R}^2$.

A popular subspace clustering method ([5, 12, 17]) is based on spectral clustering, which relies on an affinity matrix that measures how likely any pair of points belong to the same subspace.

Spectral clustering [15, 18] is a generic clustering method that groups a set of points into clusters based on their connectivity. The point connectivity is given as an $N \times N$ affinity matrix A with A_{ij} close to 1 if point i is close to point j and close to zero if they are far away. The quality of the affinity matrix is very important for obtaining good clustering results. The affinity matrix for spectral subspace clustering will be described below.

The spectral clustering algorithm proceeds by computing the matrix $S = L^{-1/2}AL^{-1/2}$, where L is a diagonal matrix with L_{ii} as the sum of row i of A. It then computes the k-leading eigenvectors of S and obtains points in $\mathbb{R}^k$ from the eigenvectors. The points are then clustered by k-means or other clustering algorithm. The number k of clusters is assumed to be given. The spectral clustering algorithm is described in detail in Figure 13.2.

Affinity Matrix for Subspace Clustering. A common practice ([5, 12, 17]) before computing the affinity matrix is to normalize the points to unit length, as shown in Figure 13.1, right. Then the following affinity measure

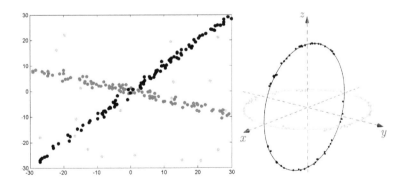

FIGURE 13.1
Left, two subspaces in 2D, Right. two 2D subspaces in 3D. The points in both 2D subspaces have been normalized to unit length. Due to noise, the points may not lie exactly on the subspace. One can observe that the angular distance finds the correct neighbors in most places except at the plane intersections.

Input: A set of points $x_1, \ldots, x_N \in \mathbb{R}^D$ and the number k of clusters.
1. Construct the affinity matrix $A \in \mathbb{R}^{N \times N}$.
2. Construct the diagonal matrix L, with $L_{ii} = \sum_{j=1}^{N} A_{ij}$.
3. .Compute $S = L^{-1/2} A L^{-1/2}$.
4. Compute the matrix $U = (u_1, \ldots, u_k) \in \mathbb{R}^{n \times k}$ containing the leading k eigenvectors of S.
5. Treat each row of U as point in $\mathbb{R}^k$ and normalize them to unit length.
6. Cluster the n points of $\mathbb{R}^k$ into k clusters by k-means or other algorithm.
7. Assign the points x_i to their corresponding clusters.

FIGURE 13.2
The spectral clustering algorithm [15].

based on the angle between the vectors has been proposed in [12]

$$A_{ij} = \left(\frac{x_i^T x_j}{\|x_i\|_2 \|x_j\|_2} \right)^{2\alpha}, \tag{13.1}$$

where α is a tuning parameter, and the value $\alpha = 4$ has been used in [12].

Fig 13.1, right shows two linear subspaces, where all points have been normalized. It is intuitive to find that the points tend to lie in the same subspace as their neighbors in angular distance except those near the intersection of the subspaces.

13.3   Scalable Subspace Clustering by Swendsen–Wang Cuts

This section presents a novel subspace clustering algorithm that formulates the subspace clustering problem as a MAP estimation of a posterior probability in a Bayesian framework and uses the Swendsen-Wang Cuts algorithm [2] for sampling and optimization.

A subspace clustering solution can be represented as a partition (labeling) $\pi : \{1, ..., N\} \to \{1, ..., M\}$ of the input points $x_1, \ldots, x_N \in \mathbb{R}^D$. The number $M \leq N$ is the maximum number of allowed clusters.

In this section, it is assumed that an affinity matrix A is given, representing the likelihood for any pair of points to belong to the same subspace. One form of A has been given in (13.1) and another one will be given in subsection 13.3.3.

13.3.1   Posterior probability

A posterior probability will be used to evaluate the quality of any partition π. A good partition can then be obtained by maximizing the posterior probability in the space of all possible partitions. The posterior probability is defined in a Bayesian framework as

$$p(\pi) \propto \exp[-E_{data}(\pi) - E_{prior}(\pi)].$$

The normalizing constant is irrelevant in the optimization since it will cancel out in the acceptance probability.

The data term $E_{data}(\pi)$ is based on the fact that the subspaces are assumed to be linear. Given the current partition (labeling) π, for each label l an affine subspace L_l is fitted in a least squares sense through all points with label l. Denote the distance of a point x with label l to the linear space L_l as $d(x, L_l)$. Then the data term is

$$E_{data}(\pi) = \sum_{l=1}^{M} \sum_{i, \pi(i)=l} d(x_i, L_l) \tag{13.2}$$

The prior term $E_{prior}(\pi)$ is set to encourage tightly connected points to stay in the same cluster.

$$E_{prior}(\pi) = -\rho \sum_{<i,j> \in E, \pi(i) \neq \pi(j)} \log(1 - A_{ij}), \tag{13.3}$$

where ρ is a parameter controlling the strength of the prior term. It will be clear in the next subsection that this prior is exactly the Potts model (13.4) that would have A_{ij} as the edge weights in the original SW algorithm.

13.3.2   Overview of the Swendsen–Wang Cuts algorithm

The precursor of the Swenden–Wang Cuts algorithm is the Swendsen-Wang (SW) method [20], which is a Markov Chain Monte Carlo (MCMC) algorithm for sampling partitions (labelings) $\pi : V \to \{1, ..., N\}$ of a given graph $G =< V, E >$. The probability distribution over the space of partitions is the Ising/Potts model [16]

$$p(\pi) = \frac{1}{Z} \exp[- \sum_{<i,j> \in E} \beta_{ij} \delta(\pi(i) \neq \pi(j)). \tag{13.4}$$

where $\beta_{ij} > 0, \forall < i, j > \in E$ and $N = |V|$.

The SW algorithm addresses the slow mixing problem of the Gibbs Sampler [10], which changes the label of a single node in one step. Instead, the SW algorithm constructs clusters of same label vertices in a random graph and flips together the label of all nodes in each cluster. The random graph is obtained by turning off (removing) all graph edges $e \in E$ between nodes with different labels and removing each of the remaining edges $< i, j > \in E$ with probability $e^{-\beta_{ij}}$. While the original SW method was developed originally for

Ising and Potts models, the Swendsen-Wang Cuts (SWC) method [2] generalized SW to arbitrary posterior probabilities defined on graph partitions.

The SWC method relies on a weighted adjacency graph $G =< V, E >$
where each edge weight $q_e, e =< i, j >\in E$ encodes an estimate of the probability that the two end nodes i, j belong to the same partition label. The
idea of the SWC method is to construct a random graph in a similar manner
to the SW but based on the edge weights, then select one connected component at random and accept a label flip of all nodes in that component with a
probability that is based on the posterior probabilities of the before and after
states and the graph edge weights.

This algorithm was proved to simulate ergodic and reversible Markov chain
jumps in the space of graph partitions and is applicable to arbitrary posterior
probabilities or energy functions.

From [2], there are different versions of the Swendsen-Wang Cut algorithm. The SWC-1 algorithm is used in this chapter, and is summarized in
Figure 13.3.

Input: Graph $G =< V, E >$, with weights q_e, $\forall e \in E$, and posterior prob-
ability $p(\pi|I)$.
Initialize: A partition $\pi : V \to \{1, ..., N\}$ by random clustering
For $t = 1, \ldots T$, current state π,

1. Find $E(\pi) = \{< i, j >\in E, \pi(i) = \pi(j)\}$
2. For $e \in E(\pi)$, turn $\mu_e =$ off with probability $1 - q_e$.
3. $V_l = \pi^{-1}(l)$ is divided into N_l connected components $V_l = V_{l1} \cup \ldots \cup V_{ln_l}$ for $l = 1, 2, \ldots, N$.
4. Collect all the connected components in set $CP = \{V_{li} : l = 1, \ldots, N, i = 1, \ldots, N_l\}$.
5. Select a connected component $V_0 \in CP$ with probability $q(V_0|CP) = \dfrac{1}{|CP|}$, say $V_0 \subset V_l$.
6. Propose to assign V_0 a new label $c_{V_0} = l'$ with a probability $q(l'|V_0, \pi)$, thus obtaining a new state π'.
7. Accept the proposed label assignment with probability

$$\alpha(\pi \to \pi') = \min \left(1, \frac{\prod_{e \in \mathcal{C}(V_0, V_{l'} - V_0)} (1 - q_e)}{\prod_{e \in \mathcal{C}(V_0, V_l - V_0)} (1 - q_e)} \cdot \frac{q(c_{V_0} = l|V_0, \pi')}{q(c_{V_0} = l'|V_0, \pi)} \cdot \frac{p(\pi'|I)}{p(\pi|I)} \right).$$

$$(13.5)$$

end For
Output: Samples $\pi \sim p(\pi|I)$.

FIGURE 13.3
The Swendsen–Wang Cut algorithm [2].

The set of edges $\mathcal{C}(V_0, V_{l'} - V_0), \mathcal{C}(V_0, V_l - V_0)$ from (13.5) are the SW cuts defined in general as

$$\mathcal{C}(V_1, V_2) = \{< i, j > \in E, i \in V_1, j \in V_2\}$$

The SWC algorithm could automatically decide the number of clusters, however in this chapter, as in most motion segmentation algorithms, it is assumed that the number of subspaces M is known. Thus the new label for the component V_0 is sampled with uniform probability from the number M of subspaces:

$$q(c_{V_0} = l'|V_0, \pi) = 1/M.$$

13.3.3 Graph construction

Section 13.2 described a popular subspace clustering method based on spectral clustering. Spectral clustering optimizes an approximation of the normalized cut or the ratio cut [25], which are discriminative measures. In contrast, the proposed subspace clustering approach optimizes a generative model where the likelihood is based on the assumption that the subspaces are linear. It is possible that the discriminative measures are more flexible and work better when the linearity assumption is violated, and will be studied in future work.

The following affinity measure, inspired by [12], will be used in this work

$$A_{ij} = \exp(-m\frac{\theta_{ij}}{\bar{\theta}}), \quad i \neq j \qquad (13.6)$$

where θ_{ij} is based on the angle between the vectors x_i and x_j,

$$\theta_{ij} = 1 - (\frac{x_i^T x_j}{\|x_i\|_2 \|x_j\|_2})^2,$$

and $\bar{\theta}$ is the average of all θ. The parameter m is a tuning parameter to control the size of the connected components obtained by the SWC algorithm. The subspace clustering performance with respect to this parameter will be evaluated in subsection 13.4.2 for motion segmentation.

The affinity measure based on the angular information between points enables to obtain the neighborhood graph, for example based on the k-nearest neighbors. After the graph has been obtained, the affinity measure is also used to obtain the edge weights for making the data driven clustering proposals in the SWC algorithm as well as for the prior term of the posterior probability.

The graph $G = (V, E)$ has as vertices the set of points that need to be clustered. The edges E are constructed based on the proposed distance measure from (13.6). Since the distance measure is more accurate in finding the nearest neighbors (NN) from the same subspace, the graph is constructed as the k-nearest neighbor graph (kNN), where k is a given parameter.

Examples of obtained graphs will be given in subsection 13.4.2.

13.3.4   Optimization by simulated annealing

The SWC algorithm is designed for sampling the posterior probability $p(\pi)$. To use SWC for optimization, a simulated annealing scheme should be applied while running the SWC algorithm.

Simulated annealing means the probability used by the algorithm is not $p(\pi)$ but $p(\pi)^{1/T}$ where T is a "temperature" parameter that is large at the beginning of the optimization and is slowly decreased according to an annealing schedule. If the annealing schedule is slow enough, it is theoretically guaranteed [11] that the global optimum of the probability $p(\pi)$ will be found.

In reality we use a faster annealing schedule, and the final partition π will only be a local optimum. We use an annealing schedule that is controlled by three parameters: the start temperature T_{start}, the end temperature as T_{end}, and the number of iterations N^{it}. The temperature at step i is calculated as

$$T_i = \frac{T_{\text{end}}}{\log\left(\frac{i}{N}[e - \exp(\frac{T_{\text{end}}}{T_{\text{start}}})] + \exp(\frac{T_{\text{end}}}{T_{\text{start}}})\right)}, i = \overline{1, N^{it}} \qquad (13.7)$$

To better explore the probability space, we also use multiple runs with different random initializations. Then the final algorithm is shown in Figure 13.4.

Input: N points $(\boldsymbol{x}_1, \ldots, \boldsymbol{x}_N)$ from M subspaces
Construct the adjacency graph G as a k-NN graph using eq (13.6).
For $r = 1, \ldots, Q$
  Initialize the partition π as $\pi(i) = 1, \forall i$.
  For $i = 1, \ldots, N^{it}$
    1. Compute the temperature T_i using eq (13.7).
    2. Run one step of the SWC algorithm 13.3 using $p(\pi|I) = p^{1/T_i}(\pi)$ in (13.5).
  end For
  Record the clustering result π_r and the final probability $p_r = p(\pi_r)$.
end For
Output: Clustering result π_r with the largest p_r.

FIGURE 13.4
The Swendsen–Wang Cuts algorithm for subspace clustering.

13.3.5 Complexity analysis

This subsection presents an evaluation of the computation complexity of the proposed SWC-based subspace clustering algorithm. To our knowledge, the complexity of SWC has not been calculated yet in the literature.

Let N be the number of points in $\mathbb{R}^D$ that need to be clustered. The computation complexity of the proposed subspace clustering method can be broken down as follows:

- The adjacency graph construction is $O(N^2 D \log k)$ where D is the space dimension. This is because one needs to calculate the distance from each point to the other $N - 1$ points and use a heap to maintain its k-NNs.

- Each of the N^{it} iterations of the SWC algorithm involves:

  - Sampling the edges at each SWC step is $O(|E|) = O(N)$ since the k-NN graph $G =< V, E >$ has at most $2kN$ edges.
  - Constructing connected components at each SWC step is $O(|E|\alpha(|E|)) = O(N\alpha(N))$ using the disjoint set forest data structure [8, 9]. The function $\alpha(N)$ is the inverse of $f(n) = A(n, n)$ where $A(m, n)$ is the fast growing Ackerman function [1] and $\alpha(N) \leq 5$ for $N \leq 2^{2^{10^{19729}}}$.
  - Computing $E_{data}(\pi)$ involves fitting linear subspaces for each motion cluster, which is $O(D^2 N + D^3)$
  - Computing the $E_{prior}(\pi)$ is $O(N)$.

  The number of iterations is $N^{it} = 2000$, so all the SWC iterations take $O(N\alpha(N))$ time.

In conclusion, the entire algorithm complexity in terms of the number N of points is $O(N^2)$ so it scales better than spectral clustering for large problems.

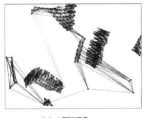  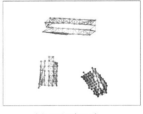

(a) 1RT2TC	(b) cars3	(c) articulated

FIGURE 13.5
Examples of SWC weighted graphs for a checkerboard (left), traffic (mid) and articulated (right) sequence. Shown are the feature point positions in the first frame. The edge intensities represent their weights from 0 (white) to 1 (black).

13.4   Application: Motion Segmentation

This section presents an application of the proposed subspace clustering algorithm to motion segmentation.

Most recent works on motion segmentation use the affine camera model, which is approximatively satisfied when the objects are far from the camera. Under the affine camera model, a point on the image plane (x, y) is related to the real world 3D point X by

$$\begin{bmatrix} x \\ y \end{bmatrix} = A \begin{bmatrix} X \\ 1 \end{bmatrix}, \tag{13.8}$$

where $A \in \mathbb{R}^{2 \times 4}$ is the affine motion matrix.

Let $t_i = (x_i^1, y_i^1, x_i^2, y_i^2, \ldots, x_i^F, y_i^F)^T, i = 1, \ldots N$ be the trajectories of tracked feature points in F frames (2D images), where N is the number of trajectories. Let the *measurement matrix* $W = [t_1, t_2, \ldots, t_N]$ be constructed by assembling the trajectories as columns.

If all trajectories undergo the same rigid motion, (13.8) implies that W can be decomposed into a *motion matrix* $M \in \mathbb{R}^{2F \times 4}$ and a *structure matrix* $S \in \mathbb{R}^{4 \times N}$ as

$$W = MS$$

$$\begin{bmatrix} x_1^1 & x_2^1 & \cdots & x_N^1 \\ y_1^1 & y_2^1 & \cdots & y_N^1 \\ \vdots & \vdots & \ddots & \vdots \\ x_1^F & x_2^F & \cdots & x_N^F \\ y_1^F & y_2^F & \cdots & y_N^F \end{bmatrix} = \begin{bmatrix} A^1 \\ \vdots \\ A^F \end{bmatrix} \begin{bmatrix} X_1 & \cdots & X_N \\ 1 & \cdots & 1 \end{bmatrix}$$

where A^f is the affine object-to-world transformation matrix at frame f. It implies that rank$(W) \leq 4$. Since the entries of the last row of S are always 1, under the affine camera model, the trajectories of feature points from a rigidly moving object reside in an affine subspace of dimension of at most 3.

In general, we are given a measurement matrix W that contains trajectories from multiple possibly nonrigid motions. The task of motion segmentation is to cluster together all trajectories coming from each motion.

A popular approach (see [5, 12, 17, 24]) is to project the trajectories to a lower dimensional space and to perform subspace clustering in that space, using spectral clustering as described in Section 13.2. These methods differ in the projection dimension D and in the affinity measure A used for spectral clustering.

13.4.1 Dimension reduction

Dimension reduction is an essential preprocessing step for obtaining a good motion segmentation. To realize this goal, the truncated SVD is often applied [5, 12, 17, 24].

To project the measurement matrix $W \in \mathbb{R}^{2F \times N}$ to $X = [x_1, ..., x_N] \in \mathbb{R}^{D \times N}$, where D is the desired projection dimension, the matrix W is decomposed via SVD as $W = U\Sigma V^T$ and the first D columns of the matrix V are chosen as X^T.

The value of D for dimension reduction is also a major concern in motion segmentation. This value has a large impact on the speed and accuracy of the final result, so it is very important to select the best dimension to perform the segmentation. The dimension of a motion is not fixed, but can vary from sequence to sequence, and since it is hard to determine the actual dimension of the mixed space when multiple motions are present, different methods may have different dimensions for projection.

The GPCA [24] suggests to project the trajectories onto a 5-dimensional space. ALC [17] chooses to use the sparsity-preserving dimension $d_{sp} = \text{argmin}_{d \geq 2D \log(2T/d)} d$ for D-dimensional subspaces. The SSC [6] and LRR [13] simply projects the trajectories to the $4M$ subspace, where M is the number of motions. Some methods [5, 12] use an exhaustive search strategy to perform the segmentation in spaces with a range of possible dimensions and pick the best result. In this chapter, we find that projecting to dimension $D = 2M + 1$ can generate good results.

The computation complexity of computing the SVD of a $m \times n$ matrix U when $m >> n$ is $O(mn^2 + n^3)$ [22]. If $n >> m$ then it is faster to compute the SVD of U^T, which takes $O(nm^2 + m^3)$.

Assuming that $2F << N$, it means that the SVD of W can be computed in $O(NF^2 + F^3)$ operations.

After projecting to the subspace of dimension $D = 2M + 1$, the SWC subspace clustering algorithm from Section 13.3 is applied and the clustering result gives the final motion segmentation result.

13.4.2 Experiments on the Hopkins 155 dataset

This subsection presents experiments with the proposed SWC-based motion segmentation algorithm on the Hopkins 155 motion database [23]. The database has been created with the goal of providing an extensive benchmark for testing feature based motion segmentation algorithms. It consists of 155 sequences of two and three motions. The ground-truth segmentation is also provided for evaluation purposes. Based on the content of the video, the sequences could be categorized into three main categories: checkerboard, traffic, and articulated sequences, with examples shown in Figure 13.6. The trajectories are extracted automatically by a tracker, so they are slightly corrupted by noise.

(a) Checkerboard   (b) Traffic   (c) Articulated

FIGURE 13.6
Sample images from some sequences of three categories in the Hopkins 155 database with ground truth superimposed.

As already mentioned, before applying the SWC algorithm, the dimension of the data is reduced from $2F$ to $D = 2M + 1$, where M is the number of motions. After the projection, the initial labeling state in the SWC algorithm has all points having the same label.

The motion segmentation results are evaluated using the misclassification error rate

$$\text{Misclassification Rate} = \frac{\# \text{ of misclassified points}}{\text{total } \# \text{ of points}} \qquad (13.9)$$

Parameter settings. The proposed motion segmentation algorithm has a number of tuning parameters that were held constant to the following values. The number of NN (nearest neighbors) for graph construction is $k = 7$, the parameter m in the affinity measure (13.6) is $m = 10$, and the prior coefficient in (13.3) is $\rho = 2.2$. The sensitivity of the average misclassification error to the parameters m and ρ is shown in Figure 13.7. The annealing parameters are $T_{\text{start}} = 1$, $T_{\text{end}} = 0.01$, $N^{it} = 2000$. The number of independent runs to obtain the most probable partition is $Q = 10$. An example of all the partition states during an SWC run is shown in Figure 13.8.

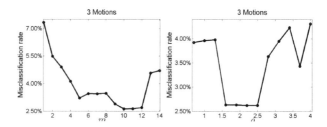

FIGURE 13.7
Dependence of the misclassification rate on the affinity parameter m (left) and prior strength ρ (right).

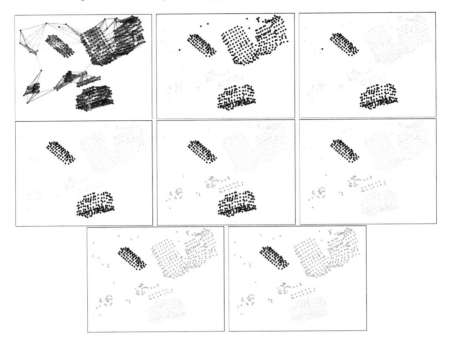

FIGURE 13.8
SWC clustering of the Hopkins 155 sequence 1R2TCR, containing $M = 3$ motions. Shown are the feature point positions in the first frame, having colors as the labeling states π obtained while running the SWC algorithm from the initial state (top left) to the final state (bottom right).

Results. The average and median misclassification errors are listed in Table 13.1. For accuracy, the results of the SWC algorithm from Table 13.1 are averaged over 10 runs and the standard deviations are shown in parentheses. In order to compare the SWC method with the state of the art methods, we also list the results of ALC [17], SC [12], SSC [6] and VC [5].

The SWC based algorithm obtains average errors that are less than twice the errors of the other methods. In our experiments we observed that the energy of the final state is usually smaller than the energy of the ground truth state. This fact indicates that the SWC algorithm is doing a good job optimizing the model but the Bayesian model is not accurate enough in its current form and needs to be improved.

The columns in Table 13.1 labeled with SC^4 and SC^{4k} are representing the misclassification errors of the SC method [12] with an affinity matrix with 4-NN and 4k-NN respectively. These errors are 13.32 and 8.59 respectively and indicate that the Spectral Clustering really needs a dense affinity matrix to work well, and it cannot be accelerated using sparse matrix operations.

TABLE 13.1

Misclassification rates (in percent) of different motion segmentation algorithms on the Hopkins 155 dataset.

Method	ALC	SC	SSC	VC	SWC (std)	SC^4	SC^{4k}	KSP
Checkerboard (2 motion)								
Average	1.55	0.85	1.12	0.67	1.25 (0.11)	10.98	5.25	1.25
Median	0.29	0.00	0.00	0.00	0.00 (0.00)	1.16	0.00	0.00
Traffic (2 motion)								
Average	1.59	0.90	0.02	0.99	1.87 (0.55)	12.85	12.48	8.65
Median	1.17	0.00	0.00	0.22	0.00 (0.00)	7.55	1.30	0.53
Articulated (2 motion)								
Average	10.70	1.71	0.62	2.94	2.15 (0.11)	11.38	12.94	18.67
Median	0.95	0.00	0.00	0.88	0.00 (0.00)	0.95	0.95	0.95
All (2 motion)								
Average	2.40	0.94	0.82	0.96	1.49 (0.19)	11.50	7.82	4.76
Median	0.43	0.00	0.00	0.00	0.00 (0.00)	2.09	0.27	0.00
Checkerboard (3 motion)								
Average	5.20	2.15	2.97	0.74	2.26 (0.04)	16.74	6.58	4.86
Median	0.67	0.47	0.27	0.21	0.67 (0.00)	14.82	0.49	1.46
Traffic (3 motion)								
Average	7.75	1.35	0.58	1.13	2.88 (0.00)	30.04	24.87	21.80
Median	0.49	0.19	0.00	0.21	0.81 (0.00)	26.82	30.91	26.36
Articulated (3 motion)								
Average	21.08	4.26	1.42	5.65	6.33 (1.88)	19.48	24.27	18.42
Median	21.08	4.26	0.00	5.65	6.33 (1.88)	19.48	24.27	18.42
All (3 motion)								
Average	6.69	2.11	2.45	1.10	2.62 (0.13)	19.55	11.25	9.00
Median	0.67	0.37	0.20	0.22	0.81 (0.00)	18.88	1.42	1.70
All sequences combined								
Average	3.37	1.20	1.24	0.99	1.75 (0.15)	13.32	8.59	5.72
Median	0.49	0.00	0.00	0.00	0.00 (0.00)	6.46	0.36	0.31

Finally, the performance of the SWC-based algorithm is compared with the KASP algorithm [26], which is a fast approximate spectral clustering and was used in place of the spectral clustering step in the SC method [12]. The data reduction parameter used was $\gamma = 10$, which still results in a $O(N^3)$ clustering algorithm. The total misclassification error is 5.72, about three times larger than the SWC method.

(a) Frame 1   (b) Frame 15   (c) Frame 30

FIGURE 13.9
Selected frames of cars10 sequence with 1000 tracked feature points.

13.4.3 Scalability experiments on large data

In order to evaluate the scalability of different algorithms, sequences with a large number of trajectories are needed. The trajectories can be generated by some optical flow algorithm, but it is difficult to obtain the ground truth segmentation and remove bad trajectories caused by occlusions. Brox et.al. [3] provided a dense segmentation for some frames in 12 sequences in the Hopkins 155 dataset[1]. From them, we picked the cars10 sequence and tracked all pixels of the first frame using the Classic+NL method [19]. There are two reasons for choosing cars10. First, it has three motions, two moving cars and the background. Second, the two moving cars are relatively large in the video, so that a large number of trajectories can be obtained from each motion.

There are 30 frames in the sequence, and 3 of them have a dense manual segmentation of all pixels. We removed trajectories that have different labels on the 3 ground truth frames. To avoid occlusion, the trajectories close to the motion boundaries were also removed. In addition, we only kept the full trajectories for clustering. Finally, we obtained around 48,000 trajectories as a pool. From the pool, different numbers N of trajectories were subsampled for evaluation. For each given N, a total of N trajectories were randomly selected from the pool such that the number of trajectories in each of the three motions was roughly the same. For example, to generate $N = 1000$ trajectories, we would randomly pick from the pool 333 trajectories from two of the motions and 334 trajectories from the third motion. If there were not enough trajectories available from one motion, we added more from the motion that has the most trajectories.

We compared the SWC method with the SC algorithm [12] discussed in Section 13.2, which is one of the fastest and most accurate algorithms [5] based on spectral clustering. We generated data containing trajectories between 1,000 and 15,000, and applied the two segmentation algorithms. Sample frames are shown in Figure 13.9. The parameters for SC were kept the

---

[1]http://lmb.informatik.uni-freiburg.de/resources/datasets/

TABLE 13.2
Average misclassification rate for the cars10 sequence (in percent).

Number of Trajectories N	SC	SWC
1000 to 15,000	2.77	0.99
24,000 to 48,000	-	1.00

same as in the original work, and those for SWC were identical with sub-section 13.4.2. The SC algorithm is implemented in MATLAB (which has optimized SVD algorithms), while the SWC code is in C++. The experiments were performed on a Windows machine with an Intel core i7-3970 CPU and 12 GB memory. We also generated data with $N = 24,000$ and $N = 48,000$ trajectories for SWC clustering. For SC, the same experiments could not be conducted because MATLAB ran out of memory.

The misclassification rate is recorded in Table 13.2 and the running time is shown on Figure 13.10. Table 13.2 shows that both methods perform well and the misclassification rate of SWC is about one third of that of the SC.

From Figure 13.10, which shows the computation time vs. the number N of trajectories, one could find that for a small number of trajectories, the SC is faster than SWC, but for more than $N = 6,000$ trajectories, the computation time of SC is greater than that of SWC, and increases much faster. We also plot the log(time) vs. log(N) and use linear regression to fit lines through the data points of the two methods. If the slope of the line is α, then the computation complexity is $O(N^\alpha)$. We observe that the slope of SC is 2.52 while the slope for SWC is 1.29, which is consistent with the complexity analysis of subsection 13.3.5.

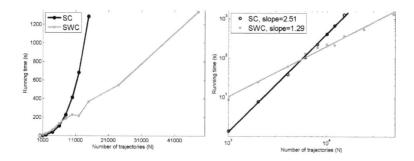

FIGURE 13.10
Left: Computation time (sec) vs number of trajectories N for SC and SWC.
Right: log-log plot of same data with the fitted regression lines.

13.5 Conclusion

This chapter presented a stochastic method for clustering points belonging to a number of linear subspaces using the Swendsen-Wang Cuts (SWC) algorithm. The posterior probability is a generative model defined in a Bayesian framework with data and likelihood parts. The graph used by the SWC algorithm for making informed relabeling proposals was defined as the k-NN graph based on an affinity matrix.

The complexity analysis showed that the proposed algorithm is $O(N^2)$ in the number N of data points that need to be clustered. This is in contrast to the spectral clustering algorithms that have $O(N^3)$ complexity.

Motion segmentation experiments on the Hopkins 155 dataset showed that the algorithm performs slightly worse than the state of the art motion segmentation algorithms based on spectral clustering. This could probably be due to the rigidity of the generative model in contrast to the flexibility of the Normalized Cut or Ratio Cut or their approximations that are optimized by the spectral clustering algorithms. Experiments also revealed that the spectral clustering is about $O(N^{2.5})$ on the data that was evaluated while the proposed SWC method is $O(N^{1.3})$.

The SWC algorithm uses many Markov Chains to better explore the partition space. These chains are independent and could be run in parallel if desired. Parallel implementations of Spectral Clustering have also been developed recently [4].

In future, we will investigate posterior probabilities based on the Normalized Cut or Ratio Cut instead of the generative model to see if the Swendsen-Wang Cuts algorithm can obtain state of the art errors using such discriminative models.

Bibliography

[1] W. Ackermann. Zum hilbertschen aufbau der reellen zahlen. *Mathematische Annalen*, 99(1):118–133, 1928.

[2] A. Barbu and S. Zhu. Generalizing Swendsen-Wang to sampling arbitrary posterior probabilities. *IEEE Transactions on Pattern Analysis and Machine Intelligence*, 27(8):1239–1253, 2005.

[3] T. Brox and J. Malik. Object segmentation by long term analysis of point trajectories. In *ECCV*, pages 282–295. 2010.

[4] W.-Y. Chen, Y. Song, H. Bai, C.-J. Lin, and E. Y. Chang. Parallel spectral clustering in distributed systems. *IEEE Transactions on Pattern Analysis and Machine Intelligence*, 33(3):568–586, 2011.

[5] L. Ding, A. Barbu, and A. Meyer-Baese. Motion segmentation by velocity clustering with estimation of subspace dimension. In *ACCV Workshop on Detection and Tracking in Challenging Environments*, 2012.

[6] E. Elhamifar and R. Vidal. Sparse subspace clustering. In *Computer Vision and Pattern Recognition (CVPR)*. IEEE, 2009.

[7] M. A. Fischler and R. C. Bolles. RANSAC random sample consensus: A paradigm for model fitting with applications to image analysis and automated cartography. *Communications of the ACM*, 26:381–395, 1981.

[8] M. Fredman and M. Saks. The cell probe complexity of dynamic data structures. In *Proceedings of the Twenty-first Annual ACM Symposium on Theory of Computing*, pages 345–354, 1989.

[9] B. A. Galler and M. J. Fisher. An improved equivalence algorithm. *Communications of the ACM*, 7(5):301–303, 1964.

[10] S. Geman and D. Geman. Stochastic relaxation, Gibbs distributions, and the Bayesian restoration of images. *IEEE Transactions on Pattern Analysis and Machine Intelligence*, 6(6):721–741, 1984.

[11] S. Kirkpatrick, C. D. Gelatt, Jr., and M. P. Vecchi. Optimization by simulated annealing. *Science*, 220(4598):671–680, 1983.

[12] F. Lauer and C. Schnörr. Spectral clustering of linear subspaces for motion segmentation. In *Proceedings of the IEEE Conference on Computer Vision (ICCV 09)*, pages 678–685, Kyoto, Japan, 2009.

[13] G. Liu, Z. Lin, and Y. Yu. Robust subspace segmentation by low-rank representation. In *Proceedings of the 27th International Conference on Machine Learning (ICML-10)*, Haifa, Israel, 2010.

[14] Y. Ma, H. Derksen, W. Hong, and J. Wright. Segmentation of multivariate mixed data via lossy coding and compression. *IEEE Transactions on Pattern Analysis and Machine Intelligence*, 29(9):1546–1562, 2007.

[15] A. Y. Ng, M. I. Jordan, and Y. Weiss. On spectral clustering: Analysis and an algorithm. In *Advances in Neural Information Processing Systems*, volume 14, pages 849–856. MIT Press, Cambridge, MA, 2001.

[16] R. B. Potts and C. Domb. Some generalized order-disorder transformations. In *Mathematical Proceedings of the Cambridge Philosophical Society*, volume 48, pages 106–109, 1952.

[17] S. Rao, R. Tron, R. Vidal, and Y. Ma. Motion segmentation in the presence of outlying, incomplete, or corrupted trajectories. *IEEE Transactions on Pattern Analysis and Machine Intelligence*, 32(10):1832–1845, 2010.

[18] J. Shi and J. Malik. Normalized cuts and image segmentation. *IEEE Transactions on Pattern Analysis and Machine Intelligence*, 22(8):888–905, 2000.

[19] D. Sun, S. Roth, and M. J. Black. Secrets of optical flow estimation and their principles. In *Computer Vision and Pattern Recognition (CVPR)*, pages 2432–2439. IEEE, 2010.

[20] R. H. Swendsen and J.-S. Wang. Nonuniversal critical dynamics in Monte Carlo simulations. *Physical Review Letters*, 58(2):86, 1987.

[21] M. E. Tipping and C. M. Bishop. Probabilistic principal component analysis. *Journal of the Royal Statistical Society: Series B (Statistical Methodology)*, 61(3):611–622, 1999.

[22] L. N. Trefethen and D. Bau III. *Numerical Linear Algebra*, volume 50. Society for Industrial and Applied Mathematics, Philadelphia, PA, 1997.

[23] R. Tron and R. Vidal. A benchmark for the comparison of 3-d motion segmentation algorithms. In *Computer Vision and Pattern Recognition (CVPR)*, pages 1–8. IEEE, 2007.

[24] R. Vidal and R. Hartley. Motion segmentation with missing data using power factorization and GPCA. In *Computer Vision and Pattern Recognition (CVPR)*, pages II–310. IEEE, 2004.

[25] U. Von Luxburg. A tutorial on spectral clustering. *Statistics and Computing*, 17(4):395–416, 2007.

[26] D. Yan, L. Huang, and M. I. Jordan. Fast approximate spectral clustering. In *15th ACM SIGKDD International Conference on Knowledge Discovery and Data Mining*, pages 907–916, 2009.

[27] J. Yan and M. Pollefeys. A general framework for motion segmentation: Independent, articulated, rigid, non-rigid, degenerate and non-degenerate. In *9th European Conference on Computer Vision (ECCV)*, pages 94–106, Austria, 2006.

14

Bayesian Inference for Logistic Regression Models Using Sequential Posterior Simulation

John Geweke

University of Technology

Garland Durham

California Polytechnic State University

Huaxin Xu

University of Technology

CONTENTS

14.1   Introduction

The multinomial logistic regression model, hereafter "logit model", is one of the most widely used models in applied statistics. It provides a straightforward link from covariates to the probabilities of discrete outcomes. More generally, it provides a workable probability distribution for discrete events, whether directly observed or not, as a function of covariates. In the latter, more general, setting it is a key component of conditional mixture models including the mixture of experts models introduced by [23] and studied by [24].

The logit model likelihood function is unimodal and globally concave, and consequently estimation by maximum likelihood is practical and reliable even in models with many outcomes and many covariates. However, it has proven less tractable in a Bayesian context, where effective posterior simulation has been a challenge. Because it also arises frequently in more complex contexts like mixture models, this is a significant impediment to the penetration of posterior simulation methods. Indeed, the multinomial probit model has proven more amenable to posterior simulation methods [1, 17] and has sometimes been used in lieu of the logit model in conditional mixture models [16]. Thus there is a need for simple and reliable posterior simulation methods for logit models.

State-of-the-art approaches to posterior simulation for logit models use combinations of likelihood function approximation and data augmentation in the context of Markov chain Monte Carlo (MCMC) algorithms: see [13, 19, 22, 29] and [28]. The last paper uses a novel representation of latent variables based on Pólya-Gamma distributions that can be applied in logit and related models, and uses this representation to develop posterior simulators that are reliable and substantially dominate alternatives with respect to computational efficiency. Going forward, we refer to the method of [28] as the PSW algorithm.

This chapter implements a sequential posterior simulator (SPS) using ideas developed in [9]. Unlike MCMC this algorithm is especially well-suited to massively parallel computation using graphics processing units (GPUs). The algorithm is highly generic; that is, the coding effort required to adapt it to a specific model is typically minimal. In particular, the algorithm is far simpler to implement for the logit models considered here than the existing MCMC algorithms mentioned in the previous paragraph. When implemented on GPUs the computational efficiency of SPS compares well with the best existing MCMC method. Moreover, SPS yields an accurate approximation of log-marginal likelihood, as well as reliable and systematic indications of the accuracy of posterior moment approximations, which existing MCMC methods do not.

Section 14.2 of the chapter describes the SPS algorithm and its implementation on GPUs. Section 14.3 provides the background for the examples taken up subsequently: specifics of the models and prior distributions, the datasets, and the various hardware and software environments used. Section 14.4 documents some of the features of models and datasets that influence computation time. Section 14.5 studies several applications of the logit model using benchmark datasets from the logit posterior simulation literature. Section 14.6 concludes with some more general observations. A quick first reading of the chapter might skim Section 14.2 and skip Section 14.4.

14.2   Sequential Posterior Simulation Algorithms for Bayesian Inference

Sequential posterior simulation (SPS) grows out of sequential Monte Carlo (SMC) methods developed over the past twenty years for state space models, in particular particle filters. Contributions leading up to the development here include [2–8], [18, 25, 26] and [27]. Despite its name, SPS is amenable to massively parallel implementation. It is well suited to GPUs, which provide massively parallel desktop computing at a cost of well under one dollar (US) per processing core. For further background on these details see [9].

14.2.1   Notation and conditions

The vectors $y_1, \ldots, y_T$ denote the data and $y_{1:t} = \{y_1, \ldots, y_t\}$. The model specifies the joint densities

$$p(y_t \mid y_{1:t-1}, \theta) \quad (t = 1, \ldots, T)$$

in which the functional form of p is specified and θ is a vector of unobservable parameters. The likelihood function is

$$p(y_{1:T} \mid \theta) = \prod_{t=1}^{T} p(y_t \mid y_{1:t-1}, \theta). \tag{14.1}$$

The model also specifies a prior distribution $p(\theta)$. The notation suppresses conditioning on the covariates x_t and treats all distributions as continuous to avoid notational clutter.

The SPS algorithm, like all posterior simulators, approximates posterior moments of the form

$$\bar{g} = \mathrm{E}\left[g(\theta) \mid y_{1:T}\right]. \tag{14.2}$$

The following conditions are sufficient for the properties of the algorithm stated here.

1. The likelihood function (14.1) is bounded.

2. The likelihood function can be evaluated directly to machine accuracy. In particular, evaluation does not entail simulation.

3. The prior distribution is proper and i.i.d. simulation of θ from this distribution is practical.

4. The prior moment $\mathrm{E}\left[g\left(\theta\right)^{2+\delta}\right]$ is finite for some $\delta > 0$.

These conditions are typically easy to check and they suffice for the purposes of this chapter. Except for condition 3 they can be weakened, but verifying these weaker conditions when the stronger stated conditions do not hold can be difficult.

Like all posterior simulation methods, the SPS algorithm generates a set of random values of θ that approximate the posterior distribution. Reflecting the evolution of the SMC literature, these random vectors are termed particles. It will prove convenient to organize the particles into J groups of N particles each, and for this reason to denote the particles θ_{jn} $(j = 1, \ldots, J; n = 1, \ldots, N)$. Evaluating a function of interest then leads to the $J \cdot N$ random values $g_{jn} = g\left(\theta_{jn}\right)$. The numerical approximation of (14.2) is

$$\overline{g}^{(J,N)} = (JN)^{-1} \sum_{j=1}^{J} \sum_{n=1}^{N} g_{jn}. \tag{14.3}$$

Reliable assessment of the numerical accuracy of this approximation is essential for SPS just as it is for all posterior simulation methods. As detailed in subsections 14.2.2 through 14.2.4, the SPS algorithm produces particles θ_{jn} that are identically distributed, independent across groups and dependent within groups. Within each group denote

$$\overline{g}_j^N = N^{-1} \sum_{n=1}^{N} g_{jn} \quad (j = 1, \ldots, J). \tag{14.4}$$

The underlying SMC theory discussed in subsection 14.2.2 guarantees

$$N^{1/2} \left(\overline{g}_j^N - \overline{g}\right) \xrightarrow{d} N\left(0, v\right) \quad (j = 1, \ldots, J). \tag{14.5}$$

The convergence (14.5) implies immediately that the numerical reliability of the SPS algorithm could be assessed based on repeated executions. But this multiplies the computational requirements many-fold. An alternative is to seek an approximation v^N of v that can be computed efficiently as a by-product of the algorithm, having the property $v^N \xrightarrow{a.s.} v$. This strategy has proven straightforward in some posterior simulation methods like importance sampling [15] but less so in others like MCMC [11].

The independence of particles across groups in the SPS algorithm implies that the reliability of the approximation $\overline{g}^{(J,N)}$ can be assessed using the same elementary approach as with repeated executions of the algorithm many-fold but without incurring the extra computational requirements. The natural approximation of v is

$$\widehat{v}^{(J,N)}\left(g\right) = \left[N/\left(J-1\right)\right] \sum_{j=1}^{J} \left(\overline{g}_j^N - \overline{g}^{(J,N)}\right)^2.$$

Define the numerical standard error of $\overline{g}^{(J,N)}$

$$\text{NSE}\left(\overline{g}^{(J,N)}\right) = \left[(JN)^{-1}\,\widehat{v}^{(J,N)}\left(g\right)\right]^{1/2}, \tag{14.6}$$

which provides a measure of variability of the moment approximation (14.3) across replications of the algorithm with fixed data. As $N \to \infty$

$$\left(J-1\right)\widehat{v}^{(J,N)}\left(g\right)/v \xrightarrow{d} \chi^2\left(J-1\right) \tag{14.7}$$

and

$$\frac{\overline{g}^{(J,N)} - \overline{g}}{\text{NSE}\left(\overline{g}^{(J,N)}\right)} \xrightarrow{d} t\left(J-1\right). \tag{14.8}$$

The relative numerical efficiency (RNE, see [15]), which approximates the population moment ratio $\text{var}\left(g\left(\theta\right) \mid y_{1:T}\right)/v$, can be obtained in a similar manner,

$$\text{RNE}\left(\overline{g}^{(J,N)}\right) = (JN)^{-1} \sum_{j=1}^{J} \sum_{n=1}^{N} \left(g_{jn} - \overline{g}^{(J,N)}\right)^2 /\widehat{v}^{(J,N)}\left(g\right). \tag{14.9}$$

RNE close to one indicates that there is little dependence amongst the particles, and that the moment approximations (14.4) and (14.3) approach the efficiency of the unattainable ideal, an i.i.d. sample from the posterior. RNE less than one indicates dependence. In this case, the moment approximations (14.4) and (14.3) are less precise than one would obtain with a hypothetical i.i.d. sample of the same size.

The detailed discussion of the SPS algorithm in subsections 14.2.2 through 14.2.4 will show that this procedure is not simply J repetitions of the algorithm with N particles each. In fact information is shared across groups, but in such a way that the independence across groups critical for (14.8) is preserved. The end of subsection 14.2.4 returns to this discussion after establishing the requisite theory. A complementary practical consideration is that when the algorithm is implemented on GPUs there are substantial returns to scale – so that, for example, a single execution with 10^n particles typically requires substantially less time than ten successive executions each with 10^{n-1} particles for n in the range of 4 to 6.

14.2.2 Non-Adaptive SPS algorithm

We start from the SMC algorithm as detailed in [6]. The algorithm generates and modifies the particles θ_{jn}, with superscripts used for further specificity at various points in the algorithm. To make the notation compact, let $\mathcal{J} = \{1, \ldots, J\}$ and $\mathcal{N} = \{1, \ldots, N\}$. The algorithm is an implementation of Bayesian learning, providing simulations from $\theta \mid y_{1:t}$ for $t = 1, 2, \ldots, T$. It processes observations, in order and in successive batches, each batch constituting a cycle of the algorithm.

The global structure of the algorithm is therefore iterative, proceeding through the sample. But it operates on many particles in exactly the same way at almost every stage, and it is this feature of the algorithm that makes it amenable to massively parallel implementations. On conventional quad-core machines and samples of typical size, one might set up the algorithm with $J = 10$ groups of 1000 particles each, and using GPUs $J = 40$ groups of 2500 particles each. (The numbers are just illustrations, to fix ideas.)

Algorithm 14.1. *(Non-adaptive)* Let $t_0, \ldots, t_L$ be fixed integers with $0 = t_0 < t_1 < \ldots < t_L = T$; these define the cycles of the algorithm. Let $\lambda_1, \ldots, \lambda_L$ be fixed vectors that parameterize transition distributions as indicated below.

1. Initialize $\ell = 0$ and let $\theta_{jn}^{(\ell)} \overset{iid}{\sim} p(\theta) \quad (j \in \mathcal{J}, n \in \mathcal{N})$

2. For $\ell = 1, \ldots, L$

    (a) Correction (C) phase, for all $j \in \mathcal{J}$ and $n \in \mathcal{N}$:

      i. $w_{jn}(t_{\ell-1}) = 1$
      ii. For $s = t_{\ell-1} + 1, \ldots, t_\ell$

$$w_{jn}(s) = w_{jn}(s-1) \cdot p\left(y_s \mid y_{1:s-1}, \theta_{jn}^{(\ell-1)}\right) \qquad (14.10)$$

      iii. $w_{jn}^{(\ell-1)} := w_{jn}(t_\ell)$

    (b) Selection (S) phase, applied independently to each group $j \in \mathcal{J}$: Using multinomial or residual sampling based on $\left\{w_{jn}^{(\ell-1)} \; (n \in \mathcal{N})\right\}$, select

$$\left\{\theta_{jn}^{(\ell,0)} \; (n \in \mathcal{N})\right\} \text{ from } \left\{\theta_{jn}^{(\ell-1)} \; (n \in \mathcal{N})\right\}$$

    (c) Mutation (M) phase, applied independently across $j \in \mathcal{J}, n \in \mathcal{N}$:

$$\theta_{jn}^{(\ell)} \sim k\left(\theta \mid y_{1:t_\ell}, \theta_{jn}^{(\ell,0)}, \lambda_\ell\right) \qquad (14.11)$$

    where the drawings are independent and the p.d.f. (14.11) satisfies the invariance condition

$$\int_\Theta k\left(\theta \mid y_{1:t_\ell}, \theta^*, \lambda_\ell\right) p\left(\theta^* \mid y_{1:t_\ell}\right) d\nu(\theta^*) = p\left(\theta \mid y_{1:t_\ell}\right)$$

3. $\theta_{jn} := \theta_{jn}^{(L)}$ ($j \in \mathcal{J}, n \in \mathcal{N}$)

The algorithm is non-adaptive because $t_0, \ldots, t_L$ and $\lambda_1, \ldots, \lambda_L$ are fixed before the algorithm starts. Going forward, it will be convenient to denote the cycle indices by $\mathcal{L} = \{1, \ldots, L\}$. At the conclusion of the algorithm, the simulation approximation of a generic posterior moment is (14.3).

The only communication between particles is in the S phase. In the C and M phases exactly the same computations are made for each particle, with no communication. This situation is ideal for GPUs, as detailed in [9]. In the S phase there is communication between particles within, but not across, the J groups. This keeps the particles in the J groups independent. Typically the fraction of computation time devoted to the S phase is minute.

For each group, $j \in \mathcal{J}$, the four regularity conditions in the previous subsection imply the assumptions of [6], Theorem 1 (for multinomial resampling) and Theorem 2 (for residual resampling). Therefore a central limit theorem(14.5) applies.

14.2.3   Adaptive SPS algorithm

In Algorithm 14.1 neither the cycles, defined by $t_1, \ldots, t_{L-1}$, nor the hyper-parameters λ_ℓ of the transition processes (14.11) depend on the particles $\{\theta_{jn}\}$. With respect to the random processes that generate these particles, these hyper-parameters are fixed – in econometric terminology, they are pre-determined with respect to $\{\theta_{jn}\}$. As a practical matter, however, one must use the knowledge of the posterior distribution inherent in the particles to choose the transition from the C phase to the S phase, and to design an effective transition distribution in the M phase. Without this feedback, it is impossible to obtain any reasonably accurate approximation $\overline{g}^{(J,N)}$ of $\overline{g}$; indeed, in all but the simplest models and smallest datasets $\overline{g}^{(J,N)}$ will otherwise be pure noise, for all intents and purposes.

The following procedure illustrates how the particles themselves can be used to choose the cycles defined by $t_1, \ldots, t_{L-1}$ and the hyper-parameters λ_ℓ of the transition processes. It is a minor modification of a procedure described in [9], that has proved effective in a number of models. It is also effective in the logit model. The algorithm requires that the user choose the number of groups, J, and the number of particles in each group, N.

Algorithm 14.2. *(Adaptive)*

1. Determine the value of t_ℓ in the C phase of cycle ℓ ($\ell \in \mathcal{L}$) as follows.

   (a) At each step s compute the effective sample size

   $$ESS(s) = \frac{\left[\sum_{j=1}^{J} \sum_{n=1}^{N} w_{jn}(s) \right]^2}{\sum_{j=1}^{J} \sum_{n=1}^{N} w_{jn}(s)^2} \tag{14.12}$$

   immediately after computing (14.10).

(b) If $ESS\left(s\right)/\left(J\cdot N\right)<0.5$ or if $s=T$ set $t_\ell=s$ and proceed to the S phase. Otherwise increment s and recompute (14.12).

2. The transition density (14.11) in the M phase of each cycle ℓ is a Metropolis Gaussian random walk, executed in steps $r=1,2,\dots$.

   (a) Initializations:

      i. $r=1$.
      ii. If $\ell=1$ then the step size scaling factor $h_{11}=0.5$.

   (b) Set RNE termination criteria:

      i. If $s<T$, $K=0.35$
      ii. If $s=T$, $K=0.9$

   (c) Execute the next Metropolis Gaussian random walk step.

      i. Compute the sample variance $V_{\ell r}$ of the particles

      $$\theta_{jn}^{(\ell,r-1)}\ (j=1,\dots,J;n=1,\dots,N),$$

      define $\Sigma_{\ell r}=h_{\ell r}\cdot V_{\ell r}$, and execute step r using a random walk Gaussian proposal density with variance matrix $\Sigma_{\ell r}$ to produce a new collection of particles $\theta_{jn}^{(\ell,r)}$ $(j=1,\dots,J;\ n=1,\dots,N)$. Let $\alpha_{\ell r}$ denote the Metropolis acceptance rate across all particles in this step.

      ii. Set
      $$\begin{aligned} h_{\ell,r+1} &= \min\left(h_{\ell r}+0.01,1.0\right) \text{ if } a_{\ell r}>0.25, \\ h_{\ell,r+1} &= \max\left(h_{\ell r}-0.01,0.1\right) \text{ if } a_{\ell r}\le0.25. \end{aligned}$$

      iii. Compute the RNE of the numerical approximation $\bar{g}^{(J,N)}$ to a test function $g^*\left(\theta\right)$. If RNE $<K$ then set $r=r+1$ and return to step 2c; otherwise set $h_{\ell+1,1}=h_{\ell,r+1}$, define $R_\ell=r$, and return to step 1.

   (d) Set $\theta_{jn}^{(\ell)}=\theta_{jn}^{(\ell,r)}$. If $s<T$ then set $h_{\ell+1,1}=h_{\ell,r+1}$ and return to step 1; otherwise set $\theta_{jn}=\theta_{jn}^{(\ell)}$, define $L=\ell$, and terminate.

At every step of the algorithm, particles are identically but not independently distributed. As the number of particles in each group $N\to\infty$ the common distribution coincides with the posterior distribution. As the number of Metropolis steps, r, in the M phase increases, dependence amongst particles decreases. The M phase terminates when the RNE criterion is satisfied, implying a specified degree of independence for the particles at the end of each cycle. The RNE criterion K ensures a specified degree of independence at the end of each cycle. The assessment of numerical accuracy is based on the comparison of different approximations in J groups of particles, and larger values of J make this assessment more reliable.

At the conclusion of the algorithm, the posterior moments of interest $E\left(g\left(\theta\right)\mid y_{1:t}\right)$ are approximated by (14.3). The asymptotic (in N) variance of the approximation is proportional to $(JN)^{-1}$, and because $K = 0.9$ in the last cycle L the factor of proportionality is approximately the posterior variance var $\left(g\left(\theta\right)\mid y_{1:T}\right)$.

The division of a given posterior sample size into a number of groups J and particles within groups N should be guided by the trade-off implied by (14.7) and the fact that values of N sufficiently small will give misleading representations of the posterior distribution. From (14.7) notice that the ratio of squared NSE from one simulation to another has an asymptotic (in N) $F\left(J-1, J-1\right)$ distribution. For $J = 8$, the ratio of NSE in two simulations will be less than 2 with probability 0.95. A good benchmark for serviceable approximation of posterior moments is $J = 10$, $N = 1000$. With implementation on GPUs much larger values can be practical, for example $J = 40$, $N = 2500$ used in Section 14.5.

14.2.4   The two-pass SPS algorithm

Algorithm 14.2 is practical and reliable in a very wide array of applications. This includes situations in which MCMC is ineffective, as illustrated in [21]. However, there is an important drawback: the algorithm has no supporting central limit theorem.

The effectiveness of the algorithm is due in no small part to the fact that the cycle definitions $\{t_\ell\}$ and parameters λ_ℓ of the M phase transition distributions are based on the particles themselves. This creates a structure of dependence amongst particles that is extremely complicated. The degree of complication stemming from the use of effective sample size in step 1b can be managed: see [8]. But the degree of complication introduced in the M phase, step 2c, is much greater. This is not addressed by any of the relevant literature, and in our view this problem is not likely to be resolved by attacking it directly anytime in the foreseeable future.

Fortunately, the problem can be solved at the cost of roughly doubling the computational burden in the following manner as proposed in [9].

Algorithm 14.3. *(Two-pass)*

1. Execute the adaptive Algorithm 14.2. Discard the particles $\{\theta_{jn}\}$. Retain the number of cycles L, values $t_0, \ldots, t_L$ that define the cycles, the number of iterations R_ℓ executed in each M phase, and the variance matrices $\lambda_\ell = \{\Sigma_{\ell r}\}$ from each M phase.

2. Execute algorithm 14.2 using t_ℓ, R_ℓ and λ_ℓ $(\ell = 1, \ldots, L)$.

Notice that in step 2 the cycle break points $t_0, \ldots, t_L$ and the variance matrices $\Sigma_{\ell r}$ are predetermined with respect to the particles generated in that step. Because they are fixed with respect to the process of random particle generation, step 2 is a specific version of Algorithm 14.1. The only change is in the notation: λ_ℓ in Algorithm 14.1 is the sequence of matrices $\{\Sigma_{\ell r}\}$ indexed by r in step 2 of Algorithm 14.3. The results in [6], and other results for SMC algorithms with fixed design parameters, now apply directly.

The software used for the work in this chapter makes it convenient to execute the two-pass algorithm. In a variety of models and applications, results using Algorithms 14.2 and 14.3 have always been similar, as illustrated in Section 14.4. Thus it is not necessary to use the two-pass algorithm exclusively, and we do not recommend doing so throughout a research project. It is prudent when SPS is first applied to a new model, because there is no central limit theorem for the one-pass algorithm (Algorithm 14.2), and one should check early for the possibility that this algorithm might be inadequate. Given that Algorithm 14.3 is available in generic SPS software, and the modest computational cost involved, it is also probably a wise step in the final stage of research before public presentation of findings.

The discussion of assessing numerical accuracy stated, in its conclusion at the end of subsection 14.2.1, that the method proposed there was not equivalent to J independent repetitions of the algorithm with N particles each. The reason is that Algorithm 14.2 uses all $J \cdot N$ particles to construct the algorithm hyper-parameters: the number of cycles L, values $t_0, \ldots, t_L$ that define the cycles, the number of iterations R_ℓ executed in each M phase, and the variance matrices $\lambda_\ell = \{\Sigma_{\ell r}\}$ from each M phase. It is more efficient and effective to construct these hyper-parameters with the larger number of particles, $J \cdot N$, than with the N particles that would be available in J independent repetitions of the algorithm with N particles each. The same argument that justifies the application of SMC theory in Algorithm 14.3 guarantees independence across groups and thereby the claims for numerical standard errors in subsection 14.2.1.

14.3   Models, Data and Software

The balance of this chapter studies the performance of the PSW and SPS algorithms in instances from the literature that have been used to assess posterior simulation approaches to Bayesian inference in the logit model. This section provides the full logit model specification in subsection 14.3.1, and describes the datasets used in subsection 14.3.2. Subsection 14.3.3 provides the details of the hardware and software used subsequently in Sections 14.4 and 14.5 to document the performance of the PSW and SPS algorithms.

14.3.1   The multinomial logit model

The multinomial logit model assigns probabilities to random variables $Y_t \in \{1, 2, \ldots, C\}$ as functions of observed $k \times 1$ covariate vectors x_t and a parameter vector θ. In the standard setup $\theta' = (\theta'_1, \ldots, \theta'_C)$ and

$$P\left(Y_t = c \mid x_t, \theta\right) = \frac{\exp\left(\theta'_c x_t\right)}{\sum_{i=1}^{C} \exp\left(\theta'_i x_t\right)} \quad (c = 1, \ldots, C; t = 1, \ldots, T). \quad (14.13)$$

There is typically a normalization $\theta_c = 0$ for some $c \in \{1, 2, \ldots, C\}$, and there could be further restrictions on θ, but these details are not important to the main points of this section.

We use the specification (14.13) of the multinomial logit model throughout. The binomial logit model is the special case $C = 2$. Going forward, denote the observed outcomes $y_t = (y_1, \ldots, y_T)$ and the full set of covariates $X = [x_1, \ldots, x_T]'$. The log odds ratio

$$\log\left[\frac{P\left(Y_t = i \mid x_t, \theta\right)}{P\left(Y_t = j \mid x_t, \theta\right)}\right] = (\theta_i - \theta_j)' x_t \quad (14.14)$$

is linear in the parameter vector θ.

The prior distribution specifies independent components

$$\theta_c \sim N\left(\mu_c, \Sigma_c\right) \quad (c = 1, \ldots, C). \quad (14.15)$$

It implies that the vectors $\theta_j - \theta_c$ $(j = 1, \ldots, C; j \neq c)$ are jointly normally distributed, with

$$\mathrm{E}\left(\theta_j - \theta_c\right) = \mu_j - \mu_c, \ \mathrm{var}\left(\theta_j - \theta_c\right) = \Sigma_j + \Sigma_c, \ \mathrm{cov}\left(\theta_j - \theta_c, \theta_i - \theta_c\right) = \Sigma_c. \quad (14.16)$$

This provides the prior distribution of the parameter vector when (14.13) is normalized by setting $\theta_c = 0$, that is, when θ_j is replaced by $\theta_j - \theta_c$ and θ_c is omitted from the parameter vector. So long as the constancy of the prior distribution (14.16) is respected, all posterior moments of the form $\mathrm{E}\left[g\left(\theta, X, y\right) \mid X, y\right]$ will be invariant with respect to normalization. While it is entirely practical to simulate from the posterior distribution of the unnormalized model, for computation it is more efficient to use the normalized model because the parameter vector is shorter, reducing both computing time and storage requirements.

If the prior distribution (14.15) is exchangeable across $c = 1, \ldots, C$ then there is no further loss of generality in specifying $\mu_c = 0$ and $\Sigma_c = \Sigma$ $(c = 1, \ldots, C)$. With one minor exception the case studies in Section 14.5 further restrict Σ to the class proposed by [30],

$$\Sigma = (gT/k)\left(X'X\right)^{-1}. \quad (14.17)$$

where k is the order of each covariate vector x_t and g is a specified hyperparameter. To interpret Σ, consider the conceptual experiment in which the prior distribution of θ is augmented with a single x_t drawn with probability T^{-1} from the set $\{x_1, \ldots, x_T\}$ and then Y is generated by (14.13). Then the

TABLE 14.1
Characteristics of datasets.

Dataset	Sample size T	Covariates k	Outcomes C	Parameters
Diabetes	768	13	2	13
Heart	270	19	2	19
Australia	690	35	2	35
Germany	1000	42	2	42
Cars	263	4	3	8
Caesarean 1	251	8	3	16
Caesarean 2	251	4	3	8
Transportation	210	9	4	27

prior distribution of the log odds ratio (14.14) has mean 0 and variance

$$\frac{1}{T}\sum_{t=1}^{T} x_t' \left[2\left(gT/k\right)\left(X'X\right)^{-1} \right] x_t = \frac{1}{T}\mathrm{tr}\sum_{t=1}^{T} x_t x_t' \left[2\left(gT/k\right)\left(X'X\right)^{-1} \right] = 2g$$

and therefore standard deviation $(2g)^{1/2}$.

The specification consisting of (14.13) and (14.15) satisfies conditions 1, 2 and 3 in subsection 14.2.1 and then (14.14) is a function of interest that satisfies condition 4. Therefore the properties of the SPS algorithm developed in subsections 14.2.2 through 14.2.4 apply here.

14.3.2   Data

We used eight different datasets to study and compare the performance of the PSW and SPS algorithms. Table 14.1 summarizes some properties of these data. The notation in the column headings is taken from subsection 14.3.1, from which the number of parameters is $k \cdot (C - 1)$.

For the binomial logit models ($C = 2$), we use the same four datasets as [28], Section 3.3. Data and documentation may be found at the University of California - Irvine Machine Learning Repository[1], using the links indicated.

- Dataset 1, "Diabetes." The outcome variable is indication for diabetes using World Health Organization criteria, from a sample of individuals of Pima Indian heritage living near Phoenix, Arizona, USA. Of the covariates, one is a constant and one is a binary indicator. Link: Pima Indians Diabetes

- Dataset 2, "Heart." The outcome is presence of heart disease. Of the covariates, one is a constant and 12 are binary indicators. $T = 270$. Link: Statlog (Heart)

---

[1] http://archive.ics.uci.edu/ml/datasets.html

- Dataset 3, "Australia." The outcome is approval or denial of an application for a credit card. Of the covariates, one is a constant and 28 are binary indicators. Link: Statlog (Australian Credit Approval)

- Dataset 4, "Germany." The outcome is approval or denial of credit. Of the covariates, one is a constant and 41 are binary indicators. $T = 1000$. Link: Statlog (German Credit Data)

For the multinomial logit models, we draw from three data sources. The first two have been used in evaluating approaches to posterior simulation and the last is typical of a simple econometric application.

- Dataset 5, "Cars." The outcome variable is the kind of car purchased (family, work or sporty). Of the covariates, one is continuous and the remainder are binary indicators. The data were used in [29] in the evaluation of latent variable approaches to posterior simulation in logit models, and are taken from the data appendix[2] of [12].

- The next two datasets derive from a common dataset described in [10], Table 1.1. The outcome variable is infection status at birth (none, Type 1, Type 2). Covariates are constructed from three binary indicators. These data ([10], Table 1.1) have been a widely used test bed for the performance of posterior simulators given severely unbalanced contingency tables, for example [14] and references therein. The data are distributed with the R statistical package.

  - Dataset 6, "Caesarean 1." The model is fully saturated, thus $2^3 = 8$ covariates. This variant of the model has been widely studied because the implicit design is severely unbalanced. One cell is empty. For the sole purpose of constructing the g prior (14.17), we supplement the covariate matrix X with a single row having an indicator in the empty cell. The likelihood function uses the actual data.

  - Dataset 7, "Caesarean 2." There are four covariates consisting of the three binary indicators and a constant term.

- Dataset 8, "Transportation." The data is a choice-based sample of mode of transportation choice (car, bus, train, air) between Sydney and Melbourne. The covariates are all continuous except for the intercept. The data (Table F21.2 of the data appendix of [20][3]) are widely used to illustrate logit choice models in econometrics.

---

[2]http://www-stat.wharton.upenn.edu/~waterman/fsw/download.html
[3]http://pages.stern.nyu.edu/~wgreene/Text/tables/tablelist5.htm

14.3.3   Hardware and software

The PSW algorithm described in [28] is implemented in the R package BayesLogit provided by the authors[4]. The R command transfers control to expertly written C code and thus execution time reflects the efficiency of fully compiled C code and not the inefficiency of interpreted R code. Except as noted in subsection 14.5.1, the code executed flawlessly without intervention. To facilitate comparison with the GPU implementation of the SPS algorithm, the execution used a single CPU (Intel Xeon 5680) and 24GB memory.

The SPS/CPU algorithm is coded in MATLAB (Edition 2012b, with the Statistics toolbox). The execution used a 12-core CPU (2× Intel Xeon E5645) and 24GB memory, exploiting multiple cores with a ratio of CPU to wall clock time of about 5.

The SPS/GPU algorithm is coded in C with the extension CUDA version 4.2. The execution used an Intel Core i7 860, 2.8 GHz CPU and one Nvidia GTX 570 GPU (480 cores).

14.4   Performance of the SPS Algorithm

The SPS algorithm can be used routinely in any model that has a bounded and directly computed likelihood function, accompanied by a proper prior distribution. It provides numerical standard errors that are reliable in the sense that they indicate correctly the likely outcome of a repeated, independent execution of the sequential posterior simulator. As a by-product, it also provides consistent (in N) approximations of log marginal likelihood and associated numerical standard error; Section 4 of [9] explains the procedure. This section illustrates these properties for the case of the multinomial logit model.

Numerical approximations of posterior moments must be accompanied by an indication of their accuracy. Even if editorial constraints often preclude accompanying each moment approximation with an indication of accuracy, decent scholarship demands that the investigator be aware of the accuracy of reported moment approximations. Moreover, the accuracy indications must themselves be interpretable and reliable.

The SPS methodology for the logit model described in Section 14.2 achieves this standard by means of a central limit theorem for posterior moment approximations accompanied by a scheme for grouping particles that leads to a simple simulation-consistent approximation of the variance in the central limit theorem. The practical consequence of these results is the numerical standard error (14.6). The underlying theory for SPS requires the two-pass procedure of Algorithm 14.3.

---

[4]http://cran.r-project.org/web/packages/BayesLogit/BayesLogit.pdf

TABLE 14.2

Reliability of one- and two-pass algorithms.

	E $(\theta_1'\bar{x} \mid$ Data$)$	E $(\theta_2'\bar{x} \mid$ Data$)$
$J = 10, N = 1000$		
Run A, Pass 1	0.6869 [.0017]	−0.3836 [.0017]
Run A, Pass 2	0.6855 [.0012]	−0.3856 [.0024]
Run B, Pass 1	0.6811 [.0014]	−0.3920 [.0017]
Run B, Pass 2	0.6847 [.0018]	−0.3877 [.0025]
Run C, Pass 1	0.6844 [.0016]	−0.3892 [.0018]
Run C, Pass 2	0.6837 [.0019]	−0.3875 [.0021]
$J = 40, N = 2500$		
Run A, Pass 1	0.6850 [.0005]	−0.3873 [.0006]
Run A, Pass 2	0.6857 [.0005]	−0.3871 [.0008]
Run B, Pass 1	0.6844 [.0004]	−0.3890 [.0006]
Run B, Pass 2	0.6849 [.0005]	−0.3885 [.0006]
Run C, Pass 1	0.6853 [.0005]	−0.3881 [.0006]
Run C, Pass 2	0.6847 [.0004]	−0.3872 [.0005]

Table 14.2 provides some evidence on these points using the multinomial logit model, the prior distribution (14.15)–(14.17) with $g/k = 1$, and the cars dataset described in subsection 14.3.2. For both small and large SPS executions (upper and lower panels, respectively) Table 14.2 indicates moment approximations for three independent executions (A, B and C) of the two-pass algorithm, and for both the first and second pass of the algorithm. The posterior moments used in the illustrations here, and subsequently, are the the log odds ratio (with respect to the outcome $Y = C$), $\theta_c'\bar{x}$ ($c = 1, \dots, C - 1$), where $\bar{x}$ is the sample mean of the covariates. For each posterior moment approximation in Table 14.2 the accompanying figure in brackets is the numerical standard error. As discussed in subsection 14.2.3, these will vary quite a bit more from one run to another when $J = 10$ than they will when $J = 40$, and this is evident in Table 14.2.

Turning first to the comparison of results at the end of Pass 1 (no formal justification for numerical standard errors) and at the end of Pass 2 (the established results for the non-adaptive algorithm discussed in subsection 14.2.2 apply) there are no unusually large differences within any run, given the numerical standard errors. That is, there is no evidence to suggest that if an investigator used Pass 1 results to anticipate what Pass 2 results would be, the investigator would be misled.

This still leaves the question of whether the numerical standard errors from a single run are a reliable indication of what the distribution of Monte Carlo approximations would be across independent runs. Comparing results for runs A, B and C, Table 14.2 provides no indication of difficulty for the large SPS executions. For the small SPS executions, there is some suggestion that variation across runs at the end of the first pass is larger than numerical standard error suggests ($E\left(\theta_1' \overline{x} \mid \text{Data}\right)$ for runs A and B).

These suggestions could be investigated more critically with scores or hundreds of runs of the SPS algorithm, but we conjecture the returns would be low and in any event there is no basis for extrapolating results across models and datasets. Most important, one cannot resort to this tactic routinely in applied work. The results here support the earlier recommendation (at the end of subsection 14.2.4) that the investigator proceed mainly using the one-pass algorithm, reserving the two-pass algorithm for checks at the start and the end of a research project.

14.5   Comparison of Algorithms

We turn now to a systematic comparison of the efficiency and reliability of the PSW and SPS algorithms in the logit model.

14.5.1   The exercise

To this end, we simulated the posterior distribution for the Cartesian product of the eight datasets described in subsection 14.3.2 and Table 14.1, and five approaches to simulation. The first approach to posterior simulation is the PSW algorithm implemented as described in subsection 14.3.3. The second approach uses the small SPS simulation ($J = 10$, $N = 1000$) with the CPU implementation described in subsection 14.3.3, and the third approach uses the GPU implementation. The fourth and fifth approaches are the same as the second and third except that they use the large SPS simulation ($J = 40$, $N = 2500$).

To complete this exercise we had to modify the multinomial logit model (last four datasets) for the PSW algorithm. The code that accompanies the algorithm requires that the vectors θ_c be independent in the prior distribution, and consequently (14.17) was modified to specify $\text{cov}\left(\theta_j - \theta_c, \theta_i - \theta_c\right) = 0$. As a consequence, posterior moments approximated by the PSW algorithm depend on the normalization employed and are never the same as those approximated by the SPS algorithms. We utilized the same normalization as in the SPS algorithms, except for the cars dataset, for which the code would not execute with this choice and we normalized instead on the second choice.

Since it is impractical to present results from all $8 \times 5 \times 5 = 200$ posterior simulations, we restrict attention to a single prior distribution for each dataset: the one producing the highest marginal likelihood. Table 14.4 provides the log-marginal likelihoods under all five prior distributions for all eight datasets, as computed using the large SPS/GPU algorithm.

TABLE 14.3
Log-marginal likelihoods, all datasets and models.

	$g/k = 1/64$	$g/k = 1/16$	$g/k = 1/4$	$g/k = 1$	$g/k = 4$
Diabetes	−405.87 [0.04]	−386.16 [0.03]	−383.31 [0.03]	−387.01 [0.04]	−392.61 [0.04]
Heart	−141.00 [0.04]	−123.36 [0.03]	−118.58 [0.04]	−124.38 [0.06]	−135.25 [0.11]
Australia	−301.87 [0.04]	−269.90 [0.05]	−267.41 [0.06]	−280.47 [0.07]	−300.20 [0.12]
Germany	−539.46 [0.06]	−535.91 [0.08]	−556.71 [0.00]	−586.66 [0.00]	−621.53 [0.00]
Cars	−263.75 [0.02]	−254.42 [0.03]	−253.62 [0.03]	−257.18 [0.03]	−262.20 [0.04]
Caesarean 1	−214.50 [0.03]	−187.19 [0.03]	−176.96 [0.02]	−177.29 [0.03]	−181.66 [0.03]
Caesarean 2	−219.20 [0.03]	−192.10 [0.03]	−180.30 [0.02]	−178.91 [0.02]	−181.42 [0.02]
Transportation	−234.58 [0.03]	−197.07 [0.04]	−176.14 [0.05]	−173.97 [0.05]	−184.24 [0.07]

Note that the accuracy of these approximations is very high, compared with existing standards for posterior simulation. The accuracy of log-marginal likelihood approximation tends to decline with increasing sample size, as detailed in [9], Section 4, and this is evident in Table 14.3. Going forward, all results pertain to the g-prior described in subsection 14.3.1 with $g/k = 1/16$ for Germany, $g/k = 1$ for Caesarean 1 and transportation, and $g/k = 1/4$ for the other five datasets.

14.5.2 Reliability

We assess the reliability of the algorithms by comparing posterior moments and log-marginal likelihood approximations for the same model. Table 14.4 provides the posterior moment approximations. The moments used are, again, the posterior expectation of the log odds ratio(s) evaluated at the sample mean $\bar{x}$, for each choice relative to the last choice. (This corresponds to the normalization used in execution.) Thus there is one moment for each of the four binomial logit datasets, two moments for the first three multinomial logit datasets, and three moments for the last multinomial logit model dataset.

The result for each moment and algorithm is presented in a block of four numbers. The first line has the simulation approximation of the posterior expectation followed by the simulation approximation of the posterior standard deviation. The second line contains [in brackets] the numerical standard error and relative numerical efficiency of the approximation. For the multinomial logit model there are multiple blocks, one for each posterior moment.

For the SPS algorithms the numerical standard error and relative numerical efficiency are the natural by-product of the results across the J groups of particles as described in subsection 14.2.1. For the PSW algorithm these are computed based on the 100 independent executions of the algorithm. The PSW approximations of the posterior expectation and standard deviation are based on a single execution.

Posterior moment approximations are consistent across the four implementations of the SPS algorithm (columns 3 through 6 of Table 14.4). For the comparable cases (the first four datasets) PSW and SPS moments are

TABLE 14.4

Posterior moments and numerical accuracy.

	PSW	($J = 10, N = 1000$)		($J = 40, N = 2500$)	
		SPS/CPU	SPS/GPU	SPS/CPU	SPS/GPU
Diabetes	−0.853 (.096)	−0.855 (.095)	−0.852 (.096)	−0.853 (.095)	−0.853 (.095)
	[.0008, 0.68]	[.0009, 1.12]	[.0009, 1.21]	[.0003, 0.99]	[.0003, 1.03]
Heart	−0.250 (.192)	−0.246 (.187)	−0.251 (.191)	−0.249 (.189)	−0.250 (.189)
	[.0021, 0.43]	[.0019, 0.96]	[.0019, 0.98]	[.0006, 0.86]	[.0006, 0.92]
Australia	−0.438, (.157)	−0.438 (.157)	−0.439 (.156)	−0.440 (.156)	−0.438 (.157)
	[.0023, 0.23]	[.0016, 0.96]	[.0016, 0.98]	[.0005, 0.97]	[.0006, 0.60]
Germany	−1.182 (.089)	−1.180 (.089)	−1.81 (.089)	−1.182 (.089)	−1.81 (.088)
	[.0010, 0.36]	[.0009, 1.00]	[.0004, 0.94]	[.0003, 0.91]	[.0004, 0.94]
Cars	−1.065 (.171)	0.684 (.156)	0.685 (.158)	0.685 (.156)	0.685 (.156)
	[.0002, 0.46]	[.0015, 1.03]	[.0017, 0.98]	[.0005, 1.12]	[.0005, 0.97]
	−0.665 (.156)	−0.386 (.195)	−0.388 (.195)	−0.387 (.194)	−0.388 (.193)
	[.0009, 0.60]	[.0021, 0.83]	[.0017, 1.29]	[.0006, 1.19]	[.0007, 0.72]
Caesarean 1	−1.975 (.241)	−2.049 (.245)	−2.052 (.248)	−2.052 (.246)	−2.052 (.246)
	[.0004, 0.26]	[.0024, 1.09]	[.0037, 0.45]	[.0008, 0.91]	[.0008, 0.96]
	−1.607 (.211)	−1.698 (.215)	−1.694 (.217)	−1.697 (.219)	−1.698 (.219)
	[.0003, 0.27]	[.0018, 1.36]	[.0024, 0.80]	[.0007, 1.07]	[.0006, 1.30]
Caesarean 2	−2.033 (.261)	−2.057 (.264)	−2.056 (.264)	−2.056 (.262)	−2.0534 (.262)
	[.0004, 0.22]	[.0027, 0.94]	[.0025, 1.32]	[.0008, 0.99]	[.0009, 0.89]
	−1.586 (.205)	−1.597 (.210)	−1.587 (.206)	−1.593 (.206)	−1.593 (.207)
	[.0003, 1.30]	[.0021, 0.98]	[.0021, 0.99]	[.0006, 1.02]	[.0006, 1.03]
Transportation	0.091 (.316)	0.130 (.321)	0.119 (.322)	0.123 (.322)	0.123 (.323)
	[.0006, 0.13]	[.0031, 1.09]	[.0030, 1.14]	[.0010, 1.01]	[.0010, 0.97]
	−0.588 (.400)	−0.416 (.380)	−0.416 (.388)	−0.421 (.386)	−0.419 (.388)
	[.0009, 0.09]	[.0020, 3.52]	[.0043, 0.82]	[.0012, 1.04]	[.0011, 1.29]
	−1.915 (.524)	−1.646 (.487)	−1.642 (.490)	−1.645 (.491)	−1.647 (.492)
	[.0013, 0.07]	[.0036, 1.84]	[.0044, 1.24]	[.0016, 0.95]	[.0019, 0.65]

also consistent with each other. As explained earlier in this section, the moments approximated by the PSW algorithm are not exactly the same as those approximated by the SPS algorithms in the last four datasets. Numerical standard errors depend on the number of iterations of the PSW algorithm and the number of particles in the SPS algorithm. For the SPS algorithm RNE clusters around 0.9 by design (Algorithm 14.2, RNE termination criterion Step 2(b)ii). Median RNE for the PSW algorithm is 0.22.

Table 14.5 compares approximations of log-marginal likelihoods across the four variants of the SPS algorithm, and there are no anomalies. (The PSW algorithm does not yield approximations of the log-marginal likelihood.) There is no evidence of unreliability of any of the algorithms in Tables 14.4 and 14.5.

14.5.3   Computational efficiency

Our comparisons are based on a single run of each of the five algorithms (PSW and four variants of SPS) for each of the eight datasets, using for each dataset one particular prior distribution chosen as indicated in subsection 14.5.1. In the case of the PSW algorithm, we used 20,000 iterations for posterior moment approximation for the first four datasets, and 21,000 for the latter four

TABLE 14.5
Log-marginal likelihoods and numerical accuracy.

	($J = 10, N = 1000$)		($J = 40, N = 2500$)	
	SPS/CPU	SPS/GPU	SPS/CPU	SPS/GPU
Diabetes	−383.15 [0.05]	−383.14 [0.17]	−383.31 [0.03]	−383.25 [0.03]
Heart	−118.29 [0.15]	−118.73 [0.14]	−118.58 [0.04]	−118.61 [0.04]
Australia	−267.25 [0.32]	−267.35 [0.19]	−267.41 [0.06]	−267.35 [0.05]
Germany	−536.05 [0.21]	−536.10 [0.18]	−535.91 [0.08]	−535.89 [0.07]
Cars	−253.57 [0.07]	−253.46 [0.10]	−253.62 [0.03]	−253.61 [0.03]
Caesarean 1	−177.06 [0.08]	−176.72 [0.13]	−176.96 [0.02]	−176.91 [0.03]
Caesarean 2	−178.89 [0.08]	−178.95 [0.03]	−178.91 [0.02]	−178.83 [0.03]
Transportation	−174.09 [0.12]	−173.92 [0.19]	−173.97 [0.05]	−173.99 [0.05]

TABLE 14.6
Clock execution time in seconds.

		($J = 10, N = 1000$)		($J = 40, N = 2500$)	
	PSW	SPS/CPU	SPS/GPU	SPS/CPU	SPS/GPU
Diabetes	14.90	106.7	6.00	739.9	26.9
Heart	9.53	140.8	13.7	923.4	73.6
Australia	41.60	1793.5	69.2	12449.9	448.7
Germany	125.59	5910.4	225.9	45263.2	1689.4
Cars	7.62	33.5	3.5	231.2	18.9
Caesarean 1	6.84	97.3	10.9	723.3	55.3
Caesarean 2	6.65	15.2	2.5	133.3	11.6
Transportation	15.10	569.7	39.7	3064.2	293.7

datasets. The entries in Table 14.6 show wall-clock time for execution on the
otherwise idle machine described in subsection 14.3.3. Execution time for the
PSW algorithm includes 1,000 burn-in iterations in all cases except Australia
and Germany, which have 5,000 burn-in iterations. Times can vary considerably, depending on the particular hardware used: for example, the SPS/CPU
algorithms were executed using a 12-core machine that utilized about 5 cores,
simultaneously, on average; and the SPS/GPU algorithms used only a single
GPU with 480 cores. The results here must be qualified by these considerations. In practice returns to additional CPU cores or additional GPUs are
substantial but less than proportionate.

Execution time also depends on memory management, clearly evident in
Table 14.6. The ratio of execution time for the SPS/CPU algorithm in the
large simulations ($J = 40, N = 2500$) to that in the small simulations ($J =
10, N = 1000$) ranges from from 8.5 (Transportation) to 16.2 (Australia).
There is no obvious pattern or source for this variation. The same ratio for the
SPS/GPU algorithm ranges from 4.48 (Diabetes) to about 7.45 (Germany and

TABLE 14.7
Computational inefficiency relative to PSW.

	($J = 10, N = 1000$)		($J = 40, N = 2500$)	
	SPS/CPU	SPS/GPU	SPS/CPU	SPS/GPU
Diabetes	9.40	0.53	6.52	0.24
Heart	14.35	1.40	9.41	0.75
Australia	28.25	1.09	19.61	0.71
Germany	45.06	1.72	34.51	1.29
Cars	5.04	0.53	3.47	0.28
Caesarean 1	8.36	0.94	6.21	0.47
Caesarean 2	3.73	0.62	3.28	0.29
Transportation	6.19	0.43	3.33	0.32

Transportation). This reflects the fact that GPU computing is more efficient to the extent that the application is intensive in arithmetic logic as opposed to flow control. Very small problems are relatively inefficient; as the number and size of particles increases, the efficiency of the SPS/GPU algorithm increases, approaching an asymptotic ratio of number and size of particles to computing time from below.

Relevant comparisons of computing time t require that we correct for the number $\widetilde{M}$ of PSW iterations or SPS particles and the relative numerical efficiency $R\widetilde{N}E$ of the algorithm. This produces an efficiency-adjusted computing time $\tilde{t} = t / \left(\widetilde{M} \cdot R\widetilde{N}E \right)$. For $R\widetilde{N}E$ we use the average of the relevant RNEs reported in Table 14.4: in the case of SPS, the averages are taken across all four variants since population RNE does not depend on the number of particles, hardware or software. This ignores variation in RNE from moment to moment and one run to the next. In the case of PSW, it also ignores dependence of RNE and number of burn-in iterations on the number of iterations used for moment approximations that arises from both practical and theoretical considerations. Therefore efficiency comparisons should be taken as indicative rather than definitive: they will vary from application to application in any event, and one will not undertake these comparisons for every (if indeed any) substantive study.

Table 14.7 provides the ratio of $\tilde{t}$ for each of the SPS algorithms to $\tilde{t}$ for the PSW algorithm, for each of the eight datasets. The SPS/CPU algorithm compares more favorably with the PSW algorithm for the small simulation exercises than for the large simulation exercises. The SPS/CPU algorithm is clearly slower than the PSW algorithm, and its disadvantage becomes more pronounced the greater the number of parameters and observations. With a single exception (Germany) the SPS/GPU algorithm is faster than the PSW algorithm for the large simulation exercises, and for the single exception it is 78% as efficient.

14.6   Conclusion

Sequential posterior simulation algorithms complement earlier approaches to the simulation approximation of posterior moments. Graphics processing units have become a convenient and cost-effective platform for scientific computing, potentially accelerating computing speed by orders of magnitude for parallel algorithms. One of the appealing features of SPS is the fact that it is well suited to GPU computing. The work reported here uses an SPS algorithm designed to exploit that potential.

The multinomial logistic regression model, the focus of this chapter, is important in applied statistics in its own right, and also as a component in mixture models, Bayesian belief networks, and machine learning. The model presents a well conditioned likelihood function that renders maximum likelihood methods straightforward, yet it has been relatively difficult to attack with posterior simulators – and hence arguably a bit of an embarrassment for applied Bayesian statisticians. Recent work by [13] [14] [22] [29] and especially, [28] has improved this state of affairs substantially, using latent variable representations specific to classes of models that include the multinomial logit.

The chapter used 8 canonical datasets to compare the performance of the SPS algorithm with the PSW algorithm of [28]. The SPS algorithm used a single GPU, the PSW algorithm a single CPU. Both were coded in C. Comparisons of efficiency were made for each of the datasets and allow for differences in the accuracy of posterior moment approximations as well as computing times. For 7 of the 8 datasets, the SPS algorithm was found to be more efficient, and more than twice as efficient for 5 of the 8 datasets. For the remaining dataset the SPS algorithm was 78% as efficient.

The SPS algorithm has other attractions that are as significant as computational efficiency. These advantages are generic, but some are more specific to the logit model than others.

1. SPS produces an accurate approximation of the log-marginal likelihood as a by-product. The latent variable algorithms, including all of those just mentioned, do not. SPS also produces accurate approximations of log-predictive likelihoods, a significant factor in time series models.

2. SPS approximations have a firm foundation in distribution theory. The algorithm produces a reliable approximation of the standard deviation in the applicable central limit theorem – again, as a by-product in the approach developed in [9]. Numerical accuracy in the latent variable methods for posterior simulation do not do this, and we are not aware of procedures for ascertaining reliable approximations of accuracy with the latent variable approaches that do not entail a many-fold increase in computing time.

3. More generally, SPS is simple to implement when the likelihood function can be evaluated in closed form. Indeed, in comparison with alternatives it

can be trivial, and this is the case for the logit model studied in this chapter. By implication, the time from conception to Bayesian implementation is greatly reduced for this class of likelihood functions.

Acknowledgment

All three authors gratefully acknowledge financial support for the research from Australian Research Council Discovery Project DP130103356.

Bibliography

[1] J. H. Albert and S. Chib. Bayesian analysis of binary and polychotomous response data. *Journal of the American Statistical Association*, 88(422):669–679, 1993.

[2] C. Andrieu, A. Doucet, and R. Holenstein. Particle Markov chain Monte Carlo methods. *Journal of the Royal Statistical Society: Series B (Statistical Methodology)*, 72(3):269–342, 2010.

[3] J. E. Baker. Adaptive selection methods for genetic algorithms. In *Proceedings of the 1st International Conference on Genetic Algorithms*, pages 101–111, Hillsdale, NJ, USA, 1985. L. Erlbaum Associates Inc.

[4] J. E. Baker. Reducing bias and inefficiency in the selection algorithm. In *Proceedings of the Second International Conference on Genetic Algorithms and Their Application*, pages 14–21, Hillsdale, NJ, USA, 1987. L. Erlbaum Associates Inc.

[5] N. Chopin. A sequential particle filter method for static models. *Biometrika*, 89(3):539–551, 2002.

[6] N. Chopin. Central limit theorem for sequential Monte Carlo methods and its application to Bayesian inference. *The Annals of Statistics*, 32(6):2385–2411, 2004.

[7] N. Chopin, P. E. Jacob, and O. Papaspiliopoulos. SMC^2: An efficient algorithm for sequential analysis of state-space models. *Journal of the Royal Statistical Society: Series B (Statistical Methodology)*, 75:397–426, 2013.

[8] P. Del Moral, A. Doucet, and A. Jasra. On adaptive resampling strategies for sequential Monte Carlo methods. *Bernoulli*, 18(1):252–278, 2012.

[9] G. Durham and J. Geweke. Adaptive sequential posterior simulators for massively parallel computing environments. Available at SSRN 2251635, 2013.

[10] L. Fahrmeir and G. Tutz. *Multivariate Statistical Modelling Based on Generalized Linear Models*. Springer Verlag, New York, 2nd edition, 2001.

[11] J. M. Flegal and G. L. Jones. Batch means and spectral variance estimators in Markov chain Monte Carlo. *The Annals of Statistics*, 38(2):1034–1070, 2010.

[12] D. P. Foster, R. A. Stine, and R. P. Waterman. *Business Analysis Using Regression - A Casebook*. Springer Verlag, New York, 1998.

[13] S. Frühwirth-Schnatter and R. Frühwirth. Auxiliary mixture sampling with applications to logistic models. *Computational Statistics & Data Analysis*, 51(7):3509–3528, 2007.

[14] S. Frühwirth-Schnatter and R. Frühwirth. Bayesian inference in the multinomial logit model. *Austrian Journal of Statistics*, 41(1):27–43, 2012.

[15] J. Geweke. Bayesian inference in econometric models using Monte Carlo integration. *Econometrica*, 57(6):1317–1339, 1989.

[16] J. Geweke and M. Keane. Smoothly mixing regressions. *Journal of Econometrics*, 138(1):252–290, 2007.

[17] J. Geweke, M. Keane, and D. Runkle. Alternative computational approaches to inference in the multinomial probit model. *The Review of Economics and Statistics*, 76(4):609–632, 1994.

[18] N. J. Gordon, D. J. Salmond, and A. F. M. Smith. Novel approach to nonlinear/non Gaussian Bayesian state estimation. *IEE Proceedings F-Radar and Signal Processing*, 140(2):107–113, Apr. 1993.

[19] R. B. Gramacy and N. G. Polson. Simulation-based regularized logistic regression. *Bayesian Analysis*, 7(3):567–590, 2012.

[20] W. H. Greene. *Econometric Analysis*. Prentice Hall, Upper Saddle River, New Jersey, 5th edition, 2003.

[21] E. P. Herbst and F. Schorfheide. Sequential Monte Carlo sampling for DSGE models. Working Papers, Federal Reserve Board, Sept. 2013.

[22] C. C. Holmes and L. Held. Bayesian auxiliary variable models for binary and multinomial regression. *Bayesian Analysis*, 1(1):145–168, 2006.

[23] R. A. Jacobs, M. I. Jordan, S. J. Nowlan, and G. E. Hinton. Adaptive mixtures of local experts. *Neural Computation*, 3(1):79–87, 1991.

[24] W. Jiang and M. A. Tanner. Hierarchical mixtures-of-experts for exponential family regression models: Approximation and maximum likelihood estimation. *The Annals of Statistics*, 27(3):987–1011, 1999.

[25] A. Kong, J. S. Liu, and W. H. Wong. Sequential imputations and Bayesian missing data problems. *Journal of the American Statistical Association*, 89(425):278–288, 1994.

[26] J. S. Liu and R. Chen. Blind deconvolution via sequential imputations. *Journal of the American Statistical Association*, 90(430):567–576, 1995.

[27] J. S. Liu and R. Chen. Sequential Monte Carlo methods for dynamic systems. *Journal of the American Statistical Association*, 93(443):1032–1044, 1998.

[28] N. G. Polson, J. G. Scott, and J. Windle. Bayesian inference for logistic models using Pólya-Gamma latent variables. *Journal of the American Statistical Association*, 108(504):1339–1349, 2013.

[29] S. L. Scott. Data augmentation, frequentist estimation, and the Bayesian analysis of multinomial logit models. *Statistical Papers*, 52(1):87–109, 2011.

[30] A. Zellner. On assessing prior distributions and Bayesian regression analysis with g-prior distributions. In P. K. Goel and A. Zellner, editors, *Bayesian Inference and Decision Techniques: Essays in Honor of Bruno de Finetti*, pages 233–243. Elsevier Science Ltd, Amsterdam, North-Holland, June 1986.

15

From Risk Analysis to Adversarial Risk Analysis

David Ríos Insua

ICMAT-CSIC

Javier Cano

University Rey Juan Carlos

Michael Pellot

Transports Metropolitans de Barcelona

Ricardo Ortega

Transports Metropolitans de Barcelona

CONTENTS

15.1   Introduction

Risk analysis may be described as a systematic analytical process for assessing, managing and communicating risks. It is performed to understand the nature of unwanted, negative consequences to human life, health, property or the environment, and to mitigate and/or eliminate them, see [1] for a review.

We adopt the classic characterisation of risk in [6], in terms of outcome scenarios, their consequences and their probability of occurrence. This entails a process to identify and evaluate the threats that a system is exposed to, which, as a result, could minimise or avoid the occurrence and impact of certain losses. Then, the negative impacts of threats can be managed and reduced to the lowest possible levels. We describe a Bayesian decision analytic framework for risk analysis.

Adversarial risk analysis (ARA), see [9], has been recently proposed to deal with threats originating from intentional actions from adversaries. Motivated by applications in counterterrorism, cybersecurity and competitive decision making, there has been a renewed interest in developing practical tools and theory for analysing the strategic calculations of intelligent opponents who must act in scenarios with random outcomes. ARA builds a Bayesian decision analytic model for one of the participants, the Defender, who then builds a forecasting model for the actions of the adversaries. We describe here a general framework for ARA focusing on the Sequential Defend-Attack model, see [8]. Since resources allocated for dealing with threats may be shared against intentional and nonintentional ones, we provide here a combined framework for standard risk analysis and adversarial risk analysis.

The structure of the chapter is as follows. In Section 15.2 we deal with standard risk analysis. We discuss its application to the fare evasion problem in Section 15.3. Section 15.4 provides a framework for adversarial risk analysis, illustrating its use in Section 15.5. We then consider risk analysis and adversarial risk analysis jointly in Section 15.6 and provide a numerical example based on the fare evasion problem in Section 15.7. We end with some discussion.

15.2   A Framework for Risk Analysis

We provide a schematic framework that formalises standard risk analysis, assessment, and management methods as in [5] or [1], adapted to the classic proposal in [6].

Figure 15.1(a) shows an *influence diagram*, see [7], that displays the simplest version of a risk analysis problem. The oval represents the costs c associated with the performance of a system under normal circumstances. The

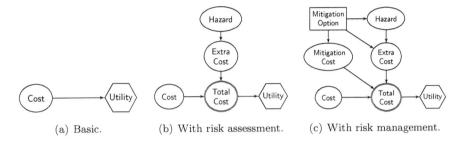

(a) Basic.   (b) With risk assessment.   (c) With risk management.

FIGURE 15.1
Risk analysis influence diagrams.

hexagon represents the net consequences in terms of the decision maker's utility function u. Costs c are uncertain and modelled through the density $\pi(c)$. The utility $u(c)$ of the cost is decreasing and, typically, nonlinear.

We globally evaluate the performance of the system through its expected utility $\psi = \int u(c)\pi(c)\mathrm{d}c$. In practice, the system owner will typically perform a risk assessment to: (1) Identify hazardous disruptive events $E_1, E_2, \ldots, E_n$, which we assume to be mutually exclusive; (2) Assess their probabilities of occurrence, $\Pr(E_j) = q_j$; and (3) Assess the (random) costs c_j conditional on the occurrence of E_j. It is convenient to let E_0 be the event for which there are no disruptions, with an associated probability q_0. Figure 15.1(b) shows the influence diagram that extends the previous formulation to include risk assessment.

Let $(q_0, q_1, \ldots, q_n)$ be the vector of event probabilities, and let $\pi_j(c)$, $j = 1, \ldots, n$ be the cost density if event E_j occurs. Then, the cost density is the mixture $\sum_{j=0}^{n} q_j\,\pi_j(c)$. Once the risk assessment is performed, the system owner estimates the expected utility $\psi_r = \sum_{j=0}^{n} q_j \int u(c)\pi_j(c)\,\mathrm{d}c$. Consider the difference $\psi - \psi_r$. This is a nonnegative quantity, as ψ describes a problem without including the costs associated with disruptive events, whereas ψ_r relies upon risk assessment. If big enough, we may reduce such difference undertaking a risk management strategy, introducing a set of choices $\mathcal{M}$ like contingency plans or insurance policies. These tend to lower the costs associated with particular disruptions and/or lower the chance of disruption, as shown in Figure 15.1(c). The risk management solution is the portfolio of countermeasures that maximise expected utility, that is, $\psi_m = \max_{m \in \mathcal{M}} \psi_r(m)$, where

$$\psi_r(m) = \sum_{j=0}^{n} q_j(m) \int u(c)\pi_j(c|m)\,\mathrm{d}c. \qquad (15.1)$$

Since risk management extends the set of choices, $\psi_m \geq \psi_r$ holds. The choice set $\mathcal{M}$ may be discrete or continuous. Additional complexity arises if there is sequential investment, if one allocates risk management resources according to a portfolio analysis, or if there are multiattribute utility functions.

15.3 Case Study: Fighting Fare Evasion in a Facility

As a case study, we consider the problem of fare evasion when accessing a physical facility through the payment of a ticket giving entrance. The figures presented in this paper have been modified from actual figures in order to protect the confidentiality of the case study provider. Therefore, *the data is realistic data but not actual data*. The service is managed by an operator who is responsible for the appropriate functioning of the facility, meeting, at the same time, customer service expectations. Because of excessive losses due to fraud, the operator studies the adoption of countermeasures, on top of already existing ones, restricted by an available budget. We display in Table 15.1 their relevant features.

The operator needs to assess the efficiency of various portfolios of countermeasures. We aim at supporting the operator in devising an optimal security portfolio.

15.3.1 Case description

We distinguish two types of customers, in terms of their attitude towards the fare system.

1. Civic customers. They pay the fare. Some might be checked by inspectors, possibly getting annoyed by that, which is an undesired consequence for the operator. In order to mitigate customer dissatisfaction, the operator launches information campaigns to make customers aware of the relevance of inspections in guaranteeing a safe and high-quality service.

2. Fare evaders. They decide not to pay individually. They risk being caught by inspectors, facing the possibility of being fined. We regard this type of evaders as 'casual'. Therefore, we do not take into account the possible consequences for them, and we only consider the relevant consequences for the operator. Some of the fare evaders will be inspected and fined, and the operator partly mitigates losses due to fare evasion.

TABLE 15.1

Relevant features of counter measures.

	Role	Features
Inspectors	Preventive/recovery	Inspect customers. Collect fines
Door guards	Preventive	Control access points
Guards	Preventive	Patrol along the facility
Doors	Preventive	New secured automatic access doors
Ticket clerks	Preventive	Current little implication in security

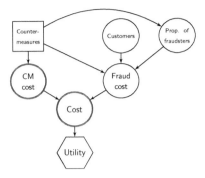

FIGURE 15.2
Influence diagram when only standard evaders are present.

Concerning the collection of fines, we simplify the problem by assuming that a fraudster caught without a ticket will pay the same (average) fine.

15.3.2 Case modelling

The problem may be seen as one of risk management, as sketched in the generic Figure 15.1(c). The "Hazard" node corresponds to the presence of evaders, whose unwillingness to pay entails a cost to the operator. To minimise such impact, the operator would deploy mitigation measures, which entail costs. All relevant costs are aggregated in the operator's utility function. A specific influence diagram is shown in Figure 15.2.

The decision node "Countermeasures" refers to the portfolio of countermeasures deployed by the operator. We have uncertainty about the proportion of fraudsters and the number of customers, from which we obtain the fraud cost. The countermeasures aim at reducing the proportion of fraud. If occurring, the inspectors are entrusted to minimise the fraud cost, through the fine system. We aim at obtaining the maximum expected utility portfolio of countermeasures.

   Assume that each inspector costs c_1, each door guard costs c_2, each guard costs c_3 and each automatic access door costs c_4 (over the relevant planning period). As ticket clerks are already hired by the company, there are no additional direct costs associated with the reassignment of their duties. However, making them switch from a passive attitude towards the fare evasion problem to a proactive one could have negative implications in terms of troubles with unions. We monetise this assuming a fixed global cost c_5 for that. Let (x_1, x_2, x_3, x_4) be, respectively, the inspectors, door guards, guards and automatic access doors deployed. We also use a binary variable $x_5 \in \{0, 1\}$, with $x_5 = 1$, indicating the involvement of clerks in observation tasks, and $x_5 = 0$, that they will preserve their operational *status quo*. b will be the budget available. Then, the feasible security countermeasure portfolios

$x = (x_1, x_2, x_3, x_4, x_5)$ will satisfy

$$c_1 x_1 + c_2 x_2 + c_3 x_3 + c_4 x_4 \leq b,$$
$$x_1, x_2, x_3, x_4 \geq 0,$$
$$x_4 \leq n_4$$
$$x_1, x_2, x_3, x_4 \text{ integer}$$
$$x_5 \in \{0, 1\},$$

where n_4 is the maximum number of access doors that may be replaced. We denote by $\mathcal{B}$ the set of feasible portfolios.

We describe now the impact of countermeasures on the fare evasion proportion. Assume that N, the number of customers in the planning period, may be modelled as a Poisson process of rate λ. Let $p(x)$ be the proportion of standard fraudsters when we implement the security plan x, and $q(x_1)$ be the proportion of customers inspected (x_2, x_3, x_4, x_5 do not serve for inspection purposes). Then, if we assume no extra cost for an annoyed customer, the operator shall have the following costs:

- The operator invests $c_1 x_1 + c_2 x_2 + c_3 x_3 + c_4 x_4$.

- The costs associated with customers are: (1) With probability $1 - p(x)$, they pay the ticket (civic customers); (2) For a standard evader, with probability $p(x)(1 - q(x_1))$, he does not pay the ticket and is not caught, therefore producing a loss of v (the cost of the ticket) to the operator; and (3) Otherwise, with probability $p(x)q(x_1)$, he does not pay the ticket, but he is caught, therefore producing an income of f (the expected income due to fines). We assume that the partitioning processes corresponding to the number of civic customers, ticket evaders or fine payers are independent with probabilities $1 - p(x)$, $p(x)(1 - q(x_1))$ and $p(x)q(x_1)$, respectively. If we denote by N_1, N_2 and N_3 the number of customers of each type, with $N = N_1 + N_2 + N_3$, then (N_1, N_2, N_3) will follow conditionally independent (given x) Poisson processes, given x, with rates $\lambda_1 = \lambda(1 - p(x))$, $\lambda_2 = \lambda p(x)(1 - q(x_1))$ and $\lambda_3 = \lambda p(x)q(x_1)$, respectively.

The increase in income for the operator associated with security plan x will be, for given N_1, N_2, N_3,

$$c_D(N_1, N_2, N_3) = 0 \cdot N_1 - v \cdot N_2 + f \cdot N_3 - (c_1 x_1 + c_2 x_2 + c_3 x_3 + c_4 x_4 + c_5 x_5).$$
(15.2)

Then, if u_D is the operator's utility function, we evaluate the security plan x through the expected utility

$$\psi(x) = \sum_{N_1, N_2, N_3} p_{N_1 x} \, p_{N_2 x} \, p_{N_3 x} \, u_D(c_D(N_1, N_2, N_3)),$$

where $p_{N_i x} = \Pr(N_i$ customers of type $i | x$ is implemented), $i = 1, 2, 3$. We would then need to find the maximum expected utility security plan subject to the constraints, through $\max_{x \in \mathcal{B}} \psi(x)$.

We discuss now the forms we have adopted for $p(x)$, $q(x_1)$ and u_D:

- Each additional resource (x_1, x_2, x_3, x_4) will have a deterrent effect, but this will be dampened as more resources are implemented. Therefore, we assume

$$p(x_1, x_2, x_3, x_4, x_5) = p_0 \exp\left(-\sum_{i=1}^{5} \gamma_i x_i\right) + p_r,$$

where γ_1, γ_2, γ_3 and γ_4 account for the fact that each additional unit of (x_1, x_2, x_3, x_4) is expected to reduce the proportion of fraudsters. γ_5 is a coefficient which accounts for the effect of having ticket clerks involved in observation tasks. $(p_0 + p_r)$ represents the fraud proportion if no additional countermeasures are deployed, i.e. if the current *status quo* is preserved, whereas p_r represents the residual proportion of fraudsters that would remain, even if infinite resources (x_1, x_2, x_3, x_4) were deployed.

- Concerning $q(x_1)$, each additional inspector adds a number of tickets to be inspected, but such increase will not be linear, as we detail below.

- The operator will be assumed to be risk averse, see [2], with respect to increase in income and, therefore, u_D will be strategically equivalent to $u_D(c_D) = -\exp(-k_D \cdot c_D)$, with $k_D > 0$.

15.3.3 Numerical results

We have assessed the required parameters with the aid of experts from a specific facility. We consider in our computations a generic facility, whose features can be regarded as representative of many others, with a single street level entrance, and a moderate daily flow of customers. We have chosen one year as our relevant planning period, since it is a sufficiently long time to observe the effect and efficiency of the measures deployed by the operator. Moreover, security budget is planned annually.

Table 15.2 displays the maximum additional investments that the operator is considering for each countermeasure, as well as their associated unit costs over the planning period. Regarding human resources, we indicate their unit annual gross salaries. We have incorporated the overall cost of installing a secured automatic access door over a whole year, including maintenance and repair, taking into account the average lifetime of a door. The available annual budget for that particular facility is 150,000 USD. With these numbers, there is a total of 84 feasible portfolios.

As the redefinition of clerk duties regards, we have estimated how much could it cost to the operator, in terms of labour troubles, the negotiation with unions over a whole year. According to the operator, such costs would amount up to, approximately, one third of their gross salary, which is 45,000 USD.

TABLE 15.2

Maximum planned investments.

Measure	Inspectors	Door guards	Guards	Automatic doors
Max	1	3	2	1
Annual cost/unit	50,000	25,000	30,000	15,000

TABLE 15.3

Number of customers.

	2008	2009	2010	2011	2012
Customers	1,007,557	1,012,002	1,045,479	996,631	1,008,370

We discuss now the assessment of the required parameters. We have first estimated the rate λ of the Poisson process, based on the data provided by the operator. We assume a diffuse, but proper, gamma prior $\lambda \sim \mathcal{G}(0.1, 0.1)$. The number of customers using the relevant facility over the last five years is shown in Table 15.3. The average number of customers is 1,014,008, with little variation. Then, a posteriori, $\lambda|data \sim \mathcal{G}(1014008.1)$. When necessary, we shall estimate such a rate through its posterior expectation $E(\lambda|data) \approx 10^6$.

With respect to the proportion of fraudsters, $p(x)$, we have estimated the current and target values at $p_0 + p_r = 0.03$ and $p_r = 0.01$, respectively. For $p_0 + p_r$, we used a beta-binomial model with a noninformative prior. Based on the data provided by the operator, we got a posterior $\mathcal{B}e(3 \cdot 10^4 + 1, 10^6 + 1)$.

For the values of the γ_i coefficients, we have assessed them through expert elicitation. Specifically, we asked the experts about the expected deterrent effect of each countermeasure when considered separately, fitting the expression for $p(x)$ and obtaining the corresponding γ_i. As an illustration, let us consider the door guards, x_2, as if they were the only countermeasure available. Using the value of 0.03 when there are no additional door guards, the experts considered that having one door guard would reduce the proportion of evasion to approximately 0.02. With this value, we fitted $\gamma_2 = 0.7$. We checked for consistency of the assessment, asking the experts about the expected reduction in the fare evasion proportion if more than one door guard were hired, obtaining consistent results. We repeated the same calculations for the other countermeasures, obtaining $\gamma_1 = 0.1$, $\gamma_3 = 0.8$ and $\gamma_4 = 0.5$. The estimation of $\gamma_5 = 0.2$ was accomplished using the only plausible values, $x_5 = \{0, 1\}$.

Regarding the proportion of inspections, the operator believes that each new inspector could contribute with a certain number of yearly inspections, as reflected in Table 15.4 for one to four inspectors. The proportion of inspections $q(x_1)$ is the ratio between such number and the number N of customers.

Finally, with the aid of experts, we have assessed the value of the risk coefficient k_D in the operator's utility function. We have used the probability equivalent method [3] to assess a few values for the utility function and then

TABLE 15.4
Expected inspections for each additional inspector.

Inspectors	1	2	3	4
Expected inspections	75,000	135,000	185,000	230,000

fit an appropriate curve through least squares, obtaining a good fit for $k_D = 5 \cdot 10^{-6}$. Other relevant parameters are the fare ticket ($v = 2$ USD) and the average fine in case somebody is caught without a valid ticket (100 USD). However, according to the facility operator, approximately only one sixth of the imposed fines are actually paid. This is equivalent to saying that the effective average fine per caught evader is, approximately, $f = 17$ USD.

We have simulated 10,000 years of operations of the facility, to identify the optimal portfolio of countermeasures. The solid line in Figure 15.3 shows the estimated expected utility of the operator for the 84 feasible portfolios.

From left to right, the portfolios on the horizontal axis begin with $x = (0,0,0,0,0)$, $x = (0,0,0,0,1)$ and so on, increasing sequentially the values in x_5, x_4, x_3, x_2 and x_1, being the last feasible portfolio under such ordering $x = (1,3,0,1,1)$. The optimal portfolio is $(0,1,0,0,0)$, corresponding to hiring just one door guard, with an estimated expected utility $\psi(x) = -1.30$, associated investment of 25,000 USD, and an expected loss for the operator of 53,149 USD (due to the investment plus the expected balance between the fraud and the collected fines, which is $-28,149$ USD). The next two portfolios with highest expected utilities are $x = (0,0,1,0,0)$, corresponding to hiring a guard, with $\psi(x) = -1.33$, associated investment of 30,000 USD and expected losses of 57,051 USD; and $x = (0,0,0,1,0)$, accounting for one automatic door, with $\psi(x) = -1.34$, associated investment of 15,000 USD and expected losses of 58,549 USD.

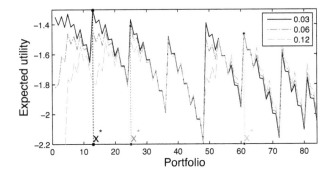

FIGURE 15.3
Expected utility for security portfolios.

TABLE 15.5

Expected utilities for representative portfolios for different standard evaders scenarios.

	$p_0 + p_r = 0.03$			$p_0 + p_r = 0.06$			$p_0 + p_r = 0.12$		
x	Invest.	$\psi(x)$	Income	x	Invest.	$\psi(x)$	x	Invest.	$\psi(x)$
$(0,1,0,0,0)$	25000	-1.30	-53149	$(0,2,0,0,0)$	50000	-1.41	$(1,1,0,0,0)$	75000	-1.47
$(1,0,0,0,0)$	50000	-1.39	-65875	$(1,0,0,0,0)$	50000	-1.50	$(1,0,0,0,0)$	50000	-1.75
$(0,3,0,0,0)$	75000	-1.48	-78625	$(0,3,0,0,0)$	75000	-1.49	$(0,3,0,0,0)$	75000	-1.51
$(0,0,2,0,0)$	60000	-1.43	-71828	$(0,0,2,0,0)$	60000	-1.47	$(0,0,2,0,0)$	60000	-1.55
$(0,0,0,1,0)$	15000	-1.34	-59264	$(0,0,0,1,0)$	15000	-1.61	$(0,0,0,1,0)$	15000	-2.32
$(0,0,0,0,1)$	15000	-1.40	-67751	$(0,0,0,0,1)$	15000	-1.79	$(0,0,0,0,1)$	15000	-2.93
$(0,3,2,1,1)$	150000	-2.19	-157227	$(0,3,2,1,1)$	150000	-2.19	$(0,3,2,1,1)$	150000	-2.19
$(0,3,2,1,0)$	150000	-2.04	-142178	$(0,3,2,1,0)$	150000	-2.03	$(0,3,2,1,0)$	150000	-2.03

The left column in Table 15.5 shows the expected utility for some other representative portfolios, together with their corresponding investments and the expected variation in operator's income (negative values correspond to losses). The first row displays the optimal portfolio. Additionally, we have also included those portfolios for which the investment is maximum in one of the countermeasures, with no investment in the other ones. For instance, the third row in Table 15.5 corresponds to a portfolio in which the operator only invests in new door guards, hiring the maximum number (3). Finally, we have also considered those portfolios which entailed highest investments, to investigate whether such investments are necessarily the most effective ones, in terms of the operator's expected utility.

As we can observe, the most worthy measures to invest in, from the operator's perspective, entail relatively small investments such as hiring a single door guard or a guard or installing a new automatic door. Investing more in these or other countermeasures is not worthy for the operator. In this respect, note that the fourth best portfolio, in terms of expected utility, is actually a 'zero additional investment' policy. On the other hand, we can observe that those portfolios entailing highest investments are definitely too expensive for the operator, although they would, undoubtedly, reduce the evasion proportion considerably.

The previous results are sensitive to variations in the fare evasion proportion. Indeed, note that the evasion proportion estimated above $(p_0 + p_r = 0.03)$ is not constant but, rather, it depends on the specific day and time considered, varying approximately between 0.005 and 0.12, according to the operator. We have repeated the previous calculations for two new scenarios, in which the evasion proportion would increase to 0.06 and 0.12, respectively. We have shown the results in Figure 15.3. The dashed-dotted line corresponds to an estimated fare evasion of 6%, whereas the dashed line corresponds to 12%.

The optimal portfolio when $p_0 + p_r = 0.06$ is $x = (0,2,0,0,0)$, corresponding to hiring two door guards, with an associated investment of 50,000 USD. Similarly, the optimal portfolio when $p_0 + p_r = 0.12$ is $x = (1,1,0,0,0)$, corresponding to hiring one door guard and one inspector, with an associated investment of 75,000 USD. We display in the middle and right columns in Table 15.5 similar results to those obtained when the evasion proportion was 3%. As we can observe, when facing higher evasion proportions, the operator needs to make higher investments in order to attain better expected utilities. When the evasion proportion becomes 12%, hiring an inspector would become crucial, as they have legal authority to impose fines, which represent the largest part of the operator's income. However, we found also that these results were quite sensitive to variations on the proportion of tickets inspected by each new inspector. Thus, it is essential that inspectors really carry out their tasks so as to ensure an effective fight against fare evasion.

15.4 A Framework for Adversarial Risk Analysis

We consider now a schematic framework for adversarial risk analysis. For illustrative purposes, and since this is the relevant model in our case study, we focus on the Sequential Defend-Attack model, see [8]. We aim at supporting a Defender facing the actions of an Attacker. Figure 15.4(a) depicts the problem graphically. It shows a coupled influence diagram (an influence diagram for each participant with several shared uncertain nodes and linking arrows), with white nodes belonging to the Defender, dark grey ones to the Attacker and, finally, a light grey node shared by both of them.

The Defender first chooses a defence $d \in \mathcal{D}$. Then, having observed it, the Attacker selects an attack $a \in \mathcal{A}$. Both $\mathcal{D}$ and $\mathcal{A}$ are assumed to be continuous here. As a result, there will be an outcome of the attack, $S \in \mathcal{S}$, which is the only uncertainty deemed relevant in the problem. The influence diagram shows explicitly that the uncertainty associated with S is dependent on the actions of both the Defender and the Attacker, $S|d, a$. Similarly, the consequences of the attack for both adversaries will depend on the outcome of the attack and their own actions.

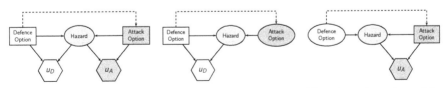

(a) Influence diagram.　(b) Defender's decision problem.　(c) Defender's analysis of Attacker's problem.

FIGURE 15.4
The Sequential Defend-Attack model.

We start by considering the Defender's problem from the decision analysis perspective: the Defender's influence diagram in Figure 15.4(b), no longer has the utility node with the Attacker's information and his decision node is perceived as a random variable. The (possibly multiattribute) consequences for the Defender are represented through $c(d, s)$. She then gets her utility $u_D(c(d, s))$, which we shall rewrite as $u_D(d, s)$. She also needs to build the probabilities $p_D(s|d, a)$, reflecting her beliefs about which outcomes are more likely when the Attacker chooses an attack a, and defensive resources d have been deployed. She then gets her expected utility given the attack a, which is

$$\psi_D(d|a) = \int u_D(d, s)\, p_D(s|d, a)\, \mathrm{d}s. \tag{15.3}$$

Suppose now that the Defender is able to build the model $p_D(a|d)$, reflecting her beliefs about which attack will be chosen by the Attacker upon seeing defence option d. Then, she may compute $\psi_D(d) = \int \psi_D(d|a)\, p_D(a|d)\, \mathrm{d}a$, and maximise $\max_{d \in \mathcal{D}} \psi_D(d)$.

The only problematic assessment is that of $p_D(a|d)$. To do so, the Defender needs to put herself into the Attacker's shoes, and try to solve his problem. Figure 15.4(c) represents the Attacker's problem, as seen by the Defender. For that, she has to assess his utility $u_A(a, s)$ and the Attacker's probabilities about success, $p_A(s|d, a)$. Then, she needs to maximise

$$a^*(d) = \max_{a \in \mathcal{A}} \int u_A(a, s)\, p_A(s|d, a)\, \mathrm{d}s.$$

However, the Defender lacks knowledge about (u_A, p_A), so she has to model her uncertainty about them through (U_A, P_A), propagating it to get the random optimal action

$$A^*(d) = \max_{a \in \mathcal{A}} \int U_A(a, s)\, P_A(s|d, a)\, \mathrm{d}s.$$

She then gets $p_D(A \leq a|d) = \Pr(A(d^*) \leq a)$. In order to get an estimate of $p_D(a|d)$, the Defender may proceed by simulation through the following steps, where K is the Monte Carlo sample size:

Algorithm 15.1 Simulating the Attacker's optimal action

```
For d ∈ 𝒟
  For k = 1 to K
    For a ∈ 𝒜
      Draw (uᵏ_A, pᵏ_A) ~ (U_A, P_A)
      Compute ψᵏ_A(a) = ∫ uᵏ_A(a, s) pᵏ_A(s|d, a) ds
    end For
    Compute aᵏ(d) = arg maxₐ∈𝒜 ψᵏ_A(a)
  end For
  Approximate p_D(a* ≤ a|d) ≈ #{1 ≤ k ≤ K : aᵏ(d) ≤ a}/K
end For
```

15.5   Case Study: Fighting Fare Colluders in a Facility

Colluders are intentional fare evaders who prepare their evasion actions in an organised manner. Irrespective of their structure and preparedness, their event flow is: (1) Some colluders eventually change their mind and decide to pay when entering the facility; (2) The rest decide not to pay; (3) Out of these, some will be inspected and fined. This will partly mitigate the losses for the operator due to fare evasion. The colluders benefit from evading the ticket fare. However, they face the possibility of being fined, in addition to having some preparation costs.

We analyse the dynamics of the Defender (the facility operator) and the Attacker (the colluders). The operator dynamics involve the same steps as in Section 15.3, and we shall use the same notation. Regarding the dynamics of colluders (He), we view the whole group as a "club" which entails M operations over the relevant planning period. We denote by (M_1, M_2, M_3) the number of aborted, successful, and failed operations, respectively. We need to take into account the following relevant considerations for them:

(a) The colluders see the security plan $x = (x_1, x_2, x_3, x_4, x_5)$.

(b) They decide the proportion r of fare evasion they will attempt.

(c) The actual proportion r' in node "Prop. of colluders" depends also on the countermeasures implemented by the operator. The colluders decide a level of evasion but, in the end, some of them will decide not to evade, say because they see more door guards than expected. Similarly, some of them, initially intending to pay the fare, eventually decide to evade, say because they see less door guards than expected.

(d) They face their operational costs. Per operation it would be: (1) With probability $(1 - r')$, the cost v of the ticket; (2) With probability $r'(1 - q_A(x_1))$, a saving of v; and (3) With probability $r'q_A(x_1)$, a cost of f (average fine cost as above), where $q_A(x_1)$ designates the probability that the Attacker gives to the fact of being inspected, if x_1 is the number of inspectors. The colluders would then face a cost/benefit balance given by:

$$c = v(M_2 - M_1) - fM_3, \tag{15.4}$$

where (M_1, M_2, M_3) come from a multinomial distribution $\mathcal{M}(M; (1 - r'), r'(1 - q_A(x_1)), r'q_A(x_1))$, with $M = M_1 + M_2 + M_3$.

(e) They get the corresponding utility, which depends on both the service savings and the costs necessary to implement their decision.

We then face an adversarial problem, whose influence diagram is shown in Figure 15.5, which is a generalised version of the standard Sequential Defend-Attack model in Figure 15.4(a).

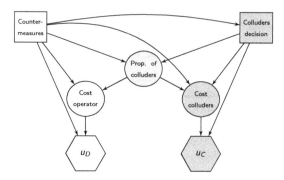

FIGURE 15.5
Influence diagram when only colluders are present.

15.5.1   The operator's problem

We sketch the problem faced by the operator in Figure 15.6(a). The colluders' decision node is perceived by the operator as a random variable, which we subsume as the perceived proportion of colluders.
The total increase in income for the operator would be:

$$c_D = -vM_2 + fM_3 - (c_1x_1 + c_2x_2 + c_3x_3 + c_4x_4 + c_5x_5). \tag{15.5}$$

If u_D is her utility, and $h(r|x)$ models her beliefs over the proportion of evasion attempts given an investment x, she has to compute the expected utility

$$\psi(x) = \int \left[\sum_{M_1, M_2, M_3} p_{M_1 M_2 M_3 x} u_D \left(-vM_2 + fM_3 - \sum_{i=1}^{5} c_i x_i \right) \right] \times h(r|x) \mathrm{d}r,$$
$$\tag{15.6}$$

where $p_{M_1 M_2 M_3 x} = \Pr(M_i$ colluders of type $i, \ i = 1, 2, 3|x$ is invested). She must then solve $\max_{x \in \mathcal{B}} \psi(x)$.

Of all the elements in (15.6), the operator will only find structural difficulties in modelling $h(r|x)$, which requires strategic thinking, as we describe below.

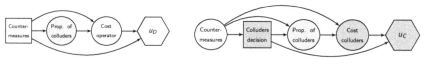

        (a) Operator's problem.   (b) Colluder's problem.

FIGURE 15.6
Fighting fare colluders.

15.5.2   The colluders' problem

We present now the problem that the colluders would solve based on the standard influence diagram reduction algorithm, see [10]. The influence diagram for the Attacker's problem, together with the involved random variables and their dependencies, is shown in Figure 15.6(b). As we can observe, the operator decision node "Countermeasures" is perceived as a random variable by the colluders. The elements required and/or involved to solve the problem are: (a) x, the security investment by the operator. We may consider $p_A(x)$, which models the Attacker's beliefs over such investments. However, this distribution will not be necessary, as the colluder sees x; (b) r is the decision made by the colluders (the proportion of fare evaders to be undertaken); (c) r' is the effective fare evasion proportion. One possible model would be $r' = r(1 - s(x))$, where $s(x)$ is the proportion of aborted fare evasion attempts with respect to the original plan. The colluders' distribution $p_A(s(x)) = p_A(s|x)$ induces the distribution $p_A(r'|r, x)$; (d) c, is the cost of the evasion operation when the effective evasion proportion is r' and the investment is x, as defined in (15.4); and (e) u_C is the utility over the consequences, which has the form $c - rc_eM$, where c_e would be the per unit (preparation of) evasion cost, i.e., it is $u_C(c-rc_eM)$, with c as in (15.4). However, in our case, the preparation costs are regarded by the operator as negligible, at least for the final "beneficiaries" of all the circulating information about when and where the inspections are being accomplished, so we shall set $c_e = 0$.

The steps needed to solve the colluders' problem are:

1. We integrate out the uncertainty over c to get the expected utility

$$\psi_C(r', r, x) = \int \left[\sum_{M_1, M_2, M_3} p_{M_1 M_2 M_3 x}\, u_C(c) \right] p_A(c|r', x)\, \mathrm{d}c,$$

   where $p_A(c|r', x)$ is the distribution over c, induced by $g_A(q_A|x_1)$, the density over q_A, given that the security investment is x_1.

2. We integrate out the uncertainty over r' to get the expected utility

$$\psi_C(r, x) = \int \psi_C(r', r, x) p_A(r'|r, x)\, \mathrm{d}r'.$$

3. We find the optimal strategy for the Attacker. This provides $r(x) = \arg\max_r \psi_C(r, x)$, the optimal planned evasion level when the security investment is x.

Note, however, that we have uncertainty about $u_C(\cdot)$, $p_A(c|\cdot)$ and $p_A(r'|\cdot)$, which we model through random utilities $U_C(\cdot)$ and probabilities $P_A(c|\cdot)$ and $P_A(r'|\cdot)$. We propagate such uncertainty as follows, for each x:

1. Compute the random expected utility

$$\Psi_C(r', r, x) = \int \left[\sum_{M_1, M_2, M_3} p_{M_1 M_2 M_3 x} \, U_C(c) \right] P_A(c|r', x) \, dc.$$

2. Compute the random expected utility

$$\Psi_C(r, x) = \int \Psi_C(r', r, x) P_A(r'|r, x) \, dr'.$$

3. Compute the random optimal alternative $R(x) = \arg\max_r \Psi_C(r, x)$.

Then, we would have an estimate of the desired distribution $h(r|x)$ in (15.6), through $p_A(R \leq r|x) = \Pr(R(x) \leq r)$. In order to estimate $R(x)$, we may proceed by simulation as outlined in Algorithm 15.2.

Algorithm 15.2 Simulating the optimal planned evasion level

For each x
    For i = 1 to K
        Sample $U_C^i(\cdot), P_A^i(c|\cdot), P_A^i(r'|\cdot)$.
        Compute

$$\Psi_C^i(r', r, x) = \int \left[\sum_{M_1, M_2, M_3} p_{M_1 M_2 M_3 x} U_C^i \Big(v(M_2 - M_1) - f M_3 \Big) \right] P_A^i(c|r', x) dc.$$

        Compute $\Psi_C^i(r, x) = \int \Psi_C^i(r', r, x) P_A^i(r'|r, x) dr'$.
        Compute $R^i(x) = \arg\max_r \Psi_C^i(r, x)$.
    end For
    Approximate $p_A(R(x) \leq r) \approx \#\{1 \leq i \leq K : R^i \leq r\}/K$.
end For

Typical assumptions would be:

- The colluders are risk prone in benefits. Therefore, their utility function is strategically equivalent to $u_C(c) = \exp(k_C \cdot c)$, $k_C > 0$. A random utility model could be $U_C(c) = \exp(k_C \cdot c)$, $k_C \sim \mathcal{U}(0, K_C)$.

- For s, we could use a beta distribution $\mathcal{B}e(\alpha_1, \beta_1)$, with s close to zero if we feel that evaders will be very committed to their plan, thus implying $\alpha_1 \ll \beta_1$. Then, $P_A \sim \mathcal{DP}(\mathcal{B}e(\alpha_1, \beta_1), \delta_1)$, a Dirichlet process with base $\mathcal{B}e(\alpha_1, \beta_1)$ and concentration parameter δ_1, see [4]. The smaller its value, the more concentrated will the Dirichlet process be, entailing less uncertainty. We could take into account here the dependence $s|x$ but we shall not model it explicitly. From it, we would derive $P_A(r'|r, x)$.

- We could consider that $q_A(x_1) \sim \mathcal{B}e(\alpha_2(x_1), \beta_2(x_1))$ with $\alpha_2(x_1)/(\alpha_2(x_1) + \beta_2(x_1)) = q(x_1)$ and small variance. Then, $G_A \sim \mathcal{DP}(\mathcal{B}e(\alpha_2(x_1), \beta_2(x_1)), \delta_2)$. From this, we would derive $P_A(c|r', x)$.

15.5.3 Numerical results

We illustrate the model when just colluders are considered. For those parameters shared with the standard evaders problem, we use the values in subsection 15.3.3. As for the specific parameters in relation with colluders, we have assessed them based on data and the aid of experts when data were not available:

- The number of colluding operations was estimated through a Poisson model, $M \sim \mathcal{P}ois(\mu)$, with a diffuse, but proper, gamma prior $\mu \sim \mathcal{G}(0.1, 0.1)$. Based on the data provided by the operator (they acknowledge around $M = 30000$ annual operations in the last five years), we got a posterior $\mu|data \sim \mathcal{G}(150000.1, 5.1)$. When necessary, we shall estimate such rate through its posterior expectation $E(\mu|data) \approx 30000$. Since the uncertainty over M is small compared with its expected value, we shall regard, for simplicity, M as constant.

- The proportion s of aborted fare evasion attempts follows a beta distribution $\mathcal{B}e(\alpha_1 = 1, \beta_1 = 9)$, which is equivalent to saying that, on average, only one out of 10 colluders will eventually change his mind when observing the preventive measures deployed by the operator. The standard deviation is, approximately, 0.1. The concentration parameter δ_1 has been set to 0.1, representing that the operator has relatively little uncertainty about the behaviour of colluders when facing possible abortions. Therefore, $P_A \sim \mathcal{DP}(\mathcal{B}e(1,9), 0.1)$.

- Regarding the operator's assessment about the colluders' beliefs on the proportion of inspections, $q_A(x_1)$, we have used a probability distribution $\mathcal{B}e(\alpha_2, \beta_2)$, whose parameters have been assessed based on the values of its first two moments. Concerning the expected value $E[q_A(x_1)] = \alpha_2/(\alpha_2 + \beta_2) = q(x_1)$, whereas for the variance σ^2 we have assumed a small value, 0.01. It is straightforward to obtain then $\alpha_2(x_1)$ and $\beta_2(x_1)$. We have also set $\delta_2 = 0.1$. Therefore, $G_A \sim \mathcal{DP}(\mathcal{B}e(\alpha_2, \beta_2), 0.1)$.

- Finally, we have assessed the value of the maximum risk coefficient in the colluders' utility function, $K_C = 10^{-5}$.

We have simulated 10,000 years of operations of the facility. The solid line in Figure 15.7 shows the estimated expected utility of the operator for all the possible portfolios x, with the same ordering as in subsection 15.3.3.

The optimal portfolio is now $x = (0,0,0,0,0)$, with $\psi(x) = -1.03$, meaning that the operator would not actually invest in additional measures. This policy entails a cost of $-55,776$ USD due to fare evasion by colluders. The next two best portfolios have very close expected utilities: $x = (0,1,0,0,0)$, corresponding to hiring a door guard, with $\psi(x) = -1.03$, associated investment of 25,000 USD and an expected loss of 32,177 USD; and $x = (1,0,0,0,0)$, corresponding to hiring an inspector, with $\psi(x) = -1.04$, an investment of 50,000 USD and an expected loss of 8,193 USD.

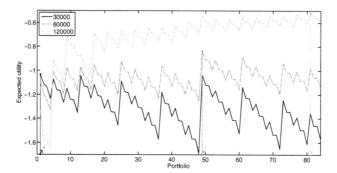

FIGURE 15.7
Operator's expected utility when colluders are present.

The left column in Table 15.6, displays additional information about other relevant portfolios, as in subsection 15.3.3. Because of the judgemental nature of several of the involved magnitudes in the case study, we have performed a sensitivity analysis to check for robustness of the results. Specifically, we study the influence of higher evasion proportions over the portfolios of measures to be adopted by the operator, considering $M = \{60000, 120000\}$. Figure 15.7 shows the expected utility for both cases, with a dashed-dotted line for $M = 60000$ and a dashed line for $M = 120000$. Additional results for other relevant portfolios are shown in the middle and right columns in Table 15.6.

As we can observe, the presence of higher proportions of colluders makes it necessary for the operator to invest in countermeasures, especially when the number of colluders gets too large. When $M = 120000$, the optimal portfolio is $x = (1, 3, 0, 0, 0)$, corresponding to hiring one inspector and three door guards, with an estimated expected utility $\psi(x) = -0.51$ and an associated investment of 125,000 USD, using over 80% of the maximum available budget. It is important to remark that, due to resource constraints, it is only possible to hire one inspector. When $M = 60000$, the optimal portfolio is already $x = (1, 0, 0, 0, 0)$, i.e. hiring just one inspector. Therefore, when the number of colluding operations increases to $M = 120000$, the operator is not able to

TABLE 15.6
Expected utilities for representative portfolios for different colluder scenarios.

	$M = 30000$			$M = 60000$			$M = 120000$	
x	Invest. $\psi(x)$	Income	x	Invest. $\psi(x)$		x	Invest. $\psi(x)$	
$(0,0,0,0,0)$	0 -1.03	-5776	$(1,0,0,0,0)$	50000 -0.84		$(1,3,0,0,0)$	125000 -0.51	
$(1,0,0,0,0)$	50000 -1.04	-7054	$(1,0,0,0,0)$	50000 -0.84		$(1,0,0,0,0)$	50000 -0.55	
$(0,3,0,0,0)$	75000 -1.18	-33052	$(0,3,0,0,0)$	75000 -0.96		$(0,3,0,0,0)$	75000 -0.64	
$(0,0,2,0,0)$	60000 -1.15	-27447	$(0,0,2,0,0)$	60000 -0.98		$(0,0,2,0,0)$	60000 -0.71	
$(0,0,0,1,0)$	15000 -1.13	-25097	$(0,0,0,1,0)$	15000 -1.23		$(0,0,0,1,0)$	15000 -1.49	
$(0,0,0,0,1)$	15000 -1.09	-17419	$(0,0,0,0,1)$	15000 -1.25		$(0,0,0,0,1)$	15000 -1.61	
$(0,3,2,1,1)$	150000 -1.69	-105135	$(0,3,2,1,1)$	150000 -1.25		$(0,3,2,1,1)$	150000 -0.69	
$(0,3,2,1,0)$	150000 -1.57	-90126	$(0,3,2,1,0)$	150000 -1.16		$(0,3,2,1,0)$	150000 -0.64	

hire more inspectors, although she acknowledges them as the most effective measure to fight fare evasion. As we can observe in Table 15.6, $x = (1, 0, 0, 0, 0)$ is the second best portfolio for this scenario, with an expected utility close to that of the optimal one. Thus, although hiring three door guards, as indicated by the optimal portfolio, will increase slightly the operator's expected utility, it is at the cost of a higher investment.

15.6 A Framework for Risk Analysis and Adversarial Risk Analysis

In many real-world cases, a given target might be at risk from threats, some of which could be casual, while others could be launched by intelligent agents, see e.g. [11]. In previous sections, we have studied separately adversarial and non-adversarial cases, but it is also important to consider when both types of threats are simultaneously present. Standard threats are related with random events. Adaptation is usually the best way to manage and minimise the negative impact of such risks to the lowest possible levels. On the other hand, intelligent threats are more challenging to deal with, since they imply the need of understanding the Attacker's motivations, preferences and abilities.

As an example, consider a critical infrastructure which is exposed to relatively frequent natural disasters, like e.g. hurricanes or floods, and, also, due to its strategic value, it is threatened by the actions of terrorist groups. The budget for new security investments is limited, and the infrastructure owner has to allocate resources in such a way that it is best protected against both types of threats. Some of the preventive measures deployed could be intended to fight against a specific threat, but others could have a multipurpose nature. The basic influence diagram for such analysis is shown in Figure 15.8

This problem would be solved in a similar way to that in Section 15.4, incorporating the uncertain node corresponding to the non-adversarial threat. The Defender's utility function would aggregate the consequences for both problems, as expressed in (15.1) and (15.3).

15.7 Case Study: Fighting Standard Evaders and Colluders Simultaneously

We consider now both types of evaders present in the fare evasion problem, as discussed in Sections 15.3 and 15.5. We join Figures 15.2 and 15.5 into a new influence diagram, shown in Figure 15.9. Note that we keep the fraud costs due to standard evaders and colluders separate, but we aggregate them in a deterministic node called "Cost".

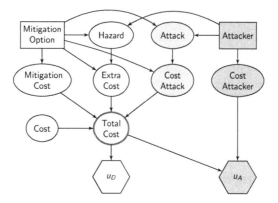

FIGURE 15.8
Influence diagram for adversarial and non-adversarial threats.

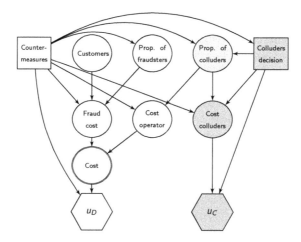

FIGURE 15.9
Influence diagram when both types of evaders are present.

15.7.1   The operator's problem

We sketch the operator problem in Figure 15.10.
We have two contributions to the total cost for the operator, arising from standard and colluding fare evaders' actions, as in (15.2) and (15.5), respectively, which get combined through

$$c_D(N_2, N_3, M_2, M_3, x) = -v(N_2 + M_2) + f(N_3 + M_3) - \sum_{i=1}^{5} c_i x_i.$$

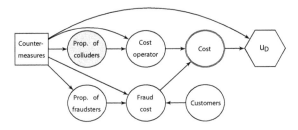

FIGURE 15.10
Operator's problem when both types of evaders are present.

Then, the operator would compute

$$\psi(x) = \int \left[\sum_{\substack{N_1,N_2,N_3 \\ M_1,M_2,M_3}} p_{M_1 M_2 M_3 x} \times p^1_{N_1 x} \, p^2_{N_2 x} \, p^3_{N_3 x} \times u_D(c_D) \right] \times h(r|x) \, dr,$$

and solve $\max_{x \in \mathcal{B}} \psi(x)$.

15.7.2 The colluders' problem

The colluders' dynamics are as in subsection 15.5.2. The standard evaders are unorganised attackers and we cannot associate with them a cost structure. Therefore, the cost faced by the evaders would be as in (15.4). The analysis in subsection 15.5.2 would be repeated to estimate the $h(r|x)$ required in subsection 15.7.1.

15.7.3 Numerical results

We deal now with the simulation of the fare evasion problem taking into account both types of evaders. We use the same values as in subsections 15.3.3 and 15.5.3. We have simulated 10,000 years of operations. The solid line in Figure 15.11 shows the estimated expected utility of the operator for all possible portfolios x when the estimated proportion of standard evaders and colluders is 3% in both cases.

The best three portfolios, in terms of their expected utilities, are $x = (1,0,0,0,0)$ (one inspector, 50,000 USD of investment, and an expected loss of 22,826 USD), with $\psi(x) = -1.12$; $x = (1,1,0,0,0)$ (one inspector plus one door guard, 75,000 USD of investment and an expected loss of 25,113 USD), with $\psi(x) = -1.14$; and $x = (0,2,0,0,0)$ (two door guards, 50,000 USD of investment and an expected loss of 27,629 USD), with $\psi(x) = -1.16$. The left column in Table 15.7 displays additional information about other relevant portfolios.

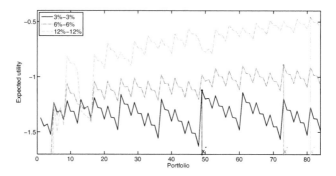

FIGURE 15.11
Operator's expected utility for both evaders.

TABLE 15.7
Expected utilities when both types of evaders are present.

$p_0 + p_r = 0.03, M = 30000$				$p_0 + p_r = 0.06, M = 60000$			$p_0 + p_r = 0.12, M = 120000$		
x	Invest.	$\psi(x)$	Income	x	Invest.	$\psi(x)$	x	Invest.	$\psi(x)$
$(1,0,0,0,0)$	50000	-1.12	-22826	$(1,2,0,0,0)$	100000	-0.89	$(1,3,0,0,0)$	125000	-0.46
$(1,0,0,0,0)$	50000	-1.12	-22826	$(1,0,0,0,0)$	50000	-0.98	$(1,0,0,0,0)$	50000	-0.75
$(0,3,0,0,0)$	75000	-1.20	-36797	$(0,3,0,0,0)$	75000	-0.98	$(0,3,0,0,0)$	75000	-0.66
$(0,0,2,0,0)$	60000	-1.22	-39409	$(0,0,2,0,0)$	60000	-1.06	$(0,0,2,0,0)$	60000	-0.81
$(0,0,0,1,0)$	15000	-1.43	-71255	$(0,0,0,1,0)$	15000	-1.82	$(0,0,0,1,0)$	15000	-3.52
$(0,0,0,0,1)$	15000	-1.45	-74147	$(0,0,0,0,1)$	15000	-2.10	$(0,0,0,0,1)$	15000	-4.19
$(0,3,2,1,1)$	150000	-1.63	-97348	$(0,3,2,1,1)$	150000	-1.20	$(0,3,2,1,1)$	150000	-0.66
$(0,3,2,1,0)$	150000	-1.51	-82303	$(0,3,2,1,0)$	150000	-1.11	$(0,3,2,1,0)$	150000	-0.61

We have also investigated the impact of higher proportions of standard
evaders and colluders. Specifically, we have considered two scenarios, when
both proportions increase to 6% and 12%, respectively. The results can be
observed in Figure 15.11 (dashed-dotted and dashed lines) and in the middle
and right columns in Table 15.7. Conclusions similar to those discussed in
subsection 15.5.3 follow for this case.

15.8 Discussion

We have provided a common framework integrating standard risk analysis
and adversarial risk analysis. As an illustration, we have used an application
in fraud detection in relation with paid access to a facility. Other related
applications could be the modelling of Internet users hacking paid websites,
or drivers forging electronic toll collection systems in highways to avoid paying
the fare. For our purpose, we have distinguished two types of evaders: those
who do not pay for the service in a casual way, and those who do not pay in

an organised manner. Traditional evaders were treated as in a standard risk analysis problem, whereas we modelled intentionality explicitly for colluders, with the aid of adversarial risk analysis. The operator of the facility aimed at deploying a portfolio of countermeasures in order to fight against the fare evasion problem.

We separated the study in three phases. First, we analysed the standard evaders alone; then we dealt with the problem when only colluders were present; and, finally, we subsumed both types of fraudsters in a single model. When problems were treated separately, we observed that the operator tended to need to invest less resources. Policies entailing little investment had higher expected utilities. This was not anymore the case when we dealt with both threats simultaneously. Then, other portfolios entailing higher investments started to being considered as suitable by the operator. From a strictly economic point of view, it would seem that this 'no investment' policy could be a valid option for this operator in some cases. However, such decision could entail negative consequences in terms of image costs, since customers could perceive that the operator is not placing sufficient resources to stop service deterioration. Moreover, an undesired side effect was observed by the operator after installing secured automatic access doors. What was originally a fraud problem, partly aggravated by obsolete access doors, has turned now into a security problem, due to more aggressive behaviour of some colluders. Occasionally, they try to 'piggyback' on fare-paying customers, violating their privacy and ending up, in extreme cases, in rough altercations. This kind of incident could have a severe impact on the image perceived by customers, increasing their feeling of insecurity. In summary, the no investment policy is likely to entail more damages than benefits, even in the short term.

Because of the judgemental nature of various magnitudes assessed throughout our analysis, we have performed sensitivity analysis in order to check for robustness of the results. We found that the most critical value was the fare evasion proportion, both for standard evaders and colluders. With this in mind, we envisaged different scenarios, with eventual increase in the fare evasion proportion, observing, consequently, increased investments.

Acknowledgments

This project has received funding from the European Union's Seventh Framework Programme for Research, Technological Development and Demonstration under grant agreement no 285223. Work has been also supported by the Spanish Ministry of Economy and Innovation program MTM2011-28983-C03-01, the Government of Madrid RIESGOS-CM program S2009/ESP-1685 the ESF-COST Action IS1304 on Expert Judgement, and the AXA-ICMAT Chair on Adversarial Risk Analysis. We are grateful to the facility experts and stakeholders for fruitful discussions about various modelling issues. We are grateful for discussion with David Banks and Jesús Ríos.

Bibliography

[1] T. Bedford and R. M. Cooke. *Probabilistic Risk Analysis: Foundations and Methods*. Cambridge University Press, UK, 2001.

[2] R. T. Clemen and T. Reilly. *Making Hard Decisions with Decision Tools*. Duxbury/Thomson Learning, 2001.

[3] P. H. Farquhar. State of the art—Utility assessment methods. *Management Science*, 30(11):1283–1300, 1984.

[4] T. S. Ferguson. A Bayesian analysis of some nonparametric problems. *The Annals of Statistics*, 1(2):209–230, 1973.

[5] Y. Y. Haimes. *Risk Modeling, Assessment, and Management*. John Wiley & Sons, New Jersey, 2009.

[6] S. Kaplan and B. J. Garrick. On the quantitative definition of risk. *Risk Analysis*, 1(1):11–27, 1981.

[7] J. Pearl. Influence diagrams–Historical and personal perspectives. *Decision Analysis*, 2(4):232–234, 2005.

[8] J. Ríos and D. Ríos Insua. Adversarial risk analysis for counterterrorism modeling. *Risk Analysis*, 32(5):894–915, 2012.

[9] D. Ríos Insua, J. Ríos, and D. Banks. Adversarial risk analysis. *Journal of the American Statistical Association*, 104(486):841–854, 2009.

[10] R. D. Shachter. Evaluating influence diagrams. *Operations Research*, 34(6):871–882, 1986.

[11] J. Zhuang and V. M. Bier. Balancing terrorism and natural disasters–Defensive strategy with endogenous attacker effort. *Operations Research*, 55(5):976–991, 2007.

16

Symmetric Power Link with Ordinal Response Model

Xun Jiang

University of Connecticut

Dipak K. Dey

University of Connecticut

CONTENTS

16.1   Introduction

Ordinal response data are often encountered in biological studies. For example, in a field survey data for the *Invasive Plant Atlas of New England* (IPANE), the abundance of *Berberis thunbergiis* is classified into 6 categories, with the higher number indicating higher coverage. In this particular data, *Berberis thunbergiis* does not exist in many locations, therefore creating extremely unbalanced data. Figure 16.1 shows the collection sites of the IPANE data across New England. Standard link functions such as logit and cloglog may not work well in this situation and a flexible link function may be needed.

Bayesian methods have been widely used in the context of ordinal response data modeling. Albert and Chib [1] used the latent variable approach to model the ordinal response data with probit and Student t links. In their model the response variable is categorized into different ordinal categories based on the scale of a latent variable and corresponding cutoff points. The link function was then introduced to adjust the behavior of the latent variable. It is shown

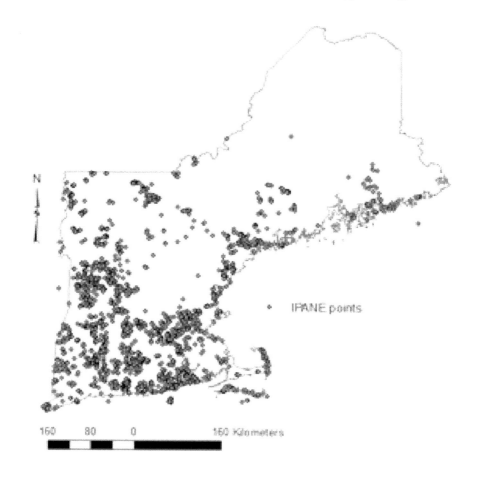

FIGURE 16.1
IPANE data collection locations.

that the symmetric power link family introduces flexiblity of skewness into
the model and performs well against link misspecifications. Here we extend
the usage of symmetric power links as introduced in [6] to ordinal response
modeling context and try to investigate the abundance of *Berberis thunbergiis*
in New England through the IPANE dataset.

The rest of this article is organized as follows. Section 16.2 introduces
the general model setup of the ordinal response data. The model with symmetric power logit (splogit) link is introduced in Section 16.3. Section 16.4
discusses the implementation of the model from a Bayesian perspective using
the Markov Chain Monte Carlo (MCMC) technique. The data analysis is carried out in Section 16.5 and various standard and flexible link function models

are compared. A cross validation study is presented in Section 16.6 to examine the predictive performance of the models. Finally Section 16.7 presents some concluding remarks.

16.2 Model Setup for Ordinal Response Data

Suppose that $y_1, ..., y_n$ are observed ordinal response data, where y_i takes one of J ordered categories. The ordinal data can be thought of as generated from a process in which a latent variable determines the category of response by crossing certain threshold values [1]. Let w_i be this continuous latent variable, and x_i be the vector of covariates. For $i = 1, ..., n$, $j = 1, ..., J$, we specify the model framework as:

$$
\begin{aligned}
y_i &= j \quad \text{if} \quad \gamma_{j-1} < w_i < \gamma_j, \qquad (16.1)\\
w_i &= x_i'\beta + \epsilon_i.
\end{aligned}
$$

Here $\epsilon_i, i = 1, ..., n$ are iid random variables with mean zero and probablity distribution corresponding to the link function in the usual generalized linear model scenario. $-\infty = \gamma_0 < \gamma_1 < ... < \gamma_{J-1} < \gamma_J = \infty$ are cutoff points that categorizes the data in to J ordered categories based on latent variables w. The cutoff points need to be standardized by scaling them with γ_J. Let F be the cdf of ϵ_i, the likelihood function of the model is clearly based on the probability distribution of random variables w and can be written as

$$
L(y|\beta, \gamma) = \prod_{i=1}^{n} \prod_{j=1}^{J} \left[F(\gamma_j - x_i'\beta) - F(\gamma_{j-1} - x_i'\beta) \right]^{I\{y_i=j\}}, \qquad (16.2)
$$

where I is the indicator function that takes 1 when $y_i = j$ and 0 otherwise. A critical problem is to choose appropriate F so that the probability of the latent variables w is properly determined by this link function to fit the oberved events $y_i = j$.

The conventional choice of F is to adopt the cdf corresponding to a standard link functions such as logit, probit or complementary log-log(cloglog). The disadvantage of such a model with fixed link function is that it lacks the ability to let the data decide what F would be better suited, therefore will lead to link misspecification. Wang and Dey [10] proposed to use a generalized extreme value (gev) distribution as an alternative flexible distribution for ϵ and let the shape parameter ξ alter the skewness of the distribution to better fit the data. Here we propose to use the symmetric power logit distribution to model ordinal response data and compare the results with other common link functions.

16.3   Ordinal Response Model with Symmetric Power Logit Link

In [6] a general class of flexible link functions based on a symmetric baseline link function and its mirror reflection was proposed. The cdf of symmetric power logit (splogit) distribution based on the logit baseline distribution F_0 is defined as

$$F(x,r) = F_0^r(\frac{x}{r})\mathbf{I}_{(0,1]}(r) + \left[1 - F_0^{\frac{1}{r}}(-rx)\right]\mathbf{I}_{(1,+\infty)}(r), \qquad (16.3)$$

where $\mathbf{I}$ is the indicator function. The idea is to introduce a power parameter r in the logit baseline in a symmetric fashion to let the skewness vary as r varies between 0 and $+\infty$. Under [2]'s measure, the splogit distribution is negatively skewed when $0 < r < 1$, positively skewed when $r > 1$, and reduces to the symmetric logit link when $r = 1$. Obviously r is the power parameter that brings flexibility in skewness for the latent variables $\boldsymbol{w}$.

16.4   Bayesian Computation

In this section, we develop the sampling scheme of the model (16.2) under the Bayesian paradigm. The full likelihood function of the model is given by equation (16.2). The parameter vector we want to estimate is $\boldsymbol{\theta} = (\boldsymbol{\beta}, \boldsymbol{\gamma}, r)'$. Assuming flat priors on $\boldsymbol{\gamma}$, normal priors on $\boldsymbol{\beta}$ and gamma priors on r, we obtain the posterior distribution on $\boldsymbol{\theta}$ as

$$\pi(\boldsymbol{\theta}|\boldsymbol{y}, \boldsymbol{X}) \quad \propto \quad L(\boldsymbol{y}|\boldsymbol{X}, \boldsymbol{\beta}, \boldsymbol{\gamma}, r)\pi(\boldsymbol{\theta}) \qquad (16.4)$$

$$\propto \quad \prod_{i=1}^{n}\prod_{j=1}^{J} \left[F(\gamma_j - \boldsymbol{x}_i'\boldsymbol{\beta}, r) - F(\gamma_{j-1} - \boldsymbol{x}_i'\boldsymbol{\beta}, r)\right]^{\mathrm{I}\{y_i=j\}}$$

$$\times \pi(\boldsymbol{\beta}|\sigma_\beta)\pi(r|\alpha_r), \qquad (16.5)$$

where $\boldsymbol{\beta} \sim \boldsymbol{N}(\boldsymbol{0}, \sigma_\beta)$ and $r \sim \mathrm{Gamma}(\alpha_r, \alpha_r)$.

Sampling from the posterior distribution of $\boldsymbol{\theta}$ needs advanced MCMC methods since there are generally no closed forms of the posterior distribution. We will use Adaptive Rejection Metropolis Sampling (ARMS) scheme embedded within Gibbs sampling to obtain the posterior samples. The Gibbs sampling method [3] samples a multivariate distribution by sampling a sequence of conditional distributions instead. It is guaranteed that under some mild regularity conditions, this sequence of samples will converge to the target multivariate distribution. The ARMS sampling method is often used within Gibbs sampling when the closed form of conditional distribution is still not available, see [4] for a detailed discussion of this sampling method. Notice that although the latent variables $\boldsymbol{w}$ are not of interest, we still need to sample it during the implemetation of MCMC. Here we establish the conditional distribution of each parameter of interest. The conditional distribution of w_i for $i = 1, ..., n$ is given as

$$w_i|\boldsymbol{\beta}, \boldsymbol{\gamma}, r, \boldsymbol{y}, \boldsymbol{X} \sim \text{splogit}(w_i, r)\text{I}(\gamma_{j-1} < w_i \leq \gamma_j), \tag{16.6}$$

which is a truncated symmetric power logistic distribution, and is easy to sample from using inverse cdf method. For $j = 1, ..., J$, the conditional distribution of γ_j is given as

$$\gamma_j|\boldsymbol{w}, \boldsymbol{y}, \gamma_{i,i \neq j} \sim \text{Uniform}(max\{max(w_i, y_i = j), \gamma_{j-1}\},$$
$$min\{min(w_i, y_i = j + 1), \gamma_{j+1}\}), \tag{16.7}$$

which is also straightforward to sample from. However, the conditional distribution of $\boldsymbol{\beta}$ is

$$\pi(\boldsymbol{\beta}|\boldsymbol{\gamma}, \boldsymbol{w}, r, \boldsymbol{y}, \boldsymbol{X}) \propto \prod_{j=1}^{J} \left[F(\gamma_j - \boldsymbol{x}_i'\boldsymbol{\beta}, r) - F(\gamma_{j-1} - \boldsymbol{x}_i'\boldsymbol{\beta}, r) \right]^{\text{I}\{y_i=j\}}$$
$$\times \exp\left[-\boldsymbol{\beta}'\boldsymbol{\beta}/(2\sigma_\beta^2) \right]. \tag{16.8}$$

Clearly (16.8) has no known closed form expression, therefore we again adopt the ARMS method to sample from this distribution. The conditional distribution of r is given as

$$\pi(r|\boldsymbol{\beta}, \boldsymbol{\gamma}, \boldsymbol{w}, \boldsymbol{y}, \boldsymbol{X}) \propto \prod_{j=1}^{J} \left[F(\gamma_j - \boldsymbol{x}_i'\boldsymbol{\beta}, r) - F(\gamma_{j-1} - \boldsymbol{x}_i'\boldsymbol{\beta}, r) \right]^{\text{I}\{y_i=j\}} r^{\alpha_r - 1}$$
$$\times \exp\left[-\alpha_r r \right]. \tag{16.9}$$

Again, no closed form expression is available for (16.9), so the ARMS method is used to conduct the sampling. [10] discussed the implementation of ordinal models under various standard link functions as well as the gev link. The implementation is similar to what we discussed in this chapter except they adopted the Metropolis-Hasting algorithm to sample from the posterior distributions when no closed form expression is available. The skewness of the gev distribution is controlled by a shape parameter ξ. Notice that the gev distribution is left skewed when $\xi > 0.306$, right skewed when $\xi < 0.306$, and reduced to loglog link when $\xi = 0$. We will implement the gev model to compare with our splogit model.

16.5   Data Analysis

In this section we apply the proposed model with a splogit link to a subset of
IPANE data. As discussed in the introduction, this particular data tracks the
coverage of species *Berberis thunbergiis* in New England area with multiple
survey locations. Several covariates of the habitat are available. One particularly interesting covariate is the maximum temperature of the warmest month.
Details on the IPANE data can be found in [7] and [5]. The ordinal response
variable coverage, which takes value from 1 to 6, is based on the percentage of
coverage of *Berberis thunbergiis* at the survey location. The response variable
coverage assumes a 1 if the species is not present in the location, while coverage
takes the value of 2 to 6 based on a certain cutoff percentage of coverage. The
higher score indicates a larger percentage of coverage. One feature of the data
is that the majority of the observed location does not have *Berberis thunbergiis*
present; as a result we have 957 out of 1789 (53.5%) of the coverage value equal
to 1. For this dataset we fit the ordinal response model discussed in previous
sections and take F to be three standard link functions, logit, cloglog, loglog,
and two flexible link functions, gev and splogit. For each model we run 100000
MCMC iterations and discard the first 20000 as a burn-in period, then only

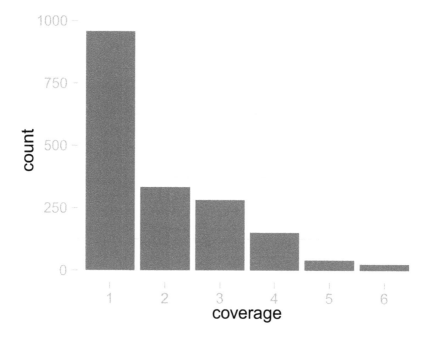

FIGURE 16.2
Bar chart of coverage of *Berberis thunbergiis*.

TABLE 16.1
Posterior mean and standard deviation (SD) of parameters under different models.

	logit		cloglog		loglog		splogit		gev	
	Mean	SD	Mean	SD	Mean	SD	Mean	SD	Mean	SD
β_0	0.76	0.07	0.73	0.04	0.07	0.04	0.09	0.08	0.08	0.04
β_1	1.97	0.07	0.86	0.03	1.26	0.04	1.31	0.08	1.25	0.04
γ_2	1.47	0.07	0.75	0.03	1.03	0.05	1.02	0.06	1.00	0.05
γ_3	2.81	0.10	1.40	0.04	2.08	0.07	2.02	0.10	1.99	0.09
γ_4	4.23	0.15	1.94	0.06	3.37	0.13	3.26	0.14	3.15	0.18
γ_5	5.29	0.22	2.23	0.07	4.39	0.21	4.25	0.21	4.02	0.29
r/ξ							2.18	0.26	−0.05	0.03
DIC	3503.7		3781.9		3492.9		3467.3		3492.7	

retain every 10th sample for the remaining 80000 iterations. Therefore, 8000 posterior samples are obtained for each model. We use flat improper priors on γ, normal priors on β with mean 0 and standard deviation 100. These non-informative (flat) and weakly informative (large standard deviation) priors are chosen to allow only the information from the data to decide the posterior estimation. For the splogit model, the prior for r is chosen to be Gamma$(2,2)$, while in gev model the prior for ξ is chosen to be Uniform$(-0.5, 0.5)$. Model comparison is based on Deviance Information Criterion (DIC) proposed by [9]. The models are implemented in R with the ARMS algorithm described in [8] and code for ARMS by [4]. [8] being the primary sampling algorithm when no closed forms are available.

We summarize the results of the parameter estimate in Table 16.1. The 8000 posterior MCMC samples converge pretty well and we calculate all of our posterior quantities and model comparison measurement based on these samples. From Table 16.1, it is clear that β_1 is significantly positive across all models, therefore we conclude that the maximum temperature of the warmest month is a significantly positive factor in determining the coverage percentage of *Berberis thunbergiis*. In other words, higher maximum temperature is associated with higher coverage percentage. The estimate of model parameters varies greatly across models except for the loglog and gev model. The fact that posterior mean of $\xi = -0.05$ explains this exception since the gev model reduces to loglog model when $\xi = 0$. On the other hand, $\xi = 0$ also indicates that the data drives the flexible gev model to a right skewed link function, which favors loglog among the standard link functions. This observation has been confirmed by observing that posterior mean of $r = 2.18 > 1$ in the splogit model. The DIC measure in the last column further confirms our observations in ξ and r. The right skewed standard model (loglog) and

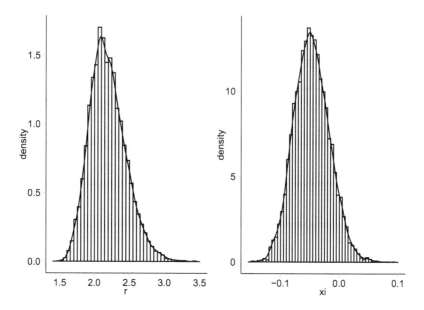

FIGURE 16.3

Posterior density plot of power parameter r and shape parameter ξ in splogit and gev models.

two flexible models with parameters skewed to the right (gev and splogit) have DICs of 3492.9, 3492.7 and 3467.3, respectively. They fit much better than the other two other models. The logit model (symmetric, with DIC 3503.7) fits better than the cloglog model (left skewed, with DIC 3781.9). Our proposed splogit model performs the best among all models with DIC advantage of 25.4 to the next closest competitor (gev model), which indicates that our proposed link function is extremely flexible and handles this case study very well. In terms of gev and loglog models, extremely close DIC value (difference is 0.2) also confirms the fact that loglog is a special case of gev from another angle.

Next we are interested in seeing how well the models fit the data by comparing predicted values against the observed ones. For observation i, $i = 1, ..., N$, we calculate the posterior probability that it will fall into category j, $j = 1, ..., J$ by the following formula:

$$P_j^{(i)} = P(y_i = j | \boldsymbol{x}_i, \boldsymbol{\theta}) = F(\gamma_j - \boldsymbol{x}_i'\boldsymbol{\beta}, r) - F(\gamma_{j-1} - \boldsymbol{x}_i'\boldsymbol{\beta}, r), \qquad (16.10)$$

where $\boldsymbol{\theta} = (\boldsymbol{\beta}, \boldsymbol{\gamma})'$ for the logit, cloglog and loglog models, $\boldsymbol{\theta} = (\boldsymbol{\beta}, \boldsymbol{\gamma}, r)'$ for the splogit model and $\boldsymbol{\theta} = (\boldsymbol{\beta}, \boldsymbol{\gamma}, \xi)'$ for the gev model, F represents the corresponding cdf based on the link function. Then the predicted value of ith observation is $\hat{y}_i = \{l, P_l^{(i)} = \max_j(P_j^{(i)})\}$. Table 16.2 summarizes the absolute

TABLE 16.2
Summary of percentage of difference between predicted response and observed response.

	logit	cloglog	loglog	splogit	gev
$prob\{|pred - obs| = 0\}$	63.0%	56.1%	62.8%	64.7%	62.8%
$prob\{|pred - obs| = 1\}$	22.4%	19.2%	21.1%	21.8%	20.7%
$prob\{|pred - obs| = 2\}$	9.7%	12.9%	10.3%	9.2%	10.7%
$prob\{|pred - obs| = 3\}$	3.2%	6.7%	4.1%	3.0%	4.1%
$prob\{|pred - obs| = 4\}$	1.2%	2.2%	1.1%	1.0%	1.1%
$prob\{|pred - obs| = 5\}$	0.4%	2.8%	0.6%	0.3%	0.6%

value of the difference between observed and predicted value, $|pred - obs|$. Obviously a model makes the correct prediction when $|pred - obs| = 0$. It is obvious from the table that among all the models considered here, the splogit model makes the highest number of correct predictions (1158 out of 1789, 64.7%), and the lowest number of predictions that miss by at least 3 (77 out of 1789, 4.3%). On the other hand, a severely misspecified model (cloglog) makes only 56.1% correct predictions and 11.7% predictions that miss by at least 3. It clearly shows the consequences of link misspecification and the necessity of flexible link functions in ordinal response data models. For each model, we also calculate the expected number of observations for each category based on posterior probabilities, and compare it with the actual observed value. The expected number of observations for category j, $j = 1, ..., 6$ is calculated as $N\bar{P}_j$, where N stands for the total number of observations and $\bar{P}_j$ is the posterior probability that an observation falls into category j calculated by averaging over MCMC samples:

$$\bar{P}_j = \sum_{i=1}^{N} P_j^{(i)} = \sum_{i=1}^{N} P(y_i = j | \boldsymbol{x}_i, \boldsymbol{\theta}). \tag{16.11}$$

The results of the expected number of observations for all models are summarized in Table 16.3. It is clear that the splogit model predicts really well on category 1, 4, 5, 6, while it overpredicts on category 2 and underpredicts on category 3. However, it is again the best model among all the models we consider here.

16.6 Cross Validation

To further assess the predictive power of our proposed model for future observations, we carry out a cross validation study. We randomly hold out 20%

TABLE 16.3
Observed counts and expected counts under different models.

	logit	cloglog	loglog	splogit	gev	observed
#{y = 1}	990	990	961	960	966	**957**
#{y = 2}	347	327	370	363	366	**333**
#{y = 3}	237	249	253	255	249	**282**
#{y = 4}	139	145	141	145	142	**151**
#{y = 5}	45	58	40	41	41	**41**
#{y = 6}	30	20	24	24	25	**25**

TABLE 16.4
Summary of training set and testing set for cross validation.

	Training Set					
	$\{y = 1\}$	$\{y = 2\}$	$\{y = 3\}$	$\{y = 4\}$	$\{y = 5\}$	$\{y = 6\}$
Count	766	263	229	121	33	19
Percentage	53.5%	18.4%	16.0%	8.5%	2.3%	1.3%
	Testing Set					
	$\{y = 1\}$	$\{y = 2\}$	$\{y = 3\}$	$\{y = 4\}$	$\{y = 5\}$	$\{y = 6\}$
Count	191	70	53	30	8	6
Percentage	53.4%	19.6%	14.8%	8.4%	2.2%	1.7%

(358 out of 1789) of the data as the testing set, and fit the models with the remaining training set. A summary of the selected training set and test set can be seen in Table 16.4. We see that the randomly selected training and testing sets reflect a similar pattern in terms of percentages for each category. We then fit the 5 models with the training set and obtain the posterior mean of parameters. The predictions of the test set are made based on new covariates and the posterior mean we obtained.

Table 16.5 summarizes the difference of predicted and observed value for the testing dataset. The splogit model again has the highest number of correct predictions (234 out of 358, 65.4%) and the lowest number of predictions missed by at least 3 (16 out of 358, 4.5%). Again splogit is the best model in terms of cross validation prediction performance. We also calculate the expected counts for each category for the testing dataset and summarize them in Table 16.6. The numbers under splogit are very close to the observed counts in each category and again are the best among all the models considered here.

TABLE 16.5
Summary of percentage of difference between predicted response and observed response for cross validation.

	logit	cloglog	loglog	splogit	gev
$prob\{|pred - obs| = 0\}$	64.0%	54.2%	62.0%	65.4%	63.1%
$prob\{|pred - obs| = 1\}$	19.6%	18.4%	21.8%	19.3%	20.7%
$prob\{|pred - obs| = 2\}$	11.5%	14.8%	10.3%	10.9%	11.2%
$prob\{|pred - obs| = 3\}$	4.2%	7.5%	4.2%	3.4%	3.4%
$prob\{|pred - obs| = 4\}$	0.6%	3.1%	1.1%	1.1%	1.4%
$prob\{|pred - obs| = 5\}$	0.3%	2.0%	0.6%	0	0.3%

TABLE 16.6
Observed counts and expected counts under different models for cross validation.

	logit	cloglog	loglog	splogit	gev	observed
$\#\{y = 1\}$	196	199	189	189	189	**191**
$\#\{y = 2\}$	69	68	75	74	76	**70**
$\#\{y = 3\}$	49	47	53	53	51	**53**
$\#\{y = 4\}$	28	24	29	29	29	**30**
$\#\{y = 5\}$	9	8	8	8	8	**8**
$\#\{y = 6\}$	6	10	5	5	5	**6**

16.7 Conclusions

We extended the symmetric power logit link in the context of ordinal response models. Although the ordinal response models adopt a latent variable approach, the role of the link function is still critical as it controls how the latent variables interact with the cutoff points. Comparing with standard link functions, the flexible splogit model is attractive as it is easy to implement and very flexible in left skewed, symmetric or right skewed scenarios. Model fittings are done in R using advanced MCMC techniques. Empirical results on IPANE data show that our proposed model performs well both in terms of model fitting and predictive power.

Bibliography

[1] J. H. Albert and S. Chib. Bayesian analysis of binary and polychotomous response data. *Journal of the American Statistical Association*, 88(422):669–679, 1993.

[2] B. C. Arnold and R. A. Groeneveld. Measuring skewness with respect to the mode. *The American Statistician*, 49(1):34–38, 1995.

[3] A. E. Gelfand and A. F. M. Smith. Sampling-based approaches to calculating marginal densities. *Journal of the American Statistical Association*, 85(410):398–409, 1990.

[4] W. R. Gilks, N. G. Best, and K. K. C. Tan. Adaptive rejection Metropolis sampling within Gibbs sampling. *Journal of the Royal Statistical Society: Series C (Applied Statistics)*, 44:455–472, 1995.

[5] I. Ibáñez, J. A. Silander Jr, A. M. Wilson, N. LaFleur, N. Tanaka, and I. Tsuyama. Multivariate forecasts of potential distributions of invasive plant species. *Ecological Applications*, 19(2):359–375, 2009.

[6] X. Jiang, D. K. Dey, R. Prunier, A. M. Wilson, and K. E. Holsinger. A new class of flexible link functions with application to species co-occurrence in cape floristic region. *The Annals of Applied Statistics*, 7(4):2180–2204, 2013.

[7] L. J. Mehrhoff, J. A. Silander Jr, S. A. Leicht, E. S. Mosher, and N. M. Tabak. IPANE: *Invasive Plant Atlas of New England*. Department of Ecology & Evolutionary Biology, University of Connecticut, Storrs, CT, USA, 2003.

[8] G. Petris and L. Tardella. *Original C code for ARMS by Wally Gilks. HI: Simulation from distributions supported by nested hyperplanes*, 2006. R Package version 0.3.

[9] D. J. Spiegelhalter, N. G. Best, B. P. Carlin, and A. van der Linde. Bayesian measures of model complexity and fit. *Journal of the Royal Statistical Society: Series B (Statistical Methodology)*, 64(4):583–616, 2002.

[10] X. Wang and D. K. Dey. Generalized extreme value regression for ordinal response data. *Environmental and Ecological Statistics*, 18(4):619–634, 2011.

17

Elastic Prior Shape Models of 3D Objects for Bayesian Image Analysis

Sebastian Kurtek

The Ohio State University

Qian Xie

Florida State University

CONTENTS

17.1 Introduction

There are several meanings of the word shape. Although the use of the words "shape" or "shape analysis" is very common, the definition is seldom made precise in a mathematical sense. According to the *Oxford English Dictionary*, it means "the external form or appearance of someone or something as produced by their outline". Kendall [7, 17] described shape as a mathematical property that remains unchanged under certain (shape-preserving) transformations such as rotation, translation, and global scaling. Shape analysis seeks to represent shapes as mathematical quantities, such as vectors or functions, that can be manipulated using appropriate rules and metrics. Statistical shape analysis is concerned with quantifying shape as a random object and

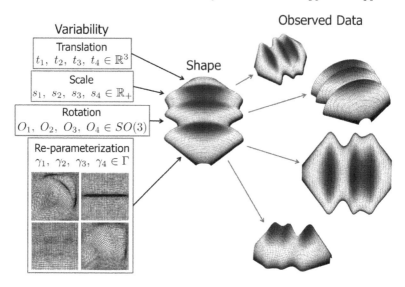

FIGURE 17.1
Statistical shape analysis of surfaces should be invariant to translation, scale,
rotation and re-parameterization.

developing tools for generating shape registrations (correspondence of points
across objects), comparisons, averages, probability models, hypothesis tests,
Bayesian estimates, and other statistical procedures on shape spaces.

Shape is an important physical property of objects that characterizes their
appearances, and can play an important role in their detection, tracking, and
recognition in images and videos. A significant part of literature has been restricted to landmark-based analysis, where shapes are represented by a coarse,
discrete sampling of the object contours. This approach is limited in that automatic detection of landmarks is not straightforward, and the ensuing shape
analysis depends heavily on the choice of landmarks. When performing shape
analysis, one usually restricts to the boundaries of objects, rather than the
whole objects, and that leads to shape analysis of surfaces in the case of 3D
images. To understand the issues and challenges in shape analysis, one has
to look at the data acquisition process. A three-dimensional image can be
captured in an arbitrary pose and zoom, which introduces a random rotation,
translation and scale of boundaries in the image volume. Therefore, any proper
metric for shape analysis should be independent of the pose and scale of the
boundaries. A visual inspection, provided in Figure 17.1, also confirms that
any rotation, translation, or scaling of a surface, while changing its coordinates
in three-dimensional space, does not change its shape.

In case of parameterized surfaces, an additional challenge arises when it
comes to invariance. Specifically, a re-parameterization of a surface (achieved

using an appropriate function $\gamma \in \Gamma$ made precise later) does not change the shape of the object. (Parameterization can be interpreted as the placement of a coordinate grid on the surface.) This is also clearly seen in Figure 17.1. This figure shows four different parameterizations (grids) on a quadrilateral surface but the overall shape of the objects remains the same. Thus, the main goal in shape analysis is to define metrics and subsequent statistical analyses, which are invariant to the introduction of an arbitrary parameterization in shape representations. For example, a metric invariant to all shape preserving transformations (translation, scale, rotation and re-parameterization) among the four surfaces shown in Figure 17.1 should be zero.

Another important issue in statistical shape analysis of 3D objects is registration (matching of points across objects). Any statistical analysis requires a registration component to help decide which point on one shape is compared to which point on the other. This fully avoids the issue of choosing discrete landmarks on the objects, because a full correspondence between the shapes is established. A registration is "good" when it matches points across shapes that have similar geometric features. One major goal in shape analysis is to solve the registration and comparison problems simultaneously under the same objective function, preferably a proper metric. This is accomplished in a Riemannian framework by finding optimal re-parameterizations of surfaces under the resulting distance functions.

As an illustrative example, consider the two closed surfaces in Figure 17.2. Structurally, these two surfaces are extremely similar; they both have two peaks but with different placements. As given, the two surfaces have uniform parameterizations, and one can clearly see that the two peaks do not match each other. The matching here is displayed by mapping the colormap from the first surface to corresponding points on the second surface. Thus, good matching is given by geometrically similar areas being shaded with similar colors. One can find the optimal re-parameterization of the second surface to match it well with the first one. We will elaborate on how to do this in later sections. This results in much better matching of structurally similar features across the two given surfaces. One can also generate comparisons of these two surfaces using geodesic paths and distances. In short, a geodesic path is the shortest path between these two objects in the space of all possible shapes. The length of this path provides the distance between them. Furthermore, the sample mean of these two surfaces is given by the midpoint of this path. When no registration is performed, the common features on these two surfaces (the two peaks) are smoothed (or averaged out) along the geodesic path. The sample mean has two peaks but they are not nearly as sharp as on the original surfaces; the resulting distance is 0.6745. This is due to misalignment of geometric features. But, when one optimally registers these surfaces, the resulting geodesic path has a much nicer structure and represents an intuitive deformation between the two given surfaces. That is, the peaks simply move from the placement on the first surface to the corresponding placement on the second surface. The geodesic distance after registration is much lower at

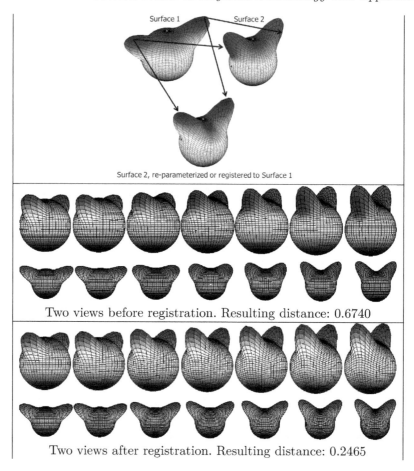

FIGURE 17.2
Optimal re-parameterizations, also referred to as registration, provide improved shape comparisons and statistics.

0.2465 and provides a more accurate measure of differences between these two surfaces. In summary, in order to generate meaningful comparisons, and representative and parsimonious models of 3D shapes one must first remove the parameterization variability, thereby optimally registering the given data.

17.1.1   Related work

Many representations have been proposed for statistical shape analysis of 3D objects. A pictorial summarization of different representations used for analysis is given in Figure 17.3. Several groups have proposed to study shapes of surfaces by embedding them in volumes and deforming the volumes [10, 15],

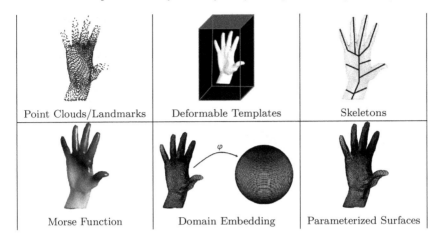

Point Clouds/Landmarks	Deformable Templates	Skeletons
Morse Function	Domain Embedding	Parameterized Surfaces

FIGURE 17.3
Different representations of shapes of 3D objects.

termed deformable templates. While pioneering, especially in the field of medical imaging, these methods are typically computationally expensive because of the high dimensionality of the resulting objects that are analyzed. An alternative approach is based on manually-generated landmarks also termed Kendall's shape analysis [7]. While this is a very popular approach in many applications, it requires a set of registered landmarks to represent the surface, which are difficult and time consuming to obtain in practice. Others have studied 3D shape variabilities using level sets [22], medial axes [2, 8], or point clouds via the iterative closest point algorithm [1].

However, the most natural representation for studying shapes of 3D objects seems to be using their boundaries, which form parameterized surfaces. As stated earlier, such a representation poses an additional issue of handling the parameterization variability. Some methods [3, 26] tackle this problem by choosing a fixed parameterization, similar to arc-length in the case of parameterized curves. A large set of papers in the literature treat the re-parameterization and statistical analysis steps as separate [4, 6]. Because in these approaches the two steps are unrelated, the computed registrations tend to be suboptimal and defining proper parameterization-invariant metrics (and statistics) is not possible. In a series of papers, Kurtek et al. [18–21] presented a comprehensive framework for parameterization-invariant shape modeling of surfaces based on the q-map representation. This was the first work that utilized a Riemannian framework and provided a truly parameterization-invariant method. A major drawback of this method is in the definition of the Riemannian metric, which does not have a clear interpretation in terms of the amount of stretching and bending needed to deform one surface into another. This issue was addressed by Jermyn et al. [13] using a novel representation of surfaces termed square-root normal fields. This is the framework we describe in the remainder of this chapter. In particular, we define a proper distance between shapes of surfaces under a simplified elastic Riemannian metric (hence the title "*elastic prior shape models*"). We define a shape space of surfaces and utilize it for statistical analysis. In particular, we describe a framework for computing summary statistics of shapes such as the mean and covariance. Using these, we learn Gaussian generative models on the shape space and provide a recipe for random sampling. Obtaining representative random samples of shapes of 3D objects is crucial in Bayesian modeling of shape data and in Bayesian image analysis.

17.1.2   Utility of shape models in Bayesian image analysis

One major utility of statistical shape models is in the form of prior information in various Bayesian approaches to image analysis. The main two examples are in image segmentation and registration [16, 28]. Segmentation refers to the process of extracting objects of interest from an image. Registration refers to determining a one-to-one pixel correspondence between two images. When shape information about the objects of interest is available *a priori*, one can incorporate it in a Bayesian segmentation or registration framework. This idea has been shown to provide improved performance in many different applications including computer vision, graphics, biometrics and medical imaging.

Many methods in literature utilize prior shape models for extracting object boundaries [5, 9, 11, 24], a process called image segmentation. The main differences lie in the chosen shape representation and metric as well as the problem formulation. Joshi et al. [16] presented a Bayesian active contour framework for extraction of planar boundaries in 2D images. They defined intrinsic prior shape models under an elastic framework and phrased the problem of boundary extraction as maximum *a posteriori* estimation: $\hat{\beta} = \underset{\beta}{\operatorname{argmin}}[E_{image}(\beta) + E_{smooth}(\beta) + E_{prior}(\beta)]$. Here, β is a curve and the minimization is carried out over all possible closed curves. In this framework, all three of the energy terms are invariant to re-parameterization of the curve. The smoothness term is invariant to translation and rotation and the prior shape term is invariant to all shape preserving transformations. The elastic shape models presented in this chapter are very useful in a similar problem formulation for extracting boundaries of objects in 3D images. Such an approach is of great interest in, for example, medical imaging where manual segmentation of anatomical structures is expensive and time consuming. We note that the elastic shape models of 3D objects presented in the following sections are analogous to the prior shape models for planar curves presented in [16].

Another important image analysis procedure is image registration, which involves matching pixels across images using an intensity-driven cost function. Image registration is especially important in medical imaging studies where images of multiple subjects must be co-registered prior to statistical analysis. Unfortunately, many medical imaging modalities such as computed tomography (CT) have low image contrast, making it difficult to distinguish different organs of interest. Thus, prior shape information is especially important in such a setting and can be used to improve registration accuracy. Again, there are many methods in literature that use such a formulation [12, 14, 23]. A recent approach [28] incorporated uncertainty in manual segmentations through a 3D elastic shape model in an image registration setting. This framework was applied to registration of fan-beam and cone-beam CT volumes, where prior shape information of the prostate was used to improve results.

17.2   Representation of Surfaces

Let $\mathcal{F}$ be the space of all smooth embeddings of a sphere in $\mathbb{R}^3$ and let Γ be the set of all diffeomorphisms from $\mathbb{S}^2$ to itself. Γ is the set of all possible re-parameterizations of a spherical surface. (We only consider closed or spherical surfaces in this chapter, but this framework easily extends beyond the spherical domain.) For a closed surface $f \in \mathcal{F}$, $f \circ \gamma$ represents a re-parameterization of this surface. Given this setup, one could adopt a standard approach and measure distances between surfaces using the $\mathbb{L}^2$ norm. Unfortunately, this framework is inappropriate for statistical shape analysis of parameterized surfaces as was previously shown in [13, 19–21]. We elaborate next. Let f_1, $f_2 \in \mathcal{F}$ be two closed surfaces and $\gamma \in \Gamma$ a re-parameterization function. Then, it is easy to see that the correspondence or registration between f_1 and f_2 is exactly the same as the correspondence between $f_1 \circ \gamma$ and $f_2 \circ \gamma$. Thus, the criterion used to measure differences between surfaces should satisfy the isometry property, $d(f_1, f_2) = d(f_1 \circ \gamma, f_2 \circ \gamma)$, for some distance function d. It is easy to show that the $\mathbb{L}^2$ norm does not satisfy this property, $\|f_1 - f_2\| \neq \|f_1 \circ \gamma - f_2 \circ \gamma\|$. That is, in this setup, a common re-parameterization of two surfaces changes the distance between them. This does not allow one to define a fully parameterization-invariant framework for shape analysis. Thus, our approach is to utilize a new representation of surfaces (and a corresponding metric) such that the isometry property is satisfied. For more detailed descriptions of these methods please refer to [13, 27].

Let $n(s) = \frac{\partial f}{\partial u}(s) \times \frac{\partial f}{\partial v}(s) \in \mathbb{R}^3$ denote the normal vector to the surface at the point $s = (u, v) \in \mathbb{S}^2$. Jermyn et al. [13] defined a mathematical representation of surfaces termed square-root normal fields (SRNFs) using a mapping $Q : f \mapsto Q(f)$ as $Q(f)(s) = q(s) = \frac{n(s)}{\sqrt{|n(s)|}}$. The space of all SRNFs is a subset of $\mathbb{L}^2(\mathbb{S}^2, \mathbb{R}^3)$ hereinafter referred to as $\mathbb{L}^2$. If a surface f is re-parameterized to $f \circ \gamma$, then its SRNF is given by $(q, \gamma) = (q \circ \gamma)\sqrt{J_\gamma}$, where J_γ is the determinant of the Jacobian of γ. For comparing shapes, we will utilize the natural $\mathbb{L}^2$ metric on the space of SRNFs. An important advantage of using the SRNF representation is that a simultaneous re-parameterization of any two surfaces does not change the distance between them. That is, for any two surfaces f_1 and f_2, represented by their SRNFs, q_1 and q_2 respectively, the isometry property is satisfied: $\|q_1 - q_2\| = \|(q_1, \gamma) - (q_2, \gamma)\|$.

Because the mapping Q is not easily invertible, we will work directly in $\mathcal{F}$, albeit with a different Riemannian metric. Actually, the Riemannian metric that we will use on $\mathcal{F}$ is the pullback of the $\mathbb{L}^2$ metric from the space of SRNFs. The expression for the pullback metric is obtained using the differential of Q, denoted by Q_*. Q_* can be used to map vectors from $T_f(\mathcal{F})$ to $T_{Q(f)}(\mathbb{L}^2)$, and is defined as (for a $w \in T_f(\mathcal{F})$):

$$Q_*(w)(s) = \frac{n_w(s)}{\sqrt{|n(s)|}} - \frac{n(s) \cdot n_w(s)}{2|n(s)|^{(5/2)}} n(s). \tag{17.1}$$

In this expression, $n_w = f_u \times w_v + w_u \times f_v$, and subscripts denote partial derivatives. Then, the induced Riemannian metric on $\mathcal{F}$ can be written as (for w_1, $w_2 \in T_f(\mathcal{F})$):

$$\langle\langle w_1, w_2 \rangle\rangle = \langle Q_*(w_1), Q_*(w_2) \rangle_{\mathbb{L}^2}$$
$$= \int_{\mathbb{S}^2} \frac{n_{w_1}(s) \cdot n_{w_1}(s)}{|n(s)|} - \frac{3(n(s) \cdot n_{w_1}(s))(n(s) \cdot n_{w_2}(s))}{4|n(s)|^3} ds.$$

With this induced metric, $\mathcal{F}$ becomes a Riemannian manifold. It turns out that this induced Riemannian metric is a special case of a general elastic metric on the space of parameterized surfaces [13]. This can be seen by rewriting the expression of this Riemannian metric in terms of perturbations of the normal vector direction (bending) and its magnitude (stretching). The main advantage of this metric over other Riemannian metrics, such as the one presented in [21], is that it provides an intuitive interpretation of the stretching and bending transformations needed to deform one surface into another. Note that the elastic metric for shape analysis of surfaces is the analog of the elastic metric used for shape analysis of curves [25]. In Figure 17.4, we provide a pictorial description of the two spaces $\mathcal{F}$ and $\mathbb{L}^2$ and how they are related.

17.2.1 Pre-shape and shape spaces

We are interested in statistical models of *shapes* of surfaces. Thus, we must ensure invariance to all shape preserving transformations including translation, scale, rotation and re-parameterization. First, note that, by definition, the SRNF representation is automatically translation invariant. We achieve scale invariance by re-scaling all surfaces to have unit area. With a slight abuse of notation we redefine $\mathcal{F}$ as the pre-shape space of surfaces.

The special orthogonal group, $SO(3)$, provides all possible rotations of a surface. For a particular rotation $O \in SO(3)$ and a surface $f \in \mathcal{F}$, the rotated surface is given by Of. The rotation and re-parameterization variabilities are removed from the representation space using equivalence classes. We define

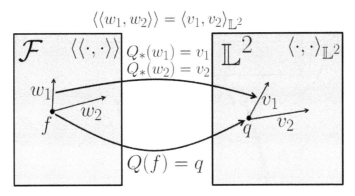

FIGURE 17.4

Q is used to map surfaces $f \in \mathcal{F}$ to corresponding SRNFs $q \in \mathbb{L}^2$. Q_* maps tangent vectors $w \in T_f(\mathcal{F})$ to corresponding tangent vectors $v \in T_q(\mathbb{L}^2)$. The Riemannian metric $(\langle \cdot, \cdot \rangle)$ on the space of SRNFs is the standard $\mathbb{L}^2$ metric. The Riemannian metric on $\mathcal{F}$ $(\langle\langle \cdot, \cdot \rangle\rangle)$ is the pullback of the $\mathbb{L}^2$ metric from the SRNF space.

the equivalence class of a surface f as $[f] = \{O(f \circ \gamma) | O \in SO(3), \gamma \in \Gamma\}$ and the set of all such equivalence classes is defined to be the shape space $\mathcal{S}$. Note that each equivalence class represents a surface shape uniquely.

17.2.2 Shape comparisons via geodesics

The next step is to define geodesic paths in the shape space $\mathcal{S}$. We remind the reader that a geodesic path is the shortest path between two shapes. This is accomplished using a path-straightening approach on $\mathcal{F}$. This is similar to the path straightening algorithm defined in [21]. The main difference is in the Riemannian metric used to compute the geodesics. Once we have an algorithm for finding geodesics in $\mathcal{F}$, we can obtain geodesic paths and geodesic distances (lengths of geodesic paths) in $\mathcal{S}$ by solving an additional optimization problem over $SO(3) \times \Gamma$. Let f_1 and f_2 denote two closed surfaces and let $\langle\langle \cdot, \cdot \rangle\rangle$ be the induced Riemannian metric on $\mathcal{F}$. Then, the geodesic distance between shapes of f_1 and f_2 is given by quantities of the following type: $d([f_1], [f_2]) =$

$$\min_{(O,\gamma) \in SO(3) \times \Gamma} \left(\min_{\substack{F : [0,1] \to \mathcal{F} \\ F(0) = f_1, \ F(1) = O(f_2 \circ \gamma)}} \left(\int_0^1 \langle\langle F_t(t), F_t(t) \rangle\rangle^{(1/2)} \, dt \right) \right).$$

(17.2)

In this equation, $F(t)$ is a path in $\mathcal{F}$ indexed by t and F_t is used to denote $\frac{dF}{dt}$. The quantity $L(F) = \int_0^1 \langle\langle F_t(t), F_t(t) \rangle\rangle^{(1/2)} dt$ provides the length of the

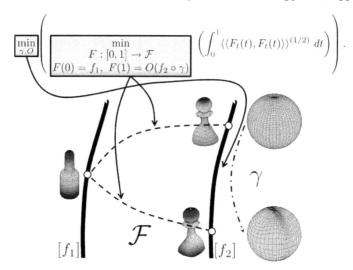

FIGURE 17.5
Given two surfaces f_1 (bottle) and f_2 (chess piece), we register them by search-ing for the optimal rotation and re-parameterization in the equivalence class of the second surface, $[f_2]$. This represents the outside minimization problem. Given a specific rotation and re-parameterization, the inside minimization problem finds the shortest path between these two surfaces. The joint solu-tion is the geodesic path in the shape space. The length of this path is the geodesic distance between the shapes of f_1 and f_2.

path F. The minimization inside the brackets, thus, is the problem of finding a geodesic path between the surfaces f_1 and $O(f_2 \circ \gamma)$, where O and γ stand for an arbitrary rotation and re-parameterization of f_2, respectively. The minimiza-tion outside the brackets seeks the optimal rotation and re-parameterization of the second surface so as to match it well with the first surface. In sim-ple words, the outside optimization solves the registration problem while the inside optimization solves for both an optimal deformation (geodesic path, F^*) and a formal distance (geodesic distance) between shapes. We provide a pictorial description of this process in Figure 17.5.

  Solving the joint optimization problem in Equation 17.2 can be compu-tationally expensive. Thus, we employ a simplification that provides an ap-proximate solution to this problem. First, we split this problem into two parts: registration (outside optimization) and deformation (inside optimization). We begin by computing the SRNF representations of f_1, $f_2 \in \mathcal{F}$ given by q_1 and q_2 and solving for the optimal registration in the SRNF space:

$$(O^*, \gamma^*) = \underset{(O,\gamma) \in SO(3) \times \Gamma}{\text{arginf}} \|q_1 - (Oq_2, \gamma)\|^2. \tag{17.3}$$

This optimization problem is solved iteratively. First, one fixes γ and searches

Pre-Shape Space	Shape Space
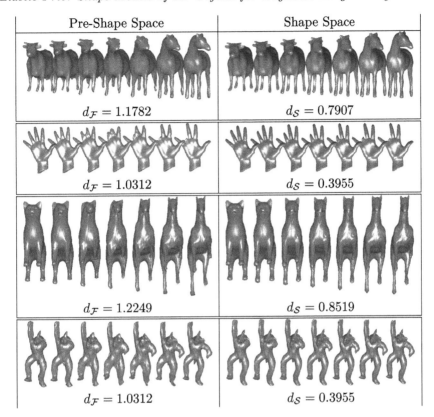	
$d_{\mathcal{F}} = 1.1782$	$d_{\mathcal{S}} = 0.7907$
$d_{\mathcal{F}} = 1.0312$	$d_{\mathcal{S}} = 0.3955$
$d_{\mathcal{F}} = 1.2249$	$d_{\mathcal{S}} = 0.8519$
$d_{\mathcal{F}} = 1.0312$	$d_{\mathcal{S}} = 0.3955$

FIGURE 17.6
Comparison of geodesics in the pre-shape space and shape space.

for an optimal rotation over $SO(3)$ using Procrustes analysis. Then, given this rotation, one searches for an optimal re-parameterization over Γ using a gradient descent algorithm [13]. This results in an optimal rotation and re-parameterization of the second surface denoted by $f_2^* = O^*(f_2 \circ \gamma^*)$. The second step is to compute the geodesic path between f_1 and f_2^* using path straightening, and to measure its length. We have found that the exact and approximate solutions are very similar in practice.

We demonstrate these ideas using several examples in Figures 17.6 and 17.7. These examples highlight improvements in registration of surfaces during the optimization over $SO(3) \times \Gamma$, by comparing corresponding geodesic paths between the same pairs of surfaces in $\mathcal{F}$ and $\mathcal{S}$. In all these experiments, we notice that the geodesic paths in the shape space are much more natural and the geodesic distances in $\mathcal{S}$ are much smaller than the geodesic distances in $\mathcal{F}$. To elaborate, let us consider the second example in Figure 17.6. The two surfaces being compared are identical except for the missing ring finger. The deformation in the pre-shape space is not intuitive because the ring finger

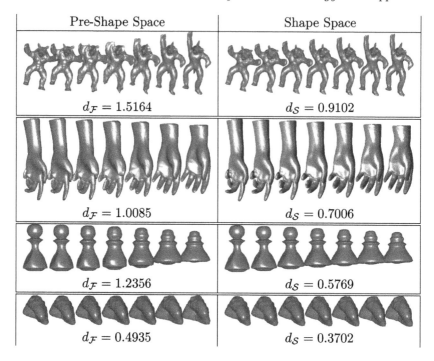

FIGURE 17.7
Comparison of geodesics in the pre-shape space and shape space. The bottom
example considers two left putamen surfaces in a medical imaging application.

grows out of the middle finger. On the contrary, the ring finger simply grows
along the geodesic path in the shape space without affecting any of the other
fingers. This more natural transformation is accompanied by a significantly
lower distance between the two surfaces ($d_{\mathcal{F}} = 1.0312$, $d_{\mathcal{S}} = 0.3955$). The last
example in Figure 17.7 considers a comparison of two left putamen subcortical
structures in the brain. While the differences in the two geodesic paths are
more subtle here, we still observe a significant decrease in the distance when
the two surfaces are optimally registered.

17.3   Shape Statistics of Surfaces

In this section, we present tools and results for computing two fundamental
shape statistics, the mean and the covariance, for a set of closed surfaces.
We then utilize these quantities to estimate a generative Gaussian model and
draw random samples from this model.

17.3.1 Estimation of the Karcher mean

We begin by defining an intrinsic mean of surfaces under the proposed elastic metric, called the Karcher mean. Let $\{f_1, f_2, \ldots, f_n\} \in \mathcal{F}$ denote a sample of closed surfaces. Also, let F_i^* denote a geodesic path (in $\mathcal{S}$) between a surface f and the ith surface in the sample, f_i. Then, the sample Karcher mean is given by $[\bar{f}] = \operatorname*{argmin}_{[f] \in \mathcal{S}} \sum_{i=1}^{n} L(F_i^*)^2$. A gradient-based approach for finding the Karcher mean is given in [7] and is repeated here for convenience. Note that the resulting Karcher mean is an entire equivalence class of surfaces, which is how we defined shape.

Algorithm 1 (Karcher Mean): Let $\bar{f}_0$ be an initial estimate of the Karcher mean. Set $j = 0$ and ϵ_1, ϵ_2 to be small positive values.
(1) For each $i = 1, \ldots, n$, compute the geodesic path between f_i and $\bar{f}_j$ to obtain F_i^*, such that $F_i^*(0) = \bar{f}_j$, $F_i^*(1) = f_i^*$.
(2) For each $i = 1, \ldots, n$, compute the shooting vector $v_i = \frac{dF_i^*}{dt}|_{t=0}$.
(3) Compute the average direction $\bar{v} = (1/n) \sum_{i=1}^{n} v_i$.
(4) If $\|\bar{v}\| < \epsilon_1$, stop. Else, update using $\bar{f}_{j+1} = \bar{f}_j + \epsilon_2 \bar{v}$.
(5) Set $j = j + 1$ and return to Step 1.

In Figure 17.8, we present several results of applying Algorithm 1 on different types of data. In the left panel, we display the sample surfaces that were used to compute the sample average. We also show the pre-shape space extrinsic mean (panel (a), no optimization over $SO(3) \times \Gamma$) and the shape space mean (panel (b)). It is clear from this figure that the averages computed in the shape space are much better representatives of the given data. The pre-shape space sample averages are often distorted, and lose most of the structure present in the given data. On the other hand, the shape space sample averages exhibit similar structures to those present in the given sample surfaces. This is due to improved alignment of geometric features. Consider Example 5 in Figure 17.8. The given surfaces are two horses and a cow. In the pre-shape space, the average of these surfaces has a distorted head and distorted legs due to severe misalignment of those parts across all sample surfaces. In the shape space, the average surface is a hybrid of the horses and cow where the head and legs are well defined. Natural shape averages are very important in subsequent tasks such as estimating the covariance and obtaining random samples from generative models.

17.3.2 Estimation of the Karcher covariance

Once the sample Karcher mean has been computed, the evaluation of the Karcher covariance is performed as follows. First, we find the shooting vectors from the estimated Karcher mean $\bar{f}$ to each of the surfaces in the sample, $\nu_i = \frac{dF_i^*}{dt}|_{t=0}$, where $i = 1, 2, \ldots, n$, and F^* denotes a geodesic path in $\mathcal{S}$. We then perform principal component analysis by applying the Gram-Schmidt

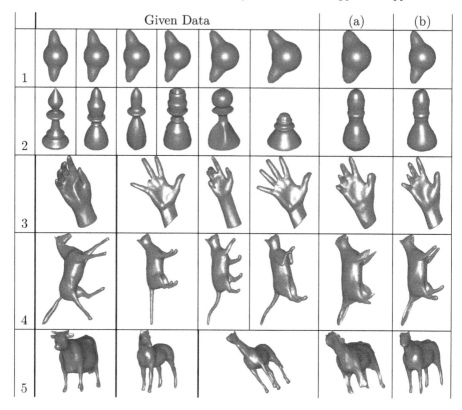

FIGURE 17.8
Comparison of sample means computed in the pre-shape space (a) and shape space (b).

procedure (under the chosen metric $\langle\langle\cdot,\cdot\rangle\rangle$), to generate an orthonormal basis $\{B_j|j=1,\ldots,k\}$, $k \leq n$, of the observed $\{\nu_i\}$. We project each of the vectors ν_i onto this orthonormal basis using $\nu_i \approx \sum_{j=1}^{k} c_{i,j}B_j$, where $c_{i,j} = \langle\langle\nu_i, B_j\rangle\rangle$. Now, each original surface can simply be represented using the coefficient vector $c_i = \{c_{i,j}\}$. Then, the sample covariance matrix can be computed in the coefficient space as $K = (1/(n-1))\sum_{i=1}^{n} c_i c_i^T \in \mathbb{R}^{k \times k}$. We can use the singular value decomposition of K to determine the principal directions of variation in the given data. For example, if $u \in \mathbb{R}^k$ corresponds to a principal singular vector of K, then the corresponding tangent vector in $T_{\bar{f}}(\mathcal{F})$ is given by $v = \sum_{j=1}^{k} u_j B_j$. One can then map this vector to a surface f using the exponential map, $\exp_{\bar{f}}(v)$. The exponential map is used to map vectors from the tangent space to the space of surfaces. Note that the exponential map must be computed under the non-standard metric introduced earlier, which is not a simple task. This can be accomplished using a tool called parallel transport, which was derived for this representation of surfaces by Xie et al.

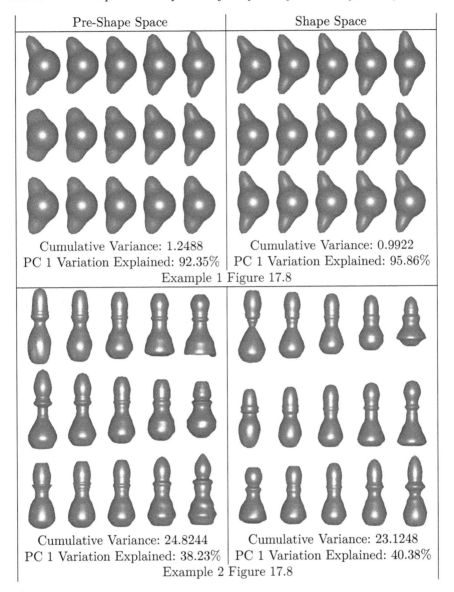

Pre-Shape Space	Shape Space
Cumulative Variance: 1.2488	Cumulative Variance: 0.9922
PC 1 Variation Explained: 92.35%	PC 1 Variation Explained: 95.86%

Example 1 Figure 17.8

Cumulative Variance: 24.8244	Cumulative Variance: 23.1248
PC 1 Variation Explained: 38.23%	PC 1 Variation Explained: 40.38%

Example 2 Figure 17.8

FIGURE 17.9
Three principal directions of variation within ±1 standard deviation of the mean for Examples 1 (top) and 2 (bottom) from Figure 17.8.

[27]. For brevity, we do not provide details here but rather refer the interested reader to that work.

In Figure 17.9, we display the three main directions of variation in the data given in Examples 1 and 2 of Figure 17.8. All the surfaces in the first

Pre-Shape Space	Shape Space
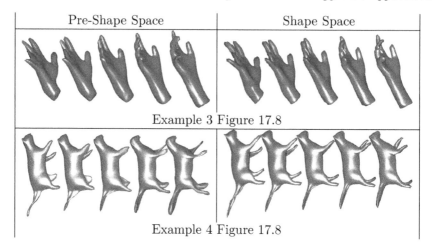

Example 3 Figure 17.8

Example 4 Figure 17.8

FIGURE 17.10
Principal direction of variation within ±1 standard deviation of the mean for
Examples 3 (top) and 4 (bottom) from Figure 17.8.

example have two peaks, but in different placements. Comparing the computed
average shapes (Example 1, Figure 17.8), we notice that the average in the
pre-shape space still has two peaks but they are smaller and wider than in
the original data. This is due to misalignment of the surfaces in their given
parameterizations. The shape space mean has the same structure as the given
data, where the two peaks are in the average placement. We computed the
covariance and performed principal component analysis in the pre-shape space
(left panel) and the shape space (right panel) for comparison. We display the
three dominant directions of variation (direction 1 in top row, direction 2 in
middle row and direction 3 in bottom row) in the given data by showing a path
starting at −1 standard deviation from the mean and ending at +1 standard
deviation from the mean, with the mean as the midpoint of this path. We
notice that almost all variation in the shape space is captured by the first
principal direction. The other two paths are nearly constant, indicating a
very low variance. This is not the case in the pre-shape space. There is fairly
large variance in both the first and second directions. Furthermore, these two
directions do not reflect natural variation in the given data. This is especially
true about the second direction where the two peaks are highly distorted when
one goes −1 standard deviation from the mean. The model in the shape space
is more compact with a cumulative variance of 0.9922 compared to 1.2488 in
the pre-shape space. The same can be observed in the second example (chess
pieces), but the differences here are more subtle.

   Figure 17.10 displays the principal direction of variation (pre-shape space
in left panel and shape space in right panel) for Examples 3 and 4 of Figure

Pre-Shape Space	Shape Space

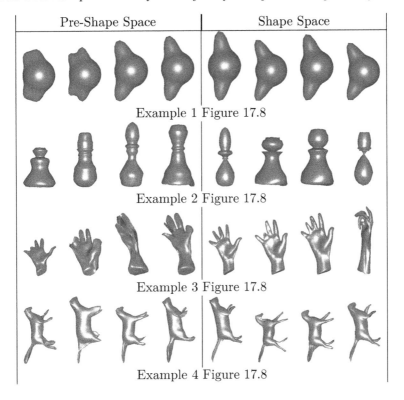

Example 1 Figure 17.8

Example 2 Figure 17.8

Example 3 Figure 17.8

Example 4 Figure 17.8

FIGURE 17.11
Four random samples from a generative Gaussian model in the pre-shape space (left panel) and the shape space (right panel).

17.8. As in the other examples, the variation around the mean shape is more natural in the shape space than in the pre-shape space.

17.3.3   Random sampling from Gaussian model

Given the mean and covariance, we can impose a Gaussian distribution in the tangent space at the mean surface. This will provide a model, which can be used to generate random sample shapes. A random tangent vector $v \in T_{\bar{f}}(\mathcal{F})$ can be sampled from the Gaussian distribution using $v = \sum_{j=1}^{k} z_j \sqrt{S_{jj}} u_j B_j$, where $z_j \overset{iid}{\sim} N(0,1)$, S_{jj} is the variance of the jth principal component, u_j is the corresponding principal singular vector and B_j is a basis element as defined in the previous section. One can then map this element of the tangent space to a surface shape using the exponential map to obtain a random shape from the Gaussian distribution. Figure 17.11 displays four random samples from a Gaussian generative model in the pre-shape space, and in the shape space. Note that in most cases, the random samples in the shape space have much better structure than those generated in the pre-shape space. This is especially evident in the top example. Here, some of the random samples in the pre-shape space have two very wide peaks or even four peaks, despite the fact that all surfaces used to estimate the model had two sharp peaks.

17.4 Summary

We have presented a comprehensive framework for shape analysis of 3D objects. This framework is based on an elastic Riemannian metric, yielding improved comparisons of shape and more parsimonious models. We have validated this framework using numerous examples and through random sampling. Statistical models of shape play a very important role in Bayesian image analysis, where they are used as prior information in numerous tasks including registration and segmentation. It is important for such models to be compact and parsimonious. They also must capture natural deformations of objects that may be present in unseen images.

Bibliography

[1] A. Almhdie, C. Léger, M. Deriche, and R. Lédée. 3D registration using a new implementation of the ICP algorithm based on a comprehensive lookup matrix: Application to medical imaging. *Pattern Recognition Letters*, 28(12):1523–1533, 2007.

[2] S. Bouix, J. C. Pruessner, D. L. Collins, and K. Siddiqi. Hippocampal shape analysis using medial surfaces. *NeuroImage*, 25:1077–1089, 2001.

[3] C. Brechbühler, G. Gerig, and O. Kübler. Parameterization of closed surfaces for 3D shape description. *Computer Vision and Image Understanding*, 61(2):154–170, 1995.

[4] J. Cates, M. Meyer, P. T. Fletcher, and R. Whitaker. Entropy-based particle systems for shape correspondence. In *Proceedings of MICCAI Workshop on Mathematical Foundations of Computational Anatomy*, pages 90–99, 2006.

[5] D. Cremers, S. J. Osher, and S. Soatto. Kernel density estimation and intrinsic alignment for shape priors in level set segmentation. *International Journal of Computer Vision*, 69(3):335–351, 2006.

[6] R. H. Davies, C. J. Twining, T. F. Cootes, and C. J. Taylor. Building 3-D statistical shape models by direct optimization. *IEEE Transactions on Medical Imaging*, 29(4):961–981, 2010.

[7] I. L. Dryden and K. V. Mardia. *Statistical Shape Analysis*. John Wiley & Sons, Chichester, U.K., 1998.

[8] K. Gorczowski, M. Styner, J. Y. Jeong, J. S. Marron, J. Piven, H. C. Hazlett, S. M. Pizer, and G. Gerig. Multi-object analysis of volume, pose, and shape using statistical discrimination. *IEEE Transactions on Pattern Analysis and Machine Intelligence*, 32(4):652–666, 2010.

[9] U. Grenander, Y. Chow, and D. M. Keenan. *Hands: A Pattern Theoretic Study of Biological Shapes*. Springer Verlag, New York, 1991.

[10] U. Grenander and M. I. Miller. Computational anatomy: An emerging discipline. *Quarterly of Applied Mathematics*, LVI(4):617–694, 1998.

[11] U. Grenander, A. Srivastava, and M. I. Miller. Asymptotic performance analysis of Bayesian target recognition. *IEEE Transactions on Information Theory*, 46(4):1658–1665, 2000.

[12] T. Hartkens, D. L. G. Hill, A. D. Castellano-Smith, D. J. Hawkes, C. R. Maurer Jr., A. J. Martin, W. A. Hall, H. Liu, and C. L. Truwit. Using points and surfaces to improve voxel-based non-rigid registration. In *Proceedings of Medical Image Computing and Computer-Assisted Intervention*, pages 565–572, 2002.

[13] I. H. Jermyn, S. Kurtek, E. Klassen, and A. Srivastava. Elastic shape matching of parameterized surfaces using square root normal fields. In *Proceedings of European Conference on Computer Vision*, pages 804–817, 2012.

[14] A. A. Joshi, D. W. Shattuck, P. M. Thompson, and R. M. Leahy. Surface-constrained volumetric brain registration using harmonic mappings. *IEEE Transactions on Medical Imaging*, 26(12):1657–1669, 2007.

[15] S. C. Joshi, M. I. Miller, and U. Grenander. On the geometry and shape of brain sub-manifolds. *Pattern Recognition and Artificial Intelligence*, 11(8):1317–1343, 1997.

[16] S. H. Joshi and A. Srivastava. Intrinsic Bayesian active contours for extraction of object boundaries in images. *International Journal of Computer Vision*, 81(3):331–355, Mar. 2009.

[17] D. G. Kendall. Shape manifolds, procrustean metrics and complex projective spaces. *Bulletin of London Mathematical Society*, 16:81–121, 1984.

[18] S. Kurtek, E. Klassen, Z. Ding, M. J. Avison, and A. Srivastava. Parameterization-invariant shape statistics and probabilistic classification of anatomical surfaces. In *Proceedings of Information Processing in Medical Imaging*, Irsee, Germany, 2011.

[19] S. Kurtek, E. Klassen, Z. Ding, S. W. Jacobson, J. L. Jacobson, M. J. Avison, and A. Srivastava. Parameterization-invariant shape comparisons of anatomical surfaces. *IEEE Transactions on Medical Imaging*, 30(3):849–858, 2011.

[20] S. Kurtek, E. Klassen, Z. Ding, and A. Srivastava. A novel Riemannian framework for shape analysis of 3D objects. In *Proceedings of IEEE Computer Vision and Pattern Recognition*, pages 1625–1632, 2010.

[21] S. Kurtek, E. Klassen, J. C. Gore, Z. Ding, and A. Srivastava. Elastic geodesic paths in shape space of parameterized surfaces. *IEEE Transactions on Pattern Analysis and Machine Intelligence*, 34(9):1717–1730, 2012.

[22] R. Malladi, J. A. Sethian, and B. C. Vemuri. A fast level set based algorithm for topology-independent shape modeling. *Journal of Mathematical Imaging and Vision*, 6(2–3):269–289, 1996.

[23] G. Postelnicu, L. Zöllei, and B. Fischl. Combined volumetric and surface registration. *IEEE Transactions on Medical Imaging*, 28(4):508–522, 2009.

[24] M. Rousson and N. Paragios. Shape priors for level set representations. In *Proceedings of European Conference on Computer Vision*, pages 78–92, 2002.

[25] A. Srivastava, E. Klassen, S. H. Joshi, and I. H. Jermyn. Shape analysis of elastic curves in Euclidean spaces. *IEEE Transactions on Pattern Analysis and Machine Intelligence*, 33(7):1415–1428, July 2011.

[26] M. Styner, I. Oguz, S. Xu, C. Brechbühler, D. Pantazis, J. Levitt, M. E. Shenton, and G. Gerig. Framework for the statistical shape analysis of brain structures using SPHARM-PDM. In *Proceedings of MICCAI Open Science Workshop*, 2006.

[27] Q. Xie, S. Kurtek, H. Le, and A. Srivastava. Parallel transport of deformations in shape space of elastic surfaces. In *IEEE International Conference on Computer Vision (ICCV)*, pages 865–872, 2013.

[28] C. Zhang, G. E. Christensen, S. Kurtek, A. Srivastava, M. J. Murphy, E. Weiss, E. Bai, and J. F. Williamson. SUPIR: Surface uncertainty-penalized, non-rigid image registration for pelvic CT imaging. In B. M. Dawant, G. E. Christensen, J. M. Fitzpatrick, and D. Rueckert, editors, *Workshop on Biomedical Image Registration*, volume 7359 of *Lecture Notes in Computer Science*, pages 236–245, 2012.

18

Multi-State Models for Disease Natural History

Amy E. Laird
Oregon Health & Science University

Rebecca A. Hubbard
University of Washington

Lurdes Y. T. Inoue
University of Washington

CONTENTS

18.1 Introduction

A variety of disease processes can be described through transitions between discrete states, such as stable or accelerated disease state in leukemia; development of AIDS-defining illnesses in HIV; or diminished lung function in asthma patients. In progressive diseases, subjects traverse disease states in only one direction while in non-progressive diseases it is possible for subjects to experience repeated occurrences of some or all states.

Studies of chronic disease often utilize longitudinal observations of a cohort of subjects to characterize natural history of disease. Subjects may be observed continuously, in which case the exact time of transition between states is known. However, in studies of human health, panel observation, in which subjects are observed and disease states assessed only at discrete time-points, is more common. In data arising from panel observation, the exact time of state transitions is unknown.

Multi-state models provide flexible tools for describing characteristics of disease processes and can be estimated under either continuous or panel observation. These models are particularly useful in the context of panel observation because they allow us to estimate the rate of disease progression even if exact transition times are not observed. If the observations are equally spaced in time then discrete time models, such as the Markov chain, can be used. If the length of time between observations is not constant, continuous time models can be employed. The most commonly used multi-state model is the Markov process model, which assumes that the probability of transition between disease states depends only on the elapsed time between observations [33]. To accommodate more complex multi-state disease processes which may feature time-varying transition probabilities, nonhomogeneous Markov models or semi-Markov models can be used (see Chapter 4 of [17]).

In this chapter, we review basic properties of some commonly used multi-state models with a focus on models that are appropriate for panel observation in biomedical/biological applications. We introduce Bayesian estimation methods for these models and demonstrate their use in two longitudinal studies of disease progression[1].

18.2   Multi-State Models: Background and Notation

In this section, we introduce notation and review a few multi-state models commonly used for describing the natural history of disease. We will generally use $\{Z(t), t \in [0, \infty)\}$ to denote a stochastic process taking a finite set of states $\mathcal{S} = \{1, 2, \ldots, m\}$. In our applications $Z(t)$ represents the disease state of a patient at time t.

---

[1]We thank Dr Jesse Fann for allowing us to use the delirium data. Delirium data collection was supported by grant No. RPG-97-035-01-PBR from the American Cancer Society and by a grant from the University of Washington Royalty Research Fund. This work was partially supported by grant RO1 CA160239 from the National Cancer Institute.

18.2.1 Markov processes

In a Markov process, the state of the process at a future time $t + s$ depends on the history of the process only through the state at present time, t. Formally, the process is *Markov* if for all $s, t \geq 0$ and for every $i, j \in \mathcal{S}$, then

$$P(Z(t+s) = j | Z(t) = i, Z(u) = z(u), 0 \leq u < s) = P(Z(t+s) = j | Z(t) = i)$$
$$\doteq p_{ij}(t, t+s).$$

Denote the initial distribution of the process by $\boldsymbol{\phi} = (\phi_1, \ldots, \phi_m)$ where $\phi_i \doteq P(Z(0) = i), \forall i \in \mathcal{S}$ and such that $\phi_i \geq 0$ for each i and $\sum_{i \in \mathcal{S}} \phi_i = 1$. Conditional on $\boldsymbol{\phi}$, the process can be characterized by the matrix of transition probabilities $\mathbf{P}(t, t+s) = [p_{ij}(t, t+s)]$ for $s, t \geq 0$. Alternatively, the process can be characterized by the matrix $\mathbf{Q}(t) = [q_{ij}(t)]$ of transition intensities, defined for $t \geq 0$ as $q_{ij}(t) \doteq \lim_{s \to 0} \frac{p_{ij}(t, t+s)}{s}$ for all $i \neq j$ with $i, j \in \mathcal{S}$ and $q_{ii}(t) \doteq -\sum_{j \neq i} q_{ij}(t)$ for all $i \in \mathcal{S}$. The $q_{ij}(\cdot)$ are also known as *cause-specific hazard functions* [31]. It follows from this definition that $\mathbf{P}(t, t) = \mathbf{I}$.

In the more general case where transition probabilities depend on both the elapsed time, s, and the chronological time, t, the process is a *nonhomogeneous Markov process*. If the transition probabilities depend only on the elapsed time s and not on the chronological time t, then the Markov process is *homogeneous* and $p_{ij}(t, t+s) \equiv p_{ij}(s)$ and $q_{ij}(t) \equiv q_{ij}$. As a consequence of the Markov property and homogeneity, the transition probability matrix $\mathbf{P}(s)$ satisfies the *Chapman-Kolmogorov equation*, that is, $p_{ij}(s) = \sum_{k \in \mathcal{S}} p_{ik}(u) p_{kj}(s - u)$, $0 < u < s$, or equivalently in matrix notation: $\mathbf{P}(s) = \mathbf{P}(u)\mathbf{P}(s - u)$, $0 < u < s$. From the Chapman-Kolmogorov equations we can derive the forward and backward equations: $\frac{d}{ds}\mathbf{P}(s) = \mathbf{Q}\mathbf{P}(s) = \mathbf{P}(s)\mathbf{Q}$, which can be solved to yield $\mathbf{P}(s) = \exp(\mathbf{Q}s) \doteq \sum_{n=0}^{\infty} \frac{\mathbf{Q}^n s^n}{n!}$, where $\mathbf{Q}^0 \equiv \mathbf{I}$. This last equation makes clear that for a homogeneous Markov process, the matrix of transition intensities and the matrix of transition probabilities give equivalent characterizations of the process.

If the domain of the Markov process is the set of nonnegative integers $\mathbb{Z}^* = \{0, 1, 2, \ldots\}$ rather than a real interval, then the process $\{Z_t, t \in \mathbb{Z}^*\}$ is called a *Markov chain* and the Markov assumption, for $t \in \mathbb{Z}^*$, reduces to

$$P(Z_{t+1} = j | Z_t = i, Z_{t-1} = z_{t-1}, \ldots, Z_0 = z_0) = P(Z_{t+1} = j | Z_t = i) \doteq p_{ij}(t).$$

A Markov chain is uniquely characterized by its initial distribution $\boldsymbol{\phi}$ and transition probability matrix $\mathbf{P}(\cdot)$. A Markov chain is *homogeneous* if the transition probabilities do not depend on chronological time so that $p_{ij}(t) \equiv p_{ij}$. For a homogeneous Markov chain, p_{ij} gives the probability of making a transition from state i to state j in one step. The matrix of n-step transition probabilities, denoted $\mathbf{P}^{(n)}$, is given by $\mathbf{P}^{(n)} = \mathbf{P}^n$, the matrix of one-step transition probabilities raised to the nth power. This follows from the discrete-time version of the Chapman-Kolmogorov equations.

Examination of the transition probability matrix $\mathbf{P}(\cdot)$ yields insights into the behavior of the Markov chain. Considering a homogeneous Markov chain,

FIGURE 18.1

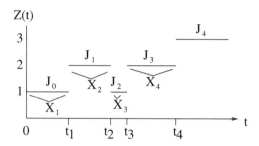

Example of a J,X process, showing relationships between $\{(J_n, X_n), \ n \geq 0\}$ and associated process $Z(\cdot)$. Under the semi-Markov assumption, the J,X process is a Markov renewal process, and $Z(\cdot)$ is the associated semi-Markov process.

if $p_{ii} = 1$, then state i is called an *absorbing state* ([6, p. 114]; [27, p. 86]). A state j is said to be *accessible* from state i if $p_{ij}^n > 0$ for some $n \geq 0$ [33, p. 168]. States of a Markov process may be classified in an analogous way by examining the transition intensity matrix, $\mathbf{Q}(\cdot)$.

18.2.2 Markov renewal processes and semi-Markov processes

The framework of Markov renewal processes in which the evolution of the process is separated into its sequence of states and of sojourn times, the length of time between two consecutive transitions, facilitates relaxation of the Markov assumption.

Consider a discrete two-dimensional stochastic process called a *J,X process*, $(J, X) = \{(J_n, X_n), \ n \geq 0\}$, where the J-process represents the states visited and the X-process represents the sojourn times in each of those states . Hence $X_n \geq 0$ and $J_n \in \mathcal{S}, \mathcal{S} = \{1, 2, \ldots, m\}$, for each $n \geq 0$, and by convention $X_0 = 0$ almost surely. The process begins in state J_0, where it remains for time X_1 before making a transition to state J_1 and, in general, remains in state J_n for time X_{n+1} before making a transition to a state J_{n+1}. The time at which the nth transition occurs is $T_n \doteq \sum_{r=1}^{n} X_r$, $n \geq 1$ and $T_0 = 0$ (see Figure 18.1).

Define the initial distribution of the process as $\boldsymbol{\phi} = (\phi_1, \ldots, \phi_m)$ where $\phi_i = P(J_0 = i)$ with $\phi_i \geq 0$ for all i and $\sum_{i \in \mathcal{S}} \phi_i = 1$. Moreover, define $N(t) = \sup\{n \geq 0 : \ T_n \leq t\}$, the number of transitions made during $[0, t]$. The semi-Markov assumption is that, for all $s \geq 0$,

$$P(J_n = j, X_n \leq s | (J_k, X_k), \ k = 0, 1, \ldots, n - 1)$$
$$= P(J_n = j, X_n \leq s | J_{n-1}, T_{n-1}, n - 1) \doteq^{(n-1)} K_{J_{n-1}j}(T_{n-1}, s)$$

for $n \geq 1$ and $j \in \mathcal{S}$, where for each i and j, the *kernel* $^{(n-1)}K_{ij}(\cdot, \cdot)$ is a real-valued function satisfying $^{(n-1)}K_{ij}(t, t + s) = 0$ for $s \leq 0$ or $t \leq 0$ and $\lim_{s \to \infty} \sum_{j \in \mathcal{S}} {}^{(n-1)}K_{ij}(t, t + s) = 1$ for each $i \in \mathcal{S}, t \geq 0$.

If this assumption holds, then $\{(J_n, X_n)\}$ is a *Markov renewal process* and the associated process $Z(t) \doteq J_{N(t)}$ is a *completely nonhomogeneous semi-Markov process* (refer to Figure 18.1). We can interpret this assumption as the statement that the future of the process depends on the entire history only through the current state, J_{n-1}, the elapsed chronological time, T_{n-1}, and the number of transitions between states, $n - 1$, that the process has made.

This very general formulation of a semi-Markov process includes several important special cases. If the kernel $[^{(n-1)}K_{ij}(\cdot,\cdot)]$ does not depend on the number of transitions, then the process is *nonhomogeneous semi-Markov* [15], and if additionally the kernel does not depend on the chronological time t, then the process is *homogeneous semi-Markov* [25, 26, 34]. We discuss the latter case in more detail. Specifically, if for all $s \geq 0$,

$$P(J_n = j, X_n \leq s | (J_k, X_k), \ k = 0, 1, \ldots, n-1) = P(J_n = j, X_n \leq s | J_{n-1})$$
$$\doteq K_{J_{n-1}j}(s),$$

where each $K_{ij}(\cdot)$ is a real-valued function satisfying $K_{ij}(s) = 0$, for $s \leq 0$ and $\lim_{s\to\infty} \sum_{j\in\mathcal{S}} K_{ij}(s) = 1$, for each $i \in \mathcal{S}$, then the associated process $Z(\cdot)$ is a homogeneous semi-Markov process. The assumption for such a process is that the future evolution depends on the history only through the current state of the process and the elapsed time in this state. This assumption is much weaker than the homogeneous Markov assumption. From this point onward we consider only homogeneous semi-Markov processes, referred to in many sources as simply semi-Markov processes, and we assume that each element of the *semi-Markov kernel* $K_{ij}(\cdot)$ is absolutely continuous. A semi-Markov process can be uniquely characterized by its initial distribution ϕ and the kernel $\mathbf{K}$.

Let us further examine the marginal process $\{J_n, n \geq 0\}$ and the process $\{X_n, n \geq 0\}$ conditional on $\{J_n, n \geq 0\}$, which we call the J- and X-processes, respectively. Using the semi-Markov assumption and the Lebesgue Monotone Convergence Theorem [32], we can show that the J-process is a homogeneous Markov chain, called the *embedded Markov chain* of the semi-Markov process that is governed by the transition probability matrix defined by $p_{ij} \doteq \lim_{s\to\infty} K_{ij}(s)$ for all $i, j \in \mathcal{S}$.

To discuss the X-process, define, for $s \geq 0$, the following functions:

$$F_{ij}(s) \doteq \begin{cases} \frac{K_{ij}(s)}{p_{ij}}, & p_{ij} > 0; \\ 1_{(s\geq 1)}, & p_{ij} = 0, \end{cases} \quad \text{for each } i, j \text{ and } \ H_i(s) \doteq \sum_{j\in\mathcal{S}} K_{ij}(s) \text{ for each } i.$$

We can show that, for $s \geq 0$, $F_{ij}(s) = P(X_n \leq s | J_{n-1} = i, J_n = j)$ and $H_i(s) = P(X_n \leq s | J_{n-1} = i)$. These are known, respectively, as the conditional and unconditional distributions of the sojourn time in state i. We note that the above definition of $F_{ij}(\cdot)$ in the case that $p_{ij} = 0$ is arbitrary. Let $f_{ij}(\cdot)$ be the density corresponding to $F_{ij}(\cdot)$ which exists given our assumption that the semi-Markov kernel is absolutely continuous.

We can express each element of the kernel in a natural way as the product of the respective transition probability and the conditional sojourn time distribution, that is, $K_{ij}(s) = F_{ij}(s) \cdot p_{ij}$ for $s \geq 0$.

We noted previously that the semi-Markov process can be uniquely characterized by $(\phi, \mathbf{K})$, and the above argument makes clear [17] that it can also be characterized by $(\phi, \mathbf{P}, \mathbf{F})$. For a process with a single absorbing state, the time to failure can be characterized by the cumulative distribution function $F(\cdot)$ or the hazard function $h(\cdot)$. By analogy, if X_n is the sojourn time in state $J_{n-1} = i$ before going to state $J_n = j$, then we can characterize the distribution of X_n by F_{ij} or equivalently by the conditional hazard function, $h_{ij}(\cdot)$, for each $i \neq j \in \mathcal{S}$, defined for $s \geq 0$ as:

$$h_{ij}(s) \doteq \lim_{\Delta s \downarrow 0} \frac{1}{\Delta s} P(s \leq X_n < s + \Delta s | J_{n-1} = i, J_n = j, X_n \geq s).$$

Hence we can alternatively characterize the semi-Markov process by $(\boldsymbol{\phi}, \mathbf{P}, \mathbf{h})$.

Finally we note the relationship between the two processes under consideration. In a semi-Markov process, the form of the conditional sojourn time has no restriction. However, a Markov process is the special case of a semi-Markov process in which the distribution of the conditional sojourn times in each state is (a) exponential, and (b) does not depend on the next state to be visited.

18.3   Estimation and Inference: A Review

In this section, we first summarize the existing approaches to estimation of multi-state processes under continuous observation and then examine available approaches to estimation under panel observation. Under panel observation, a Markov model has just enough structure so that the transition intensities can still be estimated [18]. However, in a semi-Markov model the transition intensities depend on the elapsed time in the current state, which is unknown. Estimation is therefore less tractable and methods for a general process under panel observation do not exist. Thus, we examine methods that have been developed for processes under specific assumptions for the sequence of allowed transitions, defined as state models, and under various other assumptions. As we shall see, most of the existing literature is based on maximum likelihood estimation. Thus, we end this section with a discussion of how Bayesian approaches can be implemented for estimating disease natural history.

18.3.1   Methods for continuously observed processes

When the exact time of transition between states is known, maximum likelihood (ML) estimators for transition intensities in the Markov process are given by $\hat{q}_{ij} = \frac{N_{ij}}{T_i}$, where N_{ij} is the number of transitions observed from state i to state j and T_i is the total time spent in state i [2]. Large sample properties for these estimators have been developed by [2] and [4]. Likewise, when a semi-Markov model is used for a continuously observed process, the parameters corresponding to the embedded Markov chain are estimated via the sample proportions $\hat{p}_{ij} = \frac{n_{ij}}{n_i}$, where n_{ij} is the number of observed transitions from state i to j and n_i is the total number of transitions from i to any state [3]. In the semi-Markov model, a variety of approaches to estimating the distributions of the sojourn times conditional on the sequence of visited states exist: fully parametric [42], piecewise exponential [29], or nonparametric [20, 41]. Moreover, tests have been developed to examine the Markov or semi-Markov assumption. [5] consider an illness-death model (see Figure 18.2b) and propose goodness-of-fit statistics for testing the hypotheses that the underlying process is either (1) homogeneous semi-Markov or (2) nonhomogeneous Markov, and derive asymptotic distributions of the statistics.

Parametric and semiparametric regression approaches have been used to incorporate covariates into Markov process models [37]. In a homogeneous Markov model we can specify a proportional hazards type model, $q_{jk} = q_{jk0}g(\mathbf{W}'\boldsymbol{\beta}_{jk})$, where q_{jk0} is the baseline transition intensity, $\mathbf{W}$ is a vector of covariates, and $g(\cdot)$ is a positive-valued function. Typically, $g(\cdot)$ is taken to be the exponential function to ensure positivity of the transition intensities. This model has been previously implemented in a number of applications. [21] takes this approach in modeling hepatic cancer, while [16] applies it to estimate the effect of age on rates of progression of abdominal aortic aneurysm.

In semi-Markov models, covariates may be included in a regression model via the embedded Markov chain, conditional sojourn distributions, or both. Regression modeling of the embedded Markov chain is often based on the multinomial logistic regression model. Conditional sojourn times are often modeled via the proportional hazards model. [24] give an overview of methods for modeling sojourn times that account for the presence of covariates and possible dependencies among sojourn times within a subject. They discuss the use of random effects to deal with unexplained inter-subject or temporal variability, noting that these may introduce computational difficulties and therefore should be avoided when the process is incompletely observed.

There are several classes of methods that apply to specific state models. Some methods are built on the assumption that the embedded Markov chain of the process is *ergodic*, which implies in particular that an absorbing state such as death cannot exist [33]. By contrast, other methods assume that the underlying process is progressive. [41] approach semi-Markov processes in a counting process framework and show that, under some assumptions, a progressive semi-Markov process can be transformed via a random function of the chronological time into the *multiplicative intensity model* introduced by [1]. [41] then consider a particular progressive state model and establish asymptotic properties of the estimator of the probability of being in one of the states.

Although a number of methods have addressed censoring, most have focused on right-censoring in the final state or left-censoring in the initial state [e.g., 22]. Extending these methods to a panel observation scheme has been elusive.

Finally, the Bayesian approach has been utilized to estimate Markov models [30, 35] in biomedical and biological applications. We have not identified Bayesian applications using standard semi-Markov models, however, for modeling disease processes.

18.3.2   Panel data: Markov models

In a seminal paper, [18] propose a method to estimate the instantaneous transition probabilities of a general multi-state process under panel observation assuming the process is Markov. Using the fact that the transition probability and transition intensity matrices are related via $\mathbf{P}(s) = \exp(\mathbf{Q}s) \doteq \sum_{r=0}^{\infty} \frac{\mathbf{Q}^r s^r}{r!}$ for $s \geq 0$ for a homogeneous Markov process, the authors (see [18]) propose an efficient scoring procedure to estimate $\mathbf{Q}$ via the ML method. Specifically, if subjects are observed at times $t_0, t_1, \ldots, t_m$, and if $\mathbf{Q}$ depends on $\boldsymbol{\theta}$ then the likelihood of $\boldsymbol{\theta}$ is given by $L(\boldsymbol{\theta}) = \left[\prod_{l=1}^{m} \prod_{i,j \in \mathcal{S}} p_{ij}(t_l - t_{l-1})^{n_{ijl}} \right]$ where n_{ijl} is the number of subjects who are observed in state i at t_{l-1} and state j at t_l. The closed-form expressions of $\mathbf{P}(s)$ as well as $\frac{\partial}{\partial \theta_u} \mathbf{P}(s)$ enable the application of a scoring rule involving only first derivatives to carry out inference about $\boldsymbol{\theta}$. The algorithm is extended to allow the transition rates to depend on covariates [18].

[14] extends this method to nonhomogeneous Markov models via a transformation of the time-scale. If we assume that there exists a transformation of chronological time that yields an operational time-scale, $h(t)$, on which the process is homogeneous with matrix of transition intensities $\mathbf{Q}_0$, then $\mathbf{P}(t_1, t_2) = \mathbf{P}(h(t_2) - h(t_1)) = e^{\mathbf{Q_0}(h(t_2) - h(t_1))}$. This facilitates estimation by allowing us to use standard estimation approaches for homogeneous Markov processes. Moreover, we can simultaneously estimate parameters of the time transformation, $h(t)$, and the homogeneous transition intensity matrix that exists on the transformed time-scale, $\mathbf{Q}_0$. The time transformation function represents an alternative time-scale for the process and thus must be positive and nondecreasing, that is, $h(t; \theta) > 0$ and $dh(t; \theta)/dt > 0$, for all $t > 0$. This approach has been shown to approximate nonhomogeneous processes well even if the assumption of homogeneity on the alternative time-scale is not met [14].

At their core these methods rely on the Markov assumption, and hence sojourn times in each state are modeled as exponential, where the parameters may depend on various factors. However, many disease processes are observed to progress in a way that exhibits non-constant hazard [19, 42]. Hence, methods that do not rely on the Markov assumption are needed.

18.3.3   Panel data: Progressive semi-Markov models

Under intermittent observation of subjects, the sequence of disease states and corresponding sojourn times are not necessarily known. The inherent missing information in panel data makes estimation of semi-Markov processes more challenging.

These difficulties can be overcome in some state models. In a simple progressive model with m states (see Figure 18.2a), which gives rise to chain-of-events data, events are assumed to occur in a prescribed sequence . This implies that the probability matrix of the embedded Markov chain is degenerate with $p_{ij} = 1$ for $j = i+1$, $i = 1, \ldots, m-1$. Thus, for a simple progressive

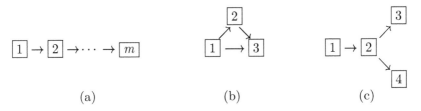

FIGURE 18.2
Examples of types of state models for which estimation for panel data is simplified:
(a) simple progressive, (b) illness-death model, (c) progressive with competing risks.

model, it remains only to estimate the sojourn time distribution in each of
these states. The assumption of absolute continuity of each conditional sojourn time distribution leads to a likelihood with convolution products of the
conditional sojourn densities $f_{i,i+1}(\cdot;\boldsymbol{\theta_i})$ and survival distributions $S_i(\cdot;\boldsymbol{\theta_i})$
for $i = 1,\ldots, m-1$. Specifically, expressing the data as in [18], or equivalently
in the "sufficient" form through $\mathbf{t}$, a vector of length $2(m-1)$, we represent
successive pairs of components for observation times preceding and following
a time of transition where each component may or may not be defined, depending on how each subject was censored. With this notation, the likelihood
of the parameters given panel observations on N subjects is given by

$$L(\boldsymbol{\theta_1},\boldsymbol{\theta_2}|\mathbf{t_1},\ldots,\mathbf{t_N}) \quad = \quad \prod_{i=1}^{N} L_1^{(1-\delta_i)\cdot(1-\epsilon_i)} \cdot L_2^{\delta_i\cdot(1-\epsilon_i)} \cdot L_3^{(1-\delta_i)\cdot\epsilon_i} \cdot L_4^{\delta_i\cdot\epsilon_i},$$

with likelihood components $L_1 = \int_{t_3}^{t_4}\int_{t_1}^{t_2} f_{12}(u_1) \cdot f_{23}(u_2 - u_1)du_1 du_2$, $L_2 =$
$\int_{t_1}^{t_4}\int_{t_1}^{u_2} f_{12}(u_1)\cdot f_{23}(u_2 - u_1)du_1 du_2$, $L_3 = \int_{t_3}^{\infty}\int_{t_1}^{t_2} f_{12}(u_1)\cdot S_{23}(u_2 - u_1)du_1 du_2$
and $L_4 = \int_{t_1}^{\infty} S_{12}(u)du$, where δ_i and ϵ_i are indicators that subject i was
not observed in states 2 and 3 respectively. That is, L_1 is the contribution
of a subject who was observed in all three states; L_2 represents a subject
observed in states 1 and 3 only; and L_3 and L_4 represent subjects who were
right-censored in state 2 and state 1 respectively.

Considering the case of a three-state simple progressive model, [8] propose a nonparametric approach to estimate the conditional distributions of
the sojourn times in states 1 and 2. Their approach, an extension of the self-
consistency algorithm of [40] for univariate survival data, involves modeling
the two sojourn time distributions as discrete random variables. Assuming
the process enters state 1 at time zero, let Y_1 and $Z = Y_1 + Y_2$ denote the
transition times into states 2 and 3 respectively, and (Y_L, Y_R, Z_L, Z_R) be the
"sufficient data" for a single realization of the process, i.e. the observation
times immediately preceding and following the two transitions. This notation
is similar to $\mathbf{t}$ introduced above. The authors choose locations of the mass
points of Y_1 and Y_2, $0 \le y_{11} < \cdots < y_{1r}$ and $0 \le y_{21} < \cdots < y_{2s}$ respectively, and note that the observation (Y_L, Y_R, Z_L, Z_R) uniquely determines a

set of "admissible values" of (y_{1j}, y_{2k}). To recast the data in this format they define α_{jk} as the indicator that (y_{1j}, y_{2k}) is an admissible value of (Y_1, Y_2). With $w_{1j} = P(Y_1 = y_{1j})$ and $w_{2k} = P(Y_2 = y_{2k})$, the likelihood is given by $L(\mathbf{w_1}, \mathbf{w_2}) = \prod_{i=1}^{N} \left(\sum_{j=1}^{r} \sum_{k=1}^{s} \alpha_{jk}^i w_{1j} w_{2k} \right)$, where $\mathbf{w_1} = (w_{11}, \ldots, w_{1r})'$ and $\mathbf{w_2} = (w_{21}, \ldots, w_{2s})'$. It can be shown that the self-consistent estimate is equivalent to that using the *EM algorithm* [9].

In parallel with [8], [13] extends the algorithm of [40] to a three-state simple progressive model, but assumes the underlying process is nonhomogeneous Markov rather than homogeneous semi-Markov. [13] additionally assumes that entry into the third state is either observed exactly or right-censored. She applies her method to the HIV application in [8] and obtains similar results except in the inference regarding the sojourn time in the infected state before developing AIDS symptoms.

Although the method of [8] avoids imposing distributional assumptions on the sojourn times in each state, it has several drawbacks. First, it implicitly assumes that each subject is observed at least once in every state. Thus, a subject who is observed in only states 1 and 3 would need to be discarded. This leads to biased estimation where the magnitude of the bias increases with the interval between observations. Second, the method involves discretizing the two sojourn times. This means that decisions must be made about the location of the mass points of Y_1 and Y_2. Though certain guidelines may be used to avoid lack of identifiability and loss of information, the particular dataset under consideration may play a big role in the choice of locations of the mass points of $(\mathbf{y_1}, \mathbf{y_2})$.

A slightly more general state model that arises in many applications is the *illness-death* model shown in Figure 18.2b . This model is useful for studying an incurable, potentially fatal disease. Beginning in a state of good health, subjects at risk may progress to illness; may die from another cause; or may die from the illness. When the third state represents death, it may be assumed that transitions to this state are observed exactly. Note that, depending on the disease process, the three states may not occur in a prescribed sequence. For example, [5] consider modeling the risk of breast cancer over time among women who may or may not have had benign breast disease. Though they assume the transition times are known exactly, in this application each of these two transitions may be subject to interval censoring. Developing methods for the illness-death model is challenging because a subject with observed trajectory $1, 1, 3$, for example, may or may not have visited state 2. Finally, another example of a progressive state model which accounts for multiple absorbing states, or competing risks, is shown in Figure 18.2c. Methods for estimation and inference exist for more general progressive state models, but are often tailored to an application and impose assumptions, specific to the application, to simplify estimation. We review a few cases below.

[11] consider a five-state progressive disease process with competing risks to model patients' natural history following kidney transplantation. The authors model sojourn times in each state as exponentiated Weibull, a parametric form

FIGURE 18.4
State model of [19] investigating cervical intraepithelial neoplasia (CIN) and human papillomavirus (HPV) infection. State 3 represents CIN diagnosis after a visit in which the patient was not infected with HPV (state 1), while state 4 represents a CIN diagnosis after a visit in which the patient was HPV infected (state 2).

FIGURE 18.3
State model of [12]. State numbers indicate (1) baseline value of creatinine clearance (CL); (2) decreased CL; (3) return to dialysis; (4) death.

that allows the conditional hazard function to be nonmonotonic. However, the assumption that subjects are observed in each visited state implies that estimation of the embedded Markov chain is trivial.

If the number of states in the progressive process is small, all possible trajectories can be accounted for when deriving the likelihood. [12] carry out this procedure for the state model shown in Figure 18.3. They assume that the times of each patient's entry into state 1 and 3 or 4 (if applicable) are known exactly. The time of entry into state 2 is interval-censored, and a patient who is not observed in state 2 may or may not have entered state 2. Similar to [11], the authors impose exponentiated Weibull sojourn times . They build the likelihood from each of the four possible trajectories using convolution products to deal with censoring of state 2 to obtain maximum likelihood estimates. The authors additionally incorporate covariates and derive a goodness-of-fit statistic to test homogeneity of the semi-Markov process. Their method for modeling progressive disease could be generalized somewhat: it could be adapted for other fairly simple state diagrams and, as they note, it could be modified to handle interval-censored absorbing states. However, numerical maximization of the likelihood is quite computationally expensive when the likelihood contributions involve more than two interval-censored times.

18.3.4 Panel data: General semi-Markov models

Several authors have developed methods for modeling non-progressive processes, but have imposed other strong assumptions. Specifically, [19] propose a method to model a non-progressive homogeneous semi-Markov process subject to interval censoring as well as misclassification of states. Their model, however, assumes that transition intensities from at least one state are duration-independent. This assumption implies the existence of a set of states for which the Markov assumption applies and allows for great simplification of the likelihood function (see Figure 18.4). Some authors have utilized semi-Markov models for nonprogressive processes with just two states as shown in Figure

(a)   (b)

FIGURE 18.5
State models considered by [28] in a study of duration of HPV infection. In the primary method, states (1) and (2) represent the uninfected and infected states respectively, as shown in (a). In the extension, the uninfected state was split into never infected (1*) and previously infected (1), as shown in (b).

18.5a. Since the embedded Markov chain has deterministic transition probabilities $p_{ij} = I_{\{i=j\}}$, $i, j \in \{1, 2\}$, the task at hand is to estimate the sojourn time distributions. [28] propose an approach to estimate the duration of HPV infection given panel observations of infection status assuming that: (1) all subjects are initially in state 1, (2) the Markov assumption is satisfied when the process is in state 1, and (3) subjects are observed at prespecified, equally-spaced, common visit times (e.g. every six months). Given these assumptions, the sojourn time in state 1 is a geometric random variable with point masses at the scheduled visit times, say $1, 2, \ldots, n_t$, where n_t is the number of possible observed time points after study entry. Let p_{12} denote the associated parameter, so that the probability of spending t units of time in state 1 before transitioning to state 2 is given by $p_{12} \cdot (1 - p_{12})^{t-1}$ for $t = 1, 2, \ldots$. The sojourn time in state 2 is a discrete random variable with point masses at these common scheduled visit times, so that the probability of spending time t in state 2 before transitioning to state 1 is $p_{21}(t)$ for $t = 1, 2, \ldots, n_t$ with $\sum_{t=1}^{n_t} p_{21}(t) = 1$. Letting $\mathbf{p} \doteq \{p_{12}; p_{21}(1), \ldots, p_{21}(n_t)\}$ and imposing the above restriction on $p_{21}(1), \ldots, p_{21}(n_t)$, each individual's likelihood contribution is $\pi_{y_0, \ldots, y_{n_t}}(\mathbf{p}) = \left\{ \left[\prod_{k=1}^{m} p_{j_{k-1}, j_k}(x_k) \right] \cdot S_{j_m}(x_m+) \right\}$ where $j_0, j_1, \ldots$ is the sequence of visited states, $x_0 = 0, x_1, x_2, \ldots$ is the sequence of corresponding sojourn times, $y_0, y_1, \ldots$ is the sequence of observed states at each time point, m is the number of states visited by visit n_t, x_m+ is the right-censored time spent in the final state, $S_{J_n}(\cdot)$ is the survival function in state J_n, and $p_{J_{n-1}J_n}(t) = P(J_n = j, X_n = t | J_{n-1} = i)$.

The authors allow for isolated missing visits assuming they occur at random (MAR assumption), but exclude subjects with two or more consecutive missing visits. Under the MAR assumption, the likelihood for N subjects is given by $L(\mathbf{p}) = \prod_{i=1}^{N} \left[\sum_{y_{n_t} \in \{0,1\}} \cdots \sum_{y_0 \in \{0,1\}} \alpha^i_{y_0, \ldots, y_{n_t}} \cdot \pi_{y_0, \ldots, y_{n_t}}(\mathbf{p}) \right]$, where $\alpha^i_{y_0, \ldots, y_{n_t}}$ is the indicator that $\{y_0, \ldots, y_{n_t}\}$ is an "admissible" observation in the sense of [8], given the possibility of missing observations at scheduled visit times. [28] maximize the log likelihood over the parameter space via a quasi-Newton algorithm.

[28] extend this method to make a distinction between subjects who are never infected during the course of follow-up and those who become infected

and later clear the infection while on the study, to allow for the possibility that previous infection influences the rate by which subjects transition into the infected state. Specifically, they consider the three-state model shown in Figure 18.5b. They assume that subjects are in state 1* at time zero, and additionally that the sojourn times in states 1 and 2 are exponential. They extend this method to relax the assumption that state 2 is Markov, but note that relaxing this assumption for state 1* would be challenging because times in the first observed state are subject to left censoring.

The primary approach of [28] is a self-consistency algorithm, similar to the method of [8]. Their approach has advantages including not imposing distributional assumptions on the sojourn time in the infected state, but has stringent assumptions about the observation scheme: since it models the process as a discrete-time semi-Markov chain, the method in its present form requires that subjects are observed at a common set of evenly-spaced visits. As a result of modeling time discretely, this method is subject to some of the same issues as [8]. [28] compare their method with that of [19], and note that their own method imposes no parametric assumptions or minimum value, above zero, on the sojourn time in the infected state. However, the discrete nature of the method, itself, imposes a minimum value on this sojourn time since there is no point mass at time zero.

Alternatively, some authors estimate semi-Markov processes by embedding a latent Markov process. This artifact simplifies the likelihood evaluation. [7] consider modeling herpes simplex virus type 2 (HSV-2), which is characterized by recurrent lesions. However, the number of lesions is not observable; only the presence or absence of lesions, known as the viral shedding status, can be ascertained. The viral shedding status itself is often asymptomatic and is therefore observed only at clinic visits, giving rise to panel data. The latent number of lesions at a point in time can be considered a *birth-death process*, a Markov process in which the states, $0, 1, 2, \ldots$ represent the size of the population at each point in time (see Figure 18.6a) . If states $1, 2, 3, \ldots$ of this homogeneous Markov process are collapsed into a single state, denoted $1+$, the corresponding process, denoted $Z(\cdot)$, is semi-Markov, as the sojourn time in state $1+$ now depends on the elapsed time in this state (Figure 18.6b). The observed viral shedding status at each point in time is an induced semi-Markov model.

(a) $\boxed{0} \leftrightarrows \boxed{1} \leftrightarrows \boxed{2} \leftrightarrows \cdots$ hidden Markov process, $W(\cdot)$

(b) $\boxed{0} \leftrightarrows \boxed{1+}$ semi-Markov process, $Z(\cdot)$

FIGURE 18.6
State models considered by [7]. The unobservable number of recurrences at time t is modeled as a birth-death process (a), while the viral shedding status is modeled as the corresponding semi-Markov process (b) defined by collapsing states $1, 2, 3, \ldots$ of the birth-death process.

[7] express the panel data likelihood via a hidden Markov model, using random effects to accommodate heterogeneity across individuals with means defined as a function of covariates and carry out Bayesian inference on the model parameters. [38] also consider a hidden Markov process which accounts for classification error in the assessment of the state at each observation time.

18.3.5   Bayesian approach to multi-state models

Our review of the literature indicates that while methods for estimating Markov models under both continuous and panel observation are well established, that is not the case for semi-Markov models, especially with panel data. Most methods are tied to particular applications and do not generalize to more complex state models. Further, they often require additional modeling assumptions to simplify the likelihood evaluation or to allow for parameter identifiability. Finally, the vast majority of methods for multi-state models are frequentist.

The Bayesian approach offers a viable alternative to estimating multi-state models. At the core of the Bayesian approach one expresses the uncertainty about all unknowns with prior distributions. Thus, Bayesian estimation of homogeneous Markov processes proceeds by assuming priors on the initial state distribution ϕ of the process as well as on the transition intensities q_{ij}. Oftentimes, we express the transition intensities as dependent on covariates via a proportional hazards formulation as discussed in subsection 18.3.1 in which case we assume priors on the regression coefficients β_{jk} instead. A similar approach can be used for nonhomogeneous Markov processes as we illustrate in subsection 18.4.1. Likewise, Bayesian estimation of continuously observed semi-Markov processes requires priors for the initial state distribution, but also for the transition probabilities of the embedded Markov chain and for the parameters governing the sojourn time distributions. Alternatively, when interest lies in estimating the effect of a given set of risk factors on the disease state transitions or sojourn times, then one assigns priors to the corresponding regression parameters. When the semi-Markov process is, however, observed intermittently, we are faced with the additional challenge that the full trajectory of the disease process and the durations in each state are unknown. One approach that has been proposed to handle this situation uses data augmentation to model the unobserved trajectories and sojourn times as latent variables, and proceeds with estimation in a Bayesian framework [23]. We illustrate this method in subsection 18.4.2.

Generally the posterior distributions of the model parameters are not available in closed form, but estimation can be accomplished using Markov chain Monte Carlo (MCMC) methods. While some computational challenges may arise, for example, in devising algorithms that allow for good mixing, the Bayesian approach offers several advantages in multi-state modeling. It allows us to incorporate available knowledge about disease processes via expert information, for example, which gives us the potential to address identifiability issues that arise given the sparsity of information in panel data. Furthermore, it provides exact inference that does not rely on asymptotic results. Although the Bayesian approach does not eliminate methodological issues arising with panel observation, it can enhance our ability to estimate models which are intractable under classical approaches.

18.4   Bayesian Multi-State Models: Applications

In this section, we illustrate the use of the Bayesian approach to multi-state modeling in the context of two applications to specific disease processes.

18.4.1   Nonhomogeneous Markov models: Modeling delirium

We first provide an illustration of nonhomogeneous Markov models using data on the course of delirium in cancer patients treated with hematopoietic stem cell transplant. Delirium is an acute neuropsychiatric condition associated with rapid onset of a change in levels of consciousness, attention, cognition, and perception. Due to the high prevalence of delirium in cancer patients, this population is of particular interest [10]. We analyzed incidence and progression of delirium in data from 90 cancer patients treated with hematopoietic stem cell transplantation collected by researchers at the Fred Hutchinson Cancer Research Center in Seattle, Washington, between 1997 and 1999. Patients were assessed for delirium using the Delirium Rating Scale (DRS) [39] on average every 2.5 days during the first thirty days following transplant. In prior studies of this cohort, delirium was treated as a dichotomous outcome and the incidence rate for delirium was assumed constant across the observation period. However, to better understand the potentially complex course of delirium we used a multi-state model that allows for both clinical and sub-clinical delirium states. The state model for the delirium disease process is presented in Figure 18.7. Sub-clinical delirium was defined as a DRS score of at least 6 and clinical delirium was defined by a DRS score of at least 12. To investigate temporal variability in onset and progression of delirium we modeled the disease via a three state nonhomogeneous model with a continuous time-scale measured in days since transplantation and using the time transformation method of [14]. Parameters of the three-state process described above can be estimated using maximum likelihood or Bayesian methods. An analysis of these data using maximum likelihood

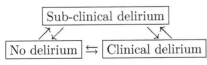

FIGURE 18.7
State model for delirium process.

methods identified a statistically significant acceleration of the disease process over time [14]. However, if we are additionally interested in allowing for subject-specific variation in the rate of acceleration of the process, maximum likelihood methods become intractable. Subject-specific variation in the rate of progression of the disease is of interest in this context because subjects included in the study are extremely heterogeneous. Variation exists in demographics, treatments, and cancer-types in these patients. Unmeasured factors and intrinsic variations in disease susceptibility and recovery may influence the course of delirium in individual patients. We can account for variation in the rate of evolution of the disease process by introducing random effects into the time-transformed nonhomogeneous Markov process. Specifically, we used a random effects model that allows for between-subject variability in the time transformation parameter. The likelihood for m subjects observed at n_i time points takes the form

$$L^{(m)}(\boldsymbol{Q}_0, \boldsymbol{\theta}) = \prod_{i=1}^{m} \left\{ P(X(u_{i1}) = x_{i1}) \prod_{j=2}^{n} \left\{ e^{\boldsymbol{Q}_0(h(u_{ij};\boldsymbol{\theta}_i) - h(u_{ij-1};\boldsymbol{\theta}_i))} \right\}_{x_{ij-1}x_{ij}} \right\},$$

where $h(u; \boldsymbol{\theta}_i)$ is a function that transforms the time-scale of the process from the observed time-scale, on which the process in nonhomogeneous, to an operational time-scale on which the process is assumed homogeneous. To allow for subject-specific variation in the rate of evolution of the process, we introduce a subject-specific time transformation parameter, $\boldsymbol{\theta}_i$.

Bayesian estimation is straightforward by introducing priors for the transition intensities and time transformation parameters. In our delirium application, we assumed independent log normal priors for the elements of $\boldsymbol{Q}_0$ and $\boldsymbol{\theta}$, such that $\pi(\log q_{0ab}) \sim \text{Normal}(\mu_q, \sigma_q^2)$, $a \neq b$ and $\pi(\log \theta_{ik}) \sim \text{Normal}(\mu, \sigma^2)$. We placed a normal prior on μ and inverse gamma prior on σ^2. Hyperparameter values for prior densities were selected according to expert information on the likelihood of observing each kind of transition. We note that normal and inverse gamma priors were chosen for convenience to guarantee that the distributions of our model parameters were defined on $\Re^{k+}$ with interpretable hyperparameters. Estimation was carried out using a hybrid Gibbs/Metropolis-Hastings MCMC simulation. We used a burn-in period of 100,000 iterations at which point chains for all parameters satisfied the Heidelberger-Welch convergence criterion at the $\alpha = 0.05$ level. We then sampled every fiftieth iteration from the chain until we had drawn 5,000 samples from the posterior distribution of the parameters.

In order to limit the dimensionality of the posterior density, we used a time transformation function with a single parameter, $h(u; \theta) = u\theta^u$. We have explored alternative functional forms for $h(u; \theta)$, including a non-parametric kernel smoother [14]. Although the non-parametric approach provides greater flexibility, a large number of finely spaced observations are required in order to successfully estimate this function. Choosing a complex functional form for $h(u; \theta)$ will lead to weak identifiability of θ if the observation scheme is sparse.

In the case of subject-specific time transformations considered here, this would require a lengthy series of observations available for each subject. Because such detailed observations were not available for this cohort, we chose a less flexible but more parsimonious representation for $h(u; \theta)$. In general, such practical considerations should guide the choice of the time transformation function.

Prior and posterior densities for transition intensities are presented in Figure 18.8. The posterior distributions for q_{13} and q_{31} are very close to their prior distributions due to the strength of the priors used for these parameters. Because few observations between these extreme states were observed, relatively stronger priors were required. Example of time transformation curves for three subjects with largest and three subjects with smallest posterior median time transformation parameters are presented in Figure 18.9. Subject-specific time transformation parameters indicate slowing of the disease process for some subjects and more rapidly evolving disease processes for others.

18.4.2   Semi-Markov models: Modeling HIV progression

We next illustrate Bayesian estimation for a semi-Markov model under panel observation. Consider the data from a retrospective cohort study of 262 patients at the Hôpital Kremlin Bicêtre and Hôpital Cœur des Yvelines in France. These patients had type A or B hemophilia and received periodic blood transfusions that were later found to be contaminated by HIV. Blood samples taken at various times allowed for intermittent retrospective assessment of HIV infection status.

[8] modeled these data using a simple progressive three-state model as depicted in Figure 18.10. They split the patients into two groups defined by the amount of blood product they had received, which we refer to as the "heavily" and "lightly" treated groups. For each of these two groups, [8] carried out separate estimation of the time to infection and the time to progression to AIDS. Of the 262 patients in the cohort, 197 were infected with HIV at the end of follow-up, 43 of whom had progressed to AIDS. All HIV infections were believed to have been caused by receiving contaminated blood. [8] divided the chronological time axis into 6-month intervals, with $Y = 1$ denoting July 1, 1978. Each interval was given a point mass, and the point masses were indexed by $1, 2, 3, \ldots$.

We carried out a Bayesian estimation of the simple progressive three-state model. The intermittent observation scheme creates a "missing" data problem since the sojourn times in each state may not be observed. We treated the true, unobservable, sojourn times as *latent variables* and used data augmentation procedures [36]. Assume, as in [8], that all subjects start in state 1 at time zero. Let the random variable $X^i \geq 0$ denote the true unobservable sojourn time in state i before proceeding to state $i+1$, where $i = 1, 2$. Let $\mathbf{X} = (X^1, X^2)$ and assume that $X^i \sim f_i$ for $i = 1, 2$, and that the densities f_1 and f_2 collectively depend on the vector of parameters $\boldsymbol{\theta}$. We observe the process periodically, giving rise to panel observations $\mathbf{Z} = (Z_0 = 1, Z_1, \ldots, Z_n)$, with $Z_i \in \{1, 2, 3\}$

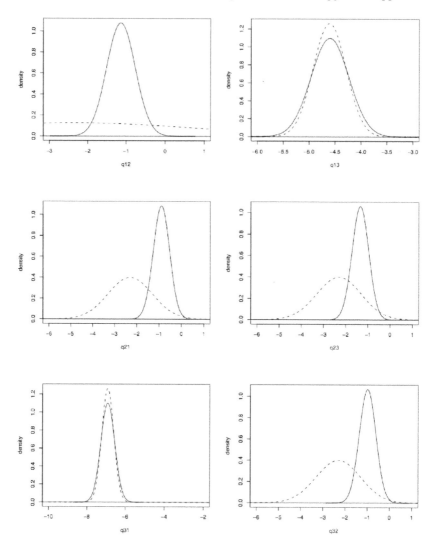

FIGURE 18.8
Prior (dashed) and posterior (solid) densities for transitions intensities (in sequence, from left to right/top to bottom: $q_{12}, q_{13}, q_{21}, q_{23}, q_{31}, q_{32}$) in time transformed random effects model for delirium data.

for each i, corresponding to observation times $\mathbf{s} = (s_0 = 0, s_1, \ldots, s_n)$. Note that in the simple progressive model, $Z_0 \leq \cdots \leq Z_n$. Note also that $\mathbf{Z}$ contains redundant information, and can be expressed equivalently as the vector $\mathbf{t} = (t_1, \ldots, t_4)$, where $t_1 = \max\{s_k : Z_k = 1\}$, $t_2 = \min\{s_k : Z_k = 2\}$, $t_3 = \max\{s_k : Z_k = 2\}$, $t_4 = \min\{s_k : Z_k = 3\}$, if each of these exists. With the above notation, the posterior and predictive equations in the data

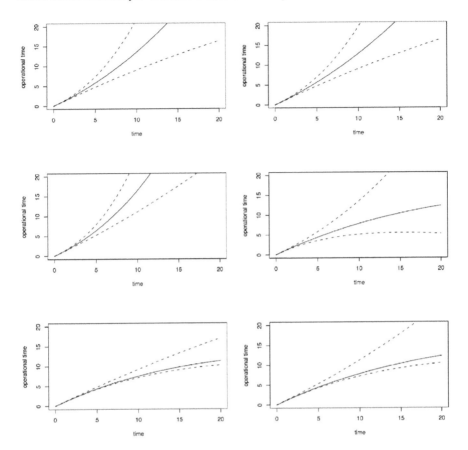

FIGURE 18.9
Estimated time transformation functions and 95% credible bands (dashed lines) for six stem cell transplantation patients with most extreme posterior median time transformation parameters from random effects time transformation model (x-axis is time, y-axis is transformed time).

$$\boxed{\text{Uninfected}} \rightarrow \boxed{\text{Infected}} \rightarrow \boxed{\text{AIDS}}$$

FIGURE 18.10
State model of [8].

augmentation algorithm [36] become

$$p(\boldsymbol{\theta}|\mathbf{t}) = \int_{\mathbf{X}} p(\boldsymbol{\theta}|\mathbf{t}, \mathbf{X})p(\mathbf{X}|\mathbf{t})d\mathbf{X} \text{ and } p(\mathbf{X}|\mathbf{t}) = \int_{\boldsymbol{\theta}} p(\mathbf{X}|\mathbf{t}, \boldsymbol{\theta})p(\boldsymbol{\theta}|\mathbf{t})d\boldsymbol{\theta}.$$

The goal is to obtain $p(\boldsymbol{\theta}|\mathbf{X}, \mathbf{t})$ and $p(\mathbf{X}|\mathbf{t}, \boldsymbol{\theta})$. The particular form of each of these expressions depends on the choice of the model for the sojourn time

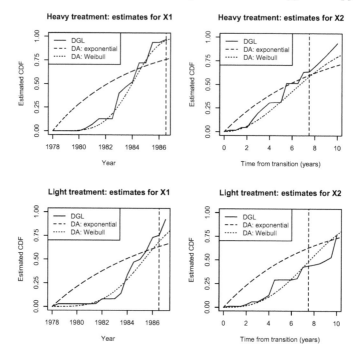

FIGURE 18.11
Estimated cumulative distribution functions (CDFs) of the sojourn times in the uninfected and infected states based on the data augmentation (DA) method. Results
are shown for heavily and lightly treated subjects (upper and lower panels, respectively), based on exponential and Weibull models of the sojourn times in each state.
Results from the method of [8] (DGL) are shown for reference. The dashed vertical lines represent the times beyond which the weights are not identifiable by the
method of [8].

distributions in each state. We complete the model by specifying priors for
parameters $\boldsymbol{\theta}$.

We applied the data augmentation approach and compared the results
to those obtained by the method of [8]. Assuming exponential and Weibull
models for the sojourn times we carried out inference about the corresponding parameters separately for the heavily and lightly treated patients . We
assumed independent noninformative uniform priors for the rate parameters
(under both exponential and Weibull models) and noninformative uniform
priors for the shape parameters (Weibull model). Figure 18.11 shows that the
Weibull model, unlike the exponential, has enough flexibility to accommodate
the shape of the hazard function in each state, as it produced an estimated
cumulative distribution function similar to that of the method of [8].

18.5   Discussion

Natural history of disease can be modeled using a variety of approaches that fall under the general framework of multi-state models, including Markov processes, nonhomogeneous Markov processes, semi-Markov processes, and hidden Markov processes. Simple approaches such as the Markov process facilitate estimation but make use of strong assumptions that may be unrealistic for certain diseases. More complex models such as semi-Markov processes may more accurately describe the disease process, but at the cost of substantially complicating estimation, especially when data arise from panel observation. In this chapter, we have discussed a few methods for estimating disease processes. In particular, we illustrated that the Bayesian approach can be used to estimate some of these more complex processes which could be intractable under the maximum likelihood framework.

Bibliography

[1] O. Aalen. Nonparametric inference for a family of counting processes. *Annals of Statistics*, 6(4):701–726, 1978.

[2] A. Albert. Estimating the infinitesimal generator of a continuous time, finite state Markov process. *Annals of Mathematical Statistics*, 33(2):727–753, 1962.

[3] T. W. Anderson and L. A. Goodman. Statistical inference about Markov chains. *Annals of Mathematical Statistics*, 28(1):89–110, 1957.

[4] P. Billingsley. *Statistical Inference for Markov Processes*. University of Chicago Press, Chicago, 1961.

[5] I.-S. Chang, Y. C. Chuang, and C. A. Hsiung. Goodness-of-fit tests for semi-Markov and Markov survival models with one intermediate state. *Scandinavian Journal of Statistics*, 28(3):505–525, 2001.

[6] C. L. Chiang. *An Introduction to Stochastic Processes and their Applications*. R.E. Krieger, New York, 1980.

[7] C. M. Crespi, W. G. Cumberland, and S. Blower. A queueing model for chronic recurrent conditions under panel observation. *Biometrics*, 61:193–198, 2005.

[8] V. De Gruttola and S. W. Lagakos. Analysis of doubly-censored survival data. *Biometrics*, 45:1–11, 1989.

[9] A. P. Dempster, N. M. Laird, and D. B. Rubin. Maximum likelihood estimation from incomplete data via the EM algorithm. *Journal of the Royal Statistical Society: Series B (Statistical Methodology)*, 39:1–22, 1977.

[10] J. R. Fann. The epidemiology of delirium: A review of studies and methodological issues. *Seminars in Clinical Neuropsychiatry*, 5(2):64–74, 2000.

[11] Y. Foucher, M. Giral, J.-P. Soulillou, and J.-P. Daures. A semi-Markov model for multistate and interval-censored data with multiple terminal events. Application in renal transplantation. *Statistics in Medicine*, 26:5381–5393, 2007.

[12] Y. Foucher, M. Giral, J.-P. Soulillou, and J.-P. Daures. A flexible semi-Markov model for interval-censored data and goodness-of-fit testing. *Statistical Methods in Medical Research*, 19:127–145, 2010.

[13] H. Frydman. A nonparametric estimation procedure for a periodically observed three-state Markov process, with application to AIDS. *Journal of the Royal Statistical Society: Series B (Statistical Methodology)*, 54(3):853–866, 1992.

[14] R. A. Hubbard, L. Y. T. Inoue, and J. R. Fann. Modeling a non-homogeneous Markov process via time transformation. *Biometrics*, 64(3):843–850, 2008.

[15] A. Iosifescu-Manu. Non-homogeneous semi-Markov processes. *Studii si Cercetari Matematice*, 24:529–533, 1972.

[16] C. H. Jackson, L. D. Sharples, S. G. Thompson, S. W. Duffy, and E. Couto. Multistate Markov models for disease progression with classification error. *Journal of the Royal Statistical Society: Series D (The Statistician)*, 52(2):193–209, 2003.

[17] J. Janssen and R. Manca. *Applied Semi-Markov Processes*. Springer Verlag, New York, 2006.

[18] J. D. Kalbfleisch and J. F. Lawless. The analysis of panel data under a Markov assumption. *Journal of the American Statistical Association*, 80(392):863–871, 1985.

[19] M. Kang and S. W. Lagakos. Statistical methods for panel data from a semi-Markov process, with application to HPV. *Biostatistics*, 8(2):863–871, 2007.

[20] E. L. Kaplan and P. Meier. Nonparametric estimation from incomplete observations. *Journal of the American Statistical Association*, 53(282):457–581, 1958.

[21] R. Kay. A Markov model for analysing cancer markers and disease states in survival studies. *Biometrics*, 42(4):855–865, 1986.

[22] S. W. Lagakos, C. J. Sommer, and M. Zelen. Semi-Markov models for partially censored data. *Biometrika*, 65(2):311–317, 1978.

[23] A. E. Laird. *Modeling a Progressive Disease Process under Panel Observation*. PhD thesis, University of Washington, 2013.

[24] J. F. Lawless and Y. T. Fong. State duration models in clinical and observational studies. *Statistics in Medicine*, 18(17-18):2365–2376, 1999.

[25] P. Lévy. Processus semi-Markoviens. *Proceedings of the International Congress of Mathematicians*, 3:416–426, 1954. Amsterdam.

[26] P. Lévy. Systèmes semi-Markoviens à au plus une infinité d'états possibles. *Proceedings of the International Congress of Mathematicians*, 2:294, 1954. Amsterdam.

[27] N. Limnios and G. Oprişan. *Semi-Markov Processes and Reliability*. Birkhäuser, Boston, 2001.

[28] C. E. Mitchell, M. G. Hudgens, C. C. King, S. Cu-Uvin, Y. Lo, A. Rompalo, J. Sobel, and J. S. Smith. Discrete-time semi-Markov modeling of human papillomavirus persistence. *Statistics in Medicine*, 30(17):2160–2170, Jul 30 2011.

[29] B. Ouhbi and N. Limnios. Nonparametric estimation for semi-Markov processes based on its hazard rate functions. *Statistical Inference for Stochastic Processes*, 2(1):151–173, 1999.

[30] S.-L. Pan, H.-M. Wu, A. M.-F. Yen, and T. H.-H. Chen. A Markov regression random-effects model for remission of functional disability in patients following a first stroke: A Bayesian approach. *Statistics in Medicine*, 26(29):5335–5353, Dec. 20 2007.

[31] R. L. Prentice, J. D. Kalbfleisch, A. V. Peterson Jr., N. Flournoy, V. T. Farewell, and N. E. Breslow. The analysis of failure times in the presence of competing risks. *Biometrics*, 34(4):541–554, 1978.

[32] R. Pyke. Markov renewal processes: Definitions and preliminary properties. *Annals of Mathematical Statistics*, 32(4):1231–1242, 1961.

[33] S. M. Ross. *Stochastic Processes*. John Wiley & Sons, Berkeley, 2nd edition, 1996.

[34] W. L. Smith. Regenerative stochastic processes. *Proceedings of the Royal Society of London: Series A*, 232:6–31, 1955.

[35] M. J. Sweeting, D. De Angelis, and O. O. Aalen. Bayesian back-calculation using a multi-state model with application to HIV. *Statistics in Medicine*, 24(24):3991–4007, Dec. 30 2005.

[36] M. A. Tanner. *Tools for Statistical Inference: Observed Data and Data Augmentation Methods*. Springer Verlag, Heidelberg, 1991.

[37] T. M. Therneau and P. M. Grambsch. *Modeling Survival Data: Extending the Cox Model*. Springer Verlag, New York, 2000.

[38] A. C. Titman and L. D. Sharples. Semi-Markov models with phase-type sojourn distributions. *Biometrics*, 66(3):742–752, 2010.

[39] P. T. Trzepacz, R. W. Baker, and J. Greenhouse. A symptom rating scale for delirium. *Psychiatry Research*, 23(1):89–97, 1988.

[40] B. W. Turnbull. The empirical distribution function with arbitrarily gropued, censored and truncated data. *Journal of the Royal Statistical Society: Series B (Statistical Methodology)*, 38(3):290–295, 1976.

[41] J. G. Voelkel and J. Crowley. Nonparametric inference for a class of semi-Markov processes with censored observations. *Annals of Statistics*, 12(1):142–160, 1984.

[42] G. H. Weiss and M. Zelen. A semi-Markov model for clinical trials. *Journal of Applied Probability*, 2(2):269–285, 1965.

19

Priors on Hypergraphical Models via Simplicial Complexes

Simón Lunagómez

Harvard University

Sayan Mukherjee

Duke University

Robert Wolpert

Duke University

CONTENTS

19.1 Introduction

It is common to model the joint probability distribution of a family of n random variables $\{X_1, \ldots, X_n\}$ in two stages: First to specify the *conditional dependence structure* of the distribution, then to specify details of the conditional distributions of the variables within that structure [3, 7]. The structure may be summarized in a variety of ways in the form of a graph $\mathcal{G} = (\mathcal{V}, \mathcal{E})$ whose vertices $\mathcal{V} = \{1, \ldots, n\}$ index the variables $\{X_i\}$ and whose edges $\mathcal{E} \subseteq \mathcal{V} \times \mathcal{V}$ in some way encode conditional dependence.

We follow the Hammersley–Clifford approach [2, 12], in which $(i, j) \in \mathcal{E}$ if and only if the conditional distribution of X_i given all other variables $\{X_k : k \neq i\}$ depends on X_j, i.e., differs from the conditional distribution of X_i given $\{X_k : k \neq i, j\}$. In this case the distribution is said to be Markov with respect to the graph. One can show that this graph is symmetric or *undirected*, i.e., all the elements of $\mathcal{E}$ are unordered pairs.

Our primary goal is the construction of informative prior distributions on undirected graphs, motivated by the problem of Bayesian inference of the dependence structure of families of observed random variables. As a side benefit, our approach also yields estimates of the conditional distributions given the graph. The model space of undirected graphs grows quickly with the dimension of $\{X_1, \ldots, X_n\}$ (there are $2^{n(n-1)/2}$ undirected graphs on n vertices) and is difficult to parametrize. We propose a novel parametrization and a simple, flexible family of prior distributions on $\mathcal{G}$ and on Markov probability distributions with respect to $\mathcal{G}$ [7]; this parametrization is based on computing the intersection pattern of a system of convex sets in $\mathbb{R}^d$. The novelty and main contribution of this chapter is structural inference for graphical models, specifically, the proposed representation of graph spaces allows for flexible prior distributions and new Markov chain Monte Carlo (MCMC) algorithms.

The problem of simultaneous inference of the graph structure and marginal distributions in a Bayesian framework was developed in [11, 27, 31] and approximate inference to identify regions with high posterior probability were proposed in [17, 28]. Computational challenges in Monte Carlo sampling over graphs were explored and partially addressed in [27, 31]. In this chapter, our focus is on constructing interesting informative priors on graphs. We think there is need for methodology that offers both efficient exploration of the model space and a simple and flexible family of distributions on graphs that can reflect meaningful prior information.

Two cases where prior distributions on graphical models have been explored in some depth are Erdös–Rényi random graphs and Gauss-Markov graphical models. In Erdös–Rényi random graph models each of the $\binom{n}{2}$ possible undirected edges (i, j) is included in $\mathcal{E}$ with specified probability $p \in [0, 1]$), the edge inclusion probabilities can also be edge-specific p_{ij}. These models have been used to place informative priors on (decomposable) graphs [14, 19].

The difficulty in using these models for prior specification on graphs is that the number of parameters in this prior specification can be enormous if the inclusion probabilities are allowed to vary. In the special case of jointly Gaussian variables $\{X_j\}$ or when an arbitrary marginal distributions $F_j(\cdot)$ can be represented in Gaussian copula form $X_j = F_j^{-1}(\Phi(Z_j))$, the problem of inferring conditional independence reduces to a search for zeros in the precision matrix C^{-1}, the inverse of the covariance matrix C [15]. A serious limitation to this approach is that the range of conditional dependencies is limited to pairwise relations. For example, a three-dimensional model in which each pair of variables is conditionally independent given the third cannot be distinguished from a model with complete joint dependence of the three variables (we return to this example in subsection 19.5.3).

In this article we establish a novel geometric approach to parametrize spaces of graphs or hypergraphs. For any integers $n, d \in \mathbb{N}$, we show in 19.2 how to use the geometrical configuration of a set $\{v_i\}$ of n points in Euclidean space $\mathbb{R}^d$ to determine a graph $\mathcal{G} = (\mathcal{V}, \mathcal{E})$ on $\mathcal{V} = \{v_1, ..., v_n\}$. Any prior distribution on point sets $\{v_i\}$ induces a prior distribution on graphs, and sampling from the posterior distribution of graphs is reduced to sampling from spatial configurations of point sets—a standard problem in spatial modeling. Relationships between graphs and finite sets of points have arisen earlier in the fields of computational topology [8] and random geometric graphs [22]. From the former we borrow the idea of *nerves*, *i.e.*, simplicial complexes computed from intersection patterns of convex subsets of $\mathbb{R}^d$; the 1-skeletons (collection of 1-dimensional simplices) of nerves are geometric graphs.

From the random geometric graph approach we gain understanding about the induced distribution on graph features when making certain features of a geometric graph (or a subclass of hypergraphs—simplicial complexes) stochastic.

19.1.1 Graphical models

The idea behind graphical models is to represent conditional dependencies for a multivariate distribution in the form of a graph or hypergraph. We first review relevant graph theoretical concepts and then relate these concepts to factorizing distributions. Note that we use the notation developed in the statistical graphical models notation as stated in [18] which at times deviates from the graph theory literature—when this is the case we state the difference.

A *graph* $\mathcal{G}$ is an ordered pair $(\mathcal{V}, \mathcal{E})$ of a set $\mathcal{V}$ of *vertices* and a set $\mathcal{E} \subseteq \mathcal{V} \times \mathcal{V}$ of *edges*. If all edges are unordered (resp., ordered), the graph is said to be *undirected* (resp., *directed*). All graphs considered in this chapter are undirected, unless stated otherwise. A *hypergraph*, denoted $\mathcal{H}$, consists of a vertex set $\mathcal{V}$ and a collection $\mathcal{K}$ of unordered subsets of $\mathcal{V}$ (known as *hyperedges*); a graph is the special case where all the subsets are vertex pairs. A graph is *complete* if $\mathcal{E} = \mathcal{V} \times \mathcal{V}$ contains all possible edges; otherwise it is *incomplete*. A complete subgraph that is maximal with respect to inclusion is a *clique*. This definition is specific to the statistical graphical models literature; the more general graph theory literature defines a clique in a graph as a subset of its vertices such that every two vertices in the subset are connected by an edge. Denote by $\mathscr{C}(\mathcal{G})$ and $\mathscr{Q}(\mathcal{G})$, respectively, the collection of complete sets and cliques of $\mathcal{G}$. A *path* between two vertices $\{v_i, v_j\} \in \mathcal{V}$ is a sequence of edges connecting v_i to v_j. A graph such that any pair of vertices can be joined by a unique path is a *tree*. A *decomposition* of an incomplete graph $\mathcal{G} = (\mathcal{V}, \mathcal{E})$ is a partition of $\mathcal{V}$ into disjoint nonempty sets (A, B, S) such that S is complete in $\mathcal{G}$ and *separates* A and B, i.e., any path from a vertex in A to a vertex in B must pass through S. Iterative decomposition of a graph $\mathcal{G}$ such that at each step the separator S_i is minimal and the subsets A_i and B_i are nonempty generates the *prime components* of $\mathcal{G}$, the collection of subgraphs that cannot be further decomposed. If all prime components of a graph $\mathcal{G}$ are complete, then $\mathcal{G}$ is said to be *decomposable*. Any graph $\mathcal{G}$ can be represented as a tree $\mathcal{T}$ whose vertices are its prime components $\mathscr{P}(\mathcal{G})$; this is called its *junction tree* representation.

Let $\mathcal{P}$ be a probability distribution on $\mathbb{R}^n$ and $X = (X_1, \ldots, X_n)$ be a random vector with distribution $\mathcal{P}$. *Graphical modeling* is the representation of the Markov or conditional dependence structure among the components $\{X_i\}$ in the form of a graph $\mathcal{G} = (\mathcal{V}, \mathcal{E})$. Denote by $f(x)$ the joint density function of $\{X_i\}$ (or probability mass function for discrete distributions— more generally, density for an arbitrary reference measure). The distribution $\mathcal{P}$ (and hence its density $f(x)$) may depend implicitly on a vector θ of parameters, taking values in some set $\Theta_\mathcal{G}$, which in some cases will depend on the graph $\mathcal{G}$; write $\Theta = \sqcup \Theta_\mathcal{G}$ for the disjoint union of the parameter spaces for all graphs on $\mathcal{V}$.

Each vertex $v_i \in \mathcal{V}$ is associated with a variable X_i, and the edges $\mathcal{E}$ determine how the distribution factors. The density $f(x)$ for the distribution can be factored in a variety of ways associated with the graph $\mathcal{G}$ [18]. It may be factored in terms of complete sets $a \in \mathscr{C}(\mathcal{G})$:

$$f(x) = \prod_{a \in \mathscr{C}(\mathcal{G})} \phi_a(x_a \mid \theta_a), \tag{19.1a}$$

or similarly in terms of cliques $a \in \mathscr{Q}$ (assuming f is positive, according to the Hammersley-Clifford theorem); if $\mathcal{G}$ is decomposable then $f(x)$ may also be factored in junction-tree form as:

$$f(x) = \frac{\prod_{a \in \mathscr{P}(\mathcal{G})} \psi_a(x_a \mid \theta_a)}{\prod_{b \in \mathscr{S}(\mathcal{G})} \psi_b(x_b \mid \theta_b)}, \tag{19.1b}$$

where $\mathscr{P}(\mathcal{G})$ and $\mathscr{S}(\mathcal{G})$ denote the prime factors and separators of $\mathcal{G}$, respectively, and where $\psi_a(x_a \mid \theta_a)$ denotes the marginal joint density for the components x_a for prime factors $a \in \mathscr{P}(\mathcal{G})$ and $\psi_b(x_b \mid \theta_b)$ that for separators $b \in \mathscr{S}(\mathcal{G})$ [7]. In the Gaussian case, a similar factorization to Equation (19.1b) holds even for non-decomposable graphs [27].

The prior distributions required for Bayesian inference about models of the form (19.1) may be specified by giving a marginal distribution on the set of all graphs $\mathcal{G} \in \mathcal{G}_n$ on n vertices and conditional distributions on each $\Theta_{\mathcal{G}}$, the space of parameters for that graph:

$$p(\mathcal{G}, \theta) = p(\mathcal{G})\, p(\theta \mid \mathcal{G}), \qquad \mathcal{G} \in \mathcal{G}_n,\ \theta \in \Theta_{\mathcal{G}} \tag{19.2}$$

where $\theta \in \Theta_{\mathcal{G}}$ determines the parameters $\{\theta_a : a \in \mathscr{C}(\mathcal{G})\}$ or $\{\theta_a : a \in \mathscr{P}(\mathcal{G})\}$ and $\{\theta_b : b \in \mathscr{S}(\mathcal{G})\}$. [10] pursue this approach in the Gaussian case, while [7] offer a rigorous framework for specifying more general prior distributions on $\Theta_{\mathcal{G}}$. Such priors, called *hyper Markov laws*, inherit the conditional independence structure from the sampling distribution, now at the parameter level. The hyper Inverse Wishart, useful when the factors are multivariate normal, is by far the most studied hyper Markov law. Most previously studied models of the form (19.2) specify very little structure on $p(\mathcal{G})$ [10, 14, 27]— typically $p(\mathcal{G})$ is taken to be a uniform distribution on the space of decomposable (or unrestricted) graphs, or perhaps an Erdös–Rényi prior to encourage sparsity [19], with no additional structure or constraints and hence no opportunity to express prior knowledge or belief.

Two inference problems arise for the model specified in (19.2): inference of the entire joint posterior distribution of the graph and factor parameters, $(\theta, \mathcal{G})$, or inference of only the *conditional independence structure*, which entails comparing different graphs via the marginal likelihood

$$\Pr\{\mathcal{G} \mid x\} \propto \int_{\Theta_{\mathcal{G}}} f(x \mid \theta, \mathcal{G})\, p(\mathcal{G})\, p(\theta \mid \mathcal{G})\, d\theta.$$

Inference about $\mathcal{G}$ may now be viewed as a Bayesian model selection procedure [25].

19.2   Geometric Graphs

Most methodology for structural inference in graphical models either assumes little prior structure on graph space, or else represents graphs using high dimensional discrete spaces with no obvious geometry or metric. In either case prior elicitation and posterior sampling can be challenging. In this section we propose parametrizations of graph space that will be used in Section 19.3 to specify flexible prior distributions and to construct new Metropolis/Hastings MCMC algorithms with local and global moves. The key idea for this parametrization is to construct graphs and hypergraphs from intersections of convex sets in $\mathbb{R}^d$.

We illustrate the approach with an example. Fix a convex region $A \subset \mathbb{R}^d$ and let $\mathcal{V} \subset A$ be a finite set of n points. For each number $r \geq 0$, the

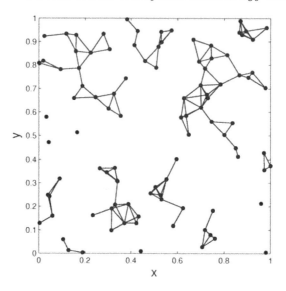

FIGURE 19.1
Proximity graph for 100 vertices and radius $r = 0.05$.

proximity graph $\text{Prox}(\mathcal{V}, r)$ (see Figure 19.1) is formed by joining every pair of (unordered) elements in $\mathcal{V}$ whose distance is $2r$ or less, *i.e.*, whose closed balls of radius r intersect. As r ranges from 0 to half the diameter of A, the graph $\text{Prox}(\mathcal{V}, r)$ ranges from the totally disconnected graph to the complete graph. This example is a particular case of a more general construction where hypergraphs can be computed from properties of intersections of classes of convex subsets in Euclidean space. The convex sets we consider are subsets of $\mathbb{R}^d$ that are simple to parametrize and compute. The key concept in our construction is the *nerve*:

Definition 19.2.1 (Nerve). *Let $F = \{A_j, \ j \in I\}$ be a finite collection of distinct nonempty convex sets. The* nerve *of F is given by*

$$Nrv(F) = \left\{ \sigma \subseteq I : \bigcap_{j \in \sigma} A_j \neq \varnothing \right\}.$$

The nerve of a family of sets uniquely determines a hypergraph. We use the following three nerves in this chapter to construct hypergraphs (for more details, see [8]).

Definition 19.2.2 (Čech Complex). *Let $\mathcal{V}$ be a finite set of points in $\mathbb{R}^d$ and $r > 0$. Denote by $\mathbb{B}^d$ the closed unit ball in $\mathbb{R}^d$. The Čech complex corresponding to $\mathcal{V}$ and r is the nerve of the sets $B_{v,r} = v + r\mathbb{B}^d$, $v \in \mathcal{V}$. This is denoted by $Nrv(\mathcal{V}, r, \check{C}ech)$.*

Definition 19.2.3 (Delaunay Triangulation). *Let* $\mathcal{V}$ *be a finite set of points in* $\mathbb{R}^d$. *The* Delaunay triangulation *corresponding to* $\mathcal{V}$ *is the nerve of the sets* $C_v = \{x \in \mathbb{R}^d : \|x - v\| \leq \|x - u\|, \ u \in \mathcal{V}\}$ *for* $v \in \mathcal{V}$. *This is denoted by* $Nrv(\mathcal{V}, Delaunay)$, *and the sets* C_v *are called* Voronoi cells.

Definition 19.2.4 (Alpha Complex). *Let* $\mathcal{V}$ *be a finite set of points in* $\mathbb{R}^d$ *and* $r > 0$. *The* Alpha complex *corresponding to* $\mathcal{V}$ *and* r *is the nerve of the sets* $B_{v,r} \cap C_v$, $v \in \mathcal{V}$. *This is denoted by* $Nrv(\mathcal{V}, r, Alpha)$.

The nerve of a family of sets is a particular class of hypergraphs known as (abstract) simplicial complexes.

Definition 19.2.5 (Abstract simplicial complex). *Let* $\mathcal{V}$ *be a finite set. A* simplicial complex *with base set* $\mathcal{V}$ *is a family* $\mathcal{K}$ *of subsets of* $\mathcal{V}$ *such that* $\tau \in \mathcal{K}$ *and* $\sigma \subseteq \tau$ *implies* $\sigma \in \mathcal{K}$. *The elements of* $\mathcal{K}$ *are called simplices, and the number of connected components of* $\mathcal{K}$ *is denoted* $\sharp(\mathcal{K})$.

19.2.1   Example

Here we use a junction tree factorization with each univariate marginal X_i associated to a point $V_i \in \mathbb{R}^d$ (the standard graphical models approach). In this case, specifying the class of sets to compute the nerve and the value for r determines a factorization for the joint density of $\{X_1, \ldots, X_n\}$. We illustrate with $n = 5$ points in Euclidean space of dimension $d = 2$.

Let $(X_1, X_2, X_3, X_4, X_5) \in \mathbb{R}^2$ be a random vector with density $f(x)$ and consider the vertex set displayed in Figure 19.2. For an Alpha complex with $r = 0.5$ the junction tree factorization (19.1b) corresponding to the graph in Figure 19.2 is

$$f(x) = \frac{\psi_{12}(x_1, x_2)\psi_{235}(x_2, x_3, x_5)\psi_4(x_4)}{\psi_2(x_2)},$$

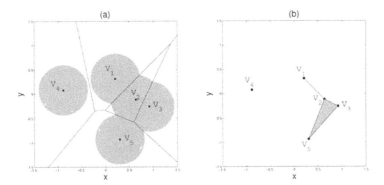

FIGURE 19.2
Top panel: (a) Alpha complex computed using $r = 0.5$ and the displayed vertex set. (b) The corresponding factorization.

we will denote the factorization as $[1, 2][2, 3, 5][4]$. In the case where the factors are potential functions rather than marginals we will use $\{\cdot\}$ instead of $[\cdot]$. Similarly, for the Čech complex and $r = 0.7$ the factorization corresponding to the graph is

$$f(x) = \frac{\psi_{1235}(x_1, x_2, x_3, x_5)\psi_{14}(x_1, x_4)}{\psi_1(x_1)}.$$

19.3 Random Geometric Graphs

In Section 19.2 we demonstrated how the geometry of a set $\mathcal{V}$ of n points in $\mathbb{R}^d$ can be used to induce a graph $\mathcal{G}$. In this section we explore the relation between prior distributions on *random* sets $\mathbf{V}$ of points in $\mathbb{R}^d$ and features of the induced distribution on graphs $\mathcal{G}$, with the goal of learning how to tailor a point process model to obtain graph distributions with desired features.

Definition 19.3.1 (Random Geometric Graph). *Fix integers $n, d \in \mathbb{N}$ and let $\mathbf{V} = (V_1, \ldots, V_n)$ be drawn from a probability distribution $\mathcal{Q}$ on $(\mathbb{R}^d)^n$. For any class $\mathcal{A}$ of convex sets in $\mathbb{R}^d$ and radius $r > 0$, the graph $\mathcal{G}(\mathbf{V}, r, \mathcal{A})$ is said to be a* Random Geometric Graph *(RGG)*.

While Definition 19.3.1 is more general than that of [22], it still cannot describe all the random graphs discussed in [23] (for example, those based on k-neighbors cannot in general be generated by nerves). For $\mathcal{A}$ we will use closed balls in $\mathbb{R}^d$ or intersections of balls and Voronoi cells; most often $\mathcal{Q}$ will be a product measure under which the $\{V_i\}$ will be n independent identically distributed (*iid*) draws from some marginal distribution $\mathcal{Q}_M$ on $\mathbb{R}^d$, such as the uniform distribution on the unit cube $[0, 1]^d$ or unit ball $\mathbb{B}^d$, but we will also explore the use of repulsive processes for $\mathbf{V}$ under which the points $\{V_i\}$ are more widely dispersed than under independence. It is clear that different choices for $\mathcal{A}$, $\mathcal{Q}$ and r will have an impact on the support of the induced RGG distribution. To make this notion precise we define feasible graphs. First,

Definition 19.3.2 (Isomorphic). *Write $\mathcal{G}_1 \cong \mathcal{G}_2$ for two graphs $\mathcal{G}_i = (\mathcal{V}_i, \mathcal{E}_i)$ and call the graphs isomorphic if there is a 1:1 mapping $\chi : \mathcal{V}_1 \to \mathcal{V}_2$ such that $(v_i, v_j) \in \mathcal{E}_1 \Leftrightarrow \left(\chi(v_i), \chi(v_j)\right) \in \mathcal{E}_2$ for all $v_i, v_j \in \mathcal{V}_1$.*

Definition 19.3.3 (Feasible Graph). *Fix numbers $d, n \in \mathbb{N}$, a class $\mathcal{A}$ of convex sets in $\mathbb{R}^d$, and a distribution $\mathcal{Q}$ on the random vectors $\mathbf{V}$ in $(\mathbb{R}^d)^n$. A graph Γ is said to be* feasible *if for some number $r > 0$,*

$$\Pr\{\mathcal{G}(\mathbf{V}, r, \mathcal{A}) \cong \Gamma\} > 0.$$

In contrast to Erdös–Rényi models, where the inclusion of graph edges are independent events, the RGG models exhibit edge dependence that depends on the metric structure of $\mathbb{R}^d$ and the class $\mathcal{A}$ of convex sets used to construct the nerves.

There is an extensive literature describing asymptotic distributions for a variety of graph features such as: subgraph counts, vertex degree, order of the largest clique, and maximum vertex degree [22]. Several results for the Delaunay triangulation, some of which generalize to the Alpha complex, are reported in [23].

Penrose [22] gives conditions which guarantee the asymptotic normality of the joint distribution of the numbers Q_j of j-simplices (edges, triads, *etc.*), for *iid* samples $\mathbf{V} = (V_1, \ldots, V_n)$ from some marginal distribution $\mathcal{Q}_M$ on $\mathbb{R}^d$, as the number $n = |\mathbf{V}|$ of vertices grows and the radius r_n shrinks.

Simulation studies suggest that the asymptotic results apply approximately for $n \geq$ 24–100. By this we mean that sometimes 24 is sufficient (the distribution of the vertices is approximately multivariate normal), and sometimes 100 may be required (distribution of the vertices far from being multivariate normal).

19.3.1   Simulation study of subgraph counts for RGGs

In this subsection we study the distribution of particular graph features as a function of the sampling distribution of the random point set $\mathbf{V}$ and contrast this with Erdös–Rényi models. Specifically we will focus on the number of edges (2-cliques) Q_2 and the number of 3-cliques Q_3.

The two spatial processes we study for $\mathcal{Q}$ are *iid* uniform draws from the unit square $[0, 1]^2$ in the plane, and dependent draws from the Matérn type III hard-core repulsive process [16], using Čech complexes with radius $r = 1/\sqrt{150} \approx 0.082$ in both cases to ensure asymptotic normality (Theorem 3.13 in [22]). In our simulations we vary both the number of variables (graph size) n and the Matérn III hard core radius ρ. Comparisons are made with an Erdös–Rényi model with a common edge inclusion parameter. Table 19.1 displays the quartiles for Q_2 and Q_3 as a function of the graph size n, hard core radius ρ, and Erdös–Rényi edge inclusion probability p.

19.4   Model Specification

19.4.1   General setting

We offer a Bayesian approach to the problem of inferring factorizations of distributions of the forms of Equation (19.1),

$$f(x) = \prod_{a \in \mathscr{C}(\mathcal{G})} \phi_a(x_a \mid \theta_a) \quad \text{or} \quad \frac{\prod_{a \in \mathscr{P}(\mathcal{G})} \psi_a(x_a \mid \theta_a)}{\prod_{b \in \mathscr{S}(\mathcal{G})} \psi_b(x_b \mid \theta_b)}.$$

TABLE 19.1

Summaries of the empirical distribution of edge and 3-clique counts for Čech complex random geometric graphs with radius $r = 0.082$, for vertex sets sampled from *iid* draws from the unit square from: a uniform distribution, a hard core process with radius $\rho = 0.035$, and from Erdös–Rényi (ER) with common edge inclusion probabilities of $p = 0.050$ and $p = 0.065$.

Graph	$\mathcal{V}$	Edges			3-Cliques		
		25%	50%	75%	25%	50%	75%
Uniform	75	161	171	182	134	160	190
Matérn (0.035)	75	154	161	170	110	124	144
ER (0.050)	75	130	138	146	6	8	11
ER (0.065)	75	172	181	189	14	18	22
Uniform	50	69	75	81	34	43	57
Matérn (0.035)	50	66	71	76	27	35	43
Matérn (0.050)	50	62	67	71	22	27	33
ER (0.050)	50	56	61	67	1	2	4
ER (0.065)	50	74	79	85	3	5	7
Uniform	20	9	12	14	1	2	4
Matérn (0.035)	20	9	11	13	1	1	3
Matérn (0.050)	20	8	10	12	0	1	2
ER (0.050)	20	8	9	11	0	0	0
ER (0.065)	20	10	12	15	0	0	1

In each case we specify the prior density function as a product

$$p(\theta, \mathcal{G}) = p(\theta \mid \mathcal{G})\, p(\mathcal{G}) \tag{19.3}$$

of a conditional hyper Markov law for $\theta \in \Theta$ and a marginal RGG law on $\mathcal{G}$. We use conventional methods to select the specific hyper Markov distribution (hyper Inverse Wishart for multivariate normal sampling distributions, for example) since our principal focus is on prior distributions for the graphs, $p(\mathcal{G})$. Every time we refer to hyper Markov laws, it will be in the strong sense according to [7]. We also present MCMC algorithms for sampling from the posterior distribution on $\mathcal{G}$, for observed data.

19.4.2   Prior specification

All the graphs in our statistical models are built from nerves constructed in Section 19.2 from a random vertex set $\mathcal{V} = \{V_i\}_{i=1}^{n} \subset \mathbb{R}^d$ and radius $r > 0$. Since the nerve construction is invariant under rigid transformations (this is, transformations that preserve angles as well as distances) of $\mathcal{V}$ or simultaneous scale changes in $\mathcal{V}$ and r, restricting the support of the prior distribution on $\mathcal{V}$ to the unit ball $\mathbb{B}^d$ does not reduce the model space:

Proposition 19.4.1. *Every feasible graph in $\mathbb{R}^d$ may be represented in the form $\mathcal{G}(\mathcal{V}, r, \mathcal{A})$ for a collection $\mathcal{V}$ of n points in the unit ball $\mathbb{B}^d$ and for $r = \frac{1}{n}$.*

Proof

Let $\mathcal{V} = \{V_1, \ldots, V_n\} \subset \mathbb{B}^d$ be a set of points and $r > 0$ a radius such that $\mathcal{G}(\mathcal{V}, r, \text{Čech})$ is the empty graph. Then the balls $V_i + r\mathbb{B}^d$ are disjoint and their union with d-dimensional volume $n\omega_d r^d$ lies wholly within the ball $(1+r)\mathbb{B}^d$ of volume $\omega_d(1+r)^d$ (where $\omega_d = \pi^{d/2}/\Gamma(1+d/2)$ is the volume of the unit ball), so $n < (1 + \frac{1}{r})^d$. ∎

Slightly stronger, the empty graph may not be attained as $\mathcal{G}(\mathcal{V}, r, \text{Čech})$ for any $r \geq 1/[(n/p_d)^{1/d} - 1]$ where p_d is the maximum spherical packing density in $\mathbb{R}^d$. For $d = 2$, this gives the asymptotically sharp bound $r < 1/\left[\sqrt{n\sqrt{12}/\pi} - 1\right]$.

19.4.3 Sampling from prior and posterior distributions

Let $\mathcal{Q}$ be a probability distribution on n-tuples in $\mathbb{R}^d$, $p(\mathcal{G})$ the induced prior distribution on graphs $\mathcal{G}(\mathbf{V}, \text{Čech})$ for $\mathbf{V} \sim \mathcal{Q}$ with $r = \frac{1}{n}$, and let $p(\theta \mid \mathcal{G})$ be a conventional hyper Markov law (see below). We wish to draw samples from the prior distribution $p(\theta, \mathcal{G})$ of Equation (19.3) and from the posterior distribution $p(\theta, \mathcal{G} \mid \mathbf{x})$, given a vector $\mathbf{x} = (x_1, ..., x_N)$ of *iid* observations $x_j \overset{\text{iid}}{\sim} f(x \mid \theta)$, using the Metropolis/Hastings approach to MCMC [13, 26].

We begin with a random walk proposal distribution in $\mathbb{B}^d$ starting at an arbitrary point $v \in \mathbb{B}^d$, that approximates the steps $\{V^{(0)}, V^{(1)}, V^{(2)}, ...\}$ of a diffusion $V^{(t)}$ on $\mathbb{B}^d$ with uniform stationary distribution and reflecting boundary conditions at the unit sphere $\partial\mathbb{B}^d$.

The random walk is conveniently parametrized in spherical coordinates with radius $\rho^{(t)} = \|V^{(t)}\|$ and Euler angles— in $d=2$ dimensions, angle $\varphi^{(t)}$— at step t. Informally, we take independent radial random walk steps such that $(\rho^{(t)})^d$ is reflecting Brownian motion on the unit interval (this ensures that the stationary distribution will be $\text{Un}(\mathbb{B}^d)$) and, conditional on the radius, angular steps from Brownian motion on the d-sphere of radius $\rho^{(t)}$.

Fix some $\eta > 0$. In $d = 2$ dimensions the reflecting random walk proposal (ρ^*, φ^*) we used for step $(t+1)$, beginning at $(\rho^{(t)}, \varphi^{(t)})$, is:

$$\rho^* = R\left([\rho^{(t)}]^2 + \zeta_\rho^{(t)} \eta\right)^{1/2}, \qquad \varphi^* = \varphi^{(t)} + \zeta_\phi^{(t)} \eta/\rho^{(t)}$$

for *iid* standard normal random variables $\left\{\zeta_\rho^{(t)}, \zeta_\phi^{(t)}\right\}$, where

$$R(x) = \left|x - 2\left\lfloor \tfrac{1}{2}(x+1)\right\rfloor\right|$$

is x reflected (as many times as necessary) to the unit interval. Similar expressions work in any dimension d, with $\rho^* = R\left([\rho^{(t)}]^d + \zeta_\rho^{(t)} \eta\right)^{1/d}$ and appropriate step sizes for the $(d-1)$ Euler angles.

For small $\eta > 0$ this diffusion-inspired random walk generates local moves under which the proposed new point (ρ^*, φ^*) is quite close to $(\rho^{(t)}, \varphi^{(t)})$ with high probability. To help escape local modes, and to simplify the proof of ergodicity below, we add the option of more dramatic "global" moves by introducing at each time step a small probability of replacing $(\rho^{(t)}, \varphi^{(t)})$ with a random draw (ρ^*, φ^*) from the uniform distribution on $\mathbb{B}^d$. Let $q(\mathcal{V}^* \mid \mathcal{V})$ denote the Lebesgue density at $\mathcal{V}^* \in (\mathbb{B}^d)^n$ of one-step of this hybrid random walk for $\mathcal{V} = (V_1, \dots, V_n)$, starting at $\mathcal{V} \in (\mathbb{B}^d)^n$.

19.4.3.1   Prior sampling

To draw sample graphs from the prior distribution, begin with $\mathbf{V}^{(0)} \sim \mathcal{Q}(d\mathbf{V})$ and, after each time step $t \geq 0$, propose a new move to $\mathbf{V}^* \sim q(\mathbf{V}^* \mid \mathbf{V}^{(t)})$. The proposed move from $\mathbf{V}^{(t)}$ (with induced graph $\mathcal{G}^{(t)} = \mathcal{G}(\mathbf{V}^{(t)})$) to $\mathbf{V}^*$ (and $\mathcal{G}^*$) is accepted (whereupon $\mathbf{V}^{(t+1)} = \mathbf{V}^*$) with probability $1 \wedge H^{(t)}$, the minimum of one and the Metropolis/Hastings ratio

$$H^{(t)} = \frac{p(\mathbf{V}^*)}{p(\mathbf{V}^{(t)})} \frac{q(\mathbf{V}^{(t)} \mid \mathbf{V}^*)}{q(\mathbf{V}^* \mid \mathbf{V}^{(t)})}.$$

Otherwise $\mathbf{V}^{(t+1)} = \mathbf{V}^{(t)}$; in either case set $t \leftarrow t{+}1$ and repeat. Note the proposal distribution $q(\cdot \mid \cdot)$ leaves the uniform distribution invariant, so $H^{(t)} \equiv 1$ for $\mathcal{Q}(d\mathbf{V}) \propto d\mathbf{V}$ and in that case every proposal is accepted.

19.4.3.2   Posterior sampling

After observing a random sample $X = \mathbf{x} = (x_1, \dots, x_N)$ from the distribution $x_j \sim f(x \mid \theta, \mathcal{G})$, let

$$f(\mathbf{x} \mid \theta, \mathcal{G}) = \prod_{i=1}^{N} f(x_i \mid \theta, \mathcal{G})$$

denote the likelihood function and

$$\mathcal{M}(\mathcal{G}) = \int_{\Theta_{\mathcal{G}}} f(\mathbf{x} \mid \theta, \mathcal{G}) \, p(\theta \mid \mathcal{G}) d\theta \tag{19.4}$$

the marginal likelihood for $\mathcal{G}$. For posterior sampling of graphs, a proposed move from $\mathbf{V}^{(t)}$ to $\mathbf{V}^*$ is accepted with probability $1 \wedge H^{(t)}$ for

$$H^{(t)} = \frac{\mathcal{M}(\mathcal{G}^*)}{\mathcal{M}(\mathcal{G}^{(t)})} \frac{p(\mathbf{V}^*)}{p(\mathbf{V}^{(t)})} \frac{q(\mathbf{V}^{(t)} \mid \mathbf{V}^*)}{q(\mathbf{V}^* \mid \mathbf{V}^{(t)})}. \tag{19.5}$$

For multivariate normal data X and hyper inverse Wishart hyper Markov law $p(\theta \mid \mathcal{G})$, $\mathcal{M}(\mathcal{G})$ from Equation (19.4) can be expressed in closed form for decomposable graphs $\mathcal{G}(\mathbf{V})$. Efficient algorithms for evaluating Equation (19.4) are still available even if this condition fails.

The model will typically be of variable dimension, since the parameter space $\Theta_{\mathcal{G}}$ for the factors may depend on the graph $\mathcal{G} = \mathcal{G}(\mathbf{V})$. Not all proposed moves of the point configuration $\mathbf{V}^{(t)} \rightsquigarrow \mathbf{V}^*$ will lead to a change in $\mathcal{G}(\mathbf{V})$; for those that do we implement reversible-jump MCMC [11, 29] using the auxiliary variable approach in [4] to simplify the book-keeping needed for non-nested moves $\Theta_{\mathcal{G}} \rightsquigarrow \Theta_{\mathcal{G}^*}$.

19.4.4   Convergence of the Markov chain

Denote by $\dot{\mathcal{G}}(n, d, \mathcal{A})$ the finite set of feasible graphs with n vertices in $\mathbb{R}^d$, *i.e.*, those generated from 1-skeletons of $\mathcal{A}$-complexes. For each $\mathcal{G} \in \dot{\mathcal{G}}(n, d, \mathcal{A})$ let $V_{\mathcal{G}} \subset (\mathbb{B}^d)^n$ denote the set of all points $\mathbf{V} = \{V_1, \ldots, V_n\} \in (\mathbb{B}^d)^n$ for which $\mathcal{G} \cong \mathcal{G}(\mathbf{V}, \frac{1}{n}, \mathcal{A})$, and set $\mu(\mathcal{G}) = \mathcal{Q}(V_{\mathcal{G}})$. Then

Proposition 19.4.2. *The sequence* $\mathcal{G}^{(t)} = \mathcal{G}(\mathbf{V}^{(t)}, \frac{1}{n}, \mathcal{A})$ *induced by the prior MCMC procedure described in subsection 19.4.3.1 samples each feasible graph* $\mathcal{G} \in \dot{\mathcal{G}}(n, d, \mathcal{A})$ *with asymptotic frequency* $\mu(\mathcal{G})$. *The posterior procedure described in subsection 19.4.3.2 samples each feasible graph with asymptotic frequency* $\mu(\mathcal{G} \mid \mathbf{x})$, *the posterior distribution of* $\mathcal{G}$ *given the data* $\mathbf{x}$ *and hyper Markov prior* $p(\theta \mid \mathcal{G})$.

Proof

Both statements follow from the Harris recurrence of the Markov chain $\mathbf{V}^{(t)}$ constructed in subsection 19.4.3. For this, it is enough to find a strictly positive lower bound for the probability of transitioning from an arbitrary point $V \in (\mathbb{B}^d)^n$ to any open neighborhood of another arbitrary point $V^* \in (\mathbb{B}^d)^n$ [26]. This follows immediately from our inclusion of the global move in which all n points $\{V_i\}$ are replaced with uniform draws from $(\mathbb{B}^d)^n$. ∎

It is interesting to note that while the sequence $\mathcal{G}^{(t)} = \mathcal{G}(\mathbf{V}^{(t)}, \frac{1}{n}, \mathcal{A})$ is a hidden Markov process, it is not itself Markovian on the finite state space $\dot{\mathcal{G}}(n, d, \mathcal{A})$; nevertheless it is ergodic, by Prop. 19.4.2.

19.5   Three Simulation Examples

We illustrate the use of the proposed parametrization in Bayesian inference for graphical models: this is done by specifying priors that encourage sparsity and the design of MCMC algorithms that allow for local as well as global moves. We offer three examples. The first example illustrates that our method works when the graph encoding the Markov structure of underlying density is contained in the space of graphs spanned by the nerve used to fit the model. In the second example, we apply our method to Gaussian Graphical Models. The third example shows that the nerve hypergraph (not just the 1-skeleton) can be used to induce different groupings in the terms of a factorization, and therefore a way to encode dependence features that go beyond pairwise relationships.

19.5.1 Example 1: $\mathcal{G}$ is in the space generated by $\mathcal{A}$

Let $(X_1, \ldots, X_{10})$ be a random vector whose distribution has factorization:

$$f_\theta(\mathbf{x}) = \tag{19.6a}$$
$$\frac{\psi_\theta(x_1, x_4, x_{10})\psi_\theta(x_1, x_8, x_{10})\psi_\theta(x_4, x_5)\psi_\theta(x_8, x_9)\psi_\theta(x_2, x_3, x_9)\psi_\theta(x_6)\psi_\theta(x_7)}{\psi_\theta(x_4)\psi_\theta(x_8)\psi_\theta(x_9)\psi_\theta(x_1, x_{10})}$$

The Markov structure of Equation (19.6a) can be encoded by the geometric graph displayed in Figure 19.3. We transform variables if necessary to achieve standard $\mathsf{Un}(0, 1)$ marginal distributions for each X_i, and model clique joint marginals with a Clayton copula [6, 21], the exchangeable multivariate model with joint distribution function

$$\Psi_\theta(\mathbf{x}_I) = \left(1 - n_I + \sum_{i \in I} x_i^{-\theta}\right)^{-1/\theta}$$

and density function

$$\psi_\theta(\mathbf{x}_I) = \theta^{n_I} \frac{\Gamma(n_I + 1/\theta)}{\Gamma(1/\theta)} \left(1 - n_I + \sum_{i \in I} x_i^{-\theta}\right)^{-n_I - 1/\theta} \left(\prod_{i \in I} x_i\right)^{-1-\theta}$$
$$\tag{19.6b}$$

on $[0, 1]^{n_I}$ for some $\theta \in \Theta = (0, \infty)$, for each clique $[v_i : i \in I]$ of size n_I.

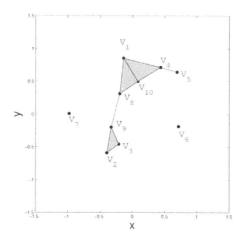

FIGURE 19.3
Geometric graph representing the model given in Equation (19.6a).

TABLE 19.2
The three models with highest estimated posterior probability. The true model is shown in bold (see Figure 19.3). Here $\theta = 4$.

Graph Topology	Posterior Probability
[1, 4, 10][1, 8, 10][4, 5][8, 9][2, 3, 9][6][7]	0.963
[1, 4, 10][1, 8, 10][4, 5][8, 9][2, 3, 9][6][5, 7]	0.021
[1, 4, 10][1, 8][4, 5][8, 9][2, 3, 9][6][7]	0.010

We drew 250 samples from the model given in Equation (19.6b) with $\theta = 4$. For inference about $\mathcal{G}$ we set $\mathcal{A} = $ Alpha and $r = 0.30$, with independent uniform prior distributions for the vertices $V_i \overset{iid}{\sim} \mathsf{Un}(\mathbb{B}^2)$ on the unit ball in the plane. We used the random walk described in subsection 19.4.3 to draw posterior samples with extra constraints applied to enforce decomposability. To estimate θ, we take a unit Exponential prior distribution $\theta \sim \mathsf{Ex}(1)$ and employ a Metropolis/Hastings approach using a symmetric random walk proposal distribution with reflecting boundary conditions at $\theta = 0$,

$$\theta^* = \left| \theta^{(t)} + \varepsilon \right|,$$

with $\varepsilon_t \sim \mathsf{Un}(-\beta, \beta)$ for fixed $\beta > 0$. We drew 1 000 samples after a burn-in period of 25 000 draws. The three models with the highest posterior probabilities are displayed in Table 19.2. The geometric graphs computed from nine posterior samples (one in every 100 draws) are shown in Figure 19.4; note that the computed nerves appear to stabilize after a few hundred iterations while the actual position of the vertex set continues to vary.

19.5.2 Example 2: Gaussian graphical model

We use our procedure to perform model selection for the Gaussian graphical model $X \sim \mathsf{No}(0, \Sigma_{\mathcal{G}})$, where $\mathcal{G}$ encodes the zeros in Σ^{-1}. We adopt a Hyper Inverse Wishart (HIW) prior distribution for $\Sigma \mid \mathcal{G}$. The marginal likelihood is given by [1]

$$\mathcal{M}(\mathbf{V}) = (2\pi)^{-nN/2} \frac{I_{\mathcal{G}(\mathbf{V})}(\delta + N, D + X^T X)}{I_{\mathcal{G}(\mathbf{V})}(\delta, D)}, \tag{19.7}$$

where

$$I_{\mathcal{G}}(\delta, D) = \int_{M^+(\mathcal{G})} |\Sigma|^{(\delta-2)/2} e^{-\frac{1}{2}<\Sigma, D>} d\Sigma$$

denotes the HIW normalizing constant. This quantity is available in closed form for weakly decomposable graphs $\mathcal{G}(\mathbf{V})$, but for our unrestricted graphs, Equation (19.7) must be approximated via simulation. For our low-dimensional examples the method of [1] suffices; for a larger number of variables we recommend that of [5]. We set $\delta = 3$ and $D = 0.4I_6 + 0.6J_6$ (I_6 and J_6 denote the identity matrix and the matrix of all ones, respectively).

We sampled 300 observations from a Multivariate Normal with mean zero and precision matrix

$$
\begin{pmatrix}
18.18 & -6.55 & 0 & 2.26 & -6.27 & 0 \\
-6.55 & 14.21 & 0 & -4.90 & 0 & 0 \\
0 & 0 & 10.47 & 0 & 0 & -3.65 \\
2.26 & -4.90 & 0 & 10.69 & 0 & 0 \\
-6.27 & 0 & 0 & 0 & 27.26 & 0 \\
0 & 0 & -3.65 & 0 & 0 & 7.41
\end{pmatrix}
\tag{19.8}
$$

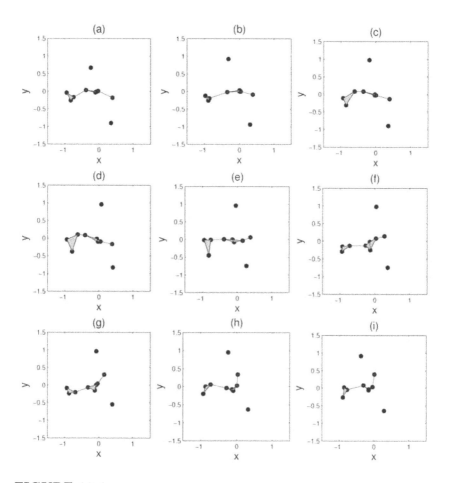

FIGURE 19.4
Geometric graphs corresponding to snapshots of posterior samples (a snapshot was taken from every 100 iterations) from model of Equation (19.6a).

TABLE 19.3
The five models with highest estimated posterior probability. The true model is shown in bold.

Graph Topology	Posterior Probability
[1,2,4][1,5][3,6]	0.152
[1,5][2,3,4][2,3,6]	0.072
[1,2,3,4,6][1,5]	0.069
[1,4][2,4][2,3,6]	0.055
[1,2,4][2,3,4][1,5][3,6]	0.052

whose conditional independence structure is given by the graph with cliques $[1, 2, 4][1, 5][3, 6]$. We fit the model described in Section 19.4 using a uniform prior for each $V_i \in \mathbb{B}^2$ and $r = 0.25$. We employed hybrid random walk proposals in which we move all five vertices $\{V_i\}$ independently according to the diffusion-inspired random walk described in subsection 19.4.3 with probability 0.85; replace one uniformly selected vertex V_i with a uniform draw from $\mathsf{Un}(\mathbb{B}^2)$ with probability 0.05; and replace all five vertices with independent uniform draws from $\mathsf{Un}(\mathbb{B}^2)$ with probability 0.10. We sampled $1,000$ observations from the posterior after a burn-in of $750,000$. Results are summarized in Table 19.3.

19.5.3 Example 3: Factorization based on nerves

While Gaussian joint distributions are determined entirely by the bivariate marginals, and so only edges appear in their complete-set factorizations (see Equation (19.1a)); more complex dependencies are possible for other distributions. The familiar example of the joint distribution of three Bernoulli variables X_1, X_2, X_3 each with mean $1/2$, with X_1 and X_2 independent but $X_3 = (X_1 - X_2)^2$ (so that $\{X_i\}$ are only *pairwise* independent) has only the complete set $\{1, 2, 3\}$ in its factorization.

Consider now a model with the continuous density function which satisfies the complete-set factorization:

$$f(x \mid \mathcal{G}, \theta) = c_{\mathcal{G}} \, \phi(x_1, x_2)\phi(x_1, x_6)\phi(x_2, x_6) \cdot \phi(x_3, x_4, x_5). \qquad (19.9a)$$

For illustration we will take each $\phi(\cdot)$ to be a Clayton copula density (see Equation (19.6b)). For simplicity we specify the same value $\theta = 4$ for each association parameter, so $f(x \mid \mathcal{G}, \theta)$ is given by Equation (19.9a) with

$$\phi(x, y) = 5 \, (x^{-4} + y^{-4} - 1)^{-9/4} \qquad (x \, y)^{-5} \qquad (19.9b)$$

$$\phi(x, y, z) = 30(x^{-4} + y^{-4} + z^{-4} - 2)^{-13/4}(x \, y \, z)^{-5}. \qquad (19.9c)$$

TABLE 19.4
Vertex set used for generating a factorization based on nerves.

Coordinate	V_1	V_2	V_3	V_4	V_5	V_6
x	-0.0936	-0.4817	0.0019	0.0930	0.2605	-0.5028
y	0.6340	0.7876	0.0055	0.0351	-0.0702	0.2839

In earlier examples we associated graphical structures (*i.e.*, edge sets) with 1-skeletons of nerves. We now associate *hyper*graphical structures (*i.e.*, abstract simplicial complexes that may include higher-order simplexes) with the entire nerves, with maximal simplices associated with complete-set factors. For example: the Alpha complex computed from the vertex set displayed in Table 19.4 with $r = 0.40$ has $\{3, 4, 5\}\{1, 2\}\{1, 6\}\{2, 6\}$ as its maximal simplices. By associating a Clayton copula to each of these hyperedges we recover the model shown in Equation (19.9a).

We use the same prior and proposal distributions constructed in subsection 19.4.3 from point distributions in $\mathbb{R}^d$; what has changed is the way the nerve is being used: as a hypergraph whose maximal hyperedges represent factors. One complicating factor is the need to evaluate the normalizing factor $c_{\mathcal{G}}$ for each graph $\mathcal{G}$ we encounter during the simulation; unavailable in closed form, we use Monte Carlo importance sampling to evaluate $c_{\mathcal{G}}$ for each new graph, and store the result to be reused when $\mathcal{G}$ recurs.

We anticipate that uniform draws $V_i \overset{\text{iid}}{\sim} \mathsf{Un}(\mathbb{B}^2)$ will give high probability to clusters of three or more points within a ball of radius r, favoring higher-dimensional features (triangles and tetrahedra) in the induced hypergraph encoding the Markov structure of $\{X_i\}$. To explore this phenomenon, we compare results for uniform draws with those from a repulsive process under which clusters of three or more points are unlikely to lie within a ball of radius r, hence favoring hypergraphs with only edges.

We began by sampling 650 observations from model Equation (19.9a) with $\mathcal{A} = $ Alpha and $r = 0.40$, with independent uniform prior distributions for the vertices $V_i \overset{\text{iid}}{\sim} \mathsf{Un}(\mathbb{B}^2)$ on the unit ball in the plane. The Metropolis/Hastings proposals for the vertex set are given by a mixture scheme:

- A random walk for each V_i, as described in subsection 19.4.3, with step size $\eta = 0.020$. This proposal is picked with probability 0.94.

- An integer $1 \le k \le 6$ is chosen uniformly and, given k, a subset of size k from $\{1, 2, 3, 4, 5, 6\}$ is sampled uniformly; the vertices corresponding to those indices are replaced with random independent draws from $\mathsf{Un}(\mathbb{B}^2)$. This proposal is picked with probability 0.06, 0.01 for each k.

For θ, we used the same standard exponential prior distribution and reflecting uniform random walk proposals described in Example 19.5.1.

TABLE 19.5
Highest posterior factorizations with uniform prior for model of Equation (19.9) (true model is shown in bold).

Maximal Simplices	Posterior Probability
{3, 4, 5}{1, 2}{2, 6}{1, 6}	0.609
{1, 2, 6}{3, 4}{4, 5}{3, 5}	0.161
{3, 5}{1, 6}{3, 4}{1, 2}{2, 6}	0.137

TABLE 19.6
Highest posterior factorizations with Strauss prior (true model is shown in bold).

Maximal Simplices	Posterior Probability
{3, 4, 5}{1, 2}{2, 6}{1, 6}	0.824
{1, 2, 6}{3, 4, 5}	0.111
{1, 2, 6}{3, 4}{3, 5}{4, 5}	0.002

Using 5 000 posterior samples after a burn-in period of 95 000 iterations, the models with highest posterior probability are summarized in Table 19.5.

To penalize higher-order simplexes we used a Strauss repulsive process [30] conditioned to have n points in $\mathbb{B}^d$ as prior distribution for the vertex set, with Lebesgue density

$$g(v) \propto \gamma^{\#\{(i,j):\ \text{dist}(v_i, v_j) < 2R\}}$$

for some $0 < \gamma \leq 1$, penalizing each pair of points closer than $2R$. Simulation results for this prior (with $R = 0.7r$ and $\gamma = 0.75$) are summarized in Table 19.6. The posterior mode is far more distinct for this prior than for the uniform prior shown in Table 19.5.

In a further experiment with $\gamma = 0.35$, the posterior was concentrated on factorizations without any triads.

19.6 Discussion

In this article we present a new parametrization of graphs by associating them with finite sets of points in $\mathbb{R}^d$. This perspective supports the design of informative prior distributions on graphs using familiar probability distributions of point sets in $\mathbb{R}^d$. It also induces new and useful Metropolis/Hastings proposal distributions in graph space that include both local and global moves. As suggested by Helly's Theorem [9] characterizing the sparsity of intersections of convex sets in $\mathbb{R}^d$, this methodology is particularly well suited for sparse graphs. The simple strategies presented here generalize easily to more detailed and subtle models for both priors and Metropolis/Hastings proposals.

An interesting feature of our approach is that the distribution on the space of graphs is modeled directly before the application of any specific hyper Markov law, in contrast to standard approaches in which it is the hyper Markov law that is used to encourage sparsity or other features on the graph. We think that working with the space of graphs explicitly opens a lot of possibilities for prior specification in graphical models, therefore, it is a perspective worth further study.

Interesting questions and extensions of this idea include: (1) achieving a deeper and more detailed understanding of the subspace of graphs spanned by different specific filtrations; (2) designing priors to control the distributions of specific features of graphs such as clique size or tree width; (3) modeling directed acyclic graphs (DAGs), and (4) concrete implementation of novel Markov structures based on nerves.

This methodology generates only graphs that are feasible for the particular filtration chosen. Although we do have some insight about which graphs can and cannot be generated by specific filtrations, a more complete and formal understanding of this aspect of the problem would be useful.

We used very simple prior distributions for the purpose of illustrating the core idea of the methodology. It is natural in this approach to incorporate tools from point processes into graphical models to define new classes of priors for graph space. Future developments in our research will involve a range of repulsive and cluster processes.

The parametrization we propose can be used to represent Markov structures on DAGs, but the strategies for obtaining such graphs from nerves will be different and will establish stronger connections between Graphical Models and computational topology.

The present work is related to that of [24] in which a nerve of convex subsets in $\mathbb{R}^d$ is used to obtain Markov structures for a distribution, an extension of the abstract tube theory of [20]. This new perspective allows for constructions that generalize the idea of junction trees. By modifying our methodology according to this framework (personal communication from H. Wynn suggests that this is feasible) we hope to fit models that factorize according to those novel Markov structures.

Another possible extension of this work is to discretize the set from which the vertex set is sampled (*e.g.*, use a grid). Such discretization may improve the behavior of the MCMC; it would also allow the use of a nonparametric form for the prior on the vertex set, leading to more flexible priors on graph space.

Acknowledgments

We are grateful to Herbert Edelsbrunner, John Harer, and Henry Wynn for helpful conversations. This work was partially supported by National Science Foundation grants DMS–0732260, DMS–0635449 and DMS-0757549, and by National Institutes of Health grants NIH R01 CA123175-01A1 and NIH P50-GM081883-02. Any opinions, findings, and conclusions or recommendations expressed in this material are those of the authors and do not necessarily reflect the views of the NSF or the NIH.

Bibliography

[1] A. Atay-Kayis and H. Massam. A Monte Carlo method for computing the marginal likelihood in nondecomposable Gaussian graphical models. *Biometrika*, 92(2):317–335, 2005.

[2] J. E. Besag. Spatial interaction and the statistical analysis of lattice systems (with discussion). *Journal of the Royal Statistical Society: Series B (Statistical Methodology)*, 36(2):192–236, 1974.

[3] J. E. Besag. Statistical analysis of non-lattice data. *Journal of the Royal Statistical Society: Series D (The Statistician)*, 24(3):179–195, 1975.

[4] S. P. Brooks, P. Giudici, and G. O. Roberts. Efficient construction of reversible jump Markov chain Monte Carlo proposal distributions. *Journal of the Royal Statistical Society: Series B (Statistical Methodology)*, 65(1):3–55, 2003.

[5] C. M. Carvalho, H. Massam, and M. West. Simulation of hyper-inverse Wishart distributions in graphical models. *Biometrika*, 94(3):647–659, 2007.

[6] D. G. Clayton. A model for association in bivariate life tables and its applications in epidemiological studies of familial tendency in chronic disease incidence. *Biometrika*, 65(1):141–151, 1978.

[7] A. P. Dawid and S. L. Lauritzen. Hyper Markov laws in the statistical analysis of decomposable graphical models. *Annals of Statistics*, 21(3):1272–1317, 1993.

[8] H. Edelsbrunner and J. Harer. Persistent homology— a survey. In J. E. Goodman, J. Pach, and R. Pollack, editors, *Surveys on Discrete and Computational Geometry: Twenty Years Later*, volume 453 of *Contemporary Mathematics*, pages 257–282. American Mathematical Society, 2008.

[9] H. Edelsbrunner and J. Harer. Lecture notes from the course 'Computational Topology'. http://www.cs.duke.edu/courses/fall06/cps296.1/, 2009.

[10] P. Giudici and P. J. Green. Decomposable graphical Gaussian model determination. *Biometrika*, 86(4):785–801, 1999.

[11] P. J. Green. Reversible jump Markov chain Monte Carlo computation and Bayesian model determination. *Biometrika*, 82(4):711–732, 1995.

[12] J. M. Hammersley and P. Clifford. Markov fields on finite graphs and lattices. Unpublished, 1971.

[13] W. K. Hastings. Monte Carlo sampling methods using Markov chains and their applications. *Biometrika*, 57(1):97–109, 1970.

[14] D. Heckerman, D. Geiger, and D. M. Chickering. Learning Bayesian networks: The combination of knowledge and statistical data. *Machine Learning*, 20(3):197–243, 1995.

[15] P. D. Hoff. Extending the rank likelihood for semiparametric copula estimation. *Annals of Applied Statistics*, 1(1):265–283, 2007.

[16] M. L. Huber and R. L. Wolpert. Likelihood-based inference for Matérn type III repulsive point processes. *Advances in Applied Probability*, 41(4):In press, 2009. Preprint available on-line at http://www.stat.duke.edu/ftp/pub/WorkingPapers/08-27.html.

[17] B. Jones, C. Carvalho, A. Dobra, C. Hans, C. K. Carter, and M. West. Experiments in stochastic computation for high-dimensional graphical models. *Statistical Science*, 20(4):388–400, 2005.

[18] S. L. Lauritzen. *Graphical Models*, volume 17 of *Oxford Statistical Science Series*. Oxford University Press, New York, NY, 1996.

[19] V. K. Mansinghka, C. Kemp, J. B. Tenenbaum, and T. L. Griffiths. Structured priors for structure learning. In L. Bertossi, A. Hunter, and T. Schaub, editors, *Uncertainty in Artificial Intelligence: Proceedings of the Twenty-second Annual Conference (UAI 2006)*, 2006.

[20] D. Q. Naiman and H. P. Wynn. Abstract tubes, improved inclusion-exclusion identities and inequalities and importance sampling. *Annals of Statistics*, 25(5):1954–1983, 1997.

[21] R. B. Nelsen. *An Introduction to Copulas*. Springer Verlag, New York, 1999.

[22] M. D. Penrose. *Random Geometric Graphs*. Oxford University Press, UK, 2003.

[23] M. D. Penrose and J. E. Yukich. Central limit theorems for some graphs in computational geometry. *Annals of Applied Probability*, 11(4):1005–1041, 2001.

[24] G. Pistone, H. Wynn, G. S. de Cabezón, and J. Q. Smith. Junction tubes and improved factorisations for Bayes nets. Unpublished, 2009.

[25] C. P. Robert. *The Bayesian Choice: From Decision-Theoretic Foundations to Computational Implementation*. Springer Verlag, New York, 2nd edition, 2001.

[26] C. P. Robert and G. Casella. *Monte Carlo Statistical Methods*. Springer Verlag, New York, 2nd edition, 2004.

[27] A. Roverato. Hyper inverse Wishart distribution for non-decomposable graphs and its application to Bayesian inference for Gaussian graphical models. *Scandinavian Journal of Statistics*, 29(3):341–411, 2002.

[28] J. G. Scott and C. M. Carvalho. Feature-inclusion stochastic search for Gaussian graphical models. *Journal of Computational and Graphical Statistics*, 17(4):790–808, 2008.

[29] S. A. Sisson. Transdimensional Markov chains: A decade of progress and future perspectives. *Journal of the American Statistical Association*, 100(471):1077–1089, 2005.

[30] D. J. Strauss. A model for clustering. *Biometrika*, 62(2):467–476, 1975.

[31] K. K. F. Wong, C. K. Carter, and R. Kohn. Hyper inverse Wishart distribution for non-decomposable graphs and its application to Bayesian inference for Gaussian graphical models. *Biometrika*, 90(4):809–830, 2003.

20

A Bayesian Uncertainty Analysis for Nonignorable Nonresponse

Balgobin Nandram

Worcester Polytechnic Institute

Namkyo Woo

Kyungpook National University

CONTENTS

20.1 Introduction

We discuss the concept of Bayesian uncertainty analysis in survey sampling for categorical data when there is nonignorable nonresponse. This is important because in sample surveys, data are typically summarized in contingency tables and there are nonresponders. In a nonignorable nonresponse model there are nonidentifiable parameters, and a sensitivity analysis is necessary to study the effects of these parameters on the parameter of interest. The sensitivity analysis is typically performed by setting the nonidentifiable parameters at various plausible values. In a Bayesian uncertainty analysis, rather than performing a sensitivity analysis, we put a prior on the nonidentifiable parameters. We illustrate Bayesian uncertainty analysis using a three-way contingency table (i.e., a single $r \times c \times u$ table) with nonignorable nonresponse.

Missing data give rise to partial classification of the sampled individuals. Thus, for each three-way table there are both item nonresponse (one of the three categories is missing and two of the three categories are missing) and unit nonresponse (all three categories are missing). One may not know how the data are missing and a model that includes some differences between the observed data and missing data (i.e., nonignorable missing data) may be preferred. For a general $r \times c \times u$ categorical table, we address the issue of estimation of the cell probabilities of the three-way table when there is possibly nonignorable nonresponse but there is really no information about ignorability. In such a situation, we would like to express a degree of uncertainty about ignorability; see [11].

We distinguish between ignorable and nonignorable nonresponse models. These are associated with the missing data mechanism; see [8]. Missing completely at random (MCAR) occurs if the missingness is independent of both the observed and the unobserved data. Missing at random (MAR) occurs, when conditional on the observed data, missingness is independent of the unobserved data. Missingness is called missing not at random (MNAR) when neither MCAR nor MAR holds. While under MCAR partially complete data are irrelevant, under MAR partially complete data are relevant. The issue with MNAR is how to fill in the missing data because they differ from the observed data. Parametric models for MCAR or MAR are called ignorable nonresponse models if the parameters of the missing data mechanism are distinct from those of the data [18]. Models for MNAR mechanisms are nonignorable nonresponse models which can also be classified into two types (pattern mixture and selection). The general difficulty with nonignorable nonresponse models is that the parameters are not identifiable (e.g., see [12]).

It is necessary to perform a sensitivity analysis for a nonignorable nonresponse model. There are at least three ways to perform a sensitivity analysis. First, as in observational studies we can introduce a latent variable to see how changes in it alter the estimate of the finite population parameters. Second, we can fit a nonignorable nonresponse model with one or more centering parameters which, when varied over a set of specified values, can provide several different models in a single family. Third, we can fit many different models (e.g., ignorable and nonignorable nonresponse models, and within the class of nonignorable models, we can fit both selection and pattern mixture models). In all three ways we can do a single Markov chain Monte Carlo (MCMC) fit; in the third case we can run a reversible jump MCMC sampler ([5]) although this is not necessary. We prefer the method to include the sensitivity parameters into a model to provide a full Bayesian uncertainty (risk) analysis [6].

Molenberghs *et al.* [9] provided examples of categorical tables in which different nonignorable nonresponse models provide the same fit to the observed data but they differ in their prediction of the unobserved counts. This implies that nonignorable nonresponse models cannot be examined using the observed data alone even if they fit well; the plausibility of the model assumptions needs to be examined carefully. The authors [9] discussed several other issues associated with missing categorical data (e.g., uniqueness and validity of estimates from nonignorable nonresponse models).

Without any information about the missing data, one does not really know how to fill in the missing cells. We need to confront both subjectivity and imprecision which are respectively due to missingness and sampling; see [10]. By fitting several plausible overspecified models, it is possible to express uncertainty about the parameters of interest. [10] discusses three types of intervals, the nonparametric (pessimistic-optimistic) interval, the interval of ignorance (missingness) and the uncertainty interval (encompassing both ignorance and imprecision). The lower bound of the nonparametric interval is set to the value when all missing data take negative (no) values and the upper bound is the value when all missing data are taken to be positive (yes). The interval of ignorance is the range of estimates that are obtained over many different plausible models, including overspecified models as well. As in the frequentist domain, the uncertainty interval is the union of the $100(1-\alpha)\%$ confidence intervals over the models under study.

Molenberghs *et al.* [10] provide a general principle for missing data within the frequentist domain. They consider the parameter space to consist of two sets (η, ν), where η is a minimal set of parameters which can be estimated when ν is specified; η is called the estimable parameter and ν the sensitivity parameter. Each specified value of ν will provide an estimate $\hat{\eta}(\nu)$. The range of these estimates over all plausible values of ν is the interval of ignorance. The union of the $100(1-\alpha)\%$ confidence intervals over all plausible values of ν gives the uncertainty interval; this latter interval is more important because it contains all sources of variability.

We can formulate the work of Molenberghs *et al.* [10] within the Bayesian framework. This will permit us to obtain a 'Bayesian uncertainty' interval for the finite population parameters to include both ignorance (subjectivity) and imprecision (sample). Starting with the two sets of parameters, we can put prior distributions on them. Thus, we can use

$$p(\eta, \nu) = p(\eta \mid \nu)p(\nu),$$

where again η is the set of parameters of interest and ν is the set of nonidentifiable parameters. The prior on ν is specified on a set of plausible values of ν. Note that if ν is specified as in a sensitivity analysis, η will be identified. Thus, one way to specify $p(\nu)$ is to put a uniform distribution over all plausible values of ν (e.g., those values specified in sensitivity analysis). This formulation, called a Bayesian uncertainty analysis, is related to the analysis of biases in observational studies; see, for example, [6] for a review. The uncertainty interval for η is the $(1-\alpha)$ credible interval (CI) for η when a prior distribution is placed on the nonidentifiable parameters.

There are two ways to model missing data from a contingency table. First, one can separate the data into a number of tables, one complete and the others partially complete. One can now fit a multinomial data model to all these tables including the missing data; one can also fit a loglinear model. See [2] for the pioneering non-Bayesian analysis for incomplete two-way tables. Second, to each categorical variable one can have a response indicator, thereby leading to a single categorical table with twice the number of categories. It is typical to fit a loglinear model to provide a nonignorable nonresponse model; see [1, 8]. Some of the interaction terms are usually omitted from the loglinear model; typically the saturated models are fitted for the categorical variables and response variables respectively and these are connected by two-way interactions.

Motivated by [4], [11] uses an expansion model to study nonignorable nonresponse binary data and [14] uses a similar model for polychotomous data. The expansion model, a nonignorable nonresponse model, degenerates into an ignorable nonresponse model (in the spirit of [3]) when a centering parameter is set to unity. This permits an expression of uncertainty about ignorability; see also [4]. However, this procedure is more appropriate for several tables (see [14]). We avoid the issue of centering parameters because we only have one categorical table. We use a simpler idea of Nandram *et al.* [15] (with a single supplemental table) and Nandram *et al.* [13] (with three supplemental tables). Nandram *et al.* [13, 15] assume an ignorable model, obtain samples of the response probabilities and use these sampled response probabilities to fit the response probabilities of a nonignorable nonresponse model while "controlling" its parameters. Of course, a possible alternative occurs when there is information about the degree of nonignorability.

The plan of this chapter is as follows. In Section 20.2, we describe the hierarchical Bayesian model. We introduce a new data distribution with a conjugate prior distribution. In Section 20.3, we illustrate Bayesian uncertainty analysis using data from the Slovenian Public Opinion survey. We use the griddy Gibbs sampler [17] to fit the model. We show how sensitive the posterior distribution of the finite population proportion is to a specified bound on one of the parameters; we study departures from the ignorable nonresponse model. Section 20.4 has concluding remarks.

20.2   The Nonignorable Nonresponse Model

For the problem of nonresponse in a three-dimensional table, we can have both item and unit nonresponse. Thus, one may consider the full data array to consist of eight tables: one for the complete data and seven supplemental tables – one for missing row information, one for missing column information, one for missing length information, one for both missing row and column, one for both missing row and length, one for both missing column and length and a table for which neither row, column nor length membership has been recorded. Throughout the chapter, we index rows by $j = 1, \ldots, r$; columns by $k = 1, \ldots, c$; lengths by $\ell = 1, \ldots, u$, and the eight tables by $t = 1, \ldots, T = 8$. We next describe the nonignorable nonresponse model (i.e., the expansion model); here it is slightly more convenient to use a pattern mixture model.

We assume that a random sample of size n is selected from a finite population of size N, there is no selection bias, and the n selected individuals can be classified into a three-way table of counts. We assume that there are three categorical variables which form the three-way table. Let $J_{it} = 1$ if the ith individual belongs to the tth table and $J_{it} = 0$ for the other seven tables (i.e., an individual belongs to only one of the eight tables). Let $I_{ijk\ell} = 1$ if the ith individual belongs to cell (j, k, ℓ) of the three-way table, and $I_{ijk\ell} = 0$ for all other cells (i.e., the ith individual belongs to only one cell of the three-way table).

Our basic model is as follows. Let

$$J_i \mid \underset{\sim}{\pi} \overset{iid}{\sim} \text{Multinomial}(1, \underset{\sim}{\pi}), I_i \mid J_{it} = 1, \underset{\sim}{p_t} \overset{iid}{\sim} \text{Multinomial}(1, \underset{\sim}{p_t}), \; t = 1, \ldots, 8.$$

Let $\psi_{jk\ell t} = \pi_t p_{jk\ell t}$. Then, because $\sum_t \pi_t = 1$ and $\sum_{jk\ell} p_{jk\ell t} = 1$ for each $t = 1, \ldots, 8$, $\sum_t \sum_{jk\ell} \pi_t p_{jk\ell t} = 1$. Let $w_{ijk\ell t} = J_{it} I_{ijk\ell}$. It follows that $\underset{\sim}{w_i} \mid \underset{\sim}{p}, \underset{\sim}{\pi} \overset{iid}{\sim} \text{Multinomial}(1, \underset{\sim}{\psi})$. Note that the vector $\underset{\sim}{\psi}$ has length of $8rcu$ and there are $8rcu - 1$ parameters subject to constraints. The parameters π_t are identifiable. However, the parameters $p_{jk\ell t}$ ($t = 1$, complete table) can be estimated, but $p_{jk\ell t}, t = 2, \ldots, 8$ (incomplete tables) cannot be estimated (i.e., the parameters $p_{jk\ell t}$ are not identifiable for $t = 2, \ldots, 8$). Clearly, if the $p_{jk\ell t}$ do not depend on t, then they will be identifiable. This latter case is the ignorable nonresponse model corresponding to the MAR mechanism. Also, note that given the constraints and the observed data, inferences for the $p_{jk\ell t}$ are independent of the π_t.

One way to reduce the effects of nonidentifiable parameters is to use a parsimonious nonignorable nonresponse model. This can be accomplished using two strategies (a) projection and (b) pooling. First, in (a) we can project $p_{jk\ell t}$ to a lower dimensional space. This can be done by expressing the $p_{jk\ell t}$ as functions of a reduced set of parameters, still one set of parameters for each of the 8 tables. Second, in (b) we can allow the reduced set of parameters to share a common distribution, as in small area estimation. This passes on the nonidentifiable effect to a smaller set of hyperparameters. Thus, (a) and (b) lead to a reduced set of nonidentifiable parameters. The specification of priors on these parameters is the Bayesian uncertainty analysis, and is the study of subjectivity. Rather than varying these nonidentifiable parameters at specified plausible values as in a formal non-Bayesian sensitivity analysis, we do so in a coherent Bayesian manner.

Our approach differs from the one in [14] which analyzes a data set arising from several $r \times c$ tables, each table corresponding to an area. Thus each table has a different probability model and there is a borrowing of strength across areas. Here a single $r \times c \times u$ contingency table is analyzed. This single table is made up of eight supplemental tables with different parameters. The parameters $p_{jk\ell t}$, $\sum_{jk\ell} p_{jk\ell t} = 1, t = 1, \ldots, 8$, are not identifiable and one strategy is to reduce this set of parameters. Henceforth, we will focus on the $2 \times 2 \times 2$ table.

20.2.1   Projection strategy

First, we consider a 2×2 categorical table. Let the first row total be x, the first column total be y and the count in $(1, 1)$ cell be z; the corresponding superpopulation proportions are p, q and θ. Note that $0 < z \leq x, y < n$ and the corresponding parameters are restricted as $0 < \theta < p, q < 1$. Then, for a random sample of n individuals, we have the joint density

$$
\begin{aligned}
&p(x, y, z \mid \theta, p, q) \\
&= \frac{n! \theta^z (p - \theta)^{x-z} (q - \theta)^{y-z} (1 - p - q + \theta)^{n-x-y+z}}{z!(x - z)!(y - z)!(n - x - y + z)!}, 0 < z \leq x, y < n.
\end{aligned}
$$

This is an extended version of the joint bivariate probability mass function (e.g., [7]). However, here we have three random variables x, y, z, and one needs to sum over z to get the bivariate binomial probability mass function of (x, y). Thus, here we have a trivariate random vector with a simple interpretation that x is the number of positives on one variable, y the number of positives on the other and z is the number of common positives (i.e., the number in the (1, 1) cell in the 2×2 table). It turns out that all three variables have binomial probability mass functions, and they are correlated because $z < x, y < n$. It is worth noting that this kind of modeling also takes care of the randomness of the marginal counts, a feature of categorical tables that is normally ignored.

Second, with a three-way table, each category having two levels, we can generalize this multinomial distribution as follows. We now assume that each level of the first category gives rise to a 2×2 table. A proportion of r individuals belong to the first table with a count of w and a proportion of $1 - r$ belong to the second table with a count of $n - w$. We relabel the counts in the two tables as (x_1, y_1, z_1) and (x_2, y_2, z_2). Thus, the joint probability mass function of $(w, x_1, y_1, z_1, x_2, y_2, z_2)$ is

$$
\begin{aligned}
&p(w, x_1, y_1, z_1, x_2, y_2, z_2 \mid r, \theta_1, p_1, q_1, \theta_2, p_2, q_2) \\
&= \frac{n!}{w!(n - w)!} r^w (1 - r)^{n-w} \\
&\quad \times \frac{w! \theta_1^{z_1} (p_1 - \theta_1)^{x_1 - z_1} (q_1 - \theta_1)^{y_1 - z_1} (1 - p_1 - q_1 + \theta_1)^{w - x_1 - y_1 + z_1}}{z_1!(x_1 - z_1)!(y_1 - z_1)!(w - x_1 - y_1 + z_1)!} \\
&\quad \times \frac{(n - w)! \theta_2^{z_2} (p_2 - \theta_2)^{x_2 - z_2} (q_2 - \theta_2)^{y_2 - z_2} (1 - p_2 - q_2 + \theta_2)^{n - w - x_2 - y_2 + z_2}}{z_2!(x_2 - z_2)!(y_2 - z_2)!(n - w - x_2 - y_2 + z_2)!},
\end{aligned}
$$

$0 < z_1 \leq x_1, y_1 < w, 0 < z_2 \leq x_2, y_2 < n - w, 0 < w < n$. Note that $0 < r < 1, 0 < \theta_{1t} < p_{1t}, q_{1t} < 1, 0 < \theta_{2t} < p_{2t}, q_{2t} < 1$. [The probabilistic model for the $r \times c \times u$ table can be developed in a similar manner.]

Again there are eight tables, and under nonignorabilty there will be eight probability mass functions with distinct parameters $(r_t, \theta_{1t}, p_{1t}, q_{1t}, \theta_{2t}, p_{2t}, q_{2t})$, $t = 1, \ldots, 8$. The number of observations falling in the tth table is n_t with $\sum_{t=1}^{8} n_t = n$ (i.e., n_t is a random variable). Also, all of these parameters are nonidentifiable, except those of the complete table. Not only these parameters are nonidentifiable, but there are missing values for the partially observed tables as well.

Finally, our nonignorable nonresponse model has a joint probability mass function of $(w_t, x_{t1}, y_{t1}, z_{t1}, x_{t2}, y_{t2}, z_{t2})$ given $(r_t, \theta_{1t}, p_{1t}, q_{1t}, \theta_{2t}, p_{2t}, q_{2t})$ and n_t, $t = 1, \ldots, 8$, are independent with probability mass functions given above. We also assume that each individual gets into the eight tables at random so that $\underline{n} \mid \underline{\pi} \sim \text{Multinomial}(n, \underline{\pi})$.

20.2.2   Pooling strategy

We discuss how to get around the nonidentifiability in the nonignorable nonresponse model using a Bayesian uncertainty analysis. This is the key issue in the current work.

We specify a joint conjugate prior distribution for $(r_t, \theta_{1t}, p_{1t}, q_{1t}, \theta_{2t}, p_{2t}, q_{2t})$ given n_t, $t = 1, \ldots, 8$, as follows. We take

$$r_t \mid \mu_0, \tau \stackrel{iid}{\sim} \text{Beta}\{\mu_0 \tau, (1 - \mu_0)\tau\}, 0 < \mu_0 < 1, \ \tau > 0,$$

$$(\theta_{st}, p_{st} - \theta_{st}, q_{st} - \theta_{st}, 1 - p_{st} - q_{st} + \theta_{st}) \mid \mu_1, \mu_2, \mu_3, \tau$$
$$\stackrel{iid}{\sim} \text{Dirichlet}\{\mu_1 \tau, (\mu_2 - \mu_1)\tau, (\mu_3 - \mu_1)\tau, (1 - \mu_2 - \mu_3 + \mu_1)\tau\},$$

$s = 1, 2$, $t = 1, \ldots, 8$, and $0 < \mu_1 < \mu_2, \mu_3 < 1$, $\tau > 0$. It is worth noting that the same parameter, τ, is used for r_t and $\theta_{st}, p_{st}, q_{st}, s = 1, 2$, thereby permitting a degree of parsimony.

We also take $\underline{\pi} \sim \text{Dirichlet}(1, \ldots, 1)$ (i.e., uniform in the simplex). This is the prior which corresponds to the cell probabilities of multinomial allocation of an individual to each of the eight tables.

It is worth noting that if $\mu_0, \mu_1, \mu_2, \mu_3$ and τ are specified, then the model will be well identified. A non-Bayesian analysis will proceed by taking various plausible values of these parameters to perform a sensitivity analysis on the finite population proportion. It is more sensible for a Bayesian to perform a Bayesian uncertainty analysis. That is, treat these parameters as hyperparameters by placing a prior on them. This will permit a study of subjectivity to provide a coherent method to obtain an uncertainty interval for the finite population proportion. But it will not work completely; an adjustment is still needed.

One way to do the adjustment is to put a bound on μ_1, say B, and take a shrinkage prior for τ (proper diffuse prior). That is, these parameters are constrained on the set, $S = \{(\mu_0, \mu_1, \mu_2, \mu_3) : 0 < \mu_0 < 1, 0 \le B \le \mu_1 < \mu_2, \mu_3 < 1\}$. We finally take

$$(\mu_0, \mu_1, \mu_2, \mu_3) \mid B \sim \text{Uniform}(S), \quad p(\tau) = 1/(1+\tau)^2, \quad \tau > 0.$$

A sensitivity analysis can proceed by fixing B at a specified value in $(0.05, 0.95)$. One needs to avoid the boundaries, otherwise there will be instability in computation (i.e., the nonidentifiability still exists). More importantly a Bayesian uncertainty analysis proceeds by putting a prior distribution on B preferably in the pessimistic-optimistic interval [a much smaller subset of $(0.05, 0.95)$], thereby making all model parameters stochastic and providing a full Bayesian model. This will help reduce boundary effects which are common in nonignorable nonresponse models (e.g., [16]).

Our nonignorable nonresponse model in which we specify B to be uniform in the pessimistic-optimistic interval is useful to perform a Bayesian uncertainty analysis. In fact, it is an important generalization and unification of a standard sensitivity analysis which proceeds by specifying plausible values (one at a time) of the nonidentifiable parameters and performing the whole analysis at each specified value.

20.3   Illustrative Example Using the Slovenian Plebiscite Data

We discuss the Slovenian Public Opinion Survey (SPOS) data. We note, however, that while [10, 19] studied a superpopulation parameter, we prefer to make inference about a finite population mean. Also, for simplicity in this illustrative example, we consider the case in which $\theta_{st} = \theta_t$, $p_{st} = p_t$ and $q_{st} = q_t, t = 1, \ldots, 8$.

20.3.1   Slovenian plebiscite data

In 1991, Slovenians voted for independence from former Yugoslavia in a plebiscite. A month before the plebiscite the Slovenian government collected data in the SPOS in which three fundamental questions were added. The three questions are: (a) Are you in favor of Slovenian's secession from Yugoslavia? (b) Will you attend the plebiscite? (c) Are you in favor of Slovenian's independence? Questions (a) and (c) are very close but different since independence would have been possible in a confederal form as well. Rubin *et al.* [19] studied these questions with emphasis on the missing data process. We will perform a Bayesian uncertainty analysis of the categorical table first presented in [10, 19]

TABLE 20.1
Data from the Slovenian Public Opinion survey (SPOS).

Secession	Attendance	Independence		
		Yes	No	DK
Yes	Yes	**1191**	8	21
	No	8	0	4
	DK	107	3	9
No	Yes	**158**	68	29
	No	7	14	3
	DK	18	43	31
DK	Yes	90	2	109
	No	1	2	25
	DK	19	8	96

NOTE: The 'don't know' category is indicated by DK [19]. The sample size (n) is 2074 and the population size (N) is 1,460,000 Slovenians; so that the sampling fraction (f) is 0.00142. The true population proportion $(P$, attend plebiscite and vote for independence) is 0.885. The complete table has 1454 Slovenians (i.e., partial nonresponse is 29.89%). The pessimistic-optimistic range is $(0.694, 0.905)$.

and in the last example of [8]. We will study inference about the finite population proportion of Slovenians who would attend the plebiscite and vote for independence.

The data from the SPOS is given in Table 20.1. Let γ denote the superpopulation proportion of Slovenians who would attend the plebiscite and vote for independence. Rubin *et al.* [19] presented six different estimates of γ. The conservative (also called pessimistic) estimate of $((1191 + 158 + 90)/2074 = 1439/2074 =)$ 0.694 assumes that every "don't know" (DK) response was really a negative (no) response. The complete-cases estimate of 0.928 uses the responses from all individuals who answered all the three questions, and the available-cases estimate of 0.929 uses everyone who answered the independence and attendance questions. A saturated ignorable nonresponse multinomial (MAR) model estimate is 0.892 with two questions and 0.883 with three questions, a value close to the true plebiscite value of 0.885. The nonignorable nonresponse model gives an estimate of 0.782, quite disappointing to Rubin *et al.* [19]. Molenberghs *et al.* [10] gave an optimistic estimate of $((1439 + 439)/2074 =)$ 0.905 which assumes that all DKs are actually yeses, $(21 + 107 + 9) + (29 + 18 + 31) + (109 + 19 + 96) = 439$; they called it the conservative estimate and gave a pessimistic-optimistic (nonparametric) interval of $(0.694, 0.905)$.

Molenberghs *et al.* [10] fit several additional models to the two-way table of attendance and independence, collapsed over secession. They started with a loglinear model of [1] and fit twelve versions of the model; nine of them with no sensitivity parameters and three as overspecified models (two of them with one sensitivity parameter and the other one with two such parameters). The range of estimates covered by the first nine models is $(0.753, 0.891)$, a considerable part of the pessimistic-optimistic range, with some nonignorable nonresponse models containing the true proportion. The intervals of ignorance of the three models are $(0.762, 0.893)$, $(0.766, 0.883)$ and $(0.694, 0.905)$ and the uncertainty intervals are $(0.744, 0.907)$, $(0.715, 0.920)$ and $(0.694, 0.905)$; all the intervals contain the true proportion except the second interval of ignorance.

20.3.2   Bayesian uncertainty analysis

Let $r_{1t} = r_t$, $r_{2t} = 1 - r_t$, $w_{1t} = w_t$, $w_{2t} = n_t - w_t$, and d_{mis} and d_{obs} denote all missing data and all observed data, respectively. Then, the joint posterior density of all parameters and missing values is

$$\pi(\theta, p, q, d_{mis}, \mu_0, \mu_1, \mu_2, \mu_3, \tau, \pi \mid d_{obs})$$

$$\propto \prod_{k=1}^{2} \prod_{t=1}^{T} \left\{ \frac{(r_{kt}\theta_t)^{z_{kt}}}{z_{kt}!} \frac{(r_{kt}(p_t - \theta_t))^{x_{kt}-z_{kt}}}{(x_{kt} - z_{kt})!} \frac{(r_{kt}(q_t - \theta_t))^{y_{kt}-z_{kt}}}{(y_{kt} - z_{kt})!} \right.$$

$$\left. \frac{(r_{kt}(1 - p_t - q_t + \theta_t))^{w_{kt}-x_{kt}-y_{kt}+z_{kt}}}{(w_{kt} - x_{kt} - y_{kt} + z_{kt})!} \right\}$$

$$\times \frac{(\Gamma(\tau))^{2T}}{(1+\tau)^2} \prod_{t=1}^{T} \left\{ \frac{\theta_t^{\mu_1\tau-1}}{\Gamma(\mu_1\tau)} \frac{(p_t - \theta_t)^{(\mu_2-\mu_1)\tau-1}}{\Gamma((\mu_2 - \mu_1)\tau)} \frac{(q_t - \theta_t)^{(\mu_3-\mu_1)\tau-1}}{\Gamma((\mu_3 - \mu_1)\tau)} \right.$$

$$\left. \frac{(1 - p_t - q_t + \theta_t)^{(1-\mu_2-\mu_3+\mu_1)\tau-1}}{\Gamma((1 - \mu_2 - \mu_3 + \mu_1)\tau)} \right\}$$

$$\times \prod_{t=1}^{T} \frac{r_t^{\mu_0\tau-1} (1-r_t)^{(1-\mu_0)\tau-1}}{\Gamma(\mu_0\tau) \Gamma((1 - \mu_0)\tau)} \times \prod_{t=1}^{T} \pi_t^{n_t}, \qquad (20.1)$$

$0 < \mu_0 < 1$, $0 < B < \mu_1 < \mu_2$, $\mu_3 < 1$ and $0 < r_t < 1$, $0 < \theta_t < p_t$, $q_t < 1$, $t = 1, \ldots, T = 8$.

We used the griddy Gibbs sampler [17] to draw samples from the joint posterior density (20.1). The joint conditional posterior distributions of the missing data have standard multinomial forms. In the joint conditional posterior density, (θ_t, p_t, q_t) are independent over t, and they have standard Dirichlet distributions. However, the joint posterior density of $(\mu_0, \mu_1, \mu_2, \mu_3, \tau)$ does not exist in a closed form. Each of these is obtained using a grid method. The detailed formulae about the conditional densities to perform the griddy Gibbs sampler are given in subsection 20.5.1, Appendix A. We monitored the griddy Gibbs sampler as follows. We drew 11,000 iterates, used 1,000 as a 'burn-in' and took each iterate thereafter. We found negligible autocorrelations among the iterates, and so it is good that 'thinning' is not needed. We have also run the Geweke test of stationarity on $\mu_o, \mu_1, \mu_2, \mu_3$ and τ and found no evidence of nonstationarity. For numerical summaries, we use the posterior mean (PM), posterior standard deviation (PSD) and CI with coverage probability 0.95.

Let P denote the finite population proportion of Slovenians who would attend the plebiscite and vote for independence. We show how to make inference about P under the nonignorable nonresponse model; the idea is the same under similar models. [10] and [19] made inference about the superpopulation mean. In subsection 20.5.2, Appendix B, we describe how to infer the finite population proportion.

First, we perform a small sensitivity analysis to investigate how inference about P changes with different values of B. [Note that this sensitivity analysis is not the Bayesian uncertainty analysis.] For τ we also compare the shrinkage prior with a data-based gamma prior which is $\tau \sim$ Gamma(185, 9). We have set B at 0.1, 0.2, 0.3, 0.4, 0.5, 0.6, 0.7, 0.8, 0.9. For the shrinkage prior, the corresponding 0.95 uncertainty intervals of P are $(0.835, 0.900)$, $(0.829, 0.897)$, $(0.832, 0.904)$, $(0.835, 0.901)$, $(0.831, 0.906)$, $(0.861, 0.910)$, $(0.859, 0.910)$, $(0.870, 0.913)$, $(0.864, 0.915)$; the intervals are similar to those for the gamma prior. All these intervals contain the true value of $P = 0.885$ and we observe that these intervals tend to move over to the right as B increases but with the smallest movement when B is within the pessimistic-optimistic range. In Figure 20.1, we also present box plots of the empirical posterior densities of P as a function of the sensitivity bound for the shrinkage prior; the box plots for the gamma prior are similar. Again, the box plots show a general increasing trend and the box plots narrow down as the pessimistic-optimsitic interval is approached. We have also obtained kernel density estimates (Parzen-Rosenblatt) of the posterior densities of the proportion for these values of B and we observe that these densities are roughly similar, unimodal and slightly skewed to the left, and they are slightly more peaked in the pessimistic-optimistic range.

Second and more importantly for the Bayesian uncertainty analysis, we set a uniform prior on the sensitivity bound, B, in the pessimistic-optimistic range. [We present results for only the shrinkage prior.] That is, we take $B \sim$ Uniform(a, b), where $a = 0.694$ and $b = 0.905$. Then, the conditional posterior density is $B \mid \mu_1 \sim$ Uniform$\{a, \min(b, \mu_1)\}$. We have now added this conditional posterior to the griddy Gibbs sampler. For inference about the proportion of Slovenians who would attend and vote for independence, we get a PM of 0.895 with a PSD of 0.011 and a 0.95 credible interval of $(0.868, 0.913)$, still informative and containing the true value. In Figure 20.2, we have plotted the estimated posterior densities of P and the hyperparameters. First, observe that the posterior density of P is unimodal and skewed to the left; thus normal theory is not appropriate, and the fact that μ_1 has informative distribution helps in the estimation of P. Also, the posterior density of B is close to the

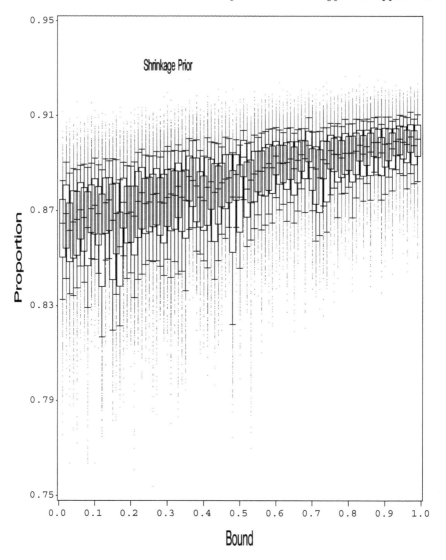

FIGURE 20.1
Boxplots of the posterior distributions of the finite population proportion as
a function of the sensitivity bound

uniform; thus as expected there is little information in the data about B. In
addition, μ_0 and μ_1 are unimodal and skewed to the left. However, although
μ_2 and μ_3 are not on the boundary at 1, they have modes close to it; thus, at
least boundary effects for μ_2 and μ_3, the more difficult parameters to estimate,
are reduced.

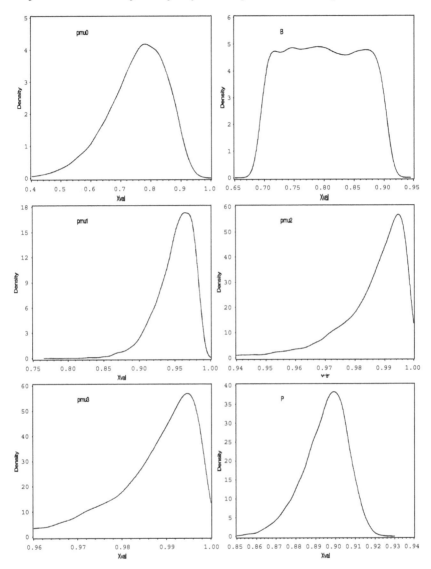

FIGURE 20.2
Plots of the estimated posterior densities of P, the hyperparameters, μ_0, μ_1, μ_2, μ_3 and, the sensitivity bound, B

20.4 Concluding Remarks

The purpose of this chapter has been to develop a methodology to analyze data from a single incomplete three-way categorical table. We have constructed a new Bayesian nonignorable nonresponse model. We have obtained a parsimonious nonignorable nonresponse model with a much reduced set of nonidentifiable parameters, each of the eight partially complete tables has a set of parameters. We allowed these parameters to share a common effect, thereby passing on the nonidentifiable effects to a manageable set of parameters. For a Bayesian uncertainty analysis, we have placed priors on these nonidentifiable hyperparameters. This allows a study of subjectivity in which an uncertainty interval (include both ignorance and imprecision) for the finite population proportion is obtained.

We have studied the Slovenian plebiscite data discussed by Rubin *et al.* [19]. We have studied the finite population proportion of Slovenians who would attend the plebiscite and vote for independence. The work in this chapter was motivated by some interesting comments made by Molenberghs *et al.* [10] about Rubin *et al.* [19]. Our model is an improvement over the one provided by Rubin *et al.* [19]. We have obtained an uncertainty interval which is not too wide and it still contains the true proportion of Slovenians who would attend the plebiscite and vote for independence.

Although our main objective is to obtain a Bayesian uncertainty analysis, we have performed a sensitivity analysis on our nonignorable nonresponse model. We observe that posterior inference about the finite population proportion of those who would attend and vote for independence is sensitive to the sensitivity bound. However, the changes in the pessimistic-optimistic interval are small. In the Bayesian uncertainty analysis (priors on nonidentifiable parameters), our nonignorable nonresponse model captures the true proportion very well as compared with the value of 0.782 [19].

Acknowledgment

We are grateful to the two referees for their interesting comments.

20.5   Appendix

20.5.1   Appendix A: Conditional posterior distributions

We state the conditional posterior densities (cpds) needed to fit the joint posterior density in (20.1) using the griddy Gibbs sampler [17].

After some algebraic manipulation in (20.1), one can show that the conditional posterior distributions of the missing data are in simple forms. Also the conditional posterior densities of $r_t, \theta_t, p_t, q_t, t = 1, \ldots, 8$ are in simple forms. The conditional posterior distributions of μ and τ are not in closed forms.

Given the data, $\underset{\sim}{d} = (\underset{\sim}{d}_{obs}, \underset{\sim}{d}_{mis})$ and μ and τ, r_t and $(\theta_t, p_t - \theta_t, q_t - \theta_t, 1 - p_t - q_t + \theta_t)$ are independent with

$$r_t \mid w_t, n_t, \mu_0, \tau, \underset{\sim}{d} \overset{ind}{\sim} \text{Beta}(w_t + \mu_0\tau, n_t - w_t + (1 - \mu_0)\tau), \quad t = 1, \ldots, 8,$$

$$(\theta_t, p_t - \theta_t, q_t - \theta_t, 1 - p_t - q_t + \theta_t) \mid \mu_0, \mu_1, \mu_2, \mu_3, \tau, \underset{\sim}{d}$$
$$\overset{ind}{\sim} \text{Dirichlet}\{z_t + \mu_1\tau, x_t - z_t + (\mu_2 - \mu_1)\tau, y_t - z_t$$
$$+ (\mu_3 - \mu_1)\tau, n_t - x_t - y_t + z_t + (1 - \mu_2 - \mu_3 + \mu_1)\tau\}.$$

For convenience, we define the geometric mean (GM) of $b_1, \ldots, b_T$ as $GM(b_1, \ldots, b_T) = (\prod_{t=1}^{T} b_t)^{1/T}, T = 8$. Let $a_1 = \mu_0\tau$, $a_2 = (1 - \mu_0)\tau$, $a_3 = \mu_1\tau$, $a_4 = (\mu_2 - \mu_1)\tau$, $a_5 = (\mu_3 - \mu_1)\tau$, $a_6 = (1 - \mu_2 - \mu_3 + \mu_1)\tau$. Also, let $G_1 = GM(r_t, t = 1, \ldots, 8)$, $G_2 = GM(1 - r_t, t = 1, \ldots, 8)$, $G_3 = GM(\theta_t, t = 1, \ldots, 8)$, $G_4 = GM(p_t - \theta_t, t = 1, \ldots, 8)$, $G_5 = GM(q_t - \theta_t, t = 1, \ldots, 8)$ and $G_6 = GM(1 - p_t - q_t + \theta_t, t = 1, \ldots, 8)$. Then,

$$p(\mu_0, \mu_1, \mu_2, \mu_3, \tau \mid \underset{\sim}{r}, \underset{\sim}{\theta}, \underset{\sim}{p}, \underset{\sim}{q}, \underset{\sim}{d})$$

$$\propto \left\{ \frac{\Gamma(\tau)^{2T}}{(1 + \tau)^2} \right\} \left\{ \prod_{s=1}^{6} \frac{G_s^{a_s}}{\Gamma(a_s)} \right\}^T, \quad 0 < \mu_0 < 1, B < \mu_1 < \mu_2, \mu_3 < 1, \tau > 0.$$

[A small adjustment is needed with a different prior for τ (e.g., a gamma prior).] Samples from this joint conditional posterior density can be obtained using the grid method.

Also, $\underset{\sim}{\pi} \mid \underset{\sim}{n} \sim \text{Dirichlet}(\underset{\sim}{n} + \underset{\sim}{1})$ where $\underset{\sim}{1}$ is a vector of ones.

We state the conditional posterior densities; see Table 20.2 for the notations. We need the conditional posterior densities for Tables s2-s8. Generally, samples can be drawn from these conditional densities using the composition method. Momentarily, we drop identifier $t = 1, \ldots, T = 8$ of the supplemental table and note that we condition on all parameters and n_t.

For Table s2, $x_1 + x_2 = x$, $y_1 + y_2 = y$ and $z_1 + z_2 = z$ are observed. Thus, the cpds are

$$z_1 \mid z, r \sim \text{Binomial}(z, r), \quad x_1 - z_1 \mid x, z, z_1 \sim \text{Binomial}(x - z, r),$$

$$y_1 - z_1 \mid y, z, z_1 \sim \text{Binomial}(y - z, r),$$

$$w - x_1 - y_1 + z_1 \mid x, y, z, x_1, y_1, z_1 \sim \text{Binomial}(n - x - y + z, r).$$

For Table s3, w, y_1 and y_2 are observed. Thus, the cpds are

$$z_1 \mid y_1, \theta, p, q \sim \text{Binomial}(y_1, \frac{\theta}{q}), x_1 - z_1 \mid z_1, \theta, p, q \sim \text{Binomial}(w - y_1, \frac{p - \theta}{1 - q}),$$

TABLE 20.2
Tables of observed and missing counts.

Table	Observed	Missing
s1	$x_k, y_k, z_k,\ k = 1, 2, w$	None
s2	$x_1 + x_2,\ y_1 + y_2,\ z_1 + z_2$	x_1, y_1, z_1, w
s3	y_1, y_2, w	x_1, x_2, z_1, z_2
s4	x_1, x_2, w	y_1, y_2, z_1, z_2
s5	w	$x_1, x_2, y_1, y_2, z_1, z_2$
s6	$x_1 + x_2$	$x_1, x_2, y_1, y_2, z_1, z_2, w$
s7	$y_1 + y_2$	$x_1, x_2, y_1, y_2, z_1, z_2, w$
s8	None	$x_k, y_k, z_k,\ k = 1, 2, w$

NOTE: Table s1 is complete; each of Tables s2, s3, s4 has one category missing; each of Tables s5, s6, s7 has two categories missing; and Table s8 has all categories missing.

$$z_2 \mid y_2, \theta, p, q \sim \text{Binomial}(y_2, \frac{\theta}{q}),\ x_2 - z_2 \mid z_2, \theta, p, q \sim \text{Binomial}(n - w - y_2, \frac{p - \theta}{1 - q}).$$

For Table s4, w, x_1 and x_2 are observed. Thus, the cpds are

$$z_1 \mid x_1, \theta, p, q \sim \text{Binomial}(x_1, \frac{\theta}{p}),\ y_1 - z_1 \mid z_1, \theta, p, q \sim \text{Binomial}(w - x_1, \frac{q - \theta}{1 - p}),$$

$$z_2 \mid x_2, \theta, p, q \sim \text{Binomial}(x_2, \frac{\theta}{p}),\ y_2 - z_2 \mid z_2, \theta, p, q \sim \text{Binomial}(n - w - x_2, \frac{q - \theta}{1 - p}).$$

For Table s5, only w is observed. Thus, the cpds are

$$(z_1, x_1 - z_1, y_1 - z_1, w - x_1 - y_1 + z_1)' \mid w, \theta, p, q$$
$$\sim \text{Multinomial}\{w, (\theta, p - \theta, q - \theta, 1 - p - q + \theta)'\},$$
$$(z_2, x_2 - z_2, y_2 - z_2, w - x_2 - y_2 + z_2)' \mid w, \theta, p, q$$
$$\sim \text{Multinomial}\{n - w, (\theta, p - \theta, q - \theta, 1 - p - q + \theta)'\}.$$

For Table s6, only $x_1 + x_2 = x$ is observed. Thus, the cpds are

$$(z_1, x_1 - z_1, z_2, x - x_1 - z_2)' \mid x, r, \theta, p, q \sim$$
$$\text{Multinomial}\{x, \frac{(r\theta, r(p - \theta), (1 - r)\theta, (1 - r)(p - \theta))'}{p}\},$$

$$(y_1 - z_1, w - x_1 - y_1 + z_1, y_2 - z_2, n - w - x + x_1 - y_2 + z_2)' \mid x_1, z_1, z_2, r, \theta, p, q \sim$$
$$\text{Multinomial}\{n - x, \frac{(r(q - \theta), r(1_{pq\theta}), (1 - r)(q - \theta), (1 - r)(1_{pq\theta}))'}{1 - p}\},$$

where $1_{pq\theta} = (1 - p - q + \theta)$.

For Table s7, only $y_1 + y_2 = y$ is observed. Thus, the cpds are

$$(z_1, y_1 - z_1, z_2, y - y_1 - z_2)' \mid y, r, \theta, p, q \sim$$
$$\text{Multinomial}\{y, \frac{(r\theta, r(q - \theta), (1 - r)\theta, (1 - r)(q - \theta))'}{q}\},$$

$$(x_1 - z_1, w - x_1 - y_1 + z_1, x_2 - z_2, n - w - x_2 - y + y_1 + z_2)' \mid y_1, z_1, z_2, r, \theta, p, q \sim$$
$$\text{Multinomial}\{n - y, \frac{(r(p - \theta), r(1_{pq\theta}), (1 - r)(p - \theta), (1 - r)(1_{pq\theta}))'}{1 - q}\},$$

where $1_{pq\theta} = (1 - p - q + \theta)$.

Finally, for Table s8, all counts are missing, and the cpds are

$$w \mid r \sim \text{Binomial}(n, r),$$
$$(z_1, x_1 - z_1, y_1 - z_1, w - x_1 - y_1 + z_1)' \mid w, \theta, p, q$$
$$\sim \text{Multinomial}\{w, (\theta, p - \theta, q - \theta, 1 - p - q + \theta)'\},$$
$$(z_2, x_2 - z_2, y_2 - z_2, w - x_2 - y_2 + z_2)' \mid w, \theta, p, q$$
$$\sim \text{Multinomial}\{n - w, (\theta, p - \theta, q - \theta, 1 - p - q + \theta)'\}.$$

20.5.2   Appendix B: Inference for the finite population proportion

Let N_t denote the total number of Slovenians responding in the tth table if all N Slovenians were sampled (i.e., a census was taken). We are assuming that there is no selection bias, and the sample is representative of the Slovenian population. Also, let $z_t = z_{1t} + z_{2t}$ and $Z_t = Z_{1t} + Z_{2t}$, $t = 1, \ldots, 8$, where Z_{1t} and Z_{2t} are the total number of nonsampled individuals (common positives) at the two levels of the first variable. So that Z_t is the total nonsample Slovenians in the tth table who would attend the plebiscite and vote for independence. [Distinguish between small z and big Z.] Then $P = \sum_{t=1}^{8} N_t / N = \{\sum_{t=1}^{8} z_t + \sum_{t=1}^{8} Z_t\}/N$.

Alternatively, using standard notation in survey sampling, we can write $P = f\bar{z}_s + (1 - f)\bar{z}_{ns}$, where $\bar{z}_s$ is the sample proportion, $\bar{z}_{ns}$ is the nonsample proportion, and $f = n/N$ is the sampling fraction. Note that both $\bar{z}_s$ and $\bar{z}_{ns}$ are unobserved; $\bar{z}_s$ is unobserved because only the values in the complete table are observed and the counts are unobserved for the seven partially complete table. Thus, given the sampled data, both $\bar{z}_s$ and $\bar{z}_{ns}$ are random variables. While $\bar{z}_s$ is obtained directly from the model fitting via the z_{t1} and z_{t2}, $\bar{z}_{ns}$ has to be predicted.

Now we show how to predict $\bar{z}_{ns}$. Let $\tilde{N} = N - n$ denote the number of nonsample individuals, $\tilde{N}_{t1} = N_{t1} - w_t$ and $\tilde{N}_{t2} = N_{t2} - (n_t - w_t)$. This gives $\tilde{N}_t = \tilde{N}_{t1} + \tilde{N}_{t2} = N_t - n_t, t = 1, \ldots, 8$. Then

$$\bar{z}_{ns} = \sum_{t=1}^{8} \left(\frac{\tilde{N}_t}{\tilde{N}}\right)\left(\frac{Z_t}{\tilde{N}_t}\right),$$

Let $\tilde{\underset{\sim}{N}} = (\tilde{N}_1, \ldots, \tilde{N}_8)'$ and $\underset{\sim}{Z} = (Z_1, \ldots, Z_8)'$. Then, under the nonignorable nonresponse model,

$$Z_{tk} \mid \theta_{tk}, \tilde{N}_{tk} \overset{ind}{\sim} \text{Binomial}(\tilde{N}_{tk}, \theta_{tk}), \; t = 1, \ldots, 8, \; k = 1, 2,$$

$$\tilde{N}_{t1} \mid \tilde{N}_t, r_t \overset{ind}{\sim} \text{Binomial}(\tilde{N}_t, r_t), \; \tilde{N}_{t2} = \tilde{N}_t - \tilde{N}_{t1}, \tilde{\underset{\sim}{N}} \mid \underset{\sim}{\pi} \sim \text{Multinomial}(N - n, \underset{\sim}{\pi}).$$

Thus, inference about $\bar{z}_{ns}$, and therefore P, can be made in a standard manner.

Bibliography

[1] S. G. Barker, W. F. Rosenberger, and R. DerSimonian. Closed-form estimates for missing counts in two-way contingency tables. *Statistics in Medicine*, 11(5):643–657, 1992.

[2] T. Chen and S. E. Fienberg. Two-dimensional contingency tables with both completely and partially cross-classified data. *Biometrics*, 30:629–642, 1974.

[3] D. Draper. Assessment and propagation of model uncertainty (with discussion). *Journal of the Royal Statistical Society: Series B (Statistical Methodology)*, 57(1):45–97, 1995.

[4] J. J. Forster and P. F. Smith. Model-based inference for categorical survey data subject to non-ignorable nonresponse. *Journal of the Royal Statistical Society: Series B (Statistical Methodology)*, 60(1):57–70, 1998.

[5] P. J. Green. Reversible jump Markov chain Monte Carlo computation and Bayesian model determination. *Biometrika*, 82(4):711–732, Dec. 1995.

[6] S. Greenland. Relaxation penalties and priors for plausible modeling of nonidentified bias sources. *Statistical Science*, 24(2):195–210, 2009.

[7] M. A. Hamdan and H. A. Al-Bayyati. A note on the bivariate Poisson distribution. *The American Statistician*, 23(4):32–33, 1969.

[8] R. J. A. Little and D. B. Rubin. *Statistical Analysis with Missing Data.* John Wiley & Sons, New York, 2nd edition, 2002.

[9] G. Molenberghs, E. Goetghebeur, S. R. Lipsitz, and M. G. Kenward. Nonrandom missingness in categorical data : Strengths and limitations. *The American Statistician*, 53(2):110–118, 1999.

[10] G. Molenberghs, M. G. Kenward, and E. Goetghebeur. Sensitivity analysis for incomplete contingency tables : The Slovenian plebiscite case. *Journal of the Royal Statistical Society: Series C (Applied Statistics)*, 50(1):15–29, 2001.

[11] B. Nandram and J. W. Choi. Hierarchical Bayesian nonresponse models for binary data from small areas with uncertainty about ignorability. *Journal of the American Statistical Association*, 97(458):381–388, 2002.

[12] B. Nandram and J. W. Choi. A Bayesian analysis of body mass index data from small domains under nonignorable nonresponse and selection. *Journal of the American Statistical Association*, 105(489):120–135, 2010.

[13] B. Nandram, L. Cox, and J. W. Choi. Bayesian analysis of nonignorable missing categorical data: An application to bone mineral density and family income. *Survey Methodology*, 31(2):213–225, 2005.

[14] B. Nandram and M. Katzoff. A hierarchical Bayesian nonresponse model for two-way categorical data from small areas with uncertainty about ignorability. *Survey Methodology*, 38(1):81–93, 2012.

[15] B. Nandram, N. Liu, J. W. Choi, and L. Cox. Bayesian nonresponse models for categorical data from small areas: An application to BMD and age. *Statistics in Medicine*, 24(7):1047–1074, 2005.

[16] T. Park and M. B. Brown. Models for categorical data with nonignorable nonresponse. *Journal of the American Statistical Association*, 89(425):44–52, 1994.

[17] C. Ritter and M. A. Tanner. Facilitating the Gibbs sampler: The Gibbs stopper and the griddy-Gibbs sampler. *Journal of the American Statistical Association.*, 87(419):861–868, 1992.

[18] D. B. Rubin. Inference and missing data. *Biometrika*, 63(3):581–592, 1976.

[19] D. B. Rubin, H. S. Stern, and V. Vehovar. Handling "Don't Know" survey responses: The case of the Slovenian plebiscite. *Journal of the American Statistical Association*, 72(431):822–828, 1995.

21

Stochastic Volatility and Realized Stochastic Volatility Models

Yasuhiro Omori

University of Tokyo

Toshiaki Watanabe

Hitotsubashi University

CONTENTS

21.1 Introduction

Asset price volatility plays an important role in financial risk management such as option pricing and Value-at-Risk. Few would dispute the fact that volatility changes over time and there has been a surge in modelling the dynamics of volatility. Many models have been proposed but they can be classified into two. One is the autoregressive conditional heteroskedasticity (ARCH) and generalized ARCH (GARCH) models proposed by [15] and [6] and their extensions. See [7] and [19] for ARCH models. The other is the stochastic volatility (SV) model proposed by [38]. While ARCH-type models can be estimated using the maximum likelihood method, it is difficult to evaluate the likelihood of SV model. There has also been a surge in developing the

estimation method for the SV model instead of the maximum likelihood estimation (see [9]) and [18]). The SV model is also becoming popular in macroe-
conometrics. For example, see [16], [25] and [33].

In this chapter, we review the SV models and the Bayesian method for the
estimation of SV models using Markov chain Monte Carlo (MCMC). In this
method, the parameters and latent volatility are sampled from their posterior distribution using MCMC. There are two efficient methods for sampling
volatility. One is the mixture sampler proposed by [26] and the other is the
block sampler proposed by [34] and [40]. We explain the block sampler. It
is a well-known phenomenon in stock markets that there is a negative correlation between today's return and tomorrow's volatility. The SV model has
been extended to the asymmetric SV (ASV) model taking account of this
phenomenon. The mixture and block samplers for the ASV model have been
proposed by [30] and [31] respectively. We also review the ASV model and its
Bayesian estimation using MCMC.

Since high-frequency data on asset prices has become available, the realized
volatility (RV), which is the sum of squared intraday returns, has recently attracted the attention of financial econometricians. The RV is, however, known
to have the bias caused by microstructure noise and non-trading hours. [36]
extends the SV model to the realized SV (RSV) model where they model the
daily return and RV jointly taking account of the bias of RV. See also [14] and
[27] for the RSV model. [20] proposes the realized GARCH model where they
apply the same idea to the GARCH model. We also review the RSV model
and its Bayesian estimation using MCMC.

We illustrate the MCMC Bayesian estimation of the SV and RSV models
by applying both models to daily returns and RV of S&P 500 stock index. We
also show that the RSV performs better than the SV model in the prediction
of the one-day-ahead volatility.

The rest of this chapter is organized as follows. Sections 21.2 and 21.3
review SV and RSV models and their estimation method. Section 21.4 estimates SV and RSV models using daily returns and RV of S&P 500. Section
21.5 concludes.

21.2   Stochastic Volatility Model and Efficient Sampler

Let p_t denote an asset price at time t and define the asset return (as a percentage) by $R_t = (\log p_t - \log p_{t-1}) \times 100$. The stochastic volatility (SV) model is

given by

$$R_t = \exp(h_t/2)\epsilon_t, \quad t = 1, \ldots, n, \tag{21.1}$$

$$h_{t+1} = \mu + \phi(h_t - \mu) + \eta_t, \quad t = 1, \ldots, n-1, \quad |\phi| < 1, \tag{21.2}$$

$$h_1 \sim \mathcal{N}(\mu, \sigma_\eta^2/(1 - \phi^2)),$$

$$\begin{pmatrix} \epsilon_t \\ \eta_t \end{pmatrix} \overset{i.i.d.}{\sim} \mathcal{N}\left(\begin{pmatrix} 0 \\ 0 \end{pmatrix}, \begin{pmatrix} 1 & \rho\sigma_\eta \\ \rho\sigma_\eta & \sigma_\eta^2 \end{pmatrix} \right),$$

where h_t is a log volatility. The log volatility is an unobserved latent variable and is assumed to follow a stationary AR(1) process as in (21.2). We let $\boldsymbol{\theta} = (\mu, \phi, \sigma_\eta^2, \rho)'$ and assume that $(\epsilon_t, \eta_t)'$ follows a bivariate normal distribution. The persistence parameter ϕ for h_t is assumed to satisfy $|\phi| < 1$ for the stationarity. In empirical studies of stock returns, we observe the volatility clustering (i.e., the high (low) volatility period tends to continue for a certain period of time) and the estimate of ϕ is usually close to one. The correlation coefficient, ρ, between R_t and h_{t+1} is often estimated to be negative, implying that the decrease (increase) in the asset return at time t is followed by the increase (decrease) in the log volatility at time $t + 1$. This is called an asymmetry or a leverage effect.

In the SV model, the state equation (21.2) is linear and Gaussian, but the measurement equation (21.1) is a nonlinear function of h_t. Thus, we are not able to obtain the likelihood function given $\boldsymbol{\theta}$ analytically and instead need to compute it numerically by Monte Carlo simulation. We take a Bayesian approach and estimate model parameters by MCMC simulation in this chapter.

Let $\pi(\boldsymbol{\theta}) = \pi(\mu)\pi(\phi)\pi(\sigma_\eta^2, \rho)$ and $f(\boldsymbol{y}, \boldsymbol{h}|\boldsymbol{\theta})$ denote the prior probability density function of the parameter $\boldsymbol{\theta}$ and the joint probability density function of $\boldsymbol{y} = (R_1, \ldots, R_n)'$ and $\boldsymbol{h} = (h_1, \ldots, h_n)'$ given $\boldsymbol{\theta}$ respectively. Then the joint posterior probability density function of $\boldsymbol{\theta}$ and $\boldsymbol{h}$ is given by

$$\pi(\boldsymbol{\theta}, \boldsymbol{h}|\boldsymbol{y})$$
$$\propto f(\boldsymbol{y}, \boldsymbol{h}|\boldsymbol{\theta})\pi(\boldsymbol{\theta})$$
$$\propto \exp\left\{ -\frac{1}{2}\sum_{t=1}^{n} h_t - \frac{1}{2}\sum_{t=1}^{n} R_t^2 \exp(-h_t) \right\}$$
$$\times \sqrt{1 - \phi^2} \times (\sigma_\eta^2)^{-\left(\frac{n}{2}\right)} \times (1 - \rho^2)^{-\frac{n-1}{2}}$$
$$\times \exp\left\{ -\frac{(1 - \phi^2)(h_1 - \mu)^2}{2\sigma_\eta^2} - \sum_{t=1}^{n-1} \frac{(h_{t+1} - (1 - \phi)\mu - \phi h_t - \bar{y}_t)^2}{2\sigma_\eta^2(1 - \rho^2)} \right\} \pi(\boldsymbol{\theta}),$$

where $\bar{y}_t = \rho\sigma_\eta \exp(-h_t/2)R_t$. We use MCMC simulation to generate random samples from the posterior distribution of $\boldsymbol{\theta}$ and conduct a statistical inference on parameters. We implement MCMC in six blocks to generate random samples from the full conditional posterior distribution.

Step 1. Initialize $\boldsymbol{\theta} = (\phi, \mu, \sigma_\eta^2, \rho)'$ and $\boldsymbol{h}$.

Step 2. Generate $\phi|\mu, \sigma_\eta^2, \rho, \boldsymbol{h}, \boldsymbol{y}$.

Step 3. Generate $\mu|\phi, \sigma_\eta^2, \rho, \boldsymbol{h}, \boldsymbol{y}$.

Step 4. Generate $(\sigma_\eta^2, \rho)|\phi, \mu, \boldsymbol{h}, \boldsymbol{y}$.

Step 5. Generate $\boldsymbol{h}|\phi, \mu, \sigma_\eta^2, \rho, \boldsymbol{y}$.

Step 6. Go to Step 2.

21.2.1   Generation of $\boldsymbol{\theta} = (\phi, \mu, \sigma_\eta^2, \rho)'$

Generation of ϕ **(Step 2).** Let $\pi(\phi)$ denote a prior probability density function of ϕ[1]. Then the conditional posterior probability density function of ϕ is given by

$$\pi(\phi|\mu, \sigma_\eta^2, \rho, \boldsymbol{h}, \boldsymbol{y}) \quad \propto \quad \pi(\phi)\sqrt{1 - \phi^2} \exp\left\{ -\frac{(\phi - \mu_\phi)^2}{2\sigma_\phi^2} \right\}, \quad (21.3)$$

where

$$\mu_\phi = \frac{\sum_{t=1}^{n-1}(h_{t+1} - \mu - \bar{y}_t)(h_t - \mu)}{\rho^2(h_1 - \mu)^2 + \sum_{t=2}^{n-1}(h_t - \mu)^2}, \quad \sigma_\phi^2 = \frac{\sigma_\eta^2(1 - \rho^2)}{\rho^2(h_1 - \mu)^2 + \sum_{t=2}^{n-1}(h_t - \mu)^2}.$$

To sample from this conditional distribution, we conduct MH algorithm as follows. Generate a candidate, $\phi^\dagger$, from the truncated normal distribution on $(-1, 1)$, $\phi^\dagger \sim \mathcal{TN}_{(-1,1)}(\mu_\phi, \sigma_\phi^2)$, and accept it with probability

$$\min\left\{ \frac{\pi(\phi^\dagger)\sqrt{1 - \phi^{\dagger 2}}}{\pi(\phi)\sqrt{1 - \phi^2}}, 1 \right\},$$

where ϕ is a current sample. If $\phi^\dagger$ is rejected, we accept ϕ as a new sample.

Generation of μ **(Step 3).** Assume that a prior distribution of μ is

$$\mu \sim \mathcal{N}(\mu_0, \sigma_0^2),$$

where μ_0 and σ_0^2 are hyper-parameters. Then the conditional posterior distribution of μ is

$$\mu|\phi, \sigma_\eta^2, \boldsymbol{h}, \boldsymbol{y} \quad \sim \quad \mathcal{N}(\mu_1, \sigma_1^2),$$

---

[1] In Section 21.4, we assume Beta distribution for $(\phi + 1)/2$.

where

$$
\mu_1 = \sigma_1^2 \left\{ \sigma_0^{-2} \mu_0 + \sigma_\eta^{-2}(1 - \phi^2) h_1 \right.
$$

$$
\left. . + (1 - \rho^2)^{-1} \sigma_\eta^{-2}(1 - \phi) \sum_{t=1}^{n-1} (h_{t+1} - \phi h_t - \bar{y}_t) \right\},
$$

$$
\sigma_1^2 = \left\{ \sigma_0^{-2} + \sigma_\eta^{-2}(1 - \phi^2) + (1 - \rho^2)^{-1} \sigma_\eta^{-2}(1 - \phi)^2 (n - 1) \right\}^{-1}.
$$

Generation of $\boldsymbol{\vartheta} = (\sigma_\eta^2, \rho)'$ (Step 4). Let $\boldsymbol{\vartheta} = (\sigma_\eta^2, \rho)'$ and $\pi(\boldsymbol{\vartheta}) = \pi(\sigma_\eta^2)\pi(\rho)$ denote the corresponding prior probability density function[2]. Then the conditional posterior probability density function of $\boldsymbol{\vartheta}$ is given by

$$
\pi(\boldsymbol{\vartheta}|\mu, \phi, \boldsymbol{h}, \boldsymbol{y})
$$

$$
\propto \pi(\boldsymbol{\vartheta}) \times (\sigma_\eta^2)^{-\frac{n}{2}}(1 - \rho^2)^{\frac{n-1}{2}}
$$

$$
\times \exp \left\{ -\frac{(1 - \phi^2)(h_1 - \mu)^2}{2\sigma_\eta^2} - \sum_{t=1}^{n-1} \frac{(h_{t+1} - (1 - \phi)\mu - \phi h_t - \bar{y}_t)^2}{2\sigma_\eta^2(1 - \rho^2)} \right\} \quad (21.4)
$$

Since the parameter space is restricted to $R = \{\boldsymbol{\vartheta} : \sigma_\eta^2 > 0, |\rho| < 1\}$, we first consider the transformation from $\boldsymbol{\vartheta}$ to $\boldsymbol{\omega} = (\omega_1, \omega_2)'$ where $\omega_1 = \log \sigma_\eta^2$, $\omega_2 = \log(1+\rho) - \log(1-\rho)$ to remove such a restriction. Under this parameterization, the parameter space is $R' = \{-\infty < \omega_i < \infty; i = 1, 2\}$. Then we compute the mode, $\hat{\boldsymbol{\omega}}$, of the conditional posterior probability density function of $\boldsymbol{\omega}$ (which we denote by $\tilde{\pi}(\boldsymbol{\omega}|\cdot)$) numerically. We conduct MH algorithm to sample $\boldsymbol{\omega}$ from its conditional posterior distribution: Generate a candidate $\boldsymbol{\omega}^\dagger \sim \mathcal{N}(\boldsymbol{\omega}_*, \Sigma_*)$ where

$$
\boldsymbol{\omega}_* = \hat{\boldsymbol{\omega}} + \Sigma_* \left. \frac{\partial \log \tilde{\pi}(\boldsymbol{\omega}|\cdot)}{\partial \boldsymbol{\omega}} \right|_{\boldsymbol{\omega}=\hat{\boldsymbol{\omega}}}, \qquad \Sigma_*^{-1} = - \left. \frac{\partial \log \tilde{\pi}(\boldsymbol{\omega}|\cdot)}{\partial \boldsymbol{\omega} \partial \boldsymbol{\omega}'} \right|_{\boldsymbol{\omega}=\hat{\boldsymbol{\omega}}},
$$

and accept it with probability

$$
\min \left\{ \frac{\tilde{\pi}(\boldsymbol{\omega}^\dagger|\cdot) f_N(\boldsymbol{\omega}|\boldsymbol{\omega}_*, \Sigma_*)}{\tilde{\pi}(\boldsymbol{\omega}|\cdot) f_N(\boldsymbol{\omega}^\dagger|\boldsymbol{\omega}_*, \Sigma_*)}, 1 \right\},
$$

where $\boldsymbol{\omega}$ is a current sample and $f_N(\cdot|\mu, \Sigma)$ denotes a probability density function of the normal distribution with mean μ and covariance matrix Σ[3]. Finally we transform the new sample of $\boldsymbol{\omega}$ back to obtain the new sample of $\boldsymbol{\vartheta}$.

---

[2]In Section 21.4, we assume inverse gamma distribution for σ_η^2 and Beta distribution for $(\rho + 1)/2$.

[3]When Σ_* is not positive definite, we use a proposal distribution $\mathcal{N}(\boldsymbol{\omega}_*, c_0 \mathbf{I}_2)$ with some large constant c_0.

When we consider the symmetric SV model where $\rho \equiv 0$, we assume the inverse gamma prior distribution for σ_η^2 ($\sigma_\eta^2 \sim \mathcal{IG}(n_0/2, S_0/2)$). Then the conditional posterior distribution of σ_η^2 is given by the inverse gamma distribution, $\sigma_\eta^2 \sim \mathcal{IG}(n_1/2, S_1/2)$ where $n_1 = n_0 + n$ and $S_1 = S_0 + (1 - \phi^2)(h_1 - \mu)^2 + \sum_{t=1}^{n-1}(h_{t+1} - (1 - \phi)\mu - \phi h_t)^2$.

21.2.2 Generation of h: Single-move sampler

The simplest way to generate h is to generate h_t given $h_{-t} = (h_1, \ldots, h_{t-1}, h_{t+1}, \ldots, h_n)'$ and θ. Consider the conditional posterior probability density function of h_t,

$$\pi(h_t | h_{-t}, \theta, y) \propto$$

$$\exp\left\{ -\frac{1}{2}h_t - \frac{1}{2}R_t^2 \exp(-h_t) - \frac{1}{2\sigma_\eta^2(1 - \rho^2)}(h_{t+1} - (1 - \phi)\mu - \phi h_t - \bar{y}_t)^2 \right\}$$

$$\times \exp\left\{ -\frac{1}{2\sigma_\eta^2(1 - \rho^2)}(h_t - (1 - \phi)\mu - \phi h_{t-1} - \bar{y}_{t-1})^2 \right\}.$$

Although we can generate a sample of h_t from its conditional posterior distribution using the rejection sampling or MH algorithm, such algorithms are known to be inefficient in the sense that they produce highly autocorrelated posterior samples. This is because the estimate of ϕ is close to one in empirical studies and hence the correlations between h_t and (h_{t+1}, h_{t-1}) are very high, which causes high sample autocorrelations of h_t. In Monte Carlo simulation, it is desirable to obtain independent random samples for exact statistical inferences and hence we wish to have as low autocorrelations as possible in posterior samples.

There are two major efficient sampling algorithms: (1) the mixture sampler which generates h all at once ([30]) and (2) the block sampler (multi-move sampler) which generates a vector of $(h_t, \ldots, h_{t+m})$ given other h_t's ([31]). We shall describe the block sampler (multi-move sampler) for the asymmetric SV model below.

21.2.3 Block sampler for h

Let $\alpha_t = h_t - \mu$, $\sigma_\epsilon = \exp(\mu/2)$, and consider the asymmetric SV model

$$\begin{aligned}
R_t &= \exp(\alpha_t/2)\epsilon_t, \quad t = 1, \ldots, n, \\
\alpha_{t+1} &= \phi\alpha_t + \eta_t, \quad t = 1, \ldots, n-1, \quad |\phi| < 1, \\
\alpha_1 &\sim \mathcal{N}(0, \sigma_\eta^2/(1 - \phi^2)),
\end{aligned}$$

where

$$\begin{pmatrix} \epsilon_t \\ \eta_t \end{pmatrix} \sim \mathcal{N}(\mathbf{0}, \boldsymbol{\Sigma}), \quad \boldsymbol{\Sigma} = \begin{pmatrix} \sigma_\epsilon^2 & \rho\sigma_\epsilon\sigma_\eta \\ \rho\sigma_\epsilon\sigma_\eta & \sigma_\eta^2 \end{pmatrix}.$$

Given $\boldsymbol{\alpha} = (\alpha_1, \ldots, \alpha_n)'$, the conditional distribution of R_t is normal with mean μ_t and variance σ_t^2 where

$$\mu_t = \begin{cases} \rho\sigma_\epsilon\sigma_\eta^{-1}(\alpha_{t+1} - \phi\alpha_t)\exp(\alpha_t/2), & t = 1, \ldots, n-1, \\ 0, & t = n, \end{cases} \quad (21.5)$$

$$\sigma_t^2 = \begin{cases} (1-\rho^2)\sigma_\epsilon^2\exp(\alpha_t), & t = 1, \ldots, n-1, \\ \sigma_\epsilon^2\exp(\alpha_n), & t = n. \end{cases} \quad (21.6)$$

and the conditional log likelihood function (excluding a constant term) is

$$l_t = -\frac{\alpha_t}{2} - \frac{(R_t - \mu_t)^2}{2\sigma_t^2}. \quad (21.7)$$

One of efficient sampling algorithms for the asymmetric SV model is a multi-move sampler or a block sampler by [31] where (1) we divide the state vector $\boldsymbol{h}$ (equivalently, $\boldsymbol{\alpha}$) into several blocks and (2) sample one block at a time given other blocks as in [34] and [40] who proposed algorithms for the symmetric SV model.

Generation of $\boldsymbol{\alpha}$. First we divide $\boldsymbol{\alpha}$ into $K+1$ blocks and let $k_{i-1} + 1$ ($i = 1, \ldots, K$) denote the starting time for the i-th block where k_i's are knots satisfying $0 = k_0 < k_1 < k_2 < \ldots < k_K < k_{K+1} = n$ [4]. Let $\boldsymbol{\alpha} = (\boldsymbol{\alpha}^{(1)}, \boldsymbol{\alpha}^{(2)}, \ldots, \boldsymbol{\alpha}^{(K+1)})$ where $\boldsymbol{\alpha}^{(i)} = (\alpha_{k_{i-1}+1}, \alpha_{k_{i-1}+2}, \ldots, \alpha_{k_i})$ is the i-th block of state variables.

For $i = 1, \ldots, K+1$, we generate $\boldsymbol{\alpha}^{(i)}$ given other blocks $\boldsymbol{\alpha}^{-(i)} = (\boldsymbol{\alpha}^{(1)}, \ldots, \boldsymbol{\alpha}^{(i-1)}, \boldsymbol{\alpha}^{(i+1)}, \ldots, \boldsymbol{\alpha}^{(K+1)})$ and the parameter $\boldsymbol{\theta}$ from its conditional posterior distribution. For simplicity, we denote $k_{i-1} = s$, $k_i = s + m$ in sampling the i-th block $\boldsymbol{\alpha}^{(i)}$ given $(\alpha_1, \ldots, \alpha_s, \alpha_{s+m+1}, \ldots, \alpha_n)$, $\boldsymbol{\theta}$ and $\boldsymbol{y}$. Since the conditional distribution of $\boldsymbol{\alpha}^{(i)} = (\alpha_{s+1}, \ldots, \alpha_{s+m})$ depends only on $(\alpha_s, \alpha_{s+m+1})$, $(R_{s+1}, \ldots, R_{s+m})$ and $\boldsymbol{\theta}$, we consider sampling from $\boldsymbol{\alpha}^{(i)}|\alpha_s, \alpha_{s+m+1}, \boldsymbol{\theta}, R_{s+1}, \ldots, R_{s+m}$.

Given $\alpha_s, \alpha_{s+m+1}, \boldsymbol{\theta}$ and $(R_{s+1}, \ldots, R_{s+m})$, $\boldsymbol{\alpha}^{(i)}$ is a linear combination of the disturbances, $\boldsymbol{u}^{(i)} = (u_s, \ldots, u_{s+m-1})'$, $(u_t = \eta_t/\sigma_\eta)$ [5] and we consider sampling $\boldsymbol{u}^{(i)}$ from its conditional posterior distribution. The logarithm of the conditional posterior probability density function of $\boldsymbol{u}^{(i)}$ is

$$\log f(\boldsymbol{u}^{(i)}|\alpha_s, \alpha_{s+m+1}, R_s, \ldots, R_{s+m}) = c_0 - \sum_{t=s}^{s+m-1} u_t^2/2 + L,$$

---

[4]One may think it would be good to fix k_i so that the size of $K+1$ blocks are almost equal ($k_i - k_{i-1} \approx c$ for some constant c for all i). However, in such a case, state variables corresponding to starting time ($k_{i-1} + 1$) or ending time (k_i) could have high autocorrelations, which results in inefficient sampling. Thus, [34] proposes the stochastic knots as follows. Generate i.i.d. uniform random variables on $(0, 1)$, $U_i \sim U(0, 1)$ ($i = 1, \ldots, K$), and set

$$k_i = \text{int}[n \times (i + U_i)/(K+2)], \quad i = 1, \ldots, K.$$

[5]Jacobian is constant.

where c_0 is a constant term and

$$
L = \begin{cases} \sum_{t=s}^{s+m} l_t - \frac{1}{2\sigma_\eta^2}(\alpha_{s+m+1} - \phi\alpha_{s+m})^2, & s+m < n, \\ \sum_{t=s}^{s+m} l_t, & s+m = n. \end{cases}
$$

Then, by Taylor expansion of the third term around $\hat{u}^{(i)}$, we obtain

$$
\begin{aligned}
& \log f(u^{(i)}|\alpha_s, \alpha_{s+m+1}, R_s, \ldots, R_{s+m}) \\
&\approx \quad c_0 - \frac{1}{2}\sum_{t=s}^{s+m-1} u_t^2 + \hat{L} + \left.\frac{\partial L}{\partial u^{(i)\prime}}\right|_{u^{(i)}=\hat{u}^{(i)}} (u^{(i)} - \hat{u}^{(i)}) \\
&\qquad + \frac{1}{2}(u^{(i)} - \hat{u}^{(i)})' \, E\left(\left.\frac{\partial L^2}{\partial u^{(i)}\partial u^{(i)\prime}}\right)\right|_{u^{(i)}=\hat{u}^{(i)}} (u^{(i)} - \hat{u}^{(i)}) \\
&= \quad c_0 - \frac{1}{2}\sum_{t=s}^{s+m-1} u_t^2 + \hat{L} + \hat{d}'(\alpha^{(i)} - \hat{\alpha}^{(i)}) - \frac{1}{2}(\alpha^{(i)} - \hat{\alpha}^{(i)})'\hat{Q}(\alpha^{(i)} - \hat{\alpha}^{(i)}) \\
&= \quad c_1 + \log g(u^{(i)}|\alpha_s, \alpha_{s+m+1}, R_s, \ldots, R_{s+m}),
\end{aligned}
$$

where c_1 is a constant term, $d = (d_{s+1}, \ldots, d_{s+m})'$,

$$
\begin{aligned}
d_t &= \frac{\partial L}{\partial \alpha_t} \\
&= -\frac{1}{2} + \frac{(R_t - \mu_t)^2}{2\sigma_t^2} + \frac{(R_t - \mu_t)}{\sigma_t^2}\frac{\partial \mu_t}{\partial \alpha_t} + \frac{(R_{t-1} - \mu_{t-1})}{\sigma_{t-1}^2}\frac{\partial \mu_{t-1}}{\partial \alpha_t} + \kappa(\alpha_t),
\end{aligned}
$$

with

$$
\frac{\partial \mu_t}{\partial \alpha_t} = \begin{cases} \frac{\rho\sigma_\epsilon}{\sigma_\eta}\left\{-\phi + \frac{\alpha_{t+1} - \phi\alpha_t}{2}\right\}\exp\left(\frac{\alpha_t}{2}\right), & t < n, \\ 0, & t = n, \end{cases} \tag{21.8}
$$

$$
\frac{\partial \mu_{t-1}}{\partial \alpha_t} = \begin{cases} 0, & t = 1, \\ \frac{\rho\sigma_\epsilon}{\sigma_\eta}\exp\left(\frac{\alpha_{t-1}}{2}\right), & t > 1. \end{cases} \tag{21.9}
$$

$$
\kappa(\alpha_t) = \begin{cases} \frac{\phi(\alpha_{t+1} - \phi\alpha_t)}{\sigma_\eta^2}, & t = s+m < n \\ 0, & \text{otherwise,} \end{cases} \tag{21.10}
$$

and

$$\mathbf{Q} = \begin{pmatrix} A_{s+1} & B_{s+2} & 0 & \cdots & 0 \\ B_{s+2} & A_{s+2} & B_{s+3} & \cdots & 0 \\ 0 & B_{s+3} & A_{s+3} & \ddots & \vdots \\ \vdots & \ddots & \ddots & \ddots & B_{s+m} \\ 0 & \cdots & 0 & B_{s+m} & A_{s+m} \end{pmatrix},$$

$$A_t = -E\left[\frac{\partial^2 L}{\partial \alpha_t^2}\right] = \frac{1}{2} + \sigma_t^{-2}\left(\frac{\partial \mu_t}{\partial \alpha_t}\right)^2 + \sigma_{t-1}^{-2}\left(\frac{\partial \mu_{t-1}}{\partial \alpha_t}\right)^2 + \kappa'(\alpha_t) \quad (21.11)$$

$$B_t = -E\left[\frac{\partial^2 L}{\partial \alpha_t \partial \alpha_{t-1}'}\right] = \sigma_{t-1}^{-2}\frac{\partial \mu_{t-1}}{\partial \alpha_{t-1}}\frac{\partial \mu_{t-1}}{\partial \alpha_t}, \quad (21.12)$$

$$\kappa'(\alpha_t) = \begin{cases} \dfrac{\phi^2}{\sigma_\eta^2}, & t = s + m < n \\ 0, & \text{otherwise.} \end{cases} \quad (21.13)$$

Further $\hat{L}, \hat{d}, \hat{\mathbf{Q}}$ are $L, d, \mathbf{Q}$ evaluated at $\boldsymbol{\alpha}^{(i)} = \hat{\boldsymbol{\alpha}}^{(i)} \ (\boldsymbol{u}^{(i)} = \hat{\boldsymbol{u}}^{(i)})$[6]. By repeating the following disturbance smoother with some initial value of $\boldsymbol{u}^{(i)}$, we obtain the mode of conditional probability density function $f(\boldsymbol{u}^{(i)}|\cdot)$.

Disturbance smoother ([31]).

1. Initialize $\hat{\boldsymbol{u}}^{(i)}$ and compute $\hat{\boldsymbol{\alpha}}^{(i)}$ recursively using $\boldsymbol{u}^{(i)} = \hat{\boldsymbol{u}}^{(i)}$.

2. Evaluate $\hat{d}_t, \hat{A}_t, \hat{B}_t$ at $\boldsymbol{\alpha}^{(i)} = \hat{\boldsymbol{\alpha}}^{(i)} \ (t = s+1, \ldots, s+m)$.

3. Compute $D_t, J_t, b_t, t = s+2, \ldots, s+m$:

$$\begin{aligned} D_t &= \hat{A}_t - D_{t-1}^{-1}\hat{B}_t^2, \quad D_{s+1} = \hat{A}_{s+1}, \\ K_t &= \sqrt{D_t}, \\ J_t &= \hat{B}_t K_{t-1}^{-1}, \quad J_{s+1} = J_{s+m+1} = 0, \\ b_t &= \hat{d}_t - J_t K_{t-1}^{-1} b_{t-1}, \quad b_{s+1} = \hat{d}_{s+1}. \end{aligned}$$

4. Define $\hat{y}_t = \hat{\gamma}_t + D_t^{-1}b_t$ where

$$\hat{\gamma}_t = \hat{\alpha}_t + K_t^{-1}J_{t+1}\hat{\alpha}_{t+1}, \quad t = s+1, \ldots, s+m.$$

5. Consider the linear Gaussian state space model

$$\begin{aligned} \hat{y}_t &= Z_t \alpha_t + \mathbf{G}_t \boldsymbol{\xi}_t, \quad t = s+1, \ldots, s+m, & (21.14) \\ \alpha_{t+1} &= \phi \alpha_t + \mathbf{H}_t \boldsymbol{\xi}_t, \quad t = s, s+1, \ldots, s+m, & (21.15) \\ \boldsymbol{\xi}_t &\sim \mathcal{N}(\mathbf{0}, \mathbf{I}_2), \end{aligned}$$

---

[6]The expectations in A_t and B_t are taken with respect to R_t given α_t and α_{t-1}.

where

$$Z_t = 1 + K_t^{-1} J_{t+1} \phi, \quad \mathbf{G}_t = K_t^{-1}[1, J_{t+1}\sigma_\eta], \quad \mathbf{H}_t = [0, \sigma_\eta].$$

Implement the Kalman filter and disturbance smoother for this linear Gaussian state space model (21.14) and (21.15), and use the smoothed estimate of $\hat{\boldsymbol{u}}^{(i)}$ to update $\hat{\boldsymbol{u}}^{(i)}$ and $\hat{\boldsymbol{\alpha}}^{(i)}$.

6. Go to 2.

To sample from $f(\boldsymbol{u}^{(i)}|\cdot)$, we implement the following simulation algorithm.

Generation of $\boldsymbol{u}^{(i)}$ ($\boldsymbol{\alpha}^{(i)}$).

1. Let $\boldsymbol{u}^{(i)}$ denote the current sample. Repeating the disturbance smoother several times, we find the mode $\hat{\boldsymbol{u}}^{(i)}$[7].

2. Using $\hat{\boldsymbol{u}}^{(i)}$, obtain the linear Gaussian state space model (21.14) and (21.15).

3. Implement a simulation smoother for the linear Gaussian state space model (21.14) and (21.15) to propose a candidate $\boldsymbol{u}^{\dagger(i)} \sim g$ for MH algorithm and accept it with probability

$$\min\left\{\frac{f(\boldsymbol{u}^{\dagger(i)}|\cdot)g(\boldsymbol{u}^{(i)}|\cdot)}{f(\boldsymbol{u}^{(i)}|\cdot)g(\boldsymbol{u}^{\dagger(i)}|\cdot)}, 1\right\}.$$

Generation of $\boldsymbol{\Sigma}$. Given $\boldsymbol{\alpha}$, we can sample $\boldsymbol{\Sigma}$ (equivalently, $(\mu, \sigma_\eta^2, \rho)$ simultaneously) as follows. Assume that the prior distribution of $\boldsymbol{\Sigma}^{-1}$ is Wishart distribution with parameters $(\nu_0, \boldsymbol{\Sigma}_0)$ ($\boldsymbol{\Sigma}^{-1} \sim \mathcal{W}(\nu_0, \boldsymbol{\Sigma}_0)$). Then the logarithm of the conditional posterior probability density function of $\boldsymbol{\Sigma}$ is

$$\log \pi(\boldsymbol{\Sigma}|\cdot) = \text{const} - \log\sigma_\eta - \frac{\alpha_1^2(1-\phi^2)}{2\sigma_\eta^2} - \log\sigma_\epsilon$$
$$- \frac{R_n^2}{2\sigma_\epsilon^2 \exp(\alpha_n)} - \frac{\nu_1+3}{2}\log|\boldsymbol{\Sigma}| - \frac{1}{2}\text{tr}\left(\boldsymbol{\Sigma}_1^{-1}\boldsymbol{\Sigma}^{-1}\right),$$

where $\nu_1 = \nu_0 + n - 1$ and

$$\boldsymbol{\Sigma}_1^{-1} = \boldsymbol{\Sigma}_0^{-1} + \sum_{t=1}^{n-1} \boldsymbol{z}_t\boldsymbol{z}_t', \quad \boldsymbol{z}_t = (R_t\exp(-\alpha_t/2), \alpha_{t+1} - \phi\alpha_t)'.$$

To sample from $\pi(\boldsymbol{\Sigma}|\cdot)$, we conduct MH algorithm. Propose a candidate

---

[7]It is not necessary to find the exact mode. The value close to the mode is enough to implement the MH algorithm.

$\Sigma^{\dagger-1} \sim \mathcal{W}(\nu_1, \Sigma_1)$ and accept it with probability

$$
\min \left\{ \frac{\sigma_\eta^{\dagger-1} \sigma_\epsilon^{\dagger-1} \exp\left\{ -\dfrac{\alpha_1^2(1-\phi^2)}{2\sigma_\eta^{\dagger 2}} - \dfrac{R_n^2}{2\sigma_\epsilon^{\dagger 2} \exp(\alpha_n)} \right\}}{\sigma_\eta^{-1} \sigma_\epsilon^{-1} \exp\left\{ -\dfrac{\alpha_1^2(1-\phi^2)}{2\sigma_\eta^2} - \dfrac{R_n^2}{2\sigma_\epsilon^2 \exp(\alpha_n)} \right\}}, 1 \right\}.
$$

After sampling $\boldsymbol{\alpha}$, we set $h_t = \alpha_t + \mu$ for $t = 1, \ldots, n$ and $\mu = \log(\sigma_\epsilon^2)$.

21.3 Realized SV Model

21.3.1 Realized volatility

Suppose that the log-price $p(s)$ follows the simple diffusion process:

$$
dp(s) = \mu(s)dt + \sigma(s)dW(s), \tag{21.16}
$$

where $W(s)$ is a standard Brownian process and $\mu(s)$ and $\sigma(s)$ are the mean and the standard deviation of $dp(s)$ respectively, which may be time-varying but are assumed to be independent of $dW(s)$. In this chapter, we call $\sigma^2(s)$ volatility although $\sigma(s)$ is usually called volatility in the finance literature.

Then, the true volatility for day t is defined as the integral of $\sigma^2(s)$ over the interval $(t-1, t)$ where $t-1$ and t represent the market closing time on day $t-1$ and t respectively, i.e.,

$$
IV_t = \int_{t-1}^{t} \sigma^2(s)ds, \tag{21.17}
$$

which is called integrated volatility.

The integrated volatility is unobservable, but if we have the intraday return data $(r_{t-1+1/n}, r_{t-1+2/n}, \ldots, r_t)$, we can estimate it as the sum of their squares

$$
RV_t = \sum_{i=1}^{n} r_{t-1+i/n}^2, \tag{21.18}
$$

which is called realized volatility (RV). RV_t will provide a consistent estimate of IV_t, i.e.,

$$
\plim_{n \to \infty} RV_t = IV_t. \tag{21.19}
$$

There are two problems in calculating RV. One of the problems is the presence of microstructure noise in real high-frequency data. The microstructure noise can be induced by various market frictions such as the discreteness of price changes, bid-ask bounces, and asymmetric information across traders, inter alia (see [10]). A growing literature attempts to study an integrated volatility estimation from microstructure noise-contaminated high-frequency data. If there presents microstructure noise, equation (21.19) may not be true. As the time interval of intraday returns used for calculating RV becomes smaller, the influence of microstructure noise on RV will increase. See [39] for the methods to mitigate the bias of RV caused by microstructure noise.

The other problem is the presence of non-trading hours. The New York Stock Exchange is open only for 9:30–16:00. It is impossible to obtain high-frequency returns for 16:00–9:30 (overnight). Since RV obtained using high-frequency returns over 6.5-hour trading period only captures the volatility during the part of the day that the market is open, we need to extend the RV to a measure of volatility for the full day. If we simply add the squares of overnight, RV may be subject to discretization error. If we neglect non-trading hours, RV has a downward bias. See [21] and [22] for the adjustment of non-trading hours.

21.3.2   Realized SV model: Definition

Let R_t, h_t, x_t denote the daily return, the logarithm of the true volatility (the integrated volatility) defined in (21.17) and the logarithm of RV, respectively, at time t. The RSV model ([36]) is given by[8].

$$R_t = \exp(h_t/2)\epsilon_t, \tag{21.20}$$

$$h_{t+1} = \mu + \phi(h_t - \mu) + \eta_t, \quad h_1 \sim N(\mu, \sigma_\eta^2/(1-\phi^2)), \tag{21.21}$$

$$x_t = \xi + h_t + u_t, \tag{21.22}$$

$$
\begin{bmatrix} \epsilon_t \\ \eta_t \\ u_t \end{bmatrix} \sim N\left(\mathbf{0}, \boldsymbol{\Sigma}\right), \quad
\boldsymbol{\Sigma} = \begin{bmatrix} 1 & \rho\sigma_\eta & 0 \\ \rho\sigma_\eta & \sigma_\eta^2 & 0 \\ 0 & 0 & \sigma_u^2 \end{bmatrix}.
$$

In the RSV model, we introduce a measurement equation (21.22) in addition to (21.20) so that we incorporate the information regarding h_t by the RV. We note that (ϵ_t, η_t) and u_t are assumed to be uncorrelated, for simplicity.

When $\xi = 0$ in (21.22), x_t is an unbiased estimator of h_t. However, as we discussed in subsection 21.3.1, it has a bias due to the microstructure noise and non-trading hours, and hence $\xi \neq 0$ is supposed to capture the bias. As observations are sampled at higher frequencies, the noise becomes more dominant in the RV. When the non-trading hours (e.g. overnight returns) are ignored, the RV underestimates the integrated volatility and ξ will be

---

[8]It can be extended to include a slope parameter, ψ, in (21.22):

$$x_t = \xi + \psi h_t + u_t.$$

However, since such an extension did not improve the forecasting performances of the volatilities in our empirical studies, we set $\psi = 1$ in this chapter.

estimated to be negative. If the bias caused by microstructure noise is positive, the estimate of ξ may be positive or negative, but captures the bias in an appropriate way in the RSV model.

21.3.3   Bayesian estimation using MCMC method

Let $\boldsymbol{\theta} = (\mu, \phi, \sigma_\eta, \rho, \xi, \sigma_u)$, $\boldsymbol{y} = \{R_t, x_t\}_{t=1}^n$, $\boldsymbol{h} = \{h_t\}_{t=1}^n$ denote the model parameters, the observations, and the latent log volatilities in RSV models. As discussed in SV models, it is difficult to compute the likelihood $f(\boldsymbol{y}|\boldsymbol{\theta})$ and hence to implement the maximum likelihood estimation. Thus we take a Bayesian approach and implement the MCMC algorithm ([36]).

In addition to Steps 1-4 in Section 21.2, we consider the following steps.

Step 5. Generate $\xi|\phi, \sigma_\eta, \rho, \mu, \sigma_u, \boldsymbol{h}, \boldsymbol{y}$.

Step 6. Generate $\sigma_u|\phi, \sigma_\eta, \rho, \mu, \xi, \boldsymbol{h}, \boldsymbol{y}$.

Step 7. Generate $\boldsymbol{h}|\boldsymbol{\theta}, \boldsymbol{y}, \xi, \sigma_u$.

Step 8. Go to Step 2.

Generation of ξ (Step 5). We assume the normal prior distribution for ξ, $\xi \sim \mathcal{N}(m_\xi, s_\xi^2)$. Then the conditional posterior distribution ξ is normal, that is, $\xi|\cdot \sim \mathcal{N}(\tilde{m}_\xi, \tilde{s}_\xi^2)$, where

$$\tilde{m}_\xi = \frac{s_\xi^2 \sum_{t=1}^n (x_t - h_t) + \sigma_u^2 m_\xi}{n s_\xi^2 + \sigma_u^2}, \quad \tilde{s}_\xi^2 = \frac{\sigma_u^2 s_\xi^2}{n s_\xi^2 + \sigma_u^2}.$$

Generation of σ_u^2 (Step 6). The prior distribution of σ_u^2 is assumed to be $\sigma_u^2 \sim \mathcal{IG}(n_u, S_u)$. Then, the posterior distribution of σ_u^2 is $\sigma_u^2|\cdot \sim \mathcal{IG}(\tilde{n}_u, \tilde{S}_u)$, where

$$\tilde{n}_u = \frac{n}{2} + n_u, \quad \tilde{S}_u = \frac{1}{2} \sum_{t=1}^n (x_t - \xi - h_t)^2 + S_u.$$

Generation of $\boldsymbol{h}$ (Step 7). As in subsection 21.2.3, we consider the multi-move sampler to generate $\boldsymbol{h}$. First rewrite (21.20)–(21.22) using $\alpha_t = h_t - \mu$, $\sigma_\epsilon = \exp(\mu/2)$, $c = \xi + \mu$ as follows.

$$R_t = \exp(\alpha_t/2)\epsilon_t, \quad t = 1, \ldots, n, \tag{21.23}$$

$$x_t = c + \alpha_t + \varepsilon_t, \quad t = 1, \ldots, n, \tag{21.24}$$

$$\alpha_{t+1} = \phi \alpha_t + \eta_t, \quad t = 0, \ldots, n-1. \tag{21.25}$$

where

$$\begin{pmatrix} \epsilon_t \\ \eta_t \\ \varepsilon_t \end{pmatrix} \sim N(\boldsymbol{0}, \boldsymbol{\Sigma}), \quad \boldsymbol{\Sigma} = \begin{pmatrix} \sigma_\epsilon^2 & \rho \sigma_\epsilon \sigma_\eta & 0 \\ \rho \sigma_\epsilon \sigma_\eta & \sigma_\eta^2 & 0 \\ 0 & 0 & \sigma_u^2 \end{pmatrix}.$$

Since the conditional distribution of R_t given $\boldsymbol{\alpha} = (\alpha_1, \ldots, \alpha_n)'$ is normal with mean μ_t and variance σ_t^2 defined in (21.5)–(21.6), logarithm of the conditional likelihood (excluding the constant term) is

$$l_t = -\frac{\alpha_t}{2} - \frac{(R_t - \mu_t)^2}{2\sigma_t^2} - \frac{(x_t - c - \alpha_t)^2}{2\sigma_u^2}. \tag{21.26}$$

Let

$$
d_t = -\frac{1}{2} + \frac{(R_t - \mu_t)^2}{2\sigma_t^2} + \frac{R_t - \mu_t}{\sigma_t^2}\frac{\partial \mu_t}{\partial \alpha_t} + \frac{R_{t-1} - \mu_{t-1}}{\sigma_{t-1}^2}\frac{\partial \mu_{t-1}}{\partial \alpha_t}
$$
$$
\quad + \frac{(x_t - c - \alpha_t)}{\sigma_u^2} + \kappa(\alpha_t),
$$
$$
A_t = \frac{1}{2} + \sigma_t^{-2}\left(\frac{\partial \mu_t}{\partial \alpha_t}\right)^2 + \sigma_{t-1}^{-2}\left(\frac{\partial \mu_{t-1}}{\partial \alpha_t}\right)^2 + \frac{1}{\sigma_u^2} + \kappa'(\alpha_t),
$$

where $\partial \mu_t/\partial \alpha_t$, $\partial \mu_{t-1}/\partial \alpha_t$, $\kappa(\alpha_t)$, $\kappa'(\alpha_t)$ are defined in (21.8)–(21.10), (21.13) respectively. Then, using B_t defined in (21.12), we sample $\boldsymbol{\alpha}^{(i)} = (\alpha_{s+1}, \ldots, \alpha_{s+m})'$ as in subsection 21.2.3.

21.4   Empirical Application

We first estimate the asymmetric SV model that consists of equations (21.1) and (21.2). The data we use is the daily returns of the S&P 500 stock index between 2000/1/4-2013/12/31. The sample size is 3,491. The daily returns are calculated as the log difference in the closing prices on two consecutive days multiplied by 100. The closing prices of the S&P 500 stock index are obtained from the Oxford–Man Institute's Realized Library ([24]). Table 21.1 summarizes the descriptive statistics of daily returns (in percentage). Mean, SE and Stdev represent the sample mean, standard error of mean and the standard deviation respectively. The mean is not significantly different from 0. LB(10) is the Ljung-Box statistic adjusted for heteroskedasticity following [13] to test the null hypothesis of no autocorrelations up to 10 lags. According to this statistic, the null hypothesis for the return is not rejected at the

TABLE 21.1
Descriptive statistics for daily returns.

Mean	SE	Stdev	LB(10)	
			Return	Squared return
0.0069	0.0221	1.3079	14.08	173.62

TABLE 21.2
Estimation result for asymmetric SV.

Parameter	Mean	Stdev	SE	95%L	95%U	CD	Inef.
ϕ	0.9681	0.0055	0.0004	0.9565	0.9778	0.36	52.79
σ_η	0.2520	0.0211	0.0023	0.2139	0.2989	0.41	120.22
ρ	−0.5248	0.0443	0.0034	−0.6092	−0.4337	0.80	57.76
μ	0.0436	0.1234	0.0024	−0.2002	0.2887	0.89	3.76

10% significance level. Thus, we use the unadjusted daily return as R_t. On the contrary, the LB statistic for the squared return, which is the proxy for volatility, is so large that the null hypothesis of no autocorrelation is rejected. This result is consistent with volatility clustering.

The prior distribution of the parameters in the asymmetric SV model is set as follows.

$$\mu \sim \mathcal{N}(0.4, 1), \quad \frac{\phi+1}{2} \sim \mathcal{B}(20, 1.5), \quad \sigma_\eta^2 \sim \mathcal{IG}(2.5, 0.025), \quad \frac{\rho+1}{2} \sim \mathcal{B}(1, 2),$$

where $\mathcal{B}(a, b)$ represents a Beta distribution. The prior means and standard deviations are $(0.86, 0.11)$, $(0.017, 0.024)$ and $(-0.33, 0.47)$ for ϕ, σ_η^2, ρ, which are based on the past empirical studies.

We use the block sampler for sampling volatility and discard the first 1,000 samples as burn-in and use the next 10,000 samples for the estimation.

Table 21.2 summarizes the estimation result of each parameter. Mean and Stdev are the sample mean and standard deviation of 10, 000 draws after burn-in. SE is the standard error of mean, which is calculated using the Parzen window because the draws using MCMC are autocorrelated. 95%L and 95%U are the lower and upper bound of Bayesian credible interval, which are calculated as the 2.5% and 97.5% percentiles of 10, 000 draws after burn-in. CD is the p-value of the convergence diagnostic (CD) statistic proposed by [17], whose asymptotic distribution is the standard normal if the sample has converged to the one from the posterior distribution. The standard error of CD statistic is also calculated using the Parzen window. According to the p-values of CD statistic, the null hypothesis of convergence to the posterior distribution is accepted for all parameters at the standard significance level. Ineff is the inefficiency factor proposed by [11]. If this is equal to 2, it implies that the number of draws must be twice as much as that of random sampling to make both standard errors equal. The inefficiency factor increases with the autocorrelation. The inefficiency factor of σ_η is rather large but affordable.

The posterior mean and 95% interval of ϕ are 0.9681 and [0.9565, 0.9778] respectively, which are close to one. This result indicates the high-persistence of volatility and is consistent with the well-known phenomenon of volatility clustering. The posterior mean and 95% interval of ρ are −0.5248 and [−0.6092, −0.4337]. The negative value of ρ is consistent with the well-known

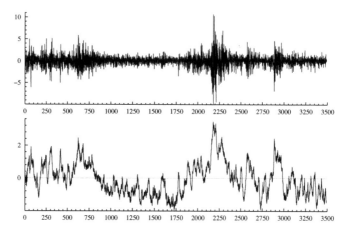

FIGURE 21.1
(a) Daily return (top) (b) log-volatility (bottom).

phenomenon of the negative correlation between today's return and tomorrow's volatility observed in stock markets. Figure 21.1 shows the return (top) and the posterior mean of log-volatiliy h_t (bottom).

Next, we estimate the RSV model that consists of equations (21.20)–(21.22). We use the log of realized kernel (RK) of S&P500 stock index obtained from the Oxford–Man Institute's Realized Library as well as the daily returns. RK is the RV calculated taking account of the bias caused by microstructure noise ([4]). Since we calculate the daily returns as the log-difference in closing prices on two consecutive days, the true volatility $\exp(h_t)$ is the volatility on the whole day. On the contrary, the RK we use is calculated using one-minuite returns only during trading hours (9:30-16:00). Hence, the RK has a downward bias. We may convert it to the one on the whole day using the method proposed by [21] and [22] but we need not do so because the downward bias of RK can be captured by the parameter ξ in the RSV model. The sample period is the same and the same prior distribution is used for the parameters ϕ, σ_η^2, ρ and μ. For ξ and σ_u^2, we set the following prior distribution.

$$\xi \sim \mathcal{N}(0,1), \quad \sigma_u^2 \sim \mathcal{IG}(2.5, 0.05).$$

We use the block sampler for sampling volatility and discard the first 1,000 samples as burn-in and use the next 10,000 samples for the estimation again.

Table 21.3 summarizes the estimation result of each parameter. According to the p-values of CD statistic, the null hypothesis of convergence to the posterior distribution is accepted for all parameters at the standard significance level. The inefficiency factors are much smaller than those of SV model. This is because the additional data x_t, i.e., the log of RK is included in the RSV model. The posterior mean and 95% interval of ξ are -0.2679 and

TABLE 21.3
Estimation result for RSV.

Parameter	Mean	Stdev	SE	95%L	95%U	CD	Inef.
ϕ	0.9611	0.0042	0.0001	0.9527	0.9691	0.75	9.42
σ_η	0.2388	0.0081	0.0005	0.2236	0.2553	0.46	34.13
ρ	−0.5260	0.0258	0.0010	−0.5757	−0.4744	0.90	15.75
μ	−0.0458	0.0934	0.0019	−0.2260	0.1377	0.73	4.13
ξ	−0.2679	0.0295	0.0016	−0.3285	−0.2121	0.58	28.54
σ_u	0.4199	0.0075	0.0003	0.4056	0.4348	0.48	11.89

$[-0.3285, -0.2121]$. The negative value of ξ implies that RK has a downward bias as we expected. Figure 21.2 shows the daily return (top), log-RK (middle) and the posterior mean of log-volatility h_t (bottom).

To compare the performance between the SV and RSV models, we calculate one-day-ahead volatility forecasts $\sigma^2_{n+1|n}$ ($n = 1500, \ldots, 3490$) using 1500 samples up to n. Define $\boldsymbol{y} = \{R_t\}^n_{t=n-1499}$ for SV model and $\boldsymbol{y} = \{R_t, x_t\}^n_{t=n-1499}$ for RSV model. Define also $\boldsymbol{h} = \{h_t\}^n_{n=n-1499}$. Then, all we have to do is to add the sampling from the following distribution between Steps 5 and 6 in Section 21.2 for SV model and between Steps 7 and 8 in subsection 21.3.3 for RSV model.

$$h_{n+1}|\boldsymbol{\theta}, \boldsymbol{h}, \boldsymbol{y}$$

It is straightforward to sample from this conditional distribution because it is

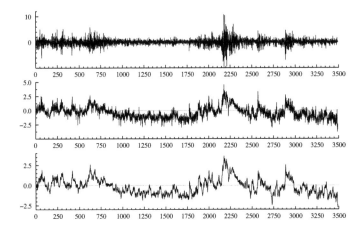

FIGURE 21.2
(a) Daily return (top) (b) log-RK (middle) (c) log-volatility (bottom).

normal with mean μ_{n+1} and variance σ_{n+1}^2 given by

$$\mu_{n+1} = \mu + \phi(h_n - \mu) + \rho\sigma_\eta R_n \exp(-h_n/2), \quad \sigma_{n+1}^2 = \sigma_\eta^2(1 - \rho^2).$$

We calcuete the one-day-ahead volatility forecast $\sigma_{n+1|n}^2$ as the average of the sample of $\exp(h_{n+1})$. We repeat this procedure from $n = 1500$ to 3490.

We need the proxy for the true volatility and loss function to compare the predictive ability of volatility. We use the following loss functions.

$$\text{MSE} = \frac{1}{1991} \sum_{n=1500}^{3490} (\sigma_{n+1|n}^2 - \hat{\sigma}_{n+1}^2)^2,$$

$$\text{QLIKE} = \frac{1}{1991} \sum_{t=1500}^{3490} \left(\ln \sigma_{n+1|n}^2 + \frac{\hat{\sigma}_{n+1}^2}{\sigma_{n+1|n}^2} \right),$$

where $\hat{\sigma}_{n+1}^2$ is the proxy for the true volatility σ_{n+1}^2. [32] shows that the above loss functions are robust in the sense that they lead to the same result as the one when the true volatility as used, if the proxy is unbiased, i.e., $E(\hat{\sigma}_{n+1}^2) = \sigma_{n+1}^2$. RK has a downward bias because it is the open-to-close volatility calculated using the intraday returns only when the market is open and the true volatility σ_{n+1}^2 is the close-to-close volatility for a whole day. Following [21], we use the RK multiplied by the following constant c for the proxy.

$$c = \frac{\sum_{t=1}^{3491} (R_t - \overline{R})^2}{\sum_{t=1}^{n} RK_t}, \tag{21.27}$$

where $(R_1, \ldots, R_{3491})$ is the daily returns in the full sample and $\overline{R}$ is the sample mean. This adjustment makes the sample mean of RK equal to the sample variance of daily returns.

TABLE 21.4
Predictive ability of one-day-ahead volatility.

	MSE	QLIKE
SV	11.3774	1.0678
RSV	10.9769	1.0072

Table 21.4 reports the result. Both loss functions have smaller values in the RSV model, showing that the RSV model performs better than the SV model in the prediction of one-day-ahead volatility.

21.5 Conclusion

This chapter reviews the SV and RSV models and the Bayesian method for the estimation of those models using MCMC. We estimate these models using the daily return and RK of S&P500 stock index.

Several extensions are possible. First, we assumed that the log-volatility h_t follows a stationary AR(1) model. [8] and [23] propose a long memory SV model where equation (21.2) in the SV model is replaced by the autoregressive fractionally integrated moving average (ARFIMA) model. Since many researchers have documented that RV follows a long memory process ([1], [2]), [35] replaces equation (21.21) in the RSV model by the ARFIMA model. Second, we assumed that the error terms (ϵ_t, η_t) in the SV model and $(\epsilon_t, \eta_t, u_t)$ in the RSV model follow normal distributions. [29] extends the SV model such that ϵ_t follows the generalized hyperbolic (GH) skew Student's t distribution and [37] extends the RSV model in the same fashion for volatility and quantile forecasts. Third, we used the log-RK for x_t in the RSV model but it is interesting to use the other realized measures such as realized range-based volatility (RRV) ([12], [28]) and jump-robust realized volatility ([3], [5]). Fourth and finally, it is important to extend the SV and RSV models to multivariate models.

Bibliography

[1] T. G. Andersen, T. Bollerslev, and H. Ebens. The distribution of realized stock return volatility. *Journal of Financial Economics*, 61(1):43–76, 2001.

[2] T. G. Andersen, T. Bollerslev, and P. Labys. Modeling and forecasting realized volatility. *Econometrica*, 71(2):579–625, 2003.

[3] T. G. Andersen, D. Dobrev, and E. Schaumburg. Jump-robust volatility estimation using nearest neighbor truncation. *Journal of Econometrics*, 169(1):75–93, 2012.

[4] O. E. Barndorff-Nielsen, P. R. Hansen, A. Lunde, and N. Shephard. Designing realized kernels to measure the ex-post variation of equity prices in the presence of noise. *Econometrica*, 76(6):1481–1536, 2008.

[5] O. E. Barndorff-Nielsen and N. Shephard. Power and bipower variation with stochastic volatility and jumps (with discussions). *Journal of Financial Econometrics*, 2(1):1–37, 2004.

[6] T. Bollerslev. Generalized autoregressive conditional heteroskedasticity. *Journal of Econometrics*, 31(3):307–327, 1986.

[7] T. Bollerslev, R. F. Engle, and D. B. Nelson. ARCH models. In R. F. Engle and D. McFadden, editors, *Handbook of Econometrics*, volume 4, pages 2959–3038, North–Holland, Amsterdam. 1994.

[8] F. J. Breidt, N. Crato, and P. de Lima. The detection and estimation of long memory in stochastic volatility. *Journal of Econometrics*, 83(1–2):325–348, 1998.

[9] C. Broto and E. Ruiz. Estimation methods for stochastic volatility models: A survey. *Journal of Economic Surveys*, 18(5):613–649, 2004.

[10] J. Y. Campbell, A. W. Lo, and A. C. MacKinlay. *The Econometrics of Financial Markets*. Princeton University Press, Princeton, 1997.

[11] S. Chib. Markov chain Monte Carlo methods: Computation and inference. In J. J. Heckman and E. Leaper, editors, *Handbook of Econometrics*, volume 5, pages 3569–3649. North-Holland, Amsterdam, 2001.

[12] K. Christensen and M. Podolskij. Realized range-based estimation of integrated variance. *Journal of Econometrics*, 141(2):323–349, 2007.

[13] F. X. Diebold. *Empirical Modeling of Exchange Rate Dynamics*, Springer Verlag, Berlin, 1988.

[14] D. Dobrev and P. Szerzen. The information content of high-frequency data for estimating equity return and forecasting risk. International Finance Discussion Papers 1005, Board of Governors of the Federal Reserve System, 2010.

[15] R. F. Engle. Autoregressive conditional heteroskedasticity with estimates of the variance of United Kingdom inflation. *Econometrica*, 50(4):987–1007, 1982.

[16] J. Fernández-Villaverde and J. Rubio-Ramírez. Estimating macroeconomic models: A likelihood approach. *Review of Economic Studies*, 74(4):1059–1087, 2007.

[17] J. Geweke. Evaluating the accuracy of sampling-based approaches to the calculation of posterior moments (with discussion). In J. M. Bernardo, J. O. Berger, A. P. Dawid, and A. F. M. Smith, editors, *Bayesian Statistics*, volume 4, pages 169–194. Oxford University Press, Oxford, 1992.

[18] E. Ghysels, A. C. Harvey, and E. Renault. Stochastic volatility. In G. S. Maddala and C. R. Rao, editors, *Handbook of Statistics*, volume 14, pages 119–191. North-Holland, Amsterdam, 1996.

[19] C. Gourieroux. *ARCH Models and Financial Applications*, Springer Verlag, New York, 1997.

[20] P. R. Hansen, Z. Huang, and H. Shek. Realized GARCH: A joint model of returns and realized measures of volatility. *Journal of Applied Econometrics*, 27(6):877–906, 2012.

[21] P. R. Hansen and A. Lunde. A forecast comparison of volatility models: Does anything beat a GARCH(1,1)? *Journal of Applied Econometrics*, 20(7):873–889, 2005.

[22] P. R. Hansen and A. Lunde. A realized variance for the whole day based on intermittent high-frequency data. *Journal of Financial Econometrics*, 3(4):525–554, 2005.

[23] A. C. Harvey. Long-memory in stochastic volatility. In J. Knight and S. E. Satchell, editors, *Forecasting Volatility in Financial Markets*, pages 307–320. Butterworth–Heinemann, London, 1998.

[24] G. Heber, A. Lunde, N. Shephard, and K. Sheppard. Oxford-Man institute's realized library, version 0.1. Oxford–Man Institute, University of Oxford, 2009.

[25] A. Justiniano and G. E. Primiceri. The time varying volatility of macroeconomic fluctuations. *American Economic Review*, 98(3):604–641, 2008.

[26] S. J. Kim, N. Shephard, and S. Chib. Stochastic volatility: Likelihood inference and comparison with ARCH models. *Review of Economic Studies*, 65(3):361–393, 1998.

[27] S. J. Koopman and M. Scharth. The analysis of stochastic volatility in the presence of daily realized measures. *Journal of Financial Econometrics*, 11(1):76–115, 2013.

[28] M. Martens and D. van Dijk. Meauring volatility with the realized range. *Journal of Econometrics*, 138(1):181–207, 2007.

[29] J. Nakajima and Y. Omori. Stochastic volatility model with leverage and asymmetrically heavy–tailed error using GH skew Student's *t*-distribution. *Computational Statistics & Data Analysis*, 56(11):3690–3704, 2012.

[30] Y. Omori, S. Chib, N. Shephard, and J. Nakajima. Stochastic volatility with leverage: Fast and efficient likelihood inference. *Journal of Econometrics*, 140(2):425–449, 2007.

[31] Y. Omori and T. Watanabe. Block sampler and posterior mode estimation for asymmetric stochastic volatility models. *Computational Statistics & Data Analysis*, 52(6):2892–2910, 2008.

[32] J. A. Patton. Volatility forecast comparison using imperfect volatility proxies. *Journal of Econometrics*, 160(1):246–256, 2011.

[33] G. E. Primiceri. Time varying structural vector autoregressions and monetary policy. *Review of Economic Studies*, 72(3):821–852, 2005.

[34] N. Shephard and M. K. Pitt. Likelihood analysis of non-Gaussian measurement time series. *Biometrika*, 84(3):653–667, 1997.

[35] S. Shirota, T. Hizu, and Y. Omori. Realized stochastic volatility with leverage and long memory. *Computational Statistics & Data Analysis*, 76(C):618–641, 2014.

[36] M. Takahashi, Y. Omori, and T. Watanabe. Estimating stochastic volatility models using daily returns and realized volatility simultaneously. *Computational Statistics & Data Analysis*, 53(6):2404–2426, 2009.

[37] M. Takahashi, T. Watanabe, and Y. Omori. Volatility and quantile forecasts of financial returns using realized stochastic volatility models with generalized hyperbolic distribution, 2014. Discussion Paper series, CIRJE-F-921, Faculty of Economics, University of Tokyo.

[38] S. J. Taylor. *Modelling Financial Time Series*, John Wiley & Sons, Chichester, 1986.

[39] M. Ubukata and T. Watanabe. Pricing Nikkei 225 options using realized volatility. *Japanese Economic Review*, 65(4):431–467, 2014.

[40] T. Watanabe and Y. Omori. A multi-move sampler for estimating non-Gaussian times series models: Comments on Shephard and Pitt (1997). *Biometrika*, 91(1):246–248, 2004.

22

Monte Carlo Methods and Zero Variance Principle

Theodore Papamarkou
University of Warwick

Antonietta Mira
Swiss Finance Institute

Mark Girolami
University of Warwick

CONTENTS

22.1   Introduction

The principle dates back to 1999, when it was introduced in the physics literature by [3]. The physical nomenclature would broadly describe ZV estimators as renormalized observables used in Monte Carlo simulation with the same mean and smaller variance than the original observables of interest. More than a decade later, [18] brought ZV to the attention of the statistical community. In statistically oriented terms, ZV is a variance reduction scheme for Markov Chain Monte Carlo (MCMC) algorithms based on a specific form of control variates.

In particular, the ZV control variates are constructed by adding, to the original function (observable), g, whose expected value with respect to a target, π, we want to estimate, a normalized version of the gradient of the log-target (that, under regularity conditions, has zero mean with respect to the target and thus provides a readily available control variate). Of course, the prevailing zero-variance terminology is a misnomer, in the sense that it is not possible to eliminate variance completely without passing from the original stochastic estimator to a deterministic one.

In this chapter we will adopt a statistical terminology and will review the ZV strategy in the context of MCMC simulation. Still, since ZV is a post-processing technique aimed at variance reduction in Monte Carlo integration, it can be applied to any sequence of either iid samples from a target distribution π, or samples that form a Markov chain having π as its stationary distribution.

Since Hamiltonian Monte Carlo and Metropolis adjusted Langevin algorithms (e.g. [19], [25]) also require computing the gradient of the log-target, zero variance and geometric exploitation of the parameter space realize a "perfect marriage" in that are attained under a unified ZV-MCMC sampling framework without exceeding the computation requirements posed by the geometric MCMC scheme alone, as highlighted in [21].

Despite ZV being relatively recent, its applicability and limitations have been studied and understood to some extent. In particular, the ZV-MCMC estimator is asymptotically unbiased under mild conditions as discussed in [18], where a CLT is also established. This practically means that ZV can be applied successfully to reduce the variance of MCMC estimators constructed by taking the ergodic average along the sample path of an irreducible Markov chain having π as its unique stationary and limiting distribution. For instance, [21] quantifies the magnitude of zero variance reduction, via Monte Carlo simulation, in posterior mean estimation in various settings such as probit and logit models and dynamical systems described by non-linear differential equations. It is demonstrated later in this chapter that ZV is robust with respect to the possibly high dimensionality of the parameter space or to strong covariance between parameters in the target distribution.

22.2 Brief Overview of Variance Reduction in MCMC

Several Monte Carlo techniques, albeit developed with varying incentives, have led indirectly to reduction in asymptotic variance of MCMC estimators. For instance, Hybrid Monte Carlo tries to avoid a random walk behavior of the Markov chain by means of auxiliary momentum variables, overrelaxation aims at accelerating MCMC convergence, data augmentation attempts to enhance feasibility and simplicity of Monte Carlo simulations. Published adaptations of these MCMC techniques acquiring the desirable byproduct of variance reduction include [1], [8], [9] [10], [27], [13] and [26]. More exhaustive lists of relevant references are available in [18] and [21].

Other MCMC methods are designed with the purpose of reducing the asymptotic Monte Carlo variance in the first place. Such variance reduction methods employ Rao-Blackwellisation ([11], [24]), Riemann approximations ([22]), basis functions ([28]), the delayed rejection strategy ([13], [27]), antithetic variables ([5], [6]), and control variates.

Control variate is a variance reduction technique originally introduced for Monte Carlo simulation [23]. Suppose we are interested in estimating the expected value of $g(\boldsymbol{\theta})$ with respect to a distribution π. The main idea behind control variates is to exploit information on another variable, say $w(\boldsymbol{\theta})$, with known expected value $E_\pi[w(\boldsymbol{\theta})]$, and correlated with the original variable of interest $g(\boldsymbol{\theta})$. Using the additional knowledge, one can easily construct a new variable defined as $\tilde{g}(\boldsymbol{\theta}) = g(\boldsymbol{\theta}) + a\left[w(\boldsymbol{\theta}) - E_\pi[w(\boldsymbol{\theta})]\right]$ (where a is a constant parameter) that, by construction, has the same mean as $g(\boldsymbol{\theta})$ and whose variance is minimized by taking $a = \dfrac{\text{Cov}_\pi[g(\boldsymbol{\theta}), w(\boldsymbol{\theta})]}{\text{Var}_\pi[w(\boldsymbol{\theta})]}$. Once a sample of observations $\boldsymbol{\theta}_1, \ldots, \boldsymbol{\theta}_n$ is obtained from π (either independently and identically distributed or from a Markov chain stationary with respect to π), one can estimate $E_\pi[g(\boldsymbol{\theta})]$ by averaging $\tilde{g}(\boldsymbol{\theta}_i)$ rather than $g(\boldsymbol{\theta}_i)$. The resulting (MC)MC estimator will have a smaller variance. The larger is the correlation between $g(\boldsymbol{\theta})$ and $w(\boldsymbol{\theta})$, the stronger is the variance reduction achieved using this control variate technique.

[4] and [14] constructed control variates for independent and general Metropolis-Hastings algorithms. [2] and [15] laid the foundations for attaining control variates via the solution of the Poisson equation, which were then extended by [7] in the context of reversible MCMC samplers.

The zero variance methodology, as introduced by [3], facilitates the construction of control variates on the basis of a Hermitian operator H and of a so-called trial function ψ. [18] exploits the ZV mechanism by using linear, quadratic or higher order polynomials as ψ, with coefficients that turn out to be functions (variance and covariance) of the log-target's gradient (see details below).

22.3 Zero Variance Methodology

The ZV principle finds its main utility in estimating, with reduced variance, the expected value of a function $g(\boldsymbol{\theta})$, $\boldsymbol{\theta} \in \mathbb{R}^{n_\theta}$, with respect to a possibly unnormalized target distribution $\pi(\boldsymbol{\theta})$:

$$E_\pi[g(\boldsymbol{\theta})] = \frac{\int g(\boldsymbol{\theta})\pi(\boldsymbol{\theta})d\boldsymbol{\theta}}{\int \pi(\boldsymbol{\theta})d\boldsymbol{\theta}}. \tag{22.1}$$

In Monte Carlo or MCMC integration, $E_\pi[g(\boldsymbol{\theta})]$ is estimated by

$$\sum_{i=1}^{n} g(\boldsymbol{\theta}_i)/n \tag{22.2}$$

The samples $\boldsymbol{\theta}_i \in \mathbb{R}^{n_\theta}$, $i = 1, 2, \ldots, n$, are either iid (Monte Carlo) from the target, $\pi(\boldsymbol{\theta})/ \int \pi(\boldsymbol{\theta})d\boldsymbol{\theta}$, or are collected (possibly after a burn-in period) along the path of an ergodic Markov chain having the target as its unique stationary and limiting distribution.

In a Bayesian context π is a posterior distribution and the Bayesian estimator of a generic parameter (under squared error loss function) is its posterior mean; thus, the typical functions of interest are of the form: $g(\boldsymbol{\theta}) = \theta_i$. In the sequel we will therefore consider functions of this type i.e. $g(\boldsymbol{\theta}) : R^{n_\theta} \to R$.

The ZV control variates $\mathbf{w} : \mathbb{R}^{n_\theta} \to \mathbb{R}^{n_a}$ are used for constructing the auxiliary function

$$\tilde{g}(\boldsymbol{\theta}) = g(\boldsymbol{\theta}) + \mathbf{a}^T \mathbf{w}(\boldsymbol{\theta}). \tag{22.3}$$

that will replace the original g in the ergodic average (22.2). Imposing the condition $E_\pi[\mathbf{w}(\boldsymbol{\theta})] = 0$ ensures that $E_\pi[\tilde{g}(\boldsymbol{\theta})] = E_\pi[g(\boldsymbol{\theta})]$.

The general auxiliary function

$$\tilde{g}(\boldsymbol{\theta}) = g(\boldsymbol{\theta}) + \frac{T[\psi(\boldsymbol{\theta})]}{\sqrt{\pi(\boldsymbol{\theta})}}, \tag{22.4}$$

is proposed in [3], where T and ψ denote the so-called trial operator and respectively. ψ is an arbitrary integrable function. T is a Hermitian operator, for all practical purposes being real symmetric, and satisfies the constraint $T[\sqrt{\pi(\boldsymbol{\theta})}] = 0$. The latter condition guarantees that $\tilde{g}(\boldsymbol{\theta})$ and $g(\boldsymbol{\theta})$ have the same mean. If the pair (T, ψ) is selected appropriately, the variance of $\tilde{g}(\boldsymbol{\theta})$ will be smaller than the one of $g(\boldsymbol{\theta})$: this is how ZV methodology provides an asymptotically unbiased estimator of $E_\pi[g(\boldsymbol{\theta})]$ with reduced Monte Carlo variability.

22.3.1 Choice of Hamiltonian operator

A variety of possible choices for the trial operator T are suggested in [3]. In the sequel we will work with the Hamiltonian operator given by

$$T[\psi(\boldsymbol{\theta})] = -\frac{1}{2}\Delta_{\boldsymbol{\theta}}[\psi(\boldsymbol{\theta})] + \frac{\psi(\boldsymbol{\theta})}{2\sqrt{\pi(\boldsymbol{\theta})}}\Delta_{\boldsymbol{\theta}}[\sqrt{\pi(\boldsymbol{\theta})}], \qquad (22.5)$$

where $\Delta_{\boldsymbol{\theta}} := \sum_{i=1}^{n_\theta} \partial^2/\partial\theta_i^2$ is the Laplace operator. As it is explained in [3], the condition $E_\pi[T[\psi(\boldsymbol{\theta})]/\sqrt{\pi(\boldsymbol{\theta})}] = 0$ is satisfied for the Hamiltonian defined in (22.5) for any integrable function $\psi(\boldsymbol{\theta})$. Thus, (22.4) simplifies and [18] proves that $\sum_{i=1}^{n} \tilde{g}(\boldsymbol{\theta}_i)/n$ is an asymptotically unbiased estimator of $E_\pi[g(\boldsymbol{\theta})]$ (under mild regularity conditions).

22.3.2 Polynomial trial functions

The highest variance reduction is achieved when the variance of $\tilde{g}$ with respect to the target π is zero ($\mathrm{Var}_\pi[\tilde{g}] = 0$). This implies a constant auxiliary function $\tilde{g}$ and therefore the equivalent condition expressed in terms of the expected value of $\tilde{g}$ is $E_\pi[\tilde{g}] = \tilde{g}$. The above equality condition leads to the fundamental equation derived from (22.4):

$$T[\psi(\boldsymbol{\theta})] = -\sqrt{\pi(\boldsymbol{\theta})}[g(\boldsymbol{\theta}) - E_\pi[g(\boldsymbol{\theta})]]. \qquad (22.6)$$

For a chosen Hermitian operator T, the optimal trial function ψ is the exact solution of the fundamental equation (22.6). Solving (22.6) analytically for ψ is not always possible since it requires the knowledge of $E_\pi[g(\boldsymbol{\theta})]$ which is what we want to estimate in the first place. Due to the infeasibility of the acquisition of the exact solution, a practical workaround relies on selecting a functional for ψ parametrized by some coefficients and then deriving the parameters of ψ by minimizing $\mathrm{Var}_\pi[\tilde{g}]$. These parameters can then be estimated via MCMC simulation.

The choice of a good functional form for ψ depends on the target density π and on g. [18] define the parametric trial function ψ to be

$$\psi(\boldsymbol{\theta}) = P(\boldsymbol{\theta})\sqrt{\pi(\boldsymbol{\theta})}, \qquad (22.7)$$

where P is a polynomial whose coefficients parametrize ψ. The choice of ψ based on polynomials is made on the basis of facilitating both the constuction of ZV and the computational feasibility of the method. [18] provide sufficient conditions under which their ZV estimator is unbiased. (22.4) and (22.7) yield the auxiliary function

$$\tilde{g}(\boldsymbol{\theta}) = g(\boldsymbol{\theta}) - \frac{1}{2}\Delta_{\boldsymbol{\theta}}[P(\boldsymbol{\theta})] + \nabla_{\boldsymbol{\theta}}[P(\boldsymbol{\theta})] \cdot \mathbf{z}(\boldsymbol{\theta}), \qquad (22.8)$$

$$\mathbf{z}(\boldsymbol{\theta}) = -\frac{1}{2}\nabla_{\boldsymbol{\theta}}[\ln(\pi(\boldsymbol{\theta}))], \qquad (22.9)$$

where $\nabla_{\boldsymbol{\theta}}$ denotes the gradient. In particular, for first degree polynomials $P(\boldsymbol{\theta}) = \mathbf{a}^T \boldsymbol{\theta}$, the auxiliary function reduces to

$$\tilde{g}(\boldsymbol{\theta}) = g(\boldsymbol{\theta}) + \mathbf{a}^T \mathbf{z}(\boldsymbol{\theta}). \tag{22.10}$$

The linear polynomial coefficients $\mathbf{a}$ which minimize $\mathrm{Var}_\pi[\tilde{g}]$ are given by

$$\mathbf{a} = -\mathrm{Var}_\pi^{-1}[z(\boldsymbol{\theta})]\mathrm{Cov}_\pi[g(\boldsymbol{\theta}), \mathbf{z}(\boldsymbol{\theta})], \tag{22.11}$$

where $\mathrm{Var}_\pi[z(\boldsymbol{\theta})]$ denotes the covariance matrix (of dimension $n_\theta \times n_\theta$) of the control variates and $\mathrm{Cov}_\pi[g(\boldsymbol{\theta}), \mathbf{z}(\boldsymbol{\theta})] = E_\pi[g(\boldsymbol{\theta}) \cdot \mathbf{z}(\boldsymbol{\theta})]$ is a vector of dimension $n_\theta \times 1$.

The coefficients $\mathbf{a}$, as provided by (22.11), are not usually available (since the variance and covariance are with respect to the target π) and one needs to replace $\mathrm{Var}_\pi[z(\boldsymbol{\theta})]$ and $\mathrm{Cov}_\pi[g(\boldsymbol{\theta}), \mathbf{z}(\boldsymbol{\theta})]$ with their MCMC estimators obtained by averaging along the path of the Markov chain having π as its unique stationary and limiting distribution.

It is noted that one way of interpreting the polynomial coefficients (22.11) is to view them as the ordinary least squares estimators of the linear regression model

$$g(\boldsymbol{\theta}) = \mathbf{a}^T \mathbf{z}(\boldsymbol{\theta}) + \boldsymbol{\epsilon}, \tag{22.12}$$

where the control variates $\mathbf{z}(\boldsymbol{\theta})$ play the role of covariates and $\boldsymbol{\epsilon}$ represents the regression's independent normal error terms.

For second degree polynomials $P(\boldsymbol{\theta}) = \mathbf{c}^T \boldsymbol{\theta} + \frac{1}{2}\boldsymbol{\theta}^T B \boldsymbol{\theta}$, the auxiliary function becomes

$$\tilde{g}(\boldsymbol{\theta}) = g(\boldsymbol{\theta}) - \frac{1}{2}tr(B) + (\mathbf{c} + B\boldsymbol{\theta})^T \mathbf{z}(\boldsymbol{\theta}), \tag{22.13}$$

where $\mathbf{c}$ and B are the quadratic polynomial coefficients, $tr(B)$ denotes the trace of B and $\mathbf{z}(\boldsymbol{\theta})$ is given by (22.9). The terms on the right hand side of (22.13) can be rearranged into the form $\mathbf{a}^T \mathbf{w}(\boldsymbol{\theta})$ where the column vectors $\mathbf{a}$, $\mathbf{w}(\boldsymbol{\theta})$ have $n_\theta(n_\theta + 3)/2$ elements each, and are defined as follows:

- $\mathbf{a} := [\mathbf{c}^T \ \mathbf{d}^T \ \mathbf{b}^T]^T$, where $\mathbf{d} := diag(B)$ is the diagonal of B and $\mathbf{b}$ is a column vector with $n_\theta(n_\theta - 1)/2$ elements, whose element in the $(2n_\theta - j)(j - 1)/2 + (i - j)$ position is the lower diagonal (i, j)-th element of B.

- $\mathbf{w} := [\mathbf{z}^T \ \mathbf{u}^T \ \mathbf{v}^T]^T$, where $\mathbf{u} := \boldsymbol{\theta} \circ \mathbf{z} - \frac{1}{2}\mathbf{1}$, with $\circ$, $\mathbf{1}$ denoting the Hadamard product and the unit vector respectively, while $\mathbf{v}$ is a column vector consisting of $n_\theta(n_\theta - 1)/2$ elements, whose element in the $(2n_\theta - j)(j - 1)/2 + (i - j)$ position equals $\theta_i z_j + \theta_j z_i$, $j \in \{1, 2, \ldots, n_\theta\}$, $i \in \{2, 3, \ldots, n_\theta\}$, $j < i$.

Similar to (22.11), the quadratic polynomial coefficients $\mathbf{a}$ which minimize $\mathrm{Var}_\pi[\tilde{g}]$ are given by

$$\mathbf{a} = -\mathrm{Var}^{-1}[\mathbf{w}(\boldsymbol{\theta})]\mathrm{Cov}[g(\boldsymbol{\theta}), \mathbf{w}(\boldsymbol{\theta})], \tag{22.14}$$

Recalling that the parameter vector of interest $\boldsymbol{\theta}$ is of length n_θ, we note that LZV results in n_θ control variates while QZV results in $\frac{n_\theta(n_\theta+3)}{2}$ control variates. Consider that we might be interested in applying the ZV strategy, for controlling the variance of MCMC estimators, only for a subset of the original parameters, for example we might disregard nuisance parameters or parameters that already exhibit small Monte Carlo variability.

A comparison between (22.11) and (22.14) reveals the computational cost of the ZV method as a function of the degree of the polynomial trial function. The precision matrix $Var^{-1}[z(\boldsymbol{\theta})]$ appearing in (22.11) is of size $n_\theta \times n_\theta$. So, in order to acquire $\text{Var}_\pi^{-1}[z(\boldsymbol{\theta})]$, it is required to compute its $n_\theta(n_\theta + 1)/2$ cells, given the symmetry of the matrix. This means that in theory the computational complexity of linear ZV is $\mathcal{O}(n_\theta^2)$. On the other hand, the size of the precision matrix $\text{Var}_\pi^{-1}[w(\boldsymbol{\theta})]$ appearing in (22.14) is $n_\theta(n_\theta + 3)/2 \times n_\theta(n_\theta + 3)/2$. Hence, it is required to compute the $n_\theta(n_\theta + 1)(n_\theta + 2)(n_\theta + 3)/2$ unique elements of $\text{Var}_\pi^{-1}[w(\boldsymbol{\theta})]$, which means that quadratic ZV requires $\mathcal{O}(n_\theta^4)$ operations. This means that implementing ZV incurs a computational penalty as the dimension of the parameter vector increases. The computational cost is not prohibitive for linear ZV, and can be circumvented in the case of quadratic ZV due to the embarrassingly parallel nature of ZV.

It can be seen from (22.8), (22.9) and (22.13) that the required component for computing polynomial trial functions is the gradient of the log-target. In a Bayesian context, the target is assumed to be a posterior distribution $\pi(\boldsymbol{\theta}|\mathbf{x}) \propto \ell(\mathbf{x}|\boldsymbol{\theta})p(\boldsymbol{\theta})$, where $\boldsymbol{\theta}$, $\mathbf{x}$ denote the corresponding model parameters and data, while ℓ and p represent the likelihood and the prior respectively. Then the zero variance control variates are equal to half the negative sum of gradients of the log-likelihood, L, and of the log-prior:

$$\mathbf{z}(\boldsymbol{\theta}) = -\frac{1}{2}\nabla_{\boldsymbol{\theta}}[\ln(\pi(\boldsymbol{\theta}|\mathbf{x}))] = -\frac{1}{2}(\nabla_{\boldsymbol{\theta}}[L(\mathbf{x}|\boldsymbol{\theta})] + \nabla_{\boldsymbol{\theta}}[\ln(p(\boldsymbol{\theta}))]). \qquad (22.15)$$

22.3.3 ZV in Langevin Monte Carlo

MCMC algorithms which use a Langevin-type proposal density require, in their implementation, the computation of $\mathbf{z}(\boldsymbol{\theta})$ which, in turn, can be used to post-process the realized Markov chain to achieve variance reduction in the spirit of the ZV strategy introduced in subsection 22.3.2. In the sequel there are a few examples of such Langevin-type algorithms that can be successfully combined with ZV.

The **Position-Dependent Manifold-Adjusted Langevin Algorithm** (PMALA) defines a Langevin diffusion with invariant distribution $\pi(\boldsymbol{\theta}|\mathbf{x})$, $\boldsymbol{\theta} \in \mathbb{R}^{n_\theta}$, on the Riemann manifold of probability densities with metric tensor $G(\boldsymbol{\theta})$. PMALA constitutes the most recent development to date in Monte Carlo sampling using Langevin diffusions (see [30] for details). It corrects a transcription error which has propagated through the literature, whereby a factor of a $1/2$ has been omitted from one of the terms (e.g. [25] and [12]).

The Metropolis-Hastings proposal mechanism of PMALA is

$$\boldsymbol{\theta}' \sim \mathcal{N}\left(\boldsymbol{\theta} + \frac{h}{2}G^{-1}(\boldsymbol{\theta})\nabla_{\boldsymbol{\theta}}[\ln(\pi(\boldsymbol{\theta}|\mathbf{x}))] + h\Gamma(\boldsymbol{\theta}), hG^{-1}(\boldsymbol{\theta})\right), \qquad (22.16)$$

$$\Gamma_i(\boldsymbol{\theta}) = \frac{1}{2}\sum_{j=1}^{n_\theta}\frac{\partial}{\partial\theta_j}G_{ij}^{-1}(\boldsymbol{\theta}), \qquad (22.17)$$

where h is the squared integration step size of the Euler approximation to the solution of the diffusion. It is apparent from (22.16) that PMALA includes calculation of the log-target's gradient, so $\mathbf{z}(\boldsymbol{\theta})$ is available for implementing ZV with polynomial trial functions ψ.

Simplified MMALA (SMMALA) assumes a manifold of constant curvature, thereby the last term in (22.16) vanishes. A further simplification arises by selecting a constant metric tensor $G(\boldsymbol{\theta}) = M$, in which case PMALA coincides with the Metropolis-Adjusted Langevin Algorithm (MALA) with preconditioning matrix M (see [25], for details). SMMALA and MALA also make use of first-order gradient information making $\mathbf{z}(\boldsymbol{\theta})$ calculable and ZV applicable.

Adaptive MCMC algorithms, which use a Langevin-type proposal density, can be unified with ZV too, since they provide the log-target's gradient in general. **Subsample-Adapting Langevin Algorithm** (SALA) is a typical example, recently introduced by [17].

Riemann Manifold Hamiltonian Monte Carlo (RMHMC) defines a Hamiltonian on the Riemann manifold of probability densities as

$$H(\boldsymbol{\theta}, \mathbf{q}) = -\ln(\pi(\boldsymbol{\theta}|\mathbf{x})) + \frac{1}{2}\ln[(2\pi)^{n_\theta}|G(\boldsymbol{\theta})|] + \frac{1}{2}\mathbf{q}^T G^{-1}(\boldsymbol{\theta})\mathbf{q}, \qquad (22.18)$$

where $\boldsymbol{\theta} \in \mathbb{R}^{n_\theta}$, $\mathbf{x}$, $G(\boldsymbol{\theta})$ and $\mathbf{q}$ are respectively the model parameters, the data, a metric tensor and the auxiliary variables; $\mathbf{q} \sim \mathcal{N}(\mathbf{0}, G(\boldsymbol{\theta}))$. The position-specific metric tensor $G(\boldsymbol{\theta})$ allows for effective transitions in RMHMC and is chosen to be the expected Fisher information (see [21] for details). Hamilton's equation for RMHMC follows from (22.18):

$$\frac{\partial H(\boldsymbol{\theta}, \mathbf{q})}{\partial\theta_i} = -\nabla_{\boldsymbol{\theta}}[\ln(\pi(\boldsymbol{\theta}|\mathbf{x}))] + \frac{1}{2}tr\left[\Omega_i(\boldsymbol{\theta})\right] - \frac{1}{2}\mathbf{q}^T\Omega_i(\boldsymbol{\theta})G^{-1}(\boldsymbol{\theta})\mathbf{q}, \quad (22.19)$$

$$\Omega_i(\boldsymbol{\theta}) = G^{-1}(\boldsymbol{\theta})\frac{\partial G(\boldsymbol{\theta})}{\partial\theta_i}. \qquad (22.20)$$

The first component of Hamilton's equation (22.19) is the negative gradient of the log-target. Storing this intermediate result allows computation of $\mathbf{z}(\boldsymbol{\theta})$ and therefore to implement ZV based on linear and quadratic polynomial trial functions.

Setting the metric tensor $G(\boldsymbol{\theta})$ to be a constant mass matrix M simplifies the covariance matrix of $\mathbf{q}$, which is then distributed as $\mathbf{q} \sim \mathcal{N}(\mathbf{0}, M)$. This way Hamiltonian Monte Carlo (HMC) can be viewed as a simplified version of RMHMC. Then $\mathbf{z}(\boldsymbol{\theta})$ is given directly by Hamilton's equation according to $2\mathbf{z}(\boldsymbol{\theta}) = \partial H(\boldsymbol{\theta}, \mathbf{q})/\partial\boldsymbol{\theta}$.

[16] recently introduced **No-U-Turn Sampler** (NUTS) and **Hamiltonian Monte Carlo with Dual Averaging** (HMCDA), which are two adaptive Hamiltonian Monte Carlo samplers that automate the choice of parameters of the involved leapfrog integration. NUTS and HMCDA also compute the gradient of the log-target and therefore provide $\mathbf{z}(\boldsymbol{\theta})$ for applying the ZV method with polynomial ψ.

The HMCDA sampler is used to sample from the multivariate target t-distribution in Section 22.6.

22.4   Quantification of Variance Reduction

Simulation-based approaches have been previously employed to quantify the reduction in variance achieved by various MCMC estimators. A generic empirical approach is to run n_c independent chains $\boldsymbol{\theta}_{,j} = \{\theta_{ij}\}$, each of length n, where $i = 1, 2, \ldots, n$, $j = 1, 2, \ldots, n_c$. The n_c means $\hat{\mu}_{g(\boldsymbol{\theta},j)} = \sum_{i=1}^{n} g(\theta_{ij})/n$ are then computed and the unbiased variance estimator of these means is derived as

$$\widehat{\text{Var}_\pi}(\hat{\mu}_{g(\boldsymbol{\theta},j)}) = \frac{1}{n_c - 1} \sum_{j=1}^{n_c} (\hat{\mu}_{g(\boldsymbol{\theta},j)} - \hat{\mu}_{g(\boldsymbol{\theta})})^2, \qquad (22.21)$$

where $\hat{\mu}_{g(\boldsymbol{\theta})} = \sum_{j=1}^{n_c} \hat{\mu}_{g(\boldsymbol{\theta},j)}/n_c$ is the average of $\hat{\mu}_{g(\boldsymbol{\theta},j)}$. Furthermore, each chain $\boldsymbol{\theta}_{,j}$ is post-processed according to the variance reduction scheme to compute its ZV counterpart i.e. instead of averaging the original function $g(\boldsymbol{\theta}_{,j})$ along the sample path of the Markov chain, the new function $\tilde{g}(\boldsymbol{\theta}_{,j})$ is obtained by adding to g the ZV control variate and is averaged to obtain $\hat{\mu}_{\tilde{g}(\boldsymbol{\theta},j)} = \sum_{i=1}^{n} \tilde{g}(\theta_{ij})/n$.

The same procedure is used for estimating the variance $\widehat{\text{Var}_\pi}(\hat{\mu}_{\tilde{g}(\boldsymbol{\theta},j)})$. Ultimately, the Variance Reduction Factor (VRF) is defined as the ratio

$$r = \frac{\widehat{\text{Var}_\pi}(\hat{\mu}_{g(\boldsymbol{\theta},j)})}{\widehat{\text{Var}_\pi}(\hat{\mu}_{\tilde{g}(\boldsymbol{\theta},j)})} = \frac{\sum_{j=1}^{n_c} (\hat{\mu}_{g(\boldsymbol{\theta},j)} - \hat{\mu}_{g(\boldsymbol{\theta})})^2}{\sum_{j=1}^{n_c} (\hat{\mu}_{\tilde{g}(\boldsymbol{\theta},j)} - \hat{\mu}_{\tilde{g}(\boldsymbol{\theta})})^2}. \qquad (22.22)$$

This approach has been used extensively in the literature, see for example [7], and is employed in the examples of this chapter. It is powerful in that it works for functionals $g(\boldsymbol{\theta})$ other than the identity.

22.5   Example 1; Normal Target

As the first example, the applicability and effectiveness of the zero variance principle on Monte Carlo samples drawn from Normal targets is assessed. The importance of the results of this example are two-fold; they illustrate how well ZV works on the most prominent target, that is on the Normal family, while at the same time they provide intuition about the asymptotic properties of ZV as a consequence of the central limit theorem.

In the sequel, the subscript π will be omitted from expectations, variances and covariances operators.

22.5.1   Univariate normal target

Assume that the target follows $\theta \sim \pi(\theta|\mu, \sigma^2) = \mathcal{N}(\mu, \sigma^2)$.

22.5.1.1   ZV with linear polynomial (LZV)

The ZV estimator $\tilde{g}(\theta)$ of a smooth function $g(\theta)$ is then given by

$$\tilde{g}(\theta) = g(\theta) + a(\theta)z(\theta), \tag{22.23}$$

where the score function for ZV with linear polynomials (LZV) is found to be

$$z(\theta) = \frac{\theta - \mu}{\sigma^2}. \tag{22.24}$$

The ZV coefficients a for the univariate Normal target are computed as

$$a = -4\sigma^2 \text{Cov}\left[g(\theta), z(\theta)\right], \tag{22.25}$$

therefore the corresponding LZV estimator follows from (22.24) and (22.25) as

$$\tilde{g}(\theta) = g(\theta) - \frac{(\theta - \mu)}{\sigma^2} \text{Cov}\left[\theta, g(\theta)\right]. \tag{22.26}$$

The equation (22.26) allows expressing the LZV estimator of any moment θ^p, $p \in \mathbb{N}$, in closed form for univariate Normal targets. For instance, the auxiliary function is set to $g(\theta) = \theta$ in order to estimate the first moment, that is the mean of the target. Then $\text{Cov}\left[\theta, \theta\right] = \text{Var}\left[\theta\right] = \sigma^2$, hence (22.26) simplifies to give the exact LZV estimator $\tilde{g}(\theta) = \mu$, which has zero variance. In a similar way, the auxiliary function is set to $g(\theta) = \theta^2$ and $g(\theta) = \theta^3$ in order to estimate the second and third order moments of the univariate Normal target.

Table 22.1 provides the LZV estimators of the second and third order moments as well as their variances. The LZV estimator of the second order moment reduces the variance by $4\mu^2\sigma^2$ in comparison to the orginal estimator. This means that when estimating the second order central moment of a

TABLE 22.1
LZV estimators and their variances for the first three moments of a univariate normal target.

$g(\theta)$	$\tilde{g}(\theta)$	$\mathrm{Var}\,[\tilde{g}(\theta)]$
θ	μ	0
θ^2	$\theta^2 - 2\mu(\theta - \mu)$	$\mathrm{Var}\,[\theta^2] - 4\mu^2\sigma^2$
θ^3	$\theta^3 - 3(\mu^2 + \sigma^2)(\theta - \mu)$	$\mathrm{Var}\,[\theta^3] - 9\sigma^2(\mu^2 + \sigma^2)$

Gaussian target $\mathcal{N}(\mu = 0, \sigma^2)$, the LZV estimator has the same variance as the original estimator. More generally, the further away the mean μ of the Normal target is from zero, the greater is the variance reduction of the LZV estimator of the second moment. The LZV estimator of the third order moment reduces the variance of the corresponding original estimator by $9\sigma^2(\mu^2 + \sigma^2)$. As the variance σ^2 of the Normal target increases, the variance of the LZV estimator of the second and third order moment reduces converging to zero variance.

Specifically in the case of $\mathcal{N}(\mu = 0, \sigma^2)$, the higher moments $E\,[\theta^p]$, $p \in \mathbb{N}$, simplify to

$$E\,[\theta^p] = \begin{cases} 0 & \text{, if } p \text{ odd,} \\ (p-1)!!\sigma^p & \text{, if } p \text{ even,} \end{cases} \qquad (22.27)$$

where $(p-1)!!$ denotes the double factorial. It thus follows that, for even p, the LZV estimator of θ^p coincides with the original estimator, that is $\tilde{g}(\theta) = g(\theta) = \theta^p$, therefore the two estimators have the same variance. On the other hand, for odd p, the LZV estimator of $E_\pi(\theta^p)$ becomes

$$\tilde{g}(\theta) = g(\theta) - p!!\sigma^{p-1}\theta, \qquad (22.28)$$

whence its variance follows as

$$\mathrm{Var}\,[\tilde{g}(\theta)] = \mathrm{Var}\,[g(\theta)] - ((p)!!\sigma^p)^2. \qquad (22.29)$$

(22.29) implies that when $\mu = 0$ and p is odd, then the LZV estimator of θ^p for the $\mathcal{N}(0, \sigma^2)$ target reduces the variance of the original estimator by $((p)!!\sigma^p)^2$.

22.5.1.2 ZV with quadratic polynomial (QZV)

Assuming a quadratic polynomial $P(\theta) = c\theta + \frac{1}{2}d\theta^2$, the ZV estimator (QZV) of $g(\theta)$ for univariate Normal target becomes

$$\tilde{g}(\theta) = g(\theta) + \mathbf{a}^T \mathbf{w}(\theta), \qquad (22.30)$$

where the QZV control variates relate to the LZV control variates $z(\theta)$ as

$$\mathbf{w}(\theta) = \begin{bmatrix} z(\theta) \\ \theta z(\theta) - \frac{1}{2} \end{bmatrix} = \frac{1}{2\sigma^2} \begin{bmatrix} \theta - \mu \\ \theta(\theta - \mu) - \sigma^2 \end{bmatrix}. \qquad (22.31)$$

The QZV coefficients $\mathbf{a}(\theta)$ are estimated by

$$\mathbf{a} = -\mathrm{Var}^{-1}[\mathbf{w}(\theta)]\mathrm{Cov}[g(\theta), \mathbf{w}(\theta)]. \qquad (22.32)$$

It follows from (22.31) that the covariance matrix of the QZV control variates is given by

$$\mathrm{Var}[\mathbf{w}(\theta)] = \frac{1}{4\sigma^2} \begin{bmatrix} 1 & \mu \\ \mu & \mu^2 + 2\sigma^2 \end{bmatrix}, \qquad (22.33)$$

therefore their precision matrix evaluates to

$$\mathrm{Var}^{-1}[\mathbf{w}(\theta)] = 2 \begin{bmatrix} \mu^2 + 2\sigma^2 & -\mu \\ -\mu & 1 \end{bmatrix}. \qquad (22.34)$$

Using (22.31), (22.32) and (22.34), it can be shown that the QZV coefficients for the univariate Normal $\mathcal{N}(\mu, \sigma^2)$ are expressed as

$$\mathbf{a} = -\frac{1}{\sigma^2} \begin{bmatrix} \mu^2 + 2\sigma^2 & -\mu \\ -\mu & 1 \end{bmatrix} \begin{bmatrix} \mathrm{Cov}[g(\theta), \theta] \\ \mathrm{Cov}[g(\theta), \theta(\theta - \mu)] \end{bmatrix}, \qquad (22.35)$$

or equivalently as

$$\mathbf{a} = -\frac{1}{\sigma^2} \begin{bmatrix} \mu^2 + 2\sigma^2 & -\mu \\ -\mu & 1 \end{bmatrix} \begin{bmatrix} E[g\theta] - \mu E[g] \\ E[g\theta^2] - \mu E[g\theta] - \sigma^2 E[g] \end{bmatrix}, \qquad (22.36)$$

where g is a more compact notation for $g(\theta)$.

Particularly for estimating the moments $E[\theta^p]$ of the univariate Normal target, the QZV coefficients follow from (22.36) as

$$\mathbf{a} = -\frac{1}{\sigma^2} \begin{bmatrix} \mu^2 + 2\sigma^2 & -\mu \\ -\mu & 1 \end{bmatrix} \begin{bmatrix} E[\theta^{p+1}] - \mu E[\theta^p] \\ E[\theta^{p+2}] - \mu E[\theta^{p+1}] - \sigma^2 E[\theta^p] \end{bmatrix}. \qquad (22.37)$$

Table 22.2 shows the QZV estimators of the first three moments for the univariate Normal target $\mathcal{N}(\mu, \sigma^2)$, as derived from (22.37) for $p = 1, 2, 3$. The QZV estimators of the first two moments coincide with the moments under estimation, thus having zero variance, while the QZV estimator of the third moment reduces the variance of the original estimator by $9\sigma^2((\mu^2 + \sigma^2)^2 + 2\mu^2\sigma^2)$. Both LZV and QZV achieve zero variance for the first moment. On the other hand, only QZV attains zero variance for the second moment. A rising pattern and conjecture is whether for ZV based on a p-degree polynomial all of the moments $E[\theta^k]$, $k \leq p$, have zero variance, at least for the Gaussian target.

TABLE 22.2
QZV estimators and their variances for the first three moments of a univariate normal target.

$g(\theta)$	$\tilde{g}(\theta)$	$\mathrm{Var}[\tilde{g}(\theta)]$
θ	μ	0
θ^2	$\mu^2 + \sigma^2$	0
θ^3	$\theta^3 - 3\mu\theta^2 - 3(\sigma^2 - \mu^2)\theta + 6\mu\sigma^2$	$\mathrm{Var}[\theta^3] - 9\sigma^2((\mu^2 + \sigma^2)^2 + 2\mu^2\sigma^2)$

22.5.2 Multivariate normal target

More generally, assume that Monte Carlo samples are drawn from a k-dimensional Normal target. The score function of the multivariate Normal target is known to be

$$\nabla_{\boldsymbol{\theta}} \pi(\boldsymbol{\theta} | \boldsymbol{\mu}, \Sigma) = -\Sigma^{-1}(\boldsymbol{\theta} - \boldsymbol{\mu}) \tag{22.38}$$

therefore the LZV control variates are calculated as

$$\mathbf{z}(\boldsymbol{\theta}) = -\frac{1}{2}\nabla_{\boldsymbol{\theta}} \pi(\boldsymbol{\theta} | \boldsymbol{\mu}, \Sigma) = \frac{1}{2}\Sigma^{-1}(\boldsymbol{\theta} - \boldsymbol{\mu}). \tag{22.39}$$

It follows from (22.39) that

$$\mathrm{Var}\left[\mathbf{z}(\boldsymbol{\theta})\right] = \frac{1}{4}\Sigma^{-1}, \tag{22.40}$$

$$\mathrm{Cov}\left[g(\boldsymbol{\theta}), \mathbf{z}(\boldsymbol{\theta})\right] = \frac{1}{2}\Sigma^{-1}E\left[g(\boldsymbol{\theta})(\boldsymbol{\theta} - \boldsymbol{\mu})\right], \tag{22.41}$$

According to (22.11), (22.40) and (22.41), the LZV coefficients for the multivariate Normal target are

$$\mathbf{a} = -2E\left[g(\boldsymbol{\theta})(\boldsymbol{\theta} - \boldsymbol{\mu})\right]. \tag{22.42}$$

It is thus seen from (22.10), (22.39) and (22.42) that the LZV estimator $\tilde{g}(\boldsymbol{\theta})$ is given by

$$\tilde{g}(\boldsymbol{\theta}) = g(\boldsymbol{\theta}) - (E\left[g(\boldsymbol{\theta})(\boldsymbol{\theta} - \boldsymbol{\mu})\right])^T \Sigma^{-1}(\boldsymbol{\theta} - \boldsymbol{\mu}), \tag{22.43}$$

or equivalently

$$\tilde{g}(\boldsymbol{\theta}) = g(\boldsymbol{\theta}) - (\mathrm{Cov}\left[g(\boldsymbol{\theta}), \boldsymbol{\theta}\right])^T \Sigma^{-1}(\boldsymbol{\theta} - \boldsymbol{\mu}). \tag{22.44}$$

It is of interest to choose the projections $g(\boldsymbol{\theta}) = \theta_i^p$, $i = 1, 2, \ldots, k$, $p \in \mathbb{N}$, in order to acquire the LZV estimator of the moments of the multivariate Normal target. For such choice of g, Equation (22.43) simplifies to

$$\tilde{g}(\boldsymbol{\theta}) = \theta_i^p - (E\left[\theta_i^p \boldsymbol{\theta}\right] - E\left[\theta_i^p\right]\boldsymbol{\mu})^T \Sigma^{-1}(\boldsymbol{\theta} - \boldsymbol{\mu}). \tag{22.45}$$

As seen from (22.45), setting $g(\boldsymbol{\theta}) = \theta_i^p$ gives the LZV estimator of the moments of the multivariate Normal distribution in closed form. This is due to the fact that the moments $E\left[\theta_i^p\right]$ and $E\left[\theta_i^p \theta_j\right]$ $j = 1, 2, \ldots, k$, are given in closed form by Theorem 1.1 in [29]. Specifically for the first, second and third order moments, the LZV estimators become respectively

$$\tilde{g}(\boldsymbol{\theta}) = \theta_i - \Sigma_{i,}\Sigma^{-1}(\boldsymbol{\theta} - \boldsymbol{\mu}), \tag{22.46}$$

$$\tilde{g}(\boldsymbol{\theta}) = \theta_i^2 - 2\mu_i\Sigma_{i,}\Sigma^{-1}(\boldsymbol{\theta} - \boldsymbol{\mu}), \tag{22.47}$$

$$\tilde{g}(\boldsymbol{\theta}) = \theta_i^3 - 3(\mu_i^2 + \Sigma_{ii})\Sigma_{i,}\Sigma^{-1}(\boldsymbol{\theta} - \boldsymbol{\mu}), \tag{22.48}$$

TABLE 22.3
LZV estimators and their variances for the first three moments of a multivariate normal target.

$g(\boldsymbol{\theta})$	$\tilde{g}(\boldsymbol{\theta})$	$\mathrm{Var}\left[\tilde{g}(\boldsymbol{\theta})\right]$
θ_i	μ_i	0
θ_i^2	$\theta_i^2 - 2\mu_i(\theta_i - \mu_i)$	$\mathrm{Var}\left[\theta_i^2\right] - 4\mu_i^2 \Sigma_{ii}$
θ_i^3	$\theta_i^3 - 3(\mu_i^2 + \Sigma_{ii})(\theta_i - \mu_i)$	$\mathrm{Var}\left[\theta_i^3\right] - 9\Sigma_{ii}(\mu_i^2 + \Sigma_{ii})$

where $\Sigma_{i,}$ denotes the i-th row of the covariance matrix Σ. It is noted that $\Sigma_{i,}\Sigma^{-1}$ is the product of the i-th row of the covariance matrix with the precision matrix. Since the covariance and the precision matrices are inverse, this product gives the i-th row of the identity matrix, that is a vector whose elements are zero apart from the element in the i-th position which equals one. Therefore, (22.46), (22.47) and (22.48) simplify further to the respective expressions of Table 22.3. This means that the LZV estimators for the first three moments of the multivariate Normal target coincide with the corresponding estimators of the moments of the univariate Normal target (see also Table 22.1). The variance reduction is thereby the same for the univariate and mutlivariate Normal targets for the first three moments.

22.6   Example 2; Multivariate t-Distribution Target

As the second example and in order to explore the reduction in variance incurred by the ZV method, Monte Carlo samples are drawn from the class of n_θ-dimensional Student-t targets $t_\nu(\mathbf{0}, \frac{\nu-2}{\nu}\Sigma)$ with ν degrees of freedom and covariance matrix

$$\Sigma = \begin{pmatrix} 1 & a^1 & a^2 & \cdots & a^{n_\theta-3} & a^{n_\theta-2} & a^{n_\theta-1} \\ a^1 & 1 & a^1 & \cdots & a^{n_\theta-4} & a^{n_\theta-3} & a^{n_\theta-2} \\ a^2 & a^1 & 1 & \cdots & a^{n_\theta-5} & a^{n_\theta-4} & a^{n_\theta-3} \\ \vdots & \vdots & \vdots & \ddots & \vdots & \vdots & \vdots \\ a^{n_\theta-3} & a^{n_\theta-4} & a^{n_\theta-5} & \cdots & 1 & a^1 & a^2 \\ a^{n_\theta-2} & a^{n_\theta-3} & a^{n_\theta-4} & \cdots & a^1 & 1 & a^1 \\ a^{n_\theta-1} & a^{n_\theta-2} & a^{n_\theta-3} & \cdots & a^2 & a^1 & 1 \end{pmatrix}, \quad (22.49)$$

for some constant $0 < a < 1$. So the elements of the i-th diagonal of the $n_\theta \times n_\theta$ covariance matrix Σ are equal to a^{i-1}, $i = 1, 2, \ldots, n_\theta$. It is worth noticing that the scale matrix $\frac{\nu-2}{\nu}\Sigma$ of the t-distribution is defined by factorizing Σ by $(\nu-2)/\nu$ so that the covariance matrix of the t-distribution is Σ.

The rationale behind the choice of the matrix (22.49) is to assess the effectiveness of ZV for varying dimensions n_θ and for varying levels of covariance.

Choosing values of a closer to 1 induces stronger covariance among the parameters, while reducing a toward 0 results in less correlated parameters.

In all subsequent runs of the multivariate Student-t example, $n_c = 1,000$ independent chains are simulated using the adaptive dual averaging scheme of Hamiltonian Monte Carlo (HMCDA). $101,000$ iterations are run for the realization of each chain, of which the first nominal $1,000$ burn-in are omitted, so $n = 100,000$ samples are retained for each chain. HMCDA's adaptation occurs only in the burn-in phase. Each chain is initialized at $\mathbf{0}$, that is at the target's $\mathbf{0}$ mean and this is why there is no need for longer burn-in. It suffices to constrain the problem in terms of the simulations' starting point and burn-in length given that the aim is to evaluate the reduction in variance attained via ZV after having achieved convergence.

The source code for the examples of the current chapter is available online at `https://github.com/UniversityofWarwick/mcmc_and_zv_book_chapter.jl`. It is written using the Julia programming language and its official MCMC package. Julia's MCMC package provides a generic engine for implementing Bayesian statistical models using a variety of Hamiltonian and Langevin Monte Carlo algorithms, including HMCDA. It further provides the machinery for computing gradients via automatic differentiation, thus taking away the computational burden from the user and facilitating use of ZV or geometric MCMC. Moreover, the ZVSimulator Julia package was used for running the simulation for the multivariate t-distribution, which can be found in `https://github.com/scidom/ZVSimulator.jl`.

For a multivariate t-distribution target, the triplet of parameters (ν, n_θ, a) will be varied in the Monte Carlo study reported below. To assess the effect of each of these three parameters on the VRFs obtained via ZV, one parameter at a time is given a range of values while the other two are fixed.

22.6.1 ZV with increasing degrees of freedom

In this section, we examine the role of the degrees of freedom ν on the performance of the ZV strategy to estimate the mean of each marginal dimension θ_i, $i = 1, 2$ of a bivariate t-distribution.

Table 22.4 shows the VRFs as computed via (22.22) for $\nu = 5, 10, 25, 50, 100$ in the case of $(n_\theta, a) = (2, 0)$

The columns labelled as L and Q correspond to the VRFs of the ZV estimators obtained by linear and quadratic polynomials, respectively. According to Table 22.4, the reduction in variance is higher for increasing ν. This observation is anticipated in theory. It is known that as $\nu \to \infty$, the t-distribution $t_\nu(\mathbf{0}, \frac{\nu-2}{\nu}\Sigma)$ converges towards the normal density $\mathcal{N}(\mathbf{0}, \Sigma)$. It can be further shown that the variance of the ZV estimator for normal targets is zero and therefore the corresponding VRF is infinite. Thus, as the t-target converges to the corresponding normal target for increasing degrees of freedom, the reduction in variance is higher.

TABLE 22.4

VRFs for 2-dimensional t-distributions $t_\nu(\mathbf{0}, \frac{\nu-2}{\nu}I)$, $\nu = 5, 10, 25, 50, 100$.

	$\nu = 5$		$\nu = 10$		$\nu = 25$		$\nu = 50$		$\nu = 100$	
	L	**Q**	**L**	**Q**	**L**	**Q**	**L**	**Q**	**L**	**Q**
θ_1	2.23	2.23	12.67	12.70	81.90	82.15	353.84	353.23	1542.51	1543.90
θ_2	2.28	2.28	13.05	13.07	94.32	94.27	370.01	370.49	1415.32	1418.72

TABLE 22.5

VRFs for t-distribution of varying dimensions.

n_θ	**min L**	**min Q**	**max L**	**max Q**
2	81.9	82.15	94.32	94.27
3	40.52	40.18	42.24	42.39
5	30.95	30.77	35.6	35.48
10	21.97	22.04	27.52	27.52

22.6.2 ZV with increasing dimensionality

To assess the effect of the target's dimensionality n_θ to ZV, (ν, a) is fixed to $(25, 0)$ while n_θ is given the values $2, 3, 5, 10$. To summarize the results, Table 22.5 shows the VRFs of the model parameters in the dimensions with the minimum (min) and maximum (max) variance reduction achieved by linear ZV (L) and quadratic ZV (Q) for each of the dimensions n_θ. Although increasing dimensionality progressively attenuates the reduction in variance induced by ZV, Table 22.5 demonstrates that ZV remains relatively effective in moderately higher dimensions.

22.6.3 ZV with strongly correlated parameters

To assess the effect of the target covariance on the ZV method, $(\nu, n_\theta) = (25, 5)$ is fixed while a is given the values 0 (corresponding to IID dimensions, that is $\Sigma = I$), 0.25, 0.5, 0.75 and 0.99. This means that five 5-dimensional Student-t targets $t_{25}(\mathbf{0}, \frac{23}{25}\Sigma)$ are considered, whose covariance matrices (22.49) have diagonals with respective values 0^i, 0.25^i, 0.5^i, 0.75^i and 0.99^i, $i = 1, 2, \ldots, 4$. Values of a closer to 0 yield nearly independent variables, while for a closer to 1 the variables become strongly correlated.

Table 22.6 displays the VRFs for these five Student-t targets. As before, L and Q refer to linear and quadratic ZV, respectively. The pattern rising from Table 22.6 suggests that stronger correlation among the model parameters leads to greater reduction in variance.

TABLE 22.6
VRFs for 5-dimensional t-distributions $t_{25}(\mathbf{0}, \frac{23}{25}\Sigma(a))$ for a = $0, 0.25, 0.5, 0.75, 0.99$.

	a = 0		a = 0.25		a = 0.5		a = 0.75		a = 0.99	
	L	**Q**	**L**	**Q**	**L**	**Q**	**L**	**Q**	**L**	**Q**
θ_1	31.57	31.50	33.04	32.88	48.22	47.91	81.44	81.73	108.94	109.27
θ_2	35.60	35.48	35.36	35.14	58.08	57.86	84.56	84.93	110.50	110.70
θ_3	31.60	31.47	31.89	31.63	57.42	57.23	89.79	90.08	112.33	112.52
θ_4	32.53	32.48	35.83	35.67	57.35	57.21	86.30	86.56	112.09	112.26
θ_5	30.95	30.77	35.70	35.70	49.26	48.89	80.34	80.46	112.23	112.33

22.7 Discussion

The ZV strategy implements a transformation on the simulated (Markov chain) Monte Carlo samples aimed at reducing the variance of the MCMC estimators computed on the transformed samples relative to the one based on the original samples.

The implementation of ZV allows for a choice of a general trial function. When this function is a linear or quadratic polynomial, LZV and QZV estimators are obtained. Choices of the trial function other than polynomials are also allowed.

In this chapter it has been shown that, if the target is normal, LZV achieves exact zero variance when estimating the first moment, while QZV achieves zero variance when estimating the second moment. It is conjectured that, k-th order polynomials will achieve ZV when estimating the k-th order moment, at least in a normal target.

Furthermore, any target which approximates the normal distribution forms a suitable candidate for implementing the ZV scheme to acquire moment estimators exhibiting nearly zero variance.

In particular, it has been verified that, for a Student-t target, the higher the degrees of freedom are, the greater is the gain in variance reduction via ZV, since the approximation to the normal becomes better and better.

More generally, the statistical properties of the underlying target determine the effectiveness of ZV. There is a fundamental understanding of how the ZV methods operate on normal or approximately normal distributions. Ongoing research attempts to exploit the ZV scheme with models based on normality assumptions. At the same time, effort is being made by the authors to assess how well ZV is anticipated to work with targets departing from normality as a function of the target's properties.

The ZV principle can be implemented to achieve variance reduction of expectations of smooth transformations, g, of the model parameters. This precludes, among others, the construction of credible intervals based on the

ZV principle since quantile functions are not smooth transformations of the parameter.

An important application of Monte Carlo methods is in the computation of normalising constants for the purpose of Bayesian model comparison. The use of ZV control variates in this setting was recently explored by [20], who observed that the thermodynamic integral was particularly amenable to the ZV approach. Analogous with exact results for ZV and Gaussian distributions, it was shown that the model evidence can be estimated exactly using the "controlled thermodynamic integral" in the class of Bayesian linear regression models with Gaussian errors.

Acknowledgments

The authors would like to thank Reza Solgi and Chris Oates for helpful comments. Theodore Papamarkou is thankful for the ASSET EU Grant 259348. Antonietta Mira would like to express her gratitude for the Swiss National Science Foundation Grant 132766 "Zero-Variance Markov chain Monte Carlo". Mark Girolami is grateful for support by an Engineering and Physical Sciences Research Council Established Research Fellowship EP/J016934/1 and a Royal Society Wolfson Research Merit Award.

Bibliography

[1] S. L. Adler. Over-relaxation method for the Monte Carlo evaluation of the partition function for multiquadratic actions. *Physical Review D*, 23:2901–2904, June 1981.

[2] S. Andradóttir, D. P. Heyman, and T. J. Ott. Variance reduction through smoothing and control variates for Markov chain simulations. *ACM Transactions on Modeling and Computer Simulation*, 3(3):167–189, July 1993.

[3] R. Assaraf and M. Caffarel. Zero-variance principle for Monte Carlo algorithms. *Physical Review Letters*, 83:4682–4685, Dec 1999.

[4] Y. F. Atchadé and F. Perron. Improving on the independent Metropolis-Hastings algorithm. *Statistica Sinica*, 15(1):3–18, January 2005.

[5] P. Barone and A. Frigessi. Improving stochastic relaxation for Gussian random fields. *Probability in the Engineering and Informational Sciences*, 4(03):369–389, 1990.

[6] R. V. Craiu and C. Lemieux. Acceleration of the multiple-try Metropolis algorithm using antithetic and stratified sampling. *Statistics and Computing*, 17(2):109–120, June 2007.

[7] P. Dellaportas and I. Kontoyiannis. Control variates for estimation based on reversible Markov chain Monte Carlo samplers. *Journal of the Royal Statistical Society: Series B (Statistical Methodology)*, 74(1):133–161, 2012.

[8] S. Duane, A. D. Kennedy, B. J. Pendleton, and D. Roweth. Hybrid Monte Carlo. *Physics Letters B*, 195(2):216 – 222, 1987.

[9] D. A. V. Dyk and X.-L. Meng. The art of data augmentation. *Journal of Computational and Graphical Statistics*, 10(1):pp. 1–50, 2001.

[10] G. Fort, E. Moulines, G. O. Roberts, and J. S. Rosenthal. On the geometric ergodicity of hybrid samplers. *Journal of Applied Probability*, 40(1):pp. 123–146, 2003.

[11] A. E. Gelfand and A. F. M. Smith. Sampling-based approaches to calculating marginal densities. *Journal of the American Statistical Association*, 85(410):398–409, 1990.

[12] M. Girolami and B. Calderhead. Riemann manifold Langevin and Hamiltonian Monte Carlo methods. *Journal of the Royal Statistical Society: Series B (Statistical Methodology)*, 73(2):123–214, 2011.

[13] P. J. Green and A. Mira. Delayed rejection in reversible jump Metropolis-Hastings. *Biometrika*, 88(4):pp. 1035–1053, 2001.

[14] H. Hammer and H. Tjemeland. Control variates for the Metropolis-Hastings algorithm. *Scandinavian Journal of Statistics*, 35(3):400–414, 2008.

[15] S. G. Henderson. *Variance Reduction via an Approximating Markov Process*. PhD thesis, Stanford University, 1997.

[16] M. D. Hoffman and A. Gelman. The No-U-Turn sampler: Adaptively setting path lengths in Hamiltonian Monte Carlo. *Journal of Machine Learning Research*, 15:1351–1381, April 2014.

[17] T. Marshall and G. O. Roberts. An adaptive approach to Langevin MCMC. *Statistics and Computing*, 22(5):1041–1057, 2012.

[18] A. Mira, R. Solgi, and D. Imparato. Zero variance Markov chain Monte Carlo for Bayesian estimators. *Statistics and Computing*, 23(5):653–662, 2013.

[19] R. M. Neal. MCMC using Hamiltonian dynamics. In S. Brooks, A. Gelman, G. L. Jones, and X.-L. Meng, editors, *Handbook of Markov Chain Monte Carlo*, pages 113–162. CRC Press, New York, 2011.

[20] C. Oates, T. Papamarkou, and M. Girolami. The controlled thermodynamic integral for Bayesian model comparison. *University of Warwick CRiSM Working Paper Series*, 2014.

[21] T. Papamarkou, A. Mira, and M. Girolami. Zero variance differential geometric Markov chain Monte Carlo algorithms. *Bayesian Analysis*, 9(1):97–128, 2014.

[22] A. Philippe and C. P. Robert. Riemann sums for MCMC estimation and convergence monitoring. *Statistics and Computing*, 11(2):103–115, April 2001.

[23] B. D. Ripley. *Stochastic Simulation*. John Wiley & Sons, USA, 1987.

[24] C. P. Robert and G. Casella. *Monte Carlo Statistical Methods*. Springer Verlag, New York, 2nd edition, 2004.

[25] G. O. Roberts and O. Stramer. Langevin diffusions and Metropolis-Hastings algorithms. *Methodology and Computing in Applied Probability*, 4(4):337–357, 2002.

[26] R. Solgi and A. Mira. A Bayesian semiparametric multiplicative error model with an application to realized volatility. *Journal of Computational and Graphical Statistics*, 22(3):558–583, 2013.

[27] L. Tierney and A. Mira. Some adaptive Monte Carlo methods for Bayesian inference. *Statistics in Medicine*, 18(17-18):2507–2515, 1999.

[28] Y. Wang. Variance reduction via basis expansion in Monte Carlo integration. In J. O. Berger, T. T. Cai, and I. M. Johnstone, editors, *Borrowing Strength: Theory Powering Applications– A Festschrift for L. D. Brown*, volume 6, pages 234–248. Institute of Mathematical Statistics, USA, 2010.

[29] C. S. Withers. The moments of the multivariate normal. *Bulletin of the Australian Mathematical Society*, 32(1):103–107, August 1985.

[30] T. Xifara, C. Sherlock, S. Livingstone, S. Byrne, and M. Girolami. Langevin diffusions and the Metropolis-adjusted Langevin algorithm. *Statistics and Probability Letters*, 91:14–19, August 2014.

23

A Flexible Class of Reduced Rank Spatial Models for Large Non-Gaussian Dataset

Rajib Paul

Western Michigan University

Casey M. Jelsema

National Institute of Environmental Health Sciences

Kwok Wai Lau

CSIRO Computational Informatics

CONTENTS

23.1 Introduction

One of the most common goals of geostatistical analysis is prediction at unobserved locations and there are multiple strategies by which to accomplish spatial prediction; two popular choices are kriging and Bayesian predictive process models. There is an extensive amount of research on these two methods and it is not possible to report all of them here, but we encourage readers to see [5], Chapter 3 for kriging and Chapters 4 and 5 of [2] for an introduction on Bayesian methods. *Kriging* is derived as the best linear unbiased predictor by minimizing the mean squared prediction error (MSPE), see [5], Chapter 3.

Any necessary parameters are then estimated and substituted into the kriging equations to obtain spatial predictions. In Bayesian analysis, one puts prior distributions on model parameters; parameter estimation and spatial prediction are then obtained through Monte Carlo Markov Chain (MCMC) sampling and posterior predictive distributions.

Using either kriging or Bayesian methods, spatial prediction requires the inverse of the spatial covariance matrix of the observations on n locations of the spatial process of interest. Obtaining this inverse requires computations of order $O(n^3)$. When n becomes larger than several thousands, the inverse computation becomes computationally prohibitive. Reduced rank approaches are very popular strategies for dealing with the computational difficulty of inverting large covariance matrix without compromising the complexity of the spatial processes. The key idea is to model the spatial covariance as $SVS'+D$, where S is a spatial basis function matrix that maps the n-dimensional spatial process to a reduced dimensional (say, r, where $r \ll n$) latent process defined over a selected r knot locations on the spatial domain under analysis. V is a $r \times r$ covariance matrix of this reduced dimensional latent process and D is a diagonal matrix that takes into account the *nugget effect,* sum of the variances of measurement errors and errors due to dimension reduction, often referred as *process error.* The resulting covariance matrix can be inverted using Sherman-Morrisson-Woodbury formula (see [6], Equation 2.15), which requires only the inversion of V. Further, under Gaussian process model assumptions, model fitting and prediction only require storing quadratic forms, hence one can avoid storing large covariance matrices. There is an extensive body of literature that addresses how to implement reduced rank spatial models for analyzing large spatial and spatio-temporal datasets. As a place to start, [24], [37], [40] and [41] were all considering environmental spatial and spatio-temporal processes from the reduced rank perspective. A few recent references are: [3], [4], [6], [7], [12], [22], [26] and [34]. For review and discussion on spatial models for large datasets, see [38] and the references therein.

While a significant amount of research has been done on Gaussian reduced rank spatial model (RRSM), very limited researches were conducted on non-Gaussian datasets. For example, [39] developed a hierarchical framework for non-Gaussian space-time processes for complicated dynamics through integro-difference equations and kernel based spectral methods. [28] developed computational techniques for spatial logistic regression for large datasets, [32] developed RRSM for Poisson count data, and [33] developed a hierarchical model that comprised of a conditional exponential family model for the data and a latent geostatistical process model for some transformed mean of the data model.

Datasets exhibiting non-Gaussian tails and/or multimodality are very common in environmental studies. Both tails can be heavier or lighter than normal, or one tail can be lighter and the other heavier. There has been a substantial amount of work done for spatial skew-Gaussian models for small to moderate datasets. Most recent work has been done by [43], where they developed a skew-Gaussian and log-Gaussian spatial model and applied on Pb (lead) concentration data collected over a region in North Iran. Some notable works in this area include, for example, [1], [23] and [44]. Readers are encouraged to consult these papers and the references there in. [43] and [44] developed Monte Carlo Expectation Maximization (MCEM) algorithm to fit their models to the data and make predictions at unobserved locations. Their models and MCEM algorithms become computationally prohibitive if one wants to apply them for large datasets. On the other hand, if one uses Gaussian RRSM when the datasets depart from Gaussianity, then likelihood based approaches, such as EM algorithm, will suffer from poor fit due to incorrect distributional assumptions, and can lead to poor estimation and predictions. Instead of assuming Gaussian distribution on the reduced dimensional latent process, we propose a model where scale mixtures of Gaussian distributions are imposed and the scale parameters are exponentially distributed random variables with mean unity. We use marginally noninformative priors for covariance matrix of the reduced-dimensional latent processes developed by [13]. As far as we are aware, this prior has not been used before for RRSMs. Through simulation studies we show that this prior facilitates computationally efficient MCMC, desirable shrinkage of eigenvalues, and better predictive power compared to previously used priors, such as inverse-Wishart distribution. This prior is marginally noninformative and specification of hyperparameters does not depend on the data like the Givens angle priors used in [19].

Our proposed model is flexible enough to handle non-Gaussian tail behaviors, and it belongs to exponential family. For the estimation of model parameters and predictions at unobserved locations, we derive a Markov Chain Monte Carlo (MCMC) algorithm. The full conditional distributions for all processes under consideration and model parameters are not available in closed form. Hence, a hybrid MCMC algorithm is constructed. Markov Chain Central Limit Theorem (MCCLT) based posterior summaries alleviate the storing problem of large MCMC generated samples to a great extent. Ergodicity of a Markov Chain does not guarantee the conditions of MCCLT. Hence, we verify the conditions required for MCCLT theoretically. See [17] for more general discussion on MCCLT and their applications. The proposed model and method are applied on several simulated datasets and the daily maximums of total column ozone data for April 22, 2012, that are publicly available from National Aeronautic and Space Administration Terra satellite.

The main contributions of this chapter are: (1) A flexible reduced rank spatial model (FRRSM) that can model the variability in non-Gaussian data through a set of latent exponential random variables with mean 1 and marginally noninformative prior on the covariance matrix of the latent reduced dimensional process; (2) Verification for the conditions of MCCLT and MCCLT based posterior summaries that alleviate the storing problem of large MCMC generated samples to a great extent.

The rest of this chapter is organized as follows: Section 23.2 provides the details of our modeling approach. Section 23.3 discusses the derived MCMC algorithm and its implementation. We verify the conditions required for MC-CLT in this section. In Section 23.4, we apply our proposed model to several simulated datasets and the daily maximums of total column ozone dataset. Section 23.5 concludes the chapter with discussion on more general applicability of our proposed method.

23.2   Flexible Reduced Rank Spatial Model

The spatial random effects (SRE) model described in [19] is a very flexible class of Gaussian geostatistical models. For our model formulation, we start with their parametrization. Let $Z \equiv \{Z(s_1), \dots, Z(s_n), s_1, \dots, s_n \in \mathcal{D}\}$ denote observations of the spatial process of interest at n locations on domain $\mathcal{D} \in \mathbb{R}^d$ with positive d–dimensional volume $|\mathcal{D}|$. In most applications, $d = 2$ (for example, latitude and longitude). Under this model assumptions, the *data model* becomes:

$$Z(s_i) = Y(s_i) + \epsilon(s_i); \; i = 1, \dots, n, \qquad (23.1)$$

where $\epsilon(s_i)$ is a zero-mean Gaussian white noise (measurement errors) with variance $\sigma^2(s_i)$, and $Y(s_i)$ is the measurement-error-free latent process (truth), which is independent of measurement errors and follows a Gaussian distribution with mean $\mu(s_i)$ and covariance $C_Y(s, s')$, for $s, s' \in \mathcal{D}$. An usual choice of $\sigma^2(s_i) = \sigma^2$, free of s_i. One can either get the information on σ^2 from the measuring instrument or estimate it along with other unknown model parameters. The goal is to make predictions for $Y(s)$, over a specified prediction grid that may include observed as well as unobserved locations. As mentioned earlier, this can be done either by (see, [5], Chapter 3) or posterior predictive distributions (see [3] and [19]). A SRE model overcomes the hurdle of inverting large dimensional matrix by specifying the distribution of Y in terms of a spatial basis function matrix S and a r – dimensional zero mean Gaussian process η over a selected knot locations $\mathcal{S} \equiv (\mathcal{S}_1, \dots, \mathcal{S}_r)$, with covariance matrix $\text{cov}(\eta) = V$, where $r \ll n$. One can write the *process model* for $Y \equiv \{Y(s_i), s_i \in \mathcal{D}\}$ as:

$$Y = \mu + S\eta + \delta, \qquad (23.2)$$

where μ is the large scale variation of the process, usually modeled in terms of linear functions of p location specific covariate matrix X and p regression coefficients β as $X\beta$. The rest of the terms in Equation (23.2) explain the small scale variation where δ is the process error that accounts for the error due to dimension reduction. Usually δ is assumed to follow another zero mean Gaussian distribution, independent of η, with covariance matrix D. Here S is a $n \times r$ matrix, that maps from n dimensional observation space to r dimensional

η space. S is a sparse, completely known matrix specified by kernel/basis functions or through splines or wavelets. Additionally, S may be specified in such a way so as to induce multiresolutional features to the model, see [34] and [6] for different choices of S matrix. For the ozone dataset under analysis, it is found that a modified bisquare function of the following form works well:

$$S(s, u_{j(l)}) = \begin{cases} \left(1 - 0.25 d^2(s, u_{j(l)})\right) & \text{for } d(s, u_{j(l)}) \leq 2 \\ 0 & \text{otherwise,} \end{cases} \tag{23.3}$$

where $u_{j(l)}$ is the jth knot locations under l^{th} resolution and

$$d(s, u_{j(l)}) = \sqrt{(s(lon) - u_{j(l)}(lon))^2/r^2_{lon(l)} + (s(lat) - u_{j(l)}(lat))^2/r^2_{lat(l)}}.$$

$s(lon)$ and $s(lat)$ denote the longitude and latitude of the location s. $r_{lon(l)}$ denotes the shortest distance along longitude between two knot locations under l^{th} resolution. $r_{lat(l)}$ denotes the same along latitude. This specification enables one to capture anisotropy along east-west and north-south directions.

Combining Equations (23.1) and (23.2):

$$Z = \mu + S\eta + \delta + \epsilon, \tag{23.4}$$

where $\epsilon \equiv \{\epsilon(s_1), \ldots, \epsilon(s_n)\}$. A common choice for D is a diagonal matrix with the diagonal elements as τ^2. Under this specification, estimating both σ^2 and τ^2 is not possible as they are not identifiable. What one can estimate is $\nu^2 = \sigma^2 + \tau^2$; ν^2 is called *nugget effect*. A common practice is to assume that the measurement error variance σ^2 is known. It can either be obtained from calibration of measuring instrument, or by some data analysis approach, see [21]. The process error variance, τ^2 is treated unknown and is estimated from MCMC samples.

Under this representation of SRE model, the inverse of spatial covariance matrices for Y and Z can be obtained using the *Sherman-Woodbury-Morrison* formula which only requires the inversion of r (much smaller than n) dimensional matrix and inverse of D, which is a diagonal matrix and its inverse is available analytically. This is the key to the success of the SRE model. See Equation 2.15 in [6] for further details.

As mentioned before, there is a vast amount of literature addressing the issues with model formulation and model fitting for Gaussian spatial processes, not much work has been done to improve accuracy for datasets that depart from Gaussian assumptions. One important issue is the tail behavior. Simple QQ plots of *detailed residuals* can easily detect departures from Gaussian distribution. Detailed residuals are defined as: $Z - X\hat{\beta}_{OLS}$, where $\hat{\beta}_{OLS} = (X'X)^{-1}X'Y$, the ordinary least squares estimate of β. In [21], while analyzing CO_2 data, the authors showed, through a QQ plot and density estimation plot, that the detailed residuals exhibit right skewness. They also discussed that simple transformations were unable to provide any improvement and analyzed with Gaussian RRSM, arguing that estimates are still best linear unbiased predictor (BLUP), and conditioning on the estimates of V and ν^2 obtained from EM algorithm. When distributional assumptions are wrong, then likelihood based approaches, such as EM algorithm, suffer and do not provide accurate estimates of model parameters and the predictions become unreliable. One can use distribution free estimation techniques, such as binned method of moments (MOM) estimation (see [6] and for more robust version [15]). But binned estimation does not provide a ready positive-definite estimate of V and requires eigenvalue adjustment. Figure (23.1(a)) shows the QQ plot of detailed residuals for log transformed daily maximum ozone data and Figure (23.1(b)) shows the QQ plot of detailed residuals for untransformed daily maximum ozone data. We can clearly see from these plots that the tails are heavier than Gaussian distribution. This is just one example; when analyzing datasets on environmental processes, one very often faces situations where one tail is lighter and the other is heavier than Gaussian distribution, or both tails are lighter/heavier than Gaussian distribution. In what follows, a flexible class of RRSMs will be proposed. The proposed model can handle these different tail behaviors with lots of flexibility. Instead of imposing Gaussian distribution on $\boldsymbol{\eta}$, scale mixtures of Gaussian distributions will be imposed, where the scale parameters follow exponential distributions with mean 1.

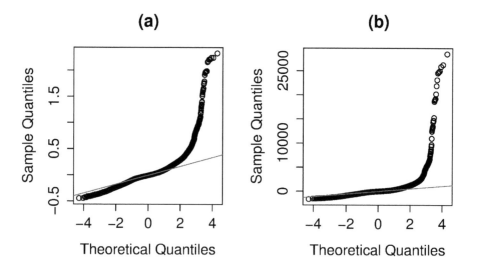

FIGURE 23.1
The QQ plots of the detailed residuals of daily total ozone maximum data. Panels (a) and (b) are respectively the detailed residuals of the log-transformed and untransformed data.

23.2.1   Proposed FRRSM

To account for non-Gaussian data, we propose a model based on equation (23.4), such that , $\boldsymbol{\eta} = \boldsymbol{A}^{\frac{1}{2}}\boldsymbol{U}$, where $\boldsymbol{U}$ is a zero mean r-dimensional Gaussian random variable with covariance matrix $\boldsymbol{V}$. The matrix $\boldsymbol{A}$ is diagonal with diagonal elements $\boldsymbol{W} = \{W_1, \ldots, W_r\}$, where $W_1, \ldots, W_r$ are identically and independently distributed exponential random variables with mean 1. Other distributional assumptions in Equation (23.2) remain the same. This distribution on $\boldsymbol{\eta}$ is motivated by a multivariate symmetric Laplace distribution. Note that this model specification does not increase the number of unknown parameters, the set of unknown parameters remains as $\{\boldsymbol{\beta}, \boldsymbol{V}, \tau^2\}$. The joint distribution of $\{\boldsymbol{Z}, \boldsymbol{U}, \boldsymbol{W}\}$ belongs to an exponential family and is log-concave. Large scale variation of the spatial process under analysis remains the same as before, $\boldsymbol{X}\boldsymbol{\beta}$.

Under our model assumptions, the latent spatial process $\boldsymbol{\eta}$ has zero mean and closed form expression for covariance matrix. Note that,

$$
\begin{aligned}
\mathrm{E}(\boldsymbol{\eta}) &= \mathrm{E}_{\boldsymbol{W},\boldsymbol{U}}(\boldsymbol{A}^{1/2}\boldsymbol{U}) & (23.5)\\
&= \mathrm{E}_{\boldsymbol{W}}(\boldsymbol{A}^{1/2})\mathrm{E}_{\boldsymbol{U}}(\boldsymbol{U}) \\
&= \left[\mathrm{E}_{W_1}(\sqrt{W_1})\mathrm{E}_{U_1}(U_1), \ldots, \mathrm{E}_{W_r}(\sqrt{W_r})\mathrm{E}_{U_r}(U_r)\right]' \\
&= \left[\mathrm{E}_{W_1}(\sqrt{W_1}) \times 0, \ldots, \mathrm{E}_{W_r}(\sqrt{W_r}) \times 0\right]' \\
&= [0, \ldots, 0]'
\end{aligned}
$$

The second equality in Equation (23.5) comes from the independence of W_j and U_j, and the fourth equality comes from the fact that means of U_j are all zero. The covariance matrix of $\boldsymbol{\eta}$ is same as $\mathrm{E}_{\boldsymbol{W},\boldsymbol{U}}(\boldsymbol{\eta}\boldsymbol{\eta}')$ and can be simplified as follows:

$$
\begin{aligned}
\mathrm{E}(\boldsymbol{\eta}\boldsymbol{\eta}') &= \mathrm{E}_{\boldsymbol{W},\boldsymbol{U}}(\boldsymbol{A}^{1/2}\boldsymbol{U}\boldsymbol{U}'\boldsymbol{A}^{1/2}) & (23.6)\\
&= \mathrm{E}_{\boldsymbol{W}}\left(\mathrm{E}_{\boldsymbol{U}|\boldsymbol{W}}(\boldsymbol{A}^{1/2}\boldsymbol{U}\boldsymbol{U}'\boldsymbol{A}^{1/2}|\boldsymbol{W})\right) \\
&= \mathrm{E}_{\boldsymbol{W}}\left(\boldsymbol{A}^{1/2}\boldsymbol{V}\boldsymbol{A}^{1/2}\right) \\
&= \mathrm{E}_{\boldsymbol{W}}
\begin{bmatrix}
W_1 V(1,1) & W_1^{1/2}V(1,2)W_2^{1/2} & \cdots & W_1^{1/2}V(1,r)W_r^{1/2} \\
\cdots & \cdots & \cdots & \cdots \\
W_r^{1/2}V(r,1)W_1^{1/2} & W_r^{1/2}V(r,2)W_2^{1/2} & \cdots & W_r V(r,r)
\end{bmatrix} \\
&=
\begin{bmatrix}
V(1,1) & \frac{\pi}{4}V(1,2) & \cdots & \frac{\pi}{4}V(1,r) \\
\cdots & \cdots & \cdots & \cdots \\
\frac{\pi}{4}V(r,1) & \frac{\pi}{4}V(r,2) & \cdots & V(r,r)
\end{bmatrix}
\end{aligned}
$$

In our notation, $V(i,j)$ denotes the $(i,j)^{th}$ element of the $\boldsymbol{V}$ matrix. The last equality in Equation (23.6) follows from the fact that W_i and W_j are independent, $E(W_i) = 1$, and $E(\sqrt{W_i}) = \sqrt{\pi}/2$.

Prior Distributions: While specifying prior distributions on model parameters, we ensure that they are noninformative or weakly informative, so that the posterior results are not sensitive to the choice of hyperparameters. We assume zero-mean *Gaussian* prior for $\boldsymbol{\beta}$ with covariance matrix $10^6 \boldsymbol{I}$.

Half-t distribution is imposed on the square root of τ^2 with hyperparameters 2 and 10^5. [9] and [13] discussed that this prior distribution is highly noninformative and has better behavior near zero compared to inverse-Gamma distributions with very small shape and scale parameters that put unnecessary prior weightage near small values. Half-t prior is usually cumbersome to deal with in MCMC implementations, but it can be specified in hierarchies of two inverse-Gamma distributions, see [13]. We assume:

$$\tau^2|a \quad \sim \quad \text{inverse-Gamma}\,(1, a/2)\,, \tag{23.7}$$

$$a \quad \sim \quad \text{inverse-Gamma}\,\left(1/2, 10^{-10}\right)\,, \tag{23.8}$$

where inverse-Gamma(b_1, b_2) denotes an inverse-Gamma distribution with shape parameter b_1 and scale parameter b_2. Under this model specification, the marginal distribution (integrating over a) of τ is $(2, 10^5)$. This prior distribution achieves a high degree of noninformativity and is better than inverse-Gamma or Uniform prior (see [13] for more details). This two-level prior specification preserves conjugacy and the full conditional distributions of τ^2 and a are inverse-Gamma.

Specifying priors for covariance matrices poses multiple challenges, see [42]. It is common in geostatistics to have only one single realization of the spatial process, hence consistent estimation of the covariance matrix is impossible. One instead hopes to obtain a covariance matrix which, while it may not be the same as the true covariance matrix, will still allow for faithful predictions, as well as reasonable interpretations. Another problem is that of specifying a noninformative prior on covariance matrices.

In the existing literature, we found at least three ways of specifying priors on the covariance matrix of the *low rank* latent process, denoted by $\boldsymbol{V}$ in this chapter. In his Gaussian process model, [12] assumed that the underlying latent process is identically and independently distributed zero-mean Gaussian distribution with a common variance parameter. This assumption compromises flexibility as the spatial covariance of the process under analysis is modeled only by basis function matrix $\boldsymbol{S}$. In the Gaussian multivariate spatial model, [3] imposed an inverse-Wishart prior for modeling cross covariances, and spatial covariance is specified by stationary Matérn covariance functions. They considered parameters in the Matérn covariance function as unknown and put priors on them. While Matérn models reduce the number of parameters, the range and smoothness parameters are somewhat delicate to deal with under Bayesian estimation as they are weakly identifiable. [36] pointed out that data cannot detect the smoothness parameter in Matérn covariance of order higher than 3. Further, a Matérn class can result in an overly smooth model. [31] overcame the smoothness problem by amending the Matérn covariance with another compactly supported covariance through tapering. [19] considered a more general covariance function and imposed priors on eigenvalues and Givens angles of V, the technique proposed by [8]. Their hyperparameters for priors on eigenvalues and Givens angles were specified using the method of moments estimator of V. The main issues with this prior specification are that Givens angles priors can slow down MCMC implementation and their priors are data dependent. If the method of moments estimate of V is unreliable (which can happen for skewed data and datasets having outliers), then specifying hyperparameters based on this will result in poor predictions.

We use the two-level hierarchical priors on the V matrix proposed by [13]. Our prior is marginally noninformative and does not depend on the data for specifying hyperparameters, furthermore our MCMC implementation is comparatively faster.

[9] noted that inverse-Gamma priors for variance parameters with small shape and scale hyperparameters give too large of weight to small values, and are not noninformative. He recommended using the Half−t prior which we used for τ^2. [13] extend this two-level hierarchy to provide a marginally non-informative prior for covariance matrices. We specify the prior on V as,

$$V|a_1,\ldots,a_r \;\sim\; \text{inverse-Wishart}\left(\lambda + r - 1, 2\lambda \text{diag}(1/a_1,\ldots,1/a_r)\right) \quad (23.9)$$

$$a_k \;\overset{ind}{\sim}\; \text{inverse-Gamma}\left(1/2, 1/B_k^2\right), \; k = 1,\ldots,r \quad (23.10)$$

When B_k takes on arbitrarily large values (e. g. $B_k = 10^5$), and $\lambda = 2$, this specification leads to the marginally noninformative prior where all variance parameters in V are Half−$t(2, 10^5)$, and all correlation parameters are uniform on $(-1, 1)$. As a result, this is a highly noninformative prior marginally, though joint noninformativity is difficult to assess and [13] suspected that joint informativity of priors on covariance matrix is hard to escape. Further, they showed that any submatrices of V belong to the same family of distributions under this prior specification, and like Givens angles priors, the multiresolutional feature is preserved. Like Givens angles priors of [19] our two-level inverse-Wishart prior takes into account the anisotropy and nonstationarity and it is more general in nature, bringing in a *nonparametric* flavor in covariance modeling.

Finally, our parameter set Ω consists of (β, τ^2, V) and latent processes U, W, and δ. The posterior distribution is unavailable in closed form and in the following section, we develop a MCMC algorithm for model fitting and predictions at unsampled locations.

23.3 MCMC Algorithm and Its Implementation

We derive the full conditional distributions for all unknown model parameters and latent processes. We block-update several parameters for better mixing and convergence, as block updating handles the correlations of Markov Chains better than individual updating. We implement our derived algorithm in R software and use the packages MASS, MCMCpack, and HI. We assume that the measurement error variance σ^2, is known, in fact, we estimate it *offline* using a data analytic approach. First we select n_0 of the observed locations, and define a small bin (2° distance in all directions from each point). Then, we calculate $\boldsymbol{\sigma}_M^2 = \left\{\sigma_1^2, \ldots, \sigma_{n_0}^2\right\}$, the variance of $\boldsymbol{Z}(\boldsymbol{s})$ within each of these bins, and compute the estimate of the measurement error variance as their median, $\hat{\sigma}^2 = \text{med}(\boldsymbol{\sigma}_M^2)$.

For hyperparameters, we set $B_k = 10^5, k = 1, \ldots, r$ and $\lambda = 2$. The full conditional distribution of $\boldsymbol{V}$ is inverse-Wishart distribution with shape parameter $\lambda + r$ and scale matrix

$$2\lambda\text{diag}(1/a_1, \ldots, 1/a_r) + \boldsymbol{U}'\boldsymbol{U}.$$

The parameters $\{a_k\}$ are sampled from its full conditional distributions, specifically, inverse-Gamma distributions with shape parameter $1 + r/2$ and rate parameter $\lambda V^{-1}(k, k) + 1/B_k^2$, where $V^{-1}(k, k)$ is the kth diagonal element of $\boldsymbol{V}^{-1}$ matrix.

The latent process, $\{W_i, i = 1, \ldots, r\}$, are linearly related, hence we block update them in our MCMC algorithm. The full conditional distribution is unavailable analytically and takes the following form:

$$[\boldsymbol{W}|\cdot] \propto \exp\left\{-0.5\boldsymbol{R}'\boldsymbol{R}\right\} \times \exp\left\{-\sum_{i=1}^{r} W_i\right\}, \tag{23.11}$$

where $\boldsymbol{R} = \boldsymbol{Z} - \boldsymbol{X}\boldsymbol{\beta} - \boldsymbol{S}\boldsymbol{A}^{\frac{1}{2}}\boldsymbol{\eta}$. This full conditional distribution is *log-concave* and we use *adaptive rejection Metropolis sampling* (ARMS) to obtain samples for $\boldsymbol{W}$. The ARMS algorithm allows samples to be obtained even from complicated densities that may not be log-concave. This is accomplished through a piecewise linear function which is an envelope for the log of the target density. Samples are drawn and either accepted, or used to update and tighten the envelope. For details see [11].

$\boldsymbol{\beta}$ and $\boldsymbol{U}$ are linearly related in our model and we sample them together from a multivariate Gaussian distribution of dimension $p+r$, with mean vector $\boldsymbol{\mu}_{\beta,U|\cdot}$ and covariance matrix $\boldsymbol{\Sigma}_{\beta,U|\cdot}$, where:

$$\boldsymbol{\Sigma}_{\beta,U|\cdot} = \begin{bmatrix} \frac{1}{\nu^2}\boldsymbol{X}'\boldsymbol{X} + 10^{-6}\boldsymbol{I}_p & \frac{1}{\nu^2}\boldsymbol{X}'\boldsymbol{S}\boldsymbol{A}^{\frac{1}{2}} \\ \frac{1}{\nu^2}\boldsymbol{A}^{\frac{1}{2}}\boldsymbol{S}'\boldsymbol{X} & \frac{1}{\nu^2}\boldsymbol{S}'\boldsymbol{A}\boldsymbol{S} + \boldsymbol{V}^{-1} \end{bmatrix}^{-1} \tag{23.12}$$

$$\boldsymbol{\mu}_{\beta,U|\cdot} = \boldsymbol{\Sigma}_{\beta,U}\begin{bmatrix} \boldsymbol{X}'\boldsymbol{Y} \\ \boldsymbol{A}^{\frac{1}{2}}\boldsymbol{S}'\boldsymbol{Y} \end{bmatrix}. \tag{23.13}$$

Recall that $\nu^2 = \sigma^2 + \tau^2$ is the nugget variance. The process error variance τ^2 is sampled from inverse-Gamma distribution with shape parameter $n/2+1$ and rate parameter $\frac{1}{2}\delta'\delta + 2/a$. The parameter a is sampled from another inverse-Gamma distribution with shape parameter $3/2$ and rate parameter $2/\tau^2 + 10^{-10}$.

The process errors, δ, for n observed locations are sampled from a multivariate Gaussian distribution with covariance matrix and mean vectors derived as:

$$\Sigma_{\delta|\cdot} = \left(\sigma^{-2} + \tau^{-2}\right)^{-1} I_n \tag{23.14}$$

$$\mu_{\delta|\cdot} = \Sigma_{\delta|\cdot}(Z - X\beta - SA^{1/2}U). \tag{23.15}$$

For each unobserved location, the process error is sampled from a zero-mean Gaussian distribution with variance τ^2.

Once we have sampled all the parameters and latent processes mentioned above, Y becomes deterministic and are obtained as follows:

$$Y_i = X_i\beta + S_i A^{1/2}U + \delta_i, \tag{23.16}$$

where X_i and S_i denote respectively the ith row of X and i^{th} row of S corresponding to Y_i, the *true* process variable at location s_i.

Multiresolutional features of two-level marginally noninformative inverse-Wishart prior: Under multiresolution model, the total number of knots r is a summation of r_l's, where r_l is the number of knots in l^{th} resolution. Then, we can write $S\eta$ as:

$$S\eta = \sum_{l=1}^{K} S_l\eta_l = \sum_{l=1}^{K} S_l A_l^{1/2} U_l, \tag{23.17}$$

where K is the total number of resolutions considered in the spatial modeling, l is the index for a particular resolution, and $A_l^{1/2}U_l$ is the parametrization of η_l in terms of W's and U's. Section 3.1 of [13] showed that the two-level inverse-Wishart-Gamma prior has the property of *"distributional invariance of subcovariance matrix."* This *invariance* property is very desirable and makes the prior even more attractive. If one considers the submatrix V_l, the $r_l \times r_l$ covariance matrix of U_l, then under this prior specification it will have the following marginal distributions:

$$V_l|a_{(1)}, \ldots, a_{(r_l)} \sim \text{inverse-Wishart}\left(\lambda + r_l - 1, 2\lambda\text{diag}(1/a_{(1)}, \ldots, 1/a_{(r_l)})\right)$$

$$a_{(k)} \overset{ind}{\sim} \text{inverse-Gamma}\left(1/2, 1/B_k^2\right), \ k = 1, \ldots, r_l.$$

This property helps the MCMC algorithm to be adaptive at different resolutions, it means that we can run the MCMC algorithm at the highest resolution and, from that, can obtain easily the parameter samples for coarser resolutions without rerunning the MCMC algorithm.

Posterior Summaries: The convergence of our Markov Chains is assessed using standard techniques, trace-plots, and Gelman and Rubin statistics, [10].

Storing simulations from an MCMC run on model parameters and latent processes requires large memory. Posterior means and standard deviations of the latent processes can be obtained sequentially within the MCMC loop. But techniques like quantile-based prediction intervals require storing all realizations of $Y(s_i)$'s after burn-in. Instead, if we can apply MCCLT, then Monte Carlo standard errors can be computed without burdening computer memory, and credible intervals of posterior means can be obtained under *normality* assumptions. Application of MCCLT on ergodic averages is not immediate; we must verify some nontrivial conditions, one of them is *geometric ergodicity*, that is, the chains converge fast enough to the target posterior at an exponential rate. For more discussion and insights on MCCLT, see [17]. The most popular posterior summary, often used, is the posterior expectation:

$$E(Y(s_i)|Z) = X_i E(\beta|Z) + S_i E(A^{1/2}U|Z) + E(\delta(s_i)|Z). \qquad (23.18)$$

It may be noted that the posterior mean is a function $g_i(\beta, W, U, \sigma^2, \tau^2)$ of $\{\beta, W, U, \sigma^2, \tau^2\}$. Since the full conditional distribution of δ is a function of $(\beta, W, U, \sigma^2, \tau^2)$ only and the measurement error variance σ^2 is assumed to be known, it is sufficient to establish the geometric ergodicity of the chain (β, W, U, τ^2). Ergodic average based on MCMC samples can be written as

$$\bar{g}_i = (1/T) \sum_{t=1}^{T} g_i(\beta^{(t)}, W^{(t)}, U^{(t)}, \tau^{2(t)}).$$

where T denotes the number of MCMC samples that are used to compute these averages and the index (t) denotes the generated sample at t^{th} MCMC iteration. The version of the CLT that we will use is:

Theorem 23.3.1. $\{\beta^{(t)}, W^{(t)}, U^{(t)}, \tau^{2(t)}, t = 0, 1, 2, \ldots\}$ *is a Harris ergodic Markov Chain with invariant probability distribution* $f(\cdot)$ *on state space* $\mathcal{X}$ *and* $E_f|g_i| < \infty$, $E_f|g_i|^{2+\epsilon} < \infty$ *for* $\epsilon > 0$. *If this Markov chain is geometrically ergodic, then:*

$$\sqrt{T}(\bar{g}_i - E(\bar{g}_i)) \to \mathcal{N}(0, \sigma_{g_i}^2),$$

where σ_{g_i} *is the* asymptotic *standard deviation.*

$\sigma_{g_i}/\sqrt{T}$ can be considered as a measure of precision of the posterior means obtained from T MCMC samples. $\sigma_{g_i}^2$ is unknown and it is estimated as follows, using the *batch mean method*, (see [17]) by splitting the generated MCMC samples into K batches, each of size J,

$$\hat{\sigma}_{g_i}^2 = \frac{J}{K-1} \sum_{k=1}^{K} (g_i^{(k)} - \bar{g}_i)^2. \qquad (23.19)$$

$g_i^{(k)}$ is the kth batch mean obtained by computing

$$g_i^{(k)} = \frac{\sum_{j=J(k-1)+1}^{kJ} g_i((\boldsymbol{\beta}^{(j)}, \boldsymbol{W}^{(j)}, \boldsymbol{U}^{(j)}, \tau^{2(j)})}{J},$$

and $T = KJ$.

Now, we verify the conditions of MCCLT on the Markov chain $\{\boldsymbol{\beta}^{(t)}, \boldsymbol{W}^{(t)}, \boldsymbol{U}^{(t)}, \tau^{2(t)}\}_{t=0}^{\infty}$ stated in Theorem (23.3.1). From the full conditional distributions that we derived in previous section, we can clearly see that all of them are *continuous*. Further, Markov chains generated from these full conditionals are *aperiodic*, *Harris recurrent*, and $\phi-$*irreducible*. In what follows, we establish the *geometric ergodicity* of this Markov chain by *drift* and *minorization* conditions, see [18].

The drift condition will be established by finding a function $V(\cdot) : \mathbb{R}^p \times \mathbb{R}^{+r} \times \mathbb{R}^r \times \mathbb{R}^+ \to \mathbb{R}^+$ of $\{\boldsymbol{\beta}, \boldsymbol{W}, \boldsymbol{U}, \tau^2\} \in \mathbb{R}^p \times \mathbb{R}^{+r} \times \mathbb{R}^r \times \mathbb{R}^+$, such that:

$$(PV)(\boldsymbol{\beta}, \boldsymbol{W}, \boldsymbol{U}, \tau^2) \leq \rho V(\boldsymbol{\beta}, \boldsymbol{W}, \boldsymbol{U}, \tau^2) + L,$$
$$\text{for all } \{\boldsymbol{\beta}, \boldsymbol{W}, \boldsymbol{U}, \tau^2\} \in \mathbb{R}^p \times \mathbb{R}^{+r} \times \mathbb{R}^r \times \mathbb{R}^+ \quad (23.20)$$

where $\rho \in [0, 1)$ and $L \in \mathbb{R}$; and

$$(PV)(\boldsymbol{\beta}, \boldsymbol{W}, \boldsymbol{U}, \tau^2) = \int_{\mathbb{R}^p} \int_{\mathbb{R}^{+r}} \int_{\mathbb{R}^r} \int_{\mathbb{R}^+} V(\boldsymbol{\beta}, \boldsymbol{W}, \boldsymbol{U}, \tau^2)$$
$$\Gamma(\boldsymbol{\beta}, \boldsymbol{W}, \boldsymbol{U}, \tau^2 | \tilde{\boldsymbol{\beta}}, \tilde{\boldsymbol{W}}, \tilde{\boldsymbol{U}}, \tilde{\tau^2}) \mathrm{d}\{\tilde{\boldsymbol{\beta}}, \tilde{\boldsymbol{W}}, \tilde{\boldsymbol{U}}, \tilde{\tau^2}\},$$
$$(23.21)$$

where $\Gamma(\boldsymbol{\beta}, \boldsymbol{W}, \boldsymbol{U}, \tau^2 | \tilde{\boldsymbol{\beta}}, \tilde{\boldsymbol{W}}, \tilde{\boldsymbol{U}}, \tilde{\tau^2})$ is the one-step transition kernel from $(\tilde{\boldsymbol{\beta}}, \tilde{\boldsymbol{W}}, \tilde{\boldsymbol{U}}, \tilde{\tau^2})$ to $(\boldsymbol{\beta}, \boldsymbol{W}, \boldsymbol{U}, \tau^2)$.

We consider $V(\boldsymbol{\beta}, \boldsymbol{W}, \boldsymbol{U}, \tau^2)$ to be $(\boldsymbol{Y} - \boldsymbol{X}\boldsymbol{\beta} - \boldsymbol{S}\boldsymbol{A}^{1/2}\boldsymbol{U})'(\boldsymbol{Y} - \boldsymbol{X}\boldsymbol{\beta} - \boldsymbol{S}\boldsymbol{A}^{1/2}\boldsymbol{U})/\tau^2$. Recall from our MCMC algorithm described in the previous section that the latent variable $\boldsymbol{Y}$ is simply a function of $\{\boldsymbol{\beta}, \boldsymbol{W}, \boldsymbol{U}, \tau^2, \sigma^2\}$, where the measurement error variance σ^2 is known. We apply Fubini's Theorem for interchanging integrals and Equation (23.21) becomes:

$$\int_{\Theta^-} \int_{\mathbb{R}^p \times \mathbb{R}^{+r} \times \mathbb{R}^r \times \mathbb{R}^+} V(\cdot) \Gamma(\boldsymbol{\beta}, \boldsymbol{W}, \boldsymbol{U}, \tau^2 | \boldsymbol{\Omega}^-, \tilde{\boldsymbol{\beta}}, \tilde{\boldsymbol{W}}, \tilde{\boldsymbol{U}}, \tilde{\tau^2})$$
$$\Gamma(\boldsymbol{\Omega}^- | \tilde{\boldsymbol{\beta}}, \tilde{\boldsymbol{W}}, \tilde{\boldsymbol{U}}, \tilde{\tau^2}) \mathrm{d}\{\tilde{\boldsymbol{\beta}}, \tilde{\boldsymbol{W}}, \tilde{\boldsymbol{U}}, \tilde{\tau^2}\} \mathrm{d}\boldsymbol{\Omega}^-$$
$$\leq \; n + 2$$
$$\leq \; \rho V(\cdot) + n + 2 \quad (23.22)$$

We denote the remaining parameters in the model except $\{\boldsymbol{\beta}, \boldsymbol{W}, \boldsymbol{U}, \tau^2\}$ by

$\mathbf{\Omega}^-$ and Θ^- is the parameter space for $\mathbf{\Omega}^-$. The full conditional distribution $\Gamma(\tau^2|\boldsymbol{\beta}, \boldsymbol{W}, \boldsymbol{U}, \mathbf{\Omega}^-, \tilde{\boldsymbol{\beta}}, \tilde{\boldsymbol{W}}, \tilde{\boldsymbol{U}}, \tilde{\tau^2})$ is an inverse-Gamma distribution with shape parameter $n/2 + 1$ and rate parameter $(\boldsymbol{Y} - \boldsymbol{X}\boldsymbol{\beta} - \boldsymbol{S}\boldsymbol{A}^{1/2}\boldsymbol{U})'(\boldsymbol{Y} - \boldsymbol{X}\boldsymbol{\beta} - \boldsymbol{S}\boldsymbol{A}^{1/2}\boldsymbol{U})/2 + 2/a$.

The innermost integral in (23.22) with respect to τ^2 becomes:

$$\frac{(n/2+1)(\boldsymbol{Y} - \boldsymbol{X}\boldsymbol{\beta} - \boldsymbol{S}\boldsymbol{A}^{1/2}\boldsymbol{U})'(\boldsymbol{Y} - \boldsymbol{X}\boldsymbol{\beta} - \boldsymbol{S}\boldsymbol{A}^{1/2}\boldsymbol{U})}{(\boldsymbol{Y} - \boldsymbol{X}\boldsymbol{\beta} - \boldsymbol{S}\boldsymbol{A}^{1/2}\boldsymbol{U})'(\boldsymbol{Y} - \boldsymbol{X}\boldsymbol{\beta} - \boldsymbol{S}\boldsymbol{A}^{1/2}\boldsymbol{U})/2 + 2/a} \quad (23.23)$$

$$\leq \quad n + 2 \quad\quad\quad\quad\quad\quad\quad\quad\quad\quad\quad\quad\quad\quad\quad\quad\quad (23.24)$$

The inequality in (23.23) follows from the fact that $2/a > 0$ and the last inequality in (23.22) holds as $V(\cdot) \geq 0$.

We establish the *minorization* condition by showing that on a set $C = \{(\boldsymbol{\beta}, \boldsymbol{W}, \boldsymbol{U}, \tau^2) : V(\cdot) < l\}$ for $l > 2L/(1-\rho)$, there exists a density q and $\varepsilon > 0$ such that for $(\tilde{\boldsymbol{\beta}}, \tilde{\boldsymbol{W}}, \tilde{\boldsymbol{U}}, \tilde{\tau^2}) \in C$ and $(\boldsymbol{\beta}, \boldsymbol{W}, \boldsymbol{U}, \tau^2) \in \mathbb{R}^p \times \mathbb{R}^{+r} \times \mathbb{R}^r \times \mathbb{R}^+$,

$$\Gamma(\boldsymbol{\beta}, \boldsymbol{W}, \boldsymbol{U}, \tau^2 | \tilde{\boldsymbol{\beta}}, \tilde{\boldsymbol{W}}, \tilde{\boldsymbol{U}}, \tilde{\tau^2}) \geq \varepsilon q(\boldsymbol{\beta}, \boldsymbol{W}, \boldsymbol{U}, \tau^2). \quad (23.25)$$

The right-hand side of Equation (23.25) is:

$$= \int_{\Theta^-} \Gamma(\boldsymbol{\beta}, \boldsymbol{W}, \boldsymbol{U}, \tau^2 | \tilde{\boldsymbol{\beta}}, \tilde{\boldsymbol{W}}, \tilde{\boldsymbol{U}}, \tilde{\tau^2}, \mathbf{\Omega}^-) \Gamma(\mathbf{\Omega}^- | \tilde{\boldsymbol{\beta}}, \tilde{\boldsymbol{W}}, \tilde{\boldsymbol{U}}, \tilde{\tau^2}) d\mathbf{\Omega}^- \quad (23.26)$$

$$= \int_{\Theta^-} \Gamma(\boldsymbol{\beta}, \boldsymbol{W}, \boldsymbol{U} | \tau^2, \tilde{\boldsymbol{\beta}}, \tilde{\boldsymbol{W}}, \tilde{\boldsymbol{U}}, \tilde{\tau^2}, \mathbf{\Omega}^-) \mathrm{IG}(n/2+1, (\boldsymbol{Y} - \boldsymbol{X}\tilde{\boldsymbol{\beta}} - \boldsymbol{S}\tilde{\boldsymbol{A}}^{1/2}\tilde{\boldsymbol{U}})'$$

$$(\boldsymbol{Y} - \boldsymbol{X}\tilde{\boldsymbol{\beta}} - \boldsymbol{S}\tilde{\boldsymbol{A}}^{1/2}\tilde{\boldsymbol{U}}) + 2/a; \tau^2) \Gamma(\mathbf{\Omega}^- | \tilde{\boldsymbol{\beta}}, \tilde{\boldsymbol{W}}, \tilde{\boldsymbol{U}}, \tilde{\tau^2}) d\mathbf{\Omega}^- \quad (23.27)$$

$$\geq \varepsilon \int_{\Theta^-} \Gamma(\boldsymbol{\beta}, \boldsymbol{W}, \boldsymbol{U} | \tau^2, \tilde{\boldsymbol{\beta}}, \tilde{\boldsymbol{W}}, \tilde{\boldsymbol{U}}, \tilde{\tau^2}, \mathbf{\Omega}^-) \mathrm{IG}(n/2+1, \tilde{\tau^2}l + 2/a; \tau^2)$$

$$\Gamma(\mathbf{\Omega}^- | \tilde{\boldsymbol{\beta}}, \tilde{\boldsymbol{W}}, \tilde{\boldsymbol{U}}, \tilde{\tau^2}) d\mathbf{\Omega}^-.$$

$\mathrm{IG}(a, b; x)$ represents an inverse-Gamma distribution with shape a and rate b, evaluated at x. In C, $(\boldsymbol{Y} - \boldsymbol{X}\tilde{\boldsymbol{\beta}} - \boldsymbol{S}\tilde{\boldsymbol{A}}^{1/2}\boldsymbol{U})'(\boldsymbol{Y} - \boldsymbol{X}\tilde{\boldsymbol{\beta}} - \boldsymbol{S}\tilde{\boldsymbol{A}}^{1/2}\boldsymbol{U}) \leq l\tilde{\tau^2}$ and the exponent term in inverse-Gamma distribution is a decreasing function of its rate parameter, hence the inequality in (23.26) holds. ε is given by:

$$\left(\frac{(\boldsymbol{Y} - \boldsymbol{X}\tilde{\boldsymbol{\beta}} - \boldsymbol{S}\tilde{\boldsymbol{A}}^{1/2}\boldsymbol{U})'(\boldsymbol{Y} - \boldsymbol{X}\tilde{\boldsymbol{\beta}} - \boldsymbol{S}\tilde{\boldsymbol{A}}^{1/2}\boldsymbol{U}) + 2/a}{\tilde{\tau^2}l + 2/a} \right)^{(n/2+1)} \quad (23.28)$$

and the remaining terms constitute $q(\cdot)$, which is a *bona fide* probability distribution. This completes the proof. For general discussion and insights on MCMC theory and the techniques and theorems used in this proof, readers are encouraged to see [16], [18], [25] and [30] and the references therein.

23.4 Application: Simulated and Real Data

In this section, we apply our model to real and simulated data. We use the simulation study to assess the validity of our model through training datasets and test datasets. Afterward, we model the daily maximum of total column ozone and obtain predictions of daily maxima over the entire globe.

23.4.1 Simulation study

To assess the performance of our FRRSM, we first apply it on simulated datasets. We generate data based on the model specified in Equation (23.4) by setting, as discussed, $\boldsymbol{\eta} = \boldsymbol{A}^{\frac{1}{2}}\boldsymbol{U}$. For observed locations we select $n = 25,600$ locations on a regular grid over the unit square. Two resolutions of knot locations were selected. The first resolution had $r_1 = 16$ knots laid out in a 4×4 grid, the second resolution had $r_2 = 64$ knots in an 8×8 grid. Within each resolution, the knots were equidistant over the unit square. For the large-scale variation, we set $\boldsymbol{X}$ to be an $n \times 3$ matrix where the first column is a vector of ones, and the second and third columns are longitude and latitude, with corresponding regression coefficients, $\boldsymbol{\beta} = (1,5,2)'$. The $\boldsymbol{S}$ matrix was specified using modified bisquare function as described in Equation (23.3). Since the knots are equidistant on a regular grid, we had $r_{lat(1)} = r_{long(1)} = 0.25$ and $r_{lat(2)} = r_{long(2)} = 0.125$. The fixed rank covariance matrix $\boldsymbol{V}$ was generated using Matérn covariance function, with range parameter $\phi = 0.25$, sill parameter $\sigma^2 = 0.02$, smoothness parameter $\kappa = 1.5$, and distance measure $\|\boldsymbol{u}_i - \boldsymbol{u}_j\|$ between knots $\boldsymbol{u}_i$ and $\boldsymbol{u}_j$ given by the Euclidean distance.

We draw a sample of $\boldsymbol{U}$ from a multivariate Gaussian distribution with mean $\boldsymbol{0}$ and covariance $\boldsymbol{V}$, as specified in the previous paragraph. To test the flexibility of the proposed model, we generate three samples for $\boldsymbol{W}$ as follows,

$$\begin{aligned} \boldsymbol{W}_1 &\sim \text{Exponential}(1), \\ \boldsymbol{W}_2 &\sim \text{Gamma}(5,1), \\ \boldsymbol{W}_3 &\sim \text{GIG}(5,1/2,1/5), \end{aligned} \qquad (23.29)$$

where $\text{GIG}(\chi, \psi, \lambda)$ is the Generalized Inverse Gaussian distribution with density:

$$f(x) = \frac{(\psi/\chi)^{\lambda/2}}{2K_\lambda(\sqrt{\psi\chi})} x^{\lambda-1} \exp\left\{ -\frac{1}{2}\left(\frac{\chi}{x} + \psi x\right) \right\}.$$

The gamma and GIG distributions were chosen in part because each contains the exponential distribution as a special case. As the model uses the exponential distribution for $\boldsymbol{W}$, good performance of the model fitting on the simulated data with the gamma and GIG distributions on $\boldsymbol{W}$ are indications of another level of flexibility of our model. We generate the measurement error ϵ from a Gaussian distribution with variance 0.05, with $\alpha = 10\%$ contamination, where the contaminated variance is 0.75. Finally, we calculate the simulated data by $\boldsymbol{Y} = \boldsymbol{X\beta} + \boldsymbol{SA}^{\frac{1}{2}}\boldsymbol{U} + \epsilon$. Quantile plots for all three simulations showed significant non-Gaussian tail behaviors.

We randomly select 5,000 observed locations and keep them aside as a test case. The remaining 20,600 observations are used for estimation and prediction (i.e. the training sample). Results are shown in Figure 23.2. The figures in the left column are histograms of $Y - \hat{Y}$ for the test data for all $\boldsymbol{W}$s mentioned in Equation (23.29). The figures in right column are analogous plots of Y vs. $\hat{Y}$ (the predictions). Most of the points lie very close to the $Y = \hat{Y}$ line, indicating that our proposed model provides excellent predictions even under contaminated data. And note that these are predictions for the *unobserved* test case locations. These results show that the proposed method is able to successfully model datasets with non-Gaussian tails. Also, note that we use the same model described in Section 23.2 with all these

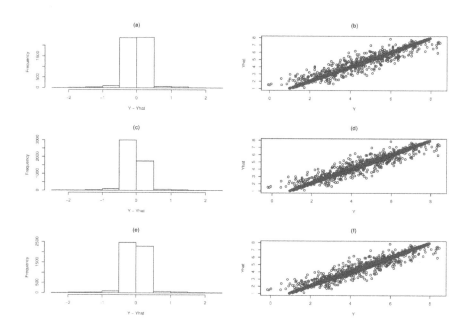

FIGURE 23.2

Simulation results when using Exponential, Gamma, and GIG distributions for $\boldsymbol{W}$. Left panel (a), (c), (e): Histograms of $Y - \hat{Y}$, correspond to the proposed model fit, obtained from 5000 test data from simulation. Right panel (b), (d), (f): Scatterplots of Y vs. $\hat{Y}$.

simulated datasets. Hence, the model is flexible enough to handle varying forms of the scale factor W even with just exponential distributions having mean 1.

Under all the three simulations for W, our model made highly accurate predictions. This indicates that the proposed model is flexible enough to account for a misspecified model for W.

We compare the performance of our FRRSM approach using the same simulated datasets with other comparable Gaussian RRSMs explained as Fixed Rank Kriging (FRK), Bayesian Gaussian process model with V matrix specified by a multiple of identity matrix (BMI), and Bayesian Gaussian model with inverse-Wishart prior (BIW) on V matrix where the hyperparameters in the scale matrix of this prior are specified using method of moments estimate of V matrix. For details on the method of moments estimation of V matrix, see [6] and [20]. In addition, we also compare the performance of our model with the EM algorithm described in [21]. In all these cases, our FRRSM outperformed other methods in terms of the absolute value of the differences between $Y - \hat{Y}$ and mean squared prediction errors or its Bayesian analog *Monte Carlo standard errors*. For the sake of space, we are not reporting all of them here, but we do compare the performance between FRRSM and BIW using daily maximum of total column ozone in the following subsection. However, readers are encouraged to see [19], the section where they discussed the limitations of FRK, BMI, and BIW models.

23.4.2 Application to daily maximum of total column Ozone data

From the Moderate Resolution Imaging Spectroradiometer (MODIS) on board NASA's Terra satellite, we obtain a large dataset of daily maximum of total column (TOM) measured in Dobson units. These data are heavily right-skewed, and are often modeled after log transformation. First, we fit our model and the BIW model to log-transformed data. Due to the flexibility evident from the simulations, we also model the data on the original scale using our proposed FRRSM. Recall that the probability plot of TOM (Figure 23.1) showed significantly non-Gaussian tail behaviors, including many outliers on the upper tails.

Many atmospheric processes display a north-south trend, but comparatively little east-west trend. The plot of TOM in Figure (23.3) shows such a trend, so we model the large-scale variation using spherical harmonics. The design matrix X is specified by Legendre polynomials (see [35]) $P_n^m(\sin(L))$ of degree $n = 80$ and order $m = 0, 1, \ldots, n$, where L is the latitude of a location.

We employ a two-resolution scheme for S, allowing the model to capture multiple scales of spatial variation, see [27] and [6]. The first resolution has $r_1 = 68$ knot locations, the second resolution has $r_2 = 200$ knot locations.

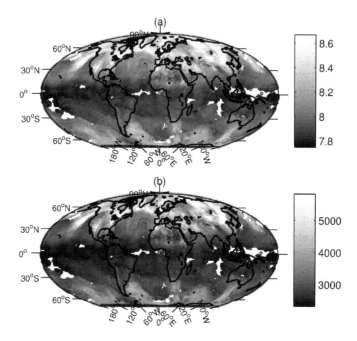

FIGURE 23.3
Map of observed data, total ozone maximum: (a) log transformed data, (b)
data on original scale, measured in Dobson unit.

Both resolutions are approximately equidistant over the entire globe. The S
matrix is based on Equation (23.3). In particular, distances between locations
and knots, $d(s_i, u_{j(l)})$, is defined using great-arc distances based on Vincenty's
formula, which is a more accurate means to compute distances on an ellipsoid
than Euclidean distances. The distance 'scaling factors' that we use for different resolutions are, for the first resolution, $r_{lon}(1) = 2725.1, r_{lat}(1) = 2700.4$,
and for the second resolution, $r_{lon}(2) = 1546.4, r_{lat}(2) = 1667.9$.

First, we perform our analysis on log scale. As described in Section 23.3, a
data analytic approach is used to estimate the measurement error variance. In
our analysis, we randomly select 20 observed locations from the dataset and
calculate the variance of all points within a $4^o \times 4^o$ bin centered around each
of the selected locations. The estimate of σ^2 is the median of these variances,
obtained as $\hat{\sigma}^2 = 0.1$. The predictions of log-TOM using Bayesian posterior
predictive distribution based on FRRSM and the Monte Carlo standard errors
are shown in Figures (23.4a) and (23.4b). We also ran a Gaussian model,
specifically the BIW model. The predictions of log-TOM from BIW model are

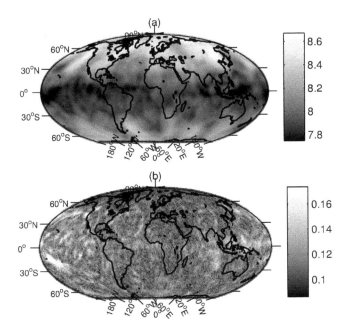

FIGURE 23.4
Maps of (a) predictions using Bayesian posterior predictive distribution and (b) Monte Carlo standard errors using the proposed FRRSM, both in log-scale.

shown in Figure (23.5a) and the corresponding Monte Carlo standard errors in Figure (23.5b). From the prediction plots, one can clearly notice that BIW model fails to capture the right tail (more smoothness is observed for the extreme observations near North Pole), where as our FRRSM produced more reliable predictions. Similar patterns are observed for the process near the South Pole.

Figure (23.6a) shows $Y - \hat{Y}$, where $\hat{Y}$ are predictions from Bayesian posterior predictive distribution of FRRSM at observed locations and Figure (23.6b) shows the same for Gaussian BIW model. One can clearly see that comparatively large magnitudes of residuals are seen in the plot for BIW model. Figure (23.6c) shows a map of the relative improvement in Monte Carlo standard errors (MCSE) for each location. Differences were taken as $(MCSE^G - MCSE^W)/MCSE^G$, where $MCSE^G$ indicates the MCSE for the Gaussian BIW model, and $MCSE^W$ is the MCSE for the proposed FRRSM. Hence, positive values indicate that the proposed method out-performed the Gaussian BIW model. We observe that there appears to be a general trend for

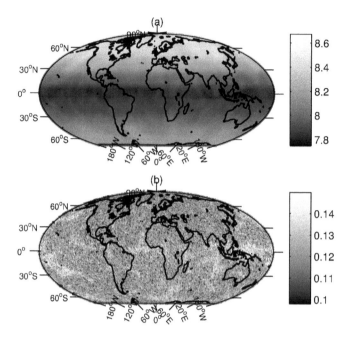

FIGURE 23.5

Maps of (a) predictions using Bayesian posterior predictive distribution and (b) Monte Carlo standard errors using Gaussian model (BIW), both in log-scale.

the proposed model to have smaller MCSEs. By the color axis, lighter shades imply the proposed method is better (smaller MCSEs), while darker shades imply the Gaussian model performed better. There appears to be more orange than green in Figure (23.6c).

Finally, we fit our FRRSM to TOM in original scale, without taking log transformation. The predictions using Bayesian posterior predictive distribution based on FRRSM and the Monte Carlo standard errors are shown in Figures (23.7a) and (23.7b), respectively. Comparing Figures (23.3b) with (23.7a), we note that our model produces reliable predictions even for highly skewed data.

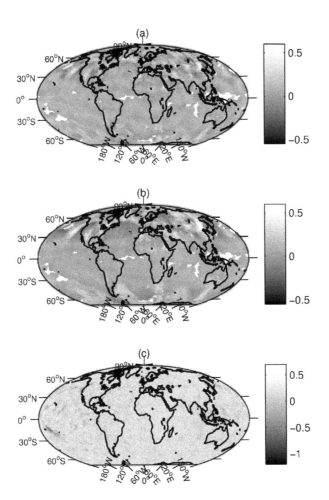

FIGURE 23.6
Maps of $Y - \hat{Y}$:(a) $\hat{Y}$ are predictions using Bayesian posterior predictive distribution of FRRSM (b)$\hat{Y}$ are predictions using Bayesian posterior predictive distribution of Gaussian BIW model. (c) Map of location-wise difference of Monte Carlo Standard Errors obtained using $(MCSE^G - MCSE^W)/MCSE^G$.

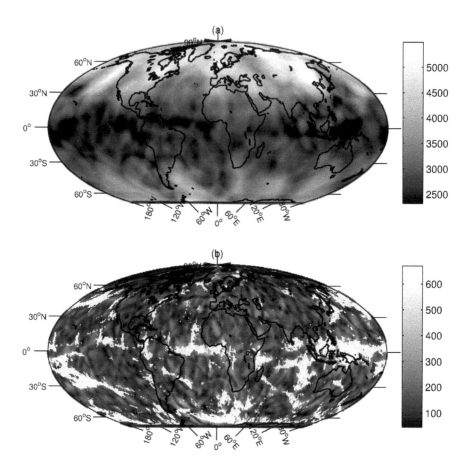

FIGURE 23.7
Maps of (a) predictions using Bayesian posterior predictive distribution and (b) Monte Carlo standard errors using the proposed FRRSM, both in original scale.

23.5   Conclusions and Discussion

In this chapter, we have proposed a new class of reduced rank spatial model. Instead of modeling the reduced dimensional latent spatial component as a multivariate Gaussian distribution, we have replaced it with a scale mixture of Gaussian distributions, where the scale parameters are independently and identically distributed exponential random variables with mean 1. This specification along with a marginally noninformative two-level inverse-Wishart prior of [13] enables the model to account for a wide variety of tail behaviors, including light or heavy tails.

All of the prior distributions that are used for FRRSM are noninformative. This setup provides more adaptivity and robust predictions compared to other data-driven prior selection models, such as the Givens angle prior of [19].

Usually, for right skewed data, analyses after taking log transformation are very common. [14] developed for large right skewed data on coal chemical composition. While, log-normal kriging is a way of modeling right skewed data, it is based on several strict assumptions and [29], in their Chapter 5, mentioned that because of its inflexibility, it is comparatively less popular. One of the big achievements of our proposed FRRSM is that it can successfully model even highly right skewed data without taking log transformation.

Finally, we have also provided a general framework for posterior summaries for reduced rank spatial models based on MCCLT and batch mean methods. The described methods and proofs can be easily adapted and extended to other types of RRSMs and alleviate the problems of storing large MCMC samples. This facilitates simple computation of prediction intervals or other similar inferential techniques.

Acknowledgments

The first author's research was supported in parts by the Faculty Research and Creative Activity Award (FRACAA), Western Michigan University and by CSIRO Computational and Simulation Science - Computational Informatics (previously known as, CSIRO Mathematics, Informatics, and Statistics - CMIS), Australia. The second author's research was supported in part by the Intramural Research Program of the NIH, National Institute of Environmental Health Sciences (Z01 ES101744).

Bibliography

[1] D. Allard and P. Naveau. A new spatial skew-normal random field model. *Communications in Statistics*, 36:1821–1834, 2007.

[2] S. Banerjee, A. E. Gelfand, and B. P. Carlin. *Hierarchical Modeling and Analysis for Spatial Data*. Boca Raton, Chapman & Hall/CRC, 2004.

[3] S. Banerjee, A. E. Gelfand, A. Finley, and H. Sang. Gaussian predictive process models for large spatial data sets. *Journal of the Royal Statistical Society: Series B (Statistical Methodology)*, 70:825–848, 2008.

[4] W. Chang, M. Haran, R. Olson, and K. Keller. Fast dimension-reduced climate model calibration. *arXiv preprint: arXiv:1303.1382*, 2013.

[5] N. Cressie. *Statistics for Spatial Data*. John Wiley & Sons, New York, 1993.

[6] N. Cressie and G. Johannesson. Fixed rank kriging for very large spatial data sets. *Journal of the Royal Statistical Society: Series B (Statistical Methodology)*, 70:209–226, 2008.

[7] N. Cressie, T. Shi, and E. L. Kang. Fixed rank filtering for spatio-temporal data. *Journal of Computational and Graphical Statistics*, 19:724–745, 2010.

[8] M. J. Daniels and R. E. Kass. Nonconjugate Bayesian estimation of covariance matrices and its use in hierarchical models. *Journal of the American Statistical Association*, 94:1254–1263, 1999.

[9] A. Gelman. Prior distributions for variance parameters in hierarchical models. *Bayesian Analysis*, 1(3):515–533, 2006.

[10] A. Gelman and D. B. Rubin. Inference from iterative simulation using multiple sequences. *Statistical Science*, 7(4):457–472, 1992.

[11] W. R. Gilks, N. G. Best, and K. K. C. Tan. Adaptive rejection Metropolis sampling within Gibbs sampling. *Journal of the Royal Statistical Society: Series C (Applied Statistics)*, 44:455–472, 1995.

[12] D. Higdon. A process-convolution approach to modelling temperatures in the north Atlantic ocean. *Environmental and Ecological Statistics*, 5:173–190, 1998.

[13] A. Huang and M. P. Wand. Simple marginally noninformative prior distributions for covariance matrices. *Bayesian Analysis*, 8(2):439–452, 2013.

[14] C. Jelsema and R. Paul. Spatial mixed effects model for compositional data with applications to coal geology. *International Journal of Coal Geology*, 114:33–43, 2013.

[15] C. M. Jelsema, R. Paul, and J. W. McKean. Robust estimation for reduced rank spatial model. Technical Report, Western Michigan University, USA, 2013.

[16] G. L. Jones. On the Markov chain central limit theorem. *Probability Surveys*, 1:299–320, 2004.

[17] G. L. Jones, M. Haran, B. S. Caffo, and R. Neath. Fixed-width output analysis for Markov chain Monte Carlo. *Journal of the American Statistical Association*, 101(476):1537–1547, 2006.

[18] G. L. Jones and J. P. Hobart. Honest exploration of intractable probability distributions via Markov chain Monte Carlo. *Statistical Science*, 16(4):312–334, 2001.

[19] E. L. Kang and N. Cressie. Bayesian inference for the spatial random effects model. *Journal of the American Statistical Association*, 106(495):972–983, 2011.

[20] E. L. Kang, N. Cressie, and T. Shi. Using temporal variability to improve spatial mapping with application to satellite data. *Canadian Journal of Statistics*, 38(2):271–289, 2010.

[21] M. Katzfuss and N. Cressie. Tutorial on fixed rank kriging (FRK) of CO2 data. Technical Report, 858, The Ohio State University, USA, 2011.

[22] M. Katzfuss and D. Hammerling. Parallel inference for massive distributed spatial data using low-rank models. *arXiv preprint arXiv:1402.1472*, 2014.

[23] H. Kim and B. K. Mallick. A Bayesian prediction using the skew-Gaussian processes. *Journal of Statistical Planning and Inference*, 120(1-2):85–101, 2004.

[24] K. V. Mardia, C. Goodall, E. J. Redfern, and F. J. Alonso. The kriged Kalman filter (with discussion). *TEST*, 7(2):217–252, 1998.

[25] S. P. Meyn and R. L. Tweedie. *Markov Chains and Stochastic Stability*. Springer Verlag, London, 1993.

[26] D. Nychka, S. Bandyopadhyay, D. Hammerling, F. Lindgren, and S. Sain. A multi-resolution Gaussian process model for the analysis of large spatial data sets. *Journal of Computational and Graphical Statistics, in press*, 2014. DOI:10.1080/10618600.2014.914946.

[27] D. Nychka, C. Wikle, and J. A. Royle. Multiresolution models for nonstationary spatial covariance functions. *Statistical Modelling*, 2(4):315–331, 2002.

[28] C. J. Paciorek. Computational techniques for spatial logistic regression with large data sets. *Computational Statistics & Data Analysis*, 51(8):3631–3653, 2007.

[29] V. Pawloswsky-Glahn and R. Olea. *Geostatistical Analysis of Compositional Data*. New York, Oxford University Press, 2004.

[30] J. S. Rosenthal. Minorization conditions and convergence rates for Markov chain Monte Carlo. *Journal of the American Statistical Association*, 90(430):558–556, 1995.

[31] H. Sang and J. Z. Huang. A full scale approximation of covariance functions for large spatial data sets. *Journal of the Royal Statistical Society: Series B (Statistical Methodology)*, 74(1):111–132, 2012.

[32] A. Sengupta and N. Cressie. Empirical hierarchical modeling for count data using the spatial random effects model. *Spatial Economic Analysis*, 8(3):389–418, 2013.

[33] A. Sengupta and N. Cressie. Hierarchical statistical modeling of big spatial datasets using the exponential family of distributions. Technical Report 870, Department of Statistics,The Ohio State University, USA, 2013.

[34] T. Shi and N. Cressie. Global statistical analysis of MISR aerosol data: A massive data product from NASA's Terra satellite. *Environmetrics*, 18(7):665–680, 2007.

[35] B. J. Smith. boa: An R package for MCMC output convergence assessment and posterior inference. *Journal of Statistical Software*, 21(11):1–37, 2007.

[36] M. L. Stein. Space-time covariance functions. *Journal of the American Statistical Association*, 100(469):310–321, 2005.

[37] J. R. Stroud, P. Muller, and B. Sanso. Dynamic models for spatiotemporal data. *Journal of the Royal Statistical Society: Series B (Statistical Methodology)*, 63(4):673–689, 2001.

[38] Y. Sun, B. Li, and M. G. Genton. Geostatistics for large datasets. In E. Porcu, J.-M. Montero, and M. Schlather, editors, *Advances and Challenges in Space-time Modelling of Natural Events*, Lecture Notes in Statistics, chapter 3, pages 55–77. Springer Verlag, Berlin Heidelberg, 2012.

[39] C. K. Wikle. A kernel-based spectral model for non-Gaussian spatiotemporal processes. *Statistical Modelling*, 2(4):299–314, 2002.

[40] C. K. Wikle and N. Cressie. A dimension-reduced approach to space-time Kalman filtering. *Biometrika*, 86(4):815–829, 1999.

[41] C. K. Wikle, R. F. Milliff, D. Nychka, and L. M. Berliner. Spatiotemporal hierarchical Bayesian modeling: Tropical ocean surface winds. *Journal of the American Statistical Association*, 96(454):382–397, 2001.

[42] R. Yang and J. O. Berger. Estimation of a covariance matrix using the reference prior. *Annals of Statistics*, 22(3):1195–1211, 1994.

[43] H. Zareifard and M. J. Khaledi. Non-Gaussian modeling of spatial data using scale mixing of a unified skew Gaussian process. *Journal of Multivariate Analysis*, 114(1):16–28, 2013.

[44] H. Zhang and A. El-Shaarawi. On spatial skew-Gaussian processes and applications. *Environmetrics*, 21(1):33–47, 2010.

24

A Bayesian Reweighting Technique for Small Area Estimation

Azizur Rahman

Charles Sturt University

Satyanshu K. Upadhyay

Banaras Hindu University

CONTENTS

24.1 Introduction

Small area estimation (SAE) has received much attention in recent decades due to increasing demand for reliable small area statistics by policy makers. To develop effective policy on various crucial social, economic and health issues at a local level, policy makers need to use accurate data on those issues collected at small area levels. SAE techniques can produce such reliable estimates for small areas (see, for example, [19, 27] and [29]). Traditionally there are direct and indirect methods for SAE and each of these methods is linked with very simple to complex theories, algorithms and models. Overall, the direct SAE method comprises a range of estimators such as the Horvitz-Thompson estimator, generalised regression estimator, modified survey estimator, etc. that are based on the survey-design and derived from very simple statistical theories and uncomplicated formulae, whereas the indirect method encompasses complex methodologies from statistics, economics and geography, among others.

The indirect SAE methods are divided into two groups: statistical and geographic approaches [21]. The statistical approach is mainly based on a range of implicit and explicit statistical models, and each of these models is widely studied to obtain small area estimates using the (empirical-) best linear unbiased prediction (E-BLUP), empirical Bayes (EB) and hierarchical Bayes (HB) techniques (see [18, 19, 21, 28] and [29]). The geographic approach is based on various spatial models and sophisticated microsimulation modelling technology, (MMT) ([22], [27]). In simple words, microsimulation modelling is a computer intensive technique that operates at the level of the individual behavioural entity, such as a person, household, firm, etc., and creates a synthetic micropopulation (i.e., a population of interest within a small area or domain) dataset of these low-level entities to produce estimates on our interest in order to draw decisions that apply to local level and higher levels of aggregation such as an entire state/country (see, for example, [2, 3, 8] and [26]).

An important issue in indirect SAE is that additional auxiliary information or covariates are needed. Once the model is chosen, whether it is statistical or geographic, its parameters are estimated using the data obtained from the sample survey. In the real world practices, sufficient representative samples are unavailable for all small areas. Depending on the type of the study and the time and money constraints, it may not be possible to conduct a sufficiently comprehensive survey to obtain an adequate sample from every small area. A basic problem with national or state level surveys is that they are not designed for efficient estimation for small areas [12]. MMT can overcome this limitation of survey data by producing reliable microdata , *i.e.*, detailed significant information of a reasonably representative micropopulation and small area estimates [23].

In the geographic method, microdata are simulated to estimate the model parameters, and the technique is robust in the sense that further aggregation or disaggregation is possible on the basis of the choice of area scales or domains. But in the statistical approach, data from different sources are used to obtain the model parameters, and it does not involve generating a small area microdata file. So, the geographic approach has an advantage over the statistical approach in that it always simulates a list-based representation of small area microdata file, which is a significant resource for various further analyses and/or model updating [23]. Another advantage is that the MMT based models are used for estimating small area effects of policy change and future estimates of socioeconomic and population attributes ([27], [36]). They are also intended to explore the links among regions and sub-regions and to project the spatial implications of inequality or economic development and policy changes at a more disaggregated level. In contrast, the statistical model based methods do not have such utility.

Although the geographic approach harks back to the microsimulation modelling ideas pioneered in the middle of last century by [17], MMT has become a popular, cost-effective and accessible method for socioeconomic policy analysis only in the recent decades. It may be further credited with the rapid development of increasingly powerful computer hardware; wider availability of individual unit record datasets [23]; and with the growing demand by policy makers [36] for SAEs at government and private sector levels including income, tax and social security benefits, income deprivation, housing stress, housing demand, care needs, etc. (see, for example, [2, 3, 7–10, 22, 24, 27, 35]). Spatially disaggregate data are not readily obtainable. Even if these data are available in some form, they suffer from severe limitations either due to lack of characteristics or geographical detail (see [10, 24] and [36]). So, reliable microdata need to be simulated using different probabilistic and deterministic methods. Synthetic reconstruction and reweighting are two common methods used in MMT [26].

Synthetic reconstruction is an old method which attempts to construct synthetic micro-populations in such a way that all variables with known benchmarks at the small area level are reproduced. There are two different ways of undertaking synthetic reconstruction — statistical data matching or fusion ([15, 33]) and iterative proportional fitting as part of the chain Monte Carlo sampling ([6, 16]). The reweighting method is a relatively new and popular method, which mainly calibrates the sampling design weights to a set of new weights based on a distance measure, using the available data at spatial scale. The reweighting approach includes a combinatorial optimisation (CO) process and the generalised regression technique called the GREGWT process. The CO iterative process selects an appropriate combination of units from the survey data that best fits the known benchmarks at small area levels using an intelligent searching optimization tool such as simulated annealing ([34, 35]). The GREGWT algorithm calibrates the weights on a dataset containing the survey returns, which are modified so that certain estimates agree with externally provided benchmarks at small areas by minimizing a truncated Chi-squared distance function with respect to the calibration equations for each small area (see [8, 23, 26] and [31]). In this case, the new weights must lie within a pre-specified boundary condition that could be constant across sample units or proportional to the sampling weights for each small area.

Creating a reliable synthetic microdata is still challenging in the geographic approach to SAE (see [2, 9, 22, 24] and [36]). Detailed discussion and comparison of various microdata generation methods are available (see [22, 23, 25]). Although a range of methods is used to generate microdata, none of these can consider the scenario of whole population at a small area level as they are not designed to encompass facts of the unobserved units (see [23, 24, 35]). So, a newly generated microdata from those approaches often leads to inaccurate data for many small areas, especially for small areas with a small-sized population. And validating outputs from a model built on such synthetic microdata is also difficult. There is no statistically robust tool yet to deal with these problems.

This chapter develops a Bayesian prediction based reweighting methodology for generating microdata set for the geographic approach of SAE. The new method takes consideration of the complete scenarios of micropopulation units at small area level and produces the statistical reliability measure of SAEs.

The plan of the chapter is as follows. Section 24.2 outlines the basic concept of the new approach with the Bayesian prediction theory. Section 24.3 illustrates the model with matrix notations and the prior and posterior distributions. Section 24.4 designs the Bayesian reweighting technique for modelling the unobserved units at small area level by determining the predictive distribution for the model. Section 24.5 provides empirical results of small area housing stress estimation. A concluding remark is presented in Section 24.6.

24.2   The Basic Concept

It is noted that in any area being sampled, a finite population usually has two parts — observed units in the sample called data and unobserved sampling units in the population (Figure 24.1).

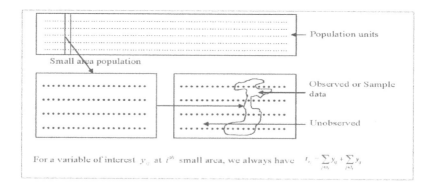

FIGURE 24.1
A diagram of new system for reweighting small area microdata.

Suppose $\mathbf{\Omega}$ represents a finite population in which Ω_i (say) is the subpopulation of small area i. Now if s_i denotes the observed sample units in the ith area then we have $s_i \cup \bar{s}_i = \Omega_i \subseteq \mathbf{\Omega} \ \forall \ i$, where $\bar{s}_i$ denotes the unobserved units in the small area population. Let y_{ij} represents a variable of interest for the jth observation in the population at i^{th} small area. Thus, we always have the estimate of population total at i^{th} small area as: $t_{y_i.} = \sum_{j \in s_i} y_{ij} + \sum_{j \in \bar{s}_i} y_{ij}$.

The main challenge in this approach of microdata simulation is to establish the linkage of observed data to the unobserved sampling units in the small area population. Essentially it is a sort of prediction problem, where a modeller tries to find a probability distribution of unobserved responses using the observed sample and the auxiliary data. The Bayesian methodology (see [1, 11, 14, 20]) can deal with such a prediction problem. This new method is principally based on the Bayesian prediction theory and Markov chain Monte Carlo (MCMC) method. The basic steps in this process of microdata simulation are as follows:

1. obtain a suitable joint prior distribution of the event under study E_i, say housing stress in the population at i^{th} small area, that is $p(E_i)$ for $\forall \ i$;
2. find the conditional distribution of unobserved sampling units, given the observed data, that is $p(y_{ij} : j \in \bar{s}_i | y_{ij} : j \in s_i)$ for $\forall \ i$;
3. derive the posterior distribution using Bayes theorem, that is, $p(\boldsymbol{\theta}|s, \boldsymbol{x}); E_i \subseteq \boldsymbol{\theta}$, where $\boldsymbol{\theta}$ is the vector of model parameters and $\boldsymbol{x}$ is an auxiliary information vector; and
4. get simulated copies of the entire population from this posterior distribution by the MCMC simulation technique.

24.2.1 The Bayesian prediction theory

The Bayesian prediction theory is very straightforward and mainly based on the Bayes theorem [4]. Let $\boldsymbol{y}$ be a set of observed sample units from a model with a joint probability density $p(\boldsymbol{y}|\boldsymbol{\theta})$, in which $\boldsymbol{\theta}$ is a set of model parameters. If the prior density of unknown parameters $\boldsymbol{\theta}$ is $g(\boldsymbol{\theta})$, the posterior density of $\boldsymbol{\theta}$ for given $\boldsymbol{y}$ can be obtained by the Bayes theorem and defined as: $p(\boldsymbol{\theta}|\boldsymbol{y}) \propto p(\boldsymbol{y}|\boldsymbol{\theta})g(\boldsymbol{\theta})$. Now, if $\bar{\boldsymbol{y}}$ is the set of unobserved units in a finite population, the predictive distribution of $\bar{\boldsymbol{y}}$ can be obtained by solving the integral

$$p(\bar{\boldsymbol{y}}|\boldsymbol{y}) \propto \int_{\boldsymbol{\theta}} p(\boldsymbol{\theta}|\boldsymbol{y})p(\bar{\boldsymbol{y}}|\boldsymbol{\theta})d\boldsymbol{\theta},$$

where $p(\bar{\boldsymbol{y}}|\boldsymbol{\theta})$ is the probability density function of unobserved units in the finite population. Further details on the Bayesian prediction theory for various linear models can be obtained in [20].

24.3   The Model

The multivariate multiple regression model is a generalization of the multiple regression model in a matrix-variate setup. This model represents the relationship between a set of values of several response variables and a single set of values of some explanatory or predictor variables. The multivariate model is used to analyze data from different situations in economics as well as in many other experimental circumstances to deal with a set of regression equations.

There are many experimental situations in real life where we need to study a set of responses from more than one dependent variable corresponding to a set of values from several independent variables. For example, in a country like Australia, if various groups of households in different states or towns are given a single set of social benefits (such as housing assistance, health subsidies, income supports, childcare benefits, etc.) to observe the responses on the living standards, then for a set of values of the predictor variables, there will be several sets of values of the response variables from various groups of households. More details on such models can be found in [20, 32].

24.3.1   Matrix notations

Suppose $\boldsymbol{y}_1, \boldsymbol{y}_2,..., \boldsymbol{y}_{n_i}$ is a set of vectors of n_i responses at i^{th} small area and $x_1, x_2,..., x_{k-1}$ is a set of $k-1$ predictor variables. Then a relation between these set of vector responses and the single set of values of predictor variables can be written as a multivariate multiple regression model when each of the n_i responses is assumed to follow the linear model, defined as $\boldsymbol{y}_{ij} = \beta_{0[j]} + \beta_{1[j]}x_{j1} + ... + \beta_{(k-1)[j]}x_{j(k-1)} + \boldsymbol{e}_{ij}$; for $\forall$ i and $j = 1, 2, ..., n_i$ where $\boldsymbol{y}_{ij}$ and $\boldsymbol{e}_{ij}$ are, respectively, the jth response at the ith small area and its associated error vectors, each of order $1 \times p$; $\beta_{.[j]}$ is a $k \times 1$ dimensional vector of regression parameters on the regression line of jth response; and $x_{j1}, x_{j2}, ..., x_{j(k-1)}$ is a single set of values of the $k-1$ predictor variables for the regression to j^{th} response. Assume $\boldsymbol{e}_{ij} \sim t_p(0, \Sigma, \nu)$. Characteristically, the Student-t distribution modelling is very useful for small sample problems and it is more robust as its limiting distribution is normal. Let $x_{j.} = [x_{j1}, x_{j2}, ..., x_{j(k-1)}]$, $\boldsymbol{y}_{.j} = [y_{j1}, y_{j2}, ..., y_{jp}]'$ and $\boldsymbol{e}_{.j} = [e_{j1}, e_{j2}, ..., e_{jp}]'$ denote the values of the explanatory, response and error variables, respectively, for the jth sample or response unit. Then for the ith small area, the multivariate model can be expressed as

$$\boldsymbol{Y}_i = \boldsymbol{X}_i\boldsymbol{\beta} + \boldsymbol{E}_i \qquad (24.1)$$

where the order of $\boldsymbol{Y}_i$ and $\boldsymbol{E}_i$ are $n_i \times p$; $\boldsymbol{\beta}$ is $k \times p$; $\boldsymbol{X}_i$ is $n_i \times k$, and the rank $(\boldsymbol{X}_i) = k$ with $n_i \geq k$ for $\forall$ i.

Now assume that the components of each row in $\boldsymbol{E}_i$ of the model are correlated, and jointly follow a multivariate Student-t distribution. Since each row of the errors matrix E_i are uncorrelated with others, the covariance of the errors matrix becomes $\frac{\nu}{\nu-2}[\Sigma \otimes I_{n_i}]$; where ν is the shape parameter of the matrix-T distribution for the errors matrix, $\otimes$ denotes the Kronecker product of two matrices Σ and I_{n_i} in which Σ is a $p \times p$ positive definite symmetric matrix and I_{n_i} is an identity matrix of order $n_i \times n_i$. Hence $E_i \sim T_{n_i p}(0, I_{n_i \times n_i}, \Sigma_{p \times p}, \nu)$ and Y_i has the following probability function

$$f(Y_i|X_i\beta, \Sigma, \nu) = C \frac{|I_{n_i}^{-1}|^{\frac{p}{2}}}{|\Sigma|^{\frac{n_i}{2}}} |I_p + \Sigma^{-1}(Y_i - X_i\beta)'(Y_i - X_i\beta)|^{-\frac{1}{2}(\nu+p+n_i-1)}$$
(24.2)

where $C = \left[\Gamma\left(\frac{1}{2}\right)\right]^{pn_i} \frac{\Gamma_p\left[\frac{1}{2}(\nu+p-1)\right]}{\Gamma_p\left[\frac{1}{2}(\nu+p+n_i-1)\right]}$ in which $\Gamma_p(.)$ is a generalized gamma function defined as: $\Gamma_d(b) = \left[\Gamma\left(\frac{1}{2}\right)\right]^{\frac{1}{2}d(d-1)} \prod_{\alpha=1}^{d} \Gamma\left(b + \frac{\alpha-d}{2}\right); b > \frac{d-1}{2}$, for the nonzero positive integers d and b, and $\alpha = 1, 2, \cdots, d$.

24.3.2 The prior and posterior distributions

Assume that the joint prior distribution of the regression matrix β and the $\frac{1}{2}p(p+1)$ distinct elements of Σ is uniform. Also assume that the elements of β and that of Σ are independently distributed. It means that if $p(\beta, \Sigma)$ is a joint prior density of β and Σ, then $p(\beta, \Sigma) = p(\beta)p(\Sigma)$. Adopting the invariance theory due to Jeffreys [34], here we can consider $p(\beta) = $ constant and $p(\Sigma) \propto |\Sigma|^{-\frac{p+1}{2}}$. Thus, the joint prior of unknown parameters is

$$p(\beta, \Sigma) = p(\beta)p(\Sigma) \propto |\Sigma|^{-\frac{p+1}{2}}.$$
(24.3)

An uniform prior distribution has been used by many researchers such as [5, 20] and [37]. Now, using the Bayes theorem, the posterior density can be defined as

$$p(\beta, \Sigma|Y_i) \propto p(Y_i|\beta, \Sigma)p(\beta, \Sigma),$$
(24.4)

where $p(Y_i|\beta, \Sigma)$ is the density of Y_i and $p(\beta, \Sigma)$ is the joint prior distribution.

Using (24.2) and (24.3) in (24.4), we get the following posterior density function

$$p(\beta, \Sigma|Y_i) = \Phi|\Sigma|^{-\frac{n_i+p+1}{2}} |I_p + \Sigma^{-1}(Y_i - X_i\beta)'(Y_i - X_i\beta)|^{-\frac{\nu+p+n_i-1}{2}}$$
(24.5)

where the normalizing constant Φ can be obtained by solving the equation

$$\Phi^{-1} = \int_\beta \int_\Sigma |\Sigma|^{-\frac{n_i+p+1}{2}} |I_p + \Sigma^{-1}(Y_i - X_i\beta)'(Y_i - X_i\beta)|^{-\frac{\nu+p+n_i-1}{2}} d\Sigma d\beta$$
(24.6)

Using the appropriate Jacobian of the transformation and then performing integrations with respect to β and Σ using the properties of the generalised beta integral and the matrix$-T$ distribution [10], the value of Φ is obtained as

$$\Phi = \frac{|\boldsymbol{X}_i'\boldsymbol{X}_i|^{\frac{p}{2}}|\boldsymbol{S}_{Y_i}|^{\frac{n_i-k}{2}}}{[\Gamma(\frac{1}{2})]^{kp}} \frac{\Gamma_p\left(\frac{\nu+n_i+p-1}{2}\right)}{\Gamma_p\left(\frac{n_i-k}{2}\right)\Gamma_p\left(\frac{\nu+p-1}{2}\right)} \quad \text{where } \boldsymbol{S}_{Y_i} = (\boldsymbol{Y}_i - \boldsymbol{X}_i\hat{\boldsymbol{\beta}})'(\boldsymbol{Y}_i - \boldsymbol{X}_i\hat{\boldsymbol{\beta}}),$$

in which $\hat{\boldsymbol{\beta}} = (\boldsymbol{X}_i'\boldsymbol{X}_i)^{-1}\boldsymbol{X}_i'\boldsymbol{Y}_i$.

24.4   Modelling Unobserved Population Units

This section demonstrates the predictive distribution for modelling unobserved population units at the ith small area.

24.4.1   The linkage model

The linkage model is set up for the unobserved population units in a small area to connect them with the observed units in the sample. Let $\bar{\boldsymbol{Y}}_i$ be the unobserved responses matrix from the model in (24.1) to the $m_i \times k$ (where $m_i = N_i - n_i; \forall\, i$) dimensional design matrix $\bar{\boldsymbol{X}}_i$. Then the linkage multivariate model for unobserved population units in the ith small area can be defined as

$$\bar{\boldsymbol{Y}}_i = \bar{\boldsymbol{X}}_i\boldsymbol{\beta} + \bar{\boldsymbol{E}}_i \tag{24.7}$$

where $\boldsymbol{\beta}$ is of order $k \times p$, and $\bar{\boldsymbol{Y}}_i$ and $\bar{\boldsymbol{E}}_i$ both are of the order $m_i \times p$.

It is assumed that $\bar{\boldsymbol{E}}_i \sim T_{m_i p}(\boldsymbol{0}, \boldsymbol{I}_{m_i \times m_i}, \boldsymbol{\Sigma}_{p \times p}, \nu)$. Since the elements in each row of the errors matrixes $\boldsymbol{E}_i$ and $\bar{\boldsymbol{E}}_i$ are correlated, and n_i rows in $\boldsymbol{E}_i$ as well as m_i rows in $\bar{\boldsymbol{E}}_i$ are uncorrelated, the joint density function of $\boldsymbol{Y}_i$ and $\bar{\boldsymbol{Y}}_i$ for $\boldsymbol{\beta}$ and $\boldsymbol{\Sigma}$ can be expressed as

$$p(\boldsymbol{Y}_i, \bar{\boldsymbol{Y}}_i|\boldsymbol{\beta}, \boldsymbol{\Sigma}) \propto |\boldsymbol{\Sigma}|^{-\frac{n_i+m_i}{2}} \left|\boldsymbol{I}_p + \boldsymbol{\Sigma}^{-1}\boldsymbol{Q}\right|^{-\frac{\nu+p+n_i+m_i-1}{2}} \tag{24.8}$$

where $\boldsymbol{Q} = (\boldsymbol{Y}_i - \boldsymbol{X}_i\boldsymbol{\beta})'(\boldsymbol{Y}_i - \boldsymbol{X}_i\boldsymbol{\beta}) + (\bar{\boldsymbol{Y}}_i - \bar{\boldsymbol{X}}_i\boldsymbol{\beta})'(\bar{\boldsymbol{Y}}_i - \bar{\boldsymbol{X}}_i\boldsymbol{\beta})$.

24.4.2   Predictive distribution

Using the prior and joint densities in (24.3) and (24.8), the joint posterior density of unknown parameters for given $\boldsymbol{Y}_i$ and $\bar{\boldsymbol{Y}}_i$ can be determined as

$$p(\boldsymbol{\beta}, \boldsymbol{\Sigma}|\boldsymbol{Y}_i, \bar{\boldsymbol{Y}}_i) \propto |\boldsymbol{\Sigma}|^{-\frac{n_i+m_i+p+1}{2}} \left|\boldsymbol{I}_p + \boldsymbol{\Sigma}^{-1}\boldsymbol{Q}\right|^{-\frac{\nu+p+n_i+m_i-1}{2}}. \tag{24.9}$$

Now the predictive distribution of the set of m_i unobserved population units at i^{th} small area can be obtained by solving the following integral

$$f(\bar{\boldsymbol{Y}}_i|\boldsymbol{Y}_i) \propto \int_{\boldsymbol{\beta}} \int_{|\boldsymbol{\Sigma}|>0} |\boldsymbol{\Sigma}|^{-\frac{n_i+m_i+p+1}{2}} \left|\boldsymbol{I}_p + \boldsymbol{\Sigma}^{-1}\boldsymbol{Q}\right|^{-\frac{\nu+p+n_i+m_i-1}{2}} d\boldsymbol{\Sigma} d\boldsymbol{\beta}. \tag{24.10}$$

Applying a matrix transformation $\boldsymbol{\Sigma}^{-1} = \boldsymbol{\Lambda}$ with the Jacobian of the transformation $|J| = \frac{d\boldsymbol{\Sigma}}{d\boldsymbol{\Lambda}} = |\boldsymbol{\Lambda}^{-1}|^{p+1}$, the density in (24.10) can be written as

$$f(\bar{\boldsymbol{Y}}_i|\boldsymbol{Y}_i) \propto \int_{\boldsymbol{\beta}} \int_{|\boldsymbol{\Lambda}|>0} |\boldsymbol{\Lambda}|^{\frac{n_i+m_i}{2}-\frac{p+1}{2}} |\boldsymbol{I}_p + \boldsymbol{\Lambda}\boldsymbol{Q}|^{-\left(\frac{n_i+m_i}{2}+\frac{\nu+p-1}{2}\right)} d\boldsymbol{\Lambda}d\boldsymbol{\beta}.$$

By integrating over $\boldsymbol{\Lambda}$ using the generalized beta integral, we get

$$f(\bar{\boldsymbol{Y}}_i|\boldsymbol{Y}_i) \propto \int_{\boldsymbol{\beta}} |\boldsymbol{Q}|^{-\frac{n_i+m_i}{2}} B_p\left(\frac{n_i+m_i}{2}, \frac{\nu+p-1}{2}\right) d\boldsymbol{\beta}. \tag{24.11}$$

Now $\boldsymbol{Q}$ can be expressed as a quadratic form in $\boldsymbol{\beta}$ as

$$\boldsymbol{Q} = \boldsymbol{R} + (\boldsymbol{\beta} - \boldsymbol{P})'\boldsymbol{M}(\boldsymbol{\beta} - \boldsymbol{P}) \tag{24.12}$$

where $\boldsymbol{R} = \boldsymbol{Y}_i'\boldsymbol{Y}_i + \bar{\boldsymbol{Y}}_i'\bar{\boldsymbol{Y}}_i - \boldsymbol{P}'\boldsymbol{M}\boldsymbol{P}$, $\boldsymbol{P} = \boldsymbol{M}^{-1}(\boldsymbol{X}_i'\boldsymbol{Y}_i + \bar{\boldsymbol{X}}_i'\bar{\boldsymbol{Y}}_i)$ and $\boldsymbol{M} = \boldsymbol{X}_i'\boldsymbol{X}_i + \bar{\boldsymbol{X}}_i'\bar{\boldsymbol{X}}_i$.

Appling (24.12) in (24.11) and integrating over $\boldsymbol{\beta}$ by using the matrix-T integral, the predictive density can be obtained as

$$f(\bar{\boldsymbol{Y}}_i|\boldsymbol{Y}_i) \propto |\boldsymbol{R}|^{-\frac{n_i+m_i-k}{2}}. \tag{24.13}$$

Here $\boldsymbol{R}$ is free from unknown parameters. To get a complete form of predictive density, we have to express $\boldsymbol{R}$ as a quadratic form of $\bar{\boldsymbol{Y}}_i$. A range of systematic mathematical operations [13] are needed to convert $\boldsymbol{R}$ into the convenient form of $\bar{\boldsymbol{Y}}_i$. Briefly, this can be achieved as follows:

$$\begin{aligned}
\boldsymbol{R} &= \boldsymbol{Y}_i'\boldsymbol{Y}_i + \bar{\boldsymbol{Y}}_i'\bar{\boldsymbol{Y}}_i - \boldsymbol{P}'\boldsymbol{M}\boldsymbol{P} \\
&= \boldsymbol{Y}_i'(\boldsymbol{I} - \boldsymbol{X}_i\boldsymbol{M}^{-1}\boldsymbol{X}_i')\boldsymbol{Y}_i + \bar{\boldsymbol{Y}}_i'\boldsymbol{H}\bar{\boldsymbol{Y}}_i - \boldsymbol{Y}_i'\boldsymbol{X}_i\boldsymbol{M}^{-1}\bar{\boldsymbol{X}}_i'\bar{\boldsymbol{Y}}_i \\
&\quad - \bar{\boldsymbol{Y}}_i'\bar{\boldsymbol{X}}_i\boldsymbol{M}^{-1}\boldsymbol{X}_i'\boldsymbol{Y}_i \tag{24.14}
\end{aligned}$$

where $\boldsymbol{H} = \boldsymbol{I} - \bar{\boldsymbol{X}}_i\boldsymbol{M}^{-1}\bar{\boldsymbol{X}}_i'$, so that $\boldsymbol{H}^{-1} = (\boldsymbol{I} - \bar{\boldsymbol{X}}_i\boldsymbol{M}^{-1}\bar{\boldsymbol{X}}_i')^{-1} = \boldsymbol{I} + \bar{\boldsymbol{X}}_i(\boldsymbol{X}_i'\boldsymbol{X}_i)^{-1}\bar{\boldsymbol{X}}_i'$.

If $G_1(\boldsymbol{M}, \boldsymbol{H}) = \boldsymbol{X}_i\boldsymbol{M}^{-1}\boldsymbol{X}_i' + \boldsymbol{X}_i\boldsymbol{M}^{-1}\bar{\boldsymbol{X}}_i\boldsymbol{H}^{-1}\bar{\boldsymbol{X}}_i\boldsymbol{M}^{-1}\boldsymbol{X}_i'$ and $G_2(\boldsymbol{M}, \boldsymbol{H}) = \boldsymbol{H}^{-1}\bar{\boldsymbol{X}}_i\boldsymbol{M}^{-1}\boldsymbol{X}_i'\boldsymbol{Y}_i$, then $\boldsymbol{R}$ can be expressed as

$$\boldsymbol{R} = \boldsymbol{Y}_i'[\boldsymbol{I} - G_1(\boldsymbol{M}, \boldsymbol{H})]\boldsymbol{Y}_i + [\bar{\boldsymbol{Y}}_i' - G_2(\boldsymbol{M}, \boldsymbol{H})]'\boldsymbol{H}[\bar{\boldsymbol{Y}}_i' - G_2(\boldsymbol{M}, \boldsymbol{H})]. \tag{24.15}$$

Now the following relationships can be established as more suitable forms

$$\begin{aligned}
G_1(\boldsymbol{M}, \boldsymbol{H}) &= \boldsymbol{X}_i\boldsymbol{M}^{-1}[\boldsymbol{X}_i' + \bar{\boldsymbol{X}}_i'\bar{\boldsymbol{X}}_i\{\boldsymbol{M}^{-1}\boldsymbol{X}_i' + (\boldsymbol{X}_i'\boldsymbol{X}_i)^{-1}\bar{\boldsymbol{X}}_i'\bar{\boldsymbol{X}}_i\boldsymbol{M}^{-1}\boldsymbol{X}_i'\}] \\
&= \boldsymbol{X}_i\boldsymbol{M}^{-1}[\boldsymbol{X}_i' + \bar{\boldsymbol{X}}_i'\bar{\boldsymbol{X}}_i\boldsymbol{U}], \tag{24.16}
\end{aligned}$$

for $\boldsymbol{U} = \boldsymbol{M}^{-1}\boldsymbol{X}_i' + (\boldsymbol{X}_i'\boldsymbol{X}_i)^{-1}\bar{\boldsymbol{X}}_i'\bar{\boldsymbol{X}}_i\boldsymbol{M}^{-1}\boldsymbol{X}_i'$.

After employing appropriate matrix operations in $\boldsymbol{U}$, we get $\boldsymbol{U} = (\boldsymbol{X}_i'\boldsymbol{X}_i)^{-1}\boldsymbol{X}_i'$. Substituting this in (24.16), we establish the form

$$G_1(\boldsymbol{M}, \boldsymbol{H}) = \boldsymbol{X}_i(\boldsymbol{X}_i'\boldsymbol{X}_i)^{-1}\boldsymbol{X}_i'. \tag{24.17}$$

Similarly, using appropriate matrix operations to other functional form, we get

$$G_2(\boldsymbol{M}, \boldsymbol{H}) = \bar{\boldsymbol{X}}_i(\boldsymbol{X}_i'\boldsymbol{X}_i)^{-1}\boldsymbol{X}_i'\boldsymbol{Y}_i. \tag{24.18}$$

Using (24.17) and (24.18) in (24.15), $\boldsymbol{R}$ can be rewritten as

$$
\begin{aligned}
\boldsymbol{R} &= \boldsymbol{Y}_i'[I - \boldsymbol{X}_i(\boldsymbol{X}_i'\boldsymbol{X}_i)^{-1}\boldsymbol{X}_i']\boldsymbol{Y}_i \\
&\quad + [\bar{\boldsymbol{Y}}_i' - \bar{\boldsymbol{X}}_i(\boldsymbol{X}_i'\boldsymbol{X}_i)^{-1}\boldsymbol{X}_i'\boldsymbol{Y}_i]'\boldsymbol{H}[\bar{\boldsymbol{Y}}_i' - \bar{\boldsymbol{X}}_i(\boldsymbol{X}_i'\boldsymbol{X}_i)^{-1}\boldsymbol{X}_i'\boldsymbol{Y}_i] \\
&= \boldsymbol{Y}_i'[I - \boldsymbol{X}_i(\boldsymbol{X}_i'\boldsymbol{X}_i)^{-1}\boldsymbol{X}_i']\boldsymbol{Y}_i + [\bar{\boldsymbol{Y}}_i' - \bar{\boldsymbol{X}}_i\hat{\boldsymbol{\beta}}]'\boldsymbol{H}[\bar{\boldsymbol{Y}}_i' - \bar{\boldsymbol{X}}_i\hat{\boldsymbol{\beta}}] \tag{24.19}
\end{aligned}
$$

Again since $\boldsymbol{S}_{Y_i} = (\boldsymbol{Y}_i - \boldsymbol{X}_i\hat{\boldsymbol{\beta}})'(\boldsymbol{Y}_i - \boldsymbol{X}_i\hat{\boldsymbol{\beta}}) = \boldsymbol{Y}_i'[I - \boldsymbol{X}_i(\boldsymbol{X}_i'\boldsymbol{X}_i)^{-1}\boldsymbol{X}_i']\boldsymbol{Y}_i$, the best convenient form of $\boldsymbol{R}$ can be accomplished as

$$\boldsymbol{R} = \boldsymbol{S}_{Y_i} + [\bar{\boldsymbol{Y}}_i' - \bar{\boldsymbol{X}}_i\hat{\boldsymbol{\beta}}]'\boldsymbol{H}[\bar{\boldsymbol{Y}}_i' - \bar{\boldsymbol{X}}_i\hat{\boldsymbol{\beta}}]. \tag{24.20}$$

Applying this form of $\boldsymbol{R}$ to (24.13) and using $m_i = N_i - n_i$, the ultimate prediction density of $\bar{\boldsymbol{Y}}_i$ conditional on $\boldsymbol{Y}_i$ for the ith area is obtained as

$$f(\bar{\boldsymbol{Y}}_i|\boldsymbol{Y}_i) = C(\boldsymbol{Y}_i, \boldsymbol{H}) \left[\boldsymbol{S}_{Y_i} + (\bar{\boldsymbol{Y}}_i - \bar{\boldsymbol{X}}_i\hat{\boldsymbol{\beta}})'\boldsymbol{H}(\bar{\boldsymbol{Y}}_i - \bar{\boldsymbol{X}}_i\hat{\boldsymbol{\beta}}) \right]^{-\frac{N_i-k}{2}}, \tag{24.21}$$

where the normalizing constant is $C(\boldsymbol{Y}_i, \boldsymbol{H}) = \dfrac{(\pi)^{-\frac{(N_i-n_i)p}{2}} \Gamma_p\left(\frac{N_i-k}{2}\right)|\boldsymbol{H}|^{-\frac{p}{2}}}{\Gamma_p\left(\frac{N_i-k}{2}\right)|\boldsymbol{S}_{Y_i}|^{\frac{n_i-k}{2}}}.$

This is a matrix-T density with location $\bar{\boldsymbol{X}}_i\hat{\boldsymbol{\beta}}$, scale factors $\boldsymbol{S}_{Y_i}$, $\boldsymbol{H}$ and shape parameter $n_i - p - k + 1$. Hence, the unobserved units matrix $\bar{\boldsymbol{Y}}_i$ has a $(N_i - n_i)p$ dimensional matrix-T distribution. The location and co-variance matrices of the predictive distribution are $\bar{\boldsymbol{X}}_i(\boldsymbol{X}_i'\boldsymbol{X}_i)^{-1}\boldsymbol{X}_i'\boldsymbol{Y}_i$ and $\frac{(n_i-p-k+1)}{(n_i-p-k-1)}[\boldsymbol{S}_{Y_i} \otimes \boldsymbol{H}]$.

24.4.3　Joint posterior based reweighting

With the result of predictive distribution in (24.21), the joint posterior density of parameters for units $\boldsymbol{Y}_i$ and $\bar{\boldsymbol{Y}}_i$ can be determined as

$$f(\boldsymbol{\beta}, \boldsymbol{\Sigma}|\boldsymbol{Y}_i\bar{\boldsymbol{Y}}_i) \cong |\boldsymbol{\Sigma}|^{-\frac{N_i+p+1}{2}} \left| \boldsymbol{I}_p + \boldsymbol{\Sigma}^{-1}\boldsymbol{Q} \right|^{-\frac{\nu+p+N_i-1}{2}} f(\bar{\boldsymbol{Y}}_i|\boldsymbol{Y}_i) \tag{24.22}$$

where $\boldsymbol{Q} = (\boldsymbol{Y}_i - \boldsymbol{X}_i\boldsymbol{\beta})'(\boldsymbol{Y}_i - \boldsymbol{X}_i\boldsymbol{\beta}) + (\bar{\boldsymbol{Y}}_i - \bar{\boldsymbol{X}}_i\boldsymbol{\beta})'(\bar{\boldsymbol{Y}}_i - \bar{\boldsymbol{X}}_i\boldsymbol{\beta})$.

Now by using the MCMC simulation to (24.22), we can obtain simulated copies of microdata for the ith small area. The key feature of this new method is that it simulates a complete scenario of the micropopulation in a small area, which means it can produce more reliable SAEs and their variance estimation. It also enables creation of the statistical reliability measures (the Bayes credible interval (BCI)) of the SAEs from MMT based models. This probabilistic approach is quite different from the deterministic approach used in GREGWT [8, 26] and the intelligent searching tool *simulated annealing* used in CO [35]. It can adopt the generalised model operated in the GREGWT to link the sample and unobserved units in the population. In contrast, from the viewpoint of CO, this method uses the MCMC simulation with a posterior density based iterative algorithm. As the joint posterior probabilities of parameters for the sample units and unobserved population units are estimated through MCMC, this microdata simulation methodology is somewhat linked with a chain Monte Carlo sampling. However, it is rather different from the multiple imputation technique advanced by Rubin [30] and others. The basic computation process here is predominantly linked with a prediction density of unobserved population units given the sample units.

One of the issues with this reweighting approach is to identify a suitable prior, and an appropriate model for linking sample data and unobserved units for each small area. These can be difficult in practice as they may vary with the real world problems. There are a range of priors including reference priors, vague priors, informative priors, non-informative priors, conjugate priors, and uniform priors, etc. to choose to fit the model. Every modelling approach would have to deal with at least few complex tasks, and perhaps they are related with suitable model selection and/or computations. The Jeffreys invariance theory [13] based prior and linkage model chosen in this methodology have worked decently by producing appropriate results.

24.5 Empirical Results

This section offers results from a MMT based model for small area estimation.

24.5.1 Data source and computational process

Data are important parts of building a MMT based spatial model for geographic approach of small area estimation. Reliable and more appropriate initially used data can generate much accurate small area estimates. A range of datasets used in MMT comes from distinctive sources and with very special formats. For example, the benchmark data used in reweighting are from the recent census of Australia. Although the census data are available in the convenient cross-tabulated format for a variety of geographic scales, they are in modified form to retain the privacy of respondents. Also, the individuals and/or households level raw data are not available at small geographic scales, due to the confidentiality agreement with respondents. As a result, some of the data files such as various Survey of Income and Housing Confidentialised Unit Record Files (SIH-CURFs) used in this study are provided by the Australian Bureau of Statistics (ABS) under special request and specific conditions. ABS is a recognised institution in Australia for reliable Census and survey datasets.

MMT based model for small area estimation has two computational stages in which the first stage simulates statistical local area (SLA) level households' synthetic weights, and the second stage simulates small area synthetic microdata and produces the ultimate small area estimates. At the first stage, the MMT uses five groups of files such as the *model file, unit records data files, auxiliary data files, benchmarks files*, and the *reweighting program*. This stage generates small area level synthetic weights for households, having a set of characteristics based on the categories of different variables selected in the model. Moreover, the *CURF files, synthetic weights* and the *CPI file* are necessary with the *model file* for further computation. The second stage computation also considers the following definition of housing stress and then determines SLA level estimates for selected region. The SAS software has been used to build the model.

24.5.2   Definition of housing stress

Typically *housing stress* describes a financial situation of households where the cost of housing – either as rental, or as a mortgage repayment – is considered to be significantly high relative to household income. Housing stress can be measured by combining two basic quantities – the income and the expenditure of the household. A household can be considered under housing stress when it is spending more than an affordable expected per cent of its household income on housing. The affordable expected cut-off point of housing expenditure can vary by the circumstance of households as well as location of dwelling. This study utilizes a common definition of housing stress known as the *30/40 rule* [24]: a household is considered to be in housing stress if it spends more than 30 per cent of its gross income on housing and the household also belongs to the bottom 40 per cent of the equivalised income distribution.

24.5.3   Estimates for private renters at small areas in Perth

Statistical local area level housing stress estimates for private renter households are presented in Table 24.1. Results reveal that the rates of housing stress for SLAs range from 11.82 to 31.76 per cent. More than eighty per cent of SLAs in Perth have more than twenty per cent of private renter households experiencing housing stress and essentially these rates are much higher. Almost all the parts of central metropolitan to north Joondalup consist of SLAs with relatively low estimates. On the other side, most of the SLAs in south west to east region, far-east region, and a few SLAs in north region are with significantly high rates. In particular, some hotspots located in these regions are Cockburn, Rockingham, Swan, Bassendean, central Stirling, Mundaring, Kwinana, Armadale and Wanneroo. It indicates a greater representation of low income households living in these areas where the house rents are also relatively high. Moreover, the statistical reliability measures confirm that most

TABLE 24.1

Proportion of private renter households in housing stress at statistical local areas in Perth, Australia.

SLA Name	Total HH	Private Renter	PE	SE	*BCI Estimate
Cambridge(T)	8556	1677	0.187	0.030	(0.129,0.245)
Claremont(T)	3482	826	0.240	0.046	(0.151,0.329)
Cottesloe(T)	2591	677	0.118	0.039	(0.041,0.195)
Mosman Park(T)	3195	948	0.279	0.045	(0.190,0.367)
Nedlands(C)	6847	1052	0.177	0.034	(0.111,0.243)
Peppermint Grove(S)	480	92	0.130	0.107	(0.079,0.340)
Perth(C)-Remainder	4169	1836	0.215	0.030	(0.157,0.273)
Subiaco(C)	7007	2642	0.248	0.026	(0.197,0.299)
Vincent(T)	11159	3618	0.190	0.020	(0.150,0.229)
Bassendean(T)	5458	1012	0.262	0.043	(0.178,0.346)
Bayswater(C)	22660	5556	0.250	0.018	(0.215,0.285)
Kalamunda(S)	17412	2127	0.246	0.029	(0.190,0.302)
Mundaring(S)	11834	1176	0.311	0.041	(0.230,0.392)
Swan(C)	31101	4670	0.254	0.020	(0.216,0.293)
Joondalup(C)-North	16187	2982	0.223	0.024	(0.176,0.269)
Joondalup(C)-South	35029	4715	0.217	0.019	(0.181,0.253)
Stirling(C)-Central	40937	10503	0.261	0.013	(0.235,0.287)
Stirling(C)-Coastal	24773	5686	0.172	0.015	(0.142,0.202)
Stirling(C)-South-Eastern	6244	1508	0.220	0.032	(0.158,0.282)
Wanneroo(C)-North-East	10735	1273	0.246	0.036	(0.175,0.317)
Wanneroo(C)-North-West	12328	2609	0.283	0.027	(0.229,0.336)
Wanneroo(C)-South	13524	1620	0.290	0.034	(0.222,0.357)
Cockburn(C)	26264	4085	0.252	0.021	(0.211,0.293)
East Fremantle(T)	2510	498	0.195	0.054	(0.089,0.301)
Fremantle(C)-Inner	316	110	0.227	0.128	(0.024,0.478)
Fremantle(C)-Remainder	10103	2246	0.232	0.028	(0.178,0.287)
Kwinana(T)	7874	1360	0.318	0.039	(0.242,0.394)
Melville(C)	34482	5775	0.212	0.017	(0.179,0.245)
Rockingham(C)	29250	6281	0.285	0.017	(0.251,0.319)
Armadale(C)	17901	2748	0.317	0.027	(0.264,0.369)
Belmont(C)	12446	3275	0.214	0.022	(0.170,0.258)
Canning(C)	27998	5929	0.253	0.018	(0.218,0.287)
Gosnells(C)	31776	5007	0.262	0.019	(0.225,0.299)
Serpentine-Jarrahdale(S)	4129	376	0.258	0.066	(0.128,0.388)
South Perth(C)	15456	4644	0.205	0.018	(0.169,0.242)
Victoria Park(T)	11874	4010	0.232	0.020	(0.193,0.272)

*Bayes credible interval (BCI) at the probability level of 0.95.

of the small area housing stress estimates are statistically accurate with fairly narrow BCIs at the desired probability level of 0.95.

24.6   Conclusion

This chapter has demonstrated a new technique to the geographic approach of small area estimation for generating spatial microdata at small area level. The technique is based on the Bayesian prediction theory and simulates complete scenario of the whole population in each small area. This approach can yield more accurate and statistically reliable small area estimates compared to the estimates from other reweighting techniques as it is evident from the values of estimated variances and the reported BCIs. Since the method depends on the predictive distribution of unobserved population units given the observed sample units; it is quite different from other reweighting approaches such as the GREGWT and the combinatorial optimisation. It is also different from the multiple imputation technique but to some extent linked with the chain Monte Carlo sampling method.

Moreover, the empirical findings have revealed that more than eighty per cent of SLAs in the Perth city have a high rate (more than 20 per cent) of private renter households in housing stress. Most of these SLAs are located in the south west, far-east, and northern parts of the city where the demand of housing is very high and the rent is also showing an increasing trend. Results have also confirmed the statistical reliability measures of small area housing stress estimates. Further research should explore applications of this approach to other social issues such as small area poverty estimates as well as other geographic regions in Australia and overseas.

Bibliography

[1] M. Aitkin. Applications of the Bayesian bootstrap in finite population inference. *Journal of Official Statistics*, 24(1):21–51, 2008.

[2] D. Ballas, G. Clarke, and J. Dewhurst. Modelling the socio-economic impacts of major job loss or gain at the local level: A spatial microsimulation framework. *Spatial Economic Analysis*, 1(1):127–146, 2006.

[3] D. Ballas, G. P. Clarke, and I. Turton. A spatial microsimulation model for social policy evaluation. In B. Boots, A. Okabe, and R. Thomas, editors, *Modelling Geographical Systems*, volume 70, pages 143–168. Springer Verlag, Netherlands, 2003.

[4] T. Bayes. An essay towards solving a problem in the doctrine of chances. *Philosophical Transactions of the Royal Society of London*, 53:370–418, 1763.

[5] J. M. Bernardo and R. Rueda. Bayesian hypothesis testing: A reference approach. *International Statistical Review*, 70(3):351–372, 2002.

[6] M. Birkin and M. Clarke. SYNTHESIS - a synthetic spatial information system for urban and regional analysis: Methods and examples. *Environment and Planning Analysis*, 20(12):1645–1671, 1988.

[7] L. Brown and A. Harding. Social modelling and public policy: Application of microsimulation modelling in Australia. *Journal of Artificial Societies and Social Simulation*, 5(1):1–14, 2002.

[8] S. F. Chin and A. Harding. Regional dimensions: Creating synthetic small-area microdata and spatial microsimulation models. Online Technical Paper–TP33, NATSEM, University of Canberra, Australia, 2006.

[9] S. F. Chin and A. Harding. Spatial MSM–NATSEM's small area household model for Australia. In A. Gupta and A. Harding, editors, *Modelling Our Future: Population Ageing Health and Aged Care*, pages 563–566. Elsevier, Oxford, 2007.

[10] S. F. Chin, A. Harding, R. Lloyd, J. McNamara, B. Phillips, and Q. N. Vu. Spatial microsimulation using synthetic small area estimates of income, tax and social security benefits. *Australian Journal of Regional Studies*, 11(3):303–335, 2005.

[11] W. A. Ericson. Subjective Bayesian models in sampling finite populations. *Journal of the Royal Statistical Society: Series B (Statistical Methodology)*, 31(2):195–233, 1969.

[12] P. Heady, P. Clarke, G. Brown, K. Ellis, D. Heasman, S. Hennell, J. Longhurst, and B. Mitchell. Model based small area estimation. Project Report, Office for National Statistics, U.K., 2003.

[13] H. Jeffreys. *The Theory of Probability*. Oxford University Press, New York, 3rd edition, 1998.

[14] A. Y. Lo. Bayesian statistical inference for sampling a finite population. *The Annals of Statistics*, 14(3):1226–1233, 1986.

[15] C. Moriarity and F. Scheuren. A note on Rubin's statistical matching using file concatenation with adjusted weights and multiple imputations. *Journal of Business and Educational Studies*, 21(1):65–73, 2003.

[16] P. Norman. Putting iterative proportional fitting on the researcher's desk. Working Paper 99/03, School of Geography, University of Leeds, U.K., 1999.

[17] G. H. Orcutt. A new type of socio-economic system. *Review of Economics and Statistics*, 39(2):116–123, 1957.

[18] D. Pfeffermann. Small area estimation - new developments and directions. *International Statistical Review*, 70(1):125–143, 2002.

[19] D. Pfeffermann. New important developments in small area estimation. *Statistical Science*, 28(1):40–68, 2013.

[20] A. Rahman. *Bayesian Predictive Inference for Some Linear Models under Student-T Errors*. VDM Verlag Press, Saarbrücken, 2008.

[21] A. Rahman. A review of small area estimation problems and methodological developments. Discussion Paper, 2008.

[22] A. Rahman. Small area estimation through spatial microsimulation models: Some methodological issues. In *2nd General Conference of the IMA*, Ottawa, June 8-10, 2009.

[23] A. Rahman. *Small Area Housing Stress Estimation in Australia: Microsimulation Modelling and Statistical Reliability*. PhD thesis, University of Canberra, Australia, 2011.

[24] A. Rahman and A. Harding. A new analysis of the characteristics of households in housing stress: Results and tools for validation. In *6th Australasian Housing Researchers' Conference (AHRC12)*, pages 1–23, South Australia, 2012. The University of Adelaide.

[25] A. Rahman and A. Harding. *Small Area Estimation and Microsimulation Modelling*. Chapman and Hall/CRC, London, 2014.

[26] A. Rahman, A. Harding, R.Tanton, and S. Liu. Methodological issues in spatial microsimulation modelling for small area estimation. *The International Journal of Microsimulation*, 3(2):3–22, 2010.

[27] A. Rahman, A. Harding, R. Tanton, and S. Liu. Simulating the characteristics of populations at the small area level: New validation techniques for a spatial microsimulation model in Australia. *Computational Statistics & Data Analysis*, 57(1):149–165, 2013.

[28] J. N. K. Rao. Some current trends in sample survey theory and methods (with discussion). *Sankhyā, The Indian Journal of Statistics: Series B*, 61(1):1–57, 1999.

[29] J. N. K. Rao. *Small Area Estimation*. John Wiley & Sons, New Jersey, 2003.

[30] D. B. Rubin. *Multiple Imputation for Nonresponse in Surveys*. John Wiley & Sons, New York, 1987.

[31] A. C. Singh and C. A. Mohl. Understanding calibration estimators in survey sampling. *Survey Methodology*, 22(2):107–115, 1996.

[32] D. C. Tiao and A. Zellner. On the Bayesian estimation of multivariate regression. *Journal of the Royal Statistical Society: Series B (Statistical Methodology)*, 26(2):277–285, 1964.

[33] M. Tranmer, A. Pickles, E. Fieldhouse, M. Elliot, A. Dale, M. Brown, D. Martin, D. Steel, and C. Gardiner. The case for small area microdata. *Journal of the Royal Statistical Society: Series A (Statistics in Society)*, 168(1):29–49, 2005.

[34] D. Voas and P. Williamson. An evaluation of the combinatorial optimisation approach to the creation of synthetic microdata. *International Journal of Population Geography*, 6(3):349–366, 2000.

[35] P. Williamson, M. Birkin, and P. Rees. The estimation of population microdata by using data from small area statistics and sample of anonymised records. *Environment and Planning Analysis*, 30(6):785–816, 1998.

[36] A. Zaidi, A. Harding, and P. Williamson. *New Frontiers in Microsimulation Modelling*. Ashgate, London, 2009.

[37] A. Zellner. *An Introduction to Bayesian Inference in Econometrics*. John Wiley & Sons, New York, 1971.

25

Empirical Bayes Methods for the Transformed Gaussian Random Field Model with Additive Measurement Errors

Vivekananda Roy

Iowa State University

Evangelos Evangelou

University of Bath

Zhengyuan Zhu

Iowa State University

CONTENTS

25.1   Introduction

If geostatistical observations are continuous but can not be modeled by the Gaussian distribution, a more appropriate model for these data may be the transformed Gaussian model. In transformed Gaussian models it is assumed that the random field of interest is a nonlinear transformation of a Gaussian random field (GRF). For example, [9] propose the Bayesian transformed Gaussian model where they use the Box-Cox family of power transformation [3] on the observations and show that prediction for unobserved random fields

can be done through posterior predictive distribution where uncertainty about
the transformation parameter is taken into account. More recently, [5] consider
maximum likelihood estimation of the parameters and a "plug-in" method of
prediction for transformed Gaussian model with Box-Cox family of transformations. Both [9] and [5] consider spatial prediction of rainfall to illustrate
their model and method of analysis. A review of the Bayesian transformed
Gaussian random fields model is given in [8]. See also [6] who discusses several
issues regarding the formulation and interpretation of transformed Gaussian
random field models, including the approximate nature of the model for positive data based on Box-Cox family of transformations, and the interpretation
of the model parameters.

In their discussion, [5] mention that for analyzing rainfall data there "must
be at least an additive measurement error" in the model, while [7] consider a
measurement error term for transformed data in their transformed Gaussian
model. An alternative approach, as suggested by [5], is to assume measurement error in the original scale of the data, *not* in the transformed scale as
done by [7]. We propose a transformed Gaussian model where an additive
measurement error term is used for the observed data, and the random fields,
after a suitable transformation, is assumed to follow Gaussian distribution. In
many practical situations this may be the more natural assumption. Unfortunately, the likelihood function is not available in closed form in this case and
as mentioned by [5], this alternative model, although "is attractive", it "raises
technical complications". In spite of the fact that the likelihood function is not
available in closed form, we show that *data augmentation* techniques can be
used for Markov chain Monte Carlo (MCMC) sampling from the target posterior density. These MCMC samples are then used for estimation of parameters
of our proposed model as well as prediction of rainfall at new locations.

The Box-Cox family of transformations is defined for *strictly positive* observations. This implies that the transformed Gaussian random variables have
restricted support and the corresponding likelihood function is not in closed
form. [5] change the observed zeros in the data to a small positive number in
order to have a closed form likelihood function. On the other hand, Stein [19]
considers Monte Carlo methods for prediction and inference in a model where
transformed observations are assumed to be a truncated Gaussian random
field. In our proposed model we consider the transformation on random effects *instead of* observed data and consider a natural extension of the Box-Cox
family of transformations for negative values of the random effect.

The so-called *full Bayesian analysis* of transformed Gaussian data requires
specification of a joint prior distribution on the Gaussian random fields parameters as well as transformation parameters, see e.g. [9]. Since a change
in the transformation parameter value results in change of location and scale
of transformed data, assuming GRF model parameters to be independent a
priori of the transformation parameter would give nonsensical results [3]. As-
signing an appropriate prior on covariance (of the transformed random field)
parameters, like the range parameter, is also not easy as the choice of prior

may influence the inference, see e.g. [4, p. 716]. Use of improper prior on correlation parameters typically results in improper posterior distribution [20, p. 224]. Thus it is difficult to specify a joint prior on all the model parameters of transformed GRF models. Here we consider an empirical Bayes (EB) approach for estimating the transformation parameter as well as the range parameter of our transformed Gaussian random field model. Our EB method avoids the difficulty of specifying a prior on the transformation parameter as well as the range parameter, which, as mentioned above, is problematic. In our EB method of analysis, we do not need to sample from the complicated nonstandard conditional distributions of these (transformation and range) parameters, which is required in the full Bayesian analysis. Further, an MCMC algorithm with updates on such parameters may not perform well in terms of mixing and convergence, see e.g. [4, p. 716]. Recently, in some simulation studies in the context of spatial generalized linear mixed models for binomial data, [18] observe that EB analysis results in estimates with less bias and variance than full Bayesian analysis. [17] uses an efficient importance sampling method based on MCMC sampling for estimating the link function parameter of a robust binary regression model, see also [11]. We use the method of [17] for estimating the transformation and range parameters of our model.

The rest of the chapter is organized as follows. Section 25.2 introduces our transformed Gaussian model with measurement error. In Section 25.3 a method based on importance sampling is described for effectively selecting the transformation parameter as well as the range parameter. Section 25.4 discusses the computation of the Bayesian predictive density function. In Section 25.5 we analyze a dataset using the proposed model and estimation procedure for constructing a continuous spatial map of rainfall amounts.

25.2   The Transformed Gaussian Model with Measurement Error

25.2.1   Model description

Let $\{Z(s), s \in \mathcal{D}\}, \mathcal{D} \in \mathbb{R}^l$ be the random field of interest. We observe a single realization from the random field with measurement errors at finite sampling locations $s_1, \ldots, s_n \in \mathcal{D}$. Let $\mathbf{y} = (y(s_1), \ldots, y(s_n))$ be the observations. We assume that the observations are sampled according to the following model,

$$Y(s) = Z(s) + \epsilon(s),$$

where we assume that $\{\epsilon(s), s \in \mathcal{D}\}$ is a process of mutually independent $N(0, \tau^2)$ random variables, which is independent of $Z(s)$. The term $\epsilon(s)$ can be interpreted as micro-scale variation, measurement error, or a combination of both. We assume that $Z(s)$, after a suitable transformation, follow a normal distribution. That is, for some family of transformations $g_\lambda(\cdot)$, $\{W(s) \equiv g_\lambda(Z(s)), s \in \mathcal{D}\}$ is assumed to be a Gaussian stochastic process with the mean function $E(W(s)) = \sum_{j=1}^{p} f_j(s)\beta_j$; $\beta = (\beta_1, \ldots, \beta_p)' \in \mathbb{R}^p$ are the unknown regression parameters, $f(s) = (f_1(s), \ldots, f_p(s))$ are known location dependent covariates, and $\text{cov}(W(s), W(u)) = \sigma^2 \rho_\theta(\|s - u\|)$, where $\|s - u\|$ denotes the Euclidean distance between s and u. Here, θ is a vector of parameters which controls the range of correlation and the smoothness/roughness of the random field.

We consider $\rho_\theta(\cdot)$ as a member of the Matérn family [15]

$$\rho(u; \phi, \kappa) = \{2^{\kappa-1}\Gamma(\kappa)\}^{-1}(u/\phi)^\kappa K_\kappa(u/\phi),$$

where $K_\kappa(\cdot)$ denotes the modified Bessel function of order κ. In this case $\theta \equiv (\phi, \kappa)$. This two-parameter family is very flexible in that the integer part of the parameter κ determines the number of times the process $W(s)$ is mean square differentiable, that is, κ controls the smoothness of the underlying process, while the parameter ϕ measures the scale (in units of distance) on which the correlation decays. We assume that κ is known and fixed and estimate ϕ using our empirical Bayes approach.

A popular choice for $g_\lambda(\cdot)$ is the Box-Cox family of power transformations [3] indexed by λ, that is,

$$g_\lambda^0(z) = \begin{cases} \frac{z^\lambda - 1}{\lambda} & \text{if } \lambda > 0 \\ \log(z) & \text{if } \lambda = 0 \end{cases}. \tag{25.1}$$

Although the above transformation holds for $\lambda < 0$, in this chapter, we assume that $\lambda \geq 0$. As mentioned in Section 25.1, the Box-Cox transformation (25.1) holds for $z > 0$. This implies that the image of the transformation is $(-1/\lambda, \infty)$, which contradicts the Gaussian assumption. Also the normalizing constant for the pdf of the transformed variable is not available in closed form. Note that the inverse transformation of (25.1) is given by

$$h_\lambda^0(w) = \begin{cases} (1 + \lambda w)^{\frac{1}{\lambda}} & \text{if } \lambda > 0 \\ \exp(w) & \text{if } \lambda = 0 \end{cases}. \tag{25.2}$$

The transformation (25.2) can be extended to the whole real line using

$$h_\lambda(w) = \begin{cases} \text{sgn}(1 + \lambda w)|1 + \lambda w|^{\frac{1}{\lambda}} & \text{if } \lambda > 0 \\ \exp(w) & \text{if } \lambda = 0, \end{cases} \tag{25.3}$$

where $\text{sgn}(x)$ denotes the sign of x, taking values -1, 0, or 1 depending on whether x is negative, zero, or positive respectively. The proposed transformation (25.3) is monotone, invertible and is continuous at $\lambda = 0$ with the inverse, the *extended Box-Cox transformation* given in [2]

$$g_\lambda(z) = \begin{cases} \frac{(\text{sgn}(z)|z|^\lambda - 1)}{\lambda} & \text{if } \lambda > 0 \\ \log(z) & \text{if } \lambda = 0 \end{cases}.$$

Let $\mathbf{w} = (w(s_1), \ldots, w(s_n))^T$, $\mathbf{z} = (z(s_1), \ldots, z(s_n))^T$, $\gamma \equiv (\lambda, \phi)$, and $\psi \equiv (\beta, \sigma^2, \tau^2)$. The reason for using different notation for (λ, ϕ) than other parameters will be clear later. Since $\{W(s) \equiv g_\lambda(Z(s)), s \in \mathcal{D}\}$ is assumed to be a Gaussian process, the joint posterior density of $(\mathbf{y}, \mathbf{w})$ is given by

$$f(\mathbf{y}, \mathbf{w}|\psi, \gamma) = (2\pi)^{-n}(\sigma\tau)^{-n} \exp\{-\frac{1}{2\tau^2} \sum_{i=1}^{n} (y_i - h_\lambda(w_i))^2\}$$

$$|R_\theta|^{-1/2} \exp\{-\frac{1}{2\sigma^2}(\mathbf{w} - F\beta)^T R_\theta^{-1}(\mathbf{w} - F\beta)\}, \tag{25.4}$$

where $\mathbf{y}, \mathbf{w} \in \mathbb{R}^n$, F is the known $n \times p$ matrix defined by $F_{ij} = f_j(s_i)$, $R_\theta \equiv H_\theta(s, s)$ is the correlation matrix with $R_{\theta,ij} = \rho_\theta(\|s_i - s_j\|)$, and $h_\lambda(\cdot)$ is defined in (25.3). The likelihood function for (ψ, γ) based on the observed data $\mathbf{y}$ is given by

$$L(\psi, \gamma|\mathbf{y}) = \int_{\mathbb{R}^n} f(\mathbf{y}, \mathbf{w}|\psi, \gamma)d\mathbf{w}. \tag{25.5}$$

25.2.2 Posterior density and MCMC

The likelihood function $L(\psi, \gamma|\mathbf{y})$ defined in (25.5) is not available in closed form. A full Bayesian analysis requires specification of a joint prior distribution on the model parameters (ψ, γ). As mentioned before, assigning a joint prior distribution is difficult in this problem. Here, we estimate γ using the method described in Section 25.3. For other model parameters ψ, we assume the following conjugate priors,

$$\beta|\sigma^2 \sim N_p(m_b, \sigma^2 V_b), \quad \sigma^2 \sim \chi^2_{ScI}(n_\sigma, a_\sigma), \text{ and } \tau^2 \sim \chi^2_{ScI}(n_\tau, a_\tau), \tag{25.6}$$

where $m_b, V_b, n_\sigma, a_\sigma, n_\tau, a_\tau$ are assumed to be known hyperparameters. (We say $W \sim \chi^2_{ScI}(n_\sigma, a_\sigma)$ if the pdf of W is $f(w) \propto w^{-(n_\sigma/2+1)} \exp(-n_\sigma a_\sigma/(2w))$.) The posterior density of ψ is given by

$$\pi_\gamma(\psi|\mathbf{y}) = \frac{L_\gamma(\psi|\mathbf{y})\pi(\psi)}{m_\gamma(\mathbf{y})}, \tag{25.7}$$

where $L_\gamma(\psi|\mathbf{y}) \equiv L(\psi, \gamma|\mathbf{y})$ is the likelihood function, and $m_\gamma(\mathbf{y}) = \int_\Omega L_\gamma(\psi|\mathbf{y})\pi(\psi)d\psi$ is the normalizing constant, with $\pi(\psi)$ being the prior on ψ and the support $\Omega = \mathbb{R}^p \times \mathbb{R}_+ \times \mathbb{R}_+$. Since the likelihood function (25.5) is not available in closed form, it is difficult to obtain MCMC sample from the posterior distribution $\pi_\gamma(\psi|\mathbf{y})$ directly using the expression in (25.7).

Here we consider the following so-called *complete* posterior density

$$\pi_\gamma(\psi, \mathbf{w}|\mathbf{y}) = \frac{f(\mathbf{y}, \mathbf{w}|\psi, \gamma)\pi(\psi)}{m_\gamma(\mathbf{y})},$$

based on the joint density $f(\mathbf{y}, \mathbf{w}|\psi, \gamma)$ defined in (25.4). Note that, integrating the complete posterior density $\pi_\gamma(\psi, \mathbf{w}|\mathbf{y})$ we get the target posterior density $\pi_\gamma(\psi|\mathbf{y})$, that is,

$$\int_{\mathbb{R}^n} \pi_\gamma(\psi, \mathbf{w}|\mathbf{y})d\mathbf{w} = \pi_\gamma(\psi|\mathbf{y}).$$

So if we can generate a Markov chain $\{\psi^{(i)}, \mathbf{w}^{(i)}\}_{i=1}^N$ with stationary density $\pi_\gamma(\psi, \mathbf{w}|\mathbf{y})$, then the marginal chain $\{\psi^{(i)}\}_{i=1}^N$ has the stationary density $\pi_\gamma(\psi|\mathbf{y})$ defined in (25.7). This is the standard technique of data augmentation and here $\mathbf{w}$ is playing the role of "latent" variables (or "missing data") [21].

Since we are using conjugate priors for (β, σ^2) in (25.6), integrating $\pi_\gamma(\psi, \mathbf{w}|\mathbf{y})$ with respect β we have

$$\pi_\gamma(\sigma^2, \tau^2, \mathbf{w}|\mathbf{y}) \propto (\sigma\tau)^{-n} \exp\{-\frac{1}{2\tau^2} \sum_{i=1}^n (y_i - h_\lambda(w_i))^2\}$$

$$\tau^{-(n_\tau+2)} \exp(-n_\tau a_\tau/2\tau^2)|\Lambda_\theta|^{-1/2} \qquad (25.8)$$

$$\exp\{-\frac{1}{2\sigma^2}(\mathbf{w} - Fm_b)^T \Lambda_\theta^{-1}(\mathbf{w} - Fm_b)\}$$

$$\sigma^{-(n_\sigma+2)} \exp(-n_\sigma a_\sigma/2\sigma^2),$$

where $\Lambda_\theta = FV_bF^T + R_\theta$.

We use a Metropolis within Gibbs algorithm for sampling from the posterior density $\pi_\gamma(\sigma^2, \tau^2, \mathbf{w}|\mathbf{y})$ given in (25.8). Note that

$$\sigma^2|\tau^2, \mathbf{w}, \mathbf{y} \sim \chi_{ScI}^2(n'_\sigma, a'_\sigma), \text{ and } \tau^2|\sigma^2, \mathbf{w}, \mathbf{y} \sim \chi_{ScI}^2(n'_\tau, a'_\tau),$$

where $n'_\sigma = n + n_\sigma$, $a'_\sigma = (n_\sigma a_\sigma + (\mathbf{w} - Fm_b)^T \Lambda_\theta^{-1}(\mathbf{w} - Fm_b))/(n + n_\sigma)$, $n'_\tau = n + n_\tau$, and $a'_\tau = (n_\tau a_\tau + \sum_{i=1}^n (y_i - h_\lambda(w_i))^2)/(n + n_\tau)$. On the other hand, the conditional distribution of $\mathbf{w}$ given $\sigma^2, \tau^2, \mathbf{y}$ is not a standard distribution. We use a Metropolis-Hastings algorithm given in [22] for sampling from this conditional distribution.

25.3 Estimation of Transformation and Correlation Parameters

Here we consider an empirical Bayes approach for estimating the transformation parameter λ and the range parameter ϕ. That is, we select that value of $\gamma \equiv (\lambda, \phi)$ which maximizes the marginal likelihood of the data $m_\gamma(\mathbf{y})$. For selecting models that are better than other models when γ varies across some set Γ, we calculate and subsequently compare the values of $B_{\gamma, \gamma_1} := m_\gamma(\mathbf{y})/m_{\gamma_1}(\mathbf{y})$, where γ_1 is a suitably chosen fixed value of (λ, ϕ). Ideally, we would like to calculate and compare B_{γ, γ_1} for a large number of values of γ. [17] used a method based on importance sampling for selecting link function parameter in a robust regression model for binary data by estimating a large family of Bayes factors. Here we apply the method of [17] to efficiently estimate B_{γ,γ_1} for a large set of possible values of γ. Recently [18] successfully used this method for estimating parameters in spatial generalized linear mixed models.

Let $f(\mathbf{y}, \mathbf{w}|\gamma) \equiv \int_{\Omega} f(\mathbf{y}, \mathbf{w}|\psi, \gamma)\pi(\psi)d\psi$. Since we are using conjugate priors for (β, σ^2) and τ^2 in (25.6), the marginal density $f(\mathbf{y}, \mathbf{w}|\gamma)$ is available in closed form. In fact, from standard Bayesian analysis of normal linear model we have

$$f(\mathbf{y}, \mathbf{w}|\gamma) \propto \{a_\tau n_\tau + (\mathbf{y} - h_\lambda(\mathbf{w}))^T(\mathbf{y} - h_\lambda(\mathbf{w}))\}^{-\frac{n_\tau + n}{2}} |\Lambda_\theta|^{-1/2}$$
$$\{a_\sigma n_\sigma + (\mathbf{w} - Fm_b)^T \Lambda_\theta^{-1}(\mathbf{w} - Fm_b)\}^{-\frac{n_\sigma + n}{2}}.$$

Note that

$$m_\gamma(\mathbf{y}) = \int_{\mathbb{R}^n} f(\mathbf{y}, \mathbf{w}|\gamma)d\mathbf{w}.$$

Let $\{(\sigma^2, \tau^2)^{(l)}, \mathbf{w}^{(l)}\}_{l=1}^N$ be the Markov chain (with stationary density $\pi_{\gamma_1}(\sigma^2, \tau^2, \mathbf{w}|\mathbf{y})$) underlying the MCMC algorithm presented in subsection 25.2.2. Then by ergodic theorem we have a simple consistent estimator of B_{γ,γ_1},

$$\frac{1}{N}\sum_{i=1}^N \frac{f(\mathbf{y}, \mathbf{w}^{(i)}|\gamma)}{f(\mathbf{y}, \mathbf{w}^{(i)}|\gamma_1)} \overset{a.s.}{\to} \int_{\mathbb{R}^n} \frac{f(\mathbf{y}, \mathbf{w}|\gamma)}{f(\mathbf{y}, \mathbf{w}|\gamma_1)}\pi_{\gamma_1}(\mathbf{w}|\mathbf{y})d\mathbf{w} = \frac{m_\gamma(\mathbf{y})}{m_{\gamma_1}(\mathbf{y})}, \qquad (25.9)$$

as $N \to \infty$, where $\pi_{\gamma_1}(\mathbf{w}|\mathbf{y}) = \int_\Omega \pi_{\gamma_1}(\psi, \mathbf{w}|\mathbf{y})d\psi$. Note that in (25.9) a single Markov chain $\{\mathbf{w}^{(l)}\}_{l=1}^N$ with stationary density $\pi_{\gamma_1}(\mathbf{w}|\mathbf{y})$ is used to estimate B_{γ,γ_1} for different values of γ. As mentioned in [17], the estimator (25.9) can be unstable and following [17] we consider the following method for estimating B_{γ,γ_1}.

Let $\gamma_1, \gamma_2, \ldots, \gamma_k \in \Gamma$ be k appropriately chosen skeleton points. Let $\{\psi_j^{(l)}, \mathbf{w}^{(j;l)}\}_{l=1}^{N_j}$ be a Markov chain with stationary density $\pi_{\gamma_j}(\psi, \mathbf{w}|\mathbf{y})$ for $j = 1\ldots, k$. Define $r_i = m_{\gamma_i}(\mathbf{y})/m_{\gamma_1}(\mathbf{y})$ for $i = 2, 3, \ldots, k$, with $r_1 = 1$. Then B_{γ,γ_1} is consistently estimated by

$$\hat{B}_{\gamma,\gamma_1} = \sum_{j=1}^k \sum_{l=1}^{N_j} \frac{f(\mathbf{y}, \mathbf{w}^{(j;l)}|\gamma)}{\sum_{i=1}^k N_i f(\mathbf{y}, \mathbf{w}^{(j;l)}|\gamma_i)/\hat{r}_i}, \qquad (25.10)$$

where $\hat{r}_1 = 1$ $\hat{r}_i$, $i = 2, 3, \ldots, k$ are consistent estimators of r_i's obtained by the "reverse logistic regression" method proposed by [12]. (See [17] for details about the above method of estimation and how to choose the skeleton point γ_i's and sample size N_i's.) The estimate of γ is obtained by maximizing (25.10), that is, $\hat{\gamma} = \arg\max_{\gamma \in \Gamma} \hat{B}_{\gamma,\gamma_1}$.

25.4   Spatial Prediction

We now discuss how we make predictions about Z_0, the values of $Z(s)$ at some locations of interest, say $(s_{01}, s_{02}, \ldots, s_{0k})$, typically a fine grid of locations covering the observed region. We use the posterior predictive distribution

$$f(\mathbf{z}_0|\mathbf{y}) = \int_\Omega \int_{\mathbb{R}^n} f_\gamma(\mathbf{z}_0|\mathbf{w}, \psi) \pi_\gamma(\psi, \mathbf{w}|\mathbf{y}) d\mathbf{w} d\psi, \qquad (25.11)$$

where $\mathbf{z}_0 = (z(s_{01}), z(s_{02}), \ldots, z(s_{0k}))$. Let

$$\mathbf{w}_0 = g_\lambda(\mathbf{z}_0) = (g_\lambda(z(s_{01})), g_\lambda(z(s_{02})), \ldots, g_\lambda(z(s_{0k}))).$$

From subsection 25.2.1, it follows that

$$(\mathbf{w}_0, \mathbf{w}|\psi) \sim \mathrm{N}_{k+n}\left(\begin{pmatrix} F_0 \beta \\ F \beta \end{pmatrix}, \sigma^2 \begin{pmatrix} H_\theta(s_0, s_0) & H_\theta(s_0, s) \\ H_\theta^T(s_0, s) & H_\theta(s, s) \end{pmatrix} \right),$$

where F_0 is the $k \times p$ matrix with $F_{0ij} = f_j(s_{0i})$, and $H_\theta(s_0, s)$ is the $k \times n$ matrix with $H_{\theta,ij}(s_0, s) = \rho_\theta(\|s_{0i} - s_j\|)$. So $\mathbf{w}_0|\mathbf{w}, \psi \sim \mathrm{N}_k(c_\gamma(\mathbf{w}, \psi), \sigma^2 D_\gamma(\mathbf{w}))$ where

$$c_\gamma(\mathbf{w}, \psi) = F_0 \beta + H_\theta(s_0, s) H_\theta^{-1}(s, s)(\mathbf{w} - F\beta),$$

and

$$D_\gamma(\psi) = H_\theta(s_0, s_0) - H_\theta(s_0, s) H_\theta^{-1}(s, s) H_\theta^T(s_0, s).$$

Suppose, we want to estimate $E(t(\mathbf{z}_0)|\mathbf{y})$ for some function t. Let $\{\psi^{(i)}, \mathbf{w}^{(i)}\}_{i=1}^N$ be a Markov chain with stationary density $\pi_{\hat\gamma}(\psi, \mathbf{w}|\mathbf{y})$, where $\hat\gamma$ is the estimate of γ obtained using the method described in Section 25.3. We then simulate $\mathbf{w}_0^{(i)}$ from $f(\mathbf{w}_0|\mathbf{w}^{(i)}, \psi^{(i)})$ for $i = 1, \ldots, N$. Finally, we calculate the following approximate minimum mean squared error predictor

$$E(t(\mathbf{z}_0)|\mathbf{y}) \approx \frac{1}{N} \sum_{i=1}^N t(h_{\hat\lambda}(\mathbf{w}_0^{(i)})).$$

We can also estimate the predictive density (25.11) using these samples $\{\mathbf{w}_0^{(i)}\}_{i=1}^N$. In particular, we can estimate the quantiles of the predictive distribution of $t(\mathbf{z}_0)$.

25.5   Example: Swiss Rainfall Data

To illustrate our model and method of analysis we apply it to a well-known example. This dataset consists of the rainfall measurements that occurred on

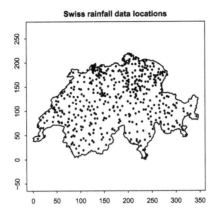

FIGURE 25.1
Sampled locations for the rainfall example.

May 8, 1986 at the 467 locations in Switzerland shown in Figure 25.1. This dataset is available in the geoR package [16] in R and it has been analyzed before using a transformed Gaussian model by [5] and [10]. The scientific objective is to construct a continuous spatial map of the average rainfall using the observed data as well as predict the proportion over the total area that the amount of rainfall exceeds a given level. The original data range from 0.5 to 585 but for our analysis these values are scaled by their geometric mean (139.663). Scaling the data helps avoid numerical overflow when computing the likelihood when the $\mathbf{w}$'s are simulated from different λ. Following [5] we use the Matérn covariance and a constant mean β in our model for analyzing the rainfall data. [5] mention that $\kappa = 1$ "gives a better fit than $\kappa = 0.5$ or $\kappa = 2$," see also [10]. We also use $\kappa = 1$ in our analysis. We estimate (λ, ϕ) using the method proposed in Section 25.3. In particular, we use the estimator (25.10) for estimating B_{γ, γ_1}. For the skeleton points we take all pairs of values of (λ, ϕ), where

$$\lambda \in \{0, 0.10, 0.20, 0.25, 0.33, 0.50, 0.67, 0.75, 1\} \text{ and } \phi \in \{10, 15, 20, 30, 50\}.$$
$$(25.12)$$

The first combination $(0, 10)$ is taken as the baseline point γ_1. MCMC samples of size 3,000 at each of the 45 skeleton points are used to estimate the Bayes factors r_i's at the skeleton points using the reverse logistic regression method. Here we use the Metropolis within Gibbs algorithm mentioned in subsection 25.2.2 for obtaining MCMC samples. These samples are taken after discarding an initial burn-in of 500 samples and keeping every 10th draw of subsequent random samples. Since the whole n-dimensional vectors $\mathbf{w}^{(i)}$'s must be stored, it is advantageous to make thinning of the MCMC sample such that saved values of $\mathbf{w}^{(i)}$ are approximately uncorrelated, see e.g. [4, p. 706]. Next we use *new* MCMC samples of size 500 corresponding to the 45 skeleton points mentioned in (25.12) to compute the Bayes factors B_{γ, γ_1} at

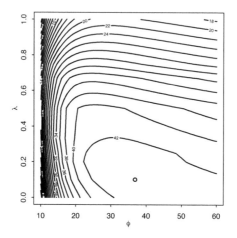

FIGURE 25.2

Contour plot of estimates of B_{γ,γ_1} for the rainfall dataset. The plot suggests $\hat{\lambda} = 0.1$ and $\hat{\phi} = 37$. Here the baseline value corresponds to $\lambda_1 = 0$ and $\phi_1 = 10$.

other points. The estimate $(\hat{\lambda}, \hat{\phi})$ is taken to be the value of (λ, ϕ) where $\hat{B}_{\gamma,\gamma_1}$ attains its maximum. Here again we collect every 10th sample after initial burn-in of 500 samples. For the entire computation it took about 70 minutes on a computer with 2.8 GHz 64-bit Intel Xeon processor and 2 Gb RAM. The computation was done using Fortran 95. Figure 25.2 shows the contour plot of the Bayes factor estimates. From the plot we see that $\hat{B}_{\gamma,\gamma_1}$ attains maximum at $\hat{\gamma} = (\hat{\lambda}, \hat{\phi}) = (0.1, 37)$. The Bayes factors for a selection of fixed λ and ϕ values are also shown in Figure 25.3.

Next, we fix λ and ϕ at their estimates and estimate β, σ^2 and τ^2, as well as the random field $\mathbf{z}$ at the observed and prediction locations. The prediction grid consists of a square grid of length and width equal to 5 kilometers. The prior hyperparameters were as follows: prior mean for β, $m_b = 0$, prior variance for β, $V_b = 100$, degrees of freedom parameter for σ^2, $n_\sigma = 1$, scale parameter for σ^2, $a_\sigma = 1$, degrees of freedom parameter for τ^2, $n_\tau = 1$, and scale parameter for τ^2, $a_\tau = 1$. A MCMC sample of size 3,000 is used for parameter estimation and prediction. Like before we discard initial 500 samples as burn-in and collected every 10th sample. Let $\{\sigma^{2(i)}, \tau^{2(i)}, \mathbf{w}^{(i)}\}_{i=1}^{N}$ be the MCMC samples (with invariant density $\pi_{\hat{\gamma}}(\sigma^2, \tau^2, \mathbf{w}|\mathbf{y})$) obtained using the Metropolis within Gibbs algorithm mentioned in subsection 25.2.2. Then we simulate $\beta^{(i)}$ from its full conditional density $\pi_{\hat{\gamma}}(\beta|\sigma^{2(i)}, \tau^{2(i)}, \mathbf{w}^{(i)}, \mathbf{y})$, which is a normal density, to obtain MCMC samples for β. This part of the algorithm took no more than 2 minutes to run on the same computer. The estimates

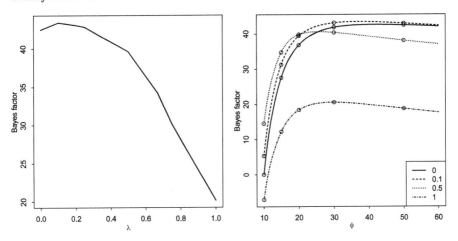

FIGURE 25.3
Estimates of B_{γ,γ_1} against λ for fixed value of $\phi = 37$ (left panel) and against ϕ for fixed values of λ (right panel). The circles show some of the skeleton points.

of posterior means of the parameter are given in Table 25.1. The standard errors of the MCMC estimators are computed using the method of overlapping batch means [13] and also given in Table 25.1. Predictions of $Z(s)$ and the corresponding prediction standard deviations are presented in Figure 25.4. Note that for the prediction, the MCMC sample is scaled back to the original scale of the data.

Using the model discussed in [5] fitted to the scaled data, the maximum likelihood estimates of the parameters for fixed $\kappa = 1$ and $\lambda = 0.5$ are $\hat{\beta} = -0.13$, $\hat{\sigma}^2 = 0.75$, $\hat{\tau}^2 = 0.05$, and $\hat{\phi} = 35.8$. These are not very different from our estimates although we emphasize that the interpretation of τ^2 in our model is different. We use the krige.conv function (used also by [5], personal communication) in the geoR package [16] in R to reproduce the prediction map of [5] and is given in Figure 25.5. From Figure 25.5, we see that the prediction map is similar to the map in Figure 25.4 obtained using our model. Note that, Figure 25.4 is prediction map for $Z(\cdot)$ whereas Figure 25.5 is prediction map for Y as done in [5]. On the other hand, if the parameter nugget τ^2 in the model of [5] is interpreted as measurement error (in the transformed scale),

TABLE 25.1
Posterior estimates of model parameters.

	β	σ^2	τ^2
Estimate	−0.23	0.74	0.05
St Error	0.00483	0.00701	0.00032

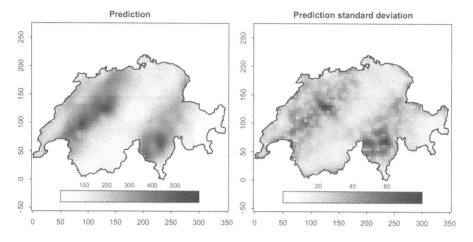

FIGURE 25.4
Maps of predictions (left panel) and prediction standard deviation (right panel) for the rainfall dataset.

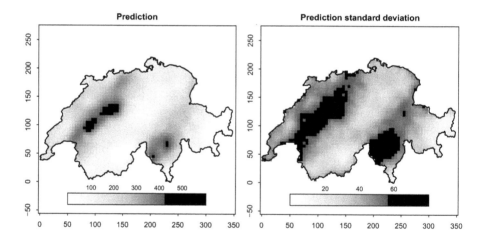

FIGURE 25.5
Maps of predictions (left panel) and prediction standard deviation (right panel) for the rainfall dataset using the model of [5].

it is not obvious how to define the target of prediction, that is, the signal part of the transformed Gaussian variables without noise when $g_\lambda(\cdot)$ is *not* the identity link function [4, p. 710]. (See also [7] who consider prediction for transformed Gaussian random fields where the parameter τ^2 is interpreted as measurement error.) When adding the error variance term to the map of the prediction variance for our model we get a similar pattern as the prediction variance plot corresponding to the model of [5] with slightly larger values. The difference in the variance is expected since our Bayesian model accounts for the uncertainty in the model parameters while the plug-in prediction in [5] does not.

Next, we consider a cross validation study to compare the performance of our model and the model of [5]. We remove 15 randomly chosen observations and predict these values using the remaining 452 observations. We repeat this procedure 31 times, each time removing 15 randomly chosen observations and predicting them with the remaining 452 data. For both our model as well as the model of [5], we keep λ and ϕ parameters fixed at their estimates when all data are observed. The average (over all 15×31 deleted observations) root mean squared error (RMSE) for our model is 7.55 and for the model of [5] is 7.48. We use 2000 samples from the predictive distribution (posterior predictive distribution in the case of our model) in order to estimate RMSE at each location. We also compute the proportion of these samples that fall below the observed (deleted) value at each of the 15×31 locations. These proportions are subtracted from 0.5 and the average of their absolute values across all locations for our model is 0.239 and for the model of [5] is 0.238. Lastly, we compute the proportions of one-sided prediction intervals that capture the observed (deleted) value. That is, we estimate the prediction intervals of the form $(-\infty, z_{0\alpha})$, where $z_{0\alpha}$ corresponds to the αth quantile of the predictive distribution. Table 25.2 shows the coverage probability of prediction intervals for different α values corresponding to our model and the model of [5]. From Table 25.2, we see that the coverage probabilities of prediction intervals for the two models are similar.

Finally, as mentioned in [5], the relative area where $Z(s) \geq c$ for some constant c is of practical significance. The proportion of locations that exceed

TABLE 25.2
Coverage probability of one-sided prediction intervals $(-\infty, z_{0\alpha})$ for different values of α (first column) corresponding to our model (second column) and the model of [5] (third column).

0.010	0.017	0.019
0.025	0.034	0.030
0.050	0.045	0.054
0.100	0.067	0.080
0.500	0.510	0.488
0.900	0.899	0.897
0.950	0.942	0.946
0.975	0.978	0.976
0.990	0.987	0.985

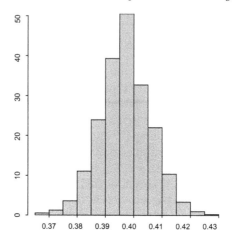

FIGURE 25.6
Histogram of random samples corresponding to the proportion of the area
with rainfall larger than 200.

the level 200 is computed using

$$\hat{E}[I(s \in \tilde{A}, Z(s) \geq 200)]/\#\tilde{A} = \frac{1}{N}\sum_{i=1}^{N} \#\{s \in \tilde{A}, h_{\hat{\lambda}}(W^{(i)}(s)) \geq 200\}/\#\tilde{A},$$

where $I(\cdot)$ is the indicator function, $\tilde{A}$ is the square grid of length and width
equal to 5 kilometers and $\{W^{(i)}(s)\}_{i=1}^{N}$ is the posterior predictive sample as
described in Section 25.4. The histogram of samples of these proportions is
shown in Figure 25.6.

25.6   Discussion

For Gaussian geostatistical models, estimation of unknown parameters as well
as minimum mean squared error prediction at unobserved location can be
done in closed form. On the other hand, many datasets in practice show non-
Gaussian behavior. Certain types of non-Gaussian random fields data may
be adequately modeled by transformed Gaussian models. In this chapter, we
present a flexible transformed Gaussian model where an additive measurement
error as well as a component representing smooth spatial variation is considered. Since specifying a joint prior distribution for all model parameters is
difficult, we consider an empirical Bayes method here. We propose an efficient
importance sampling method based on MCMC sampling for estimating the

transformation and range parameters of our model. Although, we consider an extended Box-Cox transformation in our model, other types of transformations can be used. For example, the exponential family of transformations proposed in [14], or the flexible families of transformations for binary data presented in [1] can be assumed. The method of estimating transformation parameter presented in this chapter can be used in these models also.

Acknowledgments

The authors thank two anonymous reviewers for helpful comments and valuable suggestions which led to several improvements in the manuscript.

Bibliography

[1] F. L. Aranda-Ordaz. On two families of transformations to additivity for binary response data. *Biometrika*, 68:357–363, 1981.

[2] P. J. Bickel and K. A. Doksum. An analysis of transformations revisited. *Journal of the American Statistical Association*, 76:296–311, 1981.

[3] G. E. P. Box and D. R. Cox. An analysis of transformations. *Journal of the Royal Statistical Society:* Series B *(Statistical Methodology)*, 26:211–252, 1964.

[4] O. F. Christensen. Monte Carlo maximum likelihood in model based geostatistics. *Journal of Computational and Graphical Statistics*, 13:702–718, 2004.

[5] O. F. Christensen, P. J. Diggle, and P. J. Ribeiro. Analyzing positive-valued spatial data: The transformed Gaussian model. In P. Monestiez, D. Allard, and R. Froidevaux, editors, *geoENV III-Geotatistics for Environmental Applications*, pages 287–298. Springer, 2001.

[6] V. De Oliveira. A note on the correlation structure of transformed Gaussian random fields. *Australian and New Zealand Journal of Statistics*, 45:353–366, 2003.

[7] V. De Oliveira and M. D. Ecker. Bayesian hot spot detection in the presence of a spatial trend: Application to total nitrogen concentration in Chesapeake Bay. *Environmetrics*, 13:85–101, 2002.

[8] V. De Oliveira, K. Fokianos, and B. Kedem. Bayesian transformed Gaussian random field: A review. *Japanese Journal of Applied Statistics*, 31:175–187, 2002.

[9] V. De Oliveira, B. Kedem, and D. A. Short. Bayesian prediction of transformed Gaussian random fields. *Journal of the American Statistical Association*, 92:1422–1433, 1997.

[10] P. J. Diggle, P. J. Ribeiro, and O. F. Christensen. An introduction to model-based geostatistics. In *Spatial Statistics and Computational Methods*, Lecture Notes in Statistics, pages 43–86. Springer Verlag, New York, 2003.

[11] H. Doss. Estimation of large families of Bayes factors from Markov chain output. *Statistica Sinica*, 20:537–560, 2010.

[12] C. J. Geyer. Estimating normalizing constants and reweighting mixtures in Markov chain Monte Carlo. Technical Report 568, School of Statistics, University of Minnesota, 1994.

[13] C. J. Geyer. Introduction to Markov chain Monte Carlo. In S. Brooks, A. Gelman, G. L. Jones, and X.-L. Ling, editors, *Handbook of Markov chain Monte Carlo*, pages 3–48. CRC Press, Boca Raton, FL, USA, 2011.

[14] B. F. J. Manly. Exponential data transformation. *Journal of the Royal Statistical Society: Series D (The Statistician)*, 25(1):37–42, 1976.

[15] B. Matérn. *Spatial Variation*. Springer Verlag, Berlin, 2nd edition, 1986.

[16] P. J. Ribeiro and P. J. Diggle. *geoR*, 2012. R package version 1.7-4.

[17] V. Roy. Efficient estimation of the link function parameter in a robust Bayesian binary regression model. *Computational Statistics & Data Analysis*, 73:87–102, 2014.

[18] V. Roy, E. Evangelou, and Z. Zhu. Efficient estimation and prediction for the Bayesian spatial generalized linear mixed model with flexible link functions. Technical Report, Iowa State University, 2014.

[19] M. L. Stein. Prediction and inference for truncated spatial data. *Journal of Computational and Graphical Statistics*, 1(1):91–110, 1992.

[20] M. L. Stein. *Interpolation of Spatial Data*. Springer Verlag, New York, 1999.

[21] M. A. Tanner and W. H. Wong. The calculation of posterior distributions by data augmentation (with discussion). *Journal of the American Statistical Association*, 82(398):528–550, 1987.

[22] H. Zhang. On estimation and prediction for spatial generalized linear mixed models. *Biometrics*, 58(1):129–136, 2002.

26

Mixture Kalman Filters and Beyond

Saikat Saha

Linköping University

Gustaf Hendeby

Linköping University

Fredrik Gustafsson

Linköping University

CONTENTS

26.1   Introduction

The discrete time general state-space model is a flexible framework to deal with the nonlinear and/or non-Gaussian time series problems. However, the associated (Bayesian) inference problems are often intractable. Additionally, for many applications of interest, the inference solutions are required to be recursive over time. The *particle filter* (PF) is a popular class of Monte Carlo based numerical methods to deal with such problems in real time. However, PF is known to be computationally expensive and does not scale well with the problem dimensions. If a part of the state space is analytically tractable conditioned on the remaining part, the Monte Carlo based estimation is then confined to a space of lower dimension, resulting in an estimation method known as the *Rao-Blackwellized particle filter* (RBPF).

In this chapter, we present a brief review of Rao-Blackwellized particle filtering. Especially, we outline a set of popular conditional tractable structures admitting such Rao-Blackwellization in practice. For some special and/or relatively new cases, we also provide reasonably detailed descriptions. We confine our presentation mostly to the practitioners' point of view.

26.1.1   Problem background

Let us start with the following discrete time general state-space model relating the latent state $x_k \in \mathbb{R}^{n_x}$ at time step k, to the observation $y_k \in \mathbb{R}^{n_y}$ as

$$x_k = f(x_{k-1}, w_k), \tag{26.1a}$$

$$y_k = h(x_k, e_k), \tag{26.1b}$$

where $f(x_{k-1}, w_{k-1})$ describes how the state propagates driven by the process noise w_{k-1}, and $h(x_k, e_t)$ describes how the measurements relate to the state and how the measurement is affected by noise, e_k. Note that the latent state x_k in (26.1) follows a Markovian model and the observation y_k is conditionally independent given x_k. The model (26.1) can be equivalently characterized in terms of the known state transition density $p(x_k|x_{k-1})$ and the observation likelihood $p(y_k|x_k)$. For this model, given the density for initial state (*i.e.*, , $p(x_0)$) and a stream of observations $y_{0:k} \triangleq \{y_0, y_1, \ldots, y_k\}$ up to time k, carrying the information about the state, one typical objective is to optimally estimate the sequence of posterior densities $p(x_{0:k}|y_{1:k})$ and their marginals $p(x_k|y_{1:k})$. The latter problem is generally referred to as the *filtering problem*. The filtering problem appears naturally in many applications of interest in applied science, econometrics, and engineering. Moreover, for certain applications (*e.g.*, target tracking), the above densities need to be propagated recursively in real time. The exact posterior densities $p(x_k|y_{1:k})$ (also known as filter densities) can be obtained analytically and recursively over time, only for a few special cases, including the *linear Gaussian state-*

space model (LGSSM), where the exact posterior is given by the celebrated *Kalman filter* (KF) [4, 48].

However, contrary to Kalman's linear-Gaussian assumptions (*i.e.*, , $f(\cdot)$ and $h(\cdot)$ are linear, w_k, v_k, and x_0 are Gaussian), many real world systems are nonlinear-non-Gaussian in their nature. Optimal estimates for such models can generally not be obtained in closed form and as a result, many approximative methods have been proposed. Most of these approaches are centered around approximating the state-space models with a linear-Gaussian form (*e.g.*, the extended Kalman filter, Gaussian sum filters, and nonlinear Kalman filters based on different numerical quadrature rules). Since the early 1990's, a class of sequential Monte Carlo simulation based filters, popularly known as *particle filters* have appeared to dominate this field due to their flexibility in adapting nonlinearity and/or non-Gaussianity without any *ad hoc* model assumptions. Particle filtering methods were introduced in their modern forms by the seminal contributions of Gordon *et al.* [20]. In this framework, the posterior distribution is approximated by an empirical distribution formed by (weighted) random samples, called particles, and this empirical distribution can be propagated over time recursively (see *e.g.*, [8, 15, 19, 21] for details).

Although the PF has been around for a while, it is computationally demanding and notably, it has severe limitations when scaling to higher dimensions [14, 33, 44]. As many important practical problems are high-dimensional in their nature, this poses a limitation on the general applicability of the PF. However, for many practical applications, severe nonlinearity and/or non-Gaussianity often affects only a small subset of the states. The rest of the states are often mildly nonlinear and/or non-Gaussian. Then one practical solution to the high dimensional problems can be given by first employing a particle filter to the highly nonlinear/non-Gaussian part of the state space, and conditioned on this, apply an (extended) Kalman filter (or variant thereof) to the remaining state space. This technical enabler is often known as the *mixture Kalman filter* (MKF) [10].

The mixture Kalman filter requires a special model structure, *i.e.*, , a part of the state space is required to be conditionally linear-Gaussian. Then one can conditionally apply the KF, which is analytically tractable. The immediate and foreseeable advantage is that solving part of the problem analytically leaves only a part of the state vector to be solved approximately using particle techniques, hence requiring fewer particles. However, this conditional tractability is not limited to only LGSSM (as in the MKF) and it has greater connotation in terms of efficiency improvement. When it is possible to exploit such conditional tractability, the Monte Carlo based estimation is confined to lower dimensional space. Consequently, the estimate obtained is often better and never worse than the estimate provided by the PF targeting the full state-space model. This efficiency arises as an implication of the so called *Rao-Blackwellization*. For this reason, the resulting filtering problem is also known as *Rao-Blackwellized particle filtering* [18].

26.1.2 Organization

The rest of the chapter is organized as follows. In order to make the chapter self containing, we start with brief but necessary backgrounds on PF and Rao-Blackwellization, in Section 26.2 and Section 26.3, respectively. This is followed by a detailed description on the models for mixture Kalman filter in Section 26.4. As pointed out earlier, RBPF is not only confined to the MKF and therefore we outline other popular RBPF approaches in Section 26.5. Next, we present a real life application in Section 26.6, where such model frameworks are put in place. For many practical applications, where conditional tractability is missing, the marginalization step can be carried out approximately using Monte Carlo integrations. This idea has led to the emergence of 'approximated RBPF', which is briefly discussed in Section 26.7. Finally, this is followed by the concluding remarks in Section 26.8.

26.2 Particle Filtering

In PF, the posterior distribution associated with the density $p(x_{0:k}|y_{1:k})$ is approximated by an empirical distribution represented by a set of $N(\gg 1)$ weighted particles (samples) as

$$\widehat{P}_N(dx_{0:k}|y_{1:k}) = \sum_{i=1}^{N} \widetilde{w}_k^{(i)} \delta_{x_{0:k}^{(i)}}(dx_{0:k}), \qquad (26.2)$$

where $\delta_{x_{0:k}^{(i)}}(A)$ is a Dirac measure for a given $x_{0:k}^{(i)}$ and a measurable set A. The $\widetilde{w}_k^{(i)}$ is the weight associated with each particle $x_{0:k}^{(i)}$, such that $\widetilde{w}_k^{(i)} \geq 0$ and $\sum_{i=1}^{N} \widetilde{w}_k^{(i)} = 1$. Given this representation, one can approximate the marginal distribution associated with $p(x_k|y_{1:k})$ as

$$\widehat{P}_N(dx_k|y_{1:k}) = \sum_{i=1}^{N} \widetilde{w}_k^{(i)} \delta_{x_k^{(i)}}(dx_k), \qquad (26.3)$$

and expectations of the form

$$I(g_k) = \int g_k(x_{0:k})p(x_{0:k}|y_{1:k})\,dx_{0:k} \qquad (26.4)$$

as

$$\widehat{I}_N(g_k) = \int g_k(x_{0:k})\widehat{P}_N(dx_{0:k}|y_{1:k}) \approx \sum_{i=1}^{N} \widetilde{w}_k^{(i)} g_k(x_{0:k}^{(i)}). \qquad (26.5)$$

Although the distribution $\widehat{P}_N(dx_{0:k}|y_{1:k})$ does not admit a well defined density with respect to the Lebesgue measure, the density $p(x_{0:k}|y_{1:k})$ is empirically represented as $\widehat{p}_N(x_{0:k}|y_{1:k})$ given by

$$\widehat{p}_N(x_{0:k}|y_{1:k}) = \sum_{i=1}^{N} \widetilde{w}_k^{(i)} \delta(x_{0:k} - x_{0:k}^{(i)}), \qquad (26.6)$$

where $\delta(\cdot)$ is the Dirac-delta function. The notation used in (26.6) is not mathematically rigorous; however, it is intuitively easier to follow than the more rigorous measure theoretic notations. This is especially useful when we are not concerned with theoretical convergence studies.

Now suppose at time $(k-1)$, we have a weighted particle approximation of the posterior $p(x_{0:k-1}|y_{1:k-1})$ as $\widehat{P}_N(dx_{0:k-1}|y_{1:k-1}) = \sum_{i=1}^{N} \widetilde{w}_{k-1}^{(i)} \delta_{x_{0:k-1}^{(i)}}(dx_{0:k-1})$. With the arrival of a new measurement y_k, we want to approximate $p(x_{0:k}|y_{1:k})$. This is achieved by propagating the particles in time using importance sampling and resampling steps. First, the particles are propagated to time k by sampling a new state $x_k^{(i)}$ from a proposal kernel $\pi(x_k|x_{0:k-1}^{(i)}, y_{1:k})$ and then setting $x_{0:k}^{(i)} \triangleq (x_{0:k-1}^{(i)}, x_k^{(i)})$. Since we have the posterior recursion as

$$p(x_{0:k}|y_{1:k}) \propto p(y_k|x_{0:k}, y_{1:k-1})\, p(x_k|x_{0:k-1}, y_{1:k-1})\, p(x_{0:k-1}|y_{1:k-1}) \quad (26.7)$$

and using (26.1), the corresponding (normalized) weights of the particles are updated as

$$w_k^{(i)} \propto \widetilde{w}_{k-1}^{(i)} \frac{p(y_k|x_k^{(i)})p(x_k^{(i)}|x_{k-1}^{(i)})}{\pi(x_k^{(i)}|x_{0:k-1}^{(i)}, y_{1:k})}; \quad \widetilde{w}_k^{(i)} = \frac{w_k^{(i)}}{\sum_{j=1}^{N} w_k^{(j)}}. \qquad (26.8)$$

Now, to avoid carrying trajectories with small weights, and to focus on the ones with large weights, the particles are resampled (with replacement), whenever necessary (see, *e.g.*, [17] for details). The resampled particles representing the same posterior $p(x_{0:k}|y_{1:k})$ are equally weighted. The whole procedure is repeated sequentially over time. For a more general introduction to PF, refer to, *e.g.*, [19].

26.3 Rao-Blackwellization

The foundation of the Rao-Blackwellization technique used in many estimation and filtering applications dates back to the 1940's. In the 1940's Rao [35] and Blackwell [7] showed that an estimator can be improved by making use of conditional expectations, if such structure exist in the problem formulation. They further showed how an estimator based on this knowledge can be constructed as an expected value of a conditioned estimator. This practice has over time become known as Rao-Blackwellization.

The Rao-Blackwell theorem is often used to show how the knowledge of a conditional expected value can improve the variance of an estimate, but this is only a special case of a more general theorem which shows that any convex cost-function is improved using this technique. For completeness, the Rao-Blackwell theorem is given in Theorem 26.3.1, [28, p. 51].

Theorem 26.3.1 (The Rao-Blackwell theorem [28]). *Let x be a random observable with the distribution $P_\theta \in \mathcal{P} = \{P_{\theta'} \in \Omega\}$, and let t be sufficient statistics for $\mathcal{P}^1$. Let $\hat{\theta}$ be an estimator of an estimand $g(\theta)$, and let the loss function $L(\theta, d)$ be a strictly convex function in d. Then if $\hat{\theta}$ has finite expectation and risk,*

$$R(\theta, \hat{\theta}) = E\Big(L\big(\theta, \hat{\theta}(x)\big)\Big) \quad \text{and if} \quad \hat{\theta}^{\text{RB}}(t) = E\big(\hat{\theta}(x)|t\big)$$

the risk of the estimator $\hat{\theta}^{\text{RB}}(t)$ satisfies

$$R(\theta, \hat{\theta}^{\text{RB}}) \prec R(\theta, \hat{\theta}),$$

unless $\hat{\theta}(x) = \hat{\theta}^{\text{RB}}(t)$ with probability 1, where $A \prec B$ denotes that $B - A$ is a positive definite matrix.

Proof. See [28]. □

It is worth pointing out once more that the Rao-Blackwell theorem does not only predict a reduced risk, it also shows how to construct the estimator to obtain the gain as an expected value. This makes the theorem very useful, as will be exemplified a number of times in the reminder of this chapter.

An important and often used special case of Theorem 26.3.1 describes the variance properties of a Rao-Blackwell estimator.

Corollary 1. *Let $\hat{\theta}$ be an unbiased estimator that fulfills the requirements in Theorem 26.3.1 and $\hat{\theta}^{\text{RB}}$ the Rao-Blackwell estimator, then*

$$\mathbf{C}(\hat{\theta}^{\text{RB}}) \prec \mathbf{C}(\hat{\theta})$$

unless $\hat{\theta}(x) = \hat{\theta}^{\text{RB}}(t)$ with probability 1.

Proof. To prove the corollary, assume the cost function

$$L(\theta, \hat{\theta}) = (\theta - \hat{\theta})(\theta - \hat{\theta})^T,$$

which is convex in both θ and $\hat{\theta}$, and apply Theorem 26.3.1. Finally note that $\mathbf{C}(\theta) = R(\theta, \hat{\theta}) = E\big(L(\theta, \hat{\theta})\big)$, which concludes the proof. □

---

[1]The statistic t is sufficient for θ if the conditional distribution of x given t does not depend on θ.

The covariance gain using Rao-Blackwellization is given by the separation of covariance, [24], which states

$$\mathbf{C}(x) = \mathbf{C}\big(\mathrm{E}(x|y)\big) + \mathrm{E}\big(\mathbf{C}(x|y)\big),$$

where $\mathbf{C}(x|y)$ denotes the covariance matrix of x conditioned on y. Assuming that x is the state estimate, then $\mathrm{E}(x|y)$ is the improved estimate in the Rao-Blackwell theorem. Hence, compared to the Rao-Blackwellized estimate, the regular estimate has an additional covariance term, $\mathrm{E}\big(\mathbf{C}(x|y)\big)$, that by definition is positive semi-definite. The covariance gain obtained with a Rao-Blackwellized estimator comes from eliminating this term in the covariance expression.

26.4 Models for Mixture Kalman Filters

The mixture Kalman filter was mentioned earlier as an enabling technique for PF to reasonably high dimensional problems. Since MKF uses conditional KF, it essentially requires a *conditionally linear Gaussian state-space model* (CLGSSM). This is an important class of models with many practical applications which has received extensive research attentions in the past [3, 21, 22, 29]. In the sequel, we outline two different model structures conforming to this class: the hierarchical conditionally linear Gaussian state-space model and the mixed linear/nonlinear Gaussian state-space model.

26.4.1 Hierarchical conditionally linear Gaussian model

The *hierarchical conditionally linear Gaussian model* (HCLGM) is, as the name suggests, a hierarchical model structure in two layers where a latent process c_k propagates according to a Markov kernel, independent of the second part of the latent variable x_k. Conditioned on c_k, (x_k, y_k) follows a linear Gaussian state-space model, parametrized in c_k. Mathematically the HCLGM can then be described by

$$c_{k+1} \sim p(c_{k+1}|c_k) \tag{26.9a}$$
$$x_{k+1} = f(c_k) + F(c_k)x_k + G(c_k)w_k \tag{26.9b}$$
$$y_k = h(c_k) + H(c_k)x_k + H^e(c_k)e_k, \tag{26.9c}$$

where $p(c_{k+1}|c_k)$ indicates that c_k is propagated using a Markov kernel; w_k and e_k are Gaussian process and measurement noises, respectively. A typical example of an HCLGM is a linear system with a scaled mixture of Gaussian noises (*e.g.*, Student's t noise distribution [30, 41]); here conditioned on $c_{0:k}$, the noise can be suitably represented as Gaussian and consequently KF can be applied on the conditionally linear-Gaussian part. A graphical illustration of

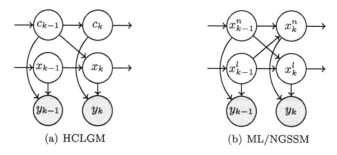

      (a) HCLGM   (b) ML/NGSSM

FIGURE 26.1

Graphical representation of dependencies in *hierarchical conditionally linear Gaussian model* (HCLGM) (26.9) and the *mixed linear/nonlinear Gaussian state-space model* (ML/NGSSM) (26.11).

the hierarchical model structure is given in Figure 26.1(a). This class of models is also known as the *conditionally dynamic linear model* (CDLM)[2] [10]. For the inference task, we can decompose the joint density $p(c_{0:k}, x_k | y_{1:k})$,

$$p(c_{0:k}, x_k | y_{1:k}) = p(c_{0:k} | y_{1:k}) \, p(x_k | c_{0:k}, y_{1:k}), \qquad (26.10)$$

where we target the propagation of $p(c_{0:k} | y_{1:k})$ using a PF and for each sequence $c_{0:k}$, the propagation of $p(x_k | c_{0:k}, y_{1:k})$ using a KF.

26.4.2   Mixed linear/nonlinear Gaussian state-space model

Another common model following a CLGSSM structure is the *mixed linear/nonlinear Gaussian state-space model* (ML/NGSSM) [43]. The ML/NGSSM is a state-space model where it is possible to divide the state vector into two parts, $x_k = \begin{pmatrix} x_k^n \\ x_k^l \end{pmatrix}$, such that conditioned on x_k^n, x_k^l assumes a linear-Gaussian state-space model,

$$x_{k+1}^n = f^n(x_k^n) + F^n(x_k^n)x_k^l + w_k^n \qquad (26.11a)$$

$$x_{k+1}^l = f^l(x_k^n) + F^l(x_k^n)x_k^l + w_k^l \qquad (26.11b)$$

$$y_k = h(x_k^n) + H(x_k^n)x_k^l + e_k, \qquad (26.11c)$$

where w_k^n, w_k^l, and e_k are mutually independent Gaussian noise processes with variance Q_k^n, Q_k^l, and R_k, respectively. Furthermore, x_0^l is assumed to follow a Gaussian distribution. The distribution of x_0^n can be arbitrary, but is assumed

---

[2]Although we restrict c_k to be a continuous variable for CDLM, in [10] CDLM is defined in a broader sense such that c_k can be either a continuous or discrete variable. The above model corresponding to a discrete c_k is popularly known as switching linear Gaussian system (or jump Markov linear system), which, according to our notations, is a special case of model (26.5.2), discussed later.

known. In [10], such a model is referred as the *partial conditionally dynamic linear model* (partial CDLM). However, [10] did not consider the influence of x_k^l in the evolution of x_k^n. This influence was first considered in [31] and was further popularized in [43]. This framework is extended to the dependent noise processes in [37]. The interconnection between the two parts of the state space, x_k^n and x_k^l is best illustrated in a graphical model as in Figure 26.1(b).

The nonlinear filtering problem can be decomposed into two sub-problems,

$$p(x_k^l, x_{0:k}^n | y_{1:k}) = p(x_{0:k}^n | y_{1:k}) \, p(x_k^l | x_{0:k}^n, y_{1:k}), \qquad (26.12)$$

where the propagation of $p(x_{0:k}^n | y_{1:k})$ is targeted using a PF and conditioned on each sequence $x_{0:k}^n$, $p(x_k^l | x_{0:k}^n, y_{1:k})$ can be propagated analytically using KF. The algorithm descriptions for this model are rather technical. The mixture Kalman filter here, when a bit simplified, works in the following way: The nonlinear part of the problem is solved using a particle filter. Attached to each particle is a KF, which estimates the (conditionally) linear part of the state. To compensate for the interactions between the linear and nonlinear parts of the state vector, a virtual measurement update must be applied to the KF each time the particle filter is propagated forward in time.

Later [25, 26] reformulated this RBPF scheme in terms of a stochastic bank of KFs, which highlighted the structural components of the solution. In essence, this RBPF scheme becomes a Kalman filter bank with stochastic pruning and weight computations provided by the PF. The algorithmic summary of this approach is given in Algorithm 26.1.

The ML/NGSSM as in (26.11) is quite common, especially in tracking and navigation applications, where the unknown state vector typically consists of position, velocity, and sometimes acceleration. When fitted into such a framework, the position typically ends up in the nonlinear part of the state space, whereas the velocity and acceleration end up in the linear part. These applications were considered in [23, 31] under the name *marginalized particle filter*.[3]

This model structure is also extensively used in robotics to address the *simultaneous localization and mapping* (SLAM) problem. Here one attempts to localize oneself in an unknown environment and at the same time, build a map of the surrounding environment. In essence, this yields a state-space model with a state comprised of the kinematic state of oneself, and the map, which grows over time. Fortunately, conditioned on the state, the map fits in a linear-Gaussian model which allows for Rao-Blackwellization. This fact is utilized in the FastSLAM algorithm which is an MKF in disguise [29].

---

[3]The reason behind that name is that the linear (or, more generally, analytical) part of the state vector is 'marginalized' (or, integrated out) in closed form to obtain the posterior distribution of the nonlinear (or, non-analytical) part.

Algorithm 26.1 Rao–Blackwellized PF [26]

For the system (26.11).

Initialization: For $i = 1, \ldots, N$, let

$$x_{0|-1}^{p(i)} \sim \pi_0^n, \qquad x_{0|-1}^{k(i)} = x_0^k, \qquad P_{0|-1}^{k(i)} = \Pi_0^l, \qquad \omega_{0|-1} = \tfrac{1}{N}.$$

Iterations: For $k = 0, \ldots$ do

- Measurement update, for each particle $i = 1, \ldots, N$ do

$$\omega_{k|k}^{(i)} \propto \mathcal{N}(y_k; \hat{y}_k^{(i)}, S_k^{(i)}) \cdot \omega_{k|k-1}^{(i)}$$
$$x_{k|k}^{(i)} = x_{k|k-1}^{(i)} + K_k^{(i)}(y_k - \hat{y}_k^{(i)})$$
$$P_{k|k}^{(i)} = P_{k|k-1}^{(i)} - K_k^{(i)} S_k^{(i)} K_k^{(i)T},$$

where

$$\hat{y}_k^{(i)} = h(x_{k|k-1}^{n(i)}) + H(x_{k|k-1}^{n(i)}) x_{k|k-1}^{l(i)}$$
$$S_k^{(i)} = H(x_{k|k-1}^{n(i)}) P_{k|k-1}^{(i)} H^T(x_{k|k-1}^{n(i)}) + R_k$$
$$K_k^{(i)} = P_{k|k-1}^{l(i)} H^T(x_{k|k-1}^{l(i)})(S_k^{(i)})^{-1}.$$

- If necessary, resample the particles using standard techniques. When selecting a particle $x_{k|k}^{n(i)}$, select the matching linear states $x_{k|k}^{l(i)}$ and $P_{k|k}^{l(i)}$, and set $\omega_{k|k}^{(i)}$ appropriately.
- Time update, for each particle $i = 1, \ldots, N$ do

$$\bar{x}_{k+1|k}^{n(i)} = f^n(x_{k|k}^n) + F^n(x_{k|k}^{n(i)}) x_{k|k}^{l(i)}$$
$$\bar{x}_{k+1|k}^{l(i)} = f^l(x_{k|k}^n) + F^l(x_{k|k}^{n(i)}) x_{k|k}^{l(i)}$$
$$\bar{P}_{k+1|k}^{n(i)} = F^n(x_{k|k}^{n(i)}) P_{k+1|k}^{l(i)} \big(F^n(x_{k|k}^{n(i)})\big)^T + Q_k^n$$
$$\bar{P}_{k+1|k}^{nl(i)} = F^n(x_{k|k}^{n(i)}) P_{k+1|k}^{l(i)} \big(F^l(x_{k|k}^{n(i)})\big)^T$$
$$\bar{P}_{k+1|k}^{l(i)} = F^l(x_{k|k}^{n(i)}) P_{k+1|k}^{l(i)} \big(F^l(x_{k|k}^{n(i)})\big)^T + Q_k^l.$$

With $\xi_{k+1}^{(i)} \sim \mathcal{N}(0, \bar{P}_{k+1|k}^{n(i)})$ do

$$x_{k+1|k}^{n(i)} = \bar{x}_{k+1|k}^{n(i)} + \xi_{k+1}^{(i)}$$
$$x_{k+1|k}^{l(i)} = \bar{x}_{k+1|k}^{l(i)} + \big(\bar{P}_{k+1|k}^{nl(i)}\big)^T \big(\bar{P}_{k+1|k}^{n(i)}\big)^{-1} \xi_{k+1}^{(i)}$$
$$P_{k+1|k}^{l(i)} = \bar{P}_{k+1|k}^{l(i)} - \big(\bar{P}_{k+1|k}^{nl(i)}\big)^T \big(\bar{P}_{k+1|k}^{n(i)}\big)^{-1} \bar{P}_{k+1|k}^{nl(i)}.$$

26.5 Beyond Mixture Kalman Filter in Practice

Note that the key to efficient Rao-Blackwellization is a (conditionally) tractable substructure, which is not necessarily limited to the *conditionally linear-Gaussian state-space model*. In this section, we present some other existing Rao-Blackwellization approaches that are popular in practice.

26.5.1 Hierarchical conditionally finite HMM

Similar to the LGSS model, when the latent state evolves according to a finite state Markov chain, the posterior of the state given the observations, can be obtained in closed form using the *hidden Markov model* (HMM) forward algorithm [34]. This model structure is known as the finite state-space HMM. Now if part of the state space follows such finite state-space HMM, it is then possible to exploit an RBPF formulation.

One such RBPF framework appears when the state space follows a finite state HMM, whose parameters are modulated by another Markovian process [18]. This model structure is denoted the *hierarchical conditionally finite HMM* (HCfHMM) and can be described as

$$p(c_k|c_{k-1}), \quad p(r_k|c_k, r_{k-1}), \quad p(y_k|c_k, r_k), \tag{26.13}$$

where r_k is a finite state-space Markov chain, whose parameter at time k depends on a Markov process c_k. Both r_k and c_k are latent and are related to the observation y_k. Note that c_k is at the top of the hierarchy and the evolution of c_k is independent of r_k. The graphical representation is shown in Figure 26.2(a). Now conditioned on c_k, r_k follows a finite state-space Markov chain of known parameters. Therefore one can implement a Rao-Blackwellized particle filter, where the PF targets only $p(c_{0:k}|y_{0:k})$ and where $p(r_k|c_{0:k}, y_{0:k})$ is analytically tractable.

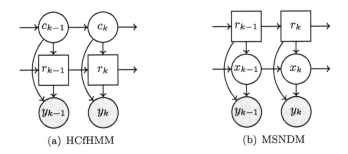

(a) HCfHMM (b) MSNDM

FIGURE 26.2
Graphical representation of the *hierarchical conditionally finite HMM* (HCfHMM) and *multiple switching nonlinear dynamics model* (MSNDM).

26.5.2 Multiple switching nonlinear dynamic models

In many practical applications of engineering and econometrics, one often deals with nonlinear dynamic systems involving both a continuous value target state and a discrete value regime variable. Such descriptions imply that the system can switch between different nonlinear dynamic regimes, where the parameters of each regime is modulated by the corresponding regime variable. The regime variable also evolves dynamically according to a finite state Markov chain. Both the target state and regime variable are latent and are related to the noisy observations. This type of model is referred to as *(Markov) regime switching nonlinear dynamic models* or *hybrid nonlinear state-space models*.

The probabilistic description of the above model can be given by

$$\Pi(r_k|r_{k-1}), \quad p_{\theta_{r_k}}(x_k|x_{k-1}, r_k), \quad p_{\theta_{r_k}}(y_k|x_k, r_k), \tag{26.14}$$

where $r_k \in S \triangleq \{1, 2, \ldots, s\}$, is a (discrete) regime indicator variable with finite number of regimes (*i.e.*, , categorical variable), $x_k \in \mathbb{R}^{n_x}$ is the (continuous) state variable. As the system can switch between different dynamic regimes, for a given regime variable $l \in S$, the corresponding dynamic regime can be characterized by a set of parameters θ_l. Both x_k and r_k are latent variables, which are related to the measurement $y_k \in \mathbb{R}^{n_y}$. The evolution of the regime variable r_k is commonly modeled by a homogeneous (time invariant) first order Markov chain with *transition probability matrix* (TPM) $\Pi = [\pi_{ij}]_{ij}$,

$$\pi_{ij} \triangleq \mathbb{P}(r_k = j|r_{k-1} = i) \quad (i, j \in S), \tag{26.15a}$$

$$\pi_{ij} \geq 0; \qquad \sum_{j=1}^{s} \pi_{ij} = 1. \tag{26.15b}$$

This model is represented graphically in Figure 26.2(b). Note that the above model can easily be generalized to the cases involving time dependent dynamics and exogenous inputs. For the model, given the densities for the initial state $\{r_0, x_0\}$ and the measurements up to a time k, our interest lies in estimating $\mathbb{P}(r_k|y_{1:k})$ and $p(x_k|y_{1:k})$ recursively. However, the above posteriors are computationally intractable. In connection to this, we note from (26.14) that conditioned on the sequence $x_{0:k}$, r_k follows a finite state hidden Markov model. As a result, $\mathbb{P}(r_k|x_{0:k}, y_{1:k})$ is analytically tractable [36]. Using this analytical substructure, it is possible to implement an efficient RBPF scheme[4]. We provide below a brief description of RBPF implementation for the same.

26.5.2.1 An RBPF implementation

The initial densities for the state and the regime variables are given by $p(x_0)$ and $\mathbb{P}(r_0) \triangleq \mathbb{P}(r_0|x_0)$, respectively, which can be arbitrary but is assumed to be known. We further assume favorable mixing conditions as in [13].

---

[4]It should be noted that the switching linear Gaussian system is a special case of this model structure, where the inference problem is popularly addressed using an alternative RBPF scheme: the propagation of $\mathbb{P}(r_{0:k}|y_{1:k})$ is targeted by a PF and conditioned on $r_{0:k}$, $p(x_k|r_{0:k}, y_{1:k})$ is obtained analytically using KF methods

Suppose that we are at time $(k - 1)$. We consider the extended target density $p(r_{k-1}, x_{0:k-1} | y_{1:k-1})$ which can be decomposed as

$$p(r_{k-1}, x_{0:k-1} | y_{1:k-1}) = p(r_{k-1} | x_{0:k-1}, y_{1:k-1}) \, p(x_{0:k-1} | y_{1:k-1}) \quad (26.16)$$

The posterior propagation of the latent state x_{k-1} can be targeted through a PF, where $p(x_{0:k-1} | y_{1:k-1})$ is represented by a set of N weighted random particles as

$$p(x_{0:k-1} | y_{1:k-1}) \approx \sum_{i=1}^{N} w_{k-1}^{(i)} \delta(x_{0:k-1} - x_{0:k-1}^{(i)}). \quad (26.17)$$

Now conditioned on $\{x_{0:k-1}, y_{1:k-1}\}$, the regime variable r_{k-1} follows a finite state-space HMM. Consequently, $p(r_{k-1} | x_{0:k-1}, y_{1:k-1})$ is analytically tractable[5], which is represented as

$$q_{k-1|k-1}^{(i)}(l) \triangleq \mathbb{P}(r_{k-1} = l | x_{0:k-1}^{(i)}, y_{1:k-1}), \quad (26.18)$$

for $l \in S$ and $i = 1, \ldots, N$. Now using (26.17) and (26.18), the extended target density in (26.16) can be represented as

$$\left[x_{0:k-1}^{(i)}, w_{k-1}^{(i)}, \{q_{k-1|k-1}^{(i)}(l)\}_{l=1}^{s} \right]_{i=1}^{N}. \quad (26.19)$$

Now having observed y_k, we want to propagate the extended target density in (26.16) to time k. This can be achieved in the following steps (a)–(d):

(a) **Prediction step for conditional HMM filter:** This is obtained as

$$q_{k|k-1}^{(i)}(l) \triangleq \mathbb{P}(r_k = l | x_{0:k-1}^{(i)}, y_{1:k-1}) = \sum_{j=1}^{s} \pi_{jl} \, q_{k-1|k-1}^{(i)}(j), \quad (l, j) \in S.$$
$$(26.20)$$

(b) **Prediction step for particle filter:** At this stage, generate N new samples $x_k^{(i)}$ from an appropriate proposal kernel as $x_k^{(i)} \sim \pi(x_k | x_{k-1}^{(i)}, y_k)$. Then set $x_{0:k}^{(i)} = \{x_{0:k-1}^{(i)}, x_k^{(i)}\}$, for $i = 1, \ldots, N$, representing the particle trajectories up to time k.

(c) **Update step for conditional HMM filter:** Noting that

$$\mathbb{P}(r_k = l | x_{0:k}, y_{1:k}) \propto p(y_k, x_k | r_k = l, x_{0:k-1}, y_{1:k-1})$$
$$\times \mathbb{P}(r_k = l | x_{0:k-1}, y_{1:k-1}), \quad (26.21)$$

---

[5]The 'favorable' mixing property ensures that $p(x_{0:k-1} | y_{1:k-1})$ can be well approximated by $p(x_{k-L:k-1} | y_{1:k-1})$, for some lag L. Consequently, $p(r_{k-1} | x_{0:k-1}, y_{1:k-1}) \approx p(r_{k-1} | x_{k-L:k-1}, y_{1:k-1})$.

we have

$$q_{k|k}^{(i)}(l) \propto p(y_k, x_k^{(i)}|r_k = l, x_{0:k-1}^{(i)}, y_{1:k-1})\, q_{k|k-1}^{(i)}(l) \tag{26.22a}$$

$$\propto p_{\theta_l}(y_k|x_k^{(i)}, r_k = l)p_{\theta_l}(x_k^{(i)}|x_{k-1}^{(i)}, r_k = l)\, q_{k|k-1}^{(i)}(l). \tag{26.22b}$$

Now defining

$$\alpha_k^{(i)}(l) \triangleq p_{\theta_l}(y_k|x_k^{(i)}, r_k = l)\, p_{\theta_l}(x_k^{(i)}|x_{k-1}^{(i)}, r_k = l)\, q_{k|k-1}^{(i)}(l) \tag{26.23}$$

we obtain

$$q_{k|k}^{(i)}(l) = \alpha_k^{(i)}(l) \,\Big/\, \sum_{j=1}^{s} \alpha_k^{(i)}(j), \tag{26.24}$$

for $l \in S$ and $i = 1, \dots, N$.

(d) **Update step for particle filter:** As the continuous state can be recursively propagated using the following relation:

$$p(x_{0:k}|y_{1:k}) \propto p(y_k, x_k|x_{0:k-1}, y_{1:k-1})\, p(x_{0:k-1}|y_{1:k-1}), \tag{26.25}$$

the corresponding weight update equation for the particle filtering is given by

$$w_k^{(i)} = \frac{p(x_k^{(i)}, y_k|x_{0:k-1}^{(i)}, y_{1:k-1})}{\pi_k(x_k^{(i)}|x_{0:k-1}^{(i)}, y_{1:k})}\tilde{w}_{k-1}^{(i)}; \quad \tilde{w}_k^{(i)} = w_k^{(i)} \,\Big/\, \sum_{j=1}^{N} w_k^{(j)}, \tag{26.26}$$

where $\{\tilde{w}_k^{(i)}\}_{i=1}^{N}$ are the normalized weights. The numerator in (26.26) can be obtained as

$$p\!\left(\begin{matrix} x_k^{(i)}, \\ y_k \end{matrix}\,\middle|\, \begin{matrix} x_{0:k-1}^{(i)}, \\ y_{1:k-1} \end{matrix}\right) = \sum_{l=1}^{s} p\!\left(\begin{matrix} y_k, \\ x_k^{(i)} \end{matrix}\,\middle|\, \begin{matrix} r_k = l, x_{0:k-1}^{(i)}, \\ y_{1:k-1} \end{matrix}\right) \mathbb{P}\!\left(r_k = l \,\middle|\, \begin{matrix} x_{0:k-1}^{(i)}, \\ y_{1:k-1} \end{matrix}\right), \tag{26.27}$$

which is basically given by the normalizing constant of (26.24). Note that the marginal density $p(x_k|y_{1:k})$ can be obtained in the particle cloud form as $p(x_k|y_{1:k}) \approx \sum_{i=1}^{N} \tilde{w}_k^{(i)}\delta(x_k - x_k^{(i)})$. The posterior probability of the regime variable can now be obtained as

$$\mathbb{P}(r_k = l|y_{0:k}) = \int \mathbb{P}(r_k = l|x_{0:k}, y_{0:k})p(x_{0:k}|y_{0:k})\, dx_{0:k} \approx \sum_{i=1}^{N} q_{k|k}^{(i)}(l)\tilde{w}_k^{(i)}. \tag{26.28}$$

For further details and algorithmic summary, please refer to [38, 39].

26.5.3 Partially observed Gaussian state-space models

Consider the following *partially observed Gaussian state-space models* (POGSSM) [2]

$$x_k = F_k x_{k-1} + G_k v_k, \tag{26.29a}$$

$$y_k = H_k x_k + H_k^e e_k, \tag{26.29b}$$

$$z_k \sim p(z_k|y_k), \tag{26.29c}$$

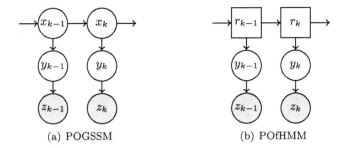

(a) POGSSM   (b) POfHMM

FIGURE 26.3
Graphical representation of a *partially observed Gaussian state-space model* (POGSSM) and a *partially observed finite HMM* (POfHMM).

where x_k is a latent (Markov) state process with initial density given by a Gaussian with mean $\hat{x}_0$ and variance P_0 as $p(x_0) = \mathcal{N}(\hat{x}_0, P_0)$; y_k and z_k are latent data process and observed data process, respectively. The processes x_k and y_k together define a standard linear Gaussian state-space model. Compared to standard LGSS model, however, one observes z_k instead of y_k. z_k is typically a quantized or censored version of y_k. For example, in a dynamic probit model, we have

$$z_k = \begin{cases} 1, & y_k > 0 \\ 0, & y_k \le 0, \end{cases} \tag{26.30}$$

whereas for a dynamic tobit model,

$$z_k = \max(y_k, 0). \tag{26.31}$$

The graphical representation of the model is shown in Figure 26.3(a).

For inference purpose, the joint density can be decomposed as

$$p(x_{0:k}, y_{0:k}|z_{0:k}) = p(x_{0:k}|y_{0:k}, z_{0:k})\, p(y_{0:k}|z_{0:k}). \tag{26.32}$$

Then the density $p(y_{0:k}|z_{0:k})$ can be targeted using a PF in the form of N weighted particles as $p(y_{0:k}|z_{0:k}) \approx \sum_{i=1}^{N} \omega_k^{(i)} \delta(y_{0:k} - y_{0:k}^{(i)})$. Now for each sequence $y_{0:k}^{(i)}$, the conditional density $p(x_{0:k}|y_{0:k}^{(i)}, z_{0:k})$ is analytically tractable using a KF. Finally, $p(x_{0:k}|z_{0:k})$ can be obtained by marginalizing over $y_{0:k}$ as

$$p(x_{0:k}|z_{0:k}) = \int p(x_{0:k}|y_{0:k}, z_{0:k})p(y_{0:k}|z_{0:k})\, dy_{0:k} \approx \sum_{i=1}^{N} \omega_k^{(i)} p(x_{0:k}|y_{0:k}^{(i)}, z_{0:k}). \tag{26.33}$$

Thus $p(x_{0:k}|z_{0:k})$ is obtained as random weighted mixture of Gaussians. For algorithmic details, model extensions and applications, please refer to the original article [2].

26.5.4   Partially observed finite HMM

Consider the following model described probabilistically as

$$r_k \sim \Pi(r_k|r_{k-1}), \qquad y_k \sim p(y_k|r_k), \qquad z_k \sim p(z_k|y_k), \qquad (26.34)$$

where the dynamics of r_k is given by a homogeneous first order finite state Markov chain. Here r_k and y_k together describe a *partially observed finite HMM* (POfHMM). However instead of y_k, z_k is actually observed. The graphical representation of the model is shown in Figure 26.3(b).

Similar to (26.5.3), the joint density can be decomposed as

$$p(r_k, y_{0:k}|z_{0:k}) = p(r_k|y_{0:k}, z_{0:k})\, p(y_{0:k}|z_{0:k}). \qquad (26.35)$$

Again, the density $p(y_{0:k}|z_{0:k})$ can be targeted using a PF in the form of N weighted particles as $p(y_{0:k}|z_{0:k}) \approx \sum_{i=1}^{N} \omega_k^{(i)} \delta(y_{0:k} - y_{0:k}^{(i)})$. Then for each sequence $y_{0:k}$, $p(r_k|y_{0:k}, z_{0:k})$ can be obtained analytically using the property of finite HMM. For details, see [2].

26.5.5   Conditionally conjugate latent process model

The key enabler here is the concept of Bayesian conjugacy. For a latent variable with known observation likelihood, if we can choose a suitable prior (represented with a parametric distribution) such that the prior and posterior belong to the same parametric family, then it is said to be conjugate to the observation likelihood. As a result, the posterior can be obtained in closed form by updating the parameters of the prior distributions. Such conjugate priors are known to exist for the members of the exponential family of distributions [6]. Using this conjugacy, one can define a latent process in such a way that both the filtering and prediction distributions are conjugate to the observation distribution in each time step. The latent process so defined is called the conjugate latent process [47].

Now if the state space can be decomposed such that part of the state space follows a conjugate latent process conditioned on the other part, one can then marginalize this part analytically. In the particle filtering framework, this leads to another Rao-Blackwellization approach. One typical application of the resulting RBPF is a noise adaptive particle filtering for a general state-space model [32, 37, 42]. A brief description of the general approach is given below.

26.5.5.1   Rao-Blackwellized noise adaptive particle filter

For many practical applications, complete knowledge of the noise processes in the state-space models is unavailable. Typically for such cases, noises are assumed to be generated from a known parametric family of distributions with unspecified parameters. In such settings, one needs to estimate the noise parameters along with the latent state using the available noisy observations.

This problem is known as noise adaptive filtering. Moreover, the noise may be non-stationary or state dependent, which prevents off-line tuning. For the above reasons, one needs a sequential method to learn the noise parameters using streamed sensor data.

When the noise distribution is a member of the exponential family, we can model the unknown noise parameter as conjugate latent process. To describe the resulting filtering problem, we first start with a description of the model (including the noise model), which is followed by the problem definition and implementation details.

Model description: Consider the following general state-space model relating a (Markovian) latent state x_k to the observation y_k at time step k as

$$x_k = f_k(x_{k-1}, u_{k-1}) + g_k(x_{k-1}, u_{k-1})v_k, \tag{26.36a}$$

$$y_k = h_k(x_k, u_k) + d_k(x_k, u_k)w_k. \tag{26.36b}$$

Here, $f_k(\cdot)$, $h_k(\cdot)$, $d_k(\cdot)$ and $g_k(\cdot)$ are nonlinear functions of the state vector x_k and the deterministic input u_k at time k. Here $d_k(\cdot)$ is assumed to be invertible whereas $g_k(\cdot)$ is not, as most motion models in practice, including those with integrators, lead to non-invertible $g_k(\cdot)$.

Noise model: We define the noise vector $e_k \triangleq [v_k^T, w_k^T]^T$ as a realization from a distribution belonging to the exponential family as

$$e_k \sim p(e_k|\theta_k) = \rho(e_k)\exp\big(\eta(\theta_k)\cdot\tau(e_k) - \phi(\theta_k)\big), \tag{26.37}$$

where θ_k is the vector of unknown parameters, $\eta(\theta_k)$ and $\phi(\theta_k)$ are vector and scalar valued functions of the parameters, respectively; $\rho(e_k)$ and $\tau(e_k)$ are scalar and vector valued functions of the realization e_k; and the symbol '$\cdot$' denotes scalar product of two vectors. Further, following [32], assume that the unknown noise parameters θ_k are slowly varying in time. This slowly varying nature can arise, *e.g.*, due to model misspecification [48].

Now for a given realization of the noise e_k, (26.37) provides the likelihood function for the unknown noise parameter θ_k. Accordingly, we can select a conjugate prior on θ_k as

$$p(\theta_k|e_{1:k-1}) = \gamma^{-1}(V_{k|k-1}, \nu_{k|k-1}) \times \exp\big(\eta(\theta_k)V_{k|k-1} - \nu_{k|k-1}\phi(\theta_k)\big), \tag{26.38}$$

where $V_{k|k-1}$ is a vector of sufficient statistics and $\nu_{k|k-1}$ is a scalar counter of the effective number of samples in the statistics. The normalization factor $\gamma(V_{k|k-1}, \nu_{k|k-1})$ is uniquely determined by the statistics $V_{k|k-1}$ and $\nu_{k|k-1}$. Then, the posterior density $p(\theta_k|e_{1:k})$ is also of the form (26.38) with statistics

$$V_{k|k} = V_{k|k-1} + \tau(e_k), \qquad \nu_{k|k} = \nu_{k|k-1} + 1. \tag{26.39}$$

This is a convenient result for recursive evaluation of sufficient statistics starting from a prior defined by V_0 and ν_0. Further, the predictive distribution of

e_k is then given by

$$p(e_k|e_{1:k-1}) = \int p(e_k|\theta_k)p(\theta_k|e_{1:k-1})\,d\theta_k = \frac{\gamma(V_{k|k}, \nu_{k|k})}{\gamma(V_{k|k-1}, \nu_{k|k-1})}\rho(e_k). \quad (26.40)$$

Now as θ_k is assumed to be slowly varying in time, to complete the state-space model, an evolution model for θ_k (*i.e.*, , $p(\theta_k|\theta_{k-1})$) is required. Given this evolution model, the predictive density of the parameter θ_{k+1} is obtained as

$$p(\theta_{k+1}|e_{1:k}) = \int p(\theta_{k+1}|\theta_k)p(\theta_k|e_{1:k})\,d\theta_k. \quad (26.41)$$

If this predictive density conforms to the specification (26.38), then the noise parameter vector follows a *conjugate latent process*. However, typically the transition model $p(\theta_{k+1}|\theta_k)$ is unknown. In [32], the evolution is defined implicitly as a Kullback-Leibler norm constraint on the time variability which leads to an exponential forgetting mechanism operating on the sufficient statistics. This keeps the predictive density $p(\theta_{k+1}|e_{1:k})$ in the same class as in (26.38) with statistics updated from (26.39) as

$$\nu_{k+1|k} = \lambda\nu_{k|k} + (1-\lambda)\nu_u, \quad (26.42a)$$
$$V_{k+1|k} = \lambda V_{k|k} + (1-\lambda)V_u, \quad (26.42b)$$

where $\lambda \in (0,\ 1]$ is a forgetting factor and the invariant measure is assumed to be also in the exponential form (26.38) with statistics ν_u, V_u. The details can be found in [32].

Problem definition: We are concerned with the evaluation of the joint posterior density $p(x_k, \theta_k|y_{1:k})$ sequentially over time. Suppose at time $(k-1)$ we have the posterior $p(x_{k-1}, \theta_{k-1}|y_{1:k-1})$. With the arrival of a new observation y_k, we want to update the posterior recursively to $p(x_k, \theta_k|y_{1:k})$.

Joint estimation of state and noise parameters: The joint posterior density at time k can be decomposed as follows:

$$p(x_{0:k}, \theta_k|y_{0:k}) = p(\theta_k|x_{0:k}, y_{0:k})\,p(x_{0:k}|y_{0:k}). \quad (26.43)$$

The posterior propagation of the latent state x_k is targeted through a PF as $p(x_{0:k}|y_{0:k}) \approx \sum_{i=1}^{n} \omega_k^{(i)}\delta(x_{0:k} - x_{0:k}^{(i)})$. Now for a given $(x_{0:k}^{(i)}, y_{0:k})$, $p(\theta_k|x_{0:k}^{(i)}, y_{0:k})$ can be considered as the posterior density of the latent parameter θ_k and can be propagated analytically using a conjugate latent process as discussed earlier. To see this, note that for a known $x_k^{(i)}$, (26.36a)–(26.36b) can be transformed into

$$e_k^{(i)} = e(x_k^{(i)}, y_k) = \begin{bmatrix} g_k^\dagger(x_{k-1}^{(i)}, u_{k-1})[x_k^{(i)} - f_k(x_{k-1}^{(i)}, u_{k-1})] \\ d_k^{-1}(x_k^{(i)}, u_k)[y_k - h_k(x_k^{(i)}, u_k)] \end{bmatrix}, \quad (26.44)$$

where $g_k^\dagger(x_{k-1}^{(i)}, u_{k-1})$ stands for the Moore-Penrose pseudo-inverse of $g_k(x_{k-1}^{(i)}, u_{k-1})$. Consequently, $p(\theta_k | x_{0:k}^{(i)}, y_{0:k}) = p(\theta_k | e_{0:k}^{(i)})$. So the parameter posterior follows a conditional conjugate process and the analytical propagation of this posterior can be obtained easily using (26.37)–(26.41).

The joint density (26.43) can be represented as

$$p(x_{0:k}, \theta_k | y_{0:k}) \approx \sum_{i=1}^{n} \omega_k^{(i)} p(\theta_k | V_{k|k}^{(i)}, \nu_{k|k}^{(i)}) \delta(x_{0:k} - x_{0:k}^{(i)}), \qquad (26.45)$$

where the statistics $\omega_k^{(i)}$, $V_{k|k}^{(i)}$, $\nu_{k|k}^{(i)}$, and $x_{0:k}^{(i)}$ are evaluated as follows. First, $x_k^{(i)}$ are sampled from a proposal density $q(x_k | x_{0:k-1}^{(i)}, y_{0:k-1})$. Second, for the known value $x_k^{(i)}$, $p(\theta_k | x_{0:k}^{(i)}, y_{0:k})$ is updated using the mapping (26.44) to $e_k^{(i)}$, and the statistics $V_k^{(i)}$ and $\nu_k^{(i)}$ are updated using (26.39). Finally, using the recursive relation $p(x_{0:k} | y_{0:k}) \propto p(y_k, x_k | x_{0:k-1}, y_{0:k-1}) p(x_{0:k-1} | y_{0:k-1})$, weights can be updated as

$$\omega_k^{(i)} \propto \frac{p(y_k, x_k | x_{0:k-1}, y_{0:k-1})}{q(x_k | x_{0:k-1}^{(i)}, y_{0:k-1})} \omega_{k-1}^{(i)},$$

where

$$p(y_k, x_k | x_{0:k-1}, y_{0:k-1}) = \int p(y_k, x_k | \theta_k, x_{0:k-1}, y_{0:k-1}) p(\theta_k | x_{0:k-1}, y_{0:k-1}) \, d\theta_k,$$

is the marginal predictive distribution of x_k and y_k. This marginal distribution can be computed by integrating out the unknown parameters, leading to the predictive distribution of x_k and y_k, and consequently e_k via (26.44). Note that the predictive distribution $p(e_k | e_{0:k-1})$ is readily available in the form of (26.40) for the exponential family. The predictor (26.46) can then be obtained using the transformation of variables in the probability density functions:

$$p(y_k, x_k | x_{0:k-1}, y_{0:k-1}) = |J(x_k, y_k)| \, p(e(x_k, y_k) | V_{k|k-1}^{(i)}, \nu_{k|k-1}^{(i)}). \qquad (26.46)$$

where $J(x_k, y_k)$ is the Jacobian of the transformation (26.44) and $p(e_k | \cdot)$ is given by (26.40). For further details and an algorithmic summary, we refer to [32].

Remark 1. *If the noise vector e_k follows a multivariate Normal distribution, with unknown mean μ_k and covariance Σ_k, a Normal-inverse-Wishart distribution defines a conjugate prior.*

Remark 2. *This approach is closely related to the works on stationary parameter estimation using RBPF by [9, 16, 45]. Note that the estimation of stationary parameters is a special case of the above approach with $\lambda = 1$, which reduces the sufficient statistics update in (26.39) to the form of [45]. For the stationary parameter, however, this approach is known to suffer from error accumulations as pointed out by [12]. Here, we note that for a sequence of $\lambda_k < 1, \forall k$, the posterior density $p(\theta_k | x_{1:k}, y_{1:k})$ and thus $p(y_k, x_k | x_{0:k-1}, y_{0:k-1})$ satisfies the exponential forgetting property that is well known to mitigate the path degeneracy problem [13].*

26.6   Terrain Navigation Problems

In this section, we outline the problem of positioning an aircraft without the aid of *global positioning systems* (GPS). GPS is vulnerable to jamming; we therefore consider the positioning by using the measurements from alternative sensors. Specifically, one can position the aircraft using only altitude measurements, measurements of the distance down to the ground, measurements of the aircraft's acceleration, and an elevation map of the visited area.

The navigation problem results in an estimation problem with linear dynamics, where the aircraft is assumed to follow a constant velocity model, and a nonlinear measurement equation in the position,

$$x_{k+1} = \begin{bmatrix} I_2 & T I_2 \\ 0 & I_2 \end{bmatrix} x_k + \begin{bmatrix} 0_2 \\ T I_2 \end{bmatrix} (w_k + a_k) \qquad (26.47a)$$

$$y_k = h^{\mathrm{map}}(x_k) + e_k, \qquad (26.47b)$$

where T is the sampling period, w_k and e_k are the independent Gaussian process and the measurement noise, respectively. The first two components of the state are the position and the next two are the velocity in two dimensional Cartesian coordinates (representing north and east). The altitude is well decoupled from these two directions, and is best handled separately. The aircraft's body accelerations a_k, are measured and enter into the system as input. The inertial sensors in an aircraft are generally good enough to provide very accurate information about accelerations. The function h^{map} relates the terrain altitude to position in the form of a digital terrain database and is highly nonlinear. The measurements are obtained by deducting the height measurement of the ground looking (on board) radar from the known altitude of the aircraft (obtained using an altimeter). Furthermore, the level of the ground can be expected to be similar in many local regions of the map, hence the position estimate tends to be a multi-modal distribution, especially after leaving the relatively uninformative planar areas. This makes the PF a natural choice when implementing this type of application (as the alternatives such as extended KF struggles with dealing multimodal distributions). However, terrain navigation is often used in systems with limited computational resources and a straightforward PF implementation is often not feasible.

Fortunately, the model that describes this navigation problem, can easily be identified as an ML/NGSSM; the Rao-Blackwellization conditioned on the position, x^n, the remaining state-space model describing the velocity, x^l, becomes linear-Gaussian. Applying a RBPF to this problem using Algorithm 26.1, is then straightforward. The following model components are identified from (26.11) as

$$f^n(x_k^n) = x_k^n \qquad F^n(x_k^n) = T I_2 \qquad h(x_k^n) = h^{\mathrm{map}}(x_k) \qquad (26.48a)$$

$$f^l(x_k^n) = 0_{2\times 1} \qquad F^l(x_k^n) = I_2 \qquad H(x_k^n) = 0, \qquad (26.48b)$$

where $h^{\mathrm{map}}(x_k)$, depends only on the position component of the state, *i.e.*, , the nonlinear part of the model. Furthermore, $w_k^n \equiv 0$ (*i.e.*, , no process noise enters directly on the nonlinear part), $w_k^l = w_k$, and e_k is the same as in the original problem formulation.

A more accurate measurement noise model would also consider secondary reflections (*e.g.*, from the tree canopies), resulting in a bi-modal Gaussian mixture representation. This could provide further structure to the problem. This application has been studied extensively in [5, 31].

26.7   Approximate Rao-Blackwellized Nonlinear Filtering

The implementation of RBPF requires that a subset of the state vector is analytically tractable conditioned on the rest of the state vector. Then one can marginalize the tractable part to obtain the posterior distribution of the non-tractable part using PF. However, for many applications, this tractability requirement is very restrictive. For example, consider the noise adaptive filtering where the noise class is not from the exponential family. For this case, the unknown noise parameters, in general, do not admit Bayesian conjugate priors. Consequently one cannot implement an RBPF for this case. For such cases, in recent times, different approximations of the RBPF schemes have been envisaged in the literature (see *e.g.*, [1, 11, 27, 40, 46]). The basic idea again is to decompose the whole state space into two (interacting) parts[6] and split the filtering problem into two nested sub-problems.

To illustrate the idea, suppose that the state vector at time k is given by $(x_k, z_k)^T$. The state vector is assumed to follow a Markovian model. Now given the sequence of observations $\{y_0, y_1, \ldots, y_k\}$, we target to evaluate $p(x_{0:k}, z_{0:k}|y_{0:k})$ recursively over k. The target distribution can be decomposed as

$$p(x_{0:k}, z_{0:k}|y_{0:k}) = p(x_{0:k}|y_{0:k})\, p(z_{0:k}|x_{0:k}, y_{0:k}) \qquad (26.49)$$

We now propose to handle each sub-problem using PFs. First using a single PF, with N_x particles, we approximate $p(x_{0:k}|y_{0:k})$. Then we run N_x conditional PFs, each with N_z particles, to estimate $p(z_{0:k}|x_{0:k}, y_{0:k})$. This is in essence an approximated RBPF, where we use local Monte Carlo approximations instead of analytical solutions. This idea was proposed in [27, 46] and also in [40] in the context of noise adaptive PF. A related but slightly different formulation is considered in [1, 11]. It is important here to emphasize that the resampling steps for the second sub-problem is local in the state space and thus can be parallelized. This is a very generic and promising approach for

---

[6]In principle, the state space can be decomposed into more than two parts to implement an approximate RBPF.

high dimensional online inference problems. On one hand, it increases the parallelism in PF and, on the other hand, it provides flexibility in implementing specialized PFs with different tailored solutions for individual sub-problems.

26.8   Concluding Remarks

The online (real time) inference problems for general state-space models do not usually, admit closed form solutions. However, thanks to the modern Monte Carlo methods, the inference aim can be achieved to an arbitrary accuracy using *particle filtering* (PF). Although, PF is a very flexible class of methods for such inference problems, it is known to suffer from the curse of dimensionality. For some models, part of the state space admits conditionally tractable structures. The possibility to exploit such conditional tractability leaves only a part of the state vector to be targeted using a PF. Consequently, Monte Carlo based estimation is confined to a space of lower dimension. This in turn leads to an efficiency improvement of the (filtering) estimator due to the implications of the *Rao-Blackwell theorem*. The resulting filtering problem is popularly known as *Rao-Blackwellized particle filtering* (RBPF).

In this chapter, we presented a brief review of such RBPF. In particular, we have outlined a range of existing RBPF models that are popular in practice and also presented some recent developments in this context. This is written mainly from the practitioners' point of view and as such, we have not included any theoretical analysis here. Our coverage is primarily aimed at providing the model structures admitting such conditional tractabilities. Nonetheless, the coverage is by no means exhaustive and partially biased towards our interest. Many other such RBPF may possibly be constructed, *e.g.*, using the exact filtering sublayer as provided in [36].

Acknowledgments

The work was supported by the projects Cooperative Localization (CoopLoc) funded by Swedish Foundation for Strategic Research (SF), CADICS funded by Swedish Research Council (VR) and the Swedish strategic research center Security Link.

Bibliography

[1] M. O. Ahmed, P. T. Bibalan, N. de Freitas, and S. Fauv. Decentralized, adaptive, look-ahead particle filtering. *arXiv:1203.2394v1*, 2012.

[2] C. Andrieu and A. Doucet. Particle filtering for partially observed Gaussian state space models. *Journal of the Royal Statistical Society: Series B (Statistical Methodology)*, 64(4):827–836, 2002.

[3] C. Andrieu and S. J. Godsill. A particle filter for model based audio source separation. In *Proceedings of the 2000 International Workshop on Independent Component Analysis and Blind Signal Separation*, Helsinki, Finland, June 2000.

[4] A. Bagchi. *Optimal Control of Stochastic Systems*. Prentice Hall Inc., New Jersey, 1993.

[5] N. Bergman, L. Ljung, and F. Gustafsson. Terrain navigation using Bayesian statistics. *Control Systems, IEEE*, 19(3):33–40, June 1999.

[6] J. M. Bernardo and A. F. M. Smith. *Bayesian Theory*. John Wiley, Chichester, U.K., 2000.

[7] D. Blackwell. Conditional expectation and unbiased sequential estimation. *The Annals of Mathematical Statistics*, 18(1):105–110, Mar. 1947.

[8] O. Cappé, S. J. Godsill, and E. Moulines. An overview of existing methods and recent advances in sequential Monte Carlo. *Proceedings of the IEEE*, 95(5):899–924, 2007.

[9] C. M. Carvalho, M. S. Johannes, H. F. Lopes, and N. G. Polson. Particle learning and smoothing. *Statistical Science*, 25(1):88–106, 2010.

[10] R. Chen and J. S. Liu. Mixture Kalman filters. *Journal of the Royal Statistical Society: Series B (Statistical Methodology)*, 62(3):493–508, 2000.

[11] T. Chen, T. B. Schön, H. Ohlsson, and L. Ljung. Decentralized particle filter with arbitrary state decomposition. *IEEE Transactions on Signal Processing*, 59(2):465–478, Feb. 2011.

[12] N. Chopin, A. Iacobucci, J. M. Marin, K. L. Mengersen, C. P. Robert, R. Ryder, and C. Schafer. On particle learning. *arXiv: 1006.0554v2*, 2010.

[13] D. Crisan and A. Doucet. A survey of convergence results on particle filtering methods for practitioners. *IEEE Transactions on Signal Processing*, 50(3):736–746, 2002.

[14] F. Daum, R. Co, and J. Huang. Curse of dimensionality and particle filters. In *Proceedings of IEEE Aerospace Conference*, Big Sky, USA, 2003.

[15] P. Del Moral. *Feynman-Kac Formulae: Genealogical and Interacting Particle Systems with Applications.* Probability and Its Applications. Springer Verlag, New York, 2004.

[16] P. M. Djuric and J. Miguez. Sequential particle filtering in the presence of additive Gaussian noise with unknown parameters. In *Proceedings of IEEE Interntaional Conference on Accoustics, Speech and Signal Processing*, pages 1621–1624, Orlando, Florida, 2002.

[17] R. Douc, O. Cappé, and E. Moulines. Comparison of resampling schemes for particle filtering. In *Proceedings of the 4th International Symposium on Image and Signal Processing and Analysis*, volume I, pages 64–69. IEEE, University of Zagreb, Croatia, 2005.

[18] A. Doucet, S. Godsill, and C. Andrieu. On sequential Monte Carlo sampling methods for Bayesian filtering. *Statistics and Computing*, 10(3):197–208, 2000.

[19] A. Doucet and A. M. Johansen. A tutorial on particle filtering and smoothing: Fiteen years later. In D. Crisan and B. Rozovsky, editors, *The Oxford Handbook of Nonlinear Filtering*, pages 656–704. Oxford University Press, Oxford, 2011.

[20] N. J. Gordon, D. J. Salmond, and A. F. M. Smith. Novel approach to nonlinear/non Gaussian Bayesian state estimation. *IEE Proceedings F-Radar and Signal Processing*, 140(2):107–113, Apr. 1993.

[21] F. Gustafsson. Particle filter theory and practice with positioning applications. *IEEE Aerospace and Electronic Systems Magazine*, 25(7):53–81, Mar. 2010.

[22] F. Gustafsson. *Statistical Sensor Fusion.* Studentlitteratur, Malmo, Sweden, 2010.

[23] F. Gustafsson, F. Gunnarson, N. Bergman, U. Forssell, J. Jansson, R. Karlsson, and P.-J. Nordlund. Particle filters for positioning, navigation, and tracking. *IEEE Transactions on Signal Processing*, 50(2):425–437, Feb. 2002.

[24] A. Gut. *An Intermediate Course in Probability.* Springer Verlag, New York, 1995.

[25] G. Hendeby. *Performance and Implementation Aspects of Nonlinear Filtering.* Dissertations no 1161, Linköping Studies in Science and Technology, Mar. 2008.

[26] G. Hendeby, R. Karlsson, and F. Gustafsson. The Rao-Blackwellized particle filter: A filter bank implementation. *EURASIP Journal on Advances in Signal Processing*, 2010(724087), 2010.

[27] A. M. Johansen, N. Whiteley, and A. Doucet. Exact approximation of Rao-Blackwellised particle filters. In *Proceedings of 16th IFAC Symposium on Systems Identification (SYSID)*, Brussels, Belgium, 2012.

[28] E. L. Lehmann. *Theory of Point Estimation*. Probability and Mathematical Statistics. John Wiley & Sons, New York, 1983.

[29] M. Montemerlo, S. Thrun, D. Koller, and W. Wegbreit. FASTSLAM: A factored solution to the simultaneous localization and mapping problem. In *Proceedings of the AAAI National Conference on Artificial Intelligence*, Edmonton, Canada, 2002.

[30] K. P. Murphy. *Machine Learning: A Probabilistic Perspective*. MIT Press, Cambridge, Massachusetts, 2012.

[31] P.-J. Nordlund. *Sequential Monte Carlo Filters and Integrated Navigation*. Licentiate Thesis no 945, Department of Electrical Engineering, Linköpings universitet, Sweden, May 2002.

[32] E. Ozkan, V. Smidl, S. Saha, C. Lundquist, and F. Gustafsson. Marginalized adaptive particle filtering for non-linear models with unknown time-varying noise parameters. *Automatica*, 49(6):1566–1575, 2013.

[33] P. B. Quang, C. Musso, and F. LeGland. An insight into the issue of dimensionality in particle filtering. In *Proceedings of 13th International Conference on Information Fusion (FUSION)*, Edinburgh, Scotland, 2010.

[34] L. R. Rabiner. A tutorial on hidden Markov models and selected applications in speech recognition. In *Proceedings of the IEEE*, volume 77(2), pages 257–286, February 1989.

[35] C. R. Rao. Information and the accuracy attainable in the estimation of statistical parameters. *Bulletin of the Calcutta Mathematical Society*, 37:81–91, 1945.

[36] B. Ristic, S. Arulampalam, and N. Gordon. *Beyond the Kalman Filter: Particle Filters for Tracking Applications*. Artech House, USA, 2004.

[37] S. Saha and F. Gustafsson. Particle filtering with dependent noise processes. *IEEE Transactions on Signal Processing*, 60(9):4497–4508, 2012.

[38] S. Saha and G. Hendeby. Online inference in Markov modulated nonlinear dynamic systems: A Rao-Blackwellized particle filtering approach. *arXiv: 1311.6486v1*, 2013.

[39] S. Saha and G. Hendeby. Rao-Blackwellized particle filter for Markov modulated nonlinear dynamic systems. In *IEEE Workshop on Statistical Signal Processing*, Gold Coast, Australia, June 2014.

[40] S. Saha, G. Hendeby, and F. Gustafsson. Noise adaptive particle filtering: A hierarchical perspective. *Working Paper (part of the work was presented at ISBA Regional Meeting and International Workshop/Conference on Bayesian Theory and Applications (IWCBTA), Varanasi)*, 2013.

[41] S. Saha, U. Orguner, and F. Gustafsson. Non-linear filtering based on observations from Student's t processes. In *Proceedings of the IEEE Aerospace Conference*, Big Sky, USA, 2012.

[42] S. Saha, E. Ozkan, F. Gustafsson, and V. Smidl. Marginalized particle filters for Bayesian estimation of Gaussian noise parameters. In *Proceedings of 13th International Conference on Information Fusion (FUSION)*, Edinburgh, Scotland, 2010.

[43] T. Schön, F. Gustafsson, and P. J. Nordlund. Marginalized particle filter for mixed linear/nonlinear state space models. *IEEE Transactions on Signal Processing*, 53(7):2279–2289, 2005.

[44] C. Snyder, T. Bengtsson, P. Bickel, and J. Anderson. Obstacles to high-dimensional particle filtering. *Monthly Weather Review*, 136(12):4629–4640, 2008.

[45] G. Storvik. Particle filters for state space models with the presence of unknown static parameters. *IEEE Transactions on Signal Processing*, 50(2):281–289, 2002.

[46] C. Vergé, C. Dubarry, P. Del Moral, and E. Moulines. On parallel implementation of sequential Monte Carlo methods: The island particle model. *arXiv:1306.3911v1*, 2013.

[47] P. Vidoni. Exponential family state space models based on a conjugate latent process. *Journal of the Royal Statistical Society: Series B (Statistical Methodology)*, 61(1):213–221, 1999.

[48] M. West and P. J. Harrison. *Bayesian Forecasting and Dynamic Models*. Springer Verlag, New York, 2nd edition, 1997.

27

Some Aspects of Bayesian Inference in Skewed Mixed Logistic Regression Models

Cristiano C. Santos

Federal University of Minas Gerais

Rosangela H. Loschi

Federal University of Minas Gerais

CONTENTS

27.1   Introduction

Mixed models arise as a parsimonious and useful strategy to model correlated data or to accommodate outliers and overdispersion produced by non-observed or latent variables. Correlated data occur, for instance, in cases where the individuals are clustered or in longitudinal studies.

    Classical inference in mixed model is not easy since the likelihood involves multiple integrals and their solutions are usually non-analytical. The Bayesian paradigm is thus a natural alternative to do inference in mixed models because the random effects can be treated as parameters to be estimated. If the posterior distributions cannot be obtained analytically, computational methods, e.g., Markov chain Monte Carlo (MCMC), can be used to approximate them. This easier way of handling the inference problem under mixed models allows the use of more realistic distributions for the random effects.

Our focus is on a mixed model for binary data. For this situation, [17] assumed a mixed logistic regression model with normal distribution for the random effects to model binary responses subject to misclassification. [20] considered the logistic regression model with random intercept and, for better detection of outliers, the Student-t and finite mixtures of normal distributions were assumed to model the random effect behavior. An empirical Bayes approach for mixed logistic regression model was presented by [11]. Another extension was presented in [16] where the skew-normal, skew-t and non-parametric prior distributions are assumed for the random effects. In the skew-normal (see [3]) case, [16] elicited a point mass probability prior for the skewness parameter, where it is assumed *a priori* that all the random effects are necessarily positive, or negative. Normal and skew-normal random effects were also considered by [19] with a more flexible class of prior distributions for the skewness parameter. Another way of modeling binary data is to consider different link functions. [5] introduced a scale mixture of multivariate normal links to model correlated binary data. An extension was considered in [12] by proposing a generalized t-link model.

One of the biggest challenges in the mixed logistic regression model is the interpretation of the parameters since it does not have the parallel interpretational features of the standard logistic model. [14] and [13] proved that the odds ratio (OR) in mixed logistic regression models depends on both fixed and random effects. Thus, under the classical approach considered in [14], the OR shall be a random quantity in a number of possible comparisons. Under normality for the random effects, [14] also discussed different measures of heterogeneity. They suggested using median (MOR) and quantile intervals (IOR) of the odds ratio. The measures MOR and IOR have nice interpretations in terms of probability. In some cases, the MOR allows a simple interpretation in terms of the well-known OR that greatly facilitates communication between the data analyst and the subject-matter researcher.

[19] generalized the results in [14] by assuming skew-normal random effects and considering posterior measures to interpret the parameters. They concluded that the posterior mean and mode can be better estimators for the OR than the posterior median. Moreover, the point estimates of the fixed effects are not substantially influenced by the prior distribution for the random effects. However, such distributions play an important role in the estimation of the OR as well as in the credible intervals for the fixed effects. For further discussions about the influence of the misspecification of the random effect distributions in the related inferences under maximum likelihood theory, see [1] and [15].

Our goal here is to review some results by [14] and [19] and to bring some new insights to the skewed logistic regression model presented in [19]. Results in [19] reveal that when a non-degenerate flat prior distribution for the skewness parameter is assumed, its estimate tends to be highly biased. The posterior uncertainty about the estimate of the skewness parameter remains high when the asymmetry in the random effects is large. Moreover, the highest posterior density (HPD) intervals show that, in general, the skewness parameter is not significantly different from zero even when the random effects are distributed according to skew-normal distribution.

In order to evaluate the effect of the prior distribution for the skewness parameter on the posterior estimates of the parameters as well as on the OR, we will perform a Monte Carlo study where we shall consider both degenerated as well as skew-normal prior distributions for the skewness parameter.

This chapter is organized as follows. In Section 27.2, we present the mixed logistic regression model with random intercept and discuss the parameter interpretation under this model. Results related to the OR, obtained by [14] assuming normal random effects as well as the extension for the independent and identically distributed (i.i.d.) skew-normal random effects, which was obtained by [19], are summarized. Section 27.3 discusses some issues related to Bayesian inference in mixed logistic regression model. In Section 27.4, we present a Monte Carlo study that shows how the prior distribution for the skewness parameter influences the posterior inferences. Section 27.5 presents conclusions and final comments.

27.2 Logistic Regression Model with Random Intercept

Suppose that the population is divided into k clusters and a sample of size n_i is selected from the ith cluster. Let y_{ij} be the response variable (y_{ij} is 1 if a success occurs and 0 otherwise) for individual j in the cluster i, $i = 1, \ldots, k$ and $j = 1, \ldots, n_i$. Let $\mathbf{x}_{ij} = (1, x_{ij1}, x_{ij2}, ..., x_{ijp})^t$ be the $(p+1) \times 1$ vector of covariates for the individual j in the ith cluster. Assume that $\mathbf{X}$ is the $N \times (p+1)$ matrix with the information related to the covariates for all $N = \sum_{i=1}^{k} n_i$ observed individuals.

Define $\boldsymbol{\gamma} = (\gamma_1, ..., \gamma_k)^t \in \mathbb{R}^k$ as the vector of random effects, where γ_i denotes the random effect for cluster i. Let $\boldsymbol{\beta} = (\beta_0, \beta_1, \beta_2, ..., \beta_p)^t \in \mathbb{R}^{p+1}$ be the vector of fixed effects. We further assume that only intercept is random. Consequently, the linear predictor can be given as $\eta_{ij} = \mathbf{x}_{ij}^t \boldsymbol{\beta} + \gamma_i$ and $\pi_{ij} = P(y_{ij} = 1|\boldsymbol{\beta}, \boldsymbol{\gamma}, \mathbf{X}) = \exp\{\eta_{ij}\}[1 + \exp\{\eta_{ij}\}]^{-1}$, i.e., $y_{ij}|\boldsymbol{\beta}, \boldsymbol{\gamma}, \mathbf{X} \overset{ind}{\sim} Ber(\pi_{ij})$. As a consequence, the likelihood function is

$$f(\mathbf{y}|\boldsymbol{\beta}, \boldsymbol{\gamma}, \mathbf{X}) = \prod_{i=1}^{k} \prod_{j=1}^{n_i} \left[\frac{\exp\{\eta_{ij}\}}{1 + \exp\{\eta_{ij}\}}\right]^{y_{ij}} \left[\frac{1}{1 + \exp\{\eta_{ij}\}}\right]^{1-y_{ij}}, \quad (27.1)$$

where $\mathbf{y} = (y_{11}, ..., y_{1n_1}, ..., y_{k1}, ..., y_{kn_k})^t$.

To complete the model specification, we must specify the prior distributions for the fixed and random effects and for the hyperparameters that index such distributions. This can become an easy task if the parameters are interpretable.

27.2.1   Parameter interpretation and Larsen et al.'s results

One of the attractive features of the standard logistic regression model is that interpretations of fixed effects are based on the OR between the highest and the lowest risk individuals. This interpretation makes easy to communicate with researchers of other areas and, in turn, it helps the elicitation of appropriate prior distributions for the fixed effects.

In mixed linear models, the fixed effects inherit the interpretational features of the parameters of the standard linear model but in mixed logistic regression models it does not happen; see [14] and [13]. [14] also exhibited the dependency of OR on both fixed and random effects. Thus under the classical approach, it will be random.

Let j_1 and j_2 be two individuals in different clusters i_1 and i_2, respectively. The OR becomes

$$OR_{i_1 j_1, i_2 j_2} = \exp\left\{(\mathbf{x}_{i_1 j_1}^t - \mathbf{x}_{i_2 j_2}^t)\boldsymbol{\beta} + \gamma_{i_1} - \gamma_{i_2}\right\}. \qquad (27.2)$$

The classical approach comparison between two individuals depends on the random effects. Thus we have an additional difficulty in the parameter interpretation due to the fact that the random effects are random quantities, which can not be directly estimated using classical methods of inference. In fact, the OR becomes a random quantity and assumes different values even for comparing individuals with same covariates belonging to different clusters. Thus, the usual interpretation for the fixed effects can be used only for comparing individuals in the same cluster. Note that if the comparison is between individuals in the same cluster, say $i_1 = i_2 = i$, having different covariates, $OR_{ij_1, ij_2} = \exp\left\{(\mathbf{x}_{ij_1}^t - \mathbf{x}_{ij_2}^t)\boldsymbol{\beta}\right\}$ depends only on the fixed effects and it is exactly same as in the usual logistic regression model. To quantify the random effects, the comparison is done assuming that two individuals, j_1 and j_2, have the same covariate vectors and are in different clusters; that is, the individual j_k belongs to cluster i_k, $k = 1, 2$. In this case, the odds ratio $OR_{i_1 j_1, i_2 j_2} = \exp\left\{\gamma_{i_1} - \gamma_{i_2}\right\}$ depends on the random effects only. The OR in (27.2) permits the comparison between the risk of individuals in two different clusters as well.

[14] proposed to interpret the OR in terms of MOR. According to [14], the MOR always provides a valid interpretation of the fixed effects because its value does not depend on the clusters under comparison. Besides these, MOR also quantifies the heterogeneity among different clusters. For general case, whenever we compare individuals with different covariates in different clusters, the MOR is defined as

$$MOR_{i_1 j_1, i_2 j_2} = \text{med}\{\exp\left\{(\mathbf{x}_{i_1 j_1}^t - \mathbf{x}_{i_2 j_2}^t)\boldsymbol{\beta} + (\gamma_{i_1} - \gamma_{i_2}) \mid \boldsymbol{\beta}, \mathbf{X}\}\}.$$

It is noticeable that the distribution of $\gamma_{i_1} - \gamma_{i_2}$ is symmetric around zero under the assumption of i.i.d. random effects. Therefore, irrespective of the clusters being compared, we shall have the same value for the MOR when individuals have the same covariates. To properly quantify the effect of clusters in these cases, [14] considered the median of $\exp\{|\gamma_{i_1} - \gamma_{i_2}|\}$, which is given by

$$MOR_{|12|} = MOR_{|i_1j_1,i_2j_2|} = \text{med}\{\exp\{|\gamma_{i_1} - \gamma_{i_2}| \mid \boldsymbol{\beta}, \mathbf{X}\}\},$$

where $|A|$ denotes the absolute value of A. Notice that in the mixed logistic regression model, influence of the covariates in the response is affected by the cluster to which the individual belongs to. [14] proposed a measure of central tendency to this effect that will not depend on the specific value assumed by the γ_i.

Let us denote now onward, $OR_{12} = OR_{i_1j_1,i_2j_2}$, $MOR_{12} = MOR_{i_1j_1,i_2j_2}$. Denote by $\phi_n(\mathbf{y} \mid \boldsymbol{\mu}, \boldsymbol{\Sigma})$, the probability density function (p.d.f.) associated to the multivariate $N_n(\boldsymbol{\mu}, \boldsymbol{\Sigma})$ distribution and by $\Phi_n(\mathbf{y} \mid \boldsymbol{\mu}, \boldsymbol{\Sigma})$, the corresponding cumulative distribution function (c.d.f.). If $\boldsymbol{\mu} = \mathbf{0}$ these functions will be denoted by $\phi_n(\mathbf{y} \mid \boldsymbol{\Sigma})$ and $\Phi_n(\mathbf{y} \mid \boldsymbol{\Sigma})$ and if $\boldsymbol{\mu} = \mathbf{0}$ and $\boldsymbol{\Sigma} = \mathbf{I}_n$, these may be denoted by $\phi_n(\mathbf{y})$ and $\Phi_n(\mathbf{y})$, respectively. For simplicity, $\phi(\mathbf{y})$ and $\Phi(\mathbf{y})$ will be used in the univariate case.

Denoting $\kappa_{12} = (\mathbf{x}_{i_1j_1}^t - \mathbf{x}_{i_2j_2}^t)\boldsymbol{\beta}$ and $W_{12} = \gamma_{i_1} - \gamma_{i_2}$, [14] showed that if $\boldsymbol{\gamma} \sim N_k(\mathbf{0}, \boldsymbol{\Sigma})$, then $W_{12} \sim N(0, \sigma_{12}^2)$ where $\sigma_{12}^2 = Var(W_{12})$. Consequently, the OR in (27.2) can be rewritten as

$$OR = \exp\{\kappa_{12} + W_{12}\}. \tag{27.3}$$

Considering this structure, we summarize the results obtained by [14] in the following lemma.

Lemma 27.2.1. *If* $\boldsymbol{\gamma} \sim N_k(\mathbf{0}, \boldsymbol{\Sigma})$ *then, given* $\mathbf{x}$, $\boldsymbol{\beta}$ *and* $\boldsymbol{\Sigma}$, *it follows that*

(i) *the random variable* OR_{12} *has lognormal distribution with location parameter* κ_{12} *and scale parameter* σ_{12}^2 *with p.d.f. given by*

$$f_{OR|\boldsymbol{\beta},\boldsymbol{\Sigma},\mathbf{x}}(t) = [\sqrt{2\pi}t\sigma_{12}]^{-1} \exp\left\{-\frac{1}{2\sigma_{12}^2}(\ln t - \kappa_{12})^2\right\}, \; t \in \mathbb{R}^+; \tag{27.4}$$

(ii) *The median of the distribution in (27.4) is*

(a) $MOR_{|12|} = \text{med}\{\exp\{|W_{12}|\}|\boldsymbol{\beta}, \boldsymbol{\Sigma}, \mathbf{x}\} = \exp\{\text{med}\{|W_{12}||\boldsymbol{\beta}, \boldsymbol{\Sigma}, \mathbf{x}\}\}$
$= \exp\{\sigma_{12}\Phi^{-1}(0, 75)\},$

(b) $\text{med}\{\exp\{-|W_{12}|\}|\boldsymbol{\beta}, \boldsymbol{\Sigma}, \mathbf{x}\} = [\text{med}\{\exp\{|W_{12}||\boldsymbol{\beta}, \boldsymbol{\Sigma}, \mathbf{X}\}\}]^{-1},$

(c) $MOR_{12} = \text{med}\{\exp\{\kappa_{12} + W_{12}|\boldsymbol{\beta}, \boldsymbol{\Sigma}, \mathbf{x}\}\} = \exp\{\kappa_{12}\}.$

The result presented in (b) in Lemma 27.2.1 is useful whenever the main interest is to compare the individual with the lowest risk with that of the highest risk.

In order to exemplify the proposal of [14], let us compare two individuals in different clusters having the same characteristics (covariates). Suppose that the observed value for the $MOR_{|12|}$ is 3.06. Thus, with probability of 50%, the chance of the individual with the highest risk to experience the event of interest is at least 3.06 times the chance of the individual with the lowest risk. Since the difference in this case is due to the different clusters to which these individuals belong the result discloses a substantial heterogeneity among the clusters. Similar inference and more intuitive interpretation can be given to the variance of the random effects using $MOR_{|12|}$.

Next, we review some of the results developed by [19], which extends some of the results given in [14].

27.2.2 Parameter interpretation under skew-normality

[19] considered the parameter interpretation under mixed logistic regression model when a skew-normal distribution [3] is assumed to model the random effect behavior. The skew-normal (SN) family was chosen because of its flexibility in fitting densities with different shapes such as the normal distribution for perfect symmetry and the half-normal distribution for strong degree of asymmetry. Besides, biological, environmental and financial data are often skewed and heavy tailed. Such behaviors are usually not well fitted by the normal distribution.

Let us assume that, given $\boldsymbol{\theta} = (\xi, \sigma^2, \lambda)$, $\gamma_{i_l} \overset{iid}{\sim} SN(\xi, \sigma^2, \lambda)$, which is a common practice in mixed model to reduce the number of parameters to be estimated. To obtain the distribution of OR_{12}, given $\mathbf{x}$, $\boldsymbol{\beta}$ and $\boldsymbol{\theta}$, it is necessary to obtain the distribution of the linear combination $W_{12} = \gamma_{i_1} - \gamma_{i_2}$. Assuming this, [19] proved that, given $\boldsymbol{\theta}$, the distribution of W_{12} has p.d.f. given by

$$f_{W_{12}|\boldsymbol{\theta}}(w) = 4\phi(w \mid 2\sigma^2) \times \Phi_2\left(\frac{w}{2\sigma^2}\boldsymbol{\epsilon}|\mathbf{I}_2 - \frac{\delta^2\boldsymbol{\epsilon}\boldsymbol{\epsilon}^t}{2}\right), \qquad (27.5)$$

where $\boldsymbol{\epsilon} = (1, -1)^t$ and $\delta = \lambda(1 + \lambda^2)^{-0.5}$. The distribution in (27.5) belongs to the unified skew-normal (SUN) family of distributions introduced by [2], which is given by

$$W_{12} \sim SUN_{1,2}\left(0, \mathbf{0}, (2\sigma^2)^{1/2}, \begin{pmatrix} \mathbf{I}_2 & \frac{\delta\boldsymbol{\epsilon}}{\sqrt{2}} \\ \frac{\delta\boldsymbol{\epsilon}^t}{\sqrt{2}} & 1 \end{pmatrix}\right).$$

Since W_{12} is the difference of two i.i.d. random variables, the distribution in (27.5) is symmetric around zero. Using this result, [19] proved the following result:

Lemma 27.2.2. *If $\gamma_{i_l} \mid \boldsymbol{\theta} \overset{i.i.d.}{\sim} SN(\xi, \sigma^2, \lambda)$ then, given $\mathbf{x}$, $\boldsymbol{\beta}$ and the hyper-parameters $\boldsymbol{\theta} = (\xi, \sigma^2, \lambda)$, it follows that*

(i) the random variable OR_{12} has a distribution with p.d.f. given by

$$f_{OR_{12}|\boldsymbol{\beta},\boldsymbol{\theta},\mathbf{x}}(r) = \frac{4}{r}\phi(\ln r|\kappa_{12}, 2\sigma^2)$$

$$\times \Phi_2\left(\frac{\delta \ln r}{2\sigma}\boldsymbol{\epsilon}\Big|\frac{\delta\kappa_{12}}{2\sigma}\boldsymbol{\epsilon}, \mathbf{I}_2 - \frac{\delta^2}{2}\boldsymbol{\epsilon}\boldsymbol{\epsilon}^t\right), \qquad (27.6)$$

for $r \in \mathbb{R}_+$, where $\mathbf{1}_k$ is the $k \times 1$ vector of ones.

(ii) The median of the distribution in (27.6) is such that

(a) $MOR_{|12|} = med\{\exp\{|W_{12}|\}|\boldsymbol{\beta},\boldsymbol{\theta},\mathbf{x}\} = \exp\{med\{|W_{12}|\boldsymbol{\beta},\boldsymbol{\theta},\mathbf{x}\}\}$

$$= \exp\left\{\Phi_{SUN}^{-1}\left(0.75 \mid 0, \mathbf{0}, (2\sigma^2)^{1/2}, \begin{pmatrix} \mathbf{I}_2 & \frac{\delta\boldsymbol{\epsilon}}{\sqrt{2}} \\ \frac{\delta\boldsymbol{\epsilon}^t}{\sqrt{2}} & 1 \end{pmatrix}\right)\right\},$$

(b) $med\{\exp\{-|W_{12}|\}|\boldsymbol{\beta},\boldsymbol{\theta},\mathbf{x}\} = [med\{\exp\{|W_{12}|\boldsymbol{\beta},\boldsymbol{\theta},\mathbf{x}\}\}]^{-1},$

(c) $MOR_{12} = med\{\exp\{\kappa_{12} + W_{12}|\boldsymbol{\beta},\boldsymbol{\theta},\mathbf{x}\} = \exp\{\kappa_{12}\},$

where Φ_{SUN} denotes the c.d.f. of the SUN distribution given in (27.5).

Results (a) and (b) in Lemma 27.2.2 do not appear in [19]. We can also note that by assuming i.i.d. SN random effects, the prior interpretation of the fixed effects through the MOR will not depend on the clusters to which the individuals belong to. Similar result was obtained by [14] in a more general setting where the random effects can have different normal distributions. However, as shown in [19], under skew-normality, if the random effects are not identically distributed, MOR_{12} will also depend on the distribution of the random effects.

The distribution in (27.6) belongs to the log canonical fundamental skew normal family of distributions. Such a family is formally defined by [18] which also discussed some of its properties and applications. Figure 27.1 shows the p.d.f. in (27.6) for different values of λ and assuming $\sigma^2 = 1$ and $\kappa_{12} = 0$. We notice that by eliciting higher values for $|\lambda|$, OR tends to be higher than assumed in Larsen et al.'s model. The results related to parameters interpretation introduced by [14] and [19] are interesting from the classical point of view and also provide a nice prior interpretation for OR, given the matrix $\mathbf{x}$, the fixed effects $\boldsymbol{\beta}$ and the hyperparameters for the prior of $\boldsymbol{\gamma}$. From the Bayesian point of view, [19] proposed to use the posterior summaries of OR in order to interpret the fixed effects. Since OR is a random quantity depending on $\boldsymbol{\beta}$ and/or $\boldsymbol{\gamma}$, it does not have a closed form but is easily approximated by computational methods, such as the MCMC algorithms. Details can be found in [19].

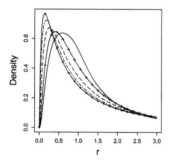

FIGURE 27.1
Prior distribution of OR for $\sigma^2 = 1$, $\kappa_{12} = 0$ and $\lambda = -8.5$ (solid line), $\lambda = -2$ ($+$) $\lambda = -1$ (dashed line), $\lambda = -0.5$ (dotdashed line), $\lambda = 0$ ($\bullet$, Larsen et al.'s model).

27.3 Bayesian Inference in Mixed Logistic Regression Models

Consider the likelihood in (27.1). To complete the model specification, we assume that $\boldsymbol{\beta} \sim N_{p+1}(\mathbf{m}, b^2 \mathbf{I}_{p+1})$ *a priori* and, given σ^2 and λ, the random effects are i.i.d. with univariate SN. That is , $\gamma_1, \ldots, \gamma_k \overset{iid}{\sim} SN(-\delta\sigma\sqrt{2/\pi}, \sigma^2, \lambda)$, where $\delta = \lambda[1 + \lambda^2]^{-1/2}$, $\mathbf{m} \in \mathbb{R}^{p+1}$ and $b^2 \in \mathbb{R}_+$. We center the SN distributions at zero to avoid nonidentifiability. We also assume a degenerate prior distribution for λ and consider $\sigma^2 \sim IG(a, d)$ where $IG(a, d)$ represents the inverted gamma distribution with $\mathrm{E}(\sigma^2) = d(a - 1)^{-1}$ and $\mathrm{V}(\sigma^2) = d[(a - 1)^2(a - 2)]^{-1}$.

From the above prior distributions and the likelihood in (27.1), we obtain the posterior distribution that does not have the closed form. Details can be found in [19]. Thus, we consider an MCMC scheme to sample from the posterior distribution for which the full conditional distributions are required (see [9], for a detailed explanation on the calculation of the full conditional distributions).

27.3.1 The posterior full conditional distributions

If the skewness parameter is fixed *a priori*, we need the full conditional distributions of the $\boldsymbol{\gamma}$, $\boldsymbol{\beta}$ and σ^2. For models where the skewness parameter is estimated, the full conditional distributions are given in [19] whenever, $\lambda \sim SN(h, \tau^2, \theta)$.

To obtain the posterior full conditional distributions of the random effects, we consider Henze's stochastic representation for the univariate SN distribution. If $T_i \sim SN(\lambda)$ then $T_i \overset{d}{=} \delta|U_i| + (1-\delta^2)^{1/2}V_i$ where U_i and V_i are i.i.d. standard normal random variables. For simplicity, let $\psi_i = |U_i|$. Since $\gamma_i \sim SN(-\delta\sigma\sqrt{2/\pi}, \sigma^2, \lambda)$, given σ^2 and λ, it follows that $\gamma_i \overset{d}{=} \delta\sigma\left(\psi_i - \sqrt{2/\pi}\right) + \sigma\left(1-\delta^2\right)^{\frac{1}{2}}V_i$.

Assuming such transformations, the posterior full conditional distributions are

$$f(\boldsymbol{\beta}|\boldsymbol{D},\boldsymbol{\gamma},\sigma^2,\boldsymbol{\psi},\lambda) \propto \left[\prod_{i=1}^{k}\prod_{j=1}^{n_i}\left[\frac{\exp\{\eta_{ij}\}}{1+\exp\{\eta_{ij}\}}\right]^{y_{ij}}\left[\frac{1}{1+\exp\{\eta_{ij}\}}\right]^{1-y_{ij}}\right]$$
$$\times \exp\left\{-\frac{(\boldsymbol{\beta}-\mathbf{m})^t(\boldsymbol{\beta}-\mathbf{m})}{2b^2}\right\},$$

$$f(\boldsymbol{\gamma}|\boldsymbol{D},\boldsymbol{\beta},\sigma^2,\boldsymbol{\psi},\lambda) \propto \left[\prod_{i=1}^{k}\prod_{j=1}^{n_i}\left[\frac{\exp\{\eta_{ij}\}}{1+\exp\{\eta_{ij}\}}\right]^{y_{ij}}\left[\frac{1}{1+\exp\{\eta_{ij}\}}\right]^{1-y_{ij}}\right]$$
$$\times \exp\left\{-\frac{\sum_{i=1}^{n}\left(\gamma_i - \sigma\delta(\psi_i - \sqrt{2/\pi})\right)^2}{2\sigma^2(1-\delta^2)}\right\},$$

$$f(\boldsymbol{\psi}|\boldsymbol{D},\boldsymbol{\beta},\boldsymbol{\gamma},\sigma^2,\lambda) \propto \prod_{i=1}^{k}\exp\left\{-\frac{\left(\psi_i - (\gamma_i + \sigma\delta\sqrt{2/\pi})\frac{\delta}{\sigma}\right)^2}{2(1-\delta^2)}\right\}\mathbf{1}_{(0,\infty)}(\psi_i),$$

$$f(\sigma^2|\boldsymbol{D},\boldsymbol{\beta},\boldsymbol{\gamma},\boldsymbol{\psi},\lambda) \propto \left(\frac{1}{\sigma^2}\right)^{\frac{k}{2}+a+1}\exp\left\{-\frac{\sum_{i=1}^{n}(\gamma_i - \sigma\delta\psi_i^*)^2}{2\sigma^2(1-\delta^2)} - \frac{d}{\sigma^2}\right\},$$

where $\psi_i^* = \psi_i - \sqrt{2/\pi}$ and $\mathbf{1}_A\{x\}$ is the indicator function assuming 1 if $x \in A$ and 0, otherwise. Notice that the posterior full conditional distribution for each component ψ_i of $\boldsymbol{\psi}$ is the normal distribution truncated below at zero, with mean $(\gamma_i + \sigma\delta\sqrt{2/\pi})\delta(\sigma)^{-1}$ and variance $1-\delta^2$. The posterior full conditional distributions for $\boldsymbol{\beta}$ and $\boldsymbol{\gamma}$ are log-concave and the adaptive rejection sampling (ARS) algorithm can be used to generate samples from these distributions. The inversion method and the adaptive rejection Metropolis sampling (ARMS) algorithm (see [10]) are used to sample from other posterior full conditional distributions.

27.3.2 Model selection

For model comparison, we consider the conditional predictive ordinate (CPO), two approaches for the deviance information criterion (DIC) and the Bayes factor (BF). The CPO statistics is a useful tool for model selection, which allows us to assess goodness of fit as well as to perform model comparisons. For the observation i, it is defined as the predictive distribution of y_i given other observations in the sample, that is, $CPO_i = f(y_i \mid \mathbf{y}_{(-i)})$. An approximation for CPO_i is given by

$$\widehat{CPO_i} = \left[\frac{1}{L} \sum_{l=1}^{L} \frac{1}{f\left(y_i | \theta^{(l)}\right)} \right]^{-1}, \tag{27.7}$$

in which $\theta^{(l)}$, $l = 1, \ldots, L$, is a sample from the posterior distribution of θ. For details, see [6–8].

The information provided by CPO_i about the model fit can be summarized using the statistic $B = n^{-1} \sum_{i=1}^{n} \log(CPO_i)$, where n is the sample size. High value of B indicates a good model. For a more detailed discussion about CPO, we refer to [6] and the references therein. Also, for simplicity, we refer the statistic B as CPO in the rest of the text.

The DIC usually fails in selecting models that includes latent variables. Some of the approaches to compute DIC in these cases can be found in [4]. We shall consider the following two approaches.

Let $D = -2\log(f(\mathbf{y}|\boldsymbol{\theta}))$ be the deviance and denote by $\bar{D}$ the posterior mean of D. Assume that $\hat{D} = -2\log(f(\mathbf{y}|\hat{\boldsymbol{\theta}}))$, in which $\hat{\boldsymbol{\theta}}$ is posterior estimate of $\boldsymbol{\theta}$. Let $pD = \bar{D} - \hat{D}$. Thus the DIC is given by

$$DIC = \bar{D} + pD = \hat{D} + 2pD = 2\bar{D} - \hat{D}. \tag{27.8}$$

The small value of DIC indicates a better model. We shall consider two posterior estimates of $\boldsymbol{\theta}$, namely the mean and the mode of the posterior marginal distributions, and the corresponding DIC's will be denoted by DIC1 and DIC2, respectively.

The BF measures the evidence in favor of a model M_1 when compared to another model M_2. High value of BF brings evidence in favor of M_1. The BF is the ratio between the predictive distributions obtained under the models being compared, that is, $BF = f_1(\mathbf{y})/f_2(\mathbf{y})$. The predictive distributions can be estimated as follows.

$$\widehat{f_i(\mathbf{y})} = \left[\frac{1}{L} \sum_{l=1}^{L} \frac{1}{f\left(\mathbf{y}|\theta^{(l)}\right)} \right]^{-1}, \tag{27.9}$$

where $\theta^{(l)}$ is defined above. The approximation in (27.9) may produce unstable estimates since $f\left(\mathbf{y}|\theta^{(l)}\right)$ can assume very small values. An alternative approach to obtain the prior predictive distribution can be found in [8]. In our case, $f(y \mid \boldsymbol{\theta})$ is the likelihood function given in (27.1) and $\boldsymbol{\theta}$ is the vector whose components are the fixed and random effects.

27.4   Monte Carlo Study

One of the conclusions drawn by [19] is that in the logistic regression model with skew-normal random effects, the skewness parameter is usually not well estimated. Since the prior distribution of random effects influences the fixed effect interpretation, our goal would be to consider different prior specifications for the skewness parameter and evaluate gain, if any, in fixing the skewness parameter (see also [16]).

We performed a Monte Carlo study where we considered 100 replications of the mixed logistic model given in (27.1) assuming the fixed effects $\beta = (\beta_0, \beta_1, \beta_2) = (-0.1, 1.2, 0.5)$. We considered 6 scenarios. The difference among the scenarios lies in the number of clusters that we considered ($k = 25$ for Scenarios 1a, 2a and 3a and $k = 100$ for Scenarios 1b, 2b and 3b) and in the distributions used to generate the random effects. In all the scenarios, the random effects γ_i are generated from skew-normal distributions with $E(\gamma_i) = 0$, $Var(\gamma_i) = 2$ with different degrees of skewness. For Scenarios 1a and 1b, we assume $\gamma_i \sim SN(0, 2, 0)$. The random effects in Scenarios 2a and 2b are generated from $SN(-1.777, 5.156, 5)$ and, in Scenarios 3a and 3b, we assumed that $\gamma_i \sim SN(-1.869, 5.493, 30)$. The covariates considered in the study are shown in Table 27.1. For data with $k = 100$ clusters, we considered the same covariates but the number of observations in each cluster was assumed to be approximately $n_i/4$, that is, we split each cluster in Table 27.1 into four.

We also fit seven different models. In all cases, we assumed informative prior distributions for the fixed effects ($\beta_i \sim N(0, 10)$, $i = 0, 1$ and 2) and, for σ^2, $\sigma^2 \sim IG(2.001, 1)$. Consequently, *a priori*, we have $E(\sigma^2) = 1$ and $V(\sigma^2) = 1,000$. We also assumed that $\gamma_i \overset{iid}{\sim} SN(-\delta\sigma\sqrt{2/\pi}, \sigma^2, \lambda)$. The models differ because we assumed different prior distributions for the skewness parameter. For models M_1 to M_5, degenerate prior distributions are assumed for λ as $\lambda = -30, -5, 0, 5, 30$, respectively. For models M_6 and M_7, we assumed non-degenerate prior distributions as $\lambda \sim N(0, 100)$ and $\lambda \sim SN(0, 257.821, 5)$, respectively. These prior distributions are shown in Figure 27.2.

For the MCMC, we generated chains of sizes 30,000 for models M_1 to M_5 and 45,000 for models M_6 and M_7. We discarded the first 5,000 samples for models M_1 to M_5 and the first 20,000 samples for models M_6 and M_7 as the burn-in period. We used a lag of 50 steps to avoid serial correlation and obtained samples of size 500 from the posterior distributions. The algorithm was implemented using the OxEdit software.

27.4.1   On the parameter estimates

In this section, we analyze the effect of the prior specification for λ on the posterior means for the fixed effects, the variance of γ_i and the skewness parameter λ.

TABLE 27.1

Covariates and number of observations in each cluster.

Cluster	n_i	X_1	X_2	Cluster	n_i	X_1	X_2
1	110	−0.429	−1.519	14	148	−1.194	0.261
2	108	−0.429	0.261	15	151	0.337	0.854
3	107	−1.194	0.261	16	148	−1.194	0.854
4	107	−0.429	0.854	17	110	1.868	0.261
5	103	−1.194	0.854	18	108	0.031	−1.519
6	98	0.031	−1.519	19	107	0.337	0.854
7	102	0.031	0.261	20	103	−0.429	0.261
8	106	0.031	0.854	21	98	−1.194	−1.519
9	203	1.868	−1.519	22	102	−0.429	0.854
10	175	1.868	0.854	23	106	0.031	0.261
11	180	−1.194	0.854	24	203	0.337	−1.519
12	156	0.337	−1.519	25	175	1.868	0.854
13	152	0.337	0.261				

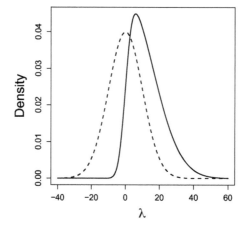

FIGURE 27.2

Prior distribution of λ under normal (dashed line) and skew-normal (solid line) distributions.

Figure 27.3 shows the mean square error (MSE) for the posterior means under all the models and for all the scenarios. As expected, the MSE tends to be smaller when the number of clusters is large and tends to be larger if the value for λ in the model is far from the true one. The model M_3 is usually the best model if the random effects are generated from a normal distribution and

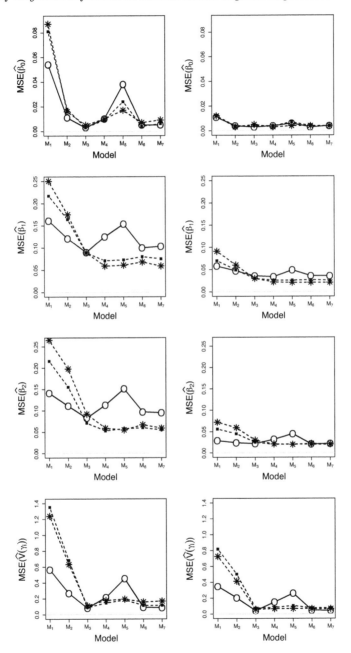

FIGURE 27.3
MSEs for the posterior means for the fixed effects and the variance of the random effects for Scenarios 1a (circle), 2a (square) and 3a (*) in the left panel and for Scenarios 1b (circle), 2b (square) and 3b (*) in the right panel.

it also works reasonably well in the other two cases. The models M_4 to M_7 are competitive models if the random effects arise from skew normal distributions with $\lambda = 5$ and 30. In these two cases, the MSEs are very close and tend to be small for all the parameters.

In conclusion, if the random effects are normally distributed, we usually obtain poor estimates due to misspecification of the prior value of λ. However, if the random effects are generated from skew-normal distributions, the model that assumes normal random effects usually works reasonably well. We also conclude that although the skewness parameter is not very well estimated, to assume a non-degenerate prior for this parameter can be a good strategy. We notice that the MSE for the fixed effects and the variance of the random effects are in general small under models M_6 and M_7.

Such conclusions are corroborated by Figure 27.4, which presents the box-plots of the posterior means of all the parameters, under all models for the Scenarios 1a, 2a and 3a. Figure 27.4 shows that the posterior estimates for β_0 are influenced by the prior specifications of the skewness parameter. In these three Scenarios, relatively better estimates are obtained for β_0 under the models M_3, M_6 and M_7.

It is also noteworthy that if the random effects are normally distributed, models M_1, M_2 and M_5 tend to overestimate the variance of the random effects, which is also observed for skew-normal random effects under models M_1 and M_2. We also notice that if the distribution of the random effects experiences strong asymmetry (Scenario 3a), the parameter λ is underestimated under the models M_6 and M_7 and, λ is overestimated in the other two Scenarios under the model M_7. Similar behavior is observed for the Scenarios with 100 clusters.

27.4.2   On the random effects estimates

As shown in previous sections, the random effects play an important role in the parameter interpretation. Here, we evaluate the effect of the prior specification for λ in the random effect estimates. We summarize the information about the quality of such estimates through the sum of the mean square error $MSE_\gamma = SSE_\gamma/k = (\gamma - \hat{\gamma})^t(\gamma - \hat{\gamma})/k$, where $\hat{\gamma}$ is the posterior mean of vector γ.

Table 27.2 shows the percentage of times in which each model provides the smallest MSE_γ. In all the scenarios, the random effects tend to be better estimated under the correct model. In general, models in which λ is fixed close to its true value also present a high percentage of times where the MSE_γ is the smallest. It is important to note that the random effects are well estimated frequently under the models M_6 and M_7, where λ is also estimated, particularly when the number of clusters is high.

Figure 27.5 shows the frequency of number of times each model provides the smallest MSE, that is, the best model. It also presents the frequency of the second best model in each case. We can notice for Scenario 1a that M_3 presents the smallest MSE in 31% of the times, and M_6 is the second best

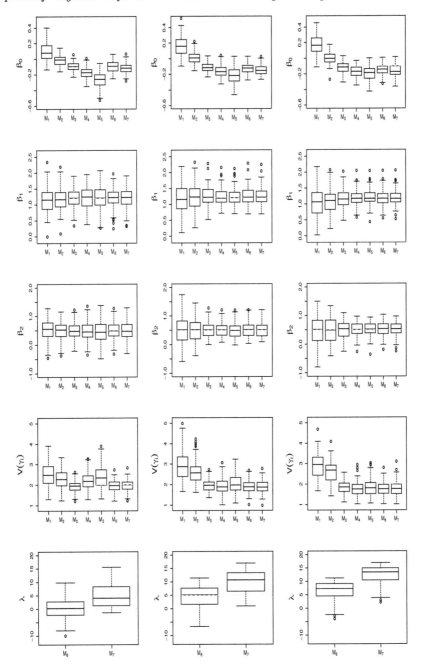

FIGURE 27.4
Boxplots of the posterior means for Scenario 1a (left), Scenario 2a (middle) and Scenario 3a (right).

TABLE 27.2

Percentage of times each model provides the smallest MSE_γ.

Scenario	Model						
	M_1	M_2	M_3	M_4	M_5	M_6	M_7
1a	1	22	**31**	13	5	13	15
1b	0	7	**37**	12	3	25	16
2a	1	6	21	**24**	**24**	11	13
2b	0	0	15	**33**	21	9	22
3a	0	3	16	20	**30**	16	15
3b	0	1	1	17	**39**	19	23

model in 48.4% of times. For Scenarios 1a and 1b, we notice that every time model M_3 is selected as the best model, the second best model is, in general, one of the models that does not fix λ *a priori*. Moreover, in Scenario 1b, when M_6 and M_7 present the smallest MSE, M_3 is usually the second best model.

Similar conclusions can be drawn when we compare the models M_4, M_6 and M_7 in Scenarios 2a and 2b and models M_5, M_6 and M_7 in Scenarios 3a and 3b. If, *a priori*, we correctly fix the value of λ, and the number of clusters is large, then the MSE for the estimates of the random effects tends to be small. Figure 27.6 presents the box-plots for the MSE_γ for all the cases under consideration. The behavior of the MSE is similar if we compare Scenarios with 25 and 100 clusters. However, the MSE is small if the number of clusters is large. We may also notice that the MSE under models M_6 and M_7 behave similar to those obtained under the model where λ is correctly specified.

In brief, for estimating either the random effects or any quantities depending on these, it is better to assume prior distributions for λ if we cannot precisely prespecify its value.

27.4.3 On the model selection

Table 27.3 shows the percentage of times that each model is selected as the best model using the CPO, DIC1, DIC2 and the Bayes factor. The importance of correctly specifying the model is mainly associated with quality of the random effects estimates and, consequently, with all the quantities depending on them such as OR.

Comparing the models M_1 to M_5 (that assume a fixed value for λ) we notice that, for Scenarios 1a and 1b, the CPO usually selects the appropriate model. However, for the other scenarios, it tends to select models that consider smaller asymmetry than the true one.

The DIC1 selects the correct model more frequently in all the Scenarios, except in the Scenario 2b. It also tends to select models in the Scenarios that consider skew-normal random effects. DIC2 does not perform well and,

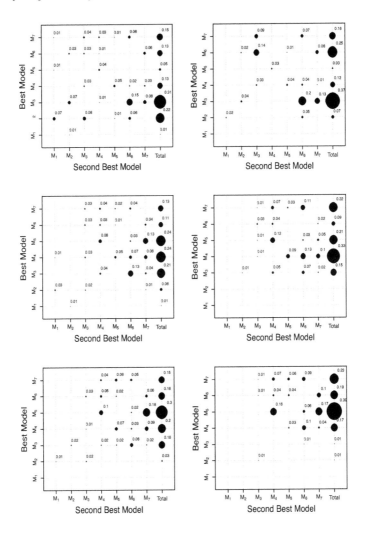

FIGURE 27.5
Percentage of times each model provides the smallest and the second smallest MSE_γ for Scenarios 1a, 2a and 3a, in the order from top, in the left panel; and for Scenarios 1b, 2b and 3b, in the order from top, in the right panel.

in many cases, it selects models in which skewness parameter is much higher than the true one.

The Bayes factor selects more frequently the appropriate model in Scenario 1b only. In Scenarios 2a, 2b, 3a and 3b, it tends to select models with small asymmetry.

580 *Current Trends in Bayesian Methodology with Applications*

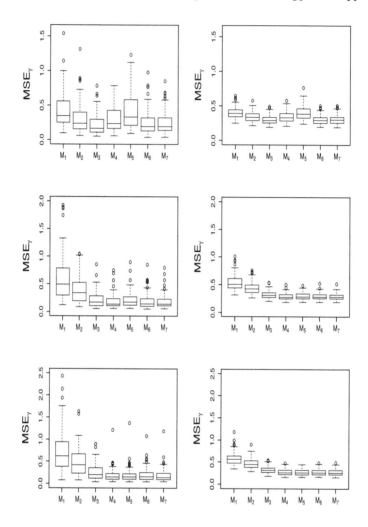

FIGURE 27.6
Boxplots of MSE_γ for Scenarios 1a, 2a and 3a, in the order from top, in the left panel; and for Scenarios 1b, 2b and 3b, in the order from top, in the right panel.

In all the cases, the models that assume non-degenerate prior distributions for λ are selected with high percentage of times. Thus, if λ could not be defined precisely in the model, then a good strategy is to estimate it. Notice that the prior distributions assumed here put small mass in values around $\lambda = 30$, which is the true value of the skewness parameter in Scenarios 3a and 3b. In

TABLE 27.3

Percentage of each model selection.

Scenario	M_1	M_2	M_3	M_4	M_5	M_6	M_7
				Model			
				CPO			
1a	7	10	**26**	15	13	17	12
1b	2	7	**30**	6	3	**30**	22
2a	8	16	**18**	15	11	17	15
2b	0	1	**34**	20	10	15	20
3a	8	2	16	**26**	14	15	19
3b	0	0	15	24	14	**27**	20
				DIC1			
1a	15	14	**23**	12	12	10	14
1b	5	10	**38**	8	8	17	14
2a	9	8	15	15	**26**	11	16
2b	1	2	10	**36**	31	13	7
3a	3	2	8	27	**34**	9	17
3b	0	0	4	13	**47**	15	21
				DIC2			
1a	9	17	13	8	**23**	16	14
1b	0	8	27	3	1	**32**	29
2a	**27**	19	11	8	12	9	14
2b	0	2	27	**31**	3	24	13
3a	**32**	14	6	7	14	12	15
3b	0	1	13	**29**	9	25	23
				Bayes Factor			
1a	16	17	12	13	8	**22**	12
1b	1	10	**37**	5	4	21	22
2a	9	6	18	11	16	19	**21**
2b	0	4	**28**	26	4	17	21
3a	3	4	13	**28**	10	23	19
3b	0	0	**22**	21	18	**22**	17

spite of this, the percentage of times the models M_5 and M_6 are selected as the best models in such Scenario is relatively high.

27.5 Final Comments

In this chapter we have reviewed the mixed logistic regression model and some aspects of inference and parameter interpretation under this model when normal and skew-normal classes of distributions are assumed to model the random effects behavior. We brought some new insights to the discussion by assuming different prior specifications for the skewness parameter. The main goal was to evaluate the influence of such prior specifications in the posterior estimates of the random and fixed effects, which has influence in the interpretation of the parameters.

We performed a Monte Carlo study and concluded that degenerate prior distributions for the skewness parameter can lead to poor estimates if they are wrongly established. If we do not have a precise information about the skewness parameter, a good strategy is to estimate it.

Acknowledgment We would like to express our gratitude to the editors, specially to Professors S.K. Upadhyay and D.K. Dey, for the invitation. We also like to thank the referees for their carefull revision of our chapter. C. C. Santos and R. H. Loschi acknowledge, respectively, CAPES (*Coordenação de Aperfeiçoamento de Pessoal de Nível Superior*) and CNPq (*Conselho Nacional de Desenvolvimento Científico e Tecnológico*) of the Ministry for Science and Technology of Brazil (CNPq-grants 301393/2013-3, 473163/2010-1, 306085/2009-7) for a partial support of their researches.

Bibliography

[1] A. Alonso, S. Litière, and G. Molemberghs. A family of tests to detect misspecifications in the random-effects structure of generalized linear mixed models. *Computational Statistics & Data Analysis*, 52:4474–4486, 2008.

[2] R. B. Arellano-Valle and A. Azzalini. On the unification of families of skew-normal distributions. *Scandinavian Journal of Statistics*, 33:561–574, 2006.

[3] A. Azzalini. A class of distributions which includes the normal ones. *Scandinavian Journal of Statistics*, 12:171–178, 1985.

[4] G. Celeux, F. Forbes, C. P. Robert, and D. M. Titterington. Deviance information criteria for missing data models. *Bayesian Analysis*, 1(4):651–673, 2006.

[5] M. H. Chen and D. K. Dey. Bayesian modeling of correlated binary responses via scale mixture of multivariate normal link function. *Sankhyā, The Indian Journal of Statistics: Series A*, 60(3):322–343, 1998.

[6] D. K. Dey, M. H. Chen, and H. Chang. Bayesian approach for nonlinear random effects models. *Biometrics*, 53:1239–1252, 1997.

[7] A. E. Gelfand. Model determination using sampling-based methods. In W. R. Gilks, S. Richardson, and D. J. Spiegelhalter, editors, *Markov Chain Monte Carlo in Pratice*, pages 145–161. Chapman & Hall, USA, 1995.

[8] A. E. Gelfand and D. K. Dey. Bayesian model choice: Asymptotics and exact calculations. *Journal of the Royal Statistical Society: Series B (Statistical Methodology)*, 56:501–514, 1994.

[9] W. R. Gilks. Full conditional distributions. In W. R. Gilks, S. Richardson, and D. J. Spiegelhalter, editors, *Markov Chain Monte Carlo in Pratice*, pages 75–88. Chapman & Hall, USA, 1995.

[10] W. R. Gilks, N. G. Best, and K. K. C. Tan. Adaptive rejection Metropolis sampling within Gibbs sampling. *Journal of the Royal Statistical Society: Series C (Applied Statistics)*, 44:455–472, 1995.

[11] T. R. Have and A. R. I. Localio. Empirical Bayes estimation of random effects parameters in mixed effects logistic regression models. *Biometrics*, 55(4):1022–1029, 1999.

[12] S. Kim, M. H. Chen, and D. K. Dey. Flexible generalized *t*-link models for binary response data. *Biometrika*, 95(1):93–106, 2008.

[13] K. Larsen and J. Merlo. Appropriate assessment of neighborhood effects on individual health: Integrating random and fixed effects in multilevel logistic regression. *American Journal of Epidemiology*, 161(1):81–88, 2005.

[14] K. Larsen, J. H. Petersen, E. Budtz-JØrgensen, and L. Endahl. Interpreting parameters in the logistic regression model with random effects. *Biometrics*, 56(3):909–914, 2000.

[15] S. Litière, A. Alonso, and G. Molenberghs. The impact of a misspecified random-effects distribution on the estimation and the performance of inferential procedures in generalized linear mixed models. *Statistics in Medicine*, 27(16):3125–3144, 2008.

[16] J. Liu and D. K. Dey. Skew random effects in multilevel binomial models: An alternative to nonparametric approach. *Statistical Modelling*, 8(3):221–241, 2008.

[17] C. D. Paulino, G. Silva, and J. A. Achcar. Bayesian analysis of correlated misclassified binary data. *Computational Statistics & Data Analysis*, 49(4):1120–1131, 2005.

[18] M. M. Queiroz, R. H. Loschi, and R. W. C. Silva. Multivariate log-skewed distributions with normal kernel and its applications. *Unpublished manuscript*, 2014.

[19] C. C. Santos, R. H. Loschi, and R. B. Arellano-Valle. Parameter interpretation in skewed logistic regression with random intercept. *Bayesian Analysis*, 8(2):381–410, 2013.

[20] A. D. P. Souza and H. S. Migon. Bayesian outlier analysis in binary regression. *Journal of Applied Statistics*, 37(8):1355–1368, 2010.

28

A Bayesian Analysis of the Solar Cycle Using Multiple Proxy Variables

David C. Stenning

University of California

David A. van Dyk

Imperial College

Yaming Yu

University of California

Vinay Kashyap

Smithsonian Astrophysical Observatory

CONTENTS

28.1   Introduction

Highly energetic solar eruptions involving bursts of radiation and discharges of plasma can eject charged particles into space and damage technological infrastructure (*e.g.* radio communications, electric power transmission, and the performance of low-Earth orbit satellites). Such "space weather" events are common during periods of high *solar activity*—a loose term that is defined only by observable proxy variables. Variations in the level of solar activity follow a roughly 11-year cyclic pattern, which is known as the *solar cycle*. Since energetic space weather events are more common near the *solar maximum*—the peak in solar activity during the 11-year cycle—there is considerable interest in predicting the timing and amplitude of future solar maxima, which has practical value in the planning of space missions. Nevertheless, predicting solar maxima remains a difficult task, with different methods yielding substantially different predictions (see [24] for an analysis of the various predictions made for the current solar cycle).

The solar cycle was first discovered by observing 11-year cyclic patterns in the average number of *sunspots* visible on the solar disk as viewed from Earth [31]. Sunspots are dark patches on the face of Sun (when viewed in optical light) that occur when intense magnetic fields inhibit convection, temporarily producing areas of reduced surface temperature. Sunspots are therefore linked to the overall magnetic activity of the Sun, and have long been a valuable proxy for solar activity. This value is partly derived from the fact that sunspot numbers (SSNs) comprise the longest uninterrupted set of observations in astronomy, with records starting in the early seventeenth century and available as monthly estimates since 1749. Correlations with solar activity have been established in other proxies such as the 10.7cm flux (i.e. the solar radio flux per unit frequency with a wavelength of 10.7cm), solar flare numbers, sunspot areas, *etc.* [18]. Solar activity can be reconstructed using radiocarbon measurements that are dated using tree-ring data [3, 27]. Still, the SSNs are the baseline for establishing properties of the solar cycle, and predictions for future cycle maxima are generally based on the SSNs.

Although the oldest SSNs were collected 265 years ago, the data are surprisingly reliable. The sunspot number is $R = k(10g + s)$, where s is the number of individual spots, g is the number of sunspot groups, and k is a factor that corrects for systematic differences between instruments and observatories. Despite advances in technology and the advent of higher resolution images of the Sun, the historical values of R are not expected to have significantly higher uncertainty for two reasons: (1) sunspot size visibility, which affects our ability to see the faintest individual spots, is limited by atmospheric conditions and that limit was reached a long time ago, and (2) sunspot groups are always counted as ten individual spots, regardless of the actual number of spots in the group.

Using this data, Waldmeier [29] determined that, within a cycle, the time for SSNs to rise to maximum is less than the time to fall to minimum. Other relations, such as the amplitude-period effect—the correlation between the duration of a cycle and the amplitude of the following cycle—were established using SSNs, and can be utilized to predict characteristics of future cycles [e.g., 16, 17, 30]. Yu *et al.* [32] analyze the SSNs to empirically derive statistically meaningful correlations between several parameters that they use to describe the solar cycle as part of building a Bayesian multilevel model that accounts for uncertainties in both the average monthly sunspot numbers, and in predicting the characteristics of future cycles. Such correlations help constrain physical models of the *solar dynamo*—the physical mechanism that generates the Sun's magnetic field—that attempt to explain the general solar cycle [25].

The statistical model of Yu *et al.* [32] uses a multilevel structure to capture complex patterns in the solar cycle. The first level of the model parameterizes the solar cycle using cycle-specific parameters, and describes the distribution of the observed monthly average SSNs around the parameterized cycle. The second level of the model incorporates relationships between the parameters of consecutive cycles, resulting in a hidden Markov model that generates characteristics of cycle $i + 1$ given the characteristics of cycle i. While this model was initially fit using observed SSNs, the model can in principle be fit using other proxies that follow the same underlying solar cycle.

A plot of the estimates of monthly average SSNs going back to the mid-nineteenth century is presented in the top panel of Figure 28.1. The cyclic pattern of SSNs that led to the discovery of the solar cycle is clearly visible. The available data extends back to January, 1749, and is maintained and made available by the Solar Influences Data Analysis Center in Belgium (http://sidc.oma.be). The bottom two panels of Figure 28.1 present the available data for two additional proxies of solar activity: (1) monthly average total sunspot areas, extending back to May, 1874, also available from the Solar Influences Data Analysis Center and (2) the monthly average 10.7cm flux, which can be obtained from the National Oceanic and Atmospheric Administration's National Geophysical Data Center (http://www.ngdc.noaa.gov/stp/solar/flux.html). Although the SSNs, sunspot areas, and the 10.7cm flux have differing temporal coverage, from the period of overlap it is clear that all three of these proxies follow a similar underlying pattern, which is the solar cycle. There are differences, however, in the cycle properties implied by the proxies. For example, the sunspot areas appear to have shorter cycle lengths and less pronounced peaks, especially in the first few observed cycles. The proxies are nonetheless highly correlated and believed to be associated with a common underlying solar cycle. Our goal is to combine information from the multiple proxies into a single omnibus estimate of the underlying cycle.

There are physical explanations as to why these proxies are correlated; the total areas of sunspots will obviously depend on the number of sunspots, and there is evidence that the 10.7cm flux values are influenced by the magnetic fields associated with sunspots [13]. Because such proxies have varying temporal coverages and may have varying cadences, it is challenging to combine them to model the underlying solar cycle in a sophisticated statistical analysis. Ideally, we would like to take advantage of the depth of high quality data that has become available in recent years, while incorporating the long record of SSNs.

Given the interest in predicting future solar maxima, diverse methodologies have been utilized including those based on i) models of the solar dynamo [e.g.,

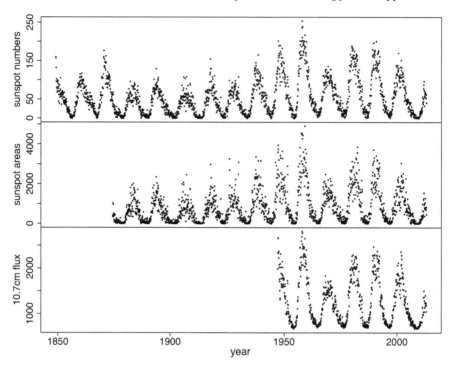

FIGURE 28.1
The observed proxies. *Top row*: monthly average sunspot numbers. *Middle row*: monthly average sunspot areas. *Bottom row*: monthly average 10.7cm flux. The roughly 11-year cycle of sunspot numbers follows the overall solar cycle, and we observe similar patterns in the sunspot areas and the 10.7cm flux.

4–7, 9], ii) measurements of geomagnetic activity that act as precursors [e.g., 15], and iii) statistical analysis of historical and current data [e.g., 2, 11, 16, 32], among others. An overview of various predictions made for the current solar cycle, cycle 24, is given in [24]. Within the solar physics community there has been debate over the amplitude of cycle 24, with different physical models producing a large range of predictions [e.g., 7, 8]. For the previous cycle, cycle 23, Kane [19] notes that among twenty predictions of the smoothed maximum sunspot number made by different researchers, only eight were within an acceptable range of the later observed value. The Bayesian method of Yu *et al.* [32] showed that there is considerable uncertainty in predicting an upcoming solar maximum using data up to the current solar minimum.

    This chapter is divided into five sections. In Section 28.2, we review the Bayesian multilevel model for the solar cycle discussed in Yu *et al.* [32]. In Section 28.3, we develop a systematic strategy for combining multiple proxies, and in particular describe how the pattern of missing data can be exploited to obtain coherent statistical inference. We present the results of fitting the solar cycle model to data that combines multiple proxies in Section 28.4. Finally, in Section 28.5 we summarize our results and discuss directions for future research.

28.2 Modeling the Solar Cycle with Sunspot Numbers

Yu *et al.* [32] propose a Bayesian multilevel model for fitting the solar cycle using monthly average SSNs as a proxy for the solar activity level. The solar cycle is parameterized with a set of cycle-specific parameters that together describe the total length, rising time, and amplitude for a given cycle. In the first level of the multilevel model, the observed SSNs are related to the parameterized solar cycle. The second level of the model incorporates a Markov structure that links parameters of consecutive cycles and encapsulates known features of the sunspot cycle in a series of sequential relations. These two stages are combined into a coherent statistical model, which is fit using Markov chain Monte Carlo methods. This structure allows for straightforward prediction of the characteristics of current cycles, even with data only extending to the beginning of the cycle. This is an important facet of the predictive capability of the Bayesian multilevel model.

28.2.1 Level one: Modeling the cycles

Figure 28.2 illustrates our parameterized model for a single solar cycle. In the figure, for cycle i, suppose $t_0^{(i)}$ is the starting time, $t_{\max}^{(i)}$ is the time of the cycle maximum, $t_1^{(i)}$ is the end time, $c^{(i)}$ is the amplitude, and $U^{[t]}$ is the "average solar activity level" at time t. Here t is recorded in units of months; although the exact number varies, there are roughly $11 \times 12 = 132$ months per cycle. Under this model, the rising phase of the cycle is described as

$$U^{[t]} = c^{(i)} \left(1 - \left(\frac{t_{\max}^{(i)} - t}{t_{\max}^{(i)} - t_0^{(i)}} \right)^{\alpha_1} \right) \quad \text{for } t < t_{\max}^{(i)}, \qquad (28.1)$$

and the declining phase as

$$U^{[t]} = c^{(i)} \left(1 - \left(\frac{t - t_{\max}^{(i)}}{t_1^{(i)} - t_{\max}^{(i)}} \right)^{\alpha_2} \right) \quad \text{for } t > t_{\max}^{(i)}, \qquad (28.2)$$

where $\alpha_1, \alpha_2 > 1$ are shape parameters assumed to be constant for all cycles. Together, (28.1) and (28.2) parameterize the solar cycle, and the curve described by (28.1) and (28.2) is the curve in Figure 28.2. An important feature

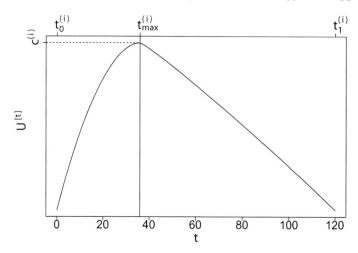

FIGURE 28.2
Parameterized form of a solar cycle. We illustrate $U^{[t]}$ with $c^{(i)} = 10, t_0^{(i)} = 0, t_{max}^{(i)} = 36, t_1^{(i)} = 120,$ $\alpha_1 = 1.9$ and $\alpha_2 = 1.1$, where $U^{[t]}$ is specified by (28.1) and (28.2).

of this parameterization is that the starting point of the the next cycle, $t_0^{(i+1)}$, is not necessarily identical to the end point of the current cycle, $t_1^{(i)}$. When two cycles overlap (i.e. when $t_1^{(i)} > t_0^{(i+1)}$), the activity level, $U^{[t]}$, is given by the sum of the contributions of the form (28.2) and (28.1) from the respective cycles; when $t_1^{(i)} < t_0^{(i+1)}$, $U^{[t]} = 0$ for $t_1^{(i)} < t < t_0^{(i+1)}$.

The observed SSNs span a total of 25 cycles that are designated cycle 0 through cycle 24. The parameters that are specific to cycle i are the set $\theta^{(i)} = (t_0^{(i)}, t_{max}^{(i)}, t_1^{(i)}, c^{(i)})$. The collection of cycle-specific parameters is contained in the set $\Theta = (\theta^{(0)}, \ldots, \theta^{(24)})$. Then, the full set of parameters that characterize the solar cycle is given by (Θ, α), where $\alpha = (\alpha_1, \alpha_2)$ does not vary from cycle to cycle and is therefore not included in Θ. With this, the distribution of the observed SSNs given Θ and the cycle-invariant parameters is

$$\sqrt{y}^{[t]} | (\Theta, \alpha, \beta, \sigma^2) \overset{\text{ind}}{\sim} N(\beta + U^{[t]}, \sigma^2), \tag{28.3}$$

where $y^{[t]}$ is the monthly average SSN at time t, and the parameter β may be regarded as a baseline. The SSNs are modeled after a square-root transformation in order to stabilize the variance. The independence assumption in (28.3) is valid since sunspots disappear or rotate over the edge of the solar disk over timescales shorter than the observed monthly average SSNs. The independence assumption is not valid when analyzing daily fluctuations since the same sunspot or group of sunspots is counted every day until it vanishes.

The decision to use monthly averages was partly motivated to avoid complex modeling of daily correlations.

28.2.2 Level two: Relationships between consecutive cycles

The evolution of the solar cycle is modeled via a Markov structure on the cycle-specific parameters $\{\theta^{(i)}, i = 0, \ldots, 24\}$. In particular, we model

$$p(\theta^{(0)}, \ldots, \theta^{(24)} | \eta) = p(\theta^{(0)} | \eta) \prod_{i=1}^{24} p(\theta^{(i)} | \theta^{(i-1)}, \eta), \qquad (28.4)$$

where η is a set of hyper-parameters that we describe below. The distribution $p(\theta^{(i)} | \theta^{(i-1)}, \eta)$ is further factored in that we model each of the components in cycle i in their temporal order within the cycle. That is, we first model the cycle's start time, $t_0^{(i)}$, given the parameters of the previous cycle, $\theta^{(i-1)}$, then model its amplitude, $c^{(i)}$, given $t_0^{(i)}$ and $\theta^{(i-1)}$, then model the time at which it reaches maximum, $t_{max}^{(i)}$, given $t_0^{(i)}$, $c^{(i)}$ and $\theta^{(i-1)}$, and finally model its end time, $t_1^{(i)}$, given $t_0^{(i)}$, $c^{(i)}$, $t_{max}^{(i)}$ and $\theta^{(i-1)}$.

Beginning with the start time of cycle i, Yu et al. [32] allowed the start time, $t_0^{(i)}$, to be different from, but dependent on, the end time of the previous cycle, $t_1^{(i-1)}$. Given $\theta^{(i-1)}$, $t_0^{(i)}$ is modeled as

$$t_0^{(i)} \mid t_1^{(i-1)} \sim t_1^{(i-1)} + N(0, \tau_0^2), \qquad (28.5)$$

where the hyper-parameter τ_0^2 regulates the time difference between $t_0^{(i)}$ and $t_1^{(i-1)}$. The conditional distribution of $t_0^{(i)}$ depends on $\theta^{(i-1)}$ only through $t_1^{(i-1)}$.

To formulate $p(c^{(i)}, t_{max}^{(i)}, t_1^{(i)} | t_0^{(i)}, \theta^{(i-1)}, \eta)$, Yu et al. [32] conducted an exploratory analysis of the observed relationships among the parameters of consecutive cycles. In particular, they fit the model described by (28.3) to each of the 25 cycles individually and used the observed correlations among the cycle-specific fitted parameters to specify the parametric form of the distribution of $\theta^{(i)}$ given $\theta^{(i-1)}$. For example, there is a positive correlation between consecutive amplitudes $c^{(i)}$ and $c^{(i-1)}$, and a negative correlation between $c^{(i)}$ and $t_0^{(i)} - t_{max}^{(i-1)}$. The predictive power of the positive correlation between consecutive amplitudes can therefore be enhanced by combining it with the negative correlation with $t_0^{(i)} - t_{max}^{(i-1)}$. This means that given $\theta^{(i-1)}$, the amplitude of cycle i, $c^{(i)}$, depends on both $c^{(i-1)}$ and $t_0^{(i)} - t_{max}^{(i-1)}$. Therefore, the distribution of $c^{(i)}$ given $\theta^{(i-1)}$ and $t_0^{(i)}$ is modeled as

$$c^{(i)} \mid (c^{(i-1)}, t_0^{(i)}, t_{max}^{(i-1)}) \sim \delta_1 + \gamma_1 \frac{c^{(i-1)}}{t_0^{(i)} - t_{max}^{(i-1)}} + N(0, \tau_1^2). \qquad (28.6)$$

Additional correlations are observed among the components of $\theta^{(i)}$. For example, Yu *et al.* [32] observe a negative correlation between the rising time of a cycle, $t_{\max}^{(i)} - t_0^{(i)}$, and the amplitude reached during the same cycle. This negative correlation was first discovered by Waldmeier [29] and is hence known as the "Waldmeier effect." This effect means that the time at which cycle i reaches a peak, $t_{\max}^{(i)}$, is dependent on the starting time and the amplitude of that cycle. The distribution of $t_{\max}^{(i)}$ given $\theta^{(i-1)}$, $t_0^{(i)}$ and $c^{(i)}$ is thus modeled as

$$t_{\max}^{(i)} \mid (t_0^{(i)}, c^{(i)}) \sim t_0^{(i)} + \delta_2 + \gamma_2 c^{(i)} + N(0, \tau_2^2). \tag{28.7}$$

Notice that $t_{\max}^{(i)}$ is conditionally independent of $\theta^{(i-1)}$.

Finally, Yu *et al.* [32] observe and incorporate a correlation between the amplitude, $c^{(i)}$, and the time-to-decline, $t_1^{(i)} - t_{\max}^{(i)}$, of that cycle. This means that $t_1^{(i)}$ depends on $c^{(i)}$ and $t_{\max}^{(i)}$. With this, the distribution of $t_1^{(i)}$ given $t_{\max}^{(i)}$ and $c^{(i)}$ is modeled as

$$t_1^{(i)} \mid (t_{\max}^{(i)}, c^{(i)}) \sim t_{\max}^{(i)} + \delta_3 + \gamma_3 c^{(i)} + N(0, \tau_3^2), \tag{28.8}$$

and $t_1^{(i)}$ is conditionally independent of $\theta^{(i-1)}$.

The relations described by (28.5) to (28.8) can be encapsulated with the Markov structure illustrated in Figure 28.3. This structure allows for straightforward prediction of the characteristics of current cycles, even with data only extending to the beginning of the cycle. This is an important facet of the predictive capability of the Bayesian multilevel model. Together, (28.5) to (28.8) define the joint distribution of $p(\theta^{(i)}|\theta^{(i-1)}, \eta)$, where $\eta = (\tau_0^2, \gamma_j, \delta_j, \tau_j^2, j = 1, 2, 3)$ are the hyper-parameters. Yu *et al.* [32] also examined correlations between non-adjacent cycles (i.e., between cycle i and cycle $i\pm2$). However, no evidence was found to suggest more than a lag-one dependence. This has important scientific implications since it suggests that the solar dynamo does not retain memory beyond one cycle. Further evidence for this property has been discovered by examining magnetic proxies of solar activity [23], and by computer simulation of the solar dynamo [20].

28.2.3 Prior distiributions

In (28.3), both β and $\log \sigma$ are given independent uniform prior distributions. To allow a wide range of cycle shapes, a uniform prior distribution on the interval $(1,3)$ is used for both α_1 and α_2. The cycle-specific parameters for cycle 0, i.e., $t_0^{(0)}$, $t_1^{(0)}$, $t_{\max}^{(0)}$, and $c^{(0)}$ are assigned non-informative uniform prior distributions, subject to physical constraints on their ranges. We consider two prior distributions on the hyper-parameters, η, namely $p(\gamma_j, \delta_j, \tau_j^2) \propto \frac{1}{\tau_j}$, $j = 1, 2, 3$, and $p(\gamma_j, \delta_j, \tau_j^2) \propto 1$, $j = 1, 2, 3$. In our numerical analyses, results are not sensitive to this choice and we therefore only report results obtained using $p(\gamma_j, \delta_j, \tau_j^2) \propto \frac{1}{\tau_j}$, $j = 1, 2, 3$.

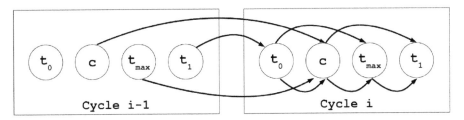

FIGURE 28.3
Markov structure relating the cycle-specific parameters $\theta^{(i-1)}$ and $\theta^{(i)}$.

28.3 Incorporating Multiple Proxies of Solar Activity

An inherent difficulty with combining multiple proxies to model the solar cycle is the varying temporal coverages of the proxies. SSNs, for example, are available as monthly estimates extending back to January, 1749, with no missing data. As technology improved, more proxies began to be observed and, like the SSNs, have been recorded up to the present. For clarity here and in our numerical illustrations, we consider three proxies: monthly average SSNs, monthly average total sunspot areas, and monthly average 10.7cm flux. Estimates of monthly average total sunspot areas extend back to May, 1874. Recordings of the 10.7cm flux began more recently, and estimates of the monthly average are available since February, 1947. It is important to note that, generally, once a proxy comes online (i.e., once a proxy begins to be recorded) it stays online and so there are no gaps in the data for individual proxies, resulting in a *monotone missing data pattern* [e.g., 22]. This pattern is readily apparent by examining the time-series of the three proxies presented in Figure 28.1, and is described in detail in subsection 28.3.2. Before we discuss our strategy for dealing with missing data, we describe how we would handle multiple proxies that were all observed over the same time period.

28.3.1 Complete-data analysis

With no missing data, we observe $Y^{[t]} = (y_1^{[t]}, y_2^{[t]}, y_3^{[t]})$ at each t, where y_1 is the monthly average SSN, y_2 is the monthly average sunspot area, and y_3 is the monthly average 10.7cm flux. Since the observed data represent monthly averages, $t = 1, \ldots, 3168$ indexes month. With this scheme, $t = 0$ corresponds to January, 1749, and $t = 3168$ corresponds to December, 2012.

The distribution of $\sqrt{y}^{[t]}$ in (28.3) can be used to model any proxy, perhaps transformed, that follows the same underlying solar cycle. In this way, (28.3) can be generalized to

$$G(Y^{[t]}) \mid (\Theta, \alpha, \beta, \sigma^2) \overset{\text{ind}}{\sim} N(\beta + U^{[t]}, \ \sigma^2), \qquad (28.9)$$

where $G(Y^{[t]})$ is a mapping from the multivariate $Y^{[t]}$ to a scalar value, and the underlying parameters are modeled in the same way as in subsections 28.2.2 and 28.2.3. Yu *et al.* [32] used $G(Y^{[t]}) = \sqrt{y_1}^{[t]}$ to obtain (28.3), but we are interested in finding a $G(Y^{[t]})$ that incorporates information from all available proxies.

The top row of Figure 28.4 displays scatterplots of the observed proxies and illustrates their strong linear correlations. Most of the variability in the data is in one linear dimension. Thus, it is appropriate to employ principal component analysis (PCA) to project the multivariate data onto the one-dimensional manifold defined by the direction of maximum variance.

Before deploying PCA, we transform the proxies to stabilize their variances, which increase with their means, see Figure 28.4. For the SSNs and sunspot areas we use the transformation $\sqrt{y_j^{[t]} + 10}$ for $j = 1, 2$. For the 10.7cm flux, we apply the transformation $\sqrt{y_3^{[t]} - \min_t(y_3^{[t]})}$. The constant offsets in the transformations were empirically chosen to improve linearity. We also normalize each transformed proxy by subtracting off its mean and dividing by its standard deviation. Subtracting off the mean is a necessary step in performing PCA, and dividing by the standard deviation controls for the differences in scale between the proxies. We denote the values of the transformed and normalized proxies at time t by $\tilde{Y}^{[t]} = (\tilde{y}_1^{[t]}, \tilde{y}_2^{[t]}, \tilde{y}_3^{[t]})$. Scatterplots of the transformed proxy data, $\tilde{Y}$, are displayed in the second row of Figure 28.4. From these plots, we note that the relationships remain linear, the correlations remain strong, but the variances are more stable.

With ω denoting the weights associated with the first principal component of the $\tilde{Y}^{[t]}$, we let $G(Y^{[t]}) = \omega^T \tilde{Y}^{[t]}$, where $G(Y^{[t]})$ is a scalar value representing the "solar activity level" at time t. The bottom row of Figure 28.4 shows that $G(Y)$ is highly correlated with each of the transformed proxies, which demonstrates its efficacy as a representation of the overall solar activity level. The first principal component accounts for 98% of the total variability in the proxies, so little information is lost in the PCA-based dimension reduction.

Once the $G(Y^{[t]})$ are obtained, they are treated as observed data with distribution given in (28.9). We use a Gibbs sampler to explore the posterior distribution, incorporating Metropolis-Hastings updates as necessary [see 32]. First, however, we must devise a scheme for handling missing observations of sunspot areas and the 10.7cm flux.

28.3.2   Multiple imputation strategy for missing data

The monthly average SSNs contain no missing data and are observed for all $\tilde{y}_1^{[t]}$, $t = 1, \ldots, 3168$. Records for the monthly average sunspot areas begin at month $t = 1505$ and contain 1664 observations, and records for the monthly

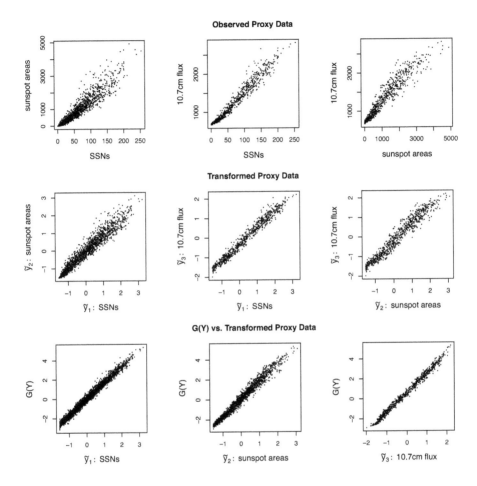

FIGURE 28.4
Transforming the proxies. The top row displays the two-dimensional scatterplots of the observed proxy data, Y. The middle row displays the two-dimensional scatterplots of the transformed proxy data, $\tilde{Y}$. The bottom row displays the computed value of $G(Y) = \omega^T \tilde{Y}$ versus each of the transformed proxies. Notice that $G(Y)$ is highly correlated with each of the transformed proxies.

average 10.7cm flux begin at month $t = 2378$ and contain 791 observations. Therefore, the SSNs are observed whenever the sunspot areas are observed, and both the SSNs and sunspot areas are observed whenever the 10.7cm flux is observed. This monotone missing data pattern (see Figure 28.5) allows the development of a straightforward strategy to account for missing data.

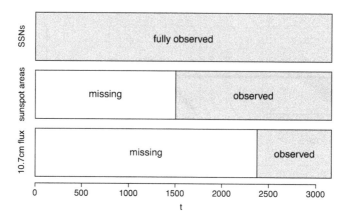

FIGURE 28.5
Illustration of the monotone missing data pattern. The solid gray bars indicate the range of observation times t for each of the three solar activity proxies: SSNs, sunspot areas, and 10.7cm flux. White bars indicate the range for which observations are missing. The SSNs are fully observed for months $t = 1, \ldots, 3168$. The sunspot areas are missing for months $t = 1, \ldots, 1504$. The 10.7cm flux is missing for months $t = 1, \ldots, 2377$. SSNs are observed whenever sunspot areas are observed, and both SSNs and sunspot areas are observed whenever the 10.7cm flux is observed. This is a monotone pattern of missing data.

Let $\tilde{Y}_{\mathrm{mis}}$ be the missing data, $\tilde{Y}_{\mathrm{mis}} = \{\tilde{y}_2^{[t]}, t = 1, \ldots, 1504; \tilde{y}_3^{[t]}, t = 1, \ldots, 2377\}$, and $\tilde{Y}_{\mathrm{obs}}$ be the observed data. A fully Bayesian strategy for handling the missing data would base inference on $p(\tilde{Y}_{\mathrm{mis}}, \Theta, \alpha, \beta, \sigma, \eta \mid \tilde{Y}_{\mathrm{obs}})$. This could be done by constructing a Gibbs sampler that at each iteration first updates $\tilde{Y}_{\mathrm{mis}} \sim p(\tilde{Y}_{\mathrm{mis}} \mid \tilde{Y}_{\mathrm{obs}}, \Theta, \alpha, \beta, \sigma, \eta)$ and then updates $(\Theta, \alpha, \beta, \sigma, \eta) \sim p(\Theta, \alpha, \beta, \sigma, \eta \mid \tilde{Y}_{\mathrm{mis}}, \tilde{Y}_{\mathrm{obs}})$ using existing computer code. This would, however, require us to specify a model for the multivariate $\tilde{Y}$ rather than simply for the univariate $G(Y)$.

Luckily, multiple imputation [e.g., 22] provides a principled way to use the univariate model and existing computer code to infer the solar cycle using multiple proxies. We first specify a separate simple local missing data model $p(\tilde{Y}_{\mathrm{mis}} \mid \tilde{Y}_{\mathrm{obs}}, \phi, \zeta)$ that incorporates the Markovian structure inherent in the data. In particular, we model

$$\tilde{y}_2^{[t]} \mid (\tilde{y}_1^{[t]}, \tilde{y}_2^{[t+1]}) \sim N(\phi_{01} + \phi_{11}\tilde{y}_1^{[t]} + \phi_{21}\tilde{y}_2^{[t+1]}, \zeta_1) \qquad (28.10)$$

for $t = 1, \ldots, 1504$, and

$$\tilde{y}_3^{[t]} \mid (\tilde{y}_1^{[t]}, \tilde{y}_2^{[t]}, \tilde{y}_3^{[t+1]}) \sim N(\phi_{02} + \phi_{12}\tilde{y}_1^{[t]} + \phi_{22}\tilde{y}_2^{[t]} + \phi_{32}\tilde{y}_3^{[t+1]}, \zeta_2) \qquad (28.11)$$

for $t = 1, \ldots, 2377$. Together, (28.10) and (28.11) define the distribution of $p(\tilde{Y}_{\text{mis}} \mid \tilde{Y}_{\text{obs}}, \phi, \varsigma)$. We fit (28.10) using only the observations for which both $\tilde{y}_1$ and $\tilde{y}_2$ are observed, and fit (28.11) using only the observations for which all three quantities are observed. With the fitted models in place, $\tilde{Y}_{\text{mis}}$ can be imputed by drawing values from $p(\tilde{Y}_{\text{mis}} \mid \tilde{Y}_{\text{obs}}, \hat{\phi}, \hat{\varsigma})$.

Multiple imputation coherently accounts for two sources of uncertainty: the uncertainty that would be present even if all proxies were observed for the same time period (i.e., if there was no missing data), and the uncertainty that arises from imputing missing data. With multiple imputation, we obtain M imputations of the missing data, $\tilde{Y}_{\text{mis}}^{(m)} \sim p(\tilde{Y}_{\text{mis}} \mid \tilde{Y}_{\text{obs}}, \hat{\phi}, \hat{\varsigma})$ and apply the complete-data analysis described in subsection 28.3.1 to each of the imputed data sets. Parameter estimates and uncertainties are based on the multiple imputation combining rules [e.g., 14, 22]. Under this scheme, we obtain M estimates, $\hat{\psi}_m$, of any particular model parameter, ψ, along with their associated variances, V_m. Since we sample from the posterior distribution to fit the Bayesian multilevel model described in Section 28.2, natural candidates for $\hat{\psi}_m$ and V_m are the posterior means and posterior variances of ψ, under each of the M imputed data sets. The multiple-imputation estimate of ψ is $\hat{\psi} = \frac{1}{M} \sum_{m=1}^{M} \hat{\psi}_m$. The estimate of the variance is a combination of the average within-imputation variance, $W = \frac{1}{M} \sum_{m=1}^{M} V_m$, and the between imputation variance, $B = \frac{1}{M-1} \sum_{m=1}^{M} (\hat{\psi}_m - \hat{\psi})^2$, and is given by $T = W + \frac{M+1}{M} B$. Interval estimates are computed from a reference t-distribution, $(\psi - \hat{\psi}) T^{-1/2} \sim t_\nu$, where the degrees of freedom is given by $\nu = (M-1) \left(1 + \frac{M}{M+1} \frac{W}{B}\right)^2$.

Our multiple imputation procedure relies upon the assumption that ς_1 and ς_2 do not vary over time. To test this assumption, we fit the local missing data model for different time periods of data separately and compared the results. We first fit (28.10) using the most recent third of the available $\tilde{y}_2$ data and obtain an estimate and standard error of ς_1. We then fit (28.10) again using the oldest third of the available $\tilde{y}_2$ data and obtain a second estimate and standard error of ς_1. We then perform an F-test for the hypothesis that ς_1 for the older time epoch is equal to ς_1 for the newer time epoch. We repeat this procedure for (28.11) using the oldest and most recent third of the $\tilde{y}_3$ data. The results are summarized in Table 28.1. We find that ς_1 appears to change over time, but that it is larger for the newer data. Since most of $\tilde{y}_2$ is observed, not imputed, we are not particularly worried about overestimating ς_1 for the older data. In addition, ς_2 does not appear to vary over time, and this is more important since, unlike $\tilde{y}_2$, the majority of the values of $\tilde{y}_3$ that go into our final analysis are imputed.

TABLE 28.1

Examining the multiple imputation assumptions. The first column displays the estimates of ζ_1 and ζ_2, along with their estimated standard errors (S.E.), using the *newest third* of the observed data for the sunspot areas and 10.7cm flux, respectively. The second column displays the same, but using the *oldest third* of the observed data for the sunspot areas and 10.7cm flux, respectively. We use an F-test of equality of variances with the null hypothesis that ζ_l for the oldest time epoch is equal to ζ_l for the newest time epoch, $l = 1, 2$. The resulting p-values are given in the third column.

	Newest Third $\hat{\zeta}_1$ (S.E.)	Oldest Third $\hat{\zeta}_2$ (S.E.)	F-test p-value
Imputing Sunspot Areas: $\tilde{y}_2$	3.88 (0.14)	3.17 (0.11)	< 0.01
Imputing 10.7cm Flux: $\tilde{y}_3$	1.37 (0.07)	1.34 (0.07)	0.74

28.4   Results

We now discuss the fit of the Bayesian multilevel model for the solar cycle. To allow for comparison, we obtain model fits using both $G(Y^{[t]}) = \sqrt{y_1}^{[t]}$ (i.e., the SSN model) and $G(Y^{[t]}) = \omega^T \tilde{Y}^{[t]}$ (i.e., the multiple-proxy model). When $G(Y^{[t]}) = \sqrt{y_1}^{[t]}$, we do not need to perform multiple imputation since there are no missing SSNs. In this case, the fitted values of all quantities are given by their posterior means and credible intervals are given by their 2.5% and 97.5% posterior quantiles. When $G(Y^{[t]}) = \omega^T \tilde{Y}^{[t]}$, we use the multiple imputation strategy described in subsection 28.3.2, with $M = 5$ imputations. The estimates for all quantities are computed by setting $\hat{\psi}_m$ and V_m equal to their posterior mean and posterior variances, respectively, and following the multiple imputation combining rules. In particular, the estimate for the solar activity level at time t is computed by setting $\hat{\psi}_m$ and V_m equal to the posterior mean and posterior variance of $U^{[t]} + \beta$. The fitted values of $U^{[t]} + \beta$ are given by the average value of the M within-imputation posterior means and a 95% interval at time t is computed from the reference t-distribution as described in subsection 28.3.2.

The fitted values of $U^{[t]} + \beta$ and associated 95% intervals for the multiple-proxy model are plotted in the top panel of Figure 28.6, for the time interval with all three proxies observed (i.e., $t = 2378, \ldots, 3168$). The data presented in this panel are the $\omega^T \tilde{Y}$ values for the given time interval. The bottom panel of Figure 28.6 shows the fitted values of $U^{[t]} + \beta$ and associated 95% intervals under the SSN model for the same time interval; the data are the observed $\sqrt{y_1}^{[t]}$ values. The solid vertical lines in both panels are the fitted

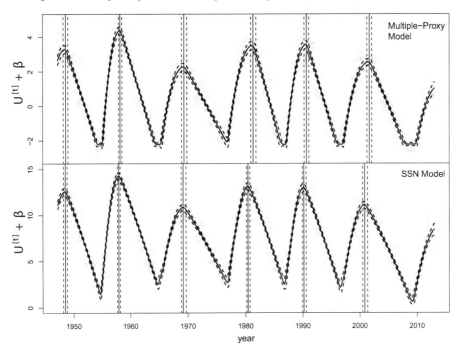

FIGURE 28.6

The fitted solar cycle. The two panels compare the fitted models and data for the multi-proxy (top) and SSN (bottom) models. They include $U^{[t]} + \beta$ (solid curves) and their 95% intervals (dashed curves), along with fitted values for $t_{\max}^{(i)}$ (solid vertical lines) and their 95% intervals (dashed vertical lines). Gray circles represent $\omega^T \tilde{Y}$ (top panel) and $\sqrt{y_1}^{[t]}$ (bottom panel). The time interval displayed covers the period when all three proxies are observed. Using multiple proxies consistently results in later fitted times for the solar maxima.

values for $t_{\max}^{(i)}$ for cycles $i = 18, \ldots, 23$, and the dashed vertical lines are their 95% intervals. The estimates of $t_{\max}^{(i)}$ under the multiple-proxy model are later than the estimates under the SSN model, although in some cases their 95% intervals overlap. We discuss further comparisons of the timing of the fitted cycles below.

To evaluate the quality of the model fits, we plotted the residuals versus time and versus the fitted values, but did not observe any patterns that might call the models into question. The plot of the residuals versus time for the SSN model did not show any significant patterns or evidence of heteroscedasticity that would lead us to question the reliability of the historical sunspot numbers. We also simulate the full time series from the posterior predictive distribution 5000 times. Figure 28.7 displays the 95% posterior predictive intervals from the simulated series along with the $\omega^T \tilde{Y}$ values for the time period when

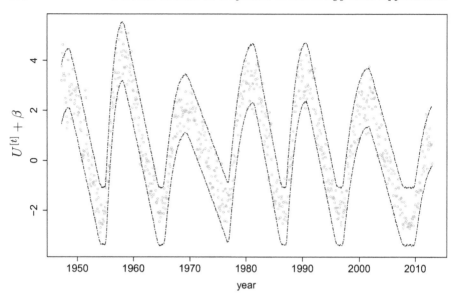

FIGURE 28.7
Posterior predictive check. The full time series is simulated from the posterior predictive distribution 5000 times, and 95% pointwise posterior predictive intervals are given by the dashed lines. Gray circles represent the observed $G(Y) = \omega^T \tilde{Y}$. The observed data are consistent with the posterior predictive distribution.

all three proxies are observed. The simulated series are consistent with the observed data.

Since solar physicists are concerned with predicting the timing of solar cycles, we can obtain fitted values and 95% intervals for the rising time of each cycle, $t_{max}^{(i)} - t_0^{(i)}$, $i = 0, \ldots, 24$, under both the SSN and multiple-proxy models. These are presented in Figure 28.8. The left panel displays the fitted values and 95% intervals for both model fits over time, and the right panel displays a scatterplot of the fitted values under the two models along with their associated 95% intervals. We do not include results for cycle 0 since this initial cycle is incomplete and has relatively large 95% intervals. The 95% intervals for cycle 24 are also larger than those of other cycles since the cycle is ongoing. Cycles 0 and 24 have fewer observed neighboring cycles than the other cycles, which also contributes to their larger 95% intervals.

Overall, the rising times do not appear to differ significantly between the two model fits. This is not the case when examining the falling times of each cycle, $t_1^{(i)} - t_{max}^{(i)}$, which are displayed in Figure 28.9. They are significantly shorter for the model fit with multiple proxies since the 95% intervals rarely intersect the 45° line plotted in the right panel. Taken together, these results suggest that fitting multiple proxies instead of only using the SSNs yields

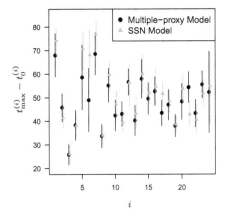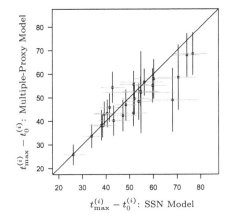

FIGURE 28.8
Fitted values and 95% intervals for the cycle rising times under the multiple-proxy and SSN models. There does not appear to be a significant difference in the rising times between the two model fits.

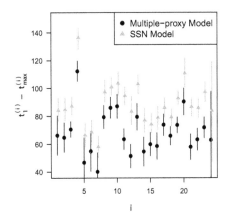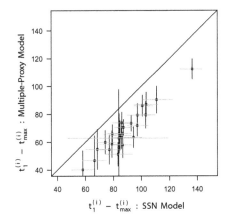

FIGURE 28.9
Fitted values and 95% intervals for the cycle falling times under the multiple-proxy and SSN models. Unlike the cycle rising times, the falling times do appear to differ between the two model fits. Specifically, the falling times are significantly shorter under the model fit with multiple proxies.

shorter overall cycle lengths, $t_1^{(i)} - t_0^{(i)}$. We display fitted values and 95% intervals for the total cycle lengths in Figure 28.10, and confirm that the model fit with multiple proxies generally has significantly shorter total cycle lengths.

There has been speculation that recent solar cycles represent a period of relatively high activity. Temmer [28] and Shapoval *et al.* [26], for example,

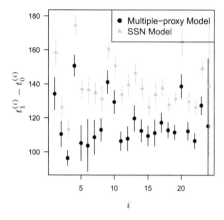
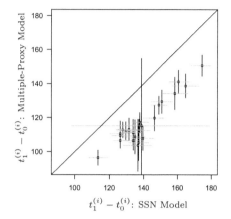

FIGURE 28.10

Fitted values and 95% intervals for the total cycle lengths under the multiple-proxy and SSN models. Following from the results displayed in Figures 28.8 and 28.9, the multiple-proxy model generally has significantly shorter cycle lengths than the SSN model has.

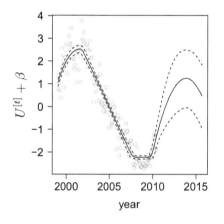
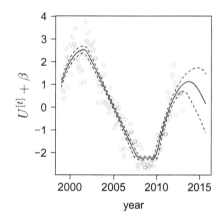

FIGURE 28.11

Prediction for cycle 24, using data up to May 2010 (left) and up to December 2012 (right). The solid curve is the fitted $U^{[t]} + \beta$, and 95% posterior (predictive) intervals are given by the dotted lines. Gray circles represent the observed $G(Y) = \omega^T \tilde{Y}$. As more data for the cycle is obtained, the uncertainty in the predictions is reduced.

suggested that the Sun was in a "low-activity" phase from around 1850 to 1915, while the Sun was in a "high-activity" phase from 1915 to the most recent solar minimum. We can evaluate this claim by obtaining an estimate and 95% interval for the mean amplitude during the high-activity phase, cycles 15 though 23, and the low-activity phase, cycles 10 through 14. Let $\bar{c}^{(j:k)}$ represent the mean amplitude from cycles j to k. Then, under the multiple-proxy model, the estimate of $\bar{c}^{(15:23)}$ is 5.01, with 95% interval $(4.86, 5.17)$, and the estimate of $\bar{c}^{(10:14)}$ is 3.86, with 95% interval $(3.70, 4.02)$. Thus, there is evidence that cycles 15 through 23 exhibit higher average solar activity than do cycles 10 through 14. Shapoval *et al.* [26] and Temmer [28] also hypothesized that cycle 24 represents a shift back to a low-activity phase. Our Bayesian approach allows for straightforward calculation of the probability that the amplitude for cycle 24 will be below $\bar{c}^{(15:23)}$: $Pr(c^{(24)} < \bar{c}^{(15:23)}) = 1$. Furthermore $Pr(c^{(24)} < \bar{c}^{(10:14)}) = 0.88$, which suggests that the amplitude for cycle 24 may be unusually low, even when compared to the low-activity regime.

The Markov structure of our model allows for straightforward prediction of the current cycle even when little data for this cycle is observed. In Figure 28.11, we show the prediction for cycle 24, using data extending up to May 2010 (left panel) and up to December 2012 (right panel). When less data is available from the current cycle, predictions rely more on the Markov structure of the model and are thus more uncertain. As data become available, the predictions are increasingly driven by the current cycle and the uncertainties diminish. The fitted solar cycle is similar in both cases, but the 95% intervals are noticeably narrower with more data. This shows that the cycle-to-cycle relationships we learn in the second stage of the model are consistent with the most recent data and we can make reasonable predictions of cycle characteristics at the start of the cycle, albeit with considerable uncertainty.

28.5 Summary and Discussion

We have carried out a fully Bayesian analysis of the solar cycle using multiple proxy variables by generalizing the model of Yu *et al.* [32]. After suitably transforming the data to stabilize the variance and increase linearity, we multiply-imputed missing data by specifying a simple local missing data model that incorporates the Markovian structure of the data. The dimensionality of each imputed data set is reduced using PCA to project the multivariate proxy observations onto a one-dimensional subspace along the direction of highest variance. In this way, we obtain a univariate summary of solar activity at each time point, allowing us to utilize the existing univariate model to infer properties of the solar cycle using multiple proxies. This approach is based on the current understanding that there is a single underlying solar cycle, and the several proxies all provide information about the cycle.

It is necessary to use the long history of the SSNs and sunspot area observations in order to learn the patterns among consecutive solar cycles. Multiple imputation is used in order to easily derive estimates based on the posterior distribution of the model parameters given all of the data. If only complete observations are used, meaning only the data with all three proxies observed, the model is overfit. This problem would be compounded in future analyses that may also include other proxies that do not extend back for more than a couple of cycles. One of our primary aims is to allow additional proxies, even if they are largely missing, so that all available data can be used in a coherent statistical framework.

We compare fits of the Bayesian multilevel model of the solar cycle based on (i) multiple proxies and (ii) the SSNs alone. We observe significant differences in the inferred cycle properties. In particular, we find that the model fit with multiple proxies has shorter falling times than the model fit with the SSNs. Since we do not find significant differences in the rising times, the shorter falling times from the multiple-proxy model also imply shorter total cycle lengths, which we also observe. Shorter cycle lengths in turn imply longer solar minima. It has been observed elsewhere that the Sun can remain in a prolonged state of minimum activity, and there is evidence that the most recent solar minimum was unusual in its depth and duration [1]. During the most recent solar minimum the 10.7cm flux was the lowest ever recorded, and physical characteristics of the solar surface and interior were unusual when compared to previous solar minima [1]. It is clear from Figure 28.6 that the most recent minimum of the fitted solar cycle has a longer duration under the multiple-proxy model. In this regard the multiple-proxy fit captures an important feature in the solar cycle that is missed by the SSN model. The use of additional proxies may further illuminate this effect.

Future work will also consider additional functional forms for the solar cycle. Recent studies have presented evidence that many solar cycles have double maxima [e.g., 10, 21], which our current parameterization does not capture. Gnevyshev [12] suggests that complex physical processes produce double peaks in all cycles, but often the gap between them is too short for the two within-cycle maxima to be distinguished. One possible explanation for double maxima that we can explore is the existence of separate cycles acting on each hemisphere of the Sun. Under this scenario the separate northern-hemisphere and southern-hemisphere cycles are parameterized by (28.1) and (28.2), with total activity being the sum over the two hemispheres and double maxima appearing when the two hemispheres reach peak activity at different times.

Our multiple-proxy Bayesian multilevel model of the solar cycle provides the flexibility needed to dynamically describe the complex structure of cycles and their varying shapes, duration, and amplitudes, while capturing the predictable way in which these features evolve over time. The effective combination of multiple imputation and PCA-based dimension reduction makes it straightforward to incorporate additional proxies, all the while taking advantage of the long history of SSN observations.

Acknowledgments

This work was supported in party by NSF grant DMS-12-09232, a British Royal Society Wolfson Research Merit Award, by a European Commission Marie-Curie Career Integration Grant, and by the STFC (UK). We thank C. Alex Young for many helpful discussions.

Bibliography

[1] S. Basu. The peculiar solar cycle 24 - where do we stand? *Journal of Physics Conference Series*, 440(1):012001, June 2013.

[2] R. E. Benestad. A review of the solar cycle length estimates. *Geophysical Research Letters*, 32(15), Aug. 2005. L:15714, doi:10.1029/2005GL023621.

[3] B. P. Bonev, K. M. Penev, and S. Sello. Long-term solar variability and the solar cycle in the 21st century. *The Astrophysical Journal Letters*, 605:L81–L84, Apr. 2004.

[4] P. Charbonneau. Babcock-Leighton models of the solar cycle: Questions and issues. *Advances in Space Research*, 39:1661–1669, 2007.

[5] P. Charbonneau and M. Dikpati. Stochastic fluctuations in a Babcock-Leighton model of the solar cycle. *The Astrophysical Journal*, 543:1027–1043, Nov. 2000.

[6] A. R. Choudhuri. Stochastic fluctuations of the solar dynamo. *Astronomy and Astrophysics*, 253:277–285, Jan. 1992.

[7] A. R. Choudhuri, P. Chatterjee, and J. Jiang. Predicting solar cycle 24 with a solar dynamo model. *Physical Review Letters*, 98(13):131103, Mar. 2007.

[8] M. Dikpati, G. de Toma, and P. A. Gilman. Predicting the strength of solar cycle 24 using a flux-transport dynamo-based tool. *Geophysical Research Letters*, 33:5102, Mar. 2006.

[9] M. Dikpati and P. A. Gilman. Simulating and predicting solar cycles using a flux-transport dynamo. *The Astrophysical Journal*, 649:498–514, Sept. 2006.

[10] K. Georgieva. Why the sunspot cycle is double peaked. *ISRN Astronomy and Astrophysics*, 2011, 2011. ID 437838, 11 pages, http://adsabs.harvard.edu/abs/2011ISRAA2011E...2G.

[11] L. A. Gil-Alana. Time series modeling of sunspot numbers using long-range cyclical dependence. *Solar Physics*, 257:371–381, July 2009.

[12] M. N. Gnevyshev. On the 11-years cycle of solar activity. *Solar Physics*, 1(1):107–120, 1967.

[13] R. A. Greenkorn. A comparison of the 10.7-cm radio flux values and the international sunspot numbers for solar activity cycles 19, 20, and 21. *Solar Physics*, 280:205–221, Sept. 2012.

[14] O. Harel and X.-H. Zhou. Multiple imputation: Review of theory, implementation and software. *Statistics in Medicine*, 26(16):3057–3077, 2007.

[15] D. H. Hathaway and R. M. Wilson. Geomagnetic activity indicates large amplitude for sunspot cycle 24. *Geophysical Research Letters*, 33(18), Sept. 2006. L:18101, doi: 10.1029/2006GL027053.

[16] D. H. Hathaway, R. M. Wilson, and E. J. Reichmann. The shape of the sunspot cycle. *Solar Physics*, 151:177–190, Apr. 1994.

[17] D. H. Hathaway, R. M. Wilson, and E. J. Reichmann. Group sunspot numbers: Sunspot cycle characteristics. *Solar Physics*, 211:357–370, Dec. 2002.

[18] H. S. Hudson. The unpredictability of the most energetic solar events. *The Astrophysical Journal Letters*, 663:L45–L48, July 2007.

[19] R. P. Kane. Did predictions of the maximum sunspot number for solar cycle 23 come true? *Solar Physics*, 202:395–406, Sept. 2001.

[20] B. B. Karak and D. Nandy. Turbulent pumping of magnetic flux reduces solar cycle memory and thus impacts predictability of the Sun's activity. *The Astrophysical Journal Letters*, 761:L13, Dec. 2012.

[21] A. Kilcik and A. Ozguc. One possible reason for double-peaked maxima in solar cycles: Is a second maximum of solar cycle 24 expected? *Solar Physics*, 289(4):1379–1386, 2014.

[22] R. J. A. Little and D. B. Rubin. *Statistical Analysis with Missing Data*. John Wiley & Sons, New York, 2nd edition, 2002.

[23] A. Muñoz-Jaramillo, M. Dasi-Espuig, L. A. Balmaceda, and E. E. DeLuca. Solar cycle propagation, memory, and prediction: Insights from a century of magnetic proxies. *The Astrophysical Journal Letters*, 767:L25, Apr. 2013.

[24] W. D. Pesnell. Solar cycle predictions (invited review). *Solar Physics*, 281(1):507–532, 2012.

[25] M. Schüssler. Are solar cycles predictable? *Astronomische Nachrichten*, 328:1087, Dec. 2007.

[26] A. Shapoval, J. L. Le Mouël, V. Courtillot, and M. Shnirman. Two regimes in the regularity of sunspot number. *The Astrophysical Journal*, 779(2):108, Dec. 2013.

[27] S. K. Solanki, I. G. Usoskin, B. Kromer, M. Schüssler, and J. Beer. Unusual activity of the Sun during recent decades compared to the previous 11,000 years. *Nature*, 431:1084–1087, Oct. 2004.

[28] M. Temmer. Statistical properties of flares and sunspots over the solar cycle. In S. R. Cranmer, J. T. Hoeksema, and J. L. Kohl, editors, *SOHO-23: Understanding a Peculiar Solar Minimum*, volume 428 of *Astronomical Society of the Pacific Conference Series*, page 161, June 2010.

[29] M. Waldmeier. Neue eigenschaften der sonnenfleckenkurve. *Astron. Mitt. Eidgen. Sternw. Zürich*, 133(105), 1935.

[30] S. Watari. Forecasting solar cycle 24 using the relationship between cycle length and maximum sunspot number. *Space Weather*, 6(12), 2008.

[31] R. Wolf. Neue untersuchungen über die periode der sonnen-flecken und ihre bedeutung. *Viertel. Natur. Ges. Bern*, 254(179), 1852.

[32] Y. Yu, D. A. van Dyk, V. L. Kashyap, and C. A. Young. A Bayesian analysis of the correlations among sunspot cycles. *Solar Physics*, 281(2):847–862, Dec. 2012.

29

Fuzzy Information, Likelihood, Bayes' Theorem, and Engineering Application

Reinhard Viertl

Vienna University of Technology

Owat Sunanta

Vienna University of Technology

CONTENTS

29.1 Introduction

In solving an engineering problem, such as developing empirical models for manufacturing processes, real data of continuous quantities are required. However, the information and/or data obtained from different sources are often clouded by different kinds of uncertainty. The available data of manufacturing problems are frequently imprecise and incomplete. To overcome such problems, fuzzy concepts along with probability theory, e.g., the Bayesian approach, can be adopted and applied to provide meaningful description that bridges the gap between real data and empirical process models.

In standard Bayesian inference, *a-priori* distributions are standard probability distributions and the observations are assumed to be numbers or vectors. Bayes' theorem formulates the transition from the *a-priori* distribution of the stochastic quantity, which describes the parameter of interest, to the *a-posteriori* distribution. In case of continuous stochastic models $X \sim f(\cdot|\theta)$,

$\theta \in \Theta$, based on observations $x_1, \cdots, x_n$ of X, the transition from an *a-priori* density to an updated information concerning the distribution of the stochastic quantity describing the parameter $\tilde{\theta}$ is given by the conditional density $\pi(\cdot|x_1, \cdots, x_n)$ of $\tilde{\theta}$, *i.e.*, Bayes' theorem:

$$\pi(\theta|x_1, \cdots, x_n) = \frac{\pi(\theta) \cdot l(\theta; x_1, \cdots, x_n)}{\int_\Theta \pi(\theta) \cdot l(\theta; x_1, \cdots, x_n) d\theta},$$

where $l(\theta; x_1, \cdots, x_n)$ is the likelihood function defined on the parameter space Θ.

However, the use of *a-priori* densities in form of standard probability densities has been criticized in some situations. Moreover, real observations from continuous quantities are not precise numbers, but, more or less, non-precise (also called fuzzy).

The first problem can be overcome by using a more general form of probability which is related to soft computing, *i.e.*, so-called fuzzy probability densities. The second problem can be solved by using general fuzzy numbers and fuzzy vectors. See also related work in [4] and [7].

29.2 Probability Based on Densities

To overcome the deficiency of standard *a-priori* distributions in Bayesian inference in some cases, fuzzy *a-priori* densities, more general forms of expressing *a-priori* information, seem to be appropriate. In order to define fuzzy densities, a special form of general fuzzy numbers is necessary.

Definition 29.2.1. *A general fuzzy number whose δ-cuts are non-empty compact intervals $[a_\delta, b_\delta]$ is called a fuzzy interval. For functions $f^*(\cdot)$ defined on a set M, whose values $f^*(x)$ are fuzzy intervals, their δ-level functions $\underline{f}_\delta(\cdot)$ and $\overline{f}_\delta(\cdot)$ are defined in the following way:*
Let $C_\delta[f^*(x)] = [a_\delta(x), b_\delta(x)]$ for all $\delta \in (0, 1]$, the lower and upper δ-level functions are standard real-valued functions defined by their values $\underline{f}_\delta(x) := a_\delta(x)$ for all $x \in M$ and $\overline{f}_\delta(x) := b_\delta(x)$ for all $x \in M$.

Fuzzy density is a fuzzy valued function $f^*(\cdot)$ defined on a measure space $(M, \mathcal{A}, \mu)$ possessing the following properties:

(i) $f^*(x)$ is a fuzzy interval for all $x \in M$;

(ii) there exists $g : M \to [0, \infty)$ which is a standard probability density on $(M, \mathcal{A}, \mu)$ where $\underline{f}_1(x) \le g(x) \le \overline{f}_1(x)$ for all $x \in M$;

(iii) all δ-level functions $\underline{f}_\delta(\cdot)$ and $\overline{f}_\delta(\cdot)$ are integrable functions with finite integral.

Probabilities of events $A \in \mathcal{A}$ based on a fuzzy probability density $f^*(\cdot)$ are defined in the following way:

Let $\mathscr{D}_\delta$ be the set of all standard probability densities $h(\cdot)$ on $(M, \mathcal{A}, \mu)$ with $\underline{f}_\delta(x) \le h(x) \le \overline{f}_\delta(x)$ for all $x \in M$. The generalized probability $P^*(A)$ is a fuzzy interval, which is determined by the following family of compact intervals $B_\delta = [a_\delta, b_\delta]$ where

$$b_\delta := \sup\{\int_A h(x)d\mu(x) : h \in \mathscr{D}_\delta\}$$
$$a_\delta := \inf\{\int_A h(x)d\mu(x) : h \in \mathscr{D}_\delta\} \qquad \text{for all } \delta \in (0,1].$$

By applying the so-called construction lemma for general fuzzy numbers [5], the characterizing function $\xi(\cdot)$ of $P^*(A)$ is given by

$$\xi(x) = \sup\{\delta.\mathbb{1}_{[a_\delta, b_\delta]}(x) : \delta \in [0,1]\} \text{ for all } x \in \mathbb{R},$$

where $\mathbb{1}_B(\cdot)$ denotes the indicator function of the set B, and $[a_0, b_0] := \mathbb{R}$.

Fuzzy probability density is a more general form of expressing *a-priori* information about the parameters θ in stochastic models $f(\cdot|\theta)$, $\theta \in \Theta$.

29.3   Fuzzy Data and Likelihood Function

Real observations of continuous stochastic quantities X are not precise numbers or vectors, whereas the measurement results are more or less non-precise, also called fuzzy. This is often the case in dealing with engineering problems. The best (to-date) mathematical description (see also [2] and [6]) of such observations is explained by means of general fuzzy numbers $x_1^*, \cdots, x_n^*$ with corresponding characterizing functions $\xi_1(\cdot), \cdots, \xi_n(\cdot)$.

Remark 3. *The fuzziness of an observation x_i^* resolves the problem in standard continuous stochastic models where observed data have zero probability.*

For a fuzzy observation x_i^* with characterizing function $\xi_i(\cdot)$ from a density $f(\cdot)$, the probability of x_i^* is given by [8]

$$Prob(x_i^*) = \int_{\mathbb{R}} \xi_i(x)f(x)d\mu(x) \ge 0.$$

This makes it possible to define the likelihood function of independent fuzzy observations $x_1^*, \cdots, x_n^*$ in a natural way:

$$l(\theta; x_1^*, \cdots, x_n^*) = \prod_{i=1}^{n} Prob(x_i^*|\theta) = \prod_{i=1}^{n} \int_{\mathbb{R}} \xi_i(x)f(x|\theta)d\mu(x) \text{ for all } \theta \in \Theta.$$

The likelihood function is, then, used in updating the *a-priori* density to yield the corresponding fuzzy *a-posteriori* density.

29.4   Generalized Bayes' Theorem

The standard Bayes' theorem has to be generalized to handle fuzzy *a-priori* density $\pi^*(\cdot)$ on the parameter space Θ and fuzzy data $x_1^*, \cdots, x_n^*$ of parametric stochastic model $X \sim f(\cdot|\theta)$, $\theta \in \Theta$. This is possible by using the δ-level functions $\underline{\pi}_\delta(\cdot)$ and $\overline{\pi}_\delta(\cdot)$ of $\pi^*(\cdot)$ along with defining the δ-level functions $\underline{\pi}_\delta(\cdot|x_1^*, \cdots, x_n^*)$ and $\overline{\pi}_\delta(\cdot|x_1^*, \cdots, x_n^*)$ of the fuzzy *a-posteriori* density in the following way:

$$\overline{\pi}_\delta(\theta|x_1^*, \cdots, x_n^*) = \frac{\overline{\pi}_\delta(\theta) \cdot l(\theta; x_1^*, \cdots, x_n^*)}{\int_\Theta \frac{\underline{\pi}_\delta(\theta) + \overline{\pi}_\delta(\theta)}{2} \cdot l(\theta; x_1^*, \cdots, x_n^*) d\theta}$$

$$\underline{\pi}_\delta(\theta|x_1^*, \cdots, x_n^*) = \frac{\underline{\pi}_\delta(\theta) \cdot l(\theta; x_1^*, \cdots, x_n^*)}{\int_\Theta \frac{\underline{\pi}_\delta(\theta) + \overline{\pi}_\delta(\theta)}{2} \cdot l(\theta; x_1^*, \cdots, x_n^*) d\theta}$$

for all $\delta \in (0, 1]$.

Remark 4. *The averaging* $\frac{\underline{\pi}_\delta(\theta) + \overline{\pi}_\delta(\theta)}{2}$ *is necessary in order to keep the sequential updating of standard Bayes' theorem.*

Based on the fuzzy *a-posteriori* density $\pi^*(\cdot|x_1^*, \cdots, x_n^*)$, the fuzzy HPD-intervals of the parameter θ, which provides valuable information on the unknown parameter θ, can be generated (see [5] for more detail on the method).

29.5   Example: Manual Lapping of Valve Discs

Lapping is a low-speed, low-pressure abrading operation that accomplishes one or more of the following results: extreme dimensional accuracy of lapped surface, *e.g.*, flat or spherical, refinement of surface finish, extremely close fit between mating surfaces, removal of minor damaged surface and subsurface layers [3]. The most intriguing aspect of lapping lies in its abrading mechanism of random loose abrasive particles, generally in form of abrasive compound, to refine and remove microscopic material. Lapping involves a number of process parameters that influence the integrity of the lapped surface. In this example, surface finish (roughness), which is a process parameter specific to manual flat lapping of valve discs (Figure 29.1), is explored.

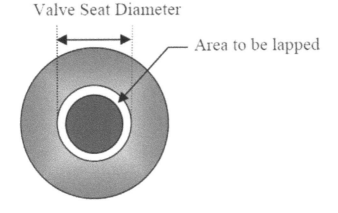

Valve Seat Diameter

Area to be lapped

Top view

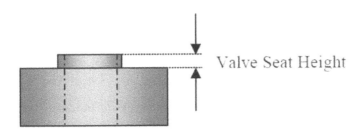

Valve Seat Height

Front/Side view

FIGURE 29.1
Seat area of a valve disc.

29.5.1 Surface finish (roughness)

Surface roughness consists of fine irregularities in the surface texture, usually including those resulting from the inherent actions of different manufacturing processes. Surface roughness can be measured by a variety of methods and instruments, including using profilometer. Surface roughness is commonly presented in terms of the arithmetic average (R_a) of the profile (absolute) deviations from the mean line within the evaluation length (Figure 29.2).

Lapping is an ultra-fine finishing process and not meant to remove a significant amount of material. Instead, lapping is a good option for correcting finished surfaces with minor imperfections. Hence, the surfaces prior to lapping must have been pre-processed to a certain level of finish. In collecting data for this example, the surfaces have already been prepared through the process of rough lapping and have surface roughness approximately between

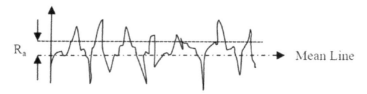

FIGURE 29.2
A profile of surface roughness (R_a).

0.00031 mm and 0.00082 mm. After lapping, based mainly on the capability of the lapping process itself and the equipment, the typical obtained surface roughness produced is approximately between 0.00005 mm and 0.00050 mm [1].

29.5.2   Fuzzy Bayesian inference

The obtained surface roughness (X) is modeled with an exponential distribution and fuzzy gamma density as *a-priori* density for the parameter $\theta \in (0, \infty)$.

$$X \sim Ex_\theta \hat{=} f(x|\theta) = \frac{1}{\theta} \exp\left[-\frac{x}{\theta}\right] \cdot \mathbb{1}_{(0,\infty)}(x)$$

The *a-priori* density $\pi(\cdot)$ for $\tilde{\theta}$ (mean value of obtained surface roughness) is assumed to be a gamma distribution. This assumption is intuitive and based on the common lapping process capability. In other words, before actual lapping of the surfaces, the anticipated (obtained) surface roughness measurements are expected to distribute with higher probability (density) around 0.00005 mm and 0.00050 mm. Different δ-level functions of the a-priori density are shown in Figure 29.3.

The obtained surface roughness data are collected and represented in the form of fuzzy observations x_i^*, i.e. $D^* = (x_1^*, x_2^*)$, with characterizing functions as depicted in Figure 29.4. The exponentially distributed likelihood function is assumed here in order to obtain conjugate distributions for *a-priori* and *a-posteriori*. The fuzzy observations are then used for updating the *a-priori* density to yield the corresponding fuzzy *a-posteriori* density $\pi^*(\cdot|D^*)$ as shown in Figure 29.5.

The collected data of obtained surface roughness (between 0.000041mm and 0.000128mm) support the prior assumption of lapping process capability. The *a-posteriori* density is more peaked than *a-priori* density, which may be interpreted in the following way: the collected data confirm the assumption prior to the experiment. As shown in Figure 29.6, a fuzzy HPD-interval of the parameter θ with 0.95 coverage probability, which is generated based on the fuzzy *a-posteriori* density $\pi^*(\cdot|D^*)$ from Figure 29.5, provides a fuzzy expected range of the obtained surface roughness.

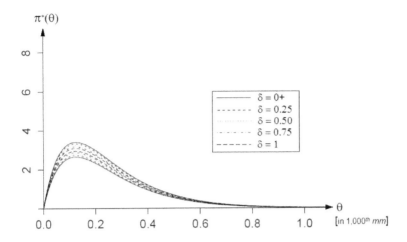

FIGURE 29.3
Fuzzy *a-priori* density.

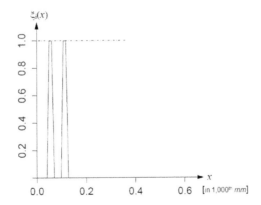

FIGURE 29.4
Fuzzy sample.

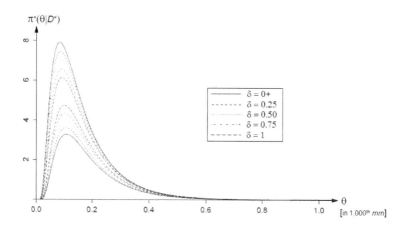

FIGURE 29.5
Fuzzy *a-posteriori* density.

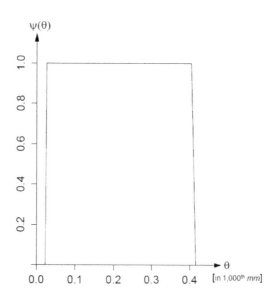

FIGURE 29.6
Fuzzy HPD interval for θ.

29.6   Final Remark

As the demand for realistic models for representing the real observations increases, generalized concepts for capturing fuzziness are necessary. A more general concept to take care of fuzzy *a-priori* information and fuzzy data of an engineering problem, as opposed to the use of standard probability, is introduced in the form of fuzzy *a-priori* densities. This concept is more suitable in modeling the prior information, which is usually uncertain, *i.e.*, fuzzy. In this case, concepts of general fuzzy numbers and fuzzy vectors along with their characterizing functions have been applied in capturing the imprecision of the collected continuous quantities. The so-called fuzzy probability densities along with the δ-level functions are used to define the fuzzy *a-posteriori* densities. The fuzzy observations are, then, used for updating the fuzzy *a-priori* density to yield the corresponding fuzzy *a-posteriori* density. As a result, Bayes' theorem is generalized to model the imprecise data.

The generalized model is equipped with explainable mathematical grounds in capturing the variability and imprecision of real observations. Fuzzy Bayesian inference provides promising results by the use of fuzzy concepts together with the probabilistic models in analyzing a critical lapping parameter. The fuzzy *a-posteriori* density and fuzzy HPD interval are useful in quantifying and broadening the understanding of the real observations from actual lapping. These results are viable for further studies in modeling flat lapping process.

Bibliography

[1] M. P. Groover. *Fundamentals of Modern Manufacturing: Materials, Processes, and Systems.* John Wiley & Sons, New Jersey, 4th edition, 2010.

[2] G. Klir and B. Yuan. *Fuzzy Sets and Fuzzy Logic – Theory and Applications.* Prentice Hall, Upper Saddle River, 1995.

[3] P. R. Lynah. *Lapping. ASM Handbook*, volume 16. ASM International, Ohio, 1990.

[4] M. Stein, M. Beer, and V. Kreinovich. Bayesian approach for inconsistent information. *Information Sciences*, 245:96–111, 2013.

[5] R. Viertl. *Statistical Methods for Fuzzy Data.* John Wiley & Sons, Chichester, 2011.

[6] O. Wolkenhauer. *Data Engineering – Fuzzy Mathematics in Systems Theory and Data Analysis.* John Wiley & Sons, New York, 2001.

[7] C. C. Yang. Fuzzy Bayesian inference. In *Proceedings of IEEE International Conference on Systems, Man, and Cybernetics*, volume 3, pages 2707–2712, Florida, USA, 2001.

[8] L. Zadeh. Probability measures of fuzzy events. *Journal of Mathematical Analysis and Applications*, 23(2):421–427, 1968.

30

Bayesian Parallel Computation for Intractable Likelihood Using Griddy-Gibbs Sampler

Nuttanan Wichitaksorn

University of Canterbury

S.T. Boris Choy

University of Sydney

CONTENTS

30.1 Introduction

The Griddy-Gibbs (GG) sampling algorithm is a Markov chain Monte Carlo (MCMC) method that was first introduced by [10]. By its name, GG is a Gibbs algorithm where each parameter can be sampled directly from its full conditional distribution with 100% acceptance rate. However, a useful and distinctive feature from the traditional Gibbs sampler is that the GG sampler can be applied to the cases where the likelihood function or the posterior density is intractable. Under the GG, the possible range of each parameter is first divided into a number of grid points. Then, parameter value at each grid point is evaluated through the full conditional density to obtain an approximate cumulative distribution function (CDF). After that, the realization of a parameter at each MCMC iteration is obtained using the inverse CDF method. The ranges of the parameters can be fine-tuned in the adaptive GG sampler to improve the performance of the algorithm. However, the use of the adaptive GG sampler in Bayesian computation has been limited, as only one parameter can be sampled at a time and each parameter has a number of grid points that need to be evaluated (though unnecessary) sequentially through its full conditional density. All this makes the algorithm very time consuming.

With the advance of computer hardware, the independently sequential computation can be replaced by parallel computation to improve the efficiency. Parallel computation is a fast growing computing environment in the last decade and has been widely used in many areas of applications. In statistics, MCMC and sequential Monte Carlo (SMC) algorithms are the two Bayesian computational methods that can benefit from parallel computing. Usually, parallel computation uses a graphics processing unit (GPU), which is an electronic circuit specially designed to process a massive amount of data in parallel. This GPU-based parallel computation has been used in phylogeny by [11] and in Bayesian mixture models by [12].

In this chapter, we propose a new approach to improve the adaptive GG sampler using GPU-based parallel computing. The proposed parallel GG sampler does not require much time and effort for computer programming. The evaluations of full conditional density at the grid points become less computationally expensive through parallel computing. The performance of the proposed sampler is illustrated in the simulation study of three statistical models. However, since the adaptive GG sampling algorithm itself samples only one parameter at a time, our proposed parallel GG sampler may be suitable for small to medium-scale problems in terms of parameter space.

In addition to the capability of dealing with the intractable likelihood function, the parallel adaptive GG sampler is also suitable for the cases where the range of a parameter is specified or known hypothetically. This is usually the case for a number of financial time series models. Even with the unknown or unspecified ranges of parameter values, the parallel adaptive GG sampler is still efficient for the posterior sampling. As a Gibbs sampling algorithm, the posterior inference including the marginal likelihood calculation can be made available immediately after the completion of random draws of the MCMC algorithm.

The organization of this chapter is as follows. Section 30.2 gives a brief introduction to the GPU-based parallel computing for Bayesian statistics and a brief review of the GG sampling algorithm and its adaptive version. Section 30.3 presents simulation studies on three statistical models and Section 30.4 contains an empirical study of a financial time series model. Finally, Section 30.5 concludes this chapter.

30.2 Parallel Computing and Griddy-Gibbs Sampler

30.2.1 Parallel computing basics

In general, the main purpose of applying GPU-based parallel computing is to enhance the time-consuming computation in many applications or algorithms where parallelism can be applied. The basic idea is to apply the GPU-based parallel computing where the independent multi-tasks can be done in parallel. In some cases, this can help speed up 20 times the computing time if the tasks have to be done sequentially. The GPU-based parallel computing arises from the use of parallel image processing. With the successful implementation of various applications in physics, engineering, and analytics, the scientific computation using GPUs has been receiving more attention in statistical computation.

The breakthrough in Bayesian computational methods in the late '80s of last century has enabled complicated real life problems to be solved using simulation-based MCMC algorithms. The beauty of the MCMC algorithms is that Bayesian statistical inferences of complicated models become accessible using these computational approaches. However, the drawback is that they are extremely computationally intensive and expensive. For example, the Gibbs sampler, the most popular MCMC algorithm, uses serial computing to obtain posterior samples for statistical inference by successive simulation from updated full conditional distributions. That is, the knowledge about a parameter or a block of parameters is updated after the other parameters have been updated in the light of data. Various methods have been proposed to increase the efficiency of the algorithm and to reduce the dependence of the simulated posterior values.

Over the past few decades, computing cost has been substantially reduced to a more affordable level while the computing speed has increased exponentially. However, MCMC algorithms still require relatively heavy computation. With the advance in computer hardware architecture, parallel computing provides a solution to speed up the MCMC algorithms. The application of parallel computation in statistics is to partition a complicated statistical problem into many smaller and manageable problems, which can be solved individually. The high-performance parallel computing performs calculations for solving these smaller problems simultaneously with a fraction of computing time compared to the serial computing.

There has been literature on parallel computing in statistical analysis. Lee *et al.* [9] demonstrate the MCMC and SMC algorithms using parallel computation with several examples. [6] introduces the massive parallel computing to the analysis of a generalized autoregressive conditional heteroscedasticity (GARCH) model using SMC algorithm while [7] uses it in stochastic volatility model. In addition, [8] illustrates the application of the parallel SMC algorithm to a logistic regression model. The implementation of the above works requires expertise in computer programming and seems suitable for some large-scale problems that can be benefited from parallel computing.

30.2.2 Griddy-Gibbs sampler

The GG sampler is an MCMC algorithm. Each model parameter is sampled directly from its univariate full conditional distribution, which is approximated at a finite number of grid points. Usually, the more grid points, the better the approximation will be. Unlike Metropolis-Hastings (MH) algorithm, this algorithm yields a 100% acceptance rate. Consider the following posterior density

$$p(\mathbf{\Theta}|\mathbf{Y}_{1:T}) \propto p(\mathbf{\Theta}) \times \prod_{t=1}^{T} p(\mathbf{Y}_t|\mathbf{\Theta}),$$

where $\mathbf{\Theta} = (\theta_1, \ldots, \theta_k)$ is the vector of model parameters, $\mathbf{Y}_{1:T}$ is the data vector or matrix, $p(\mathbf{\Theta}|\mathbf{Y}_{1:T})$ is the corresponding joint posterior density, $p(\mathbf{\Theta})$ is the prior density and $p(\mathbf{Y}_t|\mathbf{\Theta})$ is the data density. Let $\theta_j \in \mathbf{\Theta}$ denote the parameter of interest and $\mathbf{\Theta}_{-j} = (\theta_1, \ldots, \theta_{j-1}, \theta_{j+1}, \ldots, \theta_k)$. The GG sampler allows us to sample θ_j directly from the univariate full conditional density $p(\theta_j|\mathbf{\Theta}_{-j}, \mathbf{Y}_{1:T})$ regardless of the intractability of associated likelihood. The algorithm works as follows. First, the range of θ_j is divided into G grid points (they can be evenly-spaced for simplicity), say $\theta_j^1, \ldots, \theta_j^G$. Then, the full conditional density $p(\theta_j|\mathbf{\Theta}_{-j}, \mathbf{Y}_{1:T})$ is evaluated at each grid point, θ_j^g for $g = 1, \ldots, G$ and an empirical CDF of θ_j is obtained. Finally, a realization of θ_j is simulated using the inverse CDF method. In short, at a particular Gibbs sampler iteration for parameter θ_j, the algorithm is summarized below.

1. Divide the range of θ_j into G evenly-spaced grid points, $\theta_j^1, \ldots, \theta_j^G$.

2. Evaluate the full conditional density $p(\theta_j^g|\cdot)$ at each grid point to obtain weights, $w_j^1, \ldots, w_j^G$, where $w_j^g = p(\theta_g|\cdot)$ for $g = 1, \ldots, G$.

3. Normalize w_j^g to obtain $\tilde{w}_j^g = \frac{w_j^g}{\sum_{g=1}^{G} w_j^g}$ and cumulative sum $\tilde{W}_j^g = \sum_{h=1}^{g} \tilde{w}_j^h$ which is treated as the empirical CDF $F(\theta_j^g)$ where $F(\theta_j^G) = \tilde{W}_j^G = 1$.

4. Draw $u \sim U(0,1)$ and obtain a realization of θ_j, θ_j^g, using the inverse CDF method, i.e. if $F(\theta_j^{g-1}) < u \leq F(\theta_j^g)$, then set $\theta_j = \theta_j^g$.

In fact, the evaluation of $p(\theta_j|\mathbf{\Theta}_{-j}, \mathbf{Y}_{1:T})$ at each grid point θ_j^g can be done independently and simultaneously. However, in the past these evaluations were made sequentially because of the limitation on computing technology. Now, the modern parallel computing technology has made it cheaper and faster to evaluate the full conditional density using the GG sampler. From the above algorithm, one can see that each draw of an updated parameter value requires only one round-trip transfer from the host to the devices. Compared to the SMC where several round-trips are made in a sequential update, this can also save computing time. In this chapter, we adopt the parallel computing for Step 2 and the performance of the parallel GG sampler on three different models is reported in Section 30.3.

The extension to multivariate case, i.e., more than one parameter are sampled in a single block, is controversial as some do not believe that the multivariate inverse CDF exists and this is another limitation for the use of the GG sampler. For this reason, the GG sampler or parallel GG sampler may not be suitable for some large-scale problems. However, compared to the SMC where sequential updates are made through each observation, the parallel GG sampler has a smaller number of cycles. For example, with ten model parameters and 1,000 observations for a 5,000 MCMC runs (as not too many MCMC runs are necessary for the parallel GG), the parallel GG sampler takes 50,000 parameter-MCMC cycles while for the SMC with 50 Monte Carlo cycles, 250,000 observation-SMC cycles are required.

In the case where no prior knowledge is available on the range of a parameter, an initial range can be obtained from some estimation methods. For example, after a burn-in period, the mean and standard deviation of a parameter are calculated from the burn-in samples and an initial range is obtained as the mean $\pm$ 2 $\times$ standard deviation. We refer this method as an adaptive version of the GG sampler. Other adaptive GG samplers refer to some other variations. [10] indicates that the number of grid points does not have to be the same for all MCMC iterations and they call this method an adaptive grid.

The choice of grid points can be subjective and may depend on the range of a parameter. For a short range, such as (0,1), a parallel GG sampler does not require a large number of grid points; otherwise, it will make the grid to be too coarse. For a wider range, a parallel adaptive GG sampler makes the estimation easier as the range can be divided into a large number of grid points, as many as the hardware (number of cores/processors) allows.

30.3   Simulation Study

In this section, we illustrate the implementation of the parallel GG sampler and assess its performance through several examples. The models considered are a linear regression model with Student-t error, a nonlinear regression model, and a generalized autoregressive conditional heteroskedastic (GARCH) model with asymmetric Laplace error. These models are chosen because of their intractable likelihood functions. For each model, the convergence of the Markov chain is monitored by the trace plots and the simulation inefficiency factor (SIF) as in [4] is used to assess the simulation efficiency. The software package used for all computations is MATLAB (Version 2013a) with parallel computing toolbox and the hardware is a computer with two 2,496-core GPUs (though in our implementation, we use only up to 1,000 cores).

30.3.1 Linear regression model with Student-t error

Consider a linear regression model with the Student-t error distribution as follows

$$y_i = \mathbf{x}_i'\beta + \sigma\epsilon_i \text{ for } i = 1, ..., n, \tag{30.1}$$

where y_i is the scalar dependent observation i, $\mathbf{x}_i' = (1\ x_{1i}\ x_{2i})$ is a row vector of covariates, β is a column vector of length 3, σ is the scale parameter, ϵ_i is an error term which follows an independent standard Student-t, $St(0, 1, \nu)$ distribution with $\nu \in (2, 30)$ degrees of freedom. Let $\Theta = (\beta, \sigma, \nu)'$. Assuming independence across observations, the likelihood function is given by

$$L(\Theta|\text{data}) = \prod_{i=1}^{n} \frac{\Gamma\left(\frac{\nu+1}{2}\right)}{\Gamma\left(\frac{\nu}{2}\right)\sqrt{\pi\nu\sigma^2}} \left(1 + \frac{(y_i - \mathbf{x}_i'\beta)^2}{\nu\sigma^2}\right)^{-\frac{\nu+1}{2}} \tag{30.2}$$

where $\Gamma(\cdot)$ is the gamma function. A Bayesian framework is completed with the choice of a flat prior $p(\Theta) \propto \frac{1}{\sigma^2}$. Actually in the implementation, the Student-t density function can be expressed as a scale mixtures of normal representation to simplify the full conditional distributions for the Gibbs sampling algorithm, see [5]. For ν, its realizations are usually simulated using MH algorithm. However, for the sake of illustration, in this chapter we use the parallel GG sampler to draw all model parameters.

In the simulation, we set $\beta = (1, 1, -1.5)'$, $\sigma = 2$ and $\nu = 5$ and generate $n = 1,000$ observations. The values of the covariates x_{1i} and x_{2i} are generated from the uniform $U(0, 2)$ and $U(0, 1)$ distributions, respectively. With $G = 500$ grid points, we cycle 3,000 MCMC draws including 1,000 draws in the burn-in period and keep every other draw for posterior inference.

Even with only 3,000 MCMC cycles, it can be seen from Figures 30.1 and 30.2 that the Markov chains converge very fast and the posterior samples are mixed well. Table 30.1 exhibits the posterior summaries of the parameters and the SIF values. The posterior means are close to the true parameter values and the credible intervals with 0.95 coverage probability cover all the true values. The SIFs are relatively low and the computing time is around two minutes.

30.3.2 Nonlinear regression model

Bates and Watts [1] illustrate the application of a non-linear model in the experimental decay data. The data are reanalysed using a Bayesian simulation approach with a GG sampler in [10]. The non-linear model considered in these two works is

$$y_i = \beta_1(1 - \exp\{-\beta_2 x_i\}) + \epsilon_i, \text{ for } i = 1, ..., n, \tag{30.3}$$

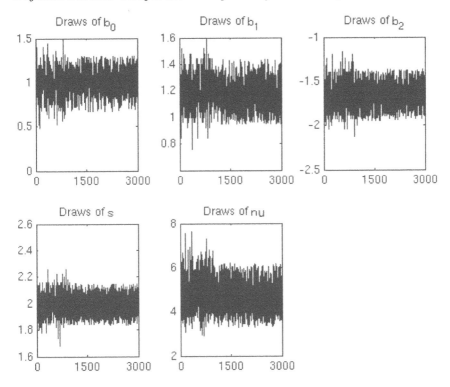

FIGURE 30.1
Convergence draws of linear regression model with Student-t error.

where x_i is the value of the covariate and $\epsilon_i \sim N(0, \sigma^2)$. In this chapter, we focus on the estimation of the regression coefficients, β_1 and β_2 in a simulation study and, therefore, without loss of generality, we set $\sigma = 1$. Let $\Theta = (\beta_1, \beta_2)'$. Assuming independence, the likelihood function is given by

$$L(\Theta | \text{data}) = \prod_{i=1}^{n} \frac{1}{\sqrt{2\pi}} \exp\left\{ -\frac{1}{2} \left(y_i - \beta_1 (1 - \exp\left\{ -\beta_2 x_i \right\}) \right)^2 \right\}. \qquad (30.4)$$

To complete the Bayesian framework, a joint prior distribution is assigned to β_1 and β_2. In the model implementation, independent vague prior distributions are assumed, i.e. $p(\Theta) \propto 1$. Since the likelihood function and hence the joint posterior distributions are intractable, MH algorithm is always adopted to obtain the posterior samples for statistical inference. However, the proposal distribution of the MH algorithm is always a multivariate distribution, or a bivariate distribution in this case, where the linear relationship among the regression coefficients is assumed through a covariance matrix. Due to the nature of non-linear relationship among the regression coefficients in this model, the MH algorithm has a very low acceptance rate and is hence inappropriate. Instead, the parallel GG sampler is used.

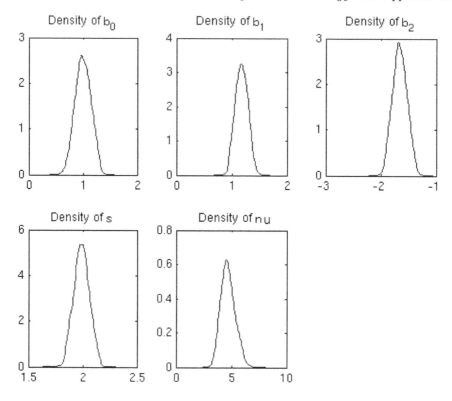

FIGURE 30.2
Posterior densities of linear regression model with Student-t error.

For illustrative purpose, we set $\beta_1 = 1$ and $\beta_2 = 1$ and generate a random sample of size $N = 1,000$ from the non-linear model discussed above. The values of the covariate, x_i, are simulated independently from the standard normal distribution. A GG sampler with $G = 500$ grid points is run to produce

TABLE 30.1
Estimation results of linear regression model with Student-t error.

$\hat{\theta}$	T.V.	Mean	S.D.	2.5%	97.5%	SIF
$\hat{\beta}_0$	1.0000	1.0089	0.1346	0.7441	1.2622	2.65
$\hat{\beta}_1$	1.0000	1.1670	0.1072	0.9774	1.3714	2.48
$\hat{\beta}_2$	−1.5000	−1.6527	0.1243	−1.8839	−1.4096	3.01
$\hat{\sigma}$	2.0000	1.9863	0.0669	1.8594	2.1164	2.05
$\hat{\nu}$	5.0000	4.6828	0.6184	3.5619	5.9728	2.11

Notes: SIF = Simulation Inefficiency Factor and Computing Time = 2.2277 mins.

TABLE 30.2
Estimation results of nonlinear regression model.

$\hat{\theta}$	T.V.	Mean	S.D.	2.5%	97.5%	SIF
$\hat{\beta}_1$	1.0000	0.9999	0.0502	0.9083	1.1011	2.46
$\hat{\beta}_2$	1.0000	0.9982	0.0182	0.9652	1.0286	2.48

Notes: SIF = Simulation Inefficiency Factor and Computing Time = 0.7070 mins.

2,000 MCMC draws, which include a burn-in period of 1,000 draws. Every 5th draw is then taken to form a random sample of size 200 from the intractable joint posterior distribution and the results are presented in Table 30.2. Even with a small sample size of 200, the posterior means of $\beta_1 = 1$ and $\beta_2 = 1$ are very close to their true values. The parallel GG sampler performed very well and the computing time is below one minute. In addition, the SIFs are reasonably low.

30.3.3   GARCH with asymmetric Laplace error

In this subsection, we consider a GARCH(1,1) model with asymmetric Laplace error distribution of the form

$$
\begin{aligned}
y_t &= \mu_y + h_t^{1/2}\epsilon_t, & (30.5)\\
h_t &= \alpha_0 + \alpha_1 h_{t-1} + \gamma_1 \epsilon_{t-1}^2, \quad \text{for } t = 2, ..., T, & (30.6)
\end{aligned}
$$

where t denotes the time index, y_t is the financial time series of return, μ_y is the conditional mean of y_t, ϵ_t is the error term that follows a standardized asymmetric Laplace distribution with location 0, variance 1 and skewness parameter $p \in (0,1)$, h_t is the conditional variance of y_t given the information up to and including time $t-1$ and α_0, α_1 and γ_1 are volatility parameters. If p is fixed, this GARCH model can yield the results for a quantile regression of a GARCH model at quantile level p. See [2] for detail. Let $\Theta = (\mu_y, \alpha_0, \alpha_1, \gamma_1, p)'$. The likelihood function is given by

$$
L(\Theta|\text{data}) = \prod_{t=2}^{T} \frac{(1-2p+2p^2)^{\frac{1}{2}}}{h_t^{1/2}} \exp\left\{ \frac{(1-2p+2p^2)^{\frac{1}{2}}(y_t-\mu_y)}{(p-I(y_t \geq \mu_y))h_t^{1/2}} \right\}. \quad (30.7)
$$

A Bayesian framework is completed with a standard normal prior distribution for μ_y and uniform prior distributions for other model parameters subject to the constraints $0 < p < 1$, $0 < \alpha_0 < 1$ and $0 < \alpha_1 + \gamma_1 < 1$ for stationarity.

In the simulation, we set $\mu_y = 0.05, p = 0.3, \alpha_0 = 0.01, \alpha_1 = 0.85$, and $\gamma_1 = 0.12$ and generate a time series of length $T = 1,000$. With the known ranges for some parameters, only $G = 100$ grid points are used in the parallel GG sampler. We iterate 3,000 MCMC cycles with the first 1,000 cycles discarded

TABLE 30.3
Estimation results of GARCH(1,1) model with asymmetric Laplace error.

$\hat{\theta}$	T.V.	Mean	S.D.	2.5%	97.5%	SIF
$\hat{\mu}_y$	0.0500	0.0457	0.0218	0.0049	0.0893	2.18
$\hat{\alpha}_0$	0.0100	0.0146	0.0109	0.0005	0.0397	2.44
$\hat{\alpha}_1$	0.8500	0.8341	0.0242	0.7820	0.8780	2.25
$\hat{\gamma}_1$	0.1200	0.1075	0.0214	0.0726	0.1542	2.15
$\hat{p}$	0.3000	0.3082	0.0139	0.2814	0.3349	2.15

Notes: SIF = Simulation Inefficiency Factor and Computing Time = 9.7355 mins.

as in the burn-in period. The remaining 2,000 cycles are used for posterior inference and the results are presented in Table 30.3.

For this complicated financial time series GARCH(1,1) model with asymmetric Laplace error distribution, the parallel GG sampler still works well and the computing time is about 10 minutes. All posterior means are close to the true parameter values and all 0.95 credible intervals contain the true parameter values. The SIF values are again very low.

30.4   Empirical Study

Durham and Geweke [6] demonstrate the applicability of parallel computing in exponential GARCH (EGARCH) models using sequential Monte Carlo methods. Here we consider an EGARCH(1,1) model of the form

$$y_t = \mu_y + \exp(v_t/2)\epsilon_t, \tag{30.8}$$

$$v_t = \alpha_1 v_{t-1} + \beta_1 \left(|\epsilon_{t-1}| - (2/\pi)^{1/2}\right) + \gamma_1 \epsilon_{t-1}, \quad \text{for } t = 2, \dots, T, \tag{30.9}$$

where y_t is the return series, v_t is the log-volatility series, $\epsilon_t \sim N(0, 1)$ is the return error term and α_1, β_1 and γ_1 are the volatility parameters. For Bayesian inference, a standard normal prior distribution is assigned to μ_y and γ_1 and a uniform $U(0, 1)$ distribution is assigned to α_1 and β_1.

Let p_t be the daily closing price of the S&P 500 index on day t. The daily percentage return is defined as $y_t = 100 \times \ln(p_t/p_{t-1})$. The return data from January 2, 1990 to March 31, 2010 are fitted using the EGARCH(1,1) model and the total number of observations is $T = 5,103$. After the burn-in of 1,000 iterations, the outputs from the subsequent 2,000 MCMC cycles are used for posterior inference. The parallel GG sampler with $G = 200$ grid points takes about 15 minutes to complete the random draws.

TABLE 30.4
Estimation results of EGARCH(1,1) model using S&P 500.

$\hat{\theta}$	Mean	S.D.	2.5%	97.5%	SIF
$\hat{\mu}_y$	0.0292	0.0085	0.0128	0.0451	2.58
$\hat{\alpha}_1$	0.9839	0.0022	0.9789	0.9873	2.76
$\hat{\beta}_1$	0.1201	0.0101	0.1013	0.1414	2.37
$\hat{\gamma}_1$	−0.0959	0.0080	−0.1117	−0.0804	2.28
log-marginal likelihood = 20,511					

Notes: SIF = Simulation Inefficiency Factor and Computing Time = 14.5470 mins.

Table 30.4 presents the posterior mean, posterior standard deviation and the credible interval with coverage probability 0.95 for each parameter of the EGARCH(1,1) model. The favorable results can be seen from the low standard deviations of the parameters. Although the conditional mean of the daily percentage return μ_y is close to zero, the credible interval shows that it is significantly different from zero. Moreover, it can be inferred that the parameter γ_1 is statistically smaller than zero and this means that the log-volatility has a negative relationship with the lag-1 value of the return. Following [3], the marginal log-likelihood is estimated to be 20,511. This value, though it may not be compared directly, is slightly higher than that of [6]. The SIFs are relatively low indicating the high efficiency of posterior draws. It is worth noting that similar results are obtained when different number of grid points and MCMC cycles are applied.

30.5   Conclusion and Discussion

GPU-based parallel computation is a fast growing computing environment in many areas including Bayesian computational statistics. However, most of the Bayesian inferences using parallel computing have been implemented through sequential Monte Carlo methods where model parameters are updated sequentially and this may be suitable for large-scale problems. This chapter is the first to revive the use of adaptive GG sampling algorithm with parallel computation under the Bayesian MCMC framework. Parallel adaptive GG sampling is suitable for (i) small to medium-scale problems where the dimensionality of model parameter space is not very high, (ii) some or all model parameters are defined on specific intervals, and (iii) the likelihood function is intractable. In addition, parallel adaptive GG sampler is relatively easy to implement and code. Since the adaptive GG sampling is a Gibbs sampling algorithm where each model parameter is directly drawn from its full conditional distribution with 100% acceptance rate, the marginal likelihood can be conveniently evaluated and immediately provided at the end of the posterior simulation. This chapter presents favourable simulation results for a linear regression model with Student-t error, a nonlinear regression model, and a GARCH(1,1) model where the model implementation relies on the adaptive GG sampling algorithm with parallel computation. Both simulation and empirical studies show very good mixing of the MCMC cycles under the parallel computing environment.

Acknowledgments

We thank Professor S.K. Upadhyay and two anonymous referees for their helpful suggestions. Useful advice and helpful discussions on GPU-based parallel computing from Paul Brouwers are fully acknowledged. This work has been originated and entirely done using the generous computing facilities at the School of Mathematics and Statistics, University of Canterbury.

Bibliography

[1] D. M. Bates and D. G. Watts. *Nonlinear Regression Analysis and Its Applications.* John Wiley & Sons, New York, 1988.

[2] C. W. S. Chen, R. Gerlach, and D. C. M. Wei. Bayesian causal effects in quantiles: Accounting for heteroscedasticity. *Computational Statistics & Data Analysis*, 53(6):1993–2007, 2009.

[3] S. Chib. Marginal likelihood from the Gibbs output. *Journal of the American Statistical Association*, 90(432):1313–1321, 1995.

[4] S. Chib. Markov chain Monte Carlo methods. In S. N. Durlauf and L. E. Blume, editors, *The New Palgrave Dictionary of Economics*, chapter 5, pages 331–338. Palgrave Macmillan, Basingstoke, 2008.

[5] S. B. Choy and A. F. M. Smith. Hierarchical models with scale mixtures of normal distribution. *TEST*, 6(1):205–221, 1997.

[6] G. Durham and J. Geweke. Adaptive sequential posterior simulators for massively parallel computing environments. Available at SSRN 2251635, 2013.

[7] A. Fulop and J. Li. Efficient learning via simulation: A marginalized resample-move approach. *Journal of Econometrics*, 176:146–161, 2013.

[8] J. Geweke, G. Durham, and H. Xu. Bayesian inference for logistic regression models using sequential posterior simulation. Working Paper, Available at SSRN 2243342, 2013.

[9] A. Lee, C. Yau, M. B. Giles, A. Doucet, and C. C. Holmes. On the utility of graphics cards to perform massively parallel simulation of advanced Monte Carlo methods. *Journal of Computational and Graphical Statistics*, 19(4):769–789, 2010.

[10] C. Ritter and M. A. Tanner. Facilitating the Gibbs sampler: The Gibbs stopper and the griddy-Gibbs sampler. *Journal of the American Statistical Association.*, 87(419):861–868, 1992.

[11] M. A. Suchard and A. Rambaut. Many-core algorithms for statistical phylogenetics. *Bioinformatics.*, 25(11):1370–1376, 2009.

[12] M. A. Suchard, Q. Wang, C. Chan, J. Frelinger, A. Cron, and M. West. Understanding GPU programming for statistical computation: Studies in massively parallel massive mixtures. *Journal of Computational and Graphical Statistics*, 19(2):419–438, 2010.

Index

Milton Keynes UK
Ingram Content Group UK Ltd.
UKHW021934071024
449327UK00022B/1806

9 780367 377625